AF252047

Progress
in Analysis and its Applications

Progress
in Analysis and its Applications

Proceedings of the 7th International ISAAC Congress

Imperial College London, UK   13 – 18 July 2009

Editors

Michael Ruzhansky
Imperial College London, UK

Jens Wirth
Imperial College London, UK

NEW JERSEY · LONDON · SINGAPORE · BEIJING · SHANGHAI · HONG KONG · TAIPEI · CHENNAI

Published by

World Scientific Publishing Co. Pte. Ltd.

5 Toh Tuck Link, Singapore 596224

USA office: 27 Warren Street, Suite 401-402, Hackensack, NJ 07601

UK office: 57 Shelton Street, Covent Garden, London WC2H 9HE

British Library Cataloguing-in-Publication Data
A catalogue record for this book is available from the British Library.

PROGRESS IN ANALYSIS AND ITS APPLICATIONS
Proceedings of the 7th International ISAAC Congress

ISBN-13 978-981-4313-16-2
ISBN-10 981-4313-16-5

Printed in Singapore by World Scientific Printers.

PREFACE

The *7th International ISAAC Congress* was held at the Imperial College London, UK, during the period 13–18 July, 2009 hosting about 550 participants. The organisation of the congress was made possible by the financial support from

- the *London Mathematical Society* (LMS),
- the *International Mathematical Union* (IMU), the Commission on Development and Exchanges and the Developing Countries Strategy Group,
- the *Engineering and Physics Research Council* (EPSRC),
- the *Oxford Centre in Collaborational and Applied Mathematics* (OCCAM),
- the *Oxford Centre for Nonlinear Partial Differential Equations* (OxPDE),
- the *Bath Institute for Complex Systems* (BICS),
- the *Imperial College London*, Strategic Fund, and
- the *Department of Mathematics, Imperial College London.*

The International Society for Analysis, its Applications and Computations (ISAAC) as well as the local organisers of the congress are grateful for the support. It allowed us to help select outstanding mathematicians from developing countries financially to enable them to participate at the meeting. We see it as one of the great successes of this congress that we could welcome participants from 59 countries from all over the world.

The congress made a significant contribution to the development of analysis internationally, and in the United Kingdom, in particular. It may be considered as the largest congress in the area of analysis that took place in the UK up to now. It provided an excellent opportunity of a large scale yet focused meeting, enabling the British mathematicians and mathematicians from many other countries to effectively communicate, to discuss a wealth of topics in analysis and partial differential equations, to start new projects, and to continue the existing cooperations. The meeting was endorsed by the *European Mathematical Society* and by the *International Mathematical*

Union.

It is already a well-established tradition to award one or several outstanding young researchers during the ISAAC congress. The ISAAC award of the 7th International ISAAC Congress was presented to

Sergey Tikhonov
(ICREA and Centre de Recerca Matemàtica, Barcelona, Spain)

for his outstanding works on criteria for convergence of Fourier series and Fourier transforms in terms of their coefficients, Fourier inequalities, function spaces, embedding theorems, polynomial inequalities, integral operators and fractional calculus.

Local organisers of the congress were Michael Ruzhansky (Imperial College) as the Chairman of the committee, together with Dan Crisan (Imperial College), Brian Davies (King's College), Jeroen Lamb, Ari Laptev, Jens Wirth and Boguslaw Zegarlinski (all from Imperial College), with further help from local post-docs and PhD students, in particular Maryam Alrashed, Laura Cattaneo, Federica Dragoni, James Inglis, Vasileos Kontis, Ioannis Papageorgiou and Muhammad Usman, who spent a lot of their time and energy in preparation of the congress week and on the registration and help desk during the meeting.

The organisers were supported and guided by the ISAAC board acting as the International Advisory Board,

H. Begehr (Berlin)	A. Berlinet (Montpellier)
B. Bojarski (Warsaw)	E. Bruning (Durban)
V. Burenkov (Padova)	O. Celebi (Istanbul)
R. Gilbert (Newark)	A. Kilbas (Minsk)
M. Lanza de Cristoforis (Padova)	M. Reissig (Freiberg)
L. Rodino (Torino)	M. Ruzhansky (London)
J. Ryan (Fayetteville)	S. Saitoh (Aveiro)
B.-W. Schulze (Potsdam)	J. Toft (Växjö)
M.W. Wong (Toronto)	Y. Xu (Louisville)
M. Yamamoto (Tokyo)	S. Zhang (Newark)

At its meeting on Wednesday 15 July 2009 the ISAAC Board has elected Michael Ruzhansky from Imperial College London as the new ISAAC President. The former president, Man Wah Wong from York University, Toronto, Canada, has completed his 4-year service. After the congress ISAAC members elected further new vice presidents. Further information can be found on the ISAAC homepage `www.mathisaac.org`

The ISAAC Board has decided the venue for the 8th International ISAAC Congress. As suggested by Victor Burenkov, it will be organised in 2011 at The People's Friendship University in Moscow in collaboration with the Steklov Mathematical Institute of the Russian Academy of Sciences.

The 7th ISAAC Congress had 12 plenary speakers and 24 parallel sessions. The local organisers estimate that during the congress week 240 hours of continuous lecturing took place, most of it parallel. The plenary talks of the congress were

- Sir John Ball (Oxford)
  The Q-tensor theory of liquid crystals
- Louis Boutet de Monvel (Paris)
  Asymptotic equivariant index of Toeplitz operators and Attiyah-Weinstein conjecture
- Brian Davies (London)
  Non-self-adjoint spectral theory
- Simon Donaldson (London)
  Asymptotic analysis and complex differential geometry
- Carlos Kenig (Chicago)
  The global behaviour of solutions to critical nonlinear dispersive and wave equations
- Vakhtang Kokilashvili (Tbilisi)
  Nonlinear harmonic analysis methods in boundary value problems
- Nicolas Lerner (Paris)
  Instability of the Cauchy-Kovalevskaya solution for a class of non-linear systems
- Paul Malliavin (Paris)
  Non-ergodicity of Euler deterministic fluid dynamics via stochastic analysis
- Vladimir Maz'ya (Liverpool and Linköping)
  Higher-order elliptic problems in non-smooth domains
- Bert-Wolfgang Schulze (Potsdam)
  Operator algebras with symbolic hierarchies on stratified spaces
- Gunther Uhlmann (Seattle)
  Visibility and invisibility
- Masahiro Yamamoto (Tokyo)
  Practise of industrial mathematics related with the manufacturing process

Two of the plenary talks were funded by institutes and therefore specially dedicated. The talk of Vladimir Maz'ya was sponsored by the Bath Institute for Complex Systems (BICS), while Masahiro Yamamoto's talk was dedicated as the OCCAM Lecture on Applied Mathematics and chaired by the Director of the Oxford Centre of Collaborational and Applied Mathematics (OCCAM), John Ockendon.

The 7th ISAAC Congress was the first congress in this series which also contained a lecture aimed at the general public and given by

- Pierre-Louis Lions (Paris)
  Analysis, models and simulations

This public lecture was made possible by the Oxford Centre for Nonlinear Partial Differential Equations (OxPDE) and also chaired by its Director, Sir John Ball.

The present volume contains the texts of a selection of invited talks delivered at the conference. As in previous years, some of the sessions or interest groups decided to publish their own volume of proceedings and are therefore excluded from this one. References to these special proceedings volumes (if already available) are given below. The sessions of the congress were

I.1. *Complex variables and potential theory,*
     organised by M. Lanza de Cristoforis, T. Aliyev, S. Plaksa and P. Tamrazov

I.2. *Differential equations: Complex and functional analytic methods, applications,*
     organised by H. Begehr, D.-Q. Dai and J. Du

I.3. *Complex analytical methods for applied sciences,*
     organised by S. Rogosin and V. Mityushev

I.4. *Zeros and gamma lines – value distributions of real and complex functions,*
     organised by G. Barsegian and G. Csordas

II.1. *Clifford and quaternion analysis,*
      organised by I. Sabadini and F. Sommen

II.2. *Analytical, geometrical and numerical methods in Clifford- and Cayley-Dickson algebras,*
      organised by W. Sprößig, V. Kisil and K. Gürlebeck

III.1. *Toeplitz operators and their applications,*
       organised by N. Vasilevski and S. Grudsky

III.2. *Reproducing kernels and related topics,*
 organised by A. Berlinet and S. Saitoh

III.3. *Modern aspects of the theory of integral transforms,*
 organised by A. Kilbas and S. Saitoh

III.4. *Spaces of differentiable functions of several real variables and applications,*
 organised by V. Burenkov and S. Samko

III.5. *Analytic and harmonic function spaces,*
 organised by R. Aulaskari, T. Kaptanoglu and J. Rättyä

IV.1. *Pseudo-differential operators,*
 organised by L. Rodino and M.W. Wong

IV.2. *Dispersive equations,*
 organised by M. Reissig and F. Hirosawa

IV.3. *Control and optimisation of nonlinear evolutionary systems,*
 organised by F. Bucci and I. Lasiecka

IV.4. *Nonlinear partial differential equations,*
 organised by V. Georgiev and T. Ozawa

IV.5. *Asymptotic and multi-scale analysis,*
 organised by V. Smyshlyaev and I. Kamotski

V.1. *Inverse problems,*
 organised by Y. Kurylev and M. Yamamoto

V.2. *Stochastic analysis,*
 organised by D. Crisan and T. Lyons

V.3. *Coercivity and functional inequalities,*
 organised by B. Zegarlinski and D. Bakry

V.4. *Dynamical systems,*
 organised by S. Luzzatto and J. Lamb

V.5. *Functional differential and difference equations,*
 organised by A. Zafer, L. Berezansky, J. Diblik

V.6. *Mathematical biology,*
 organised by R. Gilbert

VI. *Others* containing contributed talks,
 organised by the Local Organising Committee

We thank the organisers of these special sessions for their work. They spent an enormous amount of time inviting participants, arranging their sessions, providing chairmen and creating a familiar and workshop-like atmosphere within their meetings. The session organisers were also responsible for collecting contributions to this proceedings volume and for the refereeing process of the papers.

As already mentioned, there are separate proceedings publications for some sessions and also by several interest groups. We conclude this introduction with references to them.

I. Sabadini and F. Sommen will edit a special volume with contributions of Session II.1. Similarly, A. Kilbas and S. Rogosin will collect contributions of Sessions I.3 and III.3 for a publication with Cambridge Scientific Publishers, Ltd.

The proceedings of Session IV.1. will appear as publication of the special ISAAC Interest Group in Pseudo-Differential Operators (IGPDO). It will appear as *Pseudo-Differential Operators: Analysis, Applications and Computations* within the series Operator Theory: Advances and Applications, Birkhäuser, edited by L. Rodino, M.W. Wong and H. Zhu.

During the 7th ISAAC Congress a new Interest Group for Partial Differential Equations (IGPDE) was founded. Related to this, it was decided to publish selected contributions of Sessions III.6, IV.2, IV.3, IV.4, IV5 and V.1 as a special volume dedicated to new developments in the theory of partial differential equations. It will appear as *Modern Aspects of Partial Differential Equations* within the series Operator Theory: Advances and Applications, Birkhäuser, edited by M. Ruzhansky and J. Wirth.

D. Del Santo, F. Hirosawa and M. Reissig will publish contributions of Session IV.2. as special volume of *Rendiconti dell'Istituto di Matematica dell'Università di Trieste.*

Michael Ruzhansky and Jens Wirth
(Local Organising Committee)

London, UK
July 2010

CONTENTS

Preface	v
I.1. Complex variables and potential theory **(T. Aliyev, M. Lanza de Cristoforis, S. Plaksa,** **P. Tamrazov)**	1
Global mapping properties of analytic functions *C. Andreian Cazacu and D. Ghisa*	3
Global mapping properties of rational functions *C. Ballantine and D. Ghisa*	13
Microscopic behaviour of the Stokes flow in a singularly perturbed exterior domain *M. Dalla Riva*	23
Singularly perturbed loads for a nonlinear traction boundary value problem on a singularly perturbed domain *M. Dalla Riva and M. Lanza de Cristoforis*	31
Boundary behaviour of normal functions *P.V. Dovbush*	39
Spatial quasiconformal mappings and directional dilatations *A. Golberg*	45
On finite-difference smoothness of conformal mapping *O.W. Karupu*	53

Structure of non-rectifiable curves and solvability of the jump problem   59
  B.A. Kats

I.2. Differential equations: Complex and functional analytic methods, applications   65
  (H. Begehr, D.-Q. Dai, J. Du)

On some qualitative issues for the first order elliptic systems in the plane   67
  G. Akhalaia, G. Makatsaria and N. Manjavidze

Harmonic Green and Neumann representations in a triangle, quarter-disc and octo-plane   74
  H. Begehr, M.-R. Costache, S. Tappert and T. Vaitekhovic

On some classes of bicomplex pseudoanalytic functions   81
  P. Berglez

Solvability condition of the Riemann-Hilbert monodromy problem   89
  G.K. Giorgadze

A study about one kind of two dimensional integral equation of Volterra type with two interior singular lines   96
  L. Rajabova

3D rotating Navier-Stokes equations: New method of numerical analysis and families of global exact solutions   105
  A.G. Khaybullin and R.S. Saks

Phragmén-Lindelöf principle for subparabolic functions   114
  A.I. Kheyfits

I.4. Zeros and gamma lines — value distributions of real and complex functions   121
  (G. Barsegian, G. Csordas)

A universal value distribution for arbitrary meromorphic functions in a given domain   123
  G. Barsegian

A method of counting the zeros of the Riccati equation and
its application to biological and economical prognoses 129
  G. Barsegian and A. Sargsyan

A Heine-Stieltjes theorem for higher order differential operators 136
  P. Brändén

Extended Laguerre inequalities and a criterion for real zeros 143
  D.A. Cardon

New classes of stable polynomials and polynomials with real
negative roots 150
  M. Charalambides

On a general concept of order of a meromorphic function 157
  A. Fernandez Arias and J. Perez Alvares

Séries universelles construites à l'aide de la fonction zeta de Riemann 164
  A. Poirier

II.1. Clifford and quaternion analysis 171
(I. Sabadini, F. Sommen)

A higher order integral representation formula in isotonic
Clifford analysis 173
  J. Bory Reyes, H.R. Malonek, D. Peña Peña and F. Sommen

A version of Fueter's theorem in Dunkl-Clifford analysis 180
  M. Fei and S. Li

Boundary values of pluriholomorphic functions in $\mathbb{C}^2$ 188
  A. Perotti

Some properties of k-biregular function space in real Clifford analysis 195
  Y. Qiao, Y. Xie and H. Yang

**II.2. Analytical, geometrical and numerical methods
in Clifford- and Cayley-Dickson algebras** 203
(K. Gürlebeck, V. Kisil, W. Spößig)

Further results in discrete Clifford analysis   205
  N. Faustino

A note on the linear systems in quaternions   212
  S.G. Georgiev

Wavelets beyond admissibility   219
  V.V. Kisil

Itô formula for an integro-differential operator without an
associated stochastic process   226
  R. Léandre

Integral theorems in a commutative three-dimensional
harmonic algebra   232
  S.A. Plaksa and V.S. Shpakivskyi

Segre quaternions, spectral analysis and a four-dimensional
Laplace equation   240
  D.A. Pinotsis

III.2. Reproducing kernels and related topics   247
  (A. Berlinet, S. Saitoh)

Reproducing kernel Hilbert spaces and local polynomial
estimation of smooth functionals   249
  B. Abdous and A. Berlinet

Constructive reconstruction from irregular sampling in
multi-window spline-type spaces   257
  H.G. Feichtinger and D.M. Onchiş

Integral formulas on the boundary of some ball   266
  K. Fujita

Paley–Wiener spaces and their reproducing formulae   273
  J.R. Higgins

Weighted composition operators between some spaces of
analytic functions 280
    S.D. Sharma and A.K. Sharma

III.3. Modern aspects of the theory of integral transforms 287
    (A. Kilbas, S. Saitoh)

Numerical real inversion of the Laplace transform by
reproducing kernel and multiple-precision arithmetic 289
    H. Fujiwara

Some aspects of modified Kontorovitch-Lebedev integral
transforms 296
    J.M. Rappoport

**III.4. Spaces of differentiable functions of several real
variables and applications** 305
    (V. Burenkov, S. Samko)

Image normalization of WHOs in diffraction problems 307
    A. Moura Santos

Weighted estimates for the averaging integral operator and
reverse Hölder inequalities 315
    B. Opic

IV.2. Dispersive equations 323
    (F. Hirosawa, M. Reissig)

Some $L^p - L^q$ estimates for hyperbolic systems 325
    M. D'Abbico, S. Lucente and G. Taglialatela

The wave equation in the Einstein and de Sitter spacetime 332
    A. Galstyan, T. Kinoshita and K. Yagdjian

P-evolution operators with characteristics of variable
multiplicity 339
    T. Herrmann and M. Reissig

Backward uniqueness for the system of thermoelastic waves
with non-Lipschitz continuous coefficients   346
  M. Pivetta

**IV.3. Control and optimisation of nonlinear
evolutionary systems**   355
  (F. Bucci, I. Lasiecka)

Null boundary controllability of the Schrödinger equation
with a potential   357
  O. Arena and W. Littman

V.1. Inverse problems   363
  (Y. Kurylev, M. Yamamoto)

Optimal combination of data modes in inverse problems:
Maximum compatibility estimate   365
  M. Kaasalainen

Explicit and direct representations of the solutions of
nonlinear simultaneous equations   372
  M. Yamada and S. Saitoh

V.2. Stochastic analysis   379
  (D. Crisan, T. Lyons)

Information and asset pricing   381
  D.C. Brody

Solving backward stochastic differential equations using the
cubature method. Application to nonlinear pricing   389
  D. Crisan and K. Manolarakis

A spectral gap for the Brownian bridge measure on
hyperbolic spaces   398
  X. Chen, X.-M. Li and B. Wu

Individual path uniqueness of solutions of SDE   405
  A.M. Davie

Periodic homogenization with an interface 410
   M. Hairer and C. Manson

Discrete-time interest rate modelling 417
   L.P. Hughston and A. Macrina

Probabilistic representation for solutions of higher-order
elliptic equations 424
   M. Kelbert

On additive time-changes of Feller processes 431
   A. Mijatović and M. Pistorius

Statistical inference for differential equations driven by rough paths 438
   A. Papavasiliou

Constructing discrete approximations algorithms for financial
calculus from weak convergence results 445
   R.S. Tunaru

V.3. Coercivity and functional inequalities 453
   (D. Bakry, B. Zegarlinski)

Convexity along vector fields and applications to equations of
Monge-Ampére type 455
   M. Bardi and F. Dragoni

Phi-entropy inequalities and Fokker-Planck equations 463
   F. Bolley and I. Gentil

Isoperimetry for product of heavy tails distributions 470
   N. Gozlan, C. Roberto and P.-M. Samson

Hypoellipticity in infinite dimensions 479
   M. Hairer

Isoperimetry for spherically symmetric log-concave measures 485
   N. Huet

Operators on the Heisenberg group with discrete spectra   491
    J. Inglis

Liggett-type inequalities and interacting particle systems:
The Gaussian case   498
    J. Inglis, M. Neklyudov and B. Zegarlinski

Enlarging the functional space of decay estimates on semigroups   505
    C. Mouhot

The q-logarithmic Sobolev inequality in infinite dimensions   512
    I. Papageorgiou

V.4. Dynamical systems   521
    (J. Lamb, S. Luzzatto)

Homogeneous vector fields and meromorphic connections   523
    M. Abate

Period annuli and positive solutions of nonlinear boundary
value problems   530
    S. Atslega and F. Sadyrbaev

Global limit cycle bifurcations in a biomedical model   537
    V.A. Gaiko

V.5. Functional differential and difference equations   545
    (L. Berezansky, J. Diblik, A. Zafer)

Asymptotic analysis of gene regulatory networks with delay effects   547
    *Y. Nepomnyashchikh, A. Ponosov, A. Shindiapin
      and I. Shlykova*

On zero controllability of evolution equations by scalar control   554
    B. Shklyar

V.6. Mathematical biology   561
    (R. Gilbert)

New computer technologies for construction and numerical
analysis of mathematical models of molecular genetic
systems   563
*I.R. Akberdin, S.I. Fadeev, I.A. Gainova, F.V. Kazantsev,
V.K. Korolev, V.A. Likhoshvai and A.E. Medvedev*

Investigation of the acoustic properties of the cancellous bone   570
R.P. Gilbert, K. Hackl and S. Ilic

VI. Others   577
(Local Organising Committee)

The relationship between Bezoutian matrix and Newton's
matrix of divided differences   579
R.G. Airapetyan

Interpolation beyond the interval of convergence:
An extension of Erdos-Turan Theorem   585
H. Al-Attas and M.A. Bokhari

To solutions of the one non-divergent type parabolic
equation with double non-linearity   592
M. Aripov and S.A. Sadullaeva

Long-time behavior of periodic Navier-Stokes equations in
critical spaces   597
J. Benameur and R. Selmi

Regularity theory in asymptotic extensions of topological
modules and algebras   604
M.F. Hasler

Some fixed point theorems on the cone Banach spaces   612
E. Karapinar

A boundary value problem for 3D elliptic equations with
singular coefficients   619
E.T. Karimov

Numerical modelling of three-dimensional turbulent stream
of reacting gas, implying from rectangle form nozzle, on base
of $k - \varepsilon$ model of turbulence   626
   S. Khodjiev

Sequence spaces of invariant means and some matrix
transformations   630
   M. Mursaleen

Motion stabilization of a solid body with fixed point   636
   Z. Rakisheva

Session Organisers / Subeditors   643

Author Index   645

I.1. Complex variables and potential theory

Organisers: T. Aliyev, M. Lanza de Cristoforis, S. Plaksa, P. Tamrazov

This session has been devoted to recent advances in complex analysis and potential theory. It includes contributions on the global behaviour of rational functions, on the boundary behaviour of normal holomorphic functions and of Riemann mappings, estimates on the variation of conformal moduli of families of surfaces under the action of quasiconformal mappings, and on the behaviour of contour integral on nonrectifiable curves with applications to the solution of Riemann boundary value problems and contributions on the analysis of singular perturbation problems by a method alternative to those of asymptotic analysis.

Global mapping properties of analytic functions

Cabiria Andreian Cazacu

*Simion Stoilow Institute of Mathematics of Romanian Academy, P.O. Box 1-764,
RO-014700, Bucharest, Romania
E-mail: cabiria.andreian@imar.ro*

Dorin Ghisa

*York University, Glendon College, 2275 Bayview Avenue, Toronto, Canada
E-mail: dghisa@yorku.ca*

Every analytic function $f(z)$ is locally injective, except for the zeros of $f'(z)$, as well as for the multiple poles and essential singularities of $f(z)$. If E is the set of essential singularities of f, then the couple $(\widehat{\mathbb{C}} \setminus E, f)$ is a branched covering Riemann surface.

The specification of the leafs of such a surface allows one to put into evidence global mapping properties of f. Such a study has been done for rational functions in [2] and is initiated here for arbitrary analytic functions. The complexity of the problem is revealed gradually starting with some elementary examples, then introducing the concept of essential singularity of finite order and finally dealing with the general case.

The results obtained at every step are shown as representing completions to the Big Picard Theorem.

1. Introduction

It is known (see [2]) that for every rational function $f : \widehat{\mathbb{C}} \to \widehat{\mathbb{C}}$ of degree n there is a partition of $\widehat{\mathbb{C}}$ into n simply connected sets whose interiors Ω_k (fundamental domains) are mapped conformally by f onto $\widehat{\mathbb{C}} \setminus L$, where L is a slit. In fact f is conformal at every point of $\widehat{\mathbb{C}}$ except for its multiple poles and the zeros of $f'(z)$. Thus, the couple $(\widehat{\mathbb{C}}, f)$ is a branched covering Riemann surface of $\widehat{\mathbb{C}}$. We deal in this paper with transcendental analytic functions and show that they enjoy of similar properties, except that they have infinitely many fundamental domains. Since we'll make quite often reference to the Big Picard Theorem we use the abbreviation BPT for it.

2. Completion to BPT

We have proved in [4] that if (a_n) is a Blaschke sequence such that

$$\sum_{n=1}^{\infty}(1 - |a_n|) < \infty \tag{1}$$

and whose set of cluster points is a (generalized) Cantor set E, then for every point $e^{i\theta_0} \in E$, any neighborhood V of $e^{i\theta_0}$ contains infinitely many domains Ω_n (fundamental domains) which are mapped conformally by

$$B(z) = \prod_{n=1}^{\infty} \frac{\bar{a}_n}{|a_n|} \frac{a_n - z}{1 - \bar{a}_n z} \tag{2}$$

onto $\widehat{\mathbb{C}} \setminus L$, where L is a slit.

For every n,the mapping $B : \overline{\Omega}_n \to \widehat{\mathbb{C}}$ is surjective and consequently B assumes in V any value of $\widehat{\mathbb{C}}$ infinitely many times. This result comes as a completion to BPT since besides showing that in any neighborhood of the (possibly non isolated) essential singularity $e^{i\theta_0}$ the function B assumes any value infinitely many times (hence, no lacunary value), we have a precise description of the way the respective values are taken, i.e. by conformally mapping the fundamental domains onto $\widehat{\mathbb{C}} \setminus L$. Moreover, it is also stated that for this particular type of functions there is no lacunary value. Visualizations of different Blaschke product mappings can be found in [3].

It is also known (see [4]) that similar properties are valid for finite Blaschke quotients. Such a result can be generalized to infinite Blaschke quotients and it provides a completion to BPT for this type of functions too. For the next example it is known that a lacunary value exists, yet we'll be able to present a similar description of the orderly way in which every other value on the Riemann sphere is taken by the analytic function in any neighborhood of its isolated essential singularity.

The function we picked up is $w = f(z) = e^{1/z}$, which has the unique lacunary value $w = 0$. The point $z = 0$ is an essential isolated singularity for f. These are known facts, as well as some of the other affirmations we'll be making next, yet the manner in which the function f is treated as the canonical projection of a branched covering Riemann surface may be new.

Theorem 2.1. *The couple$(\widehat{\mathbb{C}} \setminus \{0\}, f)$, where$f(z) = e^{1/z}$ is a branched covering Riemann surface of $\widehat{\mathbb{C}}$ having countable many leafs, which are domains bounded by circles passing through the origin. Its group of covering transformations is an infinite cyclic group.*

Proof. We notice that with the substitution $\zeta = 1/z$ the line

$$\zeta = \xi + 2k\pi i, \xi \in R, k \in Z, \tag{3}$$

is the image of the real axis Γ_0 (as $k = 0$), or of the circle Γ_k of equation

$$|z + i/4k\pi| = 1/4|k|\pi (as \ \ k \neq 0). \tag{4}$$

The domain Ω_k bounded by two consecutive circles Γ_k and Γ_{k+1} is mapped conformally by $\zeta = 1/z$ onto the horizontal strip between $\zeta = \xi + 2k\pi i$ and $\zeta = \xi + 2(k+1)\pi i$, $\xi \in R$, which is mapped conformally by $w = e^\zeta$ onto the complex plane with a slit alongside the positive real half axis. Consequently, every domain Ω_k is mapped conformally by f onto the complex plane with a slit. We notice that the domains Ω_k contract themselves to $z = 0$ as $k \to \infty$ and the mapping $f : \overline{\Omega}_k \setminus \{0\} \to \widehat{\mathbb{C}} \setminus \{0\}$ is surjective. Moreover, for every k, the mapping f is conformal also on $\partial \Omega_k \setminus \{0\}$. $\qquad\square$

An illustration of these mappings can be seen in Fig. 1.

The function $f(z) = e^{1/z}$ is an automorphic function with respect to the infinite cyclic group generated by $U_1(z) = \frac{z}{2\pi i z + 1}$. If we denote by $U_k(z) = \frac{z}{2k\pi i z + 1}, k \in Z$, we can see easily that $U_0(z) = z, U_k^{-1}(z) = U_{-k}(z)$ and $U_k \circ U_j(z) = U_{k+j}(z)$ and $U_k(z)$ maps conformally every domain Ω_j onto the domain Ω_{k+j}. These remarks come again as a completion to BPT. The purpose of this section is to show that such a completion can be stated for much more general settings. We proceed gradually by restating first the conclusion of Theorem 2.1 for the function $\zeta(z) = \frac{1}{z - z_0}$, instead of $\zeta(z) = 1/z$.

Let us notice that the circle Γ_k mapped by $\zeta(z) = \frac{1}{z - z_0}$ on the line $\zeta = \xi + 2k\pi i$ is this time the circle of equation

$$z\bar{z} - (\frac{i}{4k\pi} + \bar{z}_0)z + (\frac{i}{4k\pi} - z_0)\bar{z} + |z_0|^2 - \frac{Imz_0}{2k\pi} = 0 \tag{5}$$

centered at $z_0 - i/4k\pi$ and passing through z_0. On the other hand, as k tends to ∞, the equation of Γ_k becomes $z\bar{z} - \bar{z}_0 z - z_0 \bar{z} + |z_0|^2 = 0$, which is $|z - z_0|^2 = 0$, i.e. $z = z_0$. Consequently, the domains Ω_k bounded by consecutive circles Γ_k and Γ_{k+1} contract themselves to the point z_0 as k tends to ∞ and each one of them is mapped conformally by the function $f(z) = Exp\{\frac{1}{z - z_0}\}$ onto $\mathbb{C}$ with a slit alongside the real positive half axis such that

$$f : \overline{\Omega}_k \setminus \{z_0\} \to \widehat{\mathbb{C}} \setminus \{0\} \tag{6}$$

is surjective. We have again a completion to BPT. The function $f(z)$ is automorphic with respect to the infinite cyclic group generated by

$$U_1(z) = [(1 + 2\pi i z_0)z - 2\pi i z_0^2]/[2\pi i z + (1 - 2\pi i z_0)]. \tag{7}$$

Now, if instead of $\zeta(z)$ we take an arbitrary meromorphic function $h(z)$, then every pole z_0 of $h(z)$ is an essential singularity for $f(z) = Exp\{h(z)\}$. We call z_0 isolated essential singularity of first order of f.

Theorem 2.2. *Let z_0 be an isolated essential singularity of the first order of $f(z) = Exp\{h(z)\}$, where z_0 is a pole of order p of h. Then any neighborhood V of z_0 is divided in p sectors by arcs starting in z_0 such that every sector contains infinitely many fundamental domains of $(\widehat{\mathbb{C}}, f)$.*

Proof. Indeed, in a small neighborhood V of z_0 we have

$$h(z) = \varphi(z)/(z - z_0)^p, \tag{8}$$

where p is the order of z_0 as a pole of $h(z)$ and $\varphi(z)$ is holomorphic in V, with $\varphi(z) \neq 0$ in V. It is easier to grasp the behavior of $f(z)$ in the neighborhood of z_0 if we study first the behavior of $g(z) = Exp\{z^p\}$ in the neighborhood of ∞. We are looking for the curves $\gamma_k^{(j)}$ which are mapped by z^p onto the lines $\zeta = \xi + 2k\pi i$, i.e. for which $Re\{z^p\} = \xi$ and $Im\{z^p\} = 2k\pi$. This last equation has the form:

$$px^{p-1}y - \frac{p(p-1)(p-2)}{3!}x^{p-3}y^3 + ... = 2k\pi \tag{9}$$

For $p = 2$ the equation is $xy = k\pi$, and it represents for k and $-k$ four hyperbola branches $\gamma_k^{(j)}$ and $\gamma_{-k}^{(j)}$, $j = 1, 2$ having as asymptotes the two coordinate axes. For $k = 0$ we get both axes. For $p = 3$ the equation is $3x^2y - y^3 = k\pi$ and it represents the family of *hyperbolas* having as asymptotes the real axis and the lines $y = \pm\sqrt{3}x$. For $k = 0$ the equation represents these three lines. Now the pattern is obvious for an arbitrary p.

If instead of $Exp\{z^p\}$ we take $Exp\{1/(z - z_0)^p\}$ the point $z = \infty$ is moved to $z = z_0$ and the curves $\gamma_k^{(j)}$ mapped by $1/(z - z_0)^p$ onto the line $\zeta = \xi + 2k\pi i$ are closed curves passing through z_0, situated between consecutive rays issuing from z_0. Since $|z - z_0|^p \leq \frac{1}{2|k|\pi}$ for z on $\gamma_k^{(j)}$ these curves contract themselves to z_0 as $|k| \to \infty$. One of the respective rays is parallel to the positive x-half axis and any two consecutive rays make between them an angle of $2\pi / p$. $\qquad\square$

An illustration of these mappings for $p = 3$ can be seen in Fig. 2.

Now we can introduce into consideration the function $\varphi(z)$ appearing in the expression of $h(z)$ in the neighborhood of z_0. The multiplication with $\varphi(z)$ has as effect a topological deformation of the curves $\gamma_k^{(j)}$ into some curves $\Gamma_k^{(j)}$, which are mapped by $h(z)$ onto the lines $\zeta = \xi + 2k\pi i$ and consequently by $f(z)$ onto the positive real half axis. All $\Gamma_k^{(j)}$ pass through z_0 and have no other common point. They contract themselves to z_0 as $k \to \infty$, hence they will be all included in V for $|k|$ greater than a given k_0.

Let us denote by $\Omega_k^{(j)}$, $j = 0, 1, ..., p-1$, the domains bounded by $\Gamma_k^{(j)}$ and $\Gamma_{k+1}^{(j)}$. It is obvious that $\Omega_k^{(j)}$ are fundamental domains and they are all included in V for $|k| > k_0$, which proves completely the theorem. There are conformal mappings $U_k^{(j)}$ with the domain and range in V mapping $\Omega_l^{(m)}$ onto $\Omega_{k+l}^{(j+m)(mod\,p)}$, $|k|$ and $|k+l|$ greater than k_0, defined by

$$U_k^{(j)}(z) = f|_{\Omega_{k+l}^{(j+m)(mod\,p)}}^{-1} \circ f(z), z \in \Omega_l^{(m)} \tag{10}$$

We notice that $h \circ U_k^{(j)}(z) = h(z)$, hence $f \circ U_k^{(j)}(z) = f(z)$, for every $z \in \Omega_l^{(m)}$ and $j = 0, 1, ..., p-1$. If all the domains involved are in V, then $U_k^{(j)} \circ U_{k'}^{(m)} = U_{k+k'}^{(j+m)(mod\,p)}$. Obviously, these are all covering transformations of $(\widehat{\mathbb{C}}, f)$, yet they might not form a group, as long as $V \neq \mathbb{C}$. However, a completion to BPT for f with the isolated essential singularity of the first order in z_0 follows again.

Suppose now that $f(z) = Exp\{g(z)\}$ and z_0 is an isolated essential singularity of the first order of $w = g(z)$. We call z_0 essential isolated singularity of second order of f, etc. By the previous section, every small neighborhood V of z_0 is divided in a number p of sectors by arcs starting in z_0 such that every sector contains infinitely many fundamental domains $\Omega_n^{(j)}$, $j = 1, 2, ..., p$, $n \in N$ of g. Each one of the domains $\Omega_n^{(j)}$ is mapped by g onto the complex plane with a slit alongside the positive real half axis. On the other hand, every strip

$$H_k = \{w \in \mathbb{C} \mid w = u + iv, u \in R, 2k\pi \leq v < 2(k+1)\pi\} \tag{11}$$

is mapped by e^w onto the complex plane with a slit alongside the real positive half axis. Then $\Omega_n^{(j)} = \cup_{k=1}^{\infty} G_{n,k}^{(j)}$, where $G_{n,k}^{(j)} = \Omega_n^{(j)} \cap g^{-1}(H_k)$ is mapped conformally by g onto the strip H_k. Finally, the function f maps every $G_{n,k}^{(j)}$ onto the slited plane.

This reasoning can be repeated for an essential singularity of any order, the conclusion being the same. This conclusion is a completion to BPT and we have:

Theorem 2.3. *Every neighborhood of an isolated essential singularity of any finite order of an analytic function can be divided in sectors containing each one of them infinitely many fundamental domains.*

3. The Main Theorem

The next theorem concerns an arbitrary essential singularity, therefore it is in some respect more general. However, it lacks specificity regarding the fundamental domains. Suppose that a is an arbitrary isolated essential singularity of a function $w = f(z)$, which is analytic in a reduced neighborhood $V \setminus \{a\}$ of a. Then $f'(z)$ is also analytic in $V \setminus \{a\}$ and has an isolated essential singularity at a. By BPT there are two possibilities:

i). $f'(z) \neq 0$ in $V \setminus \{a\}$ (we take a smaller V, if necessary), i.e. 0 is a lacunary value for f'.

ii). $f'(z)$ assumes the value 0 infinitely many times in $V \setminus \{a\}$.

In the first case, there are again two possibilities:

a). The function f has a lacunary value $l + im$.

b). There is no lacunary value for f.

Let us denote $\mathbb{Q} + im = \{r_n + im \mid r_n \in \mathbb{Q}\}$, $\mathbb{R} + im = \{x + im \mid x \in \mathbb{R}\}$ and take the pre-image by f of $\mathbb{R} + im$ in the case a) and of $\mathbb{R}$ in the case b). Let $\{z_{n,j}\}_{j=1}^{\infty}$ the pre-image by f of $r_n + im$ in the case a) and of r_n in the case b). For every n, there are infinitely many points $z_{n,j_n} \in V$ and all the sequences $\{z_{n,j_n}\}_{j_n=1}^{\infty}$ have the limit a. Thus, all the components of $f^{-1}(\mathbb{R} + im)$, respectively of $f^{-1}(\mathbb{R})$ must have the ends at a.

We can simply add a to these components and say that they pass through a. Those components containing only $z_{n,j}$ with $j = j_n$ are entirely included in V. Let us show that there are infinitely many such components. Indeed, due to the fact that $f'(z) \neq 0$ in $V \setminus \{a\}$, the function f is locally injective there. Therefore, every point $r_n + im \in \mathbb{R} + im$, respectively every point $r_n \in \mathbb{R}$ is included in an interval I_n such that $f^{-1}(I_n)$ is formed with infinitely many Jordan arcs $\gamma_{n,j}$ such that $z_{n,j_n} \in \gamma_{n,j_n}$. If I is a compact interval of $\mathbb{R} + im$, respectively of $\mathbb{R}$, then there is a finite covering of I with $f(\gamma_{n,j_n})$ and there are infinitely many arcs γ_{n,j_n} such that $f(\gamma_{n,j_n}) \cap I = \varnothing$. Performing simultaneous continuations over $\mathbb{R} + im$, respectively over $\mathbb{R}$ of the arcs γ_{n,j_n} we obtain curves Γ_{n,j_n} which must all pass through a. Indeed, for every one of them the continuation is possible as $w \to \infty$, since there are no singular points in V except for a. The only situation where the continuation stops is when $w \to l + im$. Then, again Γ_{n,j_n} approaches a and $f(\Gamma_{n,j_n})$ must be one of the intervals in which the point $l + im$ divides the line $\mathbb{R} + im$. In the case b), we have $f(\Gamma_{n,j_n}) = \mathbb{R}$.

We notice that these curves cannot intersect in a point $z_0 \in V$, $z_0 \neq a$, since we would have $f'(z_0) = 0$. They are all closed curves. If a point $z_{n,j}$ belongs to the domain bounded by such a curve, then the corresponding $\Gamma_{n,j} \setminus \{a\}$ is included in that domain. Thus these curves are divided in families of curves, as in the examples of the previous section. They border domains which are mapped by f onto half planes. We notice that in the case b) we are brought to a contradiction, since the image by f of the domain between two $\Gamma_{n,j}$ should belong simultaneously to the upper and to the lower half plane. Consequently, if $f'(z) \neq 0$, then necessarily the function f has a lacunary value. All the examples we have studied agree with this conclusion.

If 0 is not a lacunary value of f', let us denote by b_n the zeros of $f'(z)$ situated in V counted such that $(|f(b_n)|)$ is a non decreasing sequence. If for an n we have $|f(b_n)| = |f(b_{n+1})| = \ldots = |f(b_{n+k})|$, then the counting is performed such as to make possible the connections described next. We connect all $w_n = f(b_n)$ by a non self intersecting curve L as follows. If $w_1 = 0$ we connect it with w_2 by the segment between them. If $w_1 \neq 0$ and $|w_1| = |w_2|$ we connect w_1 and w_2 by the arc of the circle $|w| = |w_1|$ which does not cross the positive real axis. If $|w_1| < |w_2|$, then we draw first the segment of the ray through $w_1 = r_1 e^{i\alpha_1}$ from w_1 to $|w_2| e^{i\alpha_1}$ and then an arc of the circle centered at the origin, which does not cross the real axis, from this point to w_2.

This process can be continued indefinitely, w_{n+1} being reached at the n-th step. It is obvious that the curve L built in this way has no self intersection and passes through all the points w_n. We notice that (w_n) cannot accumulate at a finite point, since then the local inverse of $f'(z)$ would be identically zero, which is a contradiction. Consequently, L is an unbounded curve. This implies that (b_n) cannot accumulate at a point in V different of a.

The pre-image of L by f contains all the points b_n. In the neighborhood of every zero b_n of order $k_n - 1$ of $f'(z)$ we have $f(z) = w_n + (z - b_n)^{k_n} \varphi_n(z)$, where φ_n is analytic and $\varphi_n(b_n) \neq 0$, in other words, the pre-image by f of L has k_n branches issuing from b_n. There are no other self intersections of the pre-image of L, since at every point b of self intersection we must have $f'(b) = 0$, or b should be a multiple pole of f.

All those branches issuing from b_n are mapped by f on L, therefore the domains bounded by such arcs are mapped conformally by f onto the complex plane with a cut alongside the part of L determined by the image of those arcs. In other words, these are fundamental domains of f. All of them are included in V and it is obvious that their number is infinite. Thus we have:

Theorem 3.1. *Every neighborhood V of an isolated essential singularity of the analytic function f contains infinitely many fundamental domains of f.*

Obviously, the theorem applies for every isolated essential singularity of f. We notice that if f is a rational function of $f_1, f_2, \ldots, f_k$ then the essential singularities of f are exactly those of all f_j, $j = 1, 2, \ldots, k$ and the fundamental domains of f included in a small neighborhood of each one of its singularity *do not differ too much* of those of the corresponding f_j which generated that singularity. We illustrate this phenomenon by the Fig. 3, in which $f(z) = e^z e^{1/z}$.

Remark: The technique used in Theorem 3.1 allows one to study the behavior of f in the neighborhood of a non isolated essential singularity $a = \lim_{n \to \infty} a_n$, where a_n are isolated essential singularities of f. Every neighborhood V of a contains all a_n for n greater than a given k. Then V is a neighborhood of every $a_n, n > k$, hence it contains infinitely many fundamental domains of f.

Acknowledgments

The authors are grateful to Cristina Ballantine for providing computer generated graphics for this paper, as well as to Paul Gauthier for the useful comments he made on the topic of this paper during The ISAAC Congress, 2009.

References

1. Ahlfors, L.V., *Complex Analysis*, International Series in Pure and Applied Mathematics, 1979
2. Ballantine, C. and Ghisa, D. *Global Mapping Properties of Rational Functions*, in this volume
3. Ballantine, C. and Ghisa, D., *Color Visualization of Blaschke Product Mappings*, to appear in Complex Variables and Elliptic Equations
4. Ballantine, C. and Ghisa, D., *Color Visualization of Blaschke Self-Mappings of the Real Projective Plan*, to appear in Rev. Roum. Math. Pures et Appl. Vol. **54**, No.5-6, 2009
5. Barza, I. and Ghisa, D., *The Geometry of Blaschke Product Mappings*, Proceedings of the 6-th International ISAAC Congress, Ankara, 2007

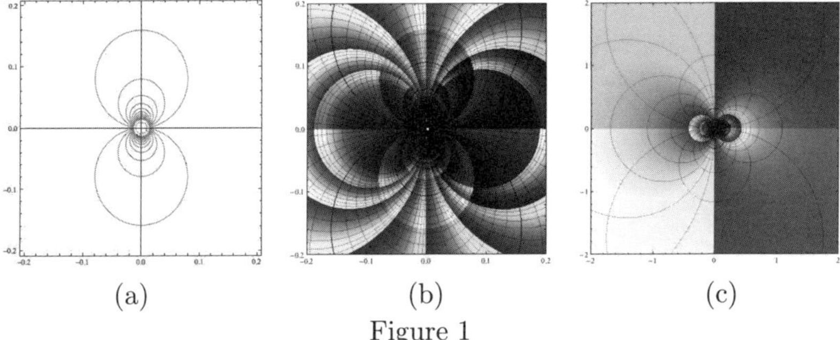

(a)   (b)   (c)

Figure 1

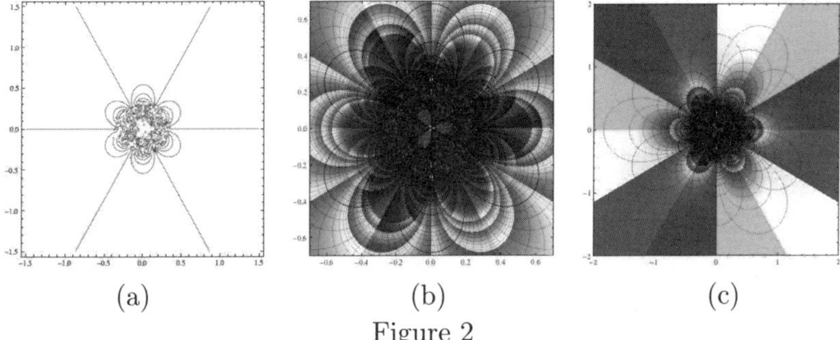

(a)   (b)   (c)

Figure 2

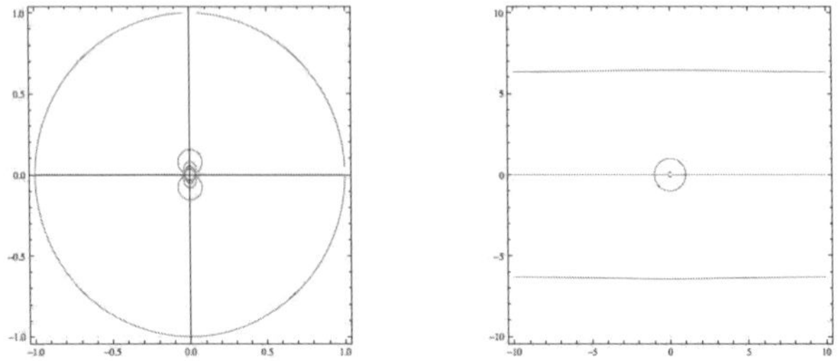

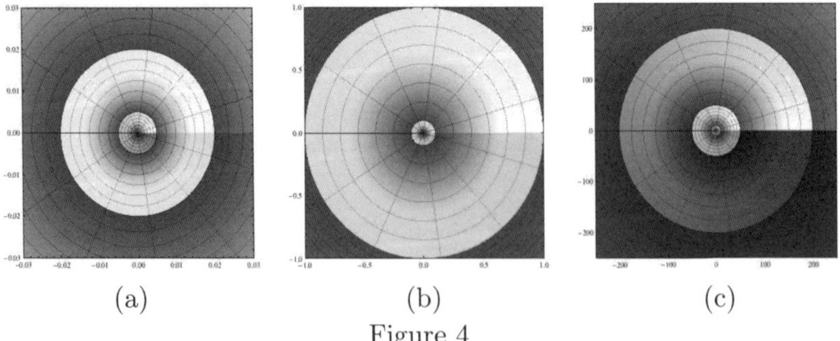

(a)   (b)

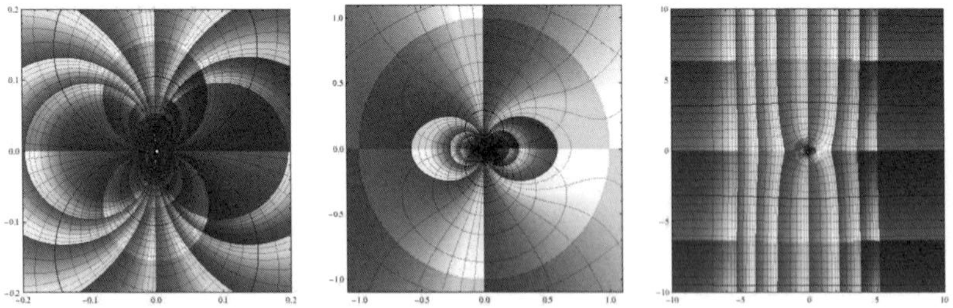

(c)   (d)   (e)

Figure 3

(a)   (b)   (c)

Figure 4

Global mapping properties of rational functions

Christina Ballantine

College of the Holy Cross, USA
E-mail: cballant@holycross.edu

Dorin Ghisa

York University, Glendon College, 2275 Bayview Avenue, Toronto, Canada
E-mail: dghisa@yorku.ca

We investigate the fundamental domains of rational functions and provide visualizations for relevant examples. The fundamental domains give a thorough understanding of the global mapping properties of the functions studied.

1. Introduction

Any rational function $f(z)$ can be viewed as the canonical projection of a branched covering Riemann surface $(\widehat{\mathbb{C}}, f)$ of the Riemann sphere $\widehat{\mathbb{C}}$. Indeed, f is locally injective in the neighborhood of every point $z \in \widehat{\mathbb{C}}$, except for the points z_k, which are solutions of the equation $f'(z) = 0$ and the points c_j which are multiple poles of f. In [4] we have studied global mapping properties of Blaschke products, showing that every Blaschke product $w = B(z)$ of degree n induces partitions of $\widehat{\mathbb{C}}$ into n sets whose interior is mapped conformally by B onto $\widehat{\mathbb{C}} \setminus L$, where L is a cut. Following [1, p. 98] we called these sets *fundamental regions* or *domains*.

The fundamental regions have played an important role in the theory of automorphic functions. In fact, a fundamental region of a group of transformations is a fundamental region of an automorphic function with respect to that group. These regions characterize the global mapping properties of automorphic functions. In this paper we show that any rational function f has similar properties. Moreover, once the fundamental regions of f are known, invariants of f can be found, *i.e.* mappings U_k of the Riemann sphere on itself such that, for every $z \in \widehat{\mathbb{C}}$, we have $f \circ U_k(z) = f(z)$. Obviously, the set of these invariants is a cyclic group of order n. They are the cover transformations (see [2, p. 37]) of $(\widehat{\mathbb{C}}, f)$ and we can extend the concept of automorphic function to such a group. Using this terminology, the main result of this paper shows that any rational function f is an automorphic function with respect to the group of cover transformations of $(\widehat{\mathbb{C}}, f)$. The proof is constructive and uses the technique of simultaneous continuations developed in [4] in order to find fundamental regions for f.

To visualize the fundamental regions, we color a set of annuli centered at the origin of the w-plane in different colors with saturation increasing counter-clockwise (*i.e.*, determined by the argument of each point) and brightness increasing outward (*i.e.*, determined by the absolute value each point) and impose the same color, saturation and brightness to the pre-image of every point in these annuli.

2. Mapping Properties of the Second Degree Rational Functions

A study of the second degree rational functions can be found in [6, p. 266]. We use Nehari's results in order to illustrate some of the mapping properties of these functions. The main result found in [6] relevant to this topic is that any mapping $w = f(z) = \dfrac{a_1 z^2 + a_2 z + a_3}{b_1 z^2 + b_2 z + b_3}$ can be written under the form

$$f(z) = S_2 \circ T \circ S_1, \tag{1}$$

where S_1 and S_2 are Möbius transformations and $\eta = T(\zeta) = \zeta^2$. Indeed, to prove this statement we only need to determine six essential parameters of the two unknown Möbius transformations S_1 and S_2 such that (1) is true, which is allays possible.

The function $\zeta = S_1(z)$ transforms the z-plane into the ζ-plane, such that a circle or a line L corresponds to the real axis from the ζ-plane. The function $\eta = T(\zeta) = \zeta^2$ transforms each one of the upper and the lower half-planes of the ζ-plane into the whole η-plane with a cut alongside the real half-axis. Finally, the function $w = S_2(\eta)$ transforms the η-plane into the w-plane and the real half-axis into an arc of a circle or a half line L'. Summing up, f maps conformally each one of the two domains determined by L onto the whole w-plane with a cut alongside L'. Thus, for such a function f, the fundamental domains can allays be taken the two domains mapped by S_1 onto the upper and the lower half planes.

In Figure 1 we illustrate the case where $f(z) = \dfrac{(1+i)z^2 + 4z + 1 - i}{(2+i)z^2 + 6z + 2 - i}$. In this case $S_1(z) = \dfrac{z+1}{z-i}$, $S_2(z) = \dfrac{z+i}{2z+i}$. The pre-image of the real axis under S_1 is the line $z = \dfrac{1+ti}{t-1}$. The image of the positive real half axis under S_2 is the semicircle of radius 0.25 centered at 0.75. The fundamental domains of f are those defined by S_1.

In Figure 1(b) we consider colored annuli centered at $(0.75, 0)$. We visualize the fundamental domains of f in Figure 1(a), by considering preimages of these annuli under f.

For an example of the case where the pre-image of the real axis under S_1 is a circle, as well as an extended version of the article, we direct the reader to the web site of the project [5].

3. Mapping properties of Blaschke Quotients

In [3] we studied the mapping properties of Blaschke quotients B of a special type, namely such that for every $z \in \widehat{\mathbb{C}}$, $B \circ h(z) = h \circ B(z)$, where $h(z) = -1/\overline{z}$. Such a rational function has the particularity that its poles and zeros appear in pairs which are opposite to each other and if z_k is a pole of order p of B, then $1/\overline{z}_k$ is a zero of order p of B and vice-versa. The point $z = 0$ is a zero or a pole of B of an odd order and therefore ∞ is a pole, respectively a zero, of the same order.

The main result of [3] shows that, for a Blaschke quotient of degree n of such a type, there is a partition of $\widehat{\mathbb{C}}$ in $2n$ simply connected sets such that the interior of each one of them is mapped conformally by B either on the open unit disc (i-set), or on the exterior of the closed unit disc (e-set). The interior of the union of an i-set and an adjacent e-set is mapped conformally by B on the Riemann sphere with a slit. The map is continuous (with respect to the spheric metric) on the borders, except for the branch points. Here, we prove that a similar property holds for any finite Blaschke quotient.

Let $B(z) = B_1(z)/B_2(z)$ be a Blaschke quotient of degree n, *i.e.* the quotient of two finite Blaschke products B_1 and B_2 of degrees n_1, respectively n_2, such that $\max\{n_1, n_2\} = n$. The function B is locally injective, except for the set of points $H_1 = \{b_1, b_2, ..., b_m\}$, which are solutions of the equation $B'(z) = 0$. Consequently, $(\widehat{\mathbb{C}}, B)$ is a branched covering Riemann surface of $\widehat{\mathbb{C}}$ having H_1 as set of branch points. In other words, $(\widehat{\mathbb{C}} \setminus H_1, B)$ is a smooth covering Riemann surface of $\widehat{\mathbb{C}}$.

Theorem 3.1. *For every Blaschke quotient B of degree n there is a partition of $\widehat{C}$ into n sets symmetric with respect to the unit circle whose interior Ω_k is mapped each one conformally by B on $\widehat{\mathbb{C}} \setminus L$, where L is a cut. Moreover, $B : \overline{\Omega}_k \to \widehat{C}$ is surjective.*

Proof. Let $H_2 = \{z_1, z_2, ..., z_n\}$ be the solutions of the equation $B(z) = e^{i\theta}$, where $\theta \in R$ has been chosen such that $H_1 \cap H_2 = \emptyset$. It is obvious that such a choice is always possible. Since the image of the unit circle by B is the unit circle, at least one of the points z_k belongs to the unit circle. Also, since $B(1/\overline{z}) = 1/\overline{B(z)}$, the solutions which are not on the unit circle, must be two by two symmetric with respect to the unit circle.

If we perform simultaneous continuation from every z_j over the unit circle (starting from $e^{i\theta}$), we obtain arcs $\gamma_{j,j'}$ starting at $z_j \in H_2$ and ending at some point $z_{j'} \in H_2$. Some of these arcs might cross each other, but this can happen only at the points in H_1 since these are the only points where the injectivity of $B(z)$ is violated.

Let $W = \{w_1, w_2, ..., w_p\}$, where $w_k = B(b_k)$, $|b_k| < 1, b_k \in H_1$ and w_k are not points of intersection of $\gamma_{j,j'}$. We connect $e^{i\theta}, w_1, ..., w_p$ by a polygonal line Γ with no self intersection and perform simultaneous continuation over Γ from all $z_j \in H_2$. The domains bounded by the pre-image of Γ and the arcs $\gamma_{j,j'}$ are mapped by B either on the unit disc (i-domains) or on the exterior of the unit circle (e-domains). Indeed, every one of these domains $\Omega_{j,j'}$ is bounded by an arc $\gamma_{j,j'}$ whose image by B is the unit circle, and by an arc having the end points in z_j and $z_{j'}$ whose image by B is a part of Γ. The previous affirmation follows from the conformal correspondence theorem (see [6, p. 154]). It is obvious that every i-domain has a symmetric e-domain with respect to the unit circle and vice-versa. An i-domain and an adjacent e-domain are always separated by an arc $\gamma_{j,j'}$ and their union to which the open $\gamma_{j,j'}$ is added as a point set constitutes a fundamental domain Ω_j of B. If we denote $L = \Gamma \cup \widetilde{\Gamma}$, where $\widetilde{\Gamma}$ is the symmetric of Γ with respect to the unit circle, then it is obvious that B maps conformally every Ω_k on $\widehat{\mathbb{C}} \setminus L$ and the mapping $B : \overline{\Omega}_k \to \widehat{\mathbb{C}}$ is surjective, which completely proves the theorem. $\qquad\qquad\square$

To illustrate, let $B_1(z) = \left(\dfrac{\overline{a_1}}{|a_1|} \dfrac{z - a_1}{\overline{a_1}z - 1} \right)^2 \cdot \dfrac{\overline{a_2}}{|a_2|} \dfrac{z - a_2}{\overline{a_2}z - 1}$ and $B_2(z) = \left(\dfrac{\overline{b}}{|b|} \dfrac{z - b}{\overline{b}z - 1} \right)^2$ with $a_1 = \dfrac{1}{4}e^{\frac{\pi i}{6}}$, $a_2 = \dfrac{1}{3}e^{-\frac{\pi i}{5}}$, $b = \dfrac{1}{2}e^{\frac{2\pi i}{3}}$. Then,

$$B(z) = \frac{B_1(z)}{B_2(z)} = \frac{e^{-\frac{4i\pi}{5}} \left(-\frac{1}{4}e^{\frac{i\pi}{6}} + z \right)^2 \left(-\frac{1}{3}e^{-\frac{i\pi}{5}} + z \right) \left(-1 + \frac{1}{2}e^{-\frac{2i\pi}{3}}z \right)^2}{\left(-\frac{1}{2}e^{\frac{2i\pi}{3}} + z \right)^2 \left(-1 + \frac{1}{4}e^{-\frac{i\pi}{6}}z \right)^2 \left(-1 + \frac{1}{3}e^{\frac{i\pi}{5}}z \right)}$$

is a Blaschke quotient of degree 5.

In Figure 2(b) we consider colored annuli centered at the origin. Figure

2(a) shows the pre-images of these annuli. For a better view, each figure shows several zoomed images.

In the next section we show that a similar property is true for any rational function.

4. Mapping Properties of Arbitrary Rational Functions

Let $w = f(z)$ be a rational function with zeros $a_1, a_2, ..., a_p$ and poles $b_1, b_2, ..., b_q$. Let α_i be the multiplicity of a_i and β_j be the multiplicity of b_j. Then, the *degree of f* is $n = \max\{u, v\}$, where $u = \alpha_1 + \alpha_2 + ... + \alpha_p$ and $v = \beta_1 + \beta_2 + ... + \beta_q$.

If $\lim\limits_{z \to \infty} f(z) = 0$, $n = v$ and $a_0 = \infty$ is said to be a zero of multiplicity $\alpha_0 = n - u$ of f. If $\lim\limits_{z \to \infty} f(z) = \infty$, $n = u$ and $b_0 = \infty$ is said to be a pole of multiplicity $\beta_0 = n - v$ of f.

Theorem 4.1. *Every rational function f of degree n defines a partition of $\widehat{\mathbb{C}}$ into n sets whose interior is mapped conformally by f on $\widehat{\mathbb{C}} \setminus L$, where L is a cut. The mapping can be analytically extended to the boundaries, except for a number $\leq n$ of common points z_j of those boundaries in the neighborhood of which f is of the form*
(i) $f(z) = w_j + (z - z_j)^k h(z)$, when $f(z_j) = w_j$,
(ii) $f(z) = (z - z_j)^{-k} h(z)$, when $f(z_j) = \infty$,
(iii) $f(z) = z^{-k} h(z)$, when $z_j = \infty$ and $f(\infty) = \infty$,
with $h(z)$ analytic and $h(z_j) \neq 0$, $k \geq 2$. In other words, $(\widehat{\mathbb{C}}, f)$ is a branched covering Riemann surface of $\widehat{\mathbb{C}}$ and the branch points are z_j.

Proof. Since $\lim\limits_{z \to a_j} f(z) = 0$, we can find a positive number r small enough such that the pre-image Γ of the circle γ_r centered at the origin and of radius r will have disjoint components Γ_j, each containing just one zero a_j. If ∞ is a zero of f, then the respective component Γ_0 must be traversed clockwise, in order for ∞ to remain on its left. We understand by the domain bounded by Γ_0 (if Γ_0 exists) that component of $\widehat{\mathbb{C}}$ defined by Γ_0 which contains ∞. For the opposite orientation of Γ_0 we have a curve containing all the other components Γ_j.

Moreover, we can choose the above r such that $f'(z) = 0$ has no solution in the closed domain bounded by Γ_j except maybe for a_j. Then, for an arbitrary $\theta \in R$, the equation $f(z) = re^{i\theta}$ has exactly α_j distinct solutions on Γ_j. Now, consider the pre-image by f of the ray inside γ_r determined by $re^{i\theta}$. In the domain bounded by Γ_j it consists of a union of α_j Jordan arcs having in common only the point a_j and connecting a_j to the solutions of $f(z) = re^{i\theta}$ on Γ_j, $j = 0, 1, 2, ..., p$ (see [1, p. 131– 133]).

Let c_k, $k = 1, 2, ..., m$, be the solutions of the equation $f'(z) = 0$ external to all Γ_j, and let $w_k = f(c_k) = r_k e^{i\theta_k}$. Suppose that $r_1 \leq r_2 \leq ... \leq r_m$. When $r_k = r_{k+1}$, then we take $\theta_k < \theta_{k+1}$, for every k. We perform simultaneous continuation starting from all a_j over a curve L from the w-plane in the following way. We take first the pre-image by f of the segment from 0 to $r_1 e^{i\theta_1}$. This is a union of arcs, α_j of which are starting in a_j, $j = 0, 1, 2, ..., p$. At least one of these arcs is connecting one of the a_j with c_1. If $r_1 = r_2$, then we take the pre-image of the shortest arc between w_1 and w_2 of the circle centered at the origin and having the radius r_1 (if $w_1 = -w_2$, we go counter-clockwise on that circle), etc. If $r_k < r_{k+1}$, we take the pre-image by f of the union of the arc of circle centered at the origin and having the radius r_k, between w_k and $r_k e^{i\theta_{k+1}}$, and the segment between this last point and w_{k+1}. After the point w_m has been reached, if f has at least one multiple pole, we take the pre-image of the ray from w_m to ∞. If f has no multiple pole, then the end of L is w_m and therefore L is a finite path. In this way we build in a few steps the path L and the simultaneous continuation over L starting from all a_j. The continuation arcs can have in common only points a_k, b_k or c_k, and all b_k and c_k are reached by several pre-image arcs. Indeed, if two such arcs meet in a point c, then they are both mapped by f on the same sub-arc of L starting in $f(c)$. One of the following four situations may happen:

a) $f(c) = 0$ and $f'(c) = 0$, hence c coincides with a multiple zero a_k. Then f has the expression (i) with $w_0 = 0$ in a neighborhood of $c = z_j$.

b) $f(c) \neq 0$ and $f'(c) = 0$, hence c coincides with a c_k. Then f has the expression (i) with $w_0 = f(c)$ in a neighborhood of $c = z_j$.

c) $f(c) = \infty$ and c is a multiple pole b_k of f. Then f has the expression (ii) in a neighborhood of $c = b_k = z_j$.

d) $c = \infty$. Then f has the expression (iii) in a neighborhood of ∞.

On the other hand, every c_k and b_k must be reached by some continuation arcs, since $f(c_k) \in L$ and $f(b_k) \in L$. More exactly, there are as many continuation arcs starting in c_k as the multiplicity of c_k as zero of the equation $f'(z) = 0$ and there are as many continuation arcs starting in b_k as the multiplicity of b_k as a pole of f. The arcs starting in simple zeros of f border exactly n bounded and/or unbounded domains Ω_k (fundamental domains) which are mapped conformally by f on the w-plane from which the curve L has been removed. This is a corollary of the boundary correspondence theorem (see [6, p. 154]). If we denote by $\overline{\Omega}_k$ the closure of Ω_k, then it is obvious that $\widehat{\mathbb{C}} = \cup_{k=1}^{n}\overline{\Omega}_k$. With the notation $A_k = \overline{\Omega}_k \setminus \cup_{j=1}^{k-1}\overline{\Omega}_j$ we have the partition in the statement of the theorem. $\qquad\square$

In Figure 3 we illustrate the theorem above using the function $f(z) = \dfrac{z^3(z+2)}{(z-i)^4(z+3-i)^3}$. In this case the saturation of the annuli increases starting at the polygonal line L.

Finally, we examine the case in which f is a polynomial of degree n. Then the unique pole of f is ∞ and it has multiplicity n. Hence, the ray from w_m to ∞ has as pre-image n infinite arcs and all the domains Ω_k are unbounded. For a polynomial $P(z) = a_0 z^n + a_1 z^{n-1} + ... + a_n$, $a_0 \neq 0$, we can describe these infinite arcs. Suppose that $\arg a_0 = \alpha$ and $\arg c_m = \beta$ and let $z_k(t)$, $t > 0$, be the parametric equation of one of these arcs. Then $P(z_k(t)) = a_0[z_k(t)]^n[1 + a_1/z_k(t) + ...]$ and $\arg P(z_k(t)) = \beta$. In other words, $\alpha + n\arg z_k(t) + o(t) = \beta + 2j\pi$, $\lim\limits_{t\to\infty} o(t) = 0$. Hence $\lim\limits_{t\to\infty}\arg z_k(t) = \dfrac{\beta-\alpha}{n} + \dfrac{2j\pi}{n}$. Thus, the arcs $z_k(t)$ tend asymptotically to the rays of slope $\dfrac{\beta-\alpha}{n} + \dfrac{2j\pi}{n}$, $j = 0,1,...,n-1$. This leads to the following theorem.

Theorem 4.2. *Every polynomial P of degree n defines a partition of $\widehat{\mathbb{C}}$ into n unbounded regions such that the interior of every region is mapped conformally by P on $\widehat{\mathbb{C}} \setminus L$, where L is a cut. The mapping can be extended analytically to L, except for a finite number of points, such that $(\widehat{\mathbb{C}}, P)$ is a branched Riemann covering of $\widehat{\mathbb{C}}$ having those points as branch points. The fundamental domains of $(\widehat{\mathbb{C}}, P)$ are bounded by arcs which tend asymptotically to n rays, every two consecutive rays forming an angle of $2\pi/n$.*

We illustrate the theorem in Figure 4 using $f(z) = \dfrac{z^7}{7} - z$.

References

1. Ahlfors, L.V., *Complex Analysis*, McGraw-Hill, 1979
2. Ahlfors, L.V., Sario, L, *Riemann Surfaces*, Princeton University Press, 1960
3. Ballantine, C. and Ghisa, D., *Color Visualization of Blaschke Self-Mappings of the Real Projective Plane*, Revue Roumaine Math. Pure Appl, 2009
4. Barza, I, Ghisa, D., *Blaschke Product Generated Covering Surfaces*, Mathematica Bohemica, 2009
5. http://math.holycross.edu/~cballant/complex/complex-functions.html
6. Nehari, Z., *Conformal Mappings*, International Series in Pure and Applied Mathematics, 1952

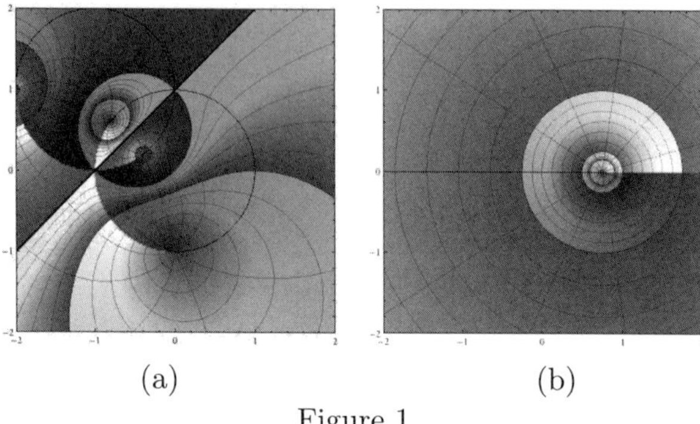

(a)   (b)

Figure 1

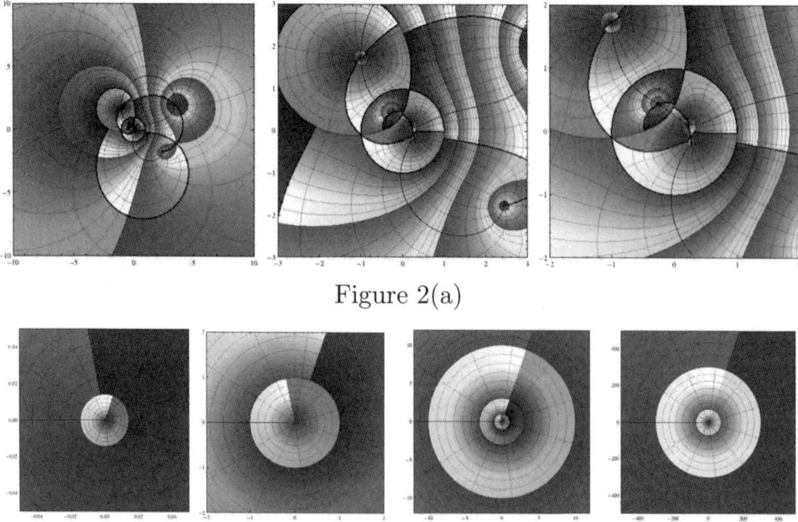

Figure 2(a)

Figure 2(b)

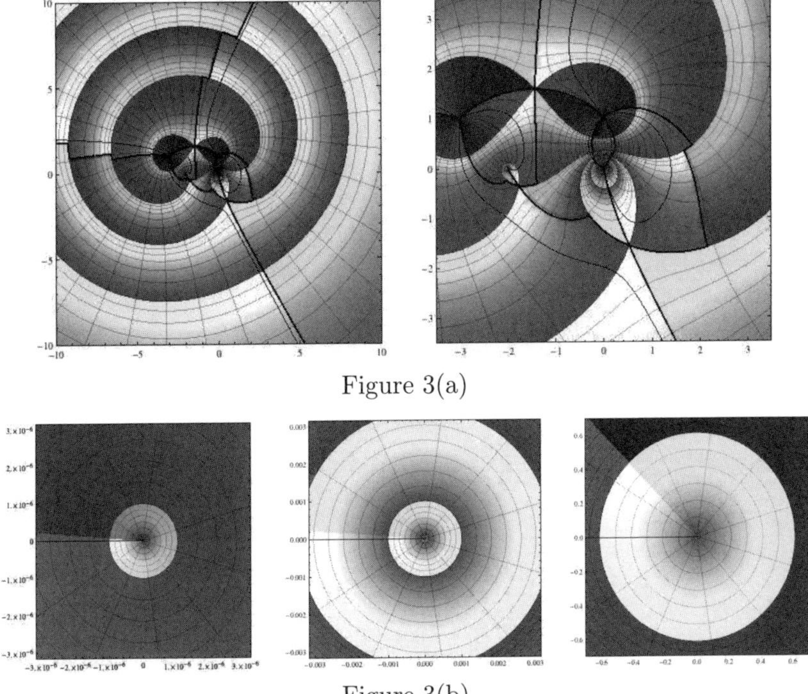

Figure 3(a)

Figure 3(b)

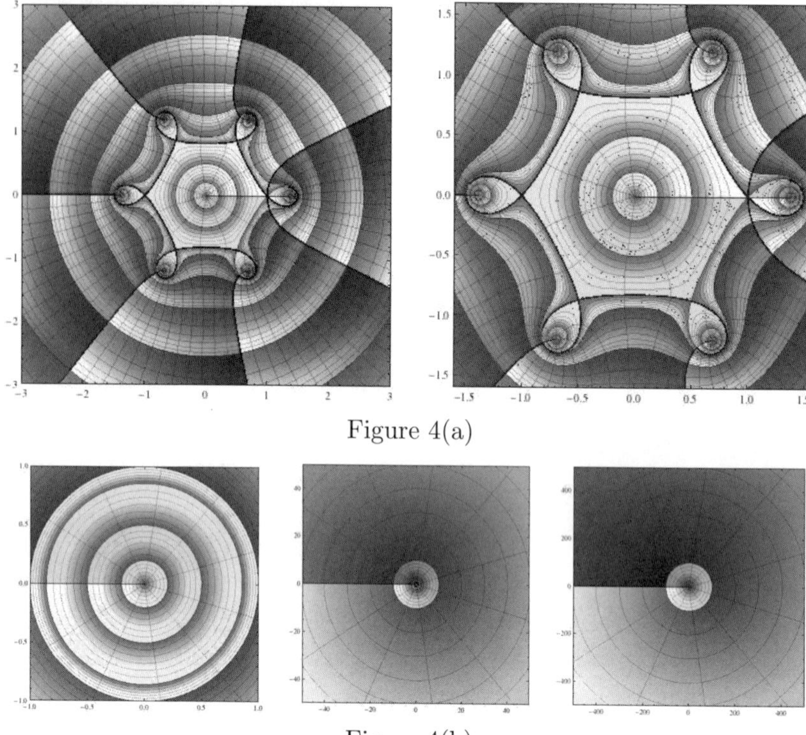

Figure 4(a)

Figure 4(b)

Microscopic behaviour of the Stokes flow in a singularly perturbed exterior domain

M. Dalla Riva

Dipartimento di Matematica Pura ed Applicata,
Università degli Studi di Padova,
Padova, 35121, Italy
E-mail: mdallari@math.unipd.it
www.unipd.it

We consider a pair of domains Ω^b and Ω^s in $\mathbb{R}^n$ and we assume that the closure of Ω^b does not intersect the closure of $\epsilon\Omega^s$ for $\epsilon \in [0, \epsilon_0[$. Then for a fixed $\epsilon \in]0, \epsilon_0[$ we consider a boundary value problem in $\mathbb{R}^n \setminus (\Omega^b \cup \epsilon\Omega^s)$ which describes the steady state Stokes flow of an incompressible viscous fluid past a body occupying the domain Ω^b and past a small impurity occupying the domain $\epsilon\Omega^s$. The unknown of the problem are the velocity field u and the pressure field p, and we impose the value of the velocity field u on the boundary both of the body and of the impurity. We assume that the boundary velocity on the impurity displays an arbitrarily strong singularity as ϵ tends to 0 and the impurity shrinks to a point. Our goal is to understand the 'microscopic' behaviour of (u, p) near the impurity when ϵ tends to 0.

Keywords: Singularly perturbed domain; boundary value problem for the Stokes system; real analytic continuation.

1. Introduction

In this paper, we present an application of a functional analytic approach to the analysis of a boundary value problem for the Stokes system in a singularly perturbed exterior domain. We first fix once for all

$$n \in \mathbb{N} \setminus \{0, 1\}, \qquad \alpha \in]0, 1[,$$

where $\mathbb{N}$ denotes the set of natural numbers including 0. Then we choose a pair of sets Ω^b and Ω^s in the n-dimensional Euclidean space $\mathbb{R}^n$ which satisfies the following condition

$$\Omega^b \text{ and } \Omega^s \text{ are open bounded connected subsets of } \mathbb{R}^n \text{ of} \qquad (1)$$

class $C^{1,\alpha}$ with connected exterior, the origin 0 of $\mathbb{R}^n$

belongs to Ω^s but not to the closure $\mathrm{cl}\Omega^b$ of Ω^b.

For the definition of the functions and sets of the usual Schauder class $C^{0,\alpha}$ or $C^{1,\alpha}$, we refer for example to Gilbarg and Trudinger, §6.2 of Ref. 1 (see also §2 of Ref. 2.) Here 'b' stands for 'body' and 's' stands for 'small impurity'. We note that condition (1) in particular implies that Ω^b and Ω^s have no holes and that there exists a real number ϵ_0 such that

$$\epsilon_0 \in]0,1[\text{ and } \mathrm{cl}\Omega^b \cap (\epsilon \mathrm{cl}\Omega^s) = \emptyset \text{ for all } \epsilon \in]0,\epsilon_0[. \tag{2}$$

Then we denote by $\Omega^e(\epsilon)$ the exterior domain defined by

$$\Omega^e(\epsilon) \equiv \mathbb{R}^n \setminus \{\mathrm{cl}\Omega^b \cup (\epsilon \mathrm{cl}\Omega^s)\} \qquad \forall \epsilon \in]0,\epsilon_0[.$$

Next we introduce a function γ such that

$$\gamma \text{ is defined from }]0,\epsilon_0[\text{ to } [0,+\infty[\text{ and } \gamma_0 \equiv \lim_{\epsilon \to 0} \gamma(\epsilon) \in [0,+\infty[. \tag{3}$$

Now let $f \in C^{1,\alpha}(\partial\Omega^s, \mathbb{R}^n)$. For $\epsilon \in]0,\epsilon_0[$ we consider the following boundary value problem in the exterior domain $\Omega^e(\epsilon)$,

$$\begin{cases} \Delta u_\epsilon - \nabla p_\epsilon = 0 & \text{in } \Omega^e(\epsilon), \\ \operatorname{div} u_\epsilon = 0 & \text{in } \Omega^e(\epsilon), \\ u_\epsilon = 0 & \text{on } \partial\Omega^b, \\ u_\epsilon(x) = \gamma(\epsilon)^{-1} f(x/\epsilon) \text{ for } x \in \epsilon\partial\Omega^s, \end{cases} \tag{4}$$

with the following decay condition,

$$\sup_{x \in \mathrm{cl}\Omega^e(\epsilon)} \left\{ |x|^{n-2} |u_\epsilon(x)|, \ |x|^{n-1} |Du_\epsilon(x)|, \ |x|^{n-1} |p_\epsilon(x)| \right\} < +\infty. \tag{5}$$

By Chang and Finn (see §4 of Ref. 3) we know that the problem in (4) admits a unique solution $(u_\epsilon, p_\epsilon) \in C^{1,\alpha}_{\mathrm{loc}}(\mathrm{cl}\Omega^e(\epsilon), \mathbb{R}^n) \times C^{0,\alpha}_{\mathrm{loc}}(\mathrm{cl}\Omega^e(\epsilon))$ which satisfy the condition in (5) (see also Varnhorn, Lemma 1.1 of Ref. 4.) Our aim is to investigate the 'microscopic' behaviour of (u_ϵ, p_ϵ) near the impurity $\epsilon\Omega^s$ as ϵ shrinks to 0. Namely, we pose the following question.

Let $\tilde{x}$ be a fixed point of $\mathbb{R}^n \setminus \Omega^s$. Let $\epsilon_{\tilde{x}} \in]0,\epsilon_0[$ be such that (6)

$\epsilon\tilde{x} \in \Omega^e(\epsilon)$ for all $\epsilon \in]0,\epsilon_{\tilde{x}}[$. What can be said of the function

which takes $\epsilon \in]0,\epsilon_{\tilde{x}}[$ to $(u_\epsilon(\epsilon\tilde{x}), p_\epsilon(\epsilon\tilde{x}))$ when ϵ is small?

Questions of this type have long been investigated with the methods of Asymptotic Analysis, which aims at giving complete asymptotic expansions of the solutions in terms of the parameter ϵ. It is perhaps difficult to provide a complete list of the contributions. Here, we mention the work of Kevorkian and Cole,[5] Van Dyke,[6] Kozlov, Maz'ya and Movchan,[7] Maz'ya, Nazarov and Plamenewskii,[8] Ward and Keller.[9] We note that, by the techniques of the Asymptotic Analysis, one can expect to obtain results which are expressed by means of regular functions of ϵ plus a reminder which is smaller than a positive known function of ϵ. Instead, the approach adopted in this paper aims to express the dependence upon ϵ in terms of real analytic functions defined in a whole open neighborhood of $\epsilon = 0$ and in terms of possibly singular but completely known function of ϵ, such as $1/\epsilon$ or $\log(\epsilon)$. Such an approach has its own advantages. For example, one could obtain asymptotic approximations which agree with those in the literature. Moreover, one would also express the dependence on ϵ in terms of convergent power series. This is the point of view adopted by Lanza de Cristoforis and his collaborators Preciso, Rossi and the author in several problems for elliptic equations and for the elliptic system of equation of linearized elastostatic (see *e.g.*, Refs. 2, 10.) In this paper we extend such methods to the analysis of boundary value problems for the Stokes system.

The paper is organized as follows. Section 2 is a section of preliminaries where we introduce some notation. In Sec. 3, we provide a representation formula for the solution of the problem in (4), (5) in terms of integral operators and real analytic operators. In Sec. 4, we prove our main Theorem 3.1.

2. Preliminaries and notation

We denote by $S_{V,n} \equiv (S_{V,n}^{i,j})_{i,j=1,\ldots,n}$, $S_{P,n} \equiv (S_{P,n}^i)_{i=1,\ldots,n}$ the functions of $\mathbb{R}^n \setminus \{0\}$ to $M_n(\mathbb{R})$ and to $\mathbb{R}^n$, respectively, defined by

$$S_{V,n}^{i,j}(x) \equiv \frac{1}{2s_n(\delta_{2,n} + (2-n))} \left\{ \left(\delta_{2,n} \log |x| + (1 - \delta_{2,n})|x|^{2-n}\right) \delta_{i,j} \right.$$

$$\left. - (\delta_{2,n} + (2-n)) \frac{x_i x_j}{|x|^n} \right\}$$

$$S_{P,n}^i(x) \equiv -\frac{1}{s_n} \frac{x_i}{|x|^n}$$

for all $i, j \in \{1, \ldots, n\}$ and all $x \in \mathbb{R}^n \setminus \{0\}$. Here $M_n(\mathbb{R})$ denotes the space of real $n \times n$-matrices, and s_n denotes the $(n-1)$ dimensional measure of the unit sphere in $\mathbb{R}^n$, and $\delta_{i,j}$ is defined by $\delta_{i,j} \equiv 0$ if $i \neq j$ and $\delta_{i,j} \equiv 1$ if $i = j$, for all $i, j \in \mathbb{N}$. Next, we denote by $\widetilde{S}_{V,n} \equiv (\widetilde{S}_{V,n}^{i,j})_{i,j=1,\ldots,n}$ the function of $\mathbb{R}^n \setminus \{0\}$ to $M_n(\mathbb{R})$ defined by

$$\widetilde{S}_{V,n}^{i,j}(x) \equiv \frac{1}{2s_n(\delta_{2,n} + (2-n))} \left(\delta_{i,j} \frac{1}{|x|^{n-2}} + (n-2) \frac{x_i x_j}{|x|^n} \right) \quad \forall x \in \mathbb{R}^n \setminus \{0\}.$$

So that $\widetilde{S}_{V,n} = S_{V,n}$ for $n \geq 3$ and $\widetilde{S}_{V,n} = (4\pi)^{-1} I$ for $n = 2$. Here I denotes the identity $n \times n$ matrix. We also find convenient to set

$$S_{V,n}^j \equiv (S_{V,n}^{i,j})_{i=1,\ldots,n}, \qquad \widetilde{S}_{V,n}^j \equiv (\widetilde{S}_{V,n}^{i,j})_{i=1,\ldots,n},$$

which we think of as column vectors for all $j = 1, \ldots, n$. For each scalar $\rho \in \mathbb{R}$ and each matrix $A \in M_n(\mathbb{R})$ we set

$$\mathcal{T}(\rho, A) \equiv -\rho I + (A + A^t),$$

where A^t denotes the transpose matrix to A. Now, let Ω be an open bounded subset of $\mathbb{R}^n$ of class $C^{1,\alpha}$. We shall denote by $\partial\Omega$ the boundary of Ω and by ν_Ω the outward unit normal to $\partial\Omega$. We set

$$w_V[\mu](x)$$
$$\equiv -\left\{ \int_{\partial\Omega} \mu^t(y)\mathcal{T}(S^j_{P,n}(x-y), D_x S^j_{V,n}(x-y))\nu_\Omega(y)\, d\sigma_y \right\}_{j=1,\ldots,n},$$
$$w_P[\mu](x) \equiv -2\operatorname{div}\left(\int_{\partial\Omega} \mu^t(y) S_{P,n}(x-y)\, \nu_\Omega(y)\, d\sigma_y \right) \qquad \forall x \in \mathbb{R}^n$$

for all $\mu \in C^{1,\alpha}(\partial\Omega, \mathbb{R}^n)$. Then we have the following Proposition 2.1 (for a proof see, *e.g.*, §5 of Ref. 11.)

Proposition 2.1. *Let Ω, $\widetilde{\Omega}$ be open bounded subsets of $\mathbb{R}^n$ with Ω of class $C^{1,\alpha}$ and $\operatorname{cl}\Omega \subset \widetilde{\Omega}$. If $\mu \in C^{1,\alpha}(\partial\Omega, \mathbb{R}^n)$, then the functions $w_V[\mu]_{|\mathbb{R}^n\setminus\operatorname{cl}\Omega}$ and $w_P[\mu]_{|\mathbb{R}^n\setminus\operatorname{cl}\Omega}$ admit a unique continuous extension to $\mathbb{R}^n \setminus \Omega$, which we denote by $w_V^-[\mu]$ and by $w_P^-[\mu]$, respectively. The maps which take μ to $w_V^-[\mu]_{|\operatorname{cl}\widetilde{\Omega}\setminus\Omega}$ and to $w_P^-[\mu]_{|\operatorname{cl}\widetilde{\Omega}\setminus\Omega}$ are linear and continuous from $C^{1,\alpha}(\partial\Omega, \mathbb{R}^n)$ to $C^{1,\alpha}(\operatorname{cl}\widetilde{\Omega} \setminus \Omega, \mathbb{R}^n)$ and to $C^{0,\alpha}(\operatorname{cl}\widetilde{\Omega} \setminus \Omega)$, respectively.*

3. Integral representation of the solution (u_ϵ, p_ϵ)

We now provide a representation formula for the solution of the problem in (4), (5) in terms of integral operators and real analytic functions. The validity of such representation is an immediate consequence of the results stated in Ref. 11. For the sake of simplicity we introduce the following notation. Let ϵ_0 be as in (2). Let γ, γ_0 be as in (3). We denote by Ψ_n the function of $\epsilon \in]0, \epsilon_0[$ defined by

$$\Psi_n[\epsilon] \equiv \left(\epsilon, \epsilon(\log\epsilon)^{\delta_{2,n}}, (\log\epsilon)^{-\delta_{2,n}}, \gamma(\epsilon) \right) \qquad \forall \epsilon \in]0, \epsilon_0[. \tag{7}$$

We note that $\Psi_n[\epsilon]$ tends to $(0, 0, 1 - \delta_{2,n}, \gamma_0)$ as ϵ tends to 0.

Theorem 3.1. *Let Ω^b, Ω^s be as in (1). Let ϵ_0 be as in (2). Let γ, γ_0 be as in (3). Let $f \in C^{1,\alpha}(\partial\Omega^s, \mathbb{R}^n)$. Let $(u_\epsilon, p_\epsilon) \in C^{1,\alpha}_{\operatorname{loc}}(\operatorname{cl}\Omega^e(\epsilon), \mathbb{R}^n) \times C^{0,\alpha}_{\operatorname{loc}}(\operatorname{cl}\Omega^e(\epsilon))$ be the unique solution of the problem in (4) which satisfy the condition in (5). Let y^b be a point of Ω^b. Then there exist $\epsilon' \in]0, \epsilon_0[$, and a neighborhood $\mathcal{U}$ of $(0, 1 - \delta_{2,n}, \gamma_0)$ in $\mathbb{R}^3$, and a real analytic operator*

$\mathcal{E} \equiv (\mathcal{E}^b, \mathcal{E}^s)$ *of* $] - \epsilon', \epsilon'[\times \mathcal{U}$ *to* $C^{1,\alpha}(\partial\Omega^b, \mathbb{R}^n) \times C^{1,\alpha}(\partial\Omega^s, \mathbb{R}^n)$ *such that* $\Psi_n(\epsilon) \in] - \epsilon', \epsilon'[\times \mathcal{U}$ *for all* $\epsilon \in]0, \epsilon'[$ *and such that*

$$u_\epsilon(x) = \frac{\epsilon^{n-2}}{\gamma(\epsilon)(\log \epsilon)^{\delta_{2,n}}} \Bigg\{ w_V[\mathcal{E}^b[\Psi_n[\epsilon]]](x) \tag{8}$$

$$+ \widetilde{S}_{V,n}(x - y^b) \int_{\partial\Omega^b} \mathcal{E}^b[\Psi_n[\epsilon]] \, d\sigma$$

$$+ \sum_{j=1}^{n(n-1)/2} \frac{Z_n^j \cdot (x - y^b)}{|x - y^b|^n} \int_{\partial\Omega^b} \mathcal{E}^b[\Psi_n[\epsilon]]^t(y) \, Z_n^j \cdot (y - y^b) \, d\sigma_y$$

$$+ \left(\widetilde{S}_{V,n}(x - y^b) - S_{V,n}(x) \right) \int_{\partial\Omega^s} \mathcal{E}^s[\Psi_n[\epsilon]] d\sigma \Bigg\}$$

$$+ \frac{\epsilon^{n-1}}{\gamma(\epsilon)} \Bigg\{ - \left(\int_{\partial\Omega^s} \mathcal{E}^s[\Psi_n[\epsilon]]^t(y) \right.$$

$$\left. \cdot \mathcal{T} \left(S_{P,n}^j(x - \epsilon y), DS_{V,n}^j(x - \epsilon y) \right) \nu_{\Omega_s}(y) \, d\sigma_y \right)_{j=1,\dots,n}$$

$$+ \sum_{j=1}^{n(n-1)/2} \frac{Z_n^j \cdot x}{|x|^n} \int_{\partial\Omega^s} \mathcal{E}^s[\Psi_n[\epsilon]]^t(y) \, Z_n^j \cdot y \, d\sigma_y \Bigg\} \qquad \forall x \in \Omega^e(\epsilon)$$

and

$$p_\epsilon(x) = \frac{\epsilon^{n-2}}{\gamma(\epsilon)(\log \epsilon)^{\delta_{2,n}}} \Bigg\{ w_P[\mathcal{E}^b[\Psi_n[\epsilon]]](x) \tag{9}$$

$$+ (1 - \delta_{2,n}) S_{P,n}^t(x - y^b) \int_{\partial\Omega^b} \mathcal{E}^b[\Psi_n[\epsilon]] \, d\sigma$$

$$+ \left(S_{P,n}^t(x - y^b) - S_{P,n}^t(x) \right) \int_{\partial\Omega^s} \mathcal{E}^s[\Psi_n[\epsilon]] \, d\sigma \Bigg\}$$

$$- \frac{2\epsilon^{n-1}}{\gamma(\epsilon)} \operatorname{div} \left(\int_{\partial\Omega^s} \mathcal{E}^s[\Psi_n[\epsilon]]^t(y) \, S_{P,n}(x - \epsilon y) \nu_\Omega(y) \, d\sigma_y \right) \quad \forall x \in \Omega^e(\epsilon)$$

for all $\epsilon \in]0, \epsilon'[$, *where* $\{Z_n^j\}_{j=1,\dots,n(n-1)/2}$ *denotes the canonical orthogonal basis of the subspace of* $M_n(\mathbb{R})$ *consisting of all skew-symmetric matrices.*

4. Microscopic behaviour of the solution (u_ϵ, p_ϵ)

We now state our main theorem where we answer to the question in (6).

Theorem 4.1. *Let* Ω^b, Ω^s *be as in (1). Let* ϵ_0 *be as in (2). Let* γ, γ_0 *be as in (3). Let* $f \in C^{1,\alpha}(\partial\Omega^s, \mathbb{R}^n)$. *Let* $(u^s, p^s) \in C^{1,\alpha}_{\mathrm{loc}}(\mathbb{R}^n \setminus \Omega^s, \mathbb{R}^n) \times C^{0,\alpha}_{\mathrm{loc}}(\mathbb{R}^n \setminus \Omega^s)$

be the unique solution of the problem

$$\Delta u^s - \nabla p^s = 0, \ \operatorname{div} u^s = 0 \quad \text{in } \mathbb{R}^n \setminus \operatorname{cl}\Omega^s, \quad u^s = f \text{ on } \partial\Omega^s$$

such that

$$\sup_{x \in \operatorname{cl}\Omega^e(\epsilon)} \left\{ |x|^{n-2}|u^s(x)|, \ |x|^{n-1}|Du^s(x)|, \ |x|^{n-1}|p^s(x)| \right\} < +\infty.$$

Let $\widetilde{\Omega}$ be an open subset of $\mathbb{R}^n \setminus \operatorname{cl}\Omega^s$.

Then there exist $\epsilon_{\widetilde{\Omega}}^m \in]0, \epsilon_0[$, and a neighborhood $\mathcal{U}_{\widetilde{\Omega}}^m$ of $(0, 1 - \delta_{2,n}, \gamma_0)$ in $\mathbb{R}^3$, and real analytic operators $U_{\widetilde{\Omega}}^{m,s}$, $U_{\widetilde{\Omega}}^{m,b}$ of $] - \epsilon_{\widetilde{\Omega}}^m, \epsilon_{\widetilde{\Omega}}^m[\times \mathcal{U}_{\widetilde{\Omega}}^m$ to $C^{1,\alpha}(\operatorname{cl}\widetilde{\Omega}, \mathbb{R}^n)$, and real analytic operators $P_{\widetilde{\Omega}}^{m,s}$, $P_{\widetilde{\Omega}}^{m,b}$ of $] - \epsilon_{\widetilde{\Omega}}^m, \epsilon_{\widetilde{\Omega}}^m[\times \mathcal{U}_{\widetilde{\Omega}}^m$ to $C^{0,\alpha}(\operatorname{cl}\widetilde{\Omega})$, such that the following statements hold.

(i) $\Psi_n(\epsilon) \in] - \epsilon_{\widetilde{\Omega}}^m, \epsilon_{\widetilde{\Omega}}^m[\times \mathcal{U}_{\widetilde{\Omega}}^m$ and $(\epsilon^{-1}\operatorname{cl}\Omega^b) \cap \operatorname{cl}\widetilde{\Omega} = \emptyset$ for all $\epsilon \in]0, \epsilon_{\widetilde{\Omega}}^m[$.

(ii) If $\epsilon \in]0, \epsilon_{\widetilde{\Omega}}^m[$ and $(u_\epsilon, p_\epsilon) \in C_{\mathrm{loc}}^{1,\alpha}(\operatorname{cl}\Omega^e(\epsilon), \mathbb{R}^n) \times C_{\mathrm{loc}}^{0,\alpha}(\operatorname{cl}\Omega^e(\epsilon))$ is the unique solution of the problem in (4) which satisfies (5), then

$$u_\epsilon(\epsilon x) = \frac{1}{\gamma(\epsilon)} U_{\widetilde{\Omega}}^{m,s}[\Psi_n[\epsilon]](x) + \frac{\epsilon^{n-2}}{\gamma(\epsilon)(\log \epsilon)^{\delta_{2,n}}} U_{\widetilde{\Omega}}^{m,b}[\Psi_n[\epsilon]](x),$$

$$p_\epsilon(\epsilon x) = \frac{1}{\epsilon\gamma(\epsilon)} P_{\widetilde{\Omega}}^{m,s}[\Psi_n[\epsilon]](x) + \frac{\epsilon^{n-2}}{\gamma(\epsilon)(\log \epsilon)^{\delta_{2,n}}} P_{\widetilde{\Omega}}^{m,b}[\Psi_n[\epsilon]](x)$$

for all $x \in \operatorname{cl}\widetilde{\Omega}$.

(iii) $U_{\widetilde{\Omega}}^{m,s}[0, 0, 1 - \delta_{2,n}, \gamma_0] = u^s_{|\operatorname{cl}\widetilde{\Omega}}$ and $P_{\widetilde{\Omega}}^{m,s}[0, 0, 1 - \delta_{2,n}, \gamma_0] = p^s_{|\operatorname{cl}\widetilde{\Omega}}$.

(iv) The limit

$$\lim_{\epsilon \to 0} \left(\gamma(\epsilon) u_\epsilon(\epsilon \, \cdot \,)_{|\operatorname{cl}\widetilde{\Omega}}, \ \epsilon\gamma(\epsilon) p_\epsilon(\epsilon \, \cdot \,)_{|\operatorname{cl}\widetilde{\Omega}} \right) = \left(u^s_{|\operatorname{cl}\widetilde{\Omega}}, \ p^s_{|\operatorname{cl}\widetilde{\Omega}} \right)$$

holds in $C^{1,\alpha}(\operatorname{cl}\widetilde{\Omega}, \mathbb{R}^n) \times C^{0,\alpha}(\operatorname{cl}\widetilde{\Omega})$.

Proof. Let $y^b \in \Omega^b$. Let ϵ', $\mathcal{U}$ be as in Theorem 3.1. Then statement (i) is verified for $\epsilon_{\widetilde{\Omega}}^m \in]0, \epsilon'[$ small enough and $\mathcal{U}_{\widetilde{\Omega}}^m \equiv \mathcal{U}$. We now define

$$U_{\widetilde{\Omega}}^{m,s}[\mathbf{e}](x) \equiv w_V^-[\mathcal{E}^s[\mathbf{e}]](x) - \left(\frac{\delta_{2,n}}{4\pi} + \frac{1}{(\log \epsilon)^{\delta_{2,n}}} S_{V,n}(x) \right) \int_{\partial\Omega^s} \mathcal{E}^s[\mathbf{e}] \, d\sigma$$

$$+ \sum_{j=1}^{n(n-1)/2} \frac{Z_n^j \cdot x}{|x|^n} \int_{\partial\Omega^s} \mathcal{E}^s[\mathbf{e}]^t(y) Z_n^j \cdot y \, d\sigma_y,$$

$$U_{\widetilde{\Omega}}^{m,b}[\mathbf{e}](x) \equiv w_V[\mathcal{E}^b[\mathbf{e}]](\epsilon x) + \widetilde{S}_{V,n}(\epsilon x - y^b)\int_{\partial\Omega^b}\mathcal{E}^b[\mathbf{e}]\,d\sigma$$

$$+ S_{V,n}(\epsilon x - y^b)\int_{\partial\Omega^b}\mathcal{E}^s[\mathbf{e}]\,d\sigma$$

$$+ \sum_{j=1}^{n(n-1)/2} \frac{Z_n^j\cdot(\epsilon x - y^b)}{|\epsilon x - y^b|^n}\int_{\partial\Omega^s}\mathcal{E}^b[\mathbf{e}]^t(y)Z_n^j\cdot(y - y^b)\,d\sigma_y\,,$$

$$P_{\widetilde{\Omega}}^{m,s}[\mathbf{e}](x) \equiv w_P^-[\mathcal{E}^s[\mathbf{e}]](x) - \frac{1}{(\log\epsilon)^{\delta_{2,n}}}S_{P,n}^t(x)\int_{\partial\Omega^s}\mathcal{E}^s[\mathbf{e}]\,d\sigma\,,$$

$$P_{\widetilde{\Omega}}^{m,b}[\mathbf{e}](x) \equiv w_P[\mathcal{E}^b[\mathbf{e}]](x)$$

$$+ S_{P,n}^t(\epsilon x - y^b)\left((1 - \delta_{2,n})\int_{\partial\Omega^b}\mathcal{E}^b[\mathbf{e}]\,d\sigma - \int_{\partial\Omega^s}\mathcal{E}^s[\mathbf{e}]\,d\sigma\right)$$

for all $x \in \mathrm{cl}\widetilde{\Omega}$ and for all $\mathbf{e} \equiv (\epsilon, \epsilon_1, \epsilon_2, \epsilon_3) \in]-\epsilon_{\widetilde{\Omega}}^m, \epsilon_{\widetilde{\Omega}}^m[\times\mathcal{U}_{\widetilde{\Omega}}^m$. Then by the integral representation formulas in (8) and in (9) we deduce the validity of statement (ii). The real analyticity of the operators $U_{\widetilde{\Omega}}^{m,s}$, $U_{\widetilde{\Omega}}^{m,b}$, $P_{\widetilde{\Omega}}^{m,s}$, $P_{\widetilde{\Omega}}^{m,b}$ follows by the real analyticity of $\mathcal{E}^s$, $\mathcal{E}^b$ (cf. Theorem 3.1), and by the standard properties of double layer Stokes potentials (cf. Proposition 2.1), and by the known properties of the nonsingular integral operators with real analytic kernel (cf. *e.g.*, Lanza, Theorem 3.7 of Ref. 10.) For a proof of statement (iii) we refer to §§ 2, 3 of Ref. 11. Statement (iv) follows immediately by (ii) and (iii). $\qquad\square$

References

1. D. Gilbarg and N. S. Trudinger, *Elliptic partial differential equations of second order*, Classics in Mathematics (Springer-Verlag, Berlin, 2001).

2. M. Dalla Riva and M. Lanza de Cristoforis, *A singularly perturbed nonlinear traction boundary value problem for linearized elastostatics. A functional analytic approach*, to appear in Analysis (Munich), (2008).

3. I.-D. Chang and R. Finn, On the solutions of a class of equations occurring in continuum mechanics, with application to the Stokes paradox *Arch. Rational Mech. Anal.* **7**, 388 (1961).

4. W. Varnhorn, *The Stokes equations*, Mathematical Research, Vol. 76 (Akademie-Verlag, Berlin, 1994).

5. J. Kevorkian and J. D. Cole, *Perturbation methods in applied mathematics*, Applied Mathematical Sciences, Vol. 34 (Springer-Verlag, New York, 1981).

6. M. Van Dyke, *Perturbation methods in fluid mechanics*, Applied Mathematics and Mechanics, Vol. 8, Applied Mathematics and Mechanics, Vol. 8 (Academic Press, New York, 1964).

7. V. Kozlov, V. Maz'ya and A. Movchan, *Asymptotic analysis of fields in multi-*

structures, Oxford Mathematical Monographs (The Clarendon Press Oxford University Press, New York, 1999).

8. V. Maz'ya, S. Nazarov and B. Plamenevskij, *Asymptotic theory of elliptic boundary value problems in singularly perturbed domains.* **I, II**, (trans. of the original in German published by Akademie Verlag 1991), Operator Theory: Advances and Applications, **111, 112** (Birkhäuser Verlag, Basel, 2000).

9. M. J. Ward and J. B. Keller, Strong localized perturbations of eigenvalue problems, *SIAM J. Appl. Math.* **53**, 770 (1993).

10. M. Lanza de Cristoforis, Asymptotic behaviour of the solutions of the Dirichlet problem for the Laplace operator in a domain with a small hole. A functional analytic approach, *Analysis (Munich)* **28**, 63 (2008).

11. M. Dalla Riva, *Stokes flow in a singularly perturbed exterior domain*, typewritten manuscript (2009).

Singularly perturbed loads for a nonlinear traction boundary value problem on a singularly perturbed domain

M. Dalla Riva and M. Lanza de Cristoforis

Dipartimento di Matematica Pura ed Applicata,
Università degli Studi di Padova,
Via Trieste 63, Padova 35121, Italy
E-mails: mdallari@math.unipd.it, mldc@math.unipd.it
www.math.unipd.it

This paper treats the equilibrium of a family of linearly elastic bodies in n dimensional Euclidean space containing a small hole of arbitrary shape of size ϵ with the traction on the hole depending nonlinearly on the displacement and depending in a singular way on ϵ. Under suitable assumptions we illustrate that this problem has a solution for small ϵ and we analyze its behaviour as ϵ tends to 0. Our approach is different from standard methods of asymptotics.

Keywords: Nonlinear traction boundary value problem; singularly perturbed domain; linearized elastostatics; real analytic continuation.

This paper is devoted to present applications of a functional analytic approach to the analysis of nonlinear traction boundary value problems for the system of equations of linearized elastostatics in a domain with a small cavity. We first introduce the problem on a domain with no cavity, and then we define a problem on a domain with a cavity. We fix once for all

$$n \in \mathbb{N} \setminus \{0, 1\}, \qquad \alpha \in]0, 1[, \qquad \omega \in]1 - (2/n), +\infty[,$$

where $\mathbb{N}$ denotes the set of natural numbers including 0. Then we choose a subset Ω^o of $\mathbb{R}^n$ satisfying the following assumption.

$$\text{It is a bounded open connected subset of } \mathbb{R}^n \text{ of class } C^{1,\alpha} \tag{1}$$

containing 0 and it has a connected exterior (and thus no holes).

For the definition of functions and sets of the usual Schauder class $C^{0,\alpha}$ or $C^{1,\alpha}$, we refer for example to Gilbarg and Trudinger [7, §6.2] (see also [3, §2]). Then we consider the map $T(\omega, \cdot)$ of $M_n(\mathbb{R})$ to $M_n(\mathbb{R})$ defined by

$$T(\omega, A) \equiv (\omega - 1)(\operatorname{tr} A)I + (A + A^t) \quad \forall A \in M_n(\mathbb{R}).$$

Here $M_n(\mathbb{R})$ denotes the set of $n \times n$ matrices with real entries, and I denotes the identity matrix, and $\operatorname{tr} A$, A^t denote the trace and the transpose matrix to A, respectively. We note that $(\omega - 1)$ plays the role of ratio between the first and second Lamé constants. Next we introduce the functions

$$a \in C^{0,\alpha}(\partial\Omega^o, M_n(\mathbb{R})), \qquad g \in C^{0,\alpha}(\partial\Omega^o, \mathbb{R}^n), \tag{2}$$

and we consider the 'unperturbed' linear traction boundary value problem

$$\begin{cases} \operatorname{div}(T(\omega, Du)) = 0 & \text{in } \Omega^o, \\ T(\omega, Du(x))\nu^o(x) = a(x)u(x) + g(x) & \forall x \in \partial\Omega^o, \end{cases} \tag{3}$$

where ν^o denotes the outward unit normal to $\partial\Omega^o$. We know that if

$$\det a(\cdot) \text{ does not vanish identically in } \partial\Omega, \tag{4}$$
$$\xi^t a(x)\xi \leq 0 \qquad \forall x \in \partial\Omega, \ \forall \xi \in \mathbb{R}^n, \tag{5}$$

then problem (3) admits a unique solution $\tilde{u} \in C^{1,\alpha}(\operatorname{cl}\Omega^o, \mathbb{R}^n)$ (cf. *e.g.*, [3, Thm. 2.2].) Next we consider Ω^i as in (1) and we take $\epsilon_0 \in]0, 1[$ such that $\epsilon \operatorname{cl}\Omega^i \subseteq \Omega^o$ for $|\epsilon| \leq \epsilon_0$, and we set $\Omega(\epsilon) \equiv \Omega^o \setminus \epsilon \operatorname{cl}\Omega^i$, and we introduce a continuous function G^i of $\partial\Omega^i \times \mathbb{R}^n$ to $\mathbb{R}^n$, and a function γ of $]0, \epsilon_0[$ to $]0, +\infty[$, and we consider the following nonlinear problem for $\epsilon \in]0, \epsilon_0[$,

$$\begin{cases} \operatorname{div}(T(\omega, Du)) = 0 & \text{in } \Omega(\epsilon), \\ -T(\omega, Du(x))\nu_{\epsilon\Omega^i}(x) = \frac{1}{\gamma(\epsilon)}G^i(x/\epsilon, u(x)) & \forall x \in \epsilon\partial\Omega^i, \\ T(\omega, Du(x))\nu^o(x) = a(x)u(x) + g(x) & \forall x \in \partial\Omega^o, \end{cases} \tag{6}$$

where $\nu_{\epsilon\Omega^i}$ denotes the outward unit normal to $\epsilon\partial\Omega^i$. We are interested in the behaviour of families of solutions $\{u(\epsilon, \cdot)\}_{\epsilon \in]0, \epsilon'[}$ of (6) for some $\epsilon' > 0$ as ϵ tends to 0. We say that the family $\{u(\epsilon, \cdot)\}_{\epsilon \in]0, \epsilon'[}$ converges in $\operatorname{cl}\Omega^o \setminus \{0\}$ to a function f defined on $\operatorname{cl}\Omega^o$ provided that

$$\lim_{\epsilon \to 0} u(\epsilon, x) = f(x) \qquad \forall x \in \operatorname{cl}\Omega^o \setminus \{0\}.$$

We say that the family $\{u(\epsilon, \cdot)\}_{\epsilon \in]0, \epsilon'[}$ converges microscopically in $\mathbb{R}^n \setminus \Omega^i$ to a function f_1 defined on $\mathbb{R}^n \setminus \Omega^i$ provided that

$$\lim_{\epsilon \to 0} u(\epsilon, \epsilon x) = f_1(x) \qquad \forall x \in \mathbb{R}^n \setminus \Omega^i.$$

We note that the function $u(\epsilon, \epsilon x)$ for $x \in \frac{1}{\epsilon}\operatorname{cl}\Omega(\epsilon)$ is a 'rescaled' version of $u(\epsilon, x)$ for $x \in \operatorname{cl}\Omega(\epsilon)$, and we think of it as the 'microscopic' version of $u(\epsilon, x)$. In our analysis, we always assume that the function γ, which determines the singularity of our boundary traction, has a prescribed limiting behaviour. We first consider the case in which the limit

$$\gamma_m \equiv \lim_{\epsilon \to 0} \gamma^{-1}(\epsilon)\epsilon(\log \epsilon)^{\delta_{2,n}} \qquad \text{exists in } \mathbb{R},$$

where $\delta_{2,n} = 1$ if $n = 2$, $\delta_{2,n} = 0$ if $n > 2$. In such a case we show that there exists a family of solutions $\{u(\epsilon,\cdot)\}_{\epsilon\in]0,\epsilon'[}$ which both converges in $\mathrm{cl}\,\Omega^o \setminus \{0\}$ and converges microscopically in $\mathbb{R}^n \setminus \Omega^i$ (cf. Theorem 1). Then we consider the case in which

$$\gamma_m \equiv \lim_{\epsilon\to 0} \gamma^{-1}(\epsilon)\epsilon(\log\epsilon)^{\delta_{2,n}} = \infty\,, \qquad \gamma_M \equiv \lim_{\epsilon\to 0}\frac{\gamma(\epsilon)}{\epsilon^{n-1}} = +\infty\,.$$

In such a case however, we consider a problem in which the right hand side of the second equation in (6) is replaced by $\frac{1}{\gamma(\epsilon)}G^i(x/\epsilon, \gamma(\epsilon)\epsilon^{-1}(\log\epsilon)^{-\delta_{2,n}}u(x))$. For a discussion on such a choice, we refer to [4]. Thus we consider the problem

$$\begin{cases} \mathrm{div}\,(T(\omega, Du)) = 0 & \text{in } \Omega(\epsilon)\,, \\ -T(\omega, Du(x))\nu_{\epsilon\Omega^i}(x) = \frac{1}{\gamma(\epsilon)}G^i(x/\epsilon, \gamma(\epsilon)\epsilon^{-1}(\log\epsilon)^{-\delta_{2,n}}u(x))\ \forall x \in \epsilon\partial\Omega^i, \\ T(\omega, Du(x))\nu^o(x) = a(x)u(x) + g(x) & \forall x \in \partial\Omega^o\,, \end{cases}$$

$$\tag{7}$$

and we show that there exists a family of solutions $\{u(\epsilon,\cdot)\}_{\epsilon\in]0,\epsilon'[}$ of (7) which converges in $\mathrm{cl}\,\Omega^o \setminus \{0\}$ and which does not necessarily converge microscopically in $\mathbb{R}^n \setminus \Omega^i$, although we show that $\left\{\frac{\gamma(\epsilon)}{\epsilon(\log\epsilon)^{\delta_{2,n}}}u(\epsilon,\cdot)\right\}_{\epsilon\in]0,\epsilon'[}$ converges microscopically in $\mathbb{R}^n \setminus \Omega^i$. Thus in a sense, we can say that $\epsilon(\log\epsilon)^{\delta_{2,n}}$ represents a microscopically critical behaviour for γ as ϵ tends to 0 (cf. Theorem 2.) Finally, we consider the case in which

$$\gamma_M \in [0, +\infty[$$

If $\gamma_M \in]0, +\infty[$, we show that there is a family of solutions $\{u(\epsilon,\cdot)\}_{\epsilon\in]0,\epsilon'[}$ of (7) which converges in $\mathrm{cl}\,\Omega^o\setminus\{0\}$ and which does not necessarily converge microscopically in $\mathbb{R}^n\setminus\Omega^i$, although we can show that $\left\{\frac{\gamma(\epsilon)}{\epsilon(\log\epsilon)^{\delta_{2,n}}}u(\epsilon,\cdot)\right\}_{\epsilon\in]0,\epsilon'[}$ converges microscopically in $\mathbb{R}^n\setminus\Omega^i$. If $\gamma_M = 0$, we show that there is a family of solutions $\{u(\epsilon,\cdot)\}_{\epsilon\in]0,\epsilon'[}$ of (7) which does not necessarily converge in $\mathrm{cl}\,\Omega^o\setminus\{0\}$ and which does not necessarily converge microscopically in $\mathbb{R}^n\setminus\Omega^i$, although we show that $\left\{\frac{\gamma(\epsilon)}{\epsilon^{n-1}}u(\epsilon,\cdot)\right\}_{\epsilon\in]0,\epsilon'[}$ converges in $\mathrm{cl}\,\Omega^o\setminus\{0\}$ and that $\left\{\frac{\gamma(\epsilon)}{\epsilon(\log\epsilon)^{\delta_{2,n}}}u(\epsilon,\cdot)\right\}_{\epsilon\in]0,\epsilon'[}$ converges microscopically in $\mathbb{R}^n \setminus \Omega^i$. Thus in a sense we can say that ϵ^{n-1} represents a critical behaviour for γ as ϵ tends to 0 (cf. Theorem 3.)

We also note that in all cases considered above, we can show that our families of solutions are unique in a local sense which we do not clarify here. However, our main interest is focused on the description of the behaviour of $u(\epsilon,\cdot)$ when ϵ is near 0, and not only on the limiting value. Actually, we pose the following two questions.

(j) Let x be a fixed point in $\mathrm{cl}\Omega^o \setminus \{0\}$. What can be said on the map $\epsilon \mapsto u(\epsilon, x)$ when ϵ is close to 0 and positive?

(jj) Let x be a fixed point in $\mathbb{R}^n \setminus \Omega^i$. What can be said on the map $\epsilon \mapsto u(\epsilon, \epsilon x)$ when ϵ is close to 0 and positive?

Questions of this type have long been investigated for linear problems with the methods of Asymptotic Analysis and of the Calculus of the Variations. Here, we mention Dal Maso and Murat [6], Kozlov, Maz'ya and Movchan [8], Maz'ya, Nazarov and Plamenewskii [10], Ozawa [11], Ward and Keller [12]. We also mention the seminal paper of Ball [1] on nonlinear elastic cavitation. For more comments, see also [3]. Our main results in this sense are Theorems 1–3 and answer questions (j), (jj) in the spirit of [9]. We now consider case $\gamma_m \in \mathbb{R}$ by the following result of [2].

Theorem 1. *Let $\gamma_m \in \mathbb{R}$. Let a satisfy (4), (5). Let the superposition operator F_{G^i} which takes $v \in C^{0,\alpha}(\partial\Omega^i, \mathbb{R}^n)$ to the function $F_{G^i}[v]$ defined by*

$$F_{G^i}[v](x) \equiv G^i(x, v(x)) \qquad \forall x \in \partial\Omega^i\,, \tag{8}$$

map $C^{0,\alpha}(\partial\Omega^i, \mathbb{R}^n)$ to itself and be real analytic. Assume that the limiting boundary value problem

$$\mathrm{div}\,(T(\omega, Du^i)) = 0 \qquad\qquad \text{in } \mathbb{R}^n \setminus \mathrm{cl}\Omega^i\,, \tag{9}$$

$$-T(\omega, Du^i(x))\nu^i(x) = \gamma_m G^i\Big(x, (1 - \delta_{2,n})u^i(x) \tag{10}$$

$$+ \frac{\delta_{2,n}}{4\pi}\frac{\omega + 2}{\omega + 1}\int_{\partial\Omega^i} T(\omega, Du^i)\nu^i\,d\sigma + \tilde{u}(0)\Big) \qquad \forall x \in \partial\Omega^i\,,$$

$$\sup_{x \in \mathbb{R}^n \setminus \Omega^i} |x|^{n-2+\delta_{2,n}}|u^{\sharp,i}(x)| < \infty, \quad \sup_{x \in \mathbb{R}^n \setminus \Omega^i} |x|^{n-1+\delta_{2,n}}|Du^{\sharp,i}(x)| < \infty, \tag{11}$$

where $\nu^i = (\nu^i_j)_{j=1,\ldots,n}$ denotes the outward unit normal to $\partial\Omega^i$, and where

$$u^{\sharp,i}(x) \equiv u^i(x) - \delta_{2,n}\Gamma_n(\omega, x)\int_{\partial\Omega^i} T(\omega, Du^i)\nu^i\,d\sigma \tag{12}$$

for all $x \in \mathbb{R}^n \setminus \Omega^i$, admits at least a solution $\tilde{u}^i \in C^{1,\alpha}_{\mathrm{loc}}(\mathbb{R}^n \setminus \Omega^i, \mathbb{R}^n)$. Here $\Gamma_n(\omega, \cdot) \equiv (\Gamma^j_{n,l}(\omega, \cdot))_{l,j=1,\ldots,n}$ denotes fundamental solution of the operator $\Delta + \omega\nabla\mathrm{div}$ in $\mathbb{R}^n$. Let $\mathcal{F}^i$ be the matrix valued function on $\partial\Omega^i$ defined by

$$\mathcal{F}^i(x) \equiv -D_\xi G^i\Big(x, (1-\delta_{2,n})\tilde{u}^i(x) + \frac{\delta_{2,n}}{4\pi}\frac{\omega + 2}{\omega + 1}\int_{\partial\Omega^i} T(\omega, D\tilde{u}^i)\nu^i\,d\sigma + \tilde{u}(0)\Big),$$

for all $x \in \partial\Omega^i$.

If $n = 2$, we assume that the matrix $I - \frac{\gamma_m(\omega+2)}{4\pi(\omega+1)} \int_{\partial\Omega^i} \mathcal{F}^i \, d\sigma$ is invertible. If $n \geq 3$ and $\gamma_m > 0$, we assume that $-\mathcal{F}^i$ satisfies assumptions (4), (5) on $\partial\Omega^i$ (while in case $\gamma_m = 0$, we do not assume that $-\mathcal{F}^i$ satisfies (4), (5).)

Then there exist $\epsilon' \in]0, \epsilon_0[$ and a family $\{u(\epsilon, \cdot)\}_{\epsilon \in]0,\epsilon'[}$ such that $u(\epsilon, \cdot)$ belongs to $C^{1,\alpha}(\mathrm{cl}\,\Omega(\epsilon), \mathbb{R}^n)$ and solves (6) for all $\epsilon \in]0, \epsilon'[$, and such that the family $\{u(\epsilon, \cdot)\}_{\epsilon \in]0,\epsilon'[}$ converges in $\mathrm{cl}\,\Omega^o \setminus \{0\}$ to $\tilde{u}$, and converges microscopically in $\mathbb{R}^n \setminus \Omega^i$ to

$$(1 - \delta_{2,n})\tilde{u}^i(\cdot) + \frac{\delta_{2,n}}{4\pi}\frac{\omega+2}{\omega+1}\int_{\partial\Omega^i} T(\omega, D\tilde{u}^i)\nu^i \, d\sigma + \tilde{u}(0) \, .$$

Moreover, the following two statements hold.

(i) Let $\tilde{\Omega}$ be a bounded open subset of $\Omega^o \setminus \{0\}$ such that $0 \notin \mathrm{cl}\,\tilde{\Omega}$. Then there exist $\epsilon_{\tilde{\Omega}} \in]0, \epsilon'[$, and an open neighborhood $\mathcal{U}_{\gamma_m}$ of $(0, 0, \gamma_m)$ in $\mathbb{R}^3$, and a real analytic operator $U_{\tilde{\Omega}}$ of $] - \epsilon_{\tilde{\Omega}}, \epsilon_{\tilde{\Omega}}[\times \mathcal{U}_{\gamma_m}$ to $C^{1,\alpha}(\mathrm{cl}\,\tilde{\Omega}, \mathbb{R}^n)$ such that

$$\tilde{\Omega} \subseteq \Omega(\epsilon) \qquad \forall \epsilon \in] - \epsilon_{\tilde{\Omega}}, \epsilon_{\tilde{\Omega}}[\, , \tag{13}$$

and such that

$$\Xi_{m,n}(\epsilon) \equiv \left(\epsilon(\log \epsilon)^{\delta_{2,n}}, \frac{\epsilon^{n-2}}{(\log \epsilon)^{\delta_{2,n}}}, \frac{\epsilon(\log \epsilon)^{\delta_{2,n}}}{\gamma(\epsilon)} \right) \in \mathcal{U}_{\gamma_m} \, ,$$

for all $\epsilon \in]0, \epsilon'[$, and such that

$$u(\epsilon, x) = U_{\tilde{\Omega}}\left[\epsilon, \Xi_{m,n}[\epsilon]\right](x) \qquad \forall x \in \mathrm{cl}\,\tilde{\Omega}, \ \forall \epsilon \in]0, \epsilon_{\tilde{\Omega}}[\, .$$

(ii) Let $\tilde{\Omega}$ be a bounded open subset of $\mathbb{R}^n \setminus \mathrm{cl}\,\Omega^i$. Then there exist $\epsilon_{\tilde{\Omega},r} \in]0, \epsilon'[$, and an open neighborhood $\mathcal{U}_{\gamma_m}$ of $(0, 0, \gamma_m)$ in $\mathbb{R}^3$, and a real analytic map $U^r_{\tilde{\Omega}}$ of $] - \epsilon_{\tilde{\Omega},r}, \epsilon_{\tilde{\Omega},r}[\times \mathcal{U}_{\gamma_m}$ to $C^{1,\alpha}(\mathrm{cl}\,\tilde{\Omega}, \mathbb{R}^n)$ such that

$$\tilde{\Omega} \subseteq \frac{1}{\epsilon}\Omega(\epsilon) \qquad \forall \epsilon \in] - \epsilon_{\tilde{\Omega},r}, \epsilon_{\tilde{\Omega},r}[\setminus \{0\} \tag{14}$$

and such that

$$u(\epsilon, \epsilon x) = U^r_{\tilde{\Omega}}\left[\epsilon, \Xi_{m,n}[\epsilon]\right](x) \qquad \forall x \in \mathrm{cl}\,\tilde{\Omega}, \ \forall \epsilon \in]0, \epsilon_{\tilde{\Omega},r}[\, .$$

We now consider case $\gamma_m = \infty$, $\gamma_M = +\infty$ by the following result of [4].

Theorem 2. Let $\gamma_m = \infty$, $\gamma_M = +\infty$. Let a satisfy (4), (5). Let the superposition operator F_{G^i} defined by (8) be real analytic in $C^{0,\alpha}(\partial\Omega^i, \mathbb{R}^n)$. Assume that the limiting boundary value problem *consisting of (9), and of*

$$-T(\omega, Du^i(x))\nu^i(x) = G^i\bigg(x, (1 - \delta_{2,n})u^i(x) \tag{15}$$

$$+ \frac{\delta_{2,n}}{4\pi}\frac{\omega+2}{\omega+1}\int_{\partial\Omega^i} T(\omega, Du^i)\nu^i \, d\sigma \bigg) \qquad \forall x \in \partial\Omega^i \, ,$$

and of (11), (12) has at least a solution $\tilde{v}^i \in C^{1,\alpha}_{\text{loc}}(\mathbb{R}^n \setminus \Omega^i, \mathbb{R}^n)$. *Let* $\mathcal{G}^i$ *be the matrix valued function on* $\partial\Omega^i$ *defined by*

$$\mathcal{G}^i(x) \equiv -D_\xi G^i\left(x, (1-\delta_{2,n})\tilde{v}^i(x) + \frac{\delta_{2,n}}{4\pi}\frac{\omega+2}{\omega+1}\int_{\partial\Omega^i} T(\omega, D\tilde{v}^i)\nu^i \, d\sigma\right), \quad (16)$$

for all $x \in \partial\Omega^i$.

If $n = 2$, *we assume that the matrix* $I - \frac{(\omega+2)}{4\pi(\omega+1)}\int_{\partial\Omega^i}\mathcal{G}^i \, d\sigma$ *is invertible. If* $n \geq 3$, *we assume that* $-\mathcal{G}^i$ *satisfies assumptions (4), (5) on* $\partial\Omega^i$.

Then there exist $\epsilon' \in]0, \epsilon_0[$ *and a family* $\{u(\epsilon, \cdot)\}_{\epsilon \in]0, \epsilon'[}$ *such that* $u(\epsilon, \cdot)$ *belongs to* $C^{1,\alpha}(\text{cl}\Omega(\epsilon), \mathbb{R}^n)$ *and solves (7) for all* $\epsilon \in]0, \epsilon'[$, *and such that the family* $\{u(\epsilon, \cdot)\}_{\epsilon \in]0, \epsilon'[}$ *converges in* $\text{cl}\Omega^o \setminus \{0\}$ *to* $\tilde{u}$, *and such that*

$$\lim_{\epsilon \to 0^+} \frac{\gamma(\epsilon)}{\epsilon(\log\epsilon)^{\delta_{2,n}}} u(\epsilon, \epsilon x) \qquad (17)$$

$$= (1 - \delta_{2,n})\tilde{v}^i(x) + \frac{\delta_{2,n}}{4\pi}\frac{\omega+2}{\omega+1}\int_{\partial\Omega^i} T(\omega, D\tilde{v}^i)\nu^i \, d\sigma \qquad \forall x \in \mathbb{R}^n \setminus \Omega^i.$$

Moreover, the following statements hold.

(i) *Let* $\tilde{\Omega}$ *be a bounded open subset of* $\Omega^o \setminus \{0\}$ *such that* $0 \notin \text{cl}\tilde{\Omega}$. *Then there exist* $\epsilon_{\tilde{\Omega}} \in]0, \epsilon'[$, *and an open neighborhood* $\mathcal{U}_1$ *of* $(0, 0, 1 - \delta_{2,n}, 0)$ *in* $\mathbb{R}^4$, *and a real analytic operator* $U_{\tilde{\Omega}}$ *of* $] - \epsilon_{\tilde{\Omega}}, \epsilon_{\tilde{\Omega}}[\times\mathcal{U}_1$ *to* $C^{1,\alpha}(\text{cl}\tilde{\Omega}, \mathbb{R}^n)$ *such that (13) holds, and such that*

$$\Xi_n(\epsilon) \equiv \left(\frac{\epsilon^{n-1}}{\gamma(\epsilon)}, \gamma(\epsilon), (\log\epsilon)^{-\delta_{2,n}}, \frac{\gamma(\epsilon)}{\epsilon(\log\epsilon)^{\delta_{2,n}}}\right) \in \mathcal{U}_1,$$

for all $\epsilon \in]0, \epsilon'[$, *and such that*

$$u(\epsilon, x) = U_{\tilde{\Omega}}\left[\epsilon, \Xi_n[\epsilon]\right](x) \qquad \forall x \in \text{cl}\tilde{\Omega}, \ \forall \epsilon \in]0, \epsilon_{\tilde{\Omega}}[.$$

(ii) *Let* $\tilde{\Omega}$ *be a bounded open subset of* $\mathbb{R}^n \setminus \text{cl}\Omega^i$. *Then there exist* $\epsilon_{\tilde{\Omega}, r} \in]0, \epsilon'[$, *and an open neighborhood* $\mathcal{U}_1$ *of* $(0, 0, 1 - \delta_{2,n}, 0)$ *in* $\mathbb{R}^4$, *and a real analytic map* $U^{r,1}_{\tilde{\Omega}}$ *of* $] - \epsilon_{\tilde{\Omega}, r}, \epsilon_{\tilde{\Omega}, r}[\times\mathcal{U}_1$ *to* $\mathbb{R}^n$, *and real analytic maps* $U^{r,j}_{\tilde{\Omega}}$ *for* $j = 2, 3$ *of* $] - \epsilon_{\tilde{\Omega}, r}, \epsilon_{\tilde{\Omega}, r}[\times\mathcal{U}_1$ *to* $C^{1,\alpha}(\text{cl}\tilde{\Omega}, \mathbb{R}^n)$ *such that (14) holds and such that*

$$u(\epsilon, \epsilon x) = \delta_{2,n}\frac{\epsilon\log\epsilon}{\gamma(\epsilon)}U^{r,1}_{\tilde{\Omega}}[\epsilon, \Xi_n[\epsilon]] + \frac{\epsilon}{\gamma(\epsilon)}U^{r,2}_{\tilde{\Omega}}[\epsilon, \Xi_n[\epsilon]](x)$$

$$+ U^{r,3}_{\tilde{\Omega}}[\epsilon, \Xi_n[\epsilon]](x) \qquad \forall x \in \text{cl}\tilde{\Omega}, \ \forall \epsilon \in]0, \epsilon_{\tilde{\Omega}, r}[.$$

We now consider case $\gamma_m = \infty$, $\gamma_M \in [0, +\infty[$ by the following result of [5].

Theorem 3. *Let* $\gamma_m = \infty$, $\gamma_M \in [0, +\infty[$. *Let* a *satisfy (4), (5). Let the superposition operator* $F_{\mathcal{G}^i}$ *defined by (8) be real analytic in* $C^{0,\alpha}(\partial\Omega^i, \mathbb{R}^n)$.

Assume that the limiting boundary value problem *consisting of (9), (15), and of*

$$\operatorname{div}\left(T(\omega, Du^o)\right) = 0 \qquad\qquad \text{in } \Omega^o\,,$$

$$T(\omega, Du^o(x))\nu^o(x) - a(x)u^o(x)$$

$$= -\sum_{l,j=1}^{n} \left(\int_{\partial\Omega^i} T_{lj}(\omega, Du^i)\nu_j^i\, d\sigma\right) T(\omega, D\Gamma_n^l(\omega, x))\nu^o(x)$$

$$+ a(x)\left\{\Gamma_n(\omega, x)\int_{\partial\Omega^i} T(\omega, Du^i)\nu^i\, d\sigma\right\} + \gamma_M g(x) \quad \forall x \in \partial\Omega^o\,,$$

where $\Gamma_n^l(\omega, \cdot) \equiv (\Gamma_{n,j}^l(\omega, \cdot))_{j=1,\dots,n}$, *and of (11), (12), has at least a solution* $(\tilde{w}^i, \tilde{w}^o)$ *in* $C^{1,\alpha}_{\mathrm{loc}}(\mathbb{R}^n \setminus \Omega^i, \mathbb{R}^n) \times C^{1,\alpha}(\mathrm{cl}\Omega^o, \mathbb{R}^n)$. *Let* $\mathcal{H}^i$ *be the matrix valued function on* $\partial\Omega^i$ *defined by the right hand side of (16) with* $\tilde{v}^i$ *replaced by* $\tilde{w}^i$.

If $n = 2$, *we assume that the matrix* $I - \frac{(\omega+2)}{4\pi(\omega+1)}\int_{\partial\Omega^i}\mathcal{H}^i\, d\sigma$ *is invertible. If* $n \geq 3$, *we assume that* $-\mathcal{H}^i$ *satisfies assumptions (4), (5) on* $\partial\Omega^i$.

Then there exist $\epsilon' \in\,]0, \epsilon_0[$ *and a family* $\{u(\epsilon, \cdot)\}_{\epsilon\in]0,\epsilon'[}$ *such that* $u(\epsilon, \cdot)$ *belongs to* $C^{1,\alpha}(\mathrm{cl}\Omega(\epsilon), \mathbb{R}^n)$ *and solves (7) for all* $\epsilon \in\,]0, \epsilon'[$, *and such that the family* $\{\frac{\gamma(\epsilon)}{\epsilon^{n-1}}u(\epsilon, \cdot)\}_{\epsilon\in]0,\epsilon'[}$ *converges in* $\mathrm{cl}\Omega^o \setminus \{0\}$ *to* $\Gamma_n(\omega, \cdot)\int_{\partial\Omega^i} T(\omega, D\tilde{w}^i)\nu^i\, d\sigma + \tilde{w}^o(\cdot)$, *and such that condition (17) with* $\tilde{v}^i$ *replaced by* $\tilde{w}^i$ *holds. Moreover, the following two statements hold.*

(i) *Let* $\tilde{\Omega}$ *be a bounded open subset of* $\Omega^o \setminus \{0\}$ *such that* $0 \notin \mathrm{cl}\tilde{\Omega}$. *Then there exist* $\epsilon_{\tilde{\Omega}} \in\,]0, \epsilon'[$, *and an open neighborhood* $\mathcal{U}_{\gamma_M}$ *in* $\mathbb{R}^2$ *of* $(\gamma_M, 1 - \delta_{2,n})$, *and a real analytic operator* $U_{\tilde{\Omega}}$ *of* $]-\epsilon_{\tilde{\Omega}}, \epsilon_{\tilde{\Omega}}[\times\mathcal{U}_{\gamma_M}$ *to* $C^{1,\alpha}(\mathrm{cl}\tilde{\Omega}, \mathbb{R}^n)$ *such that (13) holds, and such that*

$$\Xi_{M,n}(\epsilon) \equiv \left(\frac{\gamma(\epsilon)}{\epsilon^{n-1}}, (\log \epsilon)^{-\delta_{2,n}}\right) \in \mathcal{U}_{\gamma_M}$$

for all $\epsilon \in\,]0, \epsilon'[$, *and such that*

$$u(\epsilon, x) = \frac{\epsilon^{n-1}}{\gamma(\epsilon)}U_{\tilde{\Omega}}[\epsilon, \Xi_{M,n}(\epsilon)](x) \qquad \forall x \in \mathrm{cl}\tilde{\Omega}\,, \ \forall \epsilon \in\,]0, \epsilon_{\tilde{\Omega}}[\,.$$

(ii) *Let* $\tilde{\Omega}$ *be a bounded open subset of* $\mathbb{R}^n \setminus \mathrm{cl}\Omega^i$. *Then there exist* $\epsilon_{\tilde{\Omega},r} \in\,]0, \epsilon'[$, *and an open neighborhood* $\mathcal{U}_{\gamma_M}$ *in* $\mathbb{R}^2$ *of* $(\gamma_M, 1 - \delta_{2,n})$, *and a real analytic map* $U_{\tilde{\Omega}}^{r,1}$ *of* $]-\epsilon_{\tilde{\Omega},r}, \epsilon_{\tilde{\Omega},r}[\times\mathcal{U}_{\gamma_M}$ *to* $\mathbb{R}^n$, *and real analytic maps* $U_{\tilde{\Omega}}^{r,j}$ *for* $j = 2, 3$ *of* $]-\epsilon_{\tilde{\Omega},r}, \epsilon_{\tilde{\Omega},r}[\times\mathcal{U}_{\gamma_M}$ *to* $C^{1,\alpha}(\mathrm{cl}\tilde{\Omega}, \mathbb{R}^n)$ *such that*

(14) holds and such that

$$u(\epsilon, \epsilon x) = \delta_{2,n} \frac{\epsilon \log \epsilon}{\gamma(\epsilon)} U_{\tilde{\Omega}}^{r,1}[\epsilon, \Xi_{M,n}[\epsilon]] + \frac{\epsilon}{\gamma(\epsilon)} U_{\tilde{\Omega}}^{r,2}[\epsilon, \Xi_{M,n}[\epsilon]](x)$$

$$+ \frac{\epsilon^{n-1}}{\gamma(\epsilon)} U_{\tilde{\Omega}}^{r,3}[\epsilon, \Xi_{M,n}[\epsilon]](x) \qquad \forall x \in \mathrm{cl}\tilde{\Omega}, \ \forall \epsilon \in]0, \epsilon_{\tilde{\Omega},r}[.$$

References

1. J.M. Ball, *Discontinuous equilibrium solutions and cavitation in nonlinear elasticity*, Philos. Trans. Roy. Soc. London Ser. A, **306**, (1982), 557–611.
2. M. Dalla Riva and M. Lanza de Cristoforis, *Microscopically weakly singularly perturbed loads for a nonlinear traction boundary value problem. A functional analytic approach*, (2009), to appear in Complex Variables and Elliptic Equations, pp. 1–25.
3. M. Dalla Riva and M. Lanza de Cristoforis, *A singularly perturbed nonlinear traction boundary value problem for linearized elastostatics. A functional analytic approach*, (2008), to appear in Analysis (Munich), pp. 1–26.
4. M. Dalla Riva and M. Lanza de Cristoforis, *Weakly singular and microscopically hypersingular load perturbation for a nonlinear traction boundary value problem. A functional analytic approach*, (2009), submittted, pp. 1–20.
5. M. Dalla Riva and M. Lanza de Cristoforis, *Hypersingularly perturbed loads for a nonlinear traction boundary value problem. A functional analytic approach*, (2009), submitted, pp. 1–20.
6. G. Dal Maso and F. Murat, *Asymptotic behaviour and correctors for linear Dirichlet problems with simultaneously varying operators and domains* Ann. Inst. H. Poincaré Anal. Non Linéaire, **21**, (2004), 445–486.
7. D. Gilbarg and N.S. Trudinger, *Elliptic Partial Differential Equations of Second Order*, Springer Verlag, 1983.
8. V.A. Kozlov, V.G. Maz'ya and A.B. Movchan, *Asymptotic analysis of fields in multistructures*, Oxford Mathematical Monographs, The Clarendon Press, Oxford University Press, New York, 1999.
9. M. Lanza de Cristoforis, *Asymptotic behaviour of the conformal representation of a Jordan domain with a small hole, and relative capacity*, in 'Complex Analysis and Dynamical Systems', Proc. Conf. Karmiel, June 19-22, 2001, edited by M. Agranovsky, L. Karp, D. Shoikhet, and L. Zalcman, Contemp. Math., **364**, (2004) 155-167.
10. V.G. Mazya, S.A. Nazarov and B.A. Plamenewskii, *Asymptotic theory of elliptic boundary value problems in singularly perturbed domains*, **I, II**, (translation of the original in German published by Akademie Verlag 1991), Oper. Theory Adv. Appl., **111**, **112**, Birkhäuser Verlag, Basel, 2000.
11. S. Ozawa, *Electrostatic capacity and eigenvalues of the Laplacian*, J. Fac. Sci. Univ. Tokyo Sect. IA Math., **30**, (1983), 53–62.
12. M.J. Ward and J.B. Keller, *Strong localized perturbations of eigenvalue problems*, SIAM J. Appl. Math., **53**, (1993), 770–798.

Boundary behaviour of normal functions

P.V. Dovbush

*Institute of Mathematics and Computer Science of Academy of Sciences of Moldova,
5 Academy Street, MD-2028, Kishinev, Republic of Moldova
E-mail: peter.dovbush@gmail.com*

In multidimensional case we give an extension of the Lindelöf-Lehto-Virtanen theorem for normal functions and Lindelöf-Gehring-Lohwater theorem involving two paths for bounded functions.

Keywords: Normal function; boundary behavior.

1. Introduction

In the case of bounded analytic functions in the unit disc there exist various results guaranteeing the existence of an angular limit at a given boundary point. The fundamental result of this kind, due to Lindelöf, states that the existence of an asymptotic value implies the existence of an angular limit. Subsequently the hypothesis of the existence of an asymptotic value has been weakened by Gehring and Lohwater [10, Theorem 2], and by others.

Let U be the open unit disc in the complex plane $\mathbb{C}$, i.e., $U = \{\lambda \in \mathbb{C} : |\lambda| < 1\}$. A (symmetric) Stolz angle at the point $\xi \in \partial U$ is a set of the form

$$S_\beta(\xi) = \{\lambda \in U : |\arg(1 - \bar{\xi}\lambda)| < \beta\} \quad (0 < \beta < \pi/2).$$

The classical theorem of Gehring and Lohwater [10, Theorem 2, p. 165] states:

Theorem 1.1. *Let function f be analytic and bounded in U, and γ_1 and γ_2 be two Jordan arcs lying in U, and terminating at a point ξ of ∂U. If $Ref(\lambda) \to a$ as $\lambda \to \xi$ on γ_1 and $Imf(\lambda) \to b$ as $\lambda \to \xi$ on γ_2 then $f(\lambda)$ tends to $a + ib$ uniformly as $\lambda \to \xi$ inside any Stolz angle $S_\beta(\xi)$.*

The Lindelöf principle can be generalized in different directions, the two most notable of which are conditions allowing broader classes of approach curves and broader classes of functions.

The class of normal functions was introduced by Lehto and Virtanen[14] in 1957; there is now a large volume of literature on the subject (see[17] for a survey of twenty-two different characterizations of normal functions). A function f meromorphic on a simply connected domain $G \subset \mathbb{C}$ is said to be normal on G if the family $\{f(\varphi)\}$, where φ ranges over all automorphisms (one-to-one biholomorphic maps of G onto itself) of G, is a normal family.

The Lindelöf principle was improved by Lehto and Virtanen.[14]

Theorem 1.2. *If $f(\lambda)$ is meromorphic and normal function in Δ and has asymptotic value L at $\lambda_0 \in \partial\Delta$, then $f(\lambda)$ also has the angular limit L at λ_0.*

The first important several variables version of the Lindelöf principle for bounded holomorphic functions is due to Cirka.[8] Dovbush[6] established the Lindelöf principle for normal holomorphic functions of several complex variables, his approach has been further pursued in Cima and Krantz[3] (see,[11,12] among others).

The purpose of the present article is to give the extension of the Gehring–Lohwater and Lehto–Virtanen theorems which valid in bounded domains in $\mathbb{C}^n$, $n > 1$, with C^2-smooth boundary.

A domain D in $\mathbb{C}^n$ with boundary ∂D is said to have C^2-smooth boundary if there is a two times continuously differentiable function ρ defined on a neighborhood U of ∂D such that

(1) $D = \{z \in \mathbb{C}^n : \rho(z) < 0\}$;
(2) $\nabla\rho = (\partial\rho/\partial z_1, ..., \partial\rho/\partial z_n) \neq 0$ on ∂D.

Let D be a bounded domain in $\mathbb{C}^n$, $n > 1$, with C^2-smooth boundary ∂D, $T_\xi(\partial D)$ be the tangent space to ∂D at $\xi \in \partial D$, and let ν_ξ be the unit outward normal vector to ∂D at ξ. Let k_D be the infinitesimal Kobayashi metric, and let K_D be the Kobayashi metric on D. Denote by $\mathbb{C}\nu_\xi$ the complex the complex normal space.

We say that a subset $S \subset D$ is asymptotic at $\xi \in \partial D$ if $\overline{S} \cap \partial D = \{\xi\}$. Let $\pi_\xi : \mathbb{C}^n \to \mathbb{C}\nu_\xi$ be the orthogonal projection.

An open set $\mathcal{K} \subset D$ is called weakly admissible domain at $\xi \in \partial D$ if $\mathcal{K}$ is asymptotic at ξ, $\pi_\xi(\mathcal{K}) \subset \Delta$, where Δ is a triangle in $D \cap \mathbb{C}\nu_\xi$ with vertex ξ, and $\lim_{\mathcal{K} \ni z \to \xi} K_D(z, \pi_\xi(z)) = 0$.

The set

$$A_\alpha^\epsilon(\xi) = \{\, z \in D \; : |(z - \xi, \nu_\xi)| < (1 + \alpha)\delta_\xi(z), |z - \xi|^2 < \alpha\delta_\xi^{1+\epsilon}(z) \,\},$$
$$\delta_\xi(z) = \min\{p[z, \partial D], p[z, T_\xi(\partial D)]\},$$

give an example of the weakly admissible domain (see[7] for the proof). Here $\alpha, \varepsilon > 0$, $(\cdot, \cdot)$ denotes canonical hermitian product of $\mathbb{C}^n$, and p denotes the Euclidean distance in $\mathbb{C}^n$.

If $\xi \in \partial D$, $f : D \to \overline{\mathbb{C}}$, $l \in \overline{\mathbb{C}}$, we say f has weakly admissible limit l at ξ if $\lim_{\mathcal{K} \ni z \to \xi} f(z) = l$ for every weakly admissible domain $\mathcal{K}$ at ξ.

By Jordan arc we mean the homeomorphic image of $[0, 1]$. By Jordan curve we mean the homeomorphic image of ∂U; it bounds two Jordan domains. We say Ω is a Jordan domain if $\partial \Omega$ is a Jordan curve in $\mathbb{C}$.

We shall say that a curve $\gamma : [0, 1) \to D$ ending at $\xi \in \partial D$ is *special* at ξ if $\gamma_\xi = \pi_\xi \circ \gamma$ is a Jordan arc lying in $D \cap \mathbb{C}\nu_\xi$, except for the point ξ, and $\lim_{t \to 1-} K_D(\gamma(t), \pi_\xi \circ \gamma(t)) = 0$.

The class of normal functions was introduced by Lehto and Virtanen[14] in 1957; there is now a large volume of literature on the subject (see[17] for a survey of twenty-two different characterizations of normal functions). It is conceivable that the notion of normality can be generalized in various way to higher dimension. Here we adopt the following definition (see[4]):

A holomorphic function $f : D \to \mathbb{C}$ is called a normal function if there exists a constant $L > 0$ such that

$$s(f(z), f(w)) \leq L \cdot K_D(z, w) \quad \text{for all} \quad z, w \in D. \tag{1}$$

Here $s(\cdot, \cdot)$ stands for the spherical metric on the Riemann sphere $\overline{\mathbb{C}}$.

Other definition can be given: see[5] and[13] and the references therein for more work on normal functions in several variables.

2. Lindelöf's theorem

We have the following extension of the Lindelöf principle (cf.[7]):

Theorem 2.1. *Let D be a bounded domain in $\mathbb{C}^n$ with C^2-smooth boundary. Let f be a normal function on D and suppose that f has asymptotic limit L along a curve γ special at $\xi \in \partial D$. Then f has weakly admissible limit L at ξ.*

Proof. Without loss of generality, choose coordinate system in $\mathbb{C}^n$ such that $\xi = 0$, and $N_0 = \{z \in \mathbb{C}^n : z = (z_1, 0, \ldots, 0)\}$. In the following, Ω denotes a Jordan domain in $D \cap N_0$ with $0 \in \partial\Omega$, which is bounded by a circle of radius $r > 0$ about the origin 0 and $\partial D \cap N_0$. By the Riemann mapping theorem there is a conformal map F of the unit disk U onto Ω. Because F maps U onto Ω, the continuous extension (also denoted by F) must map ∂U onto $\partial\Omega$, and because F is one-to-one on ∂U, (see, e.g., [9,

Theorem 3.1, p. 13]), we may suppose that $F(1) = 0$. Since $\rho(z_1, 0, \ldots, 0)$ is twice continuously differentiate and $\rho' \neq 0$ on $\partial\Omega$ then $F'(1)$ exists and is $\neq 0$ (see, e.g., [18, Theorem 3.4, p. 51] or[19]). Furthermore, F^{-1} also has a non-zero derivative at 0. The existence of a non-zero angular derivative implies conformality (see, e.g., [9, p. 175]). Since $\lambda = F^{-1}(z_1)$ is conformal at 0 and if $|\alpha| < \pi/2$, then as $z_1 \to 0$ the image arc $F^{-1}(\{\arg(z_1) = \alpha\})$ is asymptotic to the ray $\{\arg(\lambda - 1) = \alpha + \arg((F^{-1})'(0))\}$. In other words, F^{-1} preserves angles between nontangential rays.

By Theorem 1.2 the function $f \circ F^{-1}$ has limit L along $S_\beta(1)$ for all $\beta > 0$. Since for some α, $F^{-1}(\Delta) \subset S_\alpha(1)$ we conclude that f has limit L along Δ. The assertion of the theorem now easily follows from the inequality (1). This completes the proof. $\qquad\square$

Remark 2.1. Unfortunately, E.A. Poletskiĭ and B.V. Shabat [15, p. 73] did not outline a proof of Čirka's theorem[8] in the case of domain in $\mathbb{C}^n$ with C^2-smooth boundary, because they use inequality (1.15) on p. 69 in[15] which cannot lead to the needed conclusion. They only outline a proof (unfortunately with typos) of the Lindelöf principle due to Dovbush[6] in the case of normal functions on a strictly pseudoconvex domain.

3. Lindelöf's theorem involving two paths

Let $H^\infty(D)$ be the set of all bounded holomorphic functions on D.

Theorem 3.1. *Let D be a bounded domain in $\mathbb{C}^n$ with C^2-smooth boundary, $f \in H^\infty(D)$, and let γ_1 and γ_2 be two special curves lying in D, and terminating at a point ξ of ∂D. If $Ref(\lambda) \to a$ as $\lambda \to \xi$ on γ_1 and $Imf(\lambda) \to b$ as $z \to \xi$ on γ_2 then f then f has weakly admissible limit $a + ib$ at ξ.*

Proof. We may assume that $|f(z)| < 1$ for all $z \in D$. The following is an immediate consequence of the definition and properties of the Kobayashi metric

$$|f_*(z)v| < \frac{|f_*(z)v|}{1 - |f(z)|^2} \leq k_D(z, v) \text{ for all } (z, v) \in D \times \mathbb{C}^n.$$

Here

$$f_*(z)v = \sum_{j=1}^{n} \frac{\partial f}{\partial z_j}(z)v_j.$$

Let z and w be distinct points in D. Let $\gamma : [0,1] \to X$ be a piecewise C^1 curves joining z to w in D. Then

$$|f(z) - f(w)| \leq \int_0^1 k_D(\gamma(t), \gamma'(t))dt.$$

Taking the infimum of the right-hand side over all piecewise C^1 curves γ satisfying $\gamma(0) = z$, $\gamma(1) = w$, we obtain

$$|f(z) - f(w)| \leq K_D(z,w) \quad \text{for all } z, w \in D. \tag{2}$$

We conclude from this inequality that

$$|Ref(\gamma_1(t)) - Ref(\pi_0 \circ \gamma_1(t))| \leq K_D(\gamma_1(t), \pi_0 \circ \gamma_1(t)).$$

Since curve γ_1 is special at 0 the right-hand side of the above inequality tends to zero as $t \to 1-$, and $Ref(\gamma_1(t)) \to a$ as $t \to 1-$, these implies that $Ref(\pi_0 \circ \gamma_1(t)) \to a$ as $t \to 1 -$. Arguing as above we conclude that $Imf(\pi_0 \circ \gamma_2(t)) \to b$ as $t \to 1 -$.

Since F^{-1} preserves angles between nontangential rays, by Theorem 1.1 the function $f \circ F^{-1}$ has limit $a + ib$ along $S_\beta(1)$ for all $\beta > 0$. Since for some α, $F^{-1}(\Delta) \subset S_\alpha(1)$ we conclude that f has limit $a + ib$ along Δ. The assertion of the theorem now easily follows from the inequality (2). This completes the proof. $\square$

Remark 3.1. Theorem 3.1 is a generalization of a well-known theorem of Lindelöf due to E.M. Čirka[8] (see surveys[1] and[5] for more about the Lindelöf principle), and the fact (established by Dovbush[7]) that the Lindelöf theorem holds for normal holomorphic functions makes it natural to ask whether the above theorem remains valid if the assumption that f is holomorphic and bounded in D is replaced by the hypothesis that f is holomorphic and normal in D. Using[2] one can answer this question in the negative.

References

1. Abate, M., 2004, Angular derivatives in several complex variables. - In: Real methods in complex and CR geometry, Eds. D. Zaitsev, G. Zampieri, Lect. Notes in Math. 1848, Springer, Berlin, pp. 1–47.
2. Bagemihl, F., 1962, The Lindelöf theorem and the real and imaginary parts of normal functions, Michigan Math. J., v. 9, Issue 1, pp. 15–20.
3. Cima, J.A., and Krantz, S.G., 1983, The Lindelöf principle and normal functions of several complex variables, Duke Math. J., v. 50, 303–328.
4. Dovbush, P.V., 2009, On normal and non-normal holomorphic functions on complex Banach manifolds, Ann. Scuola Norm. Sup. Pisa Cl. Sci. (5) Vol. VIII (2009), pp. 1–15

5. Dovbush, P.V., and Gavrilov, V.I., 2001, Normal mappings, Math. Montisnigri, v. XIV, pp. 5–62. (in Russian)

6. Dovbush, P.V., 1981, Lindelöf's theorem in $\mathbb{C}^n$, Vestnic Moskov. Univ. Ser. I, Mat. Meh. (6), pp. 33–36 [English transl. in Mosk. Univ. Math. Bull., 36 (1981), no. 6, pp. 41–44].

7. Dovbush, P.V., 2009, Boundary behavior of Bloch functions and normal functions, Complex Variables and Elliptic Equations.
URL: http://dx.doi.org/10.1080/17476930902999108.

8. Čirka, E.M., 1973, The Lindelöf and Fatou theorems in $\mathbb{C}^n$, Math. USSR, Sb., v. 21, pp. 619–641.

9. Garnett, J. B., and Marshall, D. E., 2005, Harmonic Measure, Cambridge University Press, Cambridge.

10. Gehring, F.W., and Lohwater, A.J., 1958, On the Lindelöf theorem, Math. Nachr., v. 19, pp. 165–170.

11. Hahn, K.T., 1984, Asymptotic behavior of normal mappings of several complex variables, Canad. J. Math. , v. 36, pp. 718–746.

12. Hahn, K.T., 1988, Non-tangential limit theorems for normal mappings, Pacific J. Math., v. 135, pp. 57–64.

13. Kwack, M.H., 1996, Families of normal maps in several variables and classical theorems in complex analysis, Lecture Notes Series 33 (Res. Inst. Math., Global Analysis Res. Center, Seoul, Korea).

14. Lehto, O., and Virtanen, K.I., 1957, Boundary behaviour and normal meromorphic functions, Acta Math., v. 97, pp. 47–65.

15. Poletskiĭ, E. A., and Shabat, B. V., Invariant metrics. - In: Several Complex Variables III, G. M. Khenkin (ed.), Springer Verlag, 1989, pp. 63–112.

16. Lappan, P., 1961, Non-normal sums and prodacts of unbounded normal functions, Mich. Math. J., v. 8, pp. 187–192.

17. Campbell, D. M., and Wickes, G., 1978, Characterizations of normal meromorphic functions, - In: Complex Analysis, Joensuu 1978, Lect. Notes in Math. 747, Springer, Berlin, pp. 55–72.

18. Pommerenke, Ch., 2002, Conformal maps at the boundary. - In: Handbook of complex analysis: Geometric function theory, Vol. 1, North-Holland, Amsterdam, pp. 37–74.

19. Warschawski, S.E., 1961, On differentiability at the boundary in conformal mapping, Proc. Amer. Math. Soc., v. 12, pp. 614–620.

Spatial quasiconformal mappings and directional dilatations

A. Golberg

*Department of Applied Mathematics, Holon Institute of Technology,
Holon, 58102, Israel
E-mail: golberga@hit.ac.il*

New geometric estimates of conformal moduli of curve and surface families
are established. The bounds are given in the terms of integrals depending on
dilatations connected with a given direction. The sharpness of the estimates is
illustrated by some examples.

Keywords: Conformal module of k-dimensional surface families; module of ring
domain; quasiconformal dilatations; directional dilatations.

1. Introduction

In geometric function theory, the quasiconformal homeomorphisms form
a natural interpolating class of mappings between the classes of bilipschitz
maps and general homeomorphisms. The main tool for investigation of such
mappings in $\mathbb{R}^n$ relies on varying and estimating conformal moduli of k-
dimensional surface families and involves more general inequalities than
quasi-invariance. This method allows to obtain, for example, the best upper
and lower bounds for various integral dilatations, which are crucial in many
problems. One of the most important applications concerns the solving
variational problems (see, e.g., Ref. 6).

The following important Kovtonyk-Ryazanov[4] inequalities estimate the
conformal moduli in the terms of integrals depending on the inner and outer
dilatations of quasiconformality and admissible metrics.

Let G be an open set in $\mathbb{R}^n$, and suppose the homeomorphism $f :
G \to \mathbb{R}^n$ belongs to FAD_k (class of mappings with finite area distortion
in dimension k) for some $1 \leq k \leq n - 1$. Then, for every family $\mathcal{S}_k$ of
k-dimensional surfaces $\mathcal{S}$ in G, f satisfies the double inequality

$$\inf \int_G \frac{\rho^n(x)}{K_O(x, f)} \, dx \leq \mathcal{M}(f(\mathcal{S}_k)) \leq \inf \int_G \varrho^n(x) K_I(x, f) \, dx, \qquad (1)$$

where the infima are taken over all $\rho \in \operatorname{extadm} \mathcal{S}_k$ and $\varrho \in \operatorname{adm} \mathcal{S}_k$, respectively.

In general case, the inequalities (1) can not be improved. However, we shall show that in the main cases $k = 1$ and $k = n - 1$ there is possible to obtain essentially better bounds. One needs for this other dilatations which depend on a given direction.

2. Moduli and quasiconformal dilatations

First recall some notions and results concerning the conformal module and quasiconformality in $\mathbb{R}^n$ starting with the definition of k-dimensional Hausdorff measure H^k, $k = 1, \ldots, n - 1$ in $\mathbb{R}^n$. For a given $E \subset \mathbb{R}^n$, put

$$H^k(E) = \sup_{r > 0} H_r^k(E),$$

where $H_r^k(E) = \Omega_k \inf \sum_i (\delta_i/2)^k$ and the infimum is taken over all countable coverings $\{E_i, i = 1, 2, \ldots\}$ of E with diameters δ_i, and Ω_k is the volume of the unit ball in $\mathbb{R}^k$.

Let $\mathcal{S}$ be a k-dimensional surface, which means that $\mathcal{S} : D_s \to \mathbb{R}^n$ is a continuous image of the closed domain $D_s \subset \mathbb{R}^k$. We denote by

$$N(\mathcal{S}, y) = \operatorname{card} \mathcal{S}^{-1}(y) = \operatorname{card} \{x \in D_s : \mathcal{S}(x) = y\}$$

the multiplicity function of the surface $\mathcal{S}$ on the points $y \in \mathbb{R}^n$. For a given Borel set $B \subseteq \mathbb{R}^n$, the k-dimensional Hausdorff area of B in $\mathbb{R}^n$ associated with the surface $\mathcal{S}$ is determined by

$$\mathcal{H}_{\mathcal{S}}(B) = \mathcal{H}_{\mathcal{S}}^k(B) = \int_B N(\mathcal{S}, y)\, dH^k y.$$

If $\rho : \mathbb{R}^n \to [0, \infty]$ is a Borel function, the integral of ρ over $\mathcal{S}$ is defined by

$$\int_{\mathcal{S}} \rho\, d\sigma_k = \int_{\mathbb{R}^n} \rho(y) N(\mathcal{S}, y)\, dH^k y.$$

Let $\mathcal{S}_k$ be a family of k-dimensional surfaces $\mathcal{S}$ in $\mathbb{R}^n$, $1 \leq k \leq n - 1$ (curves for $k = 1$). The **conformal module** of $\mathcal{S}_k$ is defined by

$$\mathcal{M}(\mathcal{S}_k) = \inf \int_{\mathbb{R}^n} \rho^n\, dx,$$

where the infimum is taken over all Borel measurable functions $\rho \geq 0$ such that $\int_{\mathcal{S}} \rho^k\, d\sigma_k \geq 1$ for every $\mathcal{S} \in \mathcal{S}_k$. We call each such ρ an **admissible function** for $\mathcal{S}_k$ ($\rho \in \operatorname{adm} \mathcal{S}_k$).

Following Ref. 4, an admissible metric $\rho \in \operatorname{adm} \mathcal{S}_k$ is said to be **extensively admissible** for $\mathcal{S}_k$ ($\rho \in \operatorname{extadm} \mathcal{S}_k$) if $\mathcal{M}(\mathcal{S}_k) \neq 0$.

A **ring domain** $\mathcal{R} \subset \mathbb{R}^n$ is a bounded domain whose complement consists of two components C_0 and C_1. The sets $F_0 = \partial C_0$ and $F_1 = \partial C_1$ are two boundary components of $\mathcal{R}$. Let us assume, for definiteness, that $\infty \in C_1$.

We say that a curve γ joins the boundary components in $\mathcal{R}$ if γ lies in $\mathcal{R}$, except for its endpoints, one of which lies on F_0 and the second on F_1. A compact set $\mathcal{S}$ is called to separate the boundary components of $\mathcal{R}$ if $\mathcal{S} \subset \mathcal{R}$ and if C_0 and C_1 are located in different components of $C\Sigma$. Denote by $\Gamma_{\mathcal{R}}$ the family of all locally rectifiable curves γ which join the boundary components of $\mathcal{R}$, and by $\Sigma_{\mathcal{R}}$ the family of all compact piecewise smooth $(n-1)$-dimensional surfaces $\mathcal{S}$ which separate the boundary components of $\mathcal{R}$.

The module of a ring domain $\mathcal{R}$ can be represented by

$$\operatorname{mod} \mathcal{R} = \left(\frac{\omega_{n-1}}{\mathcal{M}(\Gamma_{\mathcal{R}})} \right)^{\frac{1}{n-1}} = \omega_{n-1}^{\frac{1}{n-1}} \mathcal{M}(\Sigma_{\mathcal{R}}); \tag{2}$$

here ω_{n-1} is the $(n-1)$-dimension Lebesgue measure of the unit sphere S^{n-1} in $\mathbb{R}^n$. For the spherical ring $A(x_0, r, R) = \{x \in \mathbb{R}^n : 0 < r < |x - x_0| < R\}$, we have the equality $\mathcal{M}(A) = \log(R/r)$.

Let G be an open set in $\mathbb{R}^n$, $n \geq 2$. A homeomorphic mapping $f : G \to \mathbb{R}^n$ is called K-**quasiconformal**, $K \geq 1$, if $f \in W_{\mathrm{loc}}^{1,n}(G)$ and if $K_O(x, f) \leq K$ almost everywhere in G. The quantities

$$K_I(x, f) = \frac{|J_f(x)|}{l^n(f'(x))}, \qquad K_O(x, f) = \frac{\|f'(x)\|^n}{|J_f(x)|},$$

are called the **inner** and **outer** dilatations, respectively. Here (see, e.g., Ref. 5,7) $J_f(x)$ is the determinant of the Jacobian matrix,

$$\|f'(x)\| = \max_{|h|=1}\{|f'(x)h|\}, \quad l(f'(x)) = \min_{|h|=1}\{|f'(x)h|\}.$$

We also remind that a continuous mapping f satisfies (N)-property with respect to k-dimensional Hausdorff area if $\mathcal{H}_{\mathcal{S}}^k(f(B)) = 0$ whenever $\mathcal{H}_{\mathcal{S}}^k(B) = 0$. Similarly, f has (N^{-1})-property if $\mathcal{H}_{\mathcal{S}}^k(B) = 0$ whenever $\mathcal{H}_{\mathcal{S}}^k(f(B)) = 0$.

3. Directional dilatations and lower bounds for modili

Several kinds of directional dilatations have been successfully utilized for investigation of various properties of quasiconformal mappings and their generalizations in the two-dimensional case. The first attempt to extend such tool in $\mathbb{R}^n$, $n > 2$ is due to Andreian Cazacu (see, Ref. 1).

We define two another directional characteristics in $\mathbb{R}^n$, using the derivative of f in a direction $h, h \neq 0$, at x, given by $\partial_h f(x) = \lim_{t \to 0^+} (f(x + th) - f(x))t^{-1}$.

Let $f : G \to \mathbb{R}^n$ be a quasiconformal mapping. For a point $x_0 \in \mathbb{R}^n$, we define the **angular** and **normal dilatations** of the mapping f at the point x, $x \neq x_0$ **with respect to** x_0 by

$$D_f(x, x_0) = \frac{|J_f(x)|}{\ell_f^n(x, x_0)}, \qquad T_f(x, x_0) = \left(\frac{\mathcal{L}_f^n(x, x_0)}{|J_f(x)|} \right)^{\frac{1}{n-1}},$$

respectively. Here

$$\ell_f(x, x_0) = \min_{|h|=1} \frac{|\partial_h f(x)|}{|\langle h, u \rangle|}, \qquad \mathcal{L}_f(x, x_0) = \max_{|h|=1} \left(|\partial_h f(x)||\langle h, u \rangle| \right),$$

and $u = (x - x_0)/|x - x_0|$. The dilatations $D_f(x, x_0)$ and $T_f(x, x_0)$ are both measurable in G. The relations

$$l(f'(x)) \leq \ell_f(x, x_0) \leq |\partial_u f(x)| \leq \mathcal{L}_f(x, x_0) \leq ||f'(x)||,$$

being true for each x_0, imply

$$K_O^{-1}(x, f) \leq D_f(x, x_0) \leq K_I(x, f). \qquad (3)$$

The normal dilatation $T_f(x, x_0)$ has the same bounds as $D_f(x, x_0)$, since

$$K_O^{-1}(x, f) \leq K_I^{-\frac{1}{n-1}}(x, f) \leq T_f(x, x_0) \leq K_O^{\frac{1}{n-1}}(x, f) \leq K_I(x, f). \qquad (4)$$

Note that the dilatations $D_f(x, x_0)$ and $T_f(x, x_0)$ for the multi-dimensional case have been introduced in Ref. 3 and Ref. 2. Note that the angular and normal dilatations range both between 0 and ∞, while the classical dilatations are always greater than or equal to 1.

Lemma 3.1. *Let $f : G \to \mathbb{R}^n$ be a quasiconformal mapping. Suppose that $D_f(x, x_0)$ is locally integrable in the spherical ring $A = A(x_0; r, R) \subset G$. Then*

$$\mathcal{M}(f(\Gamma_A)) \geq \inf \int_A \frac{\rho^n(|x - x_0|)}{T_f^{n-1}(x, x_0)} \, dx,$$

where the infimum is taken over all $\rho \in \operatorname{extadm} \Gamma_A$.

Lemma 3.2. *Let $f : G \to \mathbb{R}^n$ be a quasiconformal mapping satisfying (N)-property with respect to $(n-1)$-dimension Hausdorff area. Suppose that $T_f(x, x_0)$ is locally integrable in the spherical ring $A = A(x_0; r, R) \subset G$. Then*

$$\mathcal{M}(f(\Sigma_A)) \geq \inf \int_A D_f^{-\frac{1}{n-1}}(x, x_0) \varrho^n \left(\frac{x - x_0}{|x - x_0|} \right) dx,$$

where the infimum is taken over all $\varrho \in \operatorname{extadm} \Sigma_A$.

4. Main results

The above Lemmas 3.1 and 3.2, together with Lemma 2.4 from Ref. 3 and Theorem 1 (Ref. 2 provide the following theorems.

Theorem 4.1. *Let $f : G \to \mathbb{R}^n$ be a quasiconformal mapping. Suppose that $D_f(x, x_0)$ is locally integrable in the spherical ring $A = A(x_0; r, R) \subset G$. Then the double inequality*

$$\inf \int_A \frac{\rho^n(|x - x_0|)}{T_f^{n-1}(x, x_0)} dx \leq \mathcal{M}(f(\Gamma_A)) \leq \inf \int_A \varrho^n(|x - x_0|) D_f(x, x_0) dx$$

is fulfilled; here the infima are taken over all $\rho \in \operatorname{extadm} \Gamma_A$ and $\varrho \in \operatorname{adm} \Gamma_A$, respectively.

Theorem 4.2. *Let $f : G \to \mathbb{R}^n$ be a quasiconformal mapping satisfying (N) and (N^{-1})-properties with respect to $(n-1)$-dimension Hausdorff area. Suppose that $T_f(x, x_0)$ is locally integrable in the spherical ring $A = A(x_0; r, R) \subset G$. Then the double inequality*

$$\inf \int_A \frac{\rho^n\left(\frac{x - x_0}{|x - x_0|}\right)}{D_f^{\frac{1}{n-1}}(x, x_0)} dx \leq \mathcal{M}(f(\Sigma_A)) \leq \inf \int_A \varrho^n\left(\frac{x - x_0}{|x - x_0|}\right) T_f(x, x_0) dx$$

is fulfilled; here the infima are taken over all $\rho \in \operatorname{extadm} \Sigma_A$ and $\varrho \in \operatorname{adm} \Sigma_A$, respectively.

Using (2) one can obtained similar estimates for the module of ring domains.

5. Illustrating examples

The following examples illustrate the sharpness of the distortion estimates provided in Theorems 4.1 and 4.2 and show that the directional dilatations are more flexible then the classical ones (cf. (3) and (4)). Without loss of generality we calculate all directional dilatations with respect to 0.

Example 5.1. Consider the radial stretching in $\mathbb{R}^n$,

$$f(x) = x(1 - \log |x|), x \neq 0, f(0) = 0.$$

Because of the radial symmetry of the mapping f, we can restrict ourselves by x and h of the form

$$x = (r, 0, 0, \ldots, 0), \ \ h = (\cos \theta, \sin \theta, 0, \ldots, 0), \ 0 < r < 1, \ 0 \leq \theta \leq 2\pi,$$

and identify the real two-dimensional plane $x_3 = \ldots = x_n = 0$ with the complex plane $z = x_1 + ix_2$. Set $\varphi(z) = f(x)$, then $\varphi_z = (1 - 2\log |z|)/2$, $\varphi_{\bar{z}} = -z/2\bar{z}$ and

$$|\partial_h \varphi| = \frac{1}{2}\left(1 + 2\log \frac{1}{|z|}\right)\left|1 - \frac{z}{\bar{z}}\frac{e^{-2i\theta}}{1 + 2\log \frac{1}{|z|}}\right|.$$

Letting $z = re^{i\psi}$, $h = e^{i\theta}$, one derives

$$|\partial_h \varphi| \leq 1 + \log \frac{1}{|z|} = \|f'(x)\|, \quad |\partial_h \varphi| \geq \log \frac{1}{|z|} = l(f'(x));$$

the equalities occur for $\psi = \theta + \pi/2$ and $\psi = \theta$, respectively. Thus, $J_f(x) = \left(1 + \log \frac{1}{r}\right)^{n-1} \log \frac{1}{r}$,

$$K_I(x, f) = \left(1 + \frac{1}{\log \frac{1}{r}}\right)^{n-1}, \quad K_O(x, f) = 1 + \frac{1}{\log \frac{1}{r}}.$$

A calculation of the directional dilatations is much more complicated. We first find the quantity $\mathcal{L}_f(x, 0)$ with respect to the origin.

$$\mathcal{L}_f(x, 0) = \max_\theta |\partial \varphi_h(r) \cos \theta| = \max_\theta \left(|\partial \varphi_z(r)| |1 + \mu(r)e^{-2i\theta}| |\cos \theta|\right)$$

$$= \frac{1 + 2\log \frac{1}{r}}{4} \max_\theta |(1 + ke^{-2i\theta})(1 + e^{-2i\theta})|,$$

where $\mu(z) = \varphi_{\bar{z}}/\varphi_z = kz/\bar{z}$, $|\varphi_z(r)| = (1 + 2\log(1/r))/2$, $k = -1/(1 + 2\log(1/r))$. A straightforward computation implies

$$\mathcal{L}_f(x, 0) = \log \frac{1}{r} \quad \text{for} \quad \log \frac{1}{r} \geq 1 + \sqrt{2},$$

$$\mathcal{L}_f(x, 0) = \frac{\left(1 + \log \frac{1}{r}\right)^2}{2\sqrt{1 + 2\log \frac{1}{r}}} \quad \text{for} \quad \log \frac{1}{r} \leq 1 + \sqrt{2}.$$

Thus,

$$T_f(x,0) = \frac{\log \frac{1}{r}}{1 + \log \frac{1}{r}}, \qquad \log \frac{1}{r} \geq 1 + \sqrt{2},$$

$$T_f(x,0) = \frac{\left(1 + \log \frac{1}{r}\right)^{\frac{n+1}{n-1}}}{2^{\frac{n}{n-1}} \left(\log \frac{1}{r}\right)^{\frac{1}{n-1}} \left(1 + 2\log \frac{1}{r}\right)^{\frac{n}{2(n-1)}}}, \qquad \log \frac{1}{r} \leq 1 + \sqrt{2}.$$

The dilatation $D_f(x,0)$ can be calculated using a technique related to functions of one complex variable and presented in Ref. 3. The result is

$$D_f(x,0) = \left(1 + \frac{1}{\log \frac{1}{r}}\right)^{n-1}.$$

Indeed,

$$\ell_f(x,0) = \min_\theta \frac{|\partial \varphi_h(r)|}{|\cos \theta|} = \min_\theta \frac{|\partial \varphi_z(r)||1 + \mu(r)e^{-2i\theta}|}{|\cos \theta|}$$

$$= \left(1 + 2\log \frac{1}{r}\right) \min_\theta \frac{|1 + ke^{-2i\theta}|}{|1 + e^{-2i\theta}|}$$

$$= \left(1 + 2\log \frac{1}{r}\right)\langle \frac{1+k}{2}, 1 \rangle = \log \frac{1}{r}.$$

The example shows that the estimate in the left-hand side of inequality (1) has been essentially improved.

Example 5.2. For the radial stretching in $\mathbb{R}^n$,

$$f(x) = x|x|^{Q-1}, \quad Q \geq 1,$$

all the dilatations have been calculated in Ref. 2,3. The results are

$$K_I(x,f) = Q, \quad K_O(x,f) = Q^{n-1}, \quad D_f(x,0) = \frac{1}{Q}, \quad T_f(x,0) = Q.$$

Example 5.3. Consider the radial stretching in $\mathbb{R}^n$

$$f(x) = x|x|^{\frac{1}{Q}-1}, \quad Q \geq 1.$$

For this mapping, the calculations similar to above imply

$$K_I(x,f) = Q^{n-1}, \quad K_O(x,f) = Q, \quad D_f(x,x_0) = Q^{n-1}.$$

The computation of $T_f(x,0)$ is splitting in two cases $Q \leq \sqrt{2}$ and $Q \geq \sqrt{2}$ and results in

$$T_f(x,0) = \frac{1}{Q}, \quad Q \leq \sqrt{2} \quad \text{and} \quad T_f(x,0) = \frac{Q^{\frac{n+1}{n-1}}}{2^{\frac{n}{n-1}}(Q^2 - 1)^{\frac{n}{2(n-1)}}}, \quad Q \geq \sqrt{2}.$$

Example 5.4. Consider the mapping $f(x) = (x_1 \cos\theta - x_2 \sin\theta, x_2 \cos\theta + x_1 \sin\theta, x_3, ..., x_n)$, $0 < |x| < 1$, $f(0) = 0$, with $x = (x_1, ..., x_n)$ and $\theta = \log(x_1^2 + x_2^2)$; this mapping is quasiconformal and preserves the volume.

A straightforward calculation given in Ref. 3 yields

$$K_I(x, f) = K_O(x, f) = (1 + \sqrt{2})^n, \quad J_f(x) = D_f(x, 0) = 1, \quad x \neq 0.$$

A calculation of the directional dilatations is much more complicated. In the two-dimensional case the dilatation $T_f(x, 0) = \mathcal{L}_f^2(x, 0)$ for every $|x| < 1$, $x \neq 0$, equals

$$\frac{3 + \sqrt[3]{7 + 4\sqrt{3}} + \sqrt[3]{7 - 4\sqrt{3}}}{\left(1 + \left(2 - \sqrt[3]{2 + \sqrt{3}} - \sqrt[3]{2 - \sqrt{3}}\right)^2\right)^2}.$$

Acknowledgments

This work is partially supported by the European Science Foundation Networking Programme "Harmonic and Complex Analysis and its Application" (HCAA).

References

1. C. Andreian Cazacu, *Some formulae on the extremal length in n-dimensional case.* Proceedings of the Romanian-Finnish Seminar on Teichmüller Spaces and Quasiconformal Mappings (Braşov, 1969), pp. 87–102. Publ. House of the Acad. of the Socialist Republic of Romania, Bucharest, 1971.
2. A. Golberg, *Directional dilatations in space*, Complex Var. Elliptic Equ., 2010, to appear.
3. V. Gutlyanskiĭ and A. Golberg, *On Lipschitz continuity of quasiconformal mappings in space*, J. Anal. Math., 2010, to appear.
4. D. Kovtonyk and V. Ryazanov, *On the theory of mappings with finite area distortion.* J. Anal. Math. 104 (2008), 291–306.
5. S. L. Krushkal, *Quasiconformal Mappings and Riemann Surfaces*, Wiley, New York, 1979.
6. S. L. Krushkal, *Variational principles in the theory of quasiconformal maps*, Handbook of Complex Analysis: Geometric Function Theory, Vol. 2 (R. Kühnau, ed.), Elsevier Science, Amsterdam etc.
7. J. Väisälä, *Lectures on n-dimensional Quasiconformal Mappings*, Springer-Verlag, 1971.

On finite-difference smoothness of conformal mapping

Olena W. Karupu

Department of Higher and Numerical Mathematics, National Aviation University,
1 Komarov ave, Kyiv, Ukraine
E-mail: karupu@ukr.net
http://www.nau.edu.ua

Some new estimates for the uniform curvilinear modulus of smoothness of arbitrary order for the function realizing conformal mapping between the domains bounded by the smooth Jordan curves are considered.

Keywords: Conformal mapping; modulus of smoothness; finite difference, smoothness.

1. Introduction

Let the finite function $w = f(z)$ be defined on a curve $\gamma \subset \mathbb{C}$. Let $(z_0, ..., z_k)$ be the collection of the points on the curve γ and let $[z_0, ..., z_k; f, z_0]$ be the finite difference of order k for the function $w = f(z)$. Let the uniform curvilinear modulus of smoothness of order k for the function $w = f(z)$ be defined (P.M. Tamrazov[1]) as

$$\tilde{\omega}_{k,N,\gamma}((z), \delta)_p = \sup_{w \in \gamma} \ \sup_{(z_0,...,z_k) \in \gamma_{w,\delta}(N)} |[z_0, ..., z_k; f, z_0]|,$$

where $\gamma_{w,\delta}(N)$ is the set of collections $(z_0, ..., z_k)$ such that curvilinear (with respect to the curve γ) distances between points $z_0, ..., z_k \in \gamma$ satisfy the condition $\rho(z_i, z_{i+1})/\rho(z_j, z_{j+1}) \leq N$, $N \in [1, \infty)$, and $\rho(z_i, w) \leq \delta(i, j = 1, ..., k)$.

In partial case when the function $w = f(z)$ is defined on the real axis and points $x_0, ..., x_k$ form the arithmetic progression we receive the real arithmetic modulus of smoothness $\omega_k(f(x), \delta)$.

Let consider a simply connected domain G in the complex plane bounded by a smooth Jordan curve Γ. Let $\tau = \tau(s)$ be the angle between the tangent to Γ and the positive real axis, $s = s(w)$ be the arc length on Γ. Let $w = \varphi(z)$ be a homeomorphism of the closed unit disk $\overline{D} = \{z : |z| \leq 1\}$

onto the closure $\overline{G}$ of the domain G, conformal in the open unit disk D. Let $z = \psi(w)$ be the function inverse to the function $w = \varphi(z)$.

Kellog in 1912 proved that if $\tau = \tau(s)$ satisfies Hölder condition with index α, $0 < \alpha < 1$, then the derivative $\varphi'(e^{i\theta})$ of the function $\varphi(z)$ on ∂D satisfies Hölder condition with the same index α. Afterwards this result was generalized in works by several authors: S.E. Warshawski, J.L. Geronimus, S. J. Alper, R.N. Kovalchuk, L I. Kolesnik.

P. M. Tamrazov[1] obtained solid reinforcement for the modulus of continuity of the function $\varphi(z)$ on $\overline{D}$. Some close problems were investigated by E.P. Dolzenko.[2-4]

P. M. Tamrazov[5] solved the problem of estimating of finite difference smoothnesses for composite function. These results gave possibility to receive generalizations and inversations of Kellog's type theorems for general moduli of smoothness of arbitrary order.

In particular, results in the terms of the uniform curvilinear and arithmetic moduli of smoothness of arbitrary order were received by author.[6-10]

2. Estimates for modulus of smoothness for the derivative of the function realizing conformal mapping of the unit disk onto the Jordan domain

The following result for the uniform curvilinear modulus of smoothness of arbitrary order for the derivative $\varphi'(z)$ of the function $\varphi(z)$ on ∂D generalizing Kellog's theorem was earlier obtained by author.[6]

Theorem 2.1. *Let modulus of smoothness $\omega_k(\tau(s), \delta)$ of arbitrary order k for the function $\tau(s)$ satisfy the condition $\omega_k(\tau(s), \delta) = O[\omega(\delta)](\delta \to 0)$, where $\omega(\delta)$ is normal majorant satisfying the condition $\int_0^l \omega(t)/t\, dt < +\infty$.*

Then the nonzero continuous on $\overline{D}$ derivative $\varphi'(z)$ of the function $\varphi(z)$ exists and the uniform curvilinear modulus of smoothness $\tilde{\omega}_{k,1,\partial D}(\varphi', \delta)$ of order k for the derivative $\varphi'(z)$ of the function $\varphi(z)$ on ∂D satisfies the conditions

$$\tilde{\omega}_{k,1,\partial D}(\arg \varphi', \delta) = O\left(\tilde{\mu}\left(\delta\right)\right) (\delta \to 0),$$

$$\tilde{\omega}_{k,1,\partial D}(\log \varphi', \delta) = O\left(\tilde{\mu} * \left(\delta\right)\right) (\delta \to 0),$$

$$\tilde{\omega}_{k,1,\partial D}(\varphi', \delta) = O\left(\tilde{\mu} * \left(\delta\right)\right) (\delta \to 0),$$

where

$$\tilde{\mu}\ (\delta) = \mu(\delta) + \delta^k \int_\delta^l \frac{\mu(t)}{t^{k+1}}\,dt,$$

$$\tilde{\mu}\ *(\delta) = \mu * (\delta) + \delta^k \int_\delta^l \frac{\mu * (t)}{t^{k+1}}\,dt,$$

$$\mu\ (\delta) = \omega_k(\tau(s),\delta) + \sum_{j=1}^{k-1}\sum_{r_1=1}^{j-1}\cdots\sum_{r_j=1}^{r_{j-1}-1} [\omega_k\,(\tau(s),\delta)]^{\frac{k-r_1}{k}}$$

$$\times \int_0^l \cdots \int_0^l \frac{[\omega_k\,(\tau(s),x_j)]^{\frac{r_j}{k}} \prod_{i=2}^{j} [\omega_k\,(\tau(s),x_{i-1})]^{\frac{r_{i-1}-r_i}{k}}}{\prod_{p=1}^{j} x_p\left(1 + \left(\frac{x_p}{x_{p-1}}\right)^{r_p}\right)}\,dx_1...dx_j,$$

$$\mu\ *(\delta) = \int_0^l \frac{\omega(x_1)}{x_1\left(1 + \left(\frac{x_1}{\delta}\right)^k\right)}\,dx_1$$

$$+ \sum_{j=1}^{k-1}\sum_{r_1=1}^{j-1}\cdots\sum_{r_j=1}^{r_{j-1}-1} \delta^{k-r_1} \int_0^l \cdots \int_0^l x_{j+1}^{r_{j-1}}\left(1 + \int_{x_{j+1}}^l \frac{\omega(y)}{y_j^{r_j+1}}\,dy\right)\left(1 + \left(\frac{x_{j+1}}{x_j}\right)^{r_{j+1}}\right)$$

$$\times \prod_{p=1}^{j}\left(1 + \int_{x_p}^l \frac{\omega(t_p)}{t_p^{r_{p-1}-r_p+1}}\,dt_p\right)\left(1 + \left(\frac{x_p}{x_{p-1}}\right)^{r_{p-1}}\right)^{-1} x_p^{r_{p-1}-r_p-1}\,dx_1...dx_{j+1}.$$

Corollary 2.1. *In partial case when modulus of smoothness $\omega_k(\tau(s),\delta)$ of order k for the function $\tau(s)$ satisfies Hölder condition $\omega_k(\tau(s),\delta) = O(\delta^\alpha)(\delta \to 0)$, $0 < \alpha < k$, then the uniform curvilinear modulus of smoothness $\tilde{\omega}_{k,1,\partial D}(\varphi',\delta)$ of the same order k for the derivative $\varphi'(z)$ of the function $\varphi(z)$ on ∂D satisfies Hölder condition with the same index α: $\tilde{\omega}_{k,1,\partial D}(\varphi',\delta) = O(\delta^\alpha)(\delta \to 0)$.*

Corollary 2.2. *In partial case when modulus of smoothness $\omega_k(\tau(s),\delta)$ of order k for the function $\tau(s)$ satisfies condition $\omega_k(\tau(s),\delta) = O(\delta^k \log 1/\delta)(\delta \to 0)$, then the uniform curvilinear modulus of smoothness $\tilde{\omega}_{k,1,\partial D}(\varphi',\delta)$ of the same order k for the derivative $\varphi'(z)$ of the function $\varphi(z)$ on ∂D satisfies the same condition $\tilde{\omega}_{k,1,\partial D}(\varphi',\delta) = O(\delta^k \log 1/\delta)(\delta \to 0)$.*

3. Estimates for modulus of smoothness for the derivative of the function realizing conformal mapping of the Jordan domain onto the unit disk

The following result for the uniform curvilinear modulus of smoothness of arbitrary order for the derivative $\psi'(w)$ of the function $\psi(w)$ on the curve Γ was earlier obtained by author.[6]

Theorem 3.1. *Let modulus of smoothness* $\omega_k(\tau(s), \delta)$ *of order* k $(k \in \mathbb{N})$ *for the function* $\tau(s)$ *satisfy the condition* $\omega_k(\tau(s), \delta) = O\left[\omega(\delta)\right](\delta \to 0)$, *where* $\omega(\delta)$ *is normal majorant satisfying the condition* $\int_0^l \omega(t)/t\, dt < +\infty$.

Then the nonzero continuous on $\overline{G}$ *derivative* $\psi'(w)$ *of the function* $\psi(w)$ *exists satisfying on* Γ *the conditions*

$$\tilde{\omega}_{k,1,\partial D}(\arg \psi', \delta) = O\left(\tilde{\eta}\left(\delta\right)\right)(\delta \to 0),$$
$$\tilde{\omega}_{k,1,\partial D}(\log \psi', \delta) = O\left(\tilde{\eta} * \left(\delta\right)\right)(\delta \to 0),$$
$$\tilde{\omega}_{k,1,\partial D}(\psi', \delta) = O\left(\tilde{\eta} * \left(\delta\right)\right)(\delta \to 0),$$

where

$$\tilde{\eta}\left(\delta\right) = \tilde{\mu}\left(\delta\right) + \delta^{1-k(k-1)/2} \int_\delta^l \frac{\tilde{\mu}\left(y\right)}{y^{k+1}} dy \left(\delta^k \int_\delta^l \frac{\mu\left(t\right)}{t^k} dt\right)^{\frac{k(k+1)/2}{k} - 1},$$

$$\tilde{\eta} * \left(\delta\right) = \tilde{\mu} * \left(\delta\right) + \delta^{1-k(k-1)/2} \int_\delta^l \frac{\tilde{\mu} * \left(y\right)}{y^{k+1}} dy \left(\delta^k \int_\delta^l \frac{\tilde{\mu}\left(t\right)}{t^k} dt\right)^{\frac{k(k+1)/2}{k} - 1}.$$

Corollary 3.1. *In partial case when modulus of smoothness* $\omega_k(\tau(s), \delta)$ *of order* k *for the function* $\tau(s)$ *satisfies Hölder condition* $\omega_k(\tau(s), \delta) = O\left(\delta^\alpha\right)(\delta \to 0)$, $0 < \alpha < k$, *then the uniform curvilinear modulus of smoothness* $\tilde{\omega}_{k,1,\Gamma}(\psi', \delta)$ *of the same order* k *for the derivative* $\psi'(w)$ *of the function* $\psi(w)$ *satisfies Hölder condition with the same index* α: $\tilde{\omega}_{k,1,\Gamma}(\psi', \delta) = O\left(\delta^\alpha\right)(\delta \to 0)$.

Corollary 3.2. *In partial case when modulus of smoothness* $\omega_k(\tau(s), \delta)$ *of order* k *for the function* $\tau(s)$ *satisfies condition* $\omega_k(\tau(s), \delta) = O\left(\delta^k \log 1/\delta\right)(\delta \to 0)$, *then the uniform curvilinear modulus of smoothness* $\tilde{\omega}_{k,1,\Gamma}(\psi', \delta)$ *for the derivative* $\psi'(w)$ *of the function* $\psi(w)$ *satisfies the same condition* $\tilde{\omega}_{k,1,\Gamma}(\psi', \delta) = O\left(\delta^k \log 1/\delta\right)(\delta \to 0)$.

4. Estimates for modulus of smoothness for the derivative of the function realizing conformal mapping between two Jordan domains

Let G_1 be the simply connected domain in the complex plane bounded by the smooth Jordan curve Γ_1 and G_2 be the simply connected domain in the complex plane bounded by the smooth Jordan curve Γ_2. Let $\tau_1(s_1)$ be the angle between the tangent to Γ_1 and the positive real axis , $s_1(\zeta)$ be the arc length on Γ_1. Let $\tau_2(s_2)$ be the angle between the tangent to Γ_2 and the positive real axis , $s_2(w)$ be the arc length on Γ_2. Let $w = f(\zeta)$ be a homeomorphism of the closure $\overline{G_1}$ of the domain G_1 onto the closure $\overline{G_2}$ of the domain G_2, conformal in the domain G_1.

Theorem 4.1. *Let moduli of smoothness $\omega_k(\tau_1(s_1), \delta)$ and $\omega_k(\tau_2(s_2), \delta)$ of order $k\,(k \in \mathbb{N})$ for the functions $\tau_1(s_1)$ and $\tau_2(s_2)$ satisfy Hölder conditions with the same index α: $\omega_k(\tau_1(s_1), \delta) = O\left(\delta^\alpha\right)(\delta \to 0)$ and $\omega_k(\tau_2(s_2), \delta) = O\left(\delta^\alpha\right)(\delta \to 0)$, $0 < \alpha < k$.*

Then the uniform curvilinear modulus of smoothness $\tilde{\omega}_{k,1,\Gamma_1}(f', \delta)$ of the same order k for the derivative of the function $f(\zeta)$ on Γ_1 satisfies Hölder condition $\tilde{\omega}_{k,1,\Gamma_1}(f', \delta) = O\left(\delta^\alpha\right)(\delta \to 0)$ with the same index α.

Proof of this theorem is based on corollary 2.1 of theorem 2.1, corollary 3.1 of theorem 3.1 and on estimates for finite difference smoothnesses of composite function.

Theorem 4.2. *Let moduli of smoothness $\omega_k(\tau_1(s_1), \delta)$ and $\omega_k(\tau_2(s_2), \delta)$ of order $k\,(k \in \mathbb{N})$ for the functions $\tau_1(s_1)$ and $\tau_2(s_2)$ satisfy conditions $\omega_k(\tau_1(s_1), \delta) = O\left(\delta^k \log 1/\delta\right)(\delta \to 0)$ and $\omega_k(\tau_2(s_2), \delta) = O\left(\delta^k \log 1/\delta\right)(\delta \to 0)$.*

Then the uniform curvilinear modulus of smoothness $\tilde{\omega}_{k,1,\Gamma_1}(f', \delta)$ of the same order k for the derivative of the function $f(\zeta)$ on Γ_1 satisfies the same condition $\tilde{\omega}_{k,1,\Gamma_1}(f', \delta) = O\left(\delta^k \log 1/\delta\right)(\delta \to 0)$.

Proof of this theorem is similar to the proof of theorem 4.1, but it is based on corollary 2.2 of theorem 2.1 and corollary 3.2 of theorem 3.1.

References

1. P. M. Tamrazov, *Smoothnesses and polynomial appoximations* (Naukova dumka, Kiev, 1975) [in Russian].
2. E.P. Dolzenko, *Dokl. Acad. Nauk SSSR* **29**, 1069 (1965) [in Russian].
3. E.P. Dolzenko, *Mat. Zametki* **60**, 176 (1996) [in Russian].
4. E.P. Dolzenko, , *Russian Acad. Nauk Dokl.* **415**, 155 (2007) [in Russian].

5. P. M. Tamrazov, *Finite difference identities and estimates for moduli of smoothness of composite functions* (Inst. Mat. of Acad. Nauk Ukraine, Kiev, 1977) [in Russian].
6. O. W. Karupu, *Ukr. Math. J.* **30**, 540 (1978) [in Russian].
7. O. W. Karupu, *Acad. Nauk Ukrain. Works Inst. Mat.* **31**, 237 (2000) [in Ukrainian].
8. O. W. Karupu, On properties of moduli of smoothness of conformal mappings, in *Complex Analysis and Potential Theory, Proc. of the Confertence Satellite to ICM 2006,* (Singapore, 2007).
9. O. W. Karupu, *Acad. Nauk Ukrain. Works Inst. Mat.* **3**, 175 (2006) [in Ukrainian].
10. O. W. Karupu, On some boundary properties of conformal mapping, in *Further progress in analysis, Proc. of the 6th International ISAAC Congress,* (Singapore, 2009).

Structure of non-rectifiable curves and solvability of the jump problem

B.A. Kats[*]

*Chair of Mathematics, Kazan State Architecture and Civil Engineering University,
Zelenaya Street, 1, Kazan, Tatarstan, 430043, Russian Federation
E-mail: katsboris877@gmail.com*

The author continues the study of the Riemann boundary value problem on non-rectifiable curves in terms of certain new characteristics of that curves.

Keywords: Jump problem; non-rectifiable curve; metric dimension.

Introduction

We consider the following boundary value problem for holomorphic functions. Let Γ be a closed Jordan curve on the complex plane $\mathbb{C}$ bounding finite domain D^+, and $D^- = \overline{\mathbb{C}} \setminus \overline{D^+}$. We seek a holomorphic in $\overline{\mathbb{C}} \setminus \Gamma$ function $\Phi(z)$ such that $\Phi(\infty) = 0$, the boundary values $\lim_{D^+ \ni z \to t} \Phi(z) \equiv \Phi^+(t)$ and $\lim_{D^- \ni z \to t} \Phi(z) \equiv \Phi^-(t)$ exist for any $t \in \Gamma$, and

$$\Phi^+(t) = G(t)\Phi^-(t) + g(t), t \in \Gamma. \tag{1}$$

This boundary value problem is called the Riemann problem. It is well known and has numerous applications in elasticity theory, hydro and aerodynamics and so on. If $G(t) \equiv 1$, then it turns to so called jump problem:

$$\Phi^+(t) - \Phi^-(t) = g(t), t \in \Gamma. \tag{2}$$

The following classical result on this problem was obtained in XIX century by Harnak, Plemelj, Sokhotskii and others (see, for instance,[1] and[2]):

–if the curve Γ is piecewise-smooth and the jump $g(t)$ satisfies the Hölder condition

$$\sup\{\frac{|f(t') - f(t'')|}{|t' - t''|^\nu} : t', t'' \in \Gamma, t' \neq t''\} \equiv h_\nu(f, \Gamma) < \infty \tag{3}$$

---

[*]The research was supported by RFBR Grants #07-01-00166-a and #09-01-12188-ofi-m.

with exponent $\nu \in (0,1]$, then a unique solution of this problem is the Cauchy integral

$$\Phi(z) = \frac{1}{2\pi i} \int_{\Gamma} \frac{g(\zeta)d\zeta}{\zeta - z}. \tag{4}$$

Below we denote $H_{\nu}(\Gamma)$ the set of all functions satisfying (3). The Riemann boundary value problem reduces to the jump problem by means of factorization.

Numerous authors studied continuity of boundary values of the Cauchy integral over non-smooth rectifiable curves during almost a century . Finally, in 1979 E. M. Dynkin[3] and T. Salimov[4] published the following important result:

– the Cauchy integral (4) over rectifiable curve Γ has boundary values $\Phi^{\pm}$ if f satisfies the Hölder condition with exponent $\nu > \frac{1}{2}$, and this bound cannot be improved in the whole class of rectifiable curves. Hence, the jump problem on non-smooth rectifiable curve is solvable if the Hölder exponent of the jump exceeds $\frac{1}{2}$, and this bound for the Hölder exponent cannot be improved.

If curve Γ is not rectifiable, then customary definition of the Cauchy integral falls, but the Riemann boundary value problem and the jump problem keep sense and applicability.

In this connection two questions arise:

– can we solve the jump problem on non-rectifiable curve without application of the Cauchy integral?

– can we define certain generalization of the Cauchy integral over non-rectifiable curve?

Clearly, the second question is closely connected with problem of definition of curvilinear integral $\int_{\Gamma} f\,dx + g\,dy$ over non-rectifiable plane curve.

In 1982 the first question was solved by the author[5] in terms of so called box dimension. It can be defined by equality

$$\mathrm{Dm}\,\Gamma = \limsup_{\varepsilon \to 0} \frac{\log N(\varepsilon, \Gamma)}{-\log \varepsilon},$$

where $N(\varepsilon, \Gamma)$ stands for the least number of disks of diameter ε covering the set Γ. For any plane continuum A we have $1 \leq \mathrm{Dm}\,A \leq 2$; if Γ is rectifiable curve, then $\mathrm{Dm}\,\Gamma = 1$. The jump problem is solvable if $g \in H_{\nu}(\Gamma)$ and $\nu > \frac{\mathrm{Dm}\,\Gamma}{2}$.

But later it was shown that the box dimension does not allow us to take into account certain important features of non-rectifiable curves. An sample of this phenomenon can be found in the paper.[6] In this connection it is of interest to study solvability and other properties of the problem in terms of other characteristics of structure of non-rectifiable curves. In the present notice we apply so called approximation dimension, which is introduced in the article.[7]

1. Approximation dimension

Let $P_1^+, P_2^+, \ldots, P_n^+, \ldots$ be a sequence of sets satisfying the following assumptions:

– for any n the set P_n^+ is open finite polygon or union of several open finite polygons, and $P_n^+ \subset P_{n+1}^+ \subset D^+$;

– if Γ_n is boundary of P_n^+ (i.e., Γ_n is closed polygonal line or union of several closed polygonal lines), then $\lim_{n\to\infty} \mathrm{dist}(\Gamma_n, \Gamma) = 0$.

We say that the sequence $\Gamma_1, \Gamma_2, \ldots, \Gamma_n, \ldots$ is increasing polygonal approximation of the curve Γ.

The definition of decreasing polygonal approximation is analogous, In this case Γ_n is boundary of set P_n^-, which is either infinite polygonal domain or union of several polygonal domains one of which is infinite, $P_{n+1}^- \subset P_n^- \subset D^-$ for any n, and, as above, $\lim_{n\to\infty} \mathrm{dist}(\Gamma_n, \Gamma) = 0$.

We call a sequence $\Gamma_1, \Gamma_2, \ldots, \Gamma_n, \ldots$ monotone approximation of Γ if it is either increasing or decreasing approximation of this curve.

We use the following notation. If γ is rectifiable curve, then $|\gamma|$ stands for its length. If a sequence of polygons $\{\Gamma_n\}$ approximates a non-rectifiable curve, then $|\Gamma_n| \to \infty$. The width $w(\delta)$ of finite domain δ is diameter of the most open disk lying in δ.

Let a sequence $\{\Gamma_n\}$ be monotone polygonal approximation of Γ. We put $\Delta_n = P_{n+1}^+ \setminus \overline{P_n^+}$ if this approximation increases, and $\Delta_n = P_n^- \setminus \overline{P_{n+1}^-}$ if it decreases, $n = 1, 2, \ldots$. As P_n^+ and P_n^- are polygonal domains, since Δ_n is either polygonal domain (maybe, double connected) or union of finite number of disjoint polygonal domains. We denote by λ_n the sum of perimeters of all connected components of Δ_n, and by ω_n the most of widths of these components.

Definition 1.1. Let $A(\Gamma)$ be set of all positive numbers q such that

$$\sum_{n=1}^{\infty} \lambda_n \omega_n^{q-1} < \infty$$

for some monotone polygonal approximation of the curve Γ. Then the value $\mathrm{Dm}_{\mathrm{ap}}\,\Gamma := \inf A(\Gamma)$ is approximation dimension of this curve.

The approximation dimension It characterizes structure of polygonal approximations of the curve.

Lemma 1.1. *(See[7]). Any plane curve Γ satisfies inequality*

$$1 \leq \mathrm{Dm_{ap}}\,\Gamma \leq \mathrm{Dm}\,\Gamma. \tag{5}$$

For any value $d \in (1,2)$ there exist a curve Γ such that $\mathrm{Dm}\,\Gamma = d$ and $\mathrm{Dm_{ap}}\,\Gamma < d$. If Γ is rectifiable curve, then $\mathrm{Dm_{ap}}\,\Gamma = 1$.

This lemma shows that $\mathrm{Dm_{ap}}\,\Gamma$ is characteristics of dimensional type, and, generally speaking, it is less than $\mathrm{Dm}\,\Gamma$. On the other hand, for a number of known self-similar fractal curves (for instance, for the von Koch snowflake) these two dimensions coincide.

2. Integration over non-rectifiable curves

As we note above, the jump problem is closely connected with problem of definition of integral over non-rectifiable curves. There exist a number of works dealing with this problem. The authors of the most of them (see, for instance, papers[8] and[9]) propose to use the Stokes formula

$$\int_{\partial D} \omega = \iint_D d\omega$$

as definition of its left side for non-rectifiable curve ∂D of null square. If 1-form ω equals to $f(t)dt$, then this definition turns into

$$\int_\Gamma f(t)dt := -\iint_{D+} \frac{\partial f^*}{\partial \bar t}\, dt d\bar t, \tag{6}$$

where f^* is certain extension of f from Γ into D^+. As shown in the papers[8] and,[9] this definition is correct under assumptions $f \in H_\nu(\Gamma)$, $\nu > \mathrm{Dm}\,\Gamma - 1$. The approximation dimension allow us to define integral over closed non-rectifiable curve in another way, and this new definition is correct under weaker assumptions.

Theorem 2.1. *If $f \in H_\nu(\Gamma), \nu > \mathrm{Dm_{ap}}\,\Gamma - 1$, and $1 < p < \mathrm{Dm_{ap}}\,\Gamma$, then for any extension f^* such that $f^*|_\Gamma = f$ and $f^* \in H_\nu(\mathbf{C})$ and for any monotone polygonal approximation $\{\Gamma_n\}$ of the curve Γ such that $\sum_{n=1}^\infty \lambda_n \omega_n^{p-1} < \infty$ there exists the limit*

$$\int_\Gamma f(t)dt := \lim_{n\to\infty} \int_{\Gamma_n} f^*(t)dt, \tag{7}$$

and meaning of this limit does not depend on the choice of the extension and polygonal approximation satisfying mentioned above assumptions.

Thus, we obtain definition (7) of the same integral, and it is correct for $\nu > \mathrm{Dm_{ap}}\,\Gamma - 1$.

3. The jump problem

We can apply the result of previous section in order to prove solvability of the jump problem (2).

Theorem 3.1. *If $g \in H_\nu(\Gamma)$ and $\nu > \frac{\mathrm{Dm_{ap}}\,\Gamma}{2}$, then the jump problem has a solution, which is representable by the Cauchy integral (4) where integration is understood in the sense of definition (7).*

The uniqueness of solution is related with the Haussdorff dimension $\mathrm{Dm_H}\,\Gamma$. According the paper,[5] we call a solution $\Phi(z)$ of the Riemann boundary value problem its $\mu-solution$ if its restrictions on domains D^+ and D^- satisfy the Hölder condition with exponent μ in closures of these domains. Then the Cauchy integral in the sense of definition (7) is $\mu-$solution of the jump problem for $\mu < \frac{2\nu - \mathrm{Dm_{ap}}\,\Gamma}{2 - \mathrm{Dm_{ap}}\,\Gamma}$. Hence, for

$$\mathrm{Dm_H}\,\Gamma - 1 < \mu < \frac{2\nu - \mathrm{Dm_{ap}}\,\Gamma}{2 - \mathrm{Dm_{ap}}\,\Gamma} \tag{8}$$

the jump problem (2) has unique $\mu-$solution under assumptions of Theorem 3.1, and this solution is representable by the generalized Cauchy integral. By virtue of E.P. Dolzhenko's theorem[10] this solution has maximal smoothness, because all other solutions can be $\mu'-$solutions only for $\mu' < \mathrm{Dm_H}\,\Gamma - 1 < \mu$.

4. The Riemann boundary value problem

The customary factorization technique (see monographs[1] and[2]) allows us to reduce the Riemann boundary value problem to the jump problem. Let us denote $\kappa := \frac{1}{2\pi}[argG]_\Gamma$, where $[argG]_\Gamma$ is increment of argument of $G(t)$ on the curve Γ, $f(t) := \log G(t)(t - z_0)^{-\kappa}$, where z_0 is a fixed point in D^+, and consider a functions $\Psi(s) := \frac{1}{2\pi i} \int_\Gamma \frac{f(\zeta)d\zeta}{\zeta - z}$ (integration in the sense (7)) and $X(z)$ equaling to $\exp \Psi(z)$ for $z \in D^+$ and to $(z - z_0)^{-\kappa} \exp \Psi(z)$ for $z \in D^-$. As a result, we obtain

Theorem 4.1. *Let G does not vanish on the curve Γ, the coefficients G and g satisfy the Hölder condition with exponent $\nu > \frac{\mathrm{Dm_{ap}}\,\Gamma}{2}$, and μ satisfies (8). Then the following propositions are valid:*
   i. if $\kappa = 0$, then function

$$\Phi_0(z) := \frac{X(z)}{2\pi i} \int_\Gamma \frac{g(\zeta)}{X^+(\zeta)} \frac{g(\zeta)d\zeta}{\zeta - z}$$

is unique $\mu-$solution of the Riemann boundary value problem (1);

    ii. if $\kappa > 0$, then general $\mu-$solution of the problem (1) is

$$\Phi(z) = \Phi_0(z) + X(z)P_\kappa(z),$$

where $P_\kappa(z)$ is arbitrary algebraic polynomial of degree less than κ;

    iii. if $\kappa < 0$, then $\Phi_0(z)$ is unique $\mu-$solution under $-\kappa$ solvability conditions.

    All integrals are understood in the sense of Theorem 2.1.

Thus, under assumptions of Theorem 4.1 the pattern of $\mu-$solvability of the Riemann boundary value problem on non-rectifiable curve coincides with the pattern of its solvability on piecewise-smooth curve.

References

1. F.D. Gakhov, *Boundary value problems*, Nauka publishers, Moscow, 1977.
2. N.I. Muskhelishvili, *Singular integral equations*, Nauka publishers, Moscow, 1962.
3. E.M. Dynkin, Smoothness of the Cauchy type integral, *Zapiski nauchn. sem. Leningr. dep. mathem. inst. AN USSR* **92** (1979), 115–133
4. T. Salimov, A direct bound for the singular Cauchy integral along a closed curve, *Nauchn. Trudy Min. vyssh. i sr. spec. obraz. Azerb. SSR, Baku* **5** (1979), 59–75.
5. B.A. Kats, The Riemann problem on closed Jordan curve, *Izvestia vuzov. Mathem.* bf 4 (1983), 68-80
6. B.A. Kats. The jump problem on non-rectifiable curves and metric dimensions, in book "FURTHER PROGRESS IN ANALYSIS, Proceedings of the 6th International ISAAC Congress, Ankara, Turkey 13 - 18 August 2007", World Scientific Publishing Co., (2009), 241-248
7. B. A. Kats. On solvability of the jump problem, *J. Math. Anal. Appl.* **356** (2009), 577-581
8. B.A. Kats. The jump problem and the integral over non-rectifiable curve, *Izvestia vuzov, Mathem.* **5** (1987), 49-57
9. J. Harrison and A. Norton. Geometric integration on fractal curves in the plane, *Indiana Univ. Math. J.* **40** (1991), 567-594
10. E.P. Dolzhenko. On "erasing" of singularities of analytic functions, *Uspekhi Math. Nauk* **18** (1963), 135-142

I.2. Differential equations: Complex and functional analytic methods, applications

Organisers: H. Begehr, D.-Q. Dai, J. Du

The session was active with 16 thirty-minute talks with topics ranging from complex boundary value problems for Cauchy-Riemann equations on Klein surfaces (C. Bolosteanu), generalized Cauchy-Riemann systems in the plane (N. Manjavidze, G. Khimshiashvili) and on Riemann surfaces (G. Giorgadze), for the Poisson equation (A. Mohammed), and for higher order complex partial differential equations (U. Aksoy, H. Begehr, O. Celebi) to bicomplex pseudoanalytic functions (P. Berglez). Other topics were volume potentials (T.S. Kalmenov), singular Volterra type integral equations (L. Rajabova, N. Rajabov), Navier-Stokes equations (R. Saks), optimization of fixed point methods (S. Graubner), superparabolic functions (A. Kheyfits), and cusped plates with deflections (N. Chinchaladze). Several proposed participants from China were not able to attend because of visa problems.

From the presentations only 6 manuscripts are published within this proceedings volume. The contributions from T.S. Kalmenov will be published in the PDE special volume.

On some qualitative issues for the first order elliptic systems in the plane

G. Akhalaia

*I. Vekua Institute of Applied Mathematics
of Tbilisi State University,
University st. 2, Tbilisi, 43, Georgia
Email: giorgi.akhalaia98@gmail.com*

G. Makatsaria

*The Patriarchate of Georgia,
Georgian University,
Chavchavadze Ave. 53a, Tbilisi, Georgia
Email: giorgi.makatsaria@gmail.com*

N. Manjavidze

*Georgian Technical University,
77 Kostava st., 0175 Tbilisi, Georgia
Email: ninomanjavidze@yahoo.com*

Keywords: Elliptic system; maximum modulus theorem; generalized Beltrami system; Q-holomorphic vector; Cauchy-Lebesgue class.

1. Introduction

The first order system of partial differential equations

$$\frac{\partial u}{\partial x} + A(x,y)\frac{\partial u}{\partial y} + B(x,y)u = 0, \tag{1}$$

where $u = (u_1, u_2, \ldots, u_{2n})$ is $2n$-component desired vector, A, B are given real $2n \times 2n$-matrices depending on two real variables x, y is called elliptic in some domain $G \subset R^2_{(x,y)}$, if

$$\det(A - \lambda I) \neq 0, \tag{2}$$

for every real λ and $(x,y) \in G$; I is an identity matrix. In other words the system (1) is elliptic if the matrix A has no real characteristic numbers in G.

The investigation of such system has a great history. Various particular cases of the system (1) were the object of investigation of Picard, Beltrami, Carleman, Bers, Vekua, Douglis, Bojarski, Hile, Begehr, D. Q. Dai and many other authors.

2. Maximum modulus theorem

In the first part of our work we study the problem of validity of the maximum modulus theorem. To this end let us mention some auxiliary explanations. Under the solution of the system (1) we mean the classical solution of the class $C^1(G) \cap C(\overline{G})$.

Denote by

$$\Lambda(A, B) \tag{3}$$

the class of all possible solutions of the system (1).

Introduce

$$\rho_u(x, y) = \left[\sum_k u_k^2(x, y) \right]^{\frac{1}{2}}, \quad (x, y) \in \overline{G} \tag{4}$$

for every u of the class (3). And now raise a question (cf. Bojarski [1]).

Is the inequality

$$\rho_u(x_0, y_0) \leq \max_{(x,y) \in \Gamma} \rho_u(x, y) \tag{5}$$

valid for arbitrary u from (3) and $(x_0, y_0) \in \overline{G}$. Γ is a boundary of the domain G. Of course, in case $n = 1$ and

$$A = \begin{pmatrix} 0 & -1 \\ 1 & 0 \end{pmatrix}, \qquad B = \begin{pmatrix} 0 & 0 \\ 0 & 0 \end{pmatrix}$$

the condition (5) if fulfilled.

Consider now $G = \{x^2 + y^2 < 1\}$, $n = 1$, A is the same matrix, $B = \begin{pmatrix} 2x & 0 \\ 2y & 0 \end{pmatrix}$ and $u = \text{column}(e^{-x^2-y^2}, 0) \in \Lambda(A, B)$.

It is evident, that $\rho_u(0, 0) = 1$ and $\rho_u(x, y) = \frac{1}{e}$, i.e. the condition (5) is not fulfilled. In this example the matrix B is not constant matrix. This example shows, that the maximum modulus theorem for minimal dimensional elliptic system is not always true. It is easy to construct the example of higher dimensional system when the condition (5) is disturbed in case A and B are constant. In fact, consider G is the same domain

$$G = \{x^2 + y^2 < 1\},$$

$$A = \begin{pmatrix} 0 & -1 & 0 & 0 \\ 1 & 0 & 0 & 0 \\ 0 & 0 & 0 & -1 \\ 0 & 0 & 1 & 0 \end{pmatrix}, \qquad B = \begin{pmatrix} -2 & 0 & 0 & 0 \\ 0 & -2 & 0 & 0 \\ -6 & 0 & -2 & 0 \\ 0 & -6 & 0 & -2 \end{pmatrix}$$

and $u = \operatorname{column}(u_1, u_2, u_3, u_4) \in \Lambda(A, B)$, where

$$u_1 = e^x(x \cos y + y \sin y), \qquad u_2 = e^x(y \cos y - x \sin y),$$
$$u_3 = 3(x^2 + y^2 - 1)e^x \cos x, \qquad u_4 = -3(x^2 + y^2 - 1)e^x \sin y.$$

It is clear, that

$$\rho_u(0, 0) = 3, \qquad \max_{(x,y)\in\Gamma} \rho_u(x, y) = e$$

and therefore the condition (5) is not fulfilled.

In case the dimension of the system(1)–(3) is minimal, i.e. when $n = 1$ and moreover, when

$$A = \begin{pmatrix} 0 & -1 \\ 1 & 0 \end{pmatrix}, \quad B = \begin{pmatrix} b_{11} & b_{12} \\ b_{21} & b_{22} \end{pmatrix}, \quad b_{kq} \in L_p(G), \quad p > 2.$$

We have with the great effort of very famous mathematicians, in some sense complete theory which is in very close connection with the theory of analytic functions of complex variable. In particular, it is well-known the following fact, that there exists the number $M \geq 1$ (depending only on the matrix B) such, that

$$\rho_u(x_0, y_0) \leq M \max_{(x,y)\in\Gamma} \rho_u(x, y) \tag{6}$$

for every $u \in \Lambda(A, B)$ and $(x_0, y_0) \in G$.

The inequality (6) is weaker than (5), but it is also very interesting problem as was noted by Bojarski in this work "General properties of the solutions of elliptic systems on the plane" in 1960.

Now we describe the sufficiently wide class of the elliptic systems (1)–(3), for which the inequality (6) as well as more strong inequality (5) holds. Consider the case of constant coefficients.

Theorem 2.1. *Let for the matrices A and B there exists the orthogonal* *matrix D such, that*

$$D^{-1}AD = \begin{pmatrix} 0 & -1 & 0 & 0 & \cdots & 0 & 0 \\ 1 & 0 & 0 & 0 & \cdots & 0 & 0 \\ 0 & 0 & 0 & -1 & \cdots & 0 & 0 \\ \hdotsfor{7} \\ 0 & 0 & 0 & 0 & \cdots & 0 & -1 \\ 0 & 0 & 0 & 0 & \cdots & 1 & 0 \end{pmatrix} \tag{7}$$

$$D^{-1}BD = \begin{pmatrix} d_{11} & -h_{11} & d_{12} & -h_{12} & \cdots & d_{1n} & -h_{1n} \\ h_{11} & d_{11} & h_{12} & d_{12} & \cdots & h_{1n} & d_{1n} \\ d_{21} & -h_{21} & d_{22} & -h_{22} & \cdots & d_{2n} & -h_{2n} \\ h_{21} & d_{21} & h_{22} & d_{22} & \cdots & h_{2n} & -d_{2n} \\ \hdotsfor{7} \\ d_{n1} & -h_{n1} & \cdots & \cdots & \cdots & d_{nn} & -h_{nn} \\ h_{n1} & -d_{n1} & \cdots & \cdots & \cdots & h_{nn} & -d_{nn} \end{pmatrix} \tag{8}$$

where d_{kp}, h_{kp}, $1 \le k \le n$, $1 \le p \le n$ are arbitrary real numbers and the constructed complex matrix

$$B_0 = \begin{pmatrix} d_{11} + ih_{11} & d_{12} + ih_{12} & \cdots & d_{1n} + ih_{1n} \\ d_{21} + ih_{21} & d_{22} + ih_{22} & \cdots & d_{2n} + ih_{2n} \\ \hdotsfor{4} \\ d_{n1} + ih_{n1} & d_{n2} + ih_{n2} & \cdots & d_{nn} + ih_{nn} \end{pmatrix} \tag{9}$$

is a normal matrix, i.e. $B_0 \overline{B}_0^T = \overline{B}_0^T B_0$. Then the inequality (5) holds for any $u \in L(A, B)$, $(x_0, y_0) \in \overline{G}$. Moreover, if the equality holds in some inner point of the domain G then the function ρ_u (but not necessarily vector-function u) is constant.

In above mentioned example, for the case $n = 2$, the conditions (7), (8) are fulfilled, but the constructed complex B_0 is not normal and therefore (5) is violated.

By the conditions (7), (8) in some sense exact class of the elliptic systems of the form (1) for which the effective general representations of the solutions through the holomorphic functions are established. These functions enable us to prove the inequality (6) in case $M = 1$. Therefore, for some definite (non-extendable) class of the generalized analytic vector-functions one gets the proof of the natural analog of the classical maximum modulus theorem.

3. The system (1)

The second part of our work is devoted to the system (1) which has the following complex form

$$w_{\bar{z}} = Q w_z, \tag{10}$$

where Q is given $n \times n$ complex matrix of the class $W_p^1(C)$, $p > 2$ and $Q(z) = 0$ outside of some circle. In this case under the solution of the system (10) we understand the so-called regular solution [4], i.e. $w(z) \in L(\overline{G})$, whose generalized derivatives $w_{\bar{z}}$, w_z belong to $L_r(G')$, $r > 2$, $G' \subset G$ is an arbitrary closed subset. (10) is to be satisfied almost everywhere in D.

Bojarski [2] assumed, that the variable matrix Q in (10) is quasi-diagonal matrix of the special form having the eigenvalues less than 1. Hile noted that what appears to be essential property of the elliptic systems of the form (10) for which one can obtain a useful extension of the analytic function theory is the self-commuting property of the matrix Q, which is

$$Q(z_1)\, Q(z_2) = Q(z_2)\, Q(z_1) \tag{11}$$

for any two points z_1, z_2. Following Hile if Q is self-commuting and if $Q(z)$ has the eigenvalues less than 1 then the system (10) is called generalized Beltrami system. The solutions of such system is called Q-holomorphic vectors.

The matrix valued function $\Phi(z)$ is a generating solution of the system (10) if it satisfies the following properties ([2]):

   (i) $\Phi(z)$ is a C^1-solution of (10) in G;

   (ii) $\Phi(z)$ is a self-commuting and commutes with Q in G;

   (iii) $\Phi(t) - \Phi(z)$ is invertible for all z, t in G, $z \neq t$;

   (iv) $\partial_z \Phi(z)$ is invertible for all z in G.

The matrix $V(t,z) = \partial_t \Phi(t)[\Phi(t) - \Phi(z)]^{-1}$ we call the generalized Cauchy kernel for the system (10).

Let now Γ be a union of simple closed non-intersecting Liapunov-smooth curves bounding finite or infinite domain; if Γ is one closed curve then G denotes the finite domain; if Γ consists of several curves then by G^+ denote the connected domain with the boundary Γ. On these curves the positive direction is chosen such, that when passing along Γ, G^+ remains left; the complement of the open set $G^+ \cup \Gamma$ till the whole plane denote by G^-.

Assume the vector $\varphi(t) \in L(\Gamma)$ is given and consider the following integral

$$\Phi(z) = \frac{1}{2\pi i} \int_\Gamma V(t,z)\, d_Q t\, \varphi(t), \tag{12}$$

where $d_Q t = I\,dt + Q\,d\bar{t}$, I is an identity matrix. It is evident, that $\Phi(z)$ is Q-holomorphic vector everywhere outside of Γ, $\Phi(\infty) = 0$.

We call the vector $\Phi(z)$ the generalized Cauchy-Lebesgue type integral for the system (10) with the jump line Γ.

The boundary values of $\Phi(z)$ on Γ are given by the formulas

$$\Phi^{\pm}(t) = \pm\frac{1}{2}\,\varphi(t) + \frac{1}{2\pi i}\int_{\Gamma} V(\tau, t)\, d_Q\tau\,\mu(\tau). \qquad (13)$$

These formulas are to be fulfilled almost everywhere on Γ, provided that $\Phi^{\pm}(t)$ are angular boundary values of the vector $\Phi(z)$ and the integral in (13) is to be understood in the sense of Cauchy principal value.

For the vector $\Phi(z)$ to be representable by the Cauchy-Lebesque type integral (12) with the jump line Γ, it is necessary and sufficient the fulfillment of the equality

$$\frac{1}{\pi i}\int_{\Gamma} V(t, t_0)\, d_Q t\big[\Phi^{+}(t) - \Phi^{-}(t)\big] = \Phi^{+}(t_0) + \Phi^{-}(t_0) \qquad (14)$$

almost everywhere on Γ.

We call the generalized Cauchy-Lebesgue type integral (12) the generalized Cauchy-Lebesgue integral in the domain G^{+} (G^{-}), if $\Phi^{+}(t) = \varphi(t)$ $(\Phi^{-}(t) = -\varphi(t))$ almost everywhere on Γ.

Theorem 3.1. *Let $\Phi(z)$ be a Q-holomorphic vector representable by the generalized Cauchy-Lebesgue type integral in G^{+} and let $\Phi^{+}(t) \in L(\Gamma)$. Then $\Phi(z)$ is representable by the generalized Cauchy-Lebesgue integral with respect to its boundary values. The analogous conclusion for the infinite domain G^{-} in case $\Phi(\infty) = 0$ is also valid.*

Introduce some classes of Q-holomorphic vectors. We say, that Q-holomorphic vector $\Phi(z)$ belongs to the class $E_p(G^{+}, Q)$ $[E_p^{-}(G^{+}, Q)]$, $p > 1$, if $\Phi(z)$ is representable by the generalized Cauchy-Lebesgue type integral with the density from the class $L_p(\Gamma)$. It follows from the Theorem 2 that every Q-holomorphic vector from $E_p(G^{\pm}, Q)$ has the angular boundary values from the class L_p and this vector is representable by the generalized Cauchy-Lebesgue integral with respect to its angular boundary values. In particular the condition that density function belongs to the class L_p, $p > 1$ is the sufficient condition for the angular boundary values to be from the class L_p.

Theorem 3.2. *Let $\Phi(z)$ be a Q-holomorphic vector representable by the generalized Cauchy-Lebesgue type integral in G^{+} (G^{-}) with the summable* *density. If the angular boundary values Φ^+ (Φ^-) belong to the class $L_p(\Gamma)$, $p > 1$ then $\Phi(z)$ belongs to the class $E_p^+(G^+, Q)$ $[E_p^-(G^-, Q)]$.*

Theorem 3.3. *Let $\Phi(z)$ be a Q-holomorphic vector representable by the generalized Cauchy-Lebesgue type integral in simple connected domain G (G may be infinite). If $\mathrm{Re}[\Phi(t)] = 0$ almost everywhere on the boundary of the domain G then $\Phi(z) = iC$, where C is a real constant vector. (In case G is infinite $C = 0$).*

The self-commutative property (11) gives us the possibility to assume that our system (10) has the triangular form. Now applying the well-known properties of the Beltrami equations and the formulas (13), (14), by means of standard methods we prove the Theorems 2,3,4.

Here we select some useful properties of above mentioned classes which are the natural classes in order to correctly pose and complete analyze the discontinuous boundary value problems for pseudo-holomorphic vectors.

We would like to thank Prof. H. Begehr and the referee for their suggestions and remarks.

References

1. Bojarski B. *General properties of the solutions of elliptic systems in the plane.* Investigation of contemporary problems of the theory of complex variables, 1960, 461–493.
2. Bojarski B. *Theory of generalized analytic vectors.* Ann. Polon. Math. **17** (1966), 281–320.
3. Hile G. N. *Function theory for generalized Beltrami systems.* Contemporary Math. **11** (1982).
4. Vekua I. *Generalized analytic functions.* Pergamon, Oxford, 1962.

Harmonic Green and Neumann representations in a triangle, quarter-disc and octo-plane

H. Begehr[1], M.-R. Costache[2], S. Tappert[1] and T. Vaitekhovic[1]

[1] *Math. Institute, FU Berlin,*
Arnimalle 3, D-12203 Berlin, Germany
E-mails: begehr@zedat.fu-berlin.de,
tappert@zedat.fu-berlin.de,
vaiteakhovich@mail.ru
[2] *Şcoala Normală Superioară Bucharest*
Department of Mathematics, Bucharest, Romania
E-mail: monica_ramona_costache@yahoo.com

The harmonic Green and Neumann functions and the Poisson kernels are explicitly constructed for a particular triangle, a quarter-disc and an octo-plane. Related representation formulas are given.

Keywords: Green function; Neumann function; Poisson kernel; representation formulas; triangle; quarter-disc; octo-plane.
Mathematics Subject Classifications 2010: 31A25, 31A30, 31A10, 35J25, 35A08

1. Introduction

The conformal invariance of the harmonic Green and Neumann functions provide a simple method to get these functions for particular domains from those of basic domains. The obstacle is often that the conformal mapping is complicated, as e.g. on the basis of the Schwarz-Christoffel formula for polygons, or even not explicitly known. There is, however, another method. It was already used in the book[1] of R. Courant and D. Hilbert for finding the harmonic Green function for a circular ring. It is based on constructing a certain meromorphic function in the entire plane having countably many simple zeros and poles by reflecting some point of the domain, the pole of the Green function, at the boundary. For the Neumann function the same set of points is used to construct a meromorphic function without zeroes having simple poles at these points. In[2] this method is used for a circular ring domain, in[3] for a particular equilateral triangle providing a parqueting of the entire plane. The results from[3] are presented here supplemented by related results for a quarter unit disc[4] and an octo plane.[5] The latter results are from two recent master theses.[4,5]

For a domain D (of the complex plane $\mathbb{C}$) having a Green $G(z,\zeta) = \frac{1}{2}G_1(z,\zeta)$ and a Neumann $N(z,\zeta) = \frac{1}{2}N_1(z,\zeta)$ function the Poisson representation[6]

$$w(z) = -\frac{1}{4\pi}\int_{\partial D} w(\zeta)\partial_{\nu_\zeta}G_1(z,\zeta)ds_\zeta - \frac{1}{\pi}\int_D w_{\zeta\bar\zeta}(\zeta)G_1(z,\zeta)d\xi d\eta,$$

and the Neumann representation[6]

$$w(z) = -\frac{1}{4\pi}\int_{\partial D}\left[w(\zeta)\partial_{\nu_\zeta}N_1(z,\zeta) - \partial_{\nu_\zeta}w(\zeta)N_1(z,\zeta)\right]ds_\zeta$$
$$-\frac{1}{\pi}\int_D w_{\zeta\bar\zeta}(\zeta)N_1(z,\zeta)d\xi d\eta$$

hold for $w \in C^2(D;\mathbb{C}) \cap C^1(\overline{D};\mathbb{C})$.

These formulas lead to solutions of the related Dirichlet and Neumann boundary value problems for the Poisson equation and solvability conditions if necessary.

In the following the Green functions $G_1(z,\zeta)$, the Poisson kernels $P(z,\zeta) = -\partial_{\nu_\zeta}G_1(z,\zeta)$ and the Neumann functions $N_1(z,\zeta)$ are listed for the three particular domains under consideration. For other examples, see e.g.[7,8]

2. Triangle

Let T be the equilateral triangle with the corner points -1, 1, $i\sqrt{3}$ in the complex plane and let $z \in T$. Denote by $\partial_1 T$ the side from 1 to $i\sqrt{3}$, by $\partial_2 T$ the one from $i\sqrt{3}$ to -1 and by $\partial_3 T$ the part on the real axis from -1 to 1. The reflection at the line $\partial_1 T$ of z leads to the point

$$z_1 = -\frac{1}{2}(1 + i\sqrt{3})\bar z + \frac{\sqrt{3}}{2}(\sqrt{3} + i).$$

The triangle T itself is reflected onto the triangle with the corners 1, $2+i\sqrt{3}$, $i\sqrt{3}$. Reflecting z_1 at the line from 1 to $2 + i\sqrt{3}$ results in

$$z_2 = -\frac{1}{2}(1 + i\sqrt{3})z + \frac{\sqrt{3}}{2}(\sqrt{3} + i).$$

For $\zeta \in T$ the points ζ_1 and ζ_2 are accordingly defined.

After three more reflections in the same direction a period arises. Therefore reflections with respect to $\partial_2 T$ are superfluous. Reflection at $\partial_3 T$ is achieved by complex conjugation. This gives another period. For defining the Green function only the half-periods

$$\omega_{m,n} = 3m + i\sqrt{3}n, \quad m, n \in \mathbb{Z},$$

are needed.

Denoting for $z, \zeta \in T$, $z \neq \zeta$,

$$G_1(z,\zeta) = \log \prod_{m+n\in 2\mathbb{Z}} \left| \frac{\zeta - \bar{z} - \omega_{m,n}}{\zeta - z - \omega_{m,n}} \frac{\zeta - z_1 - \omega_{m,n}}{\zeta - \bar{z_1} - \omega_{m,n}} \frac{\zeta - \bar{z_2} - \omega_{m,n}}{\zeta - z_2 - \omega_{m,n}} \right|^2,$$

this function turns out to be the Green function for T. Expressed by the normal derivative of the Green function the Poisson kernel for T is seen to be

$$P(z,\zeta) = 2\mathrm{Re} \sum_{\substack{m+n \\ \in 2\mathbb{Z}}} \begin{cases} \frac{\sqrt{3}+i}{z-\zeta-\omega_{m,n}} - \frac{\sqrt{3}+i}{z-\bar{\zeta}-\omega_{m,n}} + \frac{\sqrt{3}+i}{z-\zeta_1-\omega_{m,n}}, & z \in \partial_1 T, \\[2ex] \frac{\sqrt{3}-i}{z-\zeta-\omega_{m,n}} - \frac{\sqrt{3}-i}{z-\bar{\zeta}-\omega_{m,n}} + \frac{\sqrt{3}-i}{z-\zeta_2-\omega_{m,n}}, & z \in \partial_2 T, \\[2ex] \frac{i}{z-\zeta-\omega_{m,n}} + \frac{i}{z-\bar{\zeta_1}-\omega_{m,n}} - \frac{i}{z-\zeta_2-\omega_{m,n}}, & z \in \partial_3 T, \end{cases}$$

where $\zeta \in T$.

A Neumann function is

$$N_1(z,\zeta) = -\log\Big| (\zeta - z)(\zeta - z_1)(\zeta - z_2)$$

$$\times \prod_{\substack{m,n\in\mathbb{Z}, \\ 0<m^2+n^2}} \prod_{k=0}^{3} \left(\frac{\zeta - z_k}{\omega_{m,n}} - 1 \right) \left(\frac{\zeta - \bar{z_k}}{\omega_{m,n}} - 1 \right) \Big|^2,$$

Here $z_0 = z$. Its normal derivative vanishes on the boundary ∂T outside the three corner points. Therefore the Neumann problem for the Poisson equation is unconditionally solvable for the triangle T. There are other domains where this is not the case, e.g. the unit disc, see[6], the circular ring, see[2], the half ring, see[9], the half plane, see[10].

3. Quarter disc

For the quarter disc $\mathbb{D}_1 := \{|z| < 1, 0 < x, 0 < y\}$, $z = x + iy$, the Green and Neumann functions can easily be attained by mapping conformally $\mathbb{D}_1$ onto the unit disc $\mathbb{D}$. Even more simple is to apply the reflection principle. Reflecting the point z in $\mathbb{D}_1$ at the boundary leads to the reflection points $1/\bar{z}$, $\bar{z}$, $-\bar{z}$. In order to cover the entire complex plane these points are reflected at both real and imaginary axes giving $1/z$, $-1/z$, $-z$ and reflection of these points at the axes and the unit circle finally results in $-1/\overline{z}$. Observing that poles are reflected into zeroes and vice versa leads to the Green function

$$G_1(z,\zeta) = \log\left|\frac{1 - z\overline{\zeta}}{1 - z\zeta}\,\frac{\overline{\zeta} - z}{\zeta - z}\,\frac{1 + z\overline{\zeta}}{1 + z\zeta}\,\frac{\overline{\zeta} + z}{\zeta + z}\right|^2$$

for $\mathbb{D}_1$. Taking the normal derivative on the boundary $\partial\mathbb{D}_1$ gives the Poisson kernel in the form

$$P(z,\zeta) = \begin{cases} 2\mathrm{Re}\left[\frac{\zeta+z}{\zeta-z} + \frac{\zeta-z}{\zeta+z} - \frac{\overline{\zeta}+z}{\overline{\zeta}-z} - \frac{\overline{\zeta}-z}{\overline{\zeta}+z}\right], & |z| = 1,\ 0 < \mathrm{Re}\,z, \\ & \hspace{1.5em} 0 < \mathrm{Im}\,z; \\[2ex] 2\left[\frac{\zeta+\overline{\zeta}}{|\zeta-z|^2} + \frac{\zeta+\overline{\zeta}}{|1-z\zeta|^2} - \frac{\zeta+\overline{\zeta}}{|\zeta+z|^2} - \frac{\zeta+\overline{\zeta}}{|1+z\zeta|^2}\right], & z = -\overline{z},\ |z| < 1, \\ & \hspace{1.5em} \mathrm{Re}\,z = 0,\ 0 < \mathrm{Im}\,z; \\[2ex] 2i\left[\frac{\zeta-\overline{\zeta}}{|\zeta-z|^2} + \frac{\zeta-\overline{\zeta}}{|1+z\zeta|^2} - \frac{\zeta-\overline{\zeta}}{|\zeta+z|^2} - \frac{\zeta-\overline{\zeta}}{|1-z\zeta|^2}\right], & z = \overline{z},\ |z| < 1, \\ & \hspace{1.5em} 0 < \mathrm{Re}\,z,\ \mathrm{Im}\,z = 0. \end{cases}$$

Theorem 3.1. *The Dirichlet problem for the Poisson equation in the quarter disc $\mathbb{D}_1$*

$$w_{z\overline{z}} = f \ \text{ in } \ \mathbb{D}_1, \quad w = \gamma \ \text{ on } \ \partial\mathbb{D}_1,$$

with $f \in C(\mathbb{D}_1;\mathbb{C})$, $\gamma \in C(\partial\mathbb{D}_1;\mathbb{C})$, $\gamma(1) = \gamma(0) = \gamma(i) = 0$, is uniquely solvable by

$$w(z) = \frac{1}{2\pi i} \int\limits_{\substack{|\zeta|=1,\\ 0<\mathrm{Re}\zeta, 0<\mathrm{Im}\zeta}} \gamma(\zeta)\mathrm{Re}\left[\frac{\zeta+z}{\zeta-z} + \frac{\zeta-z}{\zeta+z} - \frac{\overline{\zeta}+z}{\overline{\zeta}-z} - \frac{\overline{\zeta}-z}{\overline{\zeta}+z}\right]\frac{d\zeta}{\zeta}$$

$$+ \frac{1}{2\pi i}\int_0^1 \gamma(t)\left[\frac{z-\overline{z}}{|t-z|^2} + \frac{z-\overline{z}}{|1+tz|^2} - \frac{z-\overline{z}}{|t+z|^2} - \frac{z-\overline{z}}{|1-tz|^2}\right]dt$$

$$+ \frac{1}{2\pi}\int_0^1 \gamma(it)\left[\frac{z+\overline{z}}{|z+it|^2} + \frac{z+\overline{z}}{|1+zit|^2} - \frac{z+\overline{z}}{|z-it|^2} - \frac{z+\overline{z}}{|1-zit|^2}\right]dt$$

$$- \frac{1}{\pi}\int\limits_{\mathbb{D}_1} f(\zeta)G_1(z,\zeta)d\xi d\eta.$$

A candidate for a harmonic Neumann function for $\mathbb{D}_1$ is

$$-\log\left|(\zeta - z)(\overline{\zeta} - z)(\zeta + z)(\overline{\zeta} + z)(1 - z\zeta)(1 - z\overline{\zeta})(1 + z\zeta)(1 + z\overline{\zeta})\right|^2.$$

It may altered by adding some harmonic function. Let, see[4] ,

$$N_1(z, \zeta) = 4\log |z\zeta|^2$$
$$- \log |(\zeta - z)(\overline{\zeta} - z)(\zeta + z)(\overline{\zeta} + z)(1 - z\zeta)(1 - z\overline{\zeta})(1 + z\zeta)(1 + z\overline{\zeta})|^2.$$

Then $N_1(z, \zeta)$ satisfies all properties of a Neumann function. In particular, on the whole boundary except the three corner points its normal derivative vanishes identically,

$$\partial_{\nu_z} N_1(z, \zeta) = 0 \ \text{ for } \ z \in \partial \mathbb{D}_1 \setminus \{0, 1, i\}, \ \zeta \in \mathbb{D}_1.$$

This implies that the Neumann boundary value problem for the Poisson equation in $\mathbb{D}_1$ is unconditionally (and uniquely) solvable, see.[4]

Theorem 3.2. *The Neumann problem for the Poisson equation in the quarter disc $\mathbb{D}_1$*

$$w_{z\overline{z}} = f \ \text{ in } \ \mathbb{D}_1, \quad \partial_\nu w = \gamma \ \text{ on } \ \partial \mathbb{D}_1 \setminus \{0, 1, i\},$$

with $f \in C(\mathbb{D}_1; \mathbb{C})$, $\gamma \in C(\partial \mathbb{D}_1; \mathbb{C})$, is uniquely solvable by

$$w(z) =$$
$$= \frac{1}{2\pi i} \int_{\substack{|\zeta|=1, \\ 0<\mathrm{Re}\zeta, \\ 0<\mathrm{Im}\zeta}} \gamma(\zeta) \left[2\log |z|^2 - \log |(1 - z\zeta)(1 + z\zeta)(1 - z\overline{\zeta})(1 + z\overline{\zeta})|^2 \right] \frac{d\zeta}{\zeta}$$

$$+ \frac{1}{2\pi} \int_0^1 \gamma(t) \left[2\log |zt|^2 - \log |(t - z)(t + z)(1 - zt)(1 + zt)|^2 \right] dt$$

$$- \frac{1}{2\pi} \int_0^1 \gamma(it) \left[2\log |zt|^2 - \log |(it - z)(it + z)(1 - zit)(1 + zit)|^2 \right] dt$$

$$- \frac{1}{\pi} \int_{\mathbb{D}_1} f(\zeta) N_1(z, \zeta) d\xi d\eta.$$

4. Octo plane

The simplest way to find the Green function for the octo plane

$$\mathbb{O} := \{z : 0 < \mathrm{Im}\, z < \mathrm{Re}\, z\}$$

is at first to use the reflection principle for constructing the Green function for the upper half plane

$$\mathbb{H}_+ = \{z : 0 < \mathrm{Im}\, z\}$$

giving

$$G_{1\mathbb{H}_+}(z,\zeta) = \log\left|\frac{\overline{\zeta}-z}{\zeta-z}\right|^2$$

and then combine this with the conformal map $\omega(z) = z^4$ of $\mathbb{O}$ to $\mathbb{H}_+$. Thus the Green function for $\mathbb{O}$ is

$$G_1(z,\zeta) = \log\left|\frac{\overline{\zeta}^{\,4}-z^4}{\zeta^4-z^4}\right|.$$

The Poisson kernel then is with $z = x + iy$

$$P(z,\zeta) = \begin{cases} 4ix\left[\frac{\zeta^2-\overline{\zeta}^{\,2}}{|x^2-\zeta^2|^2} - \frac{\zeta^2-\overline{\zeta}^{\,2}}{|x^2+\zeta^2|^2}\right], & z = \overline{z} = x,\ 0 < x,\ y = 0; \\[2ex] 4\sqrt{2}x\left[\frac{\zeta^2+\overline{\zeta}^{\,2}}{|2ix^2-\zeta^2|^2} - \frac{\zeta^2+\overline{\zeta}^{\,2}}{|2ix^2+\zeta^2|^2}\right], & z = (1+i)x,\ 0 < x = y. \end{cases}$$

Theorem 4.1. *The Dirichlet problem for the Poisson equation in the octo plane* $\mathbb{O}$

$$w_{z\overline{z}} = f \ \text{ in } \ \mathbb{O}, \quad w(t) = \gamma_1(t), \quad w(t+it) = \gamma_2(t) \ \text{ for } \ t \in \mathbb{R}^+,$$

with $f \in L_1(\mathbb{O}; \mathbb{C})$, $\gamma_1, \gamma_2 \in C(\mathbb{R}^+; \mathbb{C})$, $\gamma_1(0) = \gamma_2(0) = 0$, *so that* $t^{1+\delta}\gamma_1(t)$, $t^{1+\delta}\gamma_2(t)$, $z^{2+\delta}f(z)$ *are bounded for some* $0 < \delta$, *is uniquely solvable by*

$$w(z) = -\frac{1}{2\pi} \int\limits_0^{+\infty} \gamma_1(t)\mathrm{Re}\left(\frac{t}{t^2-z^2} + \frac{t}{t^2+z^2}\right) dt$$

$$+ \frac{1}{2\pi} \int\limits_0^{+\infty} \gamma_2(t)\mathrm{Re}\left(\frac{t}{2it^2-z^2} + \frac{t}{2it^2+z^2}\right) dt$$

$$- \frac{1}{\pi} \int\limits_{\mathbb{O}} f(\zeta)G_1(z,\zeta)d\xi d\eta.$$

For the Neumann function a possible choice would be

$$-\log\left|(\zeta^4 - z^4)(\overline{\zeta}^{\,4} - z^4)\right|^2.$$

In order to have a nice behavior at infinity some additional harmonic term is introduced, see[5] ,

$$N_1(z,\zeta) = -\log\left|\frac{(\zeta^4-z^4)(\overline{\zeta}^{\,4}-z^4)}{|z|^{16}|\zeta|^{16}}\right|^2.$$

Its normal derivative on $\partial\mathbb{O}$ vanishes identically with the exception of the corner point. For the solution of the Neumann boundary value problem for the Poisson equation see.[5]

Acknowledgments

This work was done while the third author was a fellow of the "Studienstiftung des Abgeordnetenhauses von Berlin" from October 2008 to September 2009.

References

1. R. Courant, D. Hilbert. *Methods of Mathematical Physics. I.* (New York: Interscience Publishers., INC, 1953).
2. T. Vaitsiakhovich. Boundary value problems for complex partial differential equations in a ring domain. Ph.D. thesis, FU Berlin, 2008: www.diss.fu-berlin.de/diss/receive/FUDISS_thesis_000000003859
3. H. Begehr, T. Vaitekhovich. Green functions, reflections, and plane parqueting. Preprint, FU Berlin, 2009.
4. M.-R. Costache. Basic boundary value problems for the Cauchy-Riemann and the Poisson equation in a quarter disc. Master thesis, Şcoala Normală Superioară Bucharest, 2009.
5. S. Tappert. Randwertprobleme in der Achtebene. Diplom thesis, FU Berlin, 2009 (in German).
6. H. Begehr. Boundary value problems in complex analysis, I, II. *F. Bol. Asoc. Mat. Venezolana,* **XII** (2005), 65–85, 217–250.
7. H. Begehr, T. Vaitekhovich. Polyharmonic Green functions for particular plane domains. Proc. 6th Congress of Roman. Mathematicians, Bucharest, 2007; Eds. L. Beznea et al. Bucharest: Publ. House Roman. Acad. Sci. **1** (2009), 119–126.
8. H. Begehr, T. Vaitekhovich. Some harmonic Robin functions in the complex plane. *J. Analysis and Appl.* **1** (2009).
9. H. Begehr, T. Vaitekhovich. Harmonic boundary value problems in half disc and half ring. *Funct. Appr.* **40.2** (2009), 251–282.
10. E. Gaertner. Basic complex boundary value problems in the upper half plane. Ph.D. thesis, FU Berlin, 2006: www.diss.fu-berlin.de/diss/receive/FUDISS_thesis_000000002129

On some classes of bicomplex pseudoanalytic functions

P. Berglez

Department of Mathematics, Graz University of Technology,
Graz, Austria
E-mail: berglez@tugraz.at

We consider certain classes of bicomplex pseudoanalytic functions using the commutative ring of bicomplex numbers $\mathbb{T} \cong Cl_{\mathbb{C}}(1,0) \cong Cl_{\mathbb{C}}(0,1)$. They obey specific bicomplex Vekua equations. We give representations for these functions using suitable differential operators acting on $\mathbb{T}$-holomorphic functions as well as on other bicomplex pseudoanalytic functions. The functions investigated here are of interest in connection with the complexified stationary Schrödinger equation for example.

Keywords: Pseudoanalytic functions; Bers-Vekua equations; bicomplex numbers; differential operators.

1. Introduction

Bicomplex numbers, also called tetranumbers, are defined as

$$\mathbb{T} := \{z_1 + i_2 z_2 \mid z_1, z_2 \in \mathbb{C}(i_1)\}$$

with $\mathbb{C}(i_k) := \{x + i_k y \mid x, y \in \mathbb{R}\}, k = 1, 2$, and the rules $i_1^2 = i_2^2 = -1$, $j^2 = 1, i_1 i_2 = i_2 i_1 = j, \ i_1 j = j i_1 = -i_2, \ i_2 j = j i_2 = -i_1$, for the imaginary units i_1, i_2, j. They form a subset of the Clifford algebra, i.e. $\mathbb{T} \cong Cl_{\mathbb{C}}(1,0) \cong Cl_{\mathbb{C}}(0,1)$ and may be given also by $\mathbb{T} = \{x_0 + i_1 x_1 + i_2 x_2 + j x_3 \mid x_0, x_1, x_2, x_3 \in \mathbb{R}\}$. This set is a commutative ring with unit and zero divisors where the set of singular elements $\mathcal{O}_2$ is given by $\mathcal{O}_2 = \{z_1 + i_2 z_2 \mid z_1^2 + z_2^2 = 0\}$ (see e.g. Refs. 1,2). For $z = z_1 + i_2 z_2 \in \mathbb{T}$ there exist three possible conjugations:

$$z^{\star} = \bar{z}_1 + i_2 \bar{z}_2, \ z^* = z_1 - i_2 z_2, \ z^{\dagger} = \bar{z}_1 - i_2 \bar{z}_2$$

A function $f : \ X \subseteq \mathbb{T} \to \mathbb{T}$ is called $\mathbb{T}$-differentiable in $z_0 \in X$ with derivative equal to $f'(z_0) \in \mathbb{T}$ if the limit

$$f'(z_0) := \lim_{z \to z_0} \frac{f(z) - f(z_0)}{z - z_0} , \ z - z_0 \notin \mathcal{O}_2 \tag{1}$$

exists. A function f is $\mathbb{T}$-holomorphic on an open set U if and only if it is $\mathbb{T}$-differentiable at each point of U.

We introduce the formal differential operators

$$D^* \equiv \partial_z = \frac{1}{2}(\partial_{z_1} - i_2\partial_{z_2})\,,\quad D \equiv \partial_{z^*} = \frac{1}{2}(\partial_{z_1} + i_2\partial_{z_2})$$

$$\partial_{z^\star} = \frac{1}{2}(\partial_{\bar{z}_1} - i_2\partial_{\bar{z}_2})\,,\quad \partial_{z\dagger} = \frac{1}{2}(\partial_{\bar{z}_1} + i_2\partial_{\bar{z}_2})$$

Here D is the generalized Cauchy-Riemann operator.

From D. Rochon[3] we get the

Lemma 1.1. *Let* $f(z) = f_0(z_1, z_2) + i_1 f_1(z_1, z_2) + i_2 f_2(z_1, z_2) + j f_3(z_1, z_2)$ *be a bicomplex function. If the derivative (1) exists, then* $f_{0,x}, f_{0,y}, f_{1,x}, f_{1,y}, f_{2,x}, f_{2,y}, f_{3,x}$ *and* $f_{3,y}$ *exist and*

$$\partial_z f(z_0) = f'(z_0) \tag{2}$$

$$\partial_{z^*} f(z_0) = 0\,,\ \partial_{z^\star} f(z_0) = 0 \quad \textit{and} \quad \partial_{z\dagger} f(z_0) = 0 \tag{3}$$

Moreover, if $f_{0,x}, f_{0,y}, f_{1,x}, f_{1,y}, f_{2,x}, f_{2,y}, f_{3,x}$ *and* $f_{3,y}$ *exist and are continuous in a neighborhood of* z_0 *and if Eqs. (3) hold, then the limit (1) exists.*

The system (3) can be considered as the corresponding Cauchy-Riemann equations a $\mathbb{T}$-holomorphic function has to obey. The 2nd and 3rd conditions in (3) mean that the complex components of a bicomplex $\mathbb{T}$-holomorphic function are holomorphic functions of two complex variables in the classic sense.

D. Rochon[3] introduced bicomplex generalizations of function theory. One of the classes he considered leads to the following bicomplex Bers-Vekua equations

$$\partial_{z^*} w = a_1 w + b_1 w^* \tag{4}$$

$$\partial_{z^\star} w = a_2 w + b_2 w^* \tag{5}$$

$$\partial_{z\dagger} w = a_3 w + b_3 w^* \tag{6}$$

Using a generating pair (F, G) (in the sense of L. Bers[4]) the coefficients $a_k, b_k,\ k = 1, 2, 3$, are given by

$$a_1 = -(F^* \partial_{z^*} G - (\partial_{z^*} F)G^*)/N\,,\quad b_1 = (F\partial_{z^*} G - (\partial_{z^*} F)G)/N$$

$$a_2 = -(F^\star \partial_{z^\star} G - (\partial_{z^\star} F)G^\star)/N\,,\quad b_2 = (F\partial_{z^\star} G - (\partial_{z^\star} F)G)/N$$

$$a_3 = -(F^\dagger \partial_{z\dagger} G - (\partial_{z\dagger} F)G^\dagger)/N\,,\quad b_3 = (F\partial_{z\dagger} G - (\partial_{z\dagger} F)G)/N$$

with $N = FG^* - F^*G$. The (F, G)-derivative $\dot{w} = d_{(F,G)}w/dz$ of a function w has the form

$$\dot{w} = \partial_z w - Aw - Bw^* \tag{7}$$

with

$$A = -(F^*\partial_z G - (\partial_z F)G^*)/N \,, \quad B = (F\partial_z G - (\partial_z F)G)/N \tag{8}$$

In this paper we shall derive different representations for the solutions of a Bers-Vekua equation of the type (4). We will investigate some interesting connections between the solutions of such a Bers-Vekua equation and different differential equations of second order. One of these equations can be considered as a generalized Bauer-Peschl equation the solutions of which may be represented by differential operators acting on $\mathbb{T}$-holomorphic functions. Other relations are obtained using the corresponding complex potentials and the (F, G)-derivative respectively. In the particular case where the functions considered here are $\mathbb{C}(i_2)$-valued functions defined in a domain $D_0 \subset \mathbb{C}(i_2)$ some of these results contain assertions obtained by K.W. Bauer.[5]

There exist a lot of investigations on the Bers-Vekua equation and its generalizations. For example V.V. Kravchenko[6] applied the complex and quaternionic Bers-Vekua theories to several problems of mathematical physics.

2. A generalized Bauer-Peschl equation

In the following we denote the set of $\mathbb{T}$-holomorphic functions by $TH(\Omega) := \{f : \Omega \subseteq \mathbb{T} \to \mathbb{T} \mid f \text{ holomorphic in } \Omega\}$ and the set of $\mathbb{T}$-antiholomorphic functions by $TH^*(\Omega) := \{f : \Omega \subseteq \mathbb{T} \to \mathbb{T} \mid f^* \text{ holomorphic in } \Omega\}$ where hereafter Ω denotes a suitable domain in $\mathbb{T}$. Now let us consider the differential equation

$$\eta^2 DD^*w - n(n+1)\, w = 0 \,, \quad n \in \mathbb{N} \tag{9}$$

with $\eta = z + z^* \neq 0$ in $\Omega \subset \mathbb{T}$.

Since $DD^* = \frac{\partial^2}{\partial z_1^2} + \frac{\partial^2}{\partial z_2^2}$ is the complex Laplacian in $\mathbb{C}^2$ Eq. (9) becomes a complexified Schrödinger equation with a special potential. There is known an extensive research on the complex Laplace equation, for example K. Fujita and M. Morimoto[7] discussed the double series expansion of holomorphic functions on the Euclidean ball, the Lie ball and the dual Lie ball. In Ref. 8 they presented integral representations for eigenfunctions of the Laplacian.

The following theorem concerning the solutions of the generalized Bauer-Peschl equation (9) was proved in Ref. 9:

Theorem 2.1.

(1) For arbitrary functions $g \in TH(\Omega)$ and $h \in TH^*(\Omega)$ the function

$$w = K_n\, g + K_n^*h \tag{10}$$

with

$$K_n g := \sum_{k=0}^{n} A_k^n\,(D^*)^k g, \quad K_n^* h := \sum_{k=0}^{n} A_k^n\, D^k h, \quad A_k^n = \frac{(-1)^{n-k}(2n-k)!}{k!(n-k)!\eta^{n-k}}$$

is a solution of (9) defined in Ω.
(2) The function

$$w = \sum_{k=0}^{n} A_k^n\,\left[(D^*)^k f + D^k f^*\right]$$

represents for each $f \in TH(\Omega)$ a solution of (9) with values in $\mathbb{C}(i_1)$.

Let u_1 and u_2 be $\mathbb{C}(i_1)$-valued functions and $u = u_1 + i_2 u_2$ be a solution of the Bers-Vekua equation

$$D\,u = (m/\eta)\,u^*, \quad m \in \mathbb{N} \tag{11}$$

Then the functions u_1 and u_2 obey the differential equations

$$\eta^2\, DD^* u_1 - m(m-1)\,u_1 = 0 \quad \text{and} \quad \eta^2\, DD^* u_2 - m(m+1)\,u_2 = 0$$

These equations are particular cases of the Bauer-Peschl equation (9). For their solutions with values in $\mathbb{C}(i_1)$ we can give the representations

$$u_1 = K_{m-1}g + K_{m-1}^* g^* \quad \text{and} \quad u_2 = K_m h + K_m^* h^*$$

with $g, h \in TH(\Omega)$. On the other hand, since $u = u_1 + i_2 u_2$ is a solution of (9) we have to choose $2g = D^* f$ and $2i_2 h = f, f \in TH(\Omega)$, and thus we can prove the following

Theorem 2.2. *For each function $f \in TH(\Omega)$ the expression*

$$u = \sum_{k=0}^{m} A_k^m\,\left(m(D^*)^k f - (m-k)D^k f^*\right) \tag{12}$$

represents a solution of (11) in Ω.

The connection between a solution u of (11) in the form (12) and its generating function f is given by

$$(D^*)^{2m} f = \eta^{-m}\left[(D^*)^m(\eta^m u)\right]$$

The pair $(\eta^m, i_2\,\eta^{-m})$ forms a generating pair in the sense of L. Bers for Eq. (11) and we can show that the corresponding Bers-Vekua equations (5) and (6) are satisfied by the function u given in (12) also.

3. Complex potentials

Proposition 3.1. *Let U be a solution of the Bers-Vekua equation*

$$DU = aU + bU^* \tag{13}$$

then the function u in the form $U = \alpha\,u$ where $\alpha \notin \mathcal{O}_2$ obeys the condition $D\alpha = a\alpha$ is a solution of

$$Du = cu^* \quad with \quad c = b\alpha^{-1}\alpha^* \tag{14}$$

Therefore it is sufficient to consider the reduced Bers-Vekua equation (14) instead of (13) in the following.

Applying the operator D^* to the Bers-Vekua equation

$$Dv = cv^* \tag{15}$$

where c is an arbitrary nonvanishing function and considering $D^*v^* = c^*v$ we get the differential equation

$$DD^*v - c^{-1}(D^*c)(Dv) - cc^*v = 0 \tag{16}$$

Now let v be a solution of Eq. (16), then the functions

$$v_1 = (v + (c^*)^{-1}D^*v^*)/2 \quad and \quad v_2 = (v - (c^*)^{-1}D^*v^*)/(2i_2)$$

are solutions of (15) again. Thus a solution of Eq. (16) may be given as $v = v_1 + i_2v_2$. Therefore the functions v_1 and v_2 are called generalized complex potentials and Eq. (16) is called the differerential equation of the complex potentials.

The following propositions can be proved by direct calculation.

Proposition 3.2. *Consider a solution w of the Bers-Vekua equation (15). Its (F,G)-derivative $\dot{w}$ according to (7) is a solution of the Bers-Vekua equation*

$$D(\dot{w}) = -B\,(\dot{w})^* \tag{17}$$

with B from to (8).

Proposition 3.3. *The differential equation of the complex potentials to Eq. (17) is*

$$DD^*u - B^{-1}(D^*B)(Du) - BB^*u = 0 \tag{18}$$

Now the functions $\dot{w}_1$ and $\dot{w}_2$ considered as solutions of (17) are solutions of (18) also and we deduce for the function $v = \dot{w}_1 + i_2\dot{w}_2$ the representation

$$v = D^*w - (c^{-1}B)\,Dw - Aw \tag{19}$$

Thus we have proved

Theorem 3.1.

(1) For each solution w of (15) the function

$$v = D^*w + (c^{-1}d)\,Dw + ew$$

with $d = -B, e = -A$, A, B according to (8) is a solution of the differential equation

$$Dv = d\,v^*$$

(2) Let w be a solution of the differential equation of the complex potentials (16) then the function v according to (19) is a solution of (18).

4. Differential operators of first order

Proposition 4.1. *Let Y be a particular $\mathbb{C}(i_1)$-valued solution and w an arbitrary solution of the Bauer-Peschl equation (9). Then the function v with $w = Yv$ is a solution of*

$$YDD^*v + (DY)D^*v + (D^*Y)Dv = 0 \tag{20}$$

Proof. By direct calculation. $\square$

Now let W be an arbitrary $\mathbb{C}(i_1)$-valued solution of

$$DD^*W + H^{-1}(DH)(D^*W) + (H^*)^{-1}(D^*H^*)\,DW = 0 \tag{21}$$

where H denotes an arbitrary suitable function nonvanishing in Ω. The function $V = D^*W$ obeys the differential equation

$$DV + H^{-1}(DH)\,V + (H^*)^{-1}(D^*H^*)\,V^* = 0$$

which with Proposition 3.1 can be transformed by $V = H^{-1}U$ into

$$DU = cU^* \quad \text{with} \quad c = -(H^*)^{-2}H\,(D^*H^*) \tag{22}$$

Thus we have proved

Proposition 4.2. *Let W be an arbitrary $\mathbb{C}(i_1)$-valued solution of Eq. (21) then U according to $U = H\,D^*W$ is a solution of (22).*

Now with Propositions 4.1 and 4.2 and using $H = Y$ we can show that the function $U = YD^*(Y^{-1}w)$ is a solution of the Bers-Vekua equation

$$DU = -Y^{-1}(D^*Y)U^* \tag{23}$$

if w denotes a solution of (9). Thus we have proved the following

Theorem 4.1. *Let w be an arbitrary solution of (9) and Y a particular $\mathbb{C}(i_1)$-valued solution of (9) then U according to*

$$U = D^*w - Y^{-1}(D^*Y)w$$

is a solution of the Bers-Vekua equation (23).

Starting from a solution w of (9) the transformation $v = \eta^n w$ according to Proposition 4.1 leads to the differential equation

$$DD^*v - (n/\eta)\,(D^*v + Dv) = 0\,, \quad n \in \mathbb{R} \tag{24}$$

In Ref. 9 the following theorem was proved:

Theorem 4.2. *Let p and q be particular solutions of (24) and v an arbitrary solution of (24), then the function V according to*

$$V = P^{-1}\left((Dp)(D^*v) - (D^*p)(Dv)\right)$$

with $P = (Dp)(D^*q) - (D^*p)(Dq) \neq 0$ is a solution of

$$DD^*V + D\left(\log\frac{P}{\eta^n(Dp)}\right)(D^*V) + D^*\left(\log\frac{P}{\eta^n(D^*p)}\right)(DV) = 0 \tag{25}$$

Now we consider a solution V of (25) where the functions p and q are assumed to be $\mathbb{C}(i_1)$-valued solutions of (24). Then in view of (25) we can use Proposition 4.2 with $H = P/(\eta^n(Dp))$ to prove the following result.

Theorem 4.3. *Let w be an arbitrary $\mathbb{C}(i_1)$-valued solution of (24) and let p and q denote two particular $\mathbb{C}(i_1)$-valued solutions of (24) with $P = (Dp)(D^*q) - (D^*p)(Dq)$. Then the function U according to*

$$U = (\eta^n(Dp))^{-1}P\,D^*\left(P^{-1}((Dp)(D^*w) - (D^*p)(Dw))\right)$$

is a solution of the Bers-Vekua equation

$$DU = P^{-1}\left((D^*p)(D^{*2}q) - (D^{*2}p)(D^*q)\right)\,U^*$$

Since the solutions of (24) can be given in an explicit way as Theorem 2.1 shows the Theorems 4.1 and 4.3 allow us to give the solutions of a large class of Bers-Vekua equations in an explicit way also.

References

1. G.B. Price, *An introduction to multicomplex spaces and functions* (Marcel Dekker, New York, 1991).
2. D. Rochon, M. Shapiro, *On algebraic properties of bicomplex and hyperbolic numbers*, An. Univ. Oradea Fasc. Mat., **11** (2004), 71–110.
3. D. Rochon, *On a relation of bicomplex pseudoanalytic function theory to the complexified stationary Schrödinger equation*, Complex Var. Elliptic Equ., **53** (6) (2008), 501–521.
4. L. Bers, *Theory of Pseudo–Analytic Functions* (New York University, 1953).
5. K.W. Bauer, *Differentialoperatoren bei verallgemeinerten Euler-Gleichungen*, Ber. Math.-Statist. Sekt. Forsch. Graz, **121** (1979), 1–17.
6. V.V. Kravchenko, *Applied pseudoanalytic function theory* (Birkhäuser, Basel, 2009).
7. K. Fujita, M. Morimoto, *On the double series expansion of holomorphic functions*, J. Math. Anal. Appl., **272** (1) (2002), 335–348.
8. K. Fujita, M. Morimoto, *Integral representation for eigenfunctions of the Laplacian*, J. Math. Soc. Japan, **51** (3) (1999), 699–713.
9. P. Berglez, *On a generalized Bauer-Peschl equation in bicomplex spaces*, to appear.

Solvability condition of the Riemann-Hilbert monodromy problem

G.K. Giorgadze

Department of Mathematics, Tbilisi State University,
Tbilisi, 0160, Georgia
E-mail: gia.giorgadze@tsu.ge
www.tsu.ge

This article explains relation between Riemann-Hilbert monodromy and Riemann-Hilbert boundary problems and gives one sufficient condition of solvability of the Riemann-Hilbert problem on the Riemann sphere in terms of partial indices corresponding to a piecewise continuous matrix function.

Keywords: Fuchsian system; monodromy representation; vector bundle; holomorphic section.

1. Riemann-Hilbert boundary problem

Let Γ be a smooth closed positively oriented loop in $\mathbb{CP}^1$ which separates $\mathbb{CP}^1$ into two connected domains U_+ and U_-. Suppose $0 \in U_+$ and $\infty \in U_-$.

Let $f : \Gamma \to GL_n(\mathbb{C})$ be a Hölder matrix-function. Find a piecewise holomorphic vector function $\Phi(t)$ in $U_+ \bigcup U_-$, which admits continuous boundary values on Γ and satisfies on Γ the boundary condition

$$\Phi^+(t) = f(t)\Phi^-(t), t \in \Gamma$$

and has finite order at ∞.

This problem is traditionally solved using the method of singular integral equations.[7] Below we will use algebraic approach to solve it.[64]

Denote by Ω the space of all Hölder loops $f : \Gamma \to GL_n(C)$. It is a Banach Lie group with respect to obvious norm and operations. Suppose

$$\Omega^+ = \{f \in \Omega : f \text{ is the boundary value of a matrix function holomorphic in } U^+\}$$

and

$\Omega^- = \{f \in \Omega : f$ is the boundary value of a matrix function holomorphic in U^- and is regular at infinity $f(\infty) = \mathbf{1}\}$.

It is known that any loop $f \in \Omega$ can be represented as

$$f(t) = f^-(t)d_K f^+(t), \tag{1}$$

where $f^\pm \in \Omega^\pm$ and d_K is a diagonal loop $d_K = diag(t^{k_1}, ..., t^{k_n})$ with condition $k_1 \geq ... \geq k_n$.

The diagonal matrix d_K will be called the *characteristic loop* of the corresponding matrix-function, $K = (k_1, k_2, ..., k_n)$ being called the *characteristic multi-index* or *partial indices* of f. Two loops $f, g \in \Omega$ will be called equivalent, if f and g have identical characteristic multi-indices.

For $K = (k_1, k_2, ..., k_n)$, denote by Ω_K the set of equivalence classes of loops Ω. The representation 1 is not unique, but if one fixes f^+ (or f^-) then f^- (respectively f^+) will be uniquely defined.

Consider the holomorphic vector bundle on $\mathbb{CP}^1$ which is obtained by covering of the Riemann sphere $\mathbb{CP}^1$ with three open sets $\{U^+, U^-, U_3\} = \mathbb{CP}^1 \setminus \{0, \infty\}$, with transition functions

$$g_{13} = h_1 : U^+ \cap U_3 \to GL_n(\mathbb{C}), g_{23} = h_2 d_K : U^- \cap U_3 \to GL_n(\mathbb{C}).$$

Let us denote this bundle by $E \to \mathbb{CP}^1$. From 1 it follows, that every holomorphic vector bundle splits into direct sum of line bundles

$$E \cong E(k_1) \oplus ... \oplus E(k_n). \tag{2}$$

The numbers $k_1,...,k_n$ are the Chern numbers of the line bundles $E(k_1),...,E(k_n)$ and satisfy the conditions $k_1 \geq ... \geq k_n$. The integer-valued vector $K = (k_1, ..., k_n) \in \mathbb{Z}^n$ is called the *splitting type* of the holomorphic vector bundle E. It defines uniquely the holomorphic type of the bundle E.

Denote by $\mathbb{E}$ the sheaf of germs of holomorphic sections of the bundle E, then the solutions of the Riemann-Hilbert boundary problem are elements of the zeroth cohomology group $H^0(\mathbb{CP}^1, \mathbb{E})$, therefore the number l of linearly independent solutions is $dim H^0(\mathbb{CP}^1, \mathbb{E})$. Since the Chern number $c_1(E)$ of the bundle E is equal to index $\det G(t)$, we obtained the known criterion of solvability of the Riemann-Hilbert boundary problem. In particular the following theorem is true:

Theorem 1.1.[6] *The Riemann-Hilbert boundary problem has solutions if and only if $c_1(E) \geq 0$ and the number l of linearly independent solutions is*

$$l = \dim H^0(\mathbb{CP}^1, \mathbb{E}) = \sum_{i=1}^{n} k_i + 1.$$

Let $l^p(\Gamma)$ be the space of Lebesgue measurable functions. All $f \in l^p_+(\Gamma)$ can be identified with functions $\hat{f}$ holomorphic in U_+. Thus $\hat{f}$ is an analytic continuation of f to U_+. Here $l^p_+(\Gamma)$ denotes the space of those holomorphic functions which are boundary values of functions from $l^p(\Gamma)$; similarly let $l^p_-(\Gamma)$ denote the space of those holomorphic functions on U_- whose extension to Γ gives an element of $l^p(\Gamma)$. Let $l^\infty(\Gamma)$ be the Banach space of Lebesgue measurable and essentially bounded functions.

Definition 1.1.[5] Factorization of a matrix-function $G \in l^\infty(U)^{n \times n}$ in the space $l^p(\Gamma)$ is its representation in the form

$$G(t) = G_+(t)\Lambda(t)G_-(t), \quad t \in \Gamma \tag{3}$$

where $\Lambda(t) = diag(t^{k_1}, ..., t^{k_n})$, $k_i \in \mathbb{Z}$, $i = 1, ..., n$, $G_+ \in l_+(\Gamma)^{n \times n}$ and $G_+(t) \in l^q_+(\Gamma)^{n \times n}$, $G_-(t) \in l^q_-(\Gamma)^{n \times n}$, and $G_-(t) \in l^p_-(\Gamma)^{n \times n}$, $\frac{1}{p} + \frac{1}{q} = 1$.

We say that G admits the *canonical* factorization in $l^p(\Gamma)$, if $k_1 = ... = k_n = 0$.

Let us consider the particular case, namely on the subspace $\mathbb{CP}^1(\Gamma)^{n \times n}$ of piecewise continuous matrix-functions. For elements of this subspace there exist the one-sided limits $G(t + 0)$ and $G(t - 0)$ for each $t \in \Gamma$. For such matrix-functions necessary and sufficient condition for the existence of Φ-factorization is given by the following theorem.

Theorem 1.2.[5] *A matrix-function $G \in PC(\Gamma)^{n \times n}$ is Φ-factorizable in the space $l_p(\Gamma)$ if and only if*

a) the matrices $G(t + 0)$ and $G(t - 0)$ are invertible for each $t \in \Gamma$;
b) for each $j = 1, ..., n$ and $t \in \Gamma$ one has

$$\frac{1}{2\pi} \arg \lambda_j(t) + \frac{1}{p} \notin \mathbb{Z}.$$

Here $\lambda_1(t), ..., \lambda_n(t)$ are eigenvalues of the matrix-function $G(t - 0)G(t + 0)^{-1}$.

If a matrix-function G is Φ-factorizable, then $\xi_j(\tau) = \frac{1}{2\pi} \arg \lambda_j(\tau)$ is a single-valued function taking values in the interval $\left(\frac{1}{p} - 1, \frac{1}{p}\right)$.

Suppose $G \in PC(\Gamma)^{n \times n}$ is moreover a piecewise constant matrix function with singular points $s_1, ..., s_m \in \Gamma$, occurring in this order on Γ. Suppose G is factorizable in the space $l_p(\Gamma)$. Let us denote $M_k = G(s_k - 0)G(s_k + 0)^{-1}$, $k = 1, ..., m$. Thus G is constant on the arc (s_k, s_{k+1}), and clearly $M_1 M_2 \cdots M_k = 1$. Suppose that the matrices are similar to the matrices $\exp(-2\pi i E_k)$ and eigenvalues of E_k belong to the interval $\left(\frac{1}{p} - 1, \frac{1}{p}\right)$, where the matrices E_k are determined uniquely up to similarity since length of that interval is 1. The numbers $\xi_1(s_k), ..., \xi_n(s_k)$ are equal to real parts of eigenvalues of E_k. This implies that for the index k one has the formula $k = \sum_{k=1}^{m} tr E_k$. Thus the matrices $E_1, ..., E_k$ depend on the space $l_p(\Gamma)$. They also depend on the choice of a logarithm of eigenvalues of the matrices M_j. Thus $G \in PC(\Gamma)^{n \times n}$ produces two m-tuples $(M_1, ..., M_m)$ and $(E_1, ..., E_m)$ of matrices.

2. Riemann-Hilbert monodromy problem

Let $s_1, ..., s_m \in \mathbb{CP}^1$ be some points, with no ∞ among them, and let

$$\rho : \pi_1(\mathbb{CP}^1 - \{s_1, ..., s_m\}, z_0) \to GL_n(\mathbb{C}) \tag{4}$$

be a representation. The problem consists in the following: for the representation ρ, find such a Fuchs system

$$df = \left(\sum_{j=1}^{m} \frac{A_j}{z - s_j} dz \right) f, \tag{5}$$

whose monodromy representation coincides with ρ. In (5), the A_j are constant matrices satisfying the condition $\sum_{j=1}^{m} A_j = 0$.

Let (7) (see below) be the regular system of differential equations which is induced by the representation (4). The fundamental matrix of solutions in a neighborhood of s_j is

$$\Phi_j(\tilde{z}) = U_j(z)(z - s_j)^{\Psi_j}(\tilde{z} - s_j)^{E_j}. \tag{6}$$

Here Ψ_j are exponents of the solution space of the system (7) and $E_j = \frac{1}{2\pi i} \ln M_j$, with eigenvalues $\mu_j^1, \mu_j^2, ..., \mu_j^n$ satisfying the conditions $0 \leq Re\mu_j^i < 1$. The numbers $\beta_j^i = \varphi_j^i + \mu_j^i$ will be called *exponents* of the solution space at the point s_j (or j-exponents).

It is known that the system (7) is Fuchsian at s_j if and only if $\det U_j(s_j) \neq 0$ and in this case in a neighborhood of s_j, one has $\omega_j = \frac{A_j}{z - s_j} dz$, where A_i is a constant matrix with eigenvalues $\beta_j^i, i = 1, ..., n$.

Using the exponents β_i^j, the condition for a regular system on $\mathbb{CP}^1$ to be Fuchs is given by the following identity: $\beta = \sum_{j=1}^{m} \sum_{i=1}^{n} \beta_i^j = 0$.

An important characteristic of the behavior of a solution of the system (7) in a neighborhood of a singular point s_j is the integer part of the real part of the number β_i^j, which equals ϕ_j^i; clearly these also influence behavior of sections of the bundle induced by the system of equations.

For a holomorphic bundle $P'_\rho \to \mathbb{CP}^1 - \{s_1, ..., s_m\}$ with connection ∇' consider its extension $P^{C,\Psi}$ to $\mathbb{CP}^1$ such that $\Psi = (0, ..., 0)$. Denote the corresponding vector bundle with connection by (E^0, ∇^0) and call it the *canonical extension*. According to (2) $E^{C,\Psi}$ decomposes into a direct sum of line bundles: $E(k_1^{C,\Psi}) \oplus \cdots \oplus E(k_n^{C,\Psi}) \to \mathbb{CP}^1$. The vector $\boldsymbol{k}^{C,\Psi} = (k_1^{C,\Psi}, ..., k_n^{C,\Psi})$ is the splitting type of the vector bundle $E^{C,\Psi}$.

Let us now formulate a condition for solvability of the 21st Hilbert problem.

Theorem 2.1.[123] *A representation ρ is realizable as a monodromy representation of a Fuchs system with given singular points $s_1, ..., s_m$ if and only if among the bundles $E^{C,\Psi} \to \mathbb{CP}^1$ there is a holomorphically trivial one, i. e. such that its splitting type is $(0, ..., 0)$.*

3. Solvability condition of the Riemann-Hilbert monodromy problem

Let

$$\frac{df}{dz} = \Omega(z) f(z) \tag{7}$$

be a system of differential equations with regular singularities, having $s_1, ..., s_m$ as singular points, and ∞ as an apparent singular point. It is known that such system has n linearly independent solutions in a neighborhood of a regular point.

Let us denote such a fundamental system of solutions by $F(\tilde{z})$. It is possible to characterize $F(\tilde{z})$ by its behavior near the singular points $s_1, ..., s_m$, using the monodromy matrices $M_1, ..., M_m$ which are determined by the matrices $E_1, ..., E_m$, and by the behavior at ∞ which is characterized by partial indices $k_1, ..., k_m$. Therefore it is said that the system (7) has the *standard form*[5] with respect to the matrices $(M_1, ..., M_m)$ and $(E_1, ..., E_m)$ satisfying the condition $M_1 \cdots M_m = 1$ such that M_k are similar to $\exp(-2\pi i E_k)$, $k = 1, ..., m$ and E_j are not resonant, with singular points $s_1, ..., s_m$ and partial indices $k_1 \geq \cdots \geq k_n$, if

i) $s_1, ..., s_m$ are the only singular points of (7), with ∞ as an apparent singular point;

ii) the monodromy group of (7) is conjugate to the subgroup of $GL_n(\mathbb{C})$ generated by the matrices $M_1, ..., M_m$;

iii) in a neighborhood U_j of the point s_j solution has the form $F(\tilde{z}) = Z_j(z)(\tilde{z} - s_j)^{E_j} C$, where $Z_j(z)$ is an analytic and invertible matrix-function on $U_j \cup \{s_j\}$ and C is a nondegenerate matrix;

iv) solution of the system in a neighborhood U_∞ of ∞ has the form $F(z) = diag(z^{k_1}, ..., z^{k_n})Z_\infty(z)C$, $z \in U_\infty$, with $Z_\infty(z)$ holomorphic and invertible on U_∞.

Theorem 3.1.[5] *Suppose $G \in PC(\Gamma)^{n \times n}$ is a piecewise constant function with jump points $s_1, ..., s_m$. Suppose G has a Φ-factorization in the space $l_p(\Gamma)$, $1 < p < \infty$, and $(M_1, ..., M_m)$, $(E_1, ..., E_m)$ are matrices associated to G on $l_p(\Gamma)$.*

Suppose there exists a system of differential equations in standard form (7) with singular points $s_1,...,s_m$ and partial indices $k_1,...,k_m$. Let $F_1(z)$, $F_2(z)$ be a fundamental system of its solutions in U_+ and $U_- \setminus \{\infty\}$.

Then there exist nondegenerate $n \times n$-matrices C_1 and C_2 such that

$$G(t) = G_+(t)\Lambda(t)G_-(t)$$

is a Φ-factorization of G in $\Lambda_p(\Gamma)$, where $\Lambda(t) = diag(t^{k_1}, ..., t^{k_n})$,

$$G_+(z) = C_1^{-1}F_1^{-1}(z), \ z \in U_+, \ G_-(z) = \Lambda^{-1}(z)F_2(z)C_2, z \in U_- \setminus \{\infty\}.$$

Let Γ be a closed simply contour as above, $s_1, ..., s_m \in \Gamma$ and $M_1, ..., M_m \in GL_n(\mathbb{C})$. We will say that the piecewise constant matrix function $G(t)$ is induced from collections $s = \{s_1, ..., s_m\}$, $M = \{M_1, ..., M_m\}$ if it is constructed in the following manner:

$$G(t) = M_j \cdot ... \cdot M_1, \ \text{if } t \in [s_j, s_{j+1}),$$

where M_j are monodromy matrices corresponding to going around small loops around the singular points s_j.

Theorem 3.2. *Let*

$$\rho : \pi_1(\mathbb{CP}^1 \setminus \{s_1, ..., s_m\}) \to GL_n(\mathbb{C}) \tag{8}$$

be a representation such that $(\rho(\gamma_1) = M_1,...,\rho(\gamma_m) = M_m)$ and $(E_1, ..., E_m)$ is admissible.

Then for the representation (8) the Riemann-Hilbert monodromy problem is solvable if $G(t)$ induced from the collection s, M above, admits a canonical factorization in $l^\alpha(\Gamma)$, for some $\alpha > 1$ sufficiently close to 1.

It is known, that for the given monodromy matrices $M_1, ..., M_m$, and singular points $s_1, ..., s_m$ there exists the system of differential equations of type

$$df = \omega f, \tag{9}$$

such that $s_1, ..., s_m$ are poles of first order for (9) and ∞ is apparent singular point, the matrices $M_1, ..., M_m$ are monodromy matrices of (9), the solution of (9) in a neighborhood of the singular point s_j has the form:

$$\Phi_j(\tilde{z}) = U_j(z)(\tilde{z} - s_j)^{E_j} C$$

where the matrix function $U_j(z)$ is invertible and analytic in the neighborhood of s_j and C is a nondegenerate matrix; in the neighborhood of ∞ the solution has the form:

$$\Phi_\infty(\tilde{z}) = diag(k_1, ..., k_n) U_\infty(z) C,$$

where $U_\infty(z)$ is analytic and invertible at ∞.[5] By theorem 3.1 the piece-wise constant matrix function $G(t)$ admits a Φ-factorization, therefore $\xi_j(\tau) = \frac{1}{2\pi} \arg \lambda_j(\tau)$ is a single-valued function taking values in the interval $\left(\frac{1}{p} - 1, \frac{1}{p} \right)$. From the factorization condition $G(t) = G_+(t)\Lambda(t)G_-(t)$ and by theorem 1.2 we have

$$G_+(z) = C_1^{-1} F_1^{-1}(z), \ z \in U_+, \ G_-(z) = \Lambda^{-1}(z) F_2(z) C_2, \ z \in U_- \setminus \{\infty\}.$$

From assumptions $G(t)$ admits a canonical factorization, i.e. $k_1 = ... = k_n = 0$. From this it follows, that ∞ is a regular point of the system (9) and conditions of Theorem 2.1 are satisfied. This precisely means that for the indicated monodromy representation the monodromy Riemann-Hilbert problem is solvable.

References

1. D.V.Anosov, A.A.Bolibruch. The Riemann-Hilbert problem, Aspects of Mathematics, Vieweg, Braunschweig, Wiesbaden, 1994
2. A.A.Bolibruch. The Riemann-Hilbert problem, Russian Math.surveys, vol. 45 (1990), N 2, 1–47.
3. A.A.Bolibruch. On the sufficient conditions for the positive resolution of the Rieamnn-Hilbert problem. Mathematical Notes, Acad. Sci. USSR. vol.51, No.2, 1992.
4. G.Giorgadze. Regular systems on Riemann surfaces. Journal of Math. Sci. 118 (5), 5347–5399, 2003.
5. T. Ehrhardt, I. Spitkovsky. Factorization of piecewise constant matrix functions and systems of linear differential equations. St. Petersburg Mathematical Journal, 2002, 13:6, 939–991
6. G. N. Khimshiashvili. Lie groups and transmission problems on Riemann surface. Contemp. Math. 1992, vol. 131, 164–178.
7. N.I.Muskhelishvili. Singular integral equations. Noordhoff, Groningen,1953.

A study about one kind of two dimensional integral equation of Volterra type with two interior singular lines

Lutfya Rajabova

Tajik Technical University
E-mail: lutfya62@mail.ru

In this study, a two dimensional integral equation of Volterra type with two singular interior lines is examined. The non-homogeneous integral equation (1) for certain values $A(a)$, $B(b)$, where $C(x,y) = -A(x)B(y)$, has always a solution and its general solutions contain arbitrary functions of one variable. For other values of $A(a)$, $B(b)$ the non-homogeneous integral equation (1) has a unique solution.

Keywords: Volterra type integral equation; interior singular lines.

Let D_0 denote the rectangle, $D_0 = \{a_0 < x < a_1,\ b_0 < y < b_1\}$ and Γ_1, Γ_2 denote straight lines, $\Gamma_1 = \{a_0 < x < a_1,\ y = b\}$, $\Gamma_2 = \{x = a,\ b_0 < y < b_1\}$, where $a_0 < a < a_1$, $b_0 < b < b_1$.

In the domain $D = D_0 \setminus \{\Gamma_1 \cup \Gamma_2\}$ we consider the two dimensional Integral Equation:

$$u(x,y) + \int_a^x \frac{A(t)u(t,y)}{|t-a|^\alpha}dt - \int_y^b \frac{B(s)u(x,s)}{|b-s|^\beta}ds$$
$$+ \int_a^x \frac{dt}{|t-a|^\alpha} \int_y^b \frac{C(t,s)u(t,s)}{|b-s|^\beta}ds = f(x,y), \quad (1)$$

where $A(x)$, $B(y)$, $C(x,y)$ are given functions in D_0, $f(x,y) \in C(\overline{D})$, $\alpha = const.$ and $\alpha > 0$, $\beta = const.$ and $\beta > 0$.

Solution of the Integral Equation (1) will be examined in the class of function $u(x,y) \in C(\overline{D_0})$ which become zero when the variables approaching Γ_1 and Γ_2 . Moreover in Γ_1, the solutions become 0 of order greater than $(\alpha - 1)$ and in Γ_2, they have a zero of order greater than $(\beta - 1)$.

Solutions of many problems having important meaning of applied characters can be figured out by the help of Integral Equation in explicit form.

For that reason, this article is dedicated to this area.

In the paper [1], the solution of a second order hyperbolic equation with two super–singular lines in the domain $\Omega = \{a < x < a_0,\ b_0 < y < b\}$ approaching infinity on singular lines are found.

The papers [2-4] are dedicated to problems of investigating two-dimensional Volterra type integral equations with two boundary singular and super-singular lines in the domain Ω, of the type:

$$u(x,y) + \lambda \int_a^x \frac{u(t,y)}{(t-a)^\alpha} dt - \mu \int_y^b \frac{u(x,s)}{(b-s)^\beta} ds$$
$$+ \delta \int_a^x \frac{dt}{(t-a)^\alpha} \int_y^b \frac{u(t,s)}{(b-s)^\beta} ds = f(x,y)$$

Problems of integral equations of type (1), with two boundary singular and super-singular lines, where $\alpha = 1,\ \beta > 1;\ \alpha > 1,\ \beta = 1;\ \alpha > 1,\ \beta < 1;\ \alpha < 1,\ \beta > 1$ are investigated in [5], [6].

Problems of finding continuous solutions of second order hyperbolic equations with two boundary singular or super-singular lines in Γ_1 and Γ_2 $(\alpha \geq 1,\ \beta \geq 1)$ in the domain Ω are reduced to considering the integral equation (1), see [7], [8].

[9] and [10] are dedicated to problems for the integral equation

$$u(x,y) + \int_a^x \frac{K_1(x,y;t)u(t,y)}{(t-a)^\alpha} dt - \int_y^b \frac{K_2(x,y;s)u(x,s)}{(b-s)^\beta} ds+$$
$$+ \int_a^x \frac{dt}{(t-a)^\alpha} \int_y^b \frac{K_3(x,y;t,s)u(t,s)}{(b-s)^\beta} ds = f(x,y),$$

in the domain Ω at $\alpha = 1,\ \beta = 1,\ \delta = -\lambda\mu;\ \alpha > 1,\ \beta > 1,\ \delta = -\lambda\mu$, where $\lambda = K_1(a,b;a),\ \mu = K_2(a,b;b),\ \delta = K_3(a,b;a,b)$; at $\alpha = 1,\ \beta = 1.\ \alpha > 1,$ $\beta > 1$ and $C_1(t,s) = C(t,s) + A(t)B(s) \neq 0$, where $A(t) = K_1(a,b;t),$ $B(s) = K_2(a,b;s),\ C(t,s) = K_3(a,b;t,s)$.

In this paper we find the solution of the two-dimensional Volterra type linear Integral Equation with interior singularity $(\alpha = 1,\ \beta = 1)$ in kernels (1).

In this case it is proved that, when the coefficients of Integral Equation mutually connected in a determined way, for certain values of $A(a)$ and $B(b)$ the homogeneous Integral Equation (1) has infinitely many linearly independent solutions, and for another certain pair of values $A(a)$ and $B(b)$ the homogeneous Integral Equation (1) not any solution except zero.

If in the domain D_0, we fix lines $x = a$ and $y = b$, the domain D_0 is divided into four domains, namely $D_1 = \{a_0 < x < a,\ b_0 < y < b\}$,

98   L. Rajabova

$D_2 = \{a < x < a_1, b_0 < y < b\}$, $D_3 = \{a_0 < x < a, b < y < b_1\}$,
$D_4 = \{a < x < a_1, b < y < b_1\}$.

If $(x, y) \in D_1$, then $a_0 < t < x < a$, $b_0 < s < y < b$ and the Integral
Equation (1) with $(\alpha = 1, \beta = 1)$ in the domain D_1 will be of the form:

$$u(x, y) - \int_x^a \frac{A(t)u(t, y)}{a - t} dt - \int_y^b \frac{B(s)u(x, s)}{b - s} ds$$
$$- \int_x^a \frac{dt}{a - t} \int_y^b \frac{C(t, s)u(t, s)}{b - s} ds = f(x, y). \quad (2)$$

If $(x, y) \in D_2$, then $a < t < x < a_1$, $b_0 < s < y < b$ and the Integral
Equation (1) in the domain D_2 will be of the form:

$$u(x, y) + \int_a^x \frac{A(t)u(t, y)}{t - a} dt - \int_y^b \frac{B(s)u(x, s)}{b - s} ds$$
$$+ \int_a^x \frac{dt}{t - a} \int_y^b \frac{C(t, s)u(t, s)}{b - s} ds = f(x, y). \quad (3)$$

If $(x, y) \in D_3$, then $a_0 < t < x < a$, $b < s < y < b_1$ and the Integral
Equation (1) will be of the form:

$$u(x, y) - \int_x^a \frac{A(t)u(t, y)}{a - t} dt + \int_b^y \frac{B(s)u(x, s)}{s - b} ds$$
$$+ \int_x^a \frac{dt}{a - t} \int_b^y \frac{C(t, s)u(t, s)}{s - b} ds = f(x, y). \quad (4)$$

If $(x, y) \in D_4$, then $a < t < x < a_1$, $b < s < y < b_1$ and the Integral
Equation (1) in the domain D_4 will be of the form:

$$u(x, y) + \int_a^x \frac{A(t)u(t, y)}{t - a} dt + \int_b^y \frac{B(s)u(x, s)}{s - b} ds$$
$$- \int_a^x \frac{dt}{t - a} \int_b^y \frac{C(t, s)u(t, s)}{s - b} ds = f(x, y). \quad (5)$$

By this way, the problem of finding the solution of the Integral Equation
(1) in D_0 is reduced to the problem of finding the solution of the Integral
Equations (2), (3), (4), (5) in D_1, D_2, D_3, D_4, respectively, that is the
theory which is investigated in [5], [6].

The following statements hold.

Theorem 1. *Let in equation (1) $A(x) \in C(\Gamma_1)$, $B(y) \in C(\Gamma_2)$, and at
the points $x = a$, $y = b$ they satisfy the Holder condition and $C(x, y) =$*

$-A(x)B(y)$, $A(a) > 0$, $B(b) > 0$. *Moreover, let* $f(x,y) \in C(\overline{D_0})$, $f(a,b) = 0$ *satisfy the asymptotic behavior on the boundaries* Γ_1 *and* Γ_2:

$$f(x,y) = o[|a - x|^{\delta_1}], \quad \delta_1 > A(a) \quad at \quad x \to a - 0,$$

$$f(x,y) = o[(x - a)^{\varepsilon}], \quad at \quad x \to a + 0,$$

$$f(x,y) = o[|b - y|^{\gamma_1}], \quad \gamma_1 > B(b) \quad at \quad y \to b - 0,$$

$$f(x,y) = o[(y - b)^{\varepsilon}], \quad at \quad y \to b + 0.$$

Then the non homogeneous Integral Equation (1) *in the class* $C(\overline{D_0})$, *approaching zero on* Γ_1 *and* Γ_2 , *is always solvable and its general solution contains four arbitrary functions of one variable and is given by the formulas*

$$u(x,y) = (a - x)^{A(a)} \exp[-W_A^{-1}(x)]\psi_1(y) + (b - y)^{B(b)} \exp[-W_B^{-1}(y)]$$

$$\times \left\{ \varphi_1(x) - \int_x^a \left(\frac{a - x}{a - t} \right)^{A(a)} \exp[W_A^{-1}(t) - W_A^{-1}(x)] \frac{A(t)}{a - t} \varphi_1(t)dt \right\}$$

$$+ K_{a,b}^{-;-}(f(x,y)), \tag{6}$$

when $(x,y) \in D_1$,

$$u(x,y) = (b - y)^{B(b)} \exp[-W_B^{-1}(y)]$$

$$\times \left\{ \varphi_2(x) - \int_a^x \exp[W_A^{+1}(x) - W_A^{+1}(t)] \left(\frac{t - a}{x - a} \right)^{A(a)} \frac{A(t)}{a - t} \varphi_2(t)dt \right\}$$

$$+ K_{a,b}^{+;-}(f(x,y)), \tag{7}$$

when $(x,y) \in D_2$,

$$u(x,y) = (a - x)^{A(a)} \exp[-W_A^{-1}(x)]\psi_2(y) + K_{a,b}^{-;+}(f(x,y)), \tag{8}$$

when $(x,y) \in D_3$,

$$u(x,y) = K_{a,b}^{+;+}(f(x,y)), \tag{9}$$

when $(x, y) \in D_4$, where

$$K_{a,b}^{-;-}(f(x,y)) = f(x,y)$$

$$+ \int_x^a \left(\frac{a-x}{a-t}\right)^{A(a)} \exp[W_A^{-1}(t) - W_A^{-1}(x)]\frac{A(t)}{a-t}f(t,y)dt$$

$$+ \int_y^b \left(\frac{b-y}{b-s}\right)^{B(b)} \exp[W_A^{-1}(s) - W_A^{-1}(y)] \cdot \frac{B(s)}{b-s}f(x,s)ds$$

$$+ \int_a^x \left(\frac{a-x}{a-t}\right)^{A(a)} \exp[W_A^{-1}(t) - W_A^{-1}(x)]\frac{A(t)}{a-t}dt$$

$$\times \int_y^b \left(\frac{b-y}{b-s}\right)^{B(b)} \exp[W_B^{-1}(s) - W_B^{-1}(y)] \cdot \frac{B(s)}{b-s}f(t,s)ds,$$

$$K_{a,b}^{+;-}(f(x,y)) = f(x,y)$$

$$- \int_a^x \left(\frac{t-a}{x-a}\right)^{A(a)} \exp[W_A^{+1}(x) - W_A^{-1}(t)]\frac{A(t)f(t,y)}{t-a}dt$$

$$+ \int_y^b \left(\frac{b-y}{b-s}\right)^{B(b)} \exp[W_B^{-1}(s) - W_B^{-1}(y)] \cdot \frac{B(s)}{b-s}f(x,s)ds$$

$$- \int_a^x \left(\frac{t-a}{x-a}\right)^{A(a)} \exp[W_A^{+1}(x) - W_A^{+1}(t)]\frac{A(t)}{t-a}dt$$

$$\times \int_y^b \left(\frac{b-y}{b-s}\right)^{B(b)} \exp[W_B^{-1}(s) - W_B^{-1}(y)] \cdot \frac{B(s)}{b-s}f(t,s)ds,$$

$$K_{a,b}^{-;+}(f(x,y)) = f(x,y)$$

$$+ \int_x^a \left(\frac{a-x}{a-t}\right)^{A(a)} \exp[W_A^{-1}(t) - W_A^{-1}(x)]\frac{A(t)}{a-t}f(t,y)dt$$

$$- \int_b^y \left(\frac{s-b}{y-b}\right)^{B(b)} \exp[W_B^{+1}(y) - W_B^{+1}(s)] \cdot \frac{B(s)f(x,s)}{b-s}ds$$

$$- \int_x^a \left(\frac{a-x}{a-t}\right)^{A(a)} \exp[W_A^{-1}(t) - W_A^{-1}(x)]\frac{A(t)}{a-t}dt$$

$$\times \int_b^y \left(\frac{s-b}{y-b}\right)^{B(b)} \exp[W_B^{+1}(y) - W_B^{+1}(s)] \cdot \frac{B(s)f(t,s)}{s-b}ds,$$

$$K_{a,b}^{+;+}(f(x,y)) = f(x,y)$$

$$- \int_a^x \left(\frac{t-a}{x-a}\right)^{A(a)} \exp[W_A^{+1}(x) - W_A^{+1}(t)] \frac{A(t)}{t-a} f(t,y)dt$$

$$- \int_b^y \left(\frac{s-b}{y-b}\right)^{B(b)} \exp[W_B^{+1}(y) - W_B^{+1}(s)] \cdot \frac{B(s)}{s-b} f(x,s)ds$$

$$+ \int_a^x \left(\frac{t-a}{x-a}\right)^{A(a)} \exp[W_A^{+1}(x) - W_A^{+1}(t)] \frac{A(t)}{t-a} dt$$

$$\times \int_b^y \left(\frac{s-b}{y-b}\right)^{B(b)} \exp[W_B^{+1}(y) - W_B^{+1}(s)] \frac{B(s)}{s-b} f(t,s)ds,$$

$$W_A^{+1}(x) = \int_a^x \frac{A(t) - A(a)}{t-a} dt, \quad W_A^{-1}(x) = \int_x^a \frac{A(a) - A(t)}{a-t} dt,$$

$$W_B^{+1}(y) = \int_b^y \frac{B(s) - B(b)}{s-b} ds, \quad W_B^{-1}(y) = \int_y^b \frac{B(b) - B(s)}{b-s} ds,$$

$\varphi_j(x)$, $\psi_j(y)$, $j = 1,2$ *are arbitrary continuous functions of the points on* Γ_1 *and* Γ_2. *Moreover at* $x \to a$, $y \to b$ $\varphi_j(x)$ *and* $\psi_j(y)$ *approach zero and their asymptotic behavior is determined by the formulas*

$$\varphi_1(x) = o[|a - x|^{\delta_2}], \quad \delta_2 > A(a), \quad at \quad x \to a,$$

$$\varphi_2(x) = o[|x - a|^{\varepsilon}], \quad \varepsilon > 0, \quad at \quad x \to a,$$

$$\psi_1(y) = o[|b - y|^{\varepsilon}], \quad at \quad y \to b,$$

$$\psi_2(y) = o[|y - b|^{\varepsilon}], \quad at \quad y \to b.$$

Theorem 2. *Let in equation* (1) $A(x) \in C(\Gamma_1)$, $B(y) \in C(\Gamma_2)$ *and at the points* $x = a$, $y = b$ *they satisfy the Holder condition and* $C(x,y) = -A(x)B(y)$, $A(a) < 0$, $B(b) < 0$. *Moreover, let* $f(x,y) \in C(\overline{D_0})$, $f(a,b) = 0$ *satisfy the asymptotic behavior on the boundaries* Γ_1 *and* Γ_2:

$$f(x,y) = o[|x - a|^{\delta_1}], \quad \delta_1 > |A(a)| \quad at \quad x \to a + 0,$$

$$f(x,y) = o[|y - b|^{\gamma_1}], \quad \gamma_1 > |B(b)| \quad at \quad y \to b + 0,$$

$$f(x,y) = o[(a - x)^{\varepsilon}], \quad \varepsilon > 0 \quad at \quad x \to a - 0,$$

$$f(x,y) = o[(b - y)^{\varepsilon}], \quad at \quad y \to b - 0.$$

Then the non homogeneous Integral Equation (1) *in the class* $C(\overline{D_0})$, *approaching zero on* Γ_1 *and* Γ_2 *,is always solvable and its general solution contains four arbitrary functions of one variable and is given by the formulas:*

$$u(x,y) = K_{a,b}^{-;-}(f(x,y)) \tag{10}$$

when $(x,y) \in D_1$,

$$u(x,y) = (x-a)^{-A(a)}\exp[W_A^{+1}(x)]\psi_1(y) + K_{a,b}^{+;-}(f(x,y)), \tag{11}$$

when $(x,y) \in D_2$,

$$u(x,y) = (y-b)^{-B(b)}\exp[W_B^{+1}(y)]$$
$$\times \left[\varphi_1(x) + \int_x^a \left(\frac{a-x}{a-t}\right)^{A(a)}\exp[W_A^{-1}(t) - W_A^{-1}(x)]\frac{A(t)}{a-t}\varphi_1(t)dt\right]$$
$$+ K_{a,b}^{-;+}(f(x,y)), \tag{12}$$

when $(x,y) \in D_3$,

$$u(x,y) = (x-a)^{-A(a)}\exp[W_A^{+1}(x)]\psi_2(y) + (y-b)^{-B(b)}\exp[W_B^{+1}(y)]$$
$$\times \left\{\varphi_2(x) - \int_a^x \left(\frac{t-a}{x-a}\right)^{A(a)}\exp[W_A^{+1}(x) - W_A^{+1}(t)]\frac{A(t)}{t-a}\varphi_2(t)dt\right\}$$
$$+ K_{a,b}^{+;+}(f(x,y)), \tag{13}$$

when $(x,y) \in D_4$, *where* $\varphi_j(x)$, $\psi_j(y)$, $j=1,2$ *are arbitrary continuous functions of the points on* Γ_1 *and* Γ_2. *Moreover at* $x \to a$, $y \to b$ $\varphi_j(x)$ *and* $\psi_j(y)$ *approach zero and their asymptotic behavior is determined by the formulas*

$$\varphi_1(x) = o[(a-x)^\varepsilon], \quad \varepsilon > 0, \quad at \quad x \to a,$$

$$\varphi_2(x) = o[(x-a)^{\delta_3}], \quad \delta_3 > |A(a)|, \quad at \quad x \to a,$$

$$\psi_1(y) = o[(b-y)^\varepsilon], \quad at \quad y \to b,$$

$$\psi_2(y) = o[(y-b)^\varepsilon], \quad at \quad y \to b.$$

Remark 1. *From formulas* (6)–(9), *it follows that, if in the Integral Equation* (1) *functions* $A(x)$, $B(y)$, $C(x,y)$, $f(x,y)$ *satisfy the conditions of theorem 1, then the solution of the Integral Equation* (1) *approaching zero* *at Γ_1 and Γ_2 and for $x \to a-0$ and $x \to a+0$, $y \to b-0$ and $y \to b+0$ it satisfies the asymptotic formulas*

$$u(x,y) = o[(a-x)^{A(a)}], x \to a-0,$$

$$u(x,y) = o[(b-y)^{\varepsilon}], \quad \text{at} \quad y \to b-0.$$

$$u(x,y) = o[(x-a)^{\varepsilon}], \quad \varepsilon > 0 \quad \text{at} \quad x \to a+0,$$

$$u(x,y) = o[(y-b)^{B(b)}], \quad \text{at} \quad y \to b+0.$$

Remark 2. *From formulas* (10)–(13), *it follows, that if in Integral Equation* (1) *the functions $A(x)$, $B(y)$, $C(x,y)$, $f(x,y)$ satisfy the conditions of theorem 2, then the solution of Integral Equation* (1) *approaching zero at Γ_1, Γ_2 and for $x \to a-0$ and $x \to a+0$, $y \to b-0$ and $y \to b+0$ satisfying the asymptotic formulas*

$$u(x,y) = o[(x-a)^{|A(a)|}], \quad \text{at} \quad x \to a+0, \quad u(x,y) = o[(a-x)^{\varepsilon}],$$

as $x \to a-0$,

$$u(x,y) = o[(y-b)^{\varepsilon}], \quad \text{at} \quad y \to b+0, \quad u(x,y) = o[(b-y)^{|B(b)|}],$$

as $y \to b-0$.

References

1. Rajabov N., An Introduction to the theory of partial differential equations with super – singular coefficients. Dushanbe, 1992.
2. Rajabov N., Rajabova L. Explicit solution of one class two dimensional linear Volterra type integral equation with two boundary singular lines // Differential Equations and its applications , Samara, 2002, 286 – 288.
3. Rajabov N.,Ronto M., Rajabova L. On some two dimentional Volterra type linear integral equations with super – singularity // Mathematical Notes, Miskolc, 2003, V. 4 ,No. 1, 65 – 76.
4. Rajabov N., Rajabova L. Investigation of one class of two dimensional Integral Equation with fixed singular Kernels, connection with Hyperbolic Equation // Russian Dokl. Academy of Sciences, 391, No. 1, 2003, 20 – 22.
5. Rajabov N., Rajabova L. To the theory of one class two dimensional, non model Volterra type Integral Equation with boundary super – singular lines in the Kernels //Russian Dokl. Academy of Sciences, 400, No. 5, 2005, 602 – 605.
6. Rajabova L. Explicit solution of a class of non model Volterra type Integral Equation with one singular and one weakly singular lines // Inter. Sci. Conf. "Differential Equations with partial derivatives and related problems of analysis and informatics", 16 – 19 November 2004, Tashkent, 4, 2004., 78–80.

7. Rajabova L. About a class of Hyperbolic Equation with singular lines. Vestnik National University, Science magazine. Dushanbe, 5, 2006, 44–51.

8. L. N. Rajabova. To theory of a class of Hyperbolic Equation with singular lines // Tajik Dokl. Academy of Sciences, 49,8,2006, 710–717.

9. L. Rajabova. About a general two dimensional Volterra type Integral Equation with singularity and supersingularity on the boundary of the domain // Proceeding Intern. Sci. Conf. "Differential Equations,function theory and applications". Novosibirsk, 2007, 458–459.

10. L. Rajabova. About a two dimensional Volterra type Integral Equation with singularities on the boundary of the domain. Vestnik Tajik State National University, Science magazine.Dushanbe,3,2007, 30–37.

3D rotating Navier-Stokes equations: New method of numerical analysis and families of global exact solutions

A.G. Khaybullin and R.S. Saks*

*Institute of Mathematics and Computer Center, Ufa Scientific Center,
Russian Academy of Sciences,
450077, Ufa, 112, Chernyshevskogo str., Russia
* E-mail: romen-saks@yandex.ru
www.umf-ovm.ucoz.ru*

The Cauchy problem for Navier-Stokes equations in frame of uniform rotation is studied. Our investigation is based on Fourier development known and unknown vector functions by eigenfunctions of the curl operator. The problem is reduced to the Cauchy problem for a system of the ordinary differential equations, which has very simple explicit form. Families of the exact global solutions of the Navier-Stokes equations are found. New method for numerical solutions of this problem was made. Some results of numerical calculation are presented in the pictures.

Keywords: Cauchy problem; Navier-Stokes equations; Fourier method; eigenvalues and eigenfunctions of curl operator.

1. Statement of the Cauchy problem

Suppose $g(x) : \mathbf{R}^3 \to \mathbf{C}^3$, 2π-periodic vector function with $div\, g = 0$, from the class $L_2(Q)$ in cube Q with the edge 2π. And let $f(x,t)$ be vector function of the same tipe for x from the class $L_2(Q)$ for any $t \geq 0$.

Problem. Find a solution (v, p) to the Navier-Stokes equation in the 3D space with uniform rotation about vertical vector e_3 at the velocity $\Omega/2$:

$$\frac{\partial v}{\partial t} - \nu\Delta v + (v \cdot \nabla)v + \Omega\,[e_3, v] = -\nabla p + f, \quad div\, v = 0, \qquad (1)$$

which satisfies the initial conditions:

$$v(x, 0) = g(x) \qquad (2)$$

and it's 2π-periodic with respect x variables. The cross product $\Omega\,[e_3, v(x)]$ is the Coriolis force in rotation system of coordinates.

This classic problem was studied by numerous authors. Please, see a book of H.P. Greenspan ("The theory of rotating fluids", 1968), the papers and revue of A. Babin, A. Mahalov, B. Nikolaenko (in Uspehi Mat. Nauk, 2003), papers of R.S. Saks (in Doklady Ac.Nauk, V.424, 2009 and in J.Theor.Math.Phys, V. 162, 2010), paper of R.S. Saks, A.G. Khaybullin (in Doklady Ac.Nauk, V.429, 2009).

2. Fourier series and modified Fourier series

The vector function $f(x) \in L_2(Q)$ is expanded in the Fourier series

$$f(x) = f_0 + \sum_{k^2=1}^{\infty} f_k \, e^{ikx}, \tag{3}$$

where

$$f_k = \frac{1}{(2\pi)^3} \int_Q f(x) e^{-ikx} dx. \tag{4}$$

This function is expanded also in the modified Fourier series:

$$f(x) = f_0 + \sum_{k^2=1}^{\infty} (\phi_k \, \widehat{k} + \phi_k^+ c_k^+ + \phi_k^- c_k^-) e^{ikx} \tag{5}$$

over the eigenfunctions of the curl operator, where

$$\phi_k = \frac{1}{(2\pi)^3} \int_Q (f(x), \widehat{k}) e^{-ikx} dx, \tag{6}$$

$$\phi_k^\pm = \frac{1}{(2\pi)^3} \int_Q (f(x), c_k^\pm) e^{-ikx} dx, \tag{7}$$

$\widehat{k} = \frac{k}{|k|}$ and vectors $c_k^\pm = a_k^\pm + i b_k^\pm$ are specified as

$$\pm \frac{\sqrt{2}}{2|k'|} \begin{pmatrix} k_2 \\ -k_1 \\ 0 \end{pmatrix} + i \frac{\sqrt{2}}{2|k||k'|} \begin{pmatrix} k_1 k_3 \\ k_2 k_3 \\ -k_1^2 - k_2^2 \end{pmatrix} \tag{8}$$

if $|k'|^2 \equiv k_1^2 + k_2^2 \neq 0$ and as

$$\pm \frac{\sqrt{2}}{2} \begin{pmatrix} \widehat{k}_3 \\ 0 \\ 0 \end{pmatrix} + i \frac{\sqrt{2}}{2} \begin{pmatrix} 0 \\ 1 \\ 0 \end{pmatrix}, \text{ if } |k'| = 0. \tag{9}$$

Vectors $u_0^j = (2\pi)^{-3/2} e_j$ and vector functions $u_k(x) = (2\pi)^{-3/2} \widehat{k} \, e^{ikx}$ if $k \neq 0$ correspond to zero eigenvalue of the curl operator, $curl \, u_k = 0$. Vector functions $u_k^+(x) = (2\pi)^{-3/2} c_k^+ e^{ikx}$ and $u_k^-(x) = (2\pi)^{-3/2} c_k^- e^{ikx}$ correspond to $|k|$ and $-|k|$ eigenvalues of the curl: $curl\, u_k^\pm = \pm|k|u_k^\pm$.

A system of these vector functions form orthonormal basis in Gilbert space $L_2(Q)$.

The vector function u_0^j and $u_k^\pm$ are also the eigenfunctions of Stokes operator S: $Su_0^j = 0, Su_k^\pm = \nu|k|^2 u_k^\pm$.

Let $f \in L_2(Q)$, then $div\, f = 0$ if and only if

$$f(x) = f_0 + \sum_{k^2=1}^{\infty} (\phi_k^+ c_k^+ + \phi_k^- c_k^-)e^{ikx}. \tag{10}$$

For a real vector function $f(x) = \overline{f(x)}$ this series has a form

$$f(x) = f_0 + 2Re \sum_{k \in M} (\phi_k^+ c_k^+ + \phi_k^- c_k^-)e^{ikx}, \tag{11}$$

where M is "positive" part of integer lattice: $M = M_1 \cup M_2 \cup M_3$, $M_1 = \{k : \, \text{k}_1 \in N, \, \text{k}_2 = \text{k}_3 = 0\}$, $M_2 = \{k : \, \text{k}_1 \in Z, \, \text{k}_2 \in N, \, \text{k}_3 = 0\}$, $M_3 = \{k : \, \text{k}_1, \text{k}_2 \in Z, \, \text{k}_3 \in N\}$.

Vector functions $\text{d}_k^\pm = Re(\phi_k^\pm c_k^\pm e^{ikx})$ are real eigenfunctions of curl operator: $curl\, d_k^\pm = \pm|k|d_k^\pm$. The flows with the velocity $d_k^\pm$ are called vortical.

Setting $\phi_k^\pm = \alpha_k^\pm + i\beta_k^\pm = |\phi_k^\pm| e^{i\theta_k^\pm}$, where $|\theta_k^\pm| \leq \pi$, we obtain

$$\text{d}_k^\pm(x) = |\phi_k^\pm| \left(\cos(kx + \theta_k^\pm)a_k^\pm - \sin(kx + \theta_k^\pm)b_k^\pm\right). \tag{12}$$

So vector-functions $d_k^+(x)$ lies in the plane formed by vectors a_k^+ and b_k^+ which are orthogonal to the wave vector k.

3. Galerkin approximations and new method of numerical analysis

We suppose that a vector function $f(x,t)$ belongs to $L_2(Q)$ and $div f = 0$ for all $t \geq 0$. Vector function $g(x)$ belongs to $L_2(Q)$ and $div g = 0$. Its Fourier series are

$$f(x,t) = f_0(t) + \sum_{k^2=1}^{\infty} (\phi_k^+(t)c_k^+ + \phi_k^-(t)c_k^-)e^{ikx}, \tag{13}$$

$$g(x) = g_0 + \sum_{k^2=1}^{\infty} (\psi_k^+ c_k^+ + \psi_k^- c_k^-)e^{ikx}. \tag{14}$$

We find the Galerkin approximations (v_l, p_l) of a solution (v, p) as:

$$v_l(x, t) = \sum_{j=1}^{3} v_0^j(t) e_j + \sum_{k^2=1}^{l} (\gamma_{k,l}^+(t) c_k^+ + \gamma_{k,l}^-(t) c_k^-) e^{ikx}, \qquad (15)$$

$$p_l(x, t) = p_0(t) + \sum_{k^2=1}^{l} p_{k,l}(t) e^{ikx}. \qquad (16)$$

We choose the functions $v_0^j(t)$, $\gamma_{k,l}^{\pm}$, $p_{k,l}(t)$ so that vector $L(v_l) + \nabla p_l - f$ is orthogonal to a linear space G_l, formed by basis vector functions u_0^j, u_k, $u_k^{\pm}$. It's equivalent to the Cauchy problems for the vector function $v_0(t)$:

$$\frac{\partial v_0}{\partial t} + \Omega\,[e_3, v_0] = f_0(t), \ \ v_0(0) = g_0; \qquad (17)$$

and a same problem for unknown functions $\gamma_{k,l}^+$ and $\gamma_{k,l}^-$:

$$\frac{\partial \gamma_{k,l}^+}{\partial t} + (\nu k^2 - i\Omega\,\widehat{k}_3 + i(v_0, k))\gamma_{k,l}^+$$

$$+ i \sum_{m^2=1}^{l} [\gamma_{k-m,l}^+(c_{k-m}^+, m) + \gamma_{k-m,l}^-(c_{k-m}^-, m)]$$

$$\times [\gamma_{m,l}^+(c_m^+, c_k^+) + \gamma_{m,l}^-(c_m^-, c_k^+)] = \phi_k^+(t), \quad (18)$$

$$\frac{\partial \gamma_{k,l}^-}{\partial t} + (\nu k^2 + i\Omega\,\widehat{k}_3 + i(v_0, k))\gamma_{k,l}^- +$$

$$+ i \sum_{m^2=1}^{l} [\gamma_{k-m,l}^+(c_{k-m}^+, m) + \gamma_{k-m,l}^-(c_{k-m}^-, m)]$$

$$\times [\gamma_{m,l}^+(c_m^+, c_k^-) + \gamma_{m,l}^-(c_m^-, c_k^-)] = \phi_k^-(t), \quad (19)$$

where $0 < |k - m|^2 \le l$ with initial conditions

$$\gamma_{k,l}^{\pm}(0) = \psi_k^{\pm}, \ 0 < |k|^2 \le l. \qquad (20)$$

Unknowns $p_{k,l}(t)$ are determined in terms of $\gamma_{k,l}^{\pm}$:

$$p_{k,l}(t) = -i\,|k|^{-1}\,(\phi_k(t) - \Omega\frac{\sqrt{2}\,|k'|}{2\,|k|}(\gamma_{k,l}^+ - \gamma_{k,l}^-) -$$

$$\sum_{m^2=1}^{l} [\gamma_{k-m,l}^+(c_{k-m}^+, m) + \gamma_{k-m,l}^-(c_{k-m}^-, m)][\gamma_{m,l}^+(c_m^+, \widehat{k}) + \gamma_{m,l}^-(c_m^-, \widehat{k})]),$$

$$(21)$$

and we suppose that $p_0(t) = 0$ for $p(x,t)$ be uniquely determined.

Principal part of the Galerkin approximation method is the problem of convergence of the sequens v_l. This problem was studied in the books of O.A. Ladyzhenskaya and R. Temam in the case of a initial boundary value problems different from the Cauchy problem. In the last case it is possible to repeat some of those inequalities.

If $f_0(t) = 0$, $g_0 = 0$ the problem (17) has only trivial solution $v_0(t) \equiv 0$. The reduced system (18), (19) will be denote by RS_l.

The system RS_l may be written in the standard form if to numerate the points k of the integer lattice and known and unknown function.

I and my graduate student A.G. Khaybullin we are working a new method for numerical analysis of the Cauchy problem. This method contain the following programms:

- calculation of the Fourier coefficients (6), (7),

- restoration of RS_l systems in LaTeX,

- numerical solution of RS_l systems with Cauchy data (using Runge-Kutta method) and others.

4. Solutions of linear Stocks-Sobolev problem

This problem is following

$$\frac{\partial v}{\partial t} - \nu \Delta v + \Omega \, [e_3, v] = -\nabla p + f, \quad div\, v = 0, \ v(x,0) = g(x). \qquad (22)$$

RS_l system

$$\frac{\partial \gamma_k^+}{\partial t} + (\nu k^2 - i\Omega \, \widehat{k}_3)\gamma_k^+ = \phi_k^+(t), \quad \frac{\partial \gamma_k^-}{\partial t} + (\nu k^2 + i\Omega \, \widehat{k}_3)\gamma_k^- = \phi_k^-(t), \ (23)$$

contain separate equations and can be solved simply under the Cauchy data (20) if $\phi_k^{\pm}(t) \in C[0,T]$. Let

$$u_k^{\pm} = c_k^{\pm} e^{ikx}, \quad \omega_k^{\pm} = \nu k^2 \mp i\Omega \, \widehat{k}_3, \quad \rho_k^{\pm}(t) = \int_0^t e^{\omega_k^{\pm}\tau} \phi_k^{\pm}(\tau) d\tau. \qquad (24)$$

1). If $f = \phi_k^+(t)\, u_k^+ + \phi_k^-(t)\, u_k^-, \quad g = \psi_k^+ u_k^+ + \psi_k^- u_k^-$ then (v_k, p_k) with

$$v_k = \left(\psi_k^+ + \rho_k^+(t)\right) u_k^+ e^{-\omega_k^+ t} + \left(\psi_k^- + \rho_k^-(t)\right) u_k^- e^{-\omega_k^- t}, \quad p_k = P(v_k) \ (25)$$

is the solution of the problem (22) on $[0,T]$ interval.

Pressure p is determined in terms of v (see (21)) and can be written explicitly.

2). If $f = Re\left(\phi_k^+(t)\, u_k^+ + \phi_k^-(t)\, u_k^-\right), g = Re\left(\psi_k^+ u_k^+ + \psi_k^- u_k^-\right)$ then $(Re\, v_k, Re\, p_k)$ is the solution of the problem (22) too.

The sum of (v_k, p_k) (and of $(Re\, v_k, Re\, p_k)$) is also the solutions of this linear problem. For nonlinear problem (1), (2) this assertion may be not correct. In general case

$$f(x,t) = \sum_{k^2=1}^{\infty} (\phi_k^+(t)c_k^+ + \phi_k^-(t)c_k^-)e^{ikx},$$

$$g(x) = \sum_{k^2=1}^{\infty} (\psi_k^+ c_k^+ + \psi_k^- c_k^-)e^{ikx}$$

and a pair (v,p) with

$$v = \sum_{k^2=1}^{\infty} \left(\psi_k^+ + \rho_k^+(t)\right) u_k^+ e^{-\omega_k^+ t} + \left(\psi_k^- + \rho_k^-(t)\right) u_k^- e^{-\omega_k^- t}, \quad p = P(v) \quad (26)$$

is a formal solution of the problem (22). If this series is convergent and superposition principle is satisfied, this solution is classical.

5. Families of global exact solutions to nonlinear problem

I've proved: 1),2) . The solution (v_k, p_k) (and $(Re\, v_k, Re\, p_k)$ also) satisfy the nonlinear problem (1), (2) with data 1) and 2).

3). If $f = \sum_{|k|=C} \phi_k^+(t)\, u_k^+,\quad g = \sum_{|k|=C} \psi_k^+ u_k^+$ a pair (v,p) with

$$v = \sum_{|k|=C} v_k^+, \quad p = \sum_{|k|=C} p_k^+ - \frac{1}{2}\Big(\sum_{|k|=C} v_k^+\Big)^2, \qquad (27)$$

$$v_k^+ = \left(\psi_k^+ + \rho_k^+(t)\right) u_k^+ e^{-\omega_k^+ t}, \quad p_k^+ = P(v_k^+),$$

is the solution of the problem (1), (2).

4). If $f = \sum_{\lambda=1}^{\infty}(\phi_{\lambda k}^+(t)\, u_{\lambda k}^+ + \phi_{\lambda k}^-(t)\, u_{\lambda k}^-),\quad g = \sum_{\lambda=1}^{\infty}(\psi_{\lambda k}^+ u_{\lambda k}^+ + \psi_{\lambda k}^- u_{\lambda k}^-)$ then a pair (v,p) with

$$v = \sum_{\lambda=1}^{\infty} \left((\psi_{\lambda k}^+ + \rho_{\lambda k}^+(t))u_{\lambda k}^+ e^{-\omega_{\lambda k}^+ t} + (\psi_{\lambda k}^- + \rho_{\lambda k}^-(t))u_{\lambda k}^- e^{-\omega_{\lambda k}^- t}\right),$$

$$p = \sum_{\lambda=1}^{\infty}(p_{\lambda k}^+ + p_{\lambda k}^-), \quad (28)$$

is a formal solution of the problem (1), (2). If each of these series is convergent and superposition principle is satisfied, this solution is classical.

6. Flows connected with fields of real curl eigenfunctions

Let P_δ be a plane defined by equation $kx = \delta$ which is orthogonal to vector k. Consider a steady flow with the velocity $d_k^\pm(x) = Re(\phi_k^\pm c_k^\pm e^{ikx})$.

On the plane P_δ this fluid flows has a constant velocity $v_\delta^\pm = d_k^\pm$ on P_δ,

$$v_k^\pm = \left|\phi_k^\pm\right|\left(\cos(\delta + \theta_k^\pm)a_k^\pm - \sin(\delta + \theta_k^\pm)b_k^\pm\right), \tag{29}$$

which is determinate by vector $\phi_k^\pm$ on complex plane. Vectors $\widehat{k}$, a_k^+, b_k^+ form orthonormal basis in $\mathbb{R}^3$ space.

If $\phi_k^\pm(t)$ is variable then it form some curve on the complex plane. Then $v_\delta^\pm(t)$ form same curve on the plane P_δ which is result of δ-rotation of the complex conjugate $\overline{\phi_k^\pm(t)}$ curve. To image the flows $d_k^\pm(x)$ it is sufficient to see the picture $v_0^\pm(t)$ on the P_0 plane or $\phi_k^\pm(t)$ on the complex plane.

7. Some results of numerical experiment

Let $\nu = 0.1$, $\Omega = 0$, $f = 0$ and the following numbers are given: ψ_k^+ for $k = (0,0,1)$, $k = (0,1,1)$, ψ_k^- for $k = (0,1,0)$. So

$$g(x) = 2Re(\psi_k^+ u_k^+)|_{k=(0,0,1)} + 2Re(\psi_k^+ u_k^+)|_{k=(0,1,1)} + 2Re(\psi_k^- u_k^-)|_{k=(0,1,0)}$$

contain 3 steady vortical flows. 8 functions $\gamma_k^\pm(t)$ appeair in calculation RS_2 system with this Cauchy data where
$k \in M' = \{(0,1,0),(0,0,1),(0,-1,1),(0,1,1)\}$.

So approximative solution v' is

$$v'(x,t) = 2Re \sum_{k \in M'} (\gamma_k^+(t)c_k^+ + \gamma_k^-(t)c_k^-)e^{ikx}. \tag{30}$$

It contain 5 new vortical flows. In the following pictures the functions $\gamma_k^\pm$ will be make out by graphics.

We repeat our calculations with $\nu = 0.01$ and you can see difference between these graphics.

Examples. In our calculations a time variable t change from 0 to 10. Cauchy data: $\psi_{4(0,0,1)}^+ = -3$, $\psi_{3(0,0,-1)}^+ = 3$, $\psi_{14(0,1,1)}^+ = 14i$, $\psi_{11(0,-1,-1)}^+ = 14i$, $\psi_{5(0,1,0)}^- = 2$, $\psi_{2(0,-1,0)}^- = -2$.

$$\nu = 0.01,\ \gamma^{+}_{4(0,0,1)}(0) = -3$$

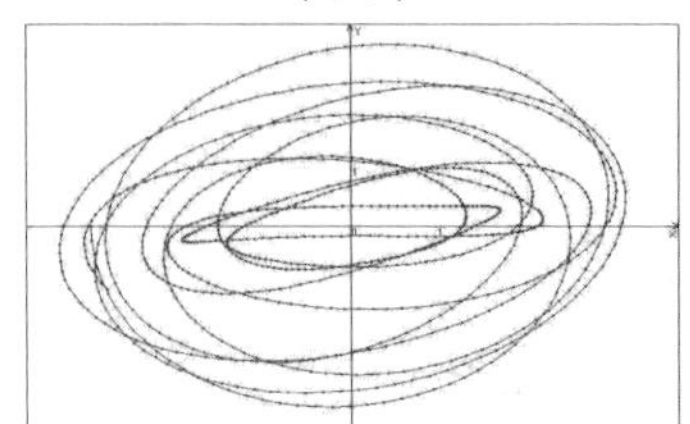

$$\nu = 0.1,\ \gamma^{+}_{4(0,0,1)}(0) = -3$$

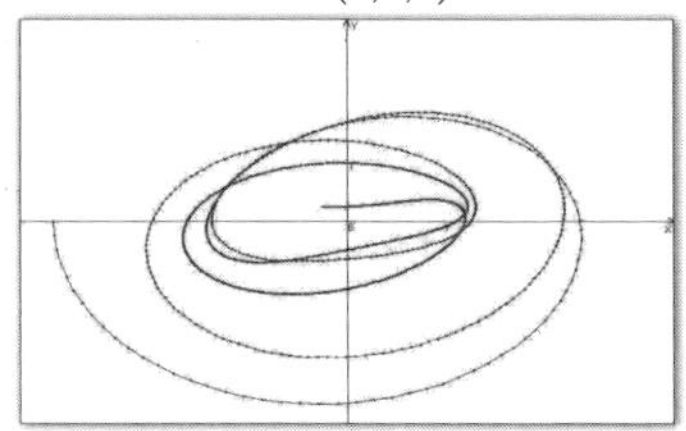

$$\nu = 0.01,\ \gamma^{+}_{5(0,1,0)}(0) = 0$$

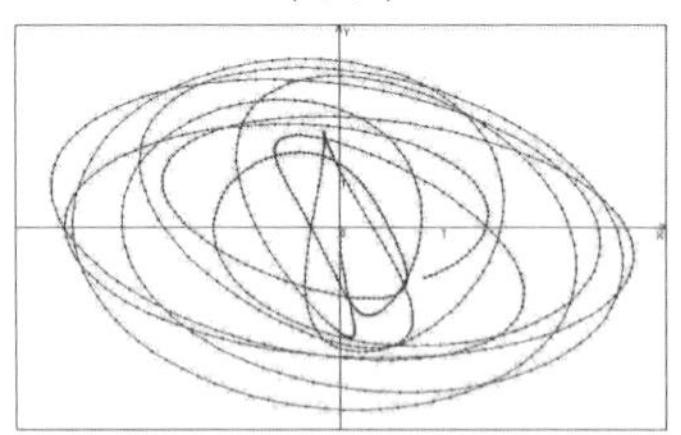

$$\nu = 0.1,\ \gamma^{+}_{5(0,1,0)}(0) = 0$$

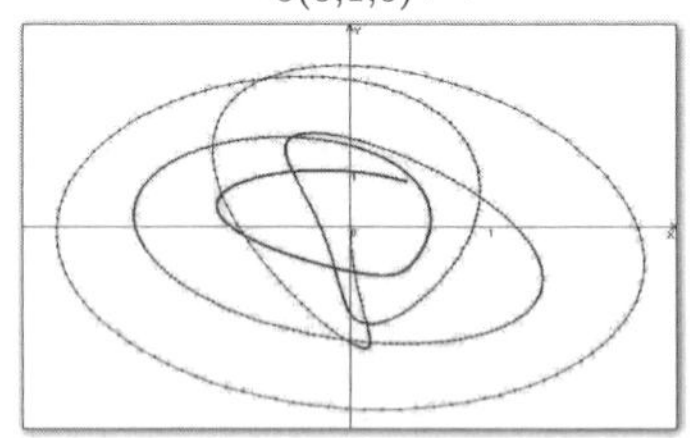

$$\nu = 0.01,\ \gamma^{+}_{12(0,-1,1)}(0) = 0$$

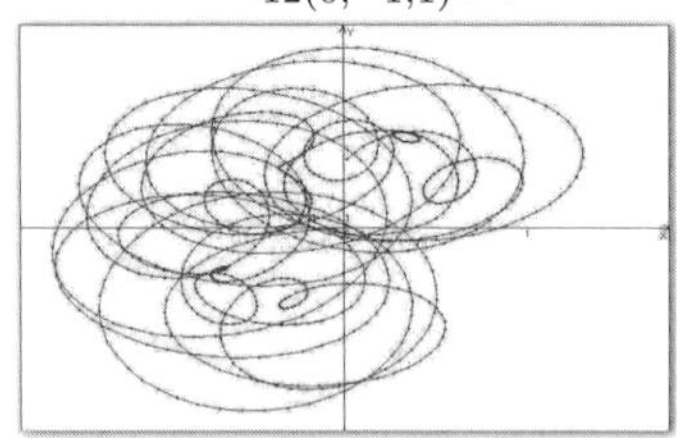

$$\nu = 0.1,\ \gamma^{+}_{12(0,-1,1)}(0) = 0$$

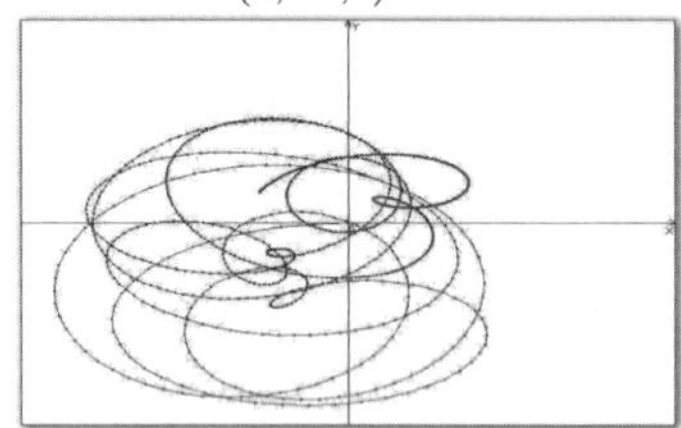

$$\nu = 0.01,\ \gamma^{+}_{14(0,1,1)}(0) = 14i$$

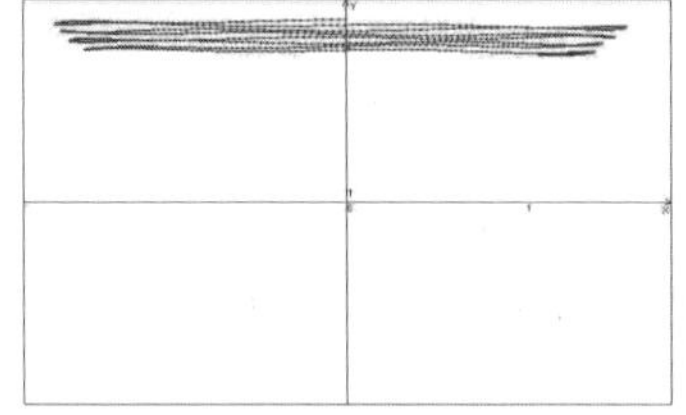

$$\nu = 0.1,\ \gamma^{+}_{14(0,1,1)}(0) = 14i$$

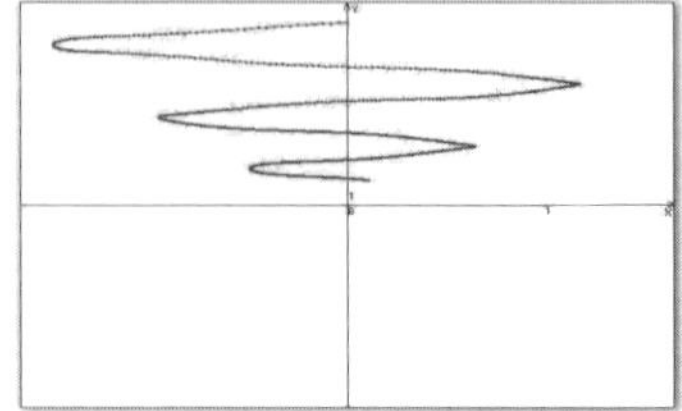

$\nu = 0.01,\ \gamma^-_{4(0,0,1)}(0) = 0$

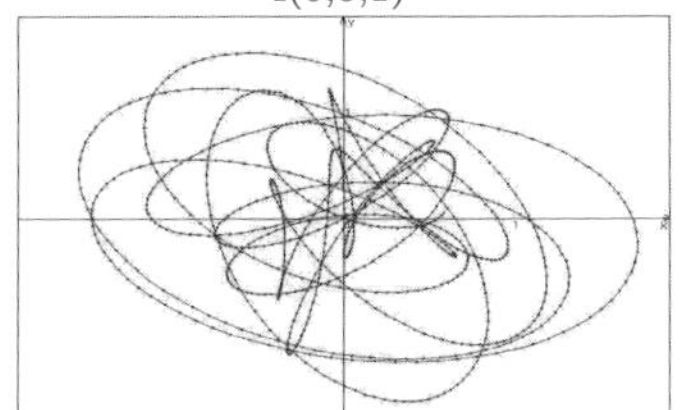

$\nu = 0.1,\ \gamma^-_{4(0,0,1)}(0) = 0$

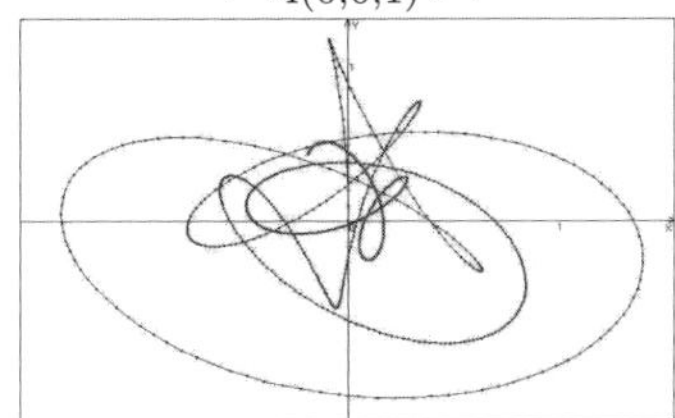

$\nu = 0.01,\ \gamma^-_{5(0,1,0)}(0) = 2$

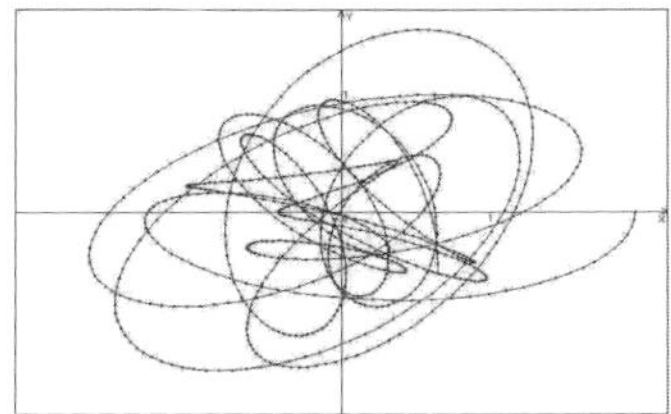

$\nu = 0.1,\ \gamma^-_{5(0,1,0)}(0) = 2$

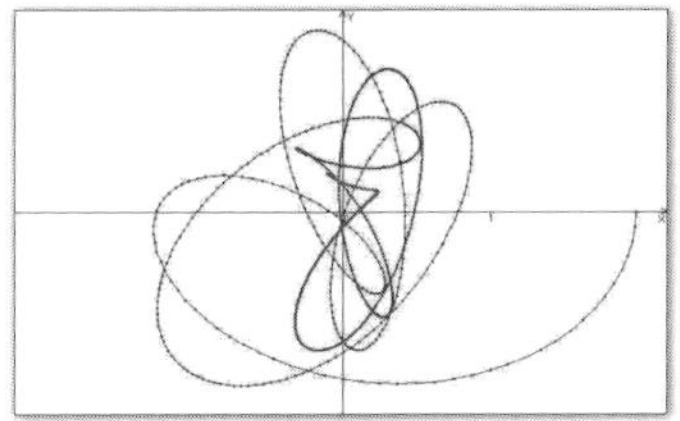

$\nu = 0.01,\ \gamma^-_{12(0,-1,1)}(0) = 0$

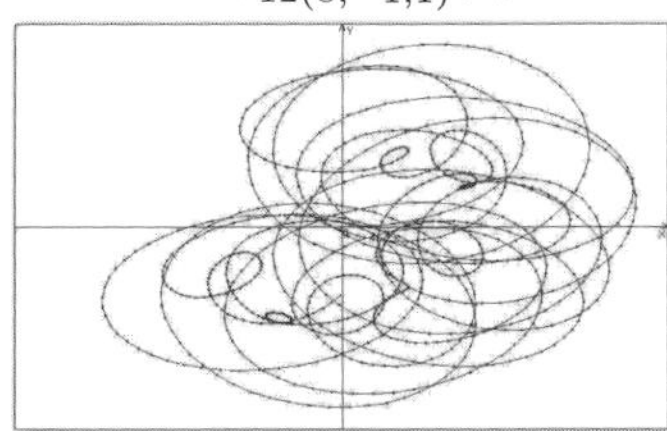

$\nu = 0.1,\ \gamma^-_{12(0,-1,1)}(0) = 0$

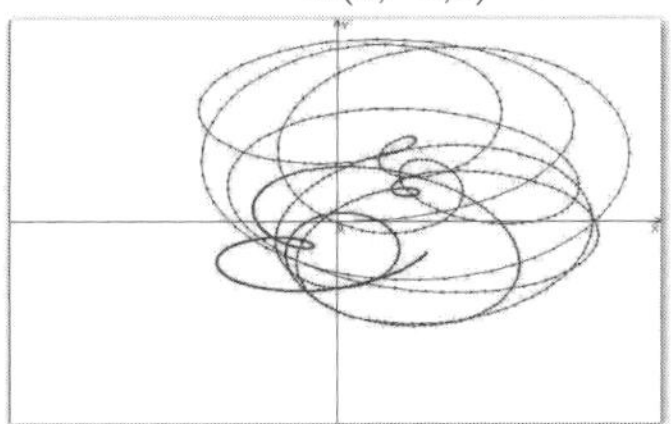

$\nu = 0.01,\ \gamma^-_{14(0,1,1)}(0) = 0$

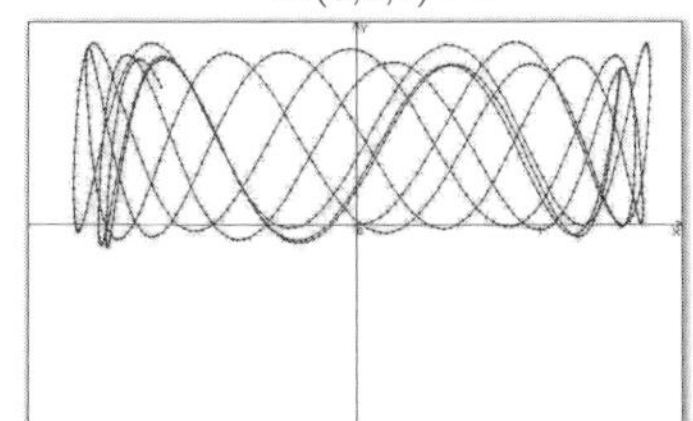

$\nu = 0.1,\ \gamma^-_{14(0,1,1)}(0) = 0$

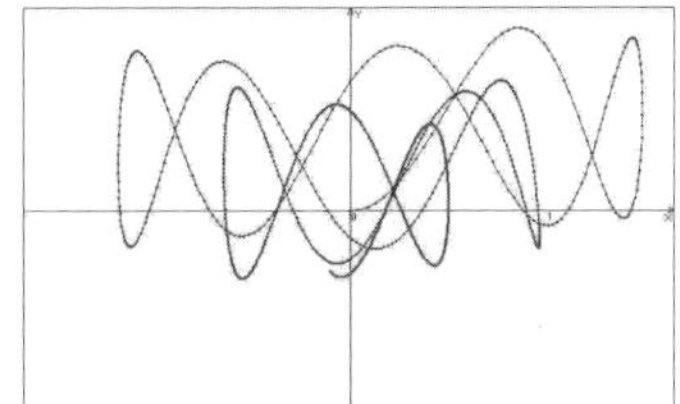

Phragmén-Lindelöf principle for subparabolic functions

A.I. Kheyfits

Graduate Center and Bronx Community College,
City University of New York
Bronx, NY, 10453, USA
E-mail: akheyfits@gc.cuny.edu

We prove Phragmén-Lindelöf theorems in the Nevanlinna integral form for subparabolic functions.

Keywords: Phragmén-Lindelöf Principle; Subparabolic Functions.

1. Introduction and Statement of Results

The following Phragmén-Lindelöf theorem is well known.[1]

Theorem A. *Let L be a uniformly parabolic operator*

$$L = \sum_{i,j=1}^{n} a_{ij}(x,t)\frac{\partial^2}{\partial x_i\,\partial x_j} + \sum_{i=1}^{n} b_i \frac{\partial}{\partial x_i} - h(x,t) - \frac{\partial}{\partial t}$$

with the coefficients a_{ij} and b_i bounded and h lower-bounded. Suppose that a function $u(x,t)$ satisfies the inequality $Lu \geq 0$ in $E = D \times (0,T)$, $0 < T < \infty$, with an unbounded $n-$dimensional domain D, and the growth condition

$$\liminf_{r \to \infty} e^{-cr^2} \left[\max_{|x|=r,\, 0 \leq t \leq T} u(x,t) \right] \leq 0, \tag{1}$$

where $|x|^2 = x_1^2 + \cdots + x_n^2$ and c is a positive constant. If $u(x,t) \leq 0$ on $\partial D \times (0,T)$ and $u(x,0) \leq 0$ for all $x \in D$, then $u \leq 0$ everywhere in E.

For subsolutions of some parabolic operators, we prove the Phragmén-Lindelöf theorem in Nevanlinna form, that is, we replace the uniform norm in growth condition (1) with a weighted integral norm. For subsolutions of time-independent Schrödinger operator a similar assertion and its applications were considered earlier.[2] To state the results, we need some notation.

Let Ω be a domain with the C^2-boundary on the unit sphere $S = S^{n-1}$ in $\mathbb{R}^n$, $n \geq 1$. The cone $K^{\Omega} = \Omega \times (0,\infty) \subset \mathbb{R}^n$ generates a "conical wedge"

$$\mathbb{W}^{\Omega} = K^{\Omega} \times (0,T) \subset \mathbb{R}^n \times \mathbb{R}_+, \; 0 < T \leq \infty.$$

Let a continuous lower-bounded function $c(x,t)$, $(x,t) \in \mathbb{W}^\Omega$, verify $c(x,t) \geq c_0$ with a constant $c_0 > -\infty$. A function $u(x,t)$, continuous together with all the derivatives involved, is called a $c-$*subparabolic function* in the domain $\mathbb{W}^\Omega$ with respect to the parabolic operator

$$L_c u(x,t) = \Delta u(x,t) - c(x,t)u(x,t) - \frac{\partial u(x,t)}{\partial t},$$

$\Delta = \Delta_x$ being the Laplacian in $\mathbb{R}^n$, if u satisfies the inequality $L_c u(x,t) \geq 0$ in $\mathbb{W}^\Omega \setminus W_0$, where W_0 is an exceptional set of zero measure in $\mathbb{R}^{n+1}$.

Let $(r,\ \theta)$ be spherical coordinates in $\mathbb{R}^n$ and δ_θ the Laplace-Beltrami operator on the unit sphere S. Denote by λ_0 and φ_0 the principal eigenvalue and the corresponding eigenfunction of the eigenvalue problem $\delta_\theta\varphi(\theta) + \lambda\varphi(\theta) = 0$, $\theta \in \Omega$, $\varphi\big|_{\partial\Omega} = 0$. We assume that $S \setminus \Omega$ is not a polar set, thus λ_0 is simple and positive and φ_0 is positive and normalized in $L^2(\Omega)$.

Our results are stated in terms of a special solution of the ordinary differential equation

$$l_q y(r) \equiv -y'' - \frac{n-1}{r}y' + \left(\frac{\lambda_0}{r^2} + \alpha^2 + q(r)\right)y(r) = 0, r > 0, \qquad (2)$$

where α is a real constant and $q \geq -\alpha^2$ a continuous function. As usual, we consider solutions of (2), which are absolutely continuous along with their first derivatives and satisfy the equation almost everywhere.

Equation (2) has a fundamental system of positive solutions $\{V(r), W(r)\}$, $0 < r < \infty$, such that $V(r)$ does not decrease as $0 < r \to \infty$, $V(0^+) \geq 0$, while $W(0^+) = \infty$ and $W(r)$ decreases. We call $V(r)$ the growing solution of (2).[3] To be specific, we norm it by $V(1) = 1$. Now we state our main result.

Theorem 1.1. *Let u be a subparabolic function in $\mathbb{W}^\Omega$ satisfying the Lindelöf boundary conditions with a constant A,*

$$\limsup_{t\to 0} u(x,t) \leq A, \text{ for all } x \in K^\Omega$$
$$\limsup_{x\to\partial(K^\Omega)} u(x,t) \leq A, \text{ for all } (x,t) \in K^\Omega \times [0,T).$$

Let V be the growing solution of equation (2) with q satisfying

$$-\alpha^2 \leq q(r) \leq \min_{\theta\in\Omega,\ 0\leq t<T} (c(x,t) - c_0).$$

If there is a constant α such that u satisfies the limit-growth condition

$$\liminf_{r\to\infty} \frac{1}{V(r)} \int_0^T e^{-(\alpha^2-c_0)t} \left(\int_\Omega u^+(r,\theta,t)\varphi_0(\theta)d\sigma(\theta)\right) dt = 0, \qquad (3)$$

where $d\sigma$ is the surface area measure on S and $a^+ = \max\{a; 0\}$, then

$$u(x, t) \le A^+ e^{-c_0 t} \quad \text{for all } (x, t) \in \mathbb{W}^\Omega. \tag{4}$$

The constant A^+ in (4) cannot be replaced by A, that is if $A < 0$, then in general we can claim only that $u(x, t) < 0$ but not $u \le A$. Moreover, the limit-growth condition (3) is precise in the class of all functions c such that $q(r) \le c(r, \theta, t) - c_0$; namely, for any $T < \infty$ there exists a positive subsolution of the equation $L_c u = 0$ with zero boundary conditions in the wedge such that the limit in (3) is positive.

The proofs are given in Section 2.

Corollary 1.1. *For the domains and operators under consideration, Theorem 1.1 implies Theorem A.*

As another corollary, let a function $u(x)$ in the cone K^Ω satisfy the time-independent Schrödinger inequality

$$\Delta u(x) - c(x)u(x) \ge 0.$$

Such a function $u(x)$ is a $c-$subparabolic function in the wedge $\mathbb{W}^\Omega$, and Theorem 1 straightforwardly implies Phragmén-Lindelöf Theorem 3.1.[2]

These results show that under the specified conditions, nontrivial solutions of parabolic inequalities in a wedge must grow at least as fast as the growing solution $V(r)$ of (2). Of course, similar Phragmén-Lindelöf theorems are valid in other unbounded domains, for example in tube domains. Let D be a bounded domain in $\mathbb{R}^p$, $p \ge 1$, with the smooth boundary and

$$\mathbb{T}^D = \mathbb{R}^n \times D = \left\{ (x, y) \ \middle| \ x \in \mathbb{R}^n, \ y \in D \right\}, \ n \ge 1.$$

The domain $\mathbb{T}^D$ generates the slab

$$\mathbb{S}^D = \mathbb{T}^D \times (0, T) \subset \mathbb{R}^n \times \mathbb{R}^p \times \mathbb{R}_+.$$

We consider $c-$subparabolic functions in $\mathbb{S}^D$, that is, classical solutions $u(x, y, t)$ of the inequality

$$L_c u \equiv \Delta_x u + \Delta_y u - c(x, y, t)u(x, y, t) - \frac{\partial u}{\partial t} \ge 0.$$

Let μ_0 and $\psi_0(y)$ denote the lowest eigenvalue and the corresponding eigenfunction of the eigenvalue problem

$$\Delta_y \psi + \mu \psi(y) = 0, \ y \in D; \ \psi(y) = 0, \ y \in \partial D,$$

and Z be a growing solution of the equation

$$y'' + (n-1)r^{-1}y' - \left(\mu_0 + \alpha^2 + q(r)\right)y = 0,\ 0 < r < \infty,$$

with a continuous function q, $0 \le q(r) \le c(x,y,t) - c_0$, $r = |x|$.

Theorem 1.2. *Let u be a subparabolic function in $\mathbb{S}^D$ satisfying*

$$\limsup_{t \to 0} u(x,y,t) \le A,\ \text{for all } (x,y) \in \mathbb{T}^D$$
$$\limsup_{(x,y) \to \mathbf{R}^n \times \partial D} u(x,y,t) \le A \text{ for all } (x,y,t) \in \mathbb{S}^D.$$

If there is a constant α such that u satisfies the limit-growth condition

$$\liminf_{r \to \infty} \frac{1}{Z(r)} \int_0^T e^{-(\alpha^2 - c_0)t} \left\{ \int_S \left[\int_D u^+(r,\theta,y,t)\psi_0(y)dy \right] d\sigma(\theta) \right\} dt = 0,$$

then $u(x,y,t) \le A^+ e^{-c_0 t}$ for all $(x,y,t) \in \mathbb{S}^D$.

2. Proofs

Proof of Theorem 1.1. First we make a few observations. Let u be a nonnegative $c-$subparabolic function and a function $c_1(x,t) \le c(x,t)$. Then

$$L_{c_1}u = L_c u + (c - c_1)u \ge 0.$$

Thus, a positive $c-$subparabolic function is c_1-subparabolic with respect to an operator L_{c_1} with any coefficient $c_1 \le c$.

Now, let u be an upper-semicontinuous subparabolic function. In an open set where $u < 0$, $L_c u^+ = 0$. At any interior point of the complementary set, where $u(x,t) \ge 0$, we have $u^+ = u$, thus in these points $L_c u^+ = L_c u \ge 0$, hence u^+ is a subparabolic function together with u.

Next we show that without loss of generality, we can assume in the proof $A = A^+ = 0$. Indeed, this is clear if $A \le 0$, hence $A^+ = 0$. Otherwise we employ a function $v_{-1}(x)$ satisfying the equation $\Delta v - q(|x|)v(x) = 0$ with the boundary values $v(x)\big|_{\partial(K^\Omega)} = -1$ in the cone K^Ω. Such function exists.[2] Now if $A > 0$, then the function $u_1(x,t) = u(x,t) + Av_{-1}(x)$ satisfies the inequality

$$L_c u_1 = L_c u - A(c - q)v_{-1} \ge 0,\ x \in \mathbb{W}^\Omega,$$

for any $q \le c$ and the boundary condition $u_1 \le A + (-1)A = 0$. Moreover,

$$\liminf_{r \to \infty} \frac{1}{V(r)} \int_0^T e^{-\alpha^2 t} \left(\int_\Omega |v_{-1}(r,\theta)|\varphi_0(\theta)d\sigma(\theta) \right) dt = 0.$$

Thus if we prove that $u_1 \le 0$, then $u(x,t) = u_1(x,t) - Av_{-1}(x) \le A$ as well, and we assume hereafter $A = 0$.

We finally remark that if a constant $\gamma \ge -c_0$, then the function $u_1 = e^{-\gamma t}u$ is a c_1−subparabolic function with $c_1 = c + \gamma \ge 0$. Therefore, we assume in the proof the function c to be everywhere nonnegative.

We use in the proof an idea of Ahlfors;[4] we have applied it earlier.[2] Fix an $R > 0$, consider the truncated cone $K_R^\Omega = K^\Omega \cap \{|x| < R\}$, and solve in the truncated wedge

$$\mathbb{W}_R^\Omega = K_R^\Omega \times [0, T)$$

the initial-boundary value problem for the operator L_q,

$$\begin{cases} L_q U(x,t) = 0, & (x,t) \in \mathbb{W}_R^\Omega \\ U(x,0) = 0, & x \in K_R^\Omega \\ U(x,t) = 0, & x \in \partial\left(K^\Omega\right) \cap \{|x| < R\} \\ U(R,\theta,t) = u^+(R,\theta,t), & \theta \in \Omega,\ 0 < t < T. \end{cases}$$

Its solution $U(x,t) \equiv U_R(x,t)$ exists.[5] As we have noticed, u^+ is a q−subparabolic, thus by the maximum principle for subparabolic functions[5]

$$u^+(x,t) \le U_R(x,t)$$

everywhere in $\mathbb{W}_R^\Omega$. Since $L_q U = 0$,

$$\int_{\mathbb{W}_R^\Omega} \chi(x,t) L_q U_R(x,t)\,dx\,dt = 0$$

for any mollifier $\chi(x,t)$, therefore, due to Green's identity,

$$\int_{\mathbb{W}_R^\Omega} U_R(x,t) L_q^* \chi(x,t)\,dx\,dt = 0. \tag{5}$$

Here $L_q^* = \Delta_x - q + \frac{\partial}{\partial t}$ is the adjoint operator. Now we set

$$\chi(x,t) = e^{-\alpha^2 t}\varphi_0(\theta)f(r),$$

where $f(r)$ is a radial test function in $(0, R)$, and evaluate

$$L_q^* \chi(x,t) = e^{-\alpha^2 t}\varphi_0(\theta)l_q f(r),$$

where l_q is the operator in (2). Computing integral (5),

$$\int_0^T \left\{ \int_0^R \left(\int_\Omega U_R(r,\theta,t) L_q^* \chi(x,t)\,d\sigma(\theta) \right) dr \right\} dt$$

$$= \int_0^T e^{-\alpha^2 t} \left\{ \int_0^R r^{n-1} l_q f(r) \left(\int_\Omega U_R(r,\theta,t)\varphi_0(\theta)d\sigma(\theta) \right) dr \right\} dt,$$

and denoting

$$y_R(r,t) = \int_\Omega U_R(r,\theta,t)\varphi_0(\theta)d\sigma(\theta) \quad \text{and} \quad Y_R(r) = \int_0^T e^{-\alpha^2 t} y_R(r,t)dt,$$

we derive the equation $\int_0^R r^{n-1} l_q f(r) Y_R(r) dr = 0$. Integrating the latter by parts, we have

$$\int_0^R r^{n-1} f(r) l_q Y_R(r) dr = 0.$$

Here f is an arbitrary test function on $(0,R)$, therefore,

$$l_q Y_R(r) = 0, \ 0 < r < R,$$

thus $Y_R(r)$ is a solution of equation (2). Noticing that the function $U_R(x,t)$ is finite at $x = 0$, we have $y_R(0,t) = 0$, so that $Y_R(0) = 0$, and finally we find $Y_R(r) = aV(r)$, where V is a growing solution of (2) and $a = a(R)$ is a constant. By setting in the latter $r = R$ we find

$$a = \frac{Y_R(R)}{V(R)}$$

and

$$Y_R(r) = \frac{Y_R(R)}{V(R)} V(r). \tag{6}$$

Since

$$Y_R(R) = \int_0^T e^{-\alpha^2 t} \left\{ \int_\Omega U_R(R,\theta,t)\varphi_0(\theta)d\sigma(\theta) \right\} dt$$

$$= \int_0^T e^{-\alpha^2 t} \left\{ \int_\Omega u^+(R,\theta,t)\varphi_0(\theta)d\sigma(\theta) \right\} dt,$$

by sending $R \to \infty$ over a subsequence while keeping $r > 0$ fixed we conclude from (6) that $Y_R(r) = 0$ for all $r > 0$. Therefore, $y_R(r,t) = 0$ for all $r > 0$ and all $0 < t < T$, hence $U_R(r,\theta,t) = 0$. Thus finally, $u^+(r,\theta,t) = 0$ for all $\theta \in \Omega$ and $0 < t < T$.

The function v_{-1} used above shows that A^+ in inequality (4) cannot be replaced by A. To prove the exactness of the growth condition of Theorem 1.1, it is sufficient to consider a q−parabolic function

$$u(r,\theta,t) = e^{\alpha^2 t}\varphi_0(\theta)V(r). \qquad \square$$

Proof of Corollary 1.1. We can assume that $q(r) = 0$, thus (2) is Bessel's equation. Replacing α with $R\alpha$ leads to

$$V_R(r) = ar^{(2-n)/2}I_\nu(\alpha R r)$$

where $\nu = (1/2)\sqrt{(n-2)^2 + 4\lambda_0}$ and a is a constant, whence

$$V_R(R) = aR^{(2-n)/2}I_\nu(\alpha R^2).$$

Choosing in Theorem 1.1 $\alpha > c$, where c is a constant in Theorem A, we immediately derive the latter from Theorem 1.1. $\qquad \square$

Proof of Theorem 1.2. The proof mimics that of Theorem 1.1. The only difference is that instead of equation (2), the function Z satisfies the equation

$$y'' + \frac{n-1}{r}y' - \left(\mu_0 + \alpha^2 + q(r)\right)y(r) = 0, \ r > 0,$$

which leads to essentially different limit growth of subparabolic functions in tube domains.[2] $\qquad \square$

Acknowledgment. The work was partially supported by a grant from the City University of New York.

References

1. M. H. Protter and H. F. Weinberger, *Maximum Principles in Differential Equations*, (Prentice-Hall, Englewood Cliffs, N.J., 1967).
2. B. Ya. Levin and A. I. Kheyfits, Asymptotic behavior of subfunctions of time-independent Schrödinger operators in *Some Topics on Value Distribution and Differentiability in Complex and P-adic Analysis*, eds. A. Escassut, W. Tutschke, C. C. Yang (Science Press, Beijing, 2008) pp. 323-397.
3. P. Hartman, *Ordinary Differential Equations*, (John Wiley & Sons, New York, London, Sydney, 1964).
4. L. Ahlfors, *Trans. Amer. Math. Soc.* **41** 1 (1937).
5. A. Friedman, *Partial Differential Equations of Parabolic Type*, (Prentice-Hall, Englewood Cliffs, N.J., 1964).

I.4. Zeros and gamma lines — value distributions of real and complex functions

Organisers: G. Barsegian and G. Csordas

This session of the 7th ISAAC Congress, July 13–18, 2009, held at the Imperial College, London, provided an excellent forum for researchers working on different aspects of value distribution theory to report their findings. The lectures covered by the participants included a wide spectrum of related topics: universal value distribution, Gamma lines, the analysis of level sets, criteria for the reality of zeros, growth of analytic funtions in unbounded open sets, perturbation of L-functions, new properties of a class of Jacobi and generalized Laguerre polynomials, quantum field theory and complex analysis, meromorphic Laguerre operators and the distribution of zeros of entire functions, generalization of the Stieltjes-Van Vleck-Bocher Theorem, spiraling Baker domains, zeros of holomorphic functions of several variables, Picard-Fuchs type equations, and number theoretic analogues of the Nevanlinna theory. Following the talks, there were several informal seminars organized. The spirited and profitable exchange of ideas have led to new collaborations. In a nutshell, this ISAAC Congress, and in particular this session, proved to be a rewarding and stimulating experience.

The speakers of this session were A. P. Singh, D. Cardon, A. Fernández, A. Hinkkanen, K. Kato, A. Prykarpatsky, G. Csordas, A. Sergeev, P. Brändén, M. Charalambides, R. Halburd, P. Gauthier and G. Barsegian.

A universal value distribution for arbitrary meromorphic functions in a given domain

Grigor Barsegian

*Institute of Mathematics, National Academy of Sciences,
Bagramian ave. 24-b, 0019 Armenia
E-mail: barseg@instmath.sci.am*

Some purely geometric results analogous to the second fundamental theorems in the classical Nevanlinna and Ahlfors theories are revealed. These analogs are valid for arbitrary analytic (meromorphic) functions in given domains in contrast to the Nevanlinna and Ahlfors theories that are applicable only for some known subclasses of functions.

Keywords: Nevanlinna theory; Ahlfors theory; Gamma-Lines.

The zeros of complex functions, generally their a-points play a pivotal role in pure and applied mathematics. They are studied in the classical Nevanlinna value distribution theory [9] and Ahlfors theory of covering surfaces [1], see also [9, chapter 13]. These theories have had an essential influence on many branches of mathematics. These theories (the one analytical the other one metric-topological) work properly only for those classes of functions that have "equidistributions": meromorphic functions in the complex plane as well as in the disks but provided that the corresponding characteristic functions grow fast enough.

As to the most applicable case, arbitrary meromorphic functions in a given domain, we had no theory and moreover, we had no idea whether there are general value distribution type regularities in this case, see the collection of open problems [6].

Surprisingly a similar regularity has been obtained long ago (1981) in [4], but it was stated there as an auxiliary result (Lemma 1) and only for the functions in the disk. In this paper we present the general case mentioned above. Also the results in [4] were not expounded as some generalities in the complex analysis. We do it in this paper.

It is easy to notice that Nevanlinna and Ahlfors theories do not have even appropriate characteristics to solve a similar problems. To see this let us take a function $w(z)$ which is meromorphic in the closure $\bar{D}$ of a given domain D with smooth boundary ∂D. Ahlfors theory covers more large classes of functions so that it is enough to consider the question in this case. Ahlfors theory works with the spherical area $A(D)$ of $w(D)$, with the spherical length $L(D)$ of $w(\partial D)$ and with the number $n(D,a)$ of a-points of w (taken with counting multiplicities) in D. Taking now the simplest function $w = z^n/n$ in the disk $D(1) := \{z|\ |z| < 1\}$ we observe that for "enough large" n the magnitudes $A(D(1))$ and $L(D(1))$ are as small as we please and the magnitude $n(D(1),0)$ is as large as we please. This means that the magnitudes A, L and n cannot be evaluated in terms of each other. Thus, to describe the value distribution of analytic functions in the given domains, in other words to describe $n(D,a)$ for a prescribed set of values a, we should deal with another set of characteristic functions or we should make use of an additional characteristic along with A and L.

We show that the role of such a characteristic plays the following magnitude:

$$K(D) := \frac{1}{2\pi} \int_{w(\partial D)} |k(s)| ds,$$

where $k(s)$ is the curvature of the curve $w(\partial D)$.

Making use the geometrically harmonious triple of characteristics $A(D)$, $L(D)$ and $K(D)$ we prove the following

Theorem 1. *Let w be a meromorphic function in $\bar{D}$. For any set of pairwise different complex values $a_1, ..., a_q$, $q \geq 1$, we have*

$$\sum_{\nu=1}^{q} |n(D,a_\nu) - A(D))| \leq K(D) + h(a_1, ..., a_q)L(D), \tag{1}$$

where $h(a_1, ..., a_q)$ is a finite positive constant depending on $a_1, ..., a_q$.

The value of $h(a_1, ..., a_q)$ will be given in the proof.

Now we show the sharpness of (1) and compare this inequality with the classical results.

Sharpness. The result is asymptotically sharp for the functions in a given domain (particularly disk) and also for the functions in the complex plane. First we consider two standard functions in the complex plane: the exponential function (entire function) and double periodic function (meromorphic function). Let us write (1) for these functions in the disks $D(r) := \{z|\ |z| < r\}$ with arbitrary $a_1, ..., a_q$ including 0 and ∞. It is easy to see that the ratio of the left and the right sides in (1) tends to 1 when $r \to \infty$. Thus the result is asymptotically sharp in the class of entire and meromorphic functions in the plane. To show the sharpness in the bounded domains we again consider the function $w = z^n$ in the disk $D(r)$. For r tending to zero we have $A(D)$ and $L(D)$ meantime $n(D(r), a_\nu) = n$ and $K(D) = n$ so that again (1) is asymptotically sharp for $q = 1$. For $q > 1$ we take $a_1 = 0, ..., a_q = q - 1$ and take the function

$$w_k = (z - 0)^{km_1} (z - 1)^{km_2} \cdots (z - q - 1)^{km_q}$$

and denote by D_ε the ε-neighbourhood of the interval $(0, q - 1)$ on x-axis. Then for ε tending to zero and k tending to infinity we have: $A(D)$ tends to zero; $L(D) = o\left(km_1 + km_2 + ... + km_q\right)$, when $k \to \infty$; $n(D, a_\nu) = km_\nu$ and $K(D)$ tends asymptotically to $km_1 + km_2 + ... + km_q$. Consequently the ratio of the left and the right sides in (1) tends to 1 when ε tends to zero and simultaneously k tends to infinity. Thus the result is asymptotically sharp in this case as well.

Comparison with the classical results and some discussions. The second fundamental theorem in Ahlfors theory asserts: for any set of pairwise different complex values $a_1, ..., a_q$, $q \geq 3$, we have

$$\sum_{\nu=1}^{q} [A(D) - n(D, a_\nu)] \leq 2A(D) + h_1(a_1, ..., a_q)L(D), \qquad (2)$$

where $h_1(a_1, ..., a_q)$ is a positive and finite constant depending on $a_1, ..., a_q$.

Notice that in (1) we deal with the modules $|n(D, a_\nu) - A(D))|$ meantime in (2) with the difference $A(D) - n(D, a_\nu)$. Due to this circumstance (1) is meaningful and describes distribution of the a-points for any function in D. Let us see when (2) is meaningful and when it describes distribution of the a-points. It is well known that (2) describes distribution of the a-points only when $L(D)$ is essentially less than $A(D)$: corresponding *exhausting surfaces* (see [1], [9], chapter 13) are regularly exhausting. As it was mentioned above this is so only for those classes of functions that have "equidistributions": for instance for meromorphic functions in the complex plane as well as in the disks $D(r)$ provided the corresponding characteristic function grows fast enough. Ahlfors' theorem does not work when we have enough powerful set of vales $a_1, ..., a_q$ such that $n(D, a_\nu)$ are essentially larger than $A(D)$. But this is quite common and important in application case (remember the simplest example z^n/n) when we deal with the functions in arbitrary domains. Corresponding Riemann surfaces have a very interesting geometry. They have one or many neighbourhoods of the algebraic branch points that look like as some thin gimlets with many coils, so that $K(D)$ is large and can majorate $\sum_{\nu=1}^{q} |n(D, a_\nu) - A(D)|$ and consequently can majorate $\sum_{\nu=1}^{q} n(D, a_\nu)$ since we assumed in this case that $A(D)$ is comparatively small.

In the case when w is meromorphic in the complex plane Miles [8] showed that

$$\sum_{\nu=1}^{q} |A(D) - n(D, a_\nu)| \le CA(D) + h_1(a_1, ..., a_q)L(D), \qquad (3)$$

where C is an absolute constant. The same result follows also from our

Theorem 1 in [4].

Proof (sketch). We deal with some non intersecting neighbourhoods of $a_1, ..., a_q$. Namely, for a given set of pairwise different complex values $a_1, ..., a_q$, $q \ge 1$, we choose $\tau(a_\nu)$ which satisfy the following
Assumption 1: if $\infty \notin a_1, ..., a_q$ then $\tau(a_\nu)$-neighbourhoods of a_ν do not intersect, that is

$$\{w|\ |w - a_\nu| < \tau(a_\nu)\} \cap \{w|\ |w - a_\mu| < \tau(a_\mu)\} = \oslash$$

for $\nu \ne \mu$, $\nu, \mu = 1, 2, ..., q$; and if $\infty \in a_1, ..., a_q$, say $a_q = \infty$, then

$$\{w|\ |w - a_\nu| < \tau(a_\nu)\} \cap \{w|\ |w - a_\mu| < \tau(a_\mu)\} = \oslash$$

for $\nu \ne \mu$, $\nu, \mu = 1, 2, ..., q - 1$.
Denote

$$G = \cup_{\nu=1}^{q}\{w|\ |w - a_\nu| < \tau(a_\nu)\}$$

when $\infty \notin a_1, ..., a_q$ and

$$G = \left\{\cup_{\nu=1}^{q-1}\{w|\ |w - a_\nu| < \tau(a_\nu)\}\right\} \cup \{w|\ |w| > \max[|a_\nu| + \tau(a_\nu)]\}.$$

Denote

$$K_X := \frac{1}{2\pi} \int_X |k(s)|ds, \quad L_X := \int_X \frac{|w'(z)|}{1 + |w(z)|^2}ds,$$

where X is a set of curves.

The above Theorem 1 is an immediate corollary of the following
Theorem 1'. *Let w be a meromorphic function in $\bar{D}$. For any set of pairwise different complex values $a_1, ..., a_q$, $q \ge 1$, we have*

$$\sum_{\nu=1}^{q} |n(D, a_\nu) - A(D))| \le K_{w(\partial D) \subset G} + h(a_1, ..., a_q)L_{w(\partial D)}, \qquad (4)$$

where $h(a_1, ..., a_q)$ is a finite positive constant depending on $a_1, ..., a_q$.

First we consider the case when $\infty \notin a_1, ..., a_q$. Following section 3 in [5] we first prove that

$$n(D, a) - A(r) = \frac{1}{2\pi} \int_{\partial D} \frac{\partial}{\partial n} \ln \frac{|w(z) - a|}{\sqrt{1 + |w(z)|^2}} ds,$$

where n is the normal vector to the boundary ∂D.

Denote by $\tau(a)$ a constant, $0 < \tau(a) < \infty$, and by $\Pi_1(D, a, \tau(a))$ and $\Pi_2(D, a, \tau(a))$ respectively $\{z|\ z \in D,\ w(z) \in \{w|\ |w - a| \leq \tau(a)\}\}$ and $\{z|\ z \in D,\ w(z) \in \{w|\ |w - a| > \tau(a)\}\}$. Then we can rewrite the above identity as

$$n(D, a) - A(r) = \frac{1}{2\pi} \int_{\partial D \subset \Pi_1(D, a, \tau(a))} \frac{\partial}{\partial n} \ln |w(z) - a| ds$$

$$+ \frac{1}{2\pi} \int_{\partial D \subset \Pi_1(D, a, \tau(a))} \frac{\partial}{\partial n} \ln \frac{1}{\sqrt{1 + |w(z)|^2}} ds$$

$$+ \frac{1}{2\pi} \int_{\partial D \subset \Pi_2(D, a, \tau(a))} \frac{\partial}{\partial n} \ln \frac{|w(z) - a|}{\sqrt{1 + |w(z)|^2}} ds = I_1 + I_2 + I_3. \qquad (5)$$

Following the proof in [2] (see also [5], section 3.2) we show then that the upper bounds I_2 and I_3 can be given in terms of $L_{w(\partial D)}$. We need now the following differential geometric result, see [7], Theorem 1.1, which, in fact, was established first in [3], Lemma 1: for any curve $\gamma := f(t) := f_1(t) + i f_2(t)$, $t \in [0, 1]$, with continuous in $[0, 1]$ derivatives $f'(t) := f'(t) + i f_2'(t)$ such that $f(t), f'(t) \neq 0$, $t \in [0, 1]$

$$\int_0^1 \left| (\arg(f(t) - a))' \right| dt \leq \int_0^1 \left| (\arg f'(t))' \right| dt + \pi.$$

Making use this inequality we can give upper bounds for I_1 in terms of $K(D) + h^*(a_1, ..., a_q)L(D)$, where $h^*(a_1, ..., a_q)$ depends on $\max_\nu |a_\nu|$, and $\min_{\nu \neq \mu} |a_\nu - a_\mu|$. Summing up all obtained upper bounds we will get Theorem $1'$, consequently Theorem 1.

References

1. L. V. Ahlfors, *Zur Theorie der Überlagerungsflächen*, Acta Soc. Sci. Fenn. **9**, n. 1 (1930), 1-40.
2. G. Barsegian, *Deficient values and the structure of covering surfaces*, Izvestia Acad. Nauk Armenii **12** (1977), 46-53.

3. G. A. Barsegian, *On geometric structure of image of disks under mappings by meromorphic functions*, Math. Sbornik **106(148)**, n. 1 (1978), 35-43.

4. G. Barsegian, *Exceptional values associated with logarithmic derivatives of meromorphic functions*, Izvestia Acad. Nauk Armenian SSR, Matematika **16**, n. 5 (1981), 408-423.

5. G. Barsegian, *Gamma-lines: on the Geometry of Real and Complex Functions*, Taylor and Francis, London, New York, 2002.

6. G. Barsegian, *A new program of investigations in analysis: Gamma-lines approaches*, 1-73, in book: *Value Distribution and Related Topics*, Editors G. Barsegian, I. Laine and C. C. Yang, Kluwer, Series: Advances in complex analysis and its applications, 2004.

7. G. Barsegian, *Some new inequalities in geometry and analysis*, Izvestia Acadmii Nauk Armenii **42**, n. 2 (2007), 17-28.

8. J. Miles, *Bounds on the ratio sup $n(r,a)/A(r)$ for meromorphic functions*, Trans. Amer.Math. Soc. **162** (1971), 383-393.

9. R. Nevanlinna, *Eindeutige analytische Funktionen*,Springer, Berlin, 1936.

A method of counting the zeros of the Riccati equation and its application to biological and economical prognoses

G. Barsegian

Institute of Mathematics, National Academy of Sciences,
Bagramian ave. 24-b, 0019 Armenia
E-mail: barseg@instmath.sci.am

A. Sargsyan

Institute of Mathematics, National Academy of Sciences,
Bagramian ave. 24-b, 0019 Armenia
E-mail: sargsyan@instmath.sci.am

A geometric method is presented for studying the zeros of the Riccati equation and some applications to bio-economical prognoses are given.

Keywords: Gamma-Lines; Riccati equation; economic growth; logistic equation; population dynamics.

Introduction: The problem and its solution for the Riccati equation

In biological and economical processes we often deal with some dates or resources we have to (or we can) control; for example, it can be a quantity of food, reproduction etc. in biology or available money, human resources etc. in economics. Assume that we have a prognostic behavior (prognostic level) for a given process. This can be some desired, expected or test level. We consider the following

Practical task: can we reach the prognostic level in a given process (say, some level of population number or economic growth) during the forthcoming time (t_1, t_2) by managing our given or available resources?

One can meet a huge amount of similar problems in many areas of applied sciences. Mathematically such a process is often described by an equation $F(y, y', t) = 0$ with a solution $y(t)$ in (t_1, t_2) and initial condition $y(t_1) = y_0$. Our resources and the possibility to affect on the process are taken into account in the coefficients of the equation. The prognostic behavior (prognostic level) can be given by a function $\tilde{y}(t)$, or, in particular, by a constant $\tilde{y}_0$.

Respectively, the above task leads to the following

Problem: Let y be a solution of $F(y, y', t) = 0$ with the given coefficients. Can we have $y(t^*) = \tilde{y}(t^*)$ for some $t^* \in (t_1, t_2)$?

Especially interesting is its particular case when we deal with constant $\tilde{y}_0$, which can be, say, a desired level we would like to reach. The question then is as follows: can our function reach $\tilde{y}_0$ during the time interval (t_1, t_2)?

In this paper we solve this problem for the Riccati equation

$$y' = a(t)y^2 + b(t)y + c(t). \tag{1}$$

In what follows we consider rather general coefficients $a(t)$, $b(t)$, $c(t)$ assuming merely that they belong to $C^1(t_1, t_2)$. A priory this should permit to study more adequate models.

Theorem 1. *Let $y(t)$ be a solution of the equation (1) in (t_1, t_2) satisfying $y(t_1) = y(t_2) = 0$. Then there is a constant K depending only on the coefficients of (1) such that*

$$t_2 - t_1 \geq \frac{\pi}{K}. \tag{2}$$

For K we have

$$K = 3(M_2 + 6M_1^2) \max\left\{\frac{4}{m_1^2}; \frac{(4M_1)^5}{m_1 m_2^4}; \frac{4}{m_2^2}\right\}, \tag{3}$$

where

$$\begin{aligned}
M_1 &:= M_1(t_1, t_2) := \sup_{t_1 < t < t_2} \max\{|a(t)|, |b(t)|, |c(t)|\}, \\
M_2 &:= M_2(t_1, t_2) := \sup_{t_1 < t < t_2} \max\{|a'(t)|, |b'(t)|, |c'(t)|\}, \\
m_1 &:= m_1(t_1, t_2) := \inf_{t_1 < t < t_2} |c(t)|, \\
m_2 &:= m_2(t_1, t_2) := \inf_{t_1 < t < t_2} |a(t)|.
\end{aligned} \tag{4}$$

From here we derive the following

Corollary 1. *Assume $y(t)$ is a solution of equation (1) in (t_1, t_2) satisfying $y(t_2) = \tilde{y}(t_2)$, where $\tilde{y}(t) \in C^1(t_1, t_2)$ is a given function. Then there is a constant K^* depending only on the coefficients of (1) and the values $y(t_1)$ and $y(t_2)$ such that*

$$\frac{(t_2 - t_1)K^*}{\pi} \geq 1. \tag{5}$$

The constant K^* is defined in terms of (4) and its precise expression is given below in (22).

Applications to the mentioned prognoses

Particular cases of the Riccati equation one can meet in many purely mathematical and applied studies, for example, in biological and economical models or processes. Among these particular cases we mention here the *logistic* equation

$$\frac{dy(t)}{dt} = ry(t)\left(1 - \frac{1}{k}y(t)\right), \qquad (6)$$

that describes the dynamics of a biological population. Here y stays for the population number, r is the population growth rate parameter and k is the carrying capacity of the environment, that arises from the limitation of resources (food, living space etc.). The process of technological innovations, adoption and diffusion of some phenomena, the demand dynamics for some product, military spending, the spread of epidemic etc. can also be modelled by (6), see papers [5], [6], [7], [8], [9], [10], [11], [12] that involve more comprehensive citations.

We should note, however, that almost all these models were studied under the additional assumption that the coefficients in (6) are constants. This assumption was taken clearly for simplicity: the cases with variable coefficients are much more complicated. In the reality the resources are changeable and hence while modelling real processes one should consider time varying coefficients.

Thanks to Theorem 1 and its Corollary 1 we are able to study much larger class of equations (Riccati instead of logistic) with variable coefficients. We proceed here following [3]. Consider a solution $y(t)$ of (1) and the desired (or prognostic) function $\tilde{y}(t)$ (or level $\tilde{y}_0$) and ask whether $y(t)$ can reach $\tilde{y}(t)$ (or $\tilde{y}_0$) at a moment t_2?

We can easily answer this question since Corollary 1 implies the following
Conclusion: *if we want to reach the level $\tilde{y}_0$ at t_2 we need to adjust our coefficients in such a way that we have (5).*

Notice that this conclusion allows to check whether we can reach the prognostic levels in different applications (despite we get it very easily).

Proofs. To prove (2) we apply a method utilized often in Gamma-lines theory [2]. The key idea is as follows. Notice first that for any $y(t) \in C^1(t_1, t_2)$ satisfying $y(t_1) = 0$, $y'(t_1) \neq 0$ and $y(t_2) = 0$, $y'(t_2) \neq 0$ due to Rolle's theorem there is a point $t^* \in C^1(t_1, t_2)$ such that $y'(t^*) = 0$. This implies that the angles $\arctan y(t_1)/y'(t_1)$ and $\arctan y(t_2)/y'(t_2)$ are equal to zero, and the angle $\arctan y(t^*)/y'(t^*)$ is equal to $\pi/2$. Hence

$$\frac{\pi}{2} \leq \int_{t_1}^{t^*} \left| \frac{d}{dt} \arctan \frac{y(t)}{y'(t)} \right| dt, \quad \frac{\pi}{2} \leq \int_{t^*}^{t_2} \left| \frac{d}{dt} \arctan \frac{y(t)}{y'(t)} \right| dt$$

and thus

$$\pi \leq \int_{t_1}^{t_2} \left| \frac{d}{dt} \arctan \frac{y(t)}{y'(t)} \right| dt. \tag{7}$$

This estimate has been applied to study zeros of solutions $y(t)$ of (1) in [3] and, in fact, it was derived during the proof of Theorem 1 there.

Arguing as in [3] we first calculate

$$\left| \frac{d}{dt} \arctan \frac{y(t)}{y'(t)} \right| \tag{8}$$

by making use the form of (1). Simple calculations reduce (8) to

$$\left| \frac{y''(t)y(t) - (y'(t))^2}{(y'(t))^2 + (y(t))^2} \right| = \frac{\left| \left(a'y^2 + b'y + c'\right) y + (2ay + b)yy' - (ay^2 + by + c)^2 \right|}{\left| (ay^2 + by + c)^2 + y^2 \right|} \tag{9}$$

In fact, it was shown in [3] that choosing y such that $|y| \geq \frac{4M_1}{m_2} \geq 1$ one can obtain

$$\left| (ay^2 + by + c)^2 + y^2 \right| \geq \frac{|a|^2 |y|^4}{4} \geq \frac{m_2^2 |y|^4}{4}. \tag{10}$$

At the same time for the numerator we have

$$|y|^4 \left| \left(\frac{a'}{y} + \frac{b'}{y^2} + \frac{c'}{y^3} \right) + \left(a + \frac{b}{y} + \frac{c}{y^2} \right) \left(2a + \frac{b}{y} \right) - \left(a + \frac{b}{y} + \frac{c}{y^2} \right)^2 \right|$$

$$\leq |y|^4 \left(3M_2 + 18M_1^2 \right). \tag{11}$$

This gives

$$\left| \frac{y''(t)y(t) - (y'(t))^2}{(y'(t))^2 + (y(t))^2} \right| \leq \frac{3M_2 + 18M_1^2}{\frac{m_2^2}{4}}, \quad |y| \geq \frac{4M_1}{m_2}. \tag{12}$$

On the other hand, for all y such that $|y| \leq \left(\frac{m_1}{4M_1}\right)^{\frac{1}{2}}$ we have

$$\left|(ay^2 + by + c)^2 + y^2\right| \geq \frac{m_1^2}{4}. \tag{13}$$

This implies

$$\left|\frac{y''(t)y(t) - (y'(t))^2}{(y'(t))^2 + (y(t))^2}\right| \leq \frac{3M_2 + 18M_1^2}{\frac{m_1^2}{4}}, \quad |y| \leq \left(\frac{m_1}{4M_1}\right)^{\frac{1}{2}}. \tag{14}$$

Finally, the straightforward computations of (9) for all $\frac{4M_1}{m_2} \leq |y| \leq \left(\frac{m_1}{4M_1}\right)^{\frac{1}{2}}$ yield

$$\left|\frac{y''(t)y(t) - (y'(t))^2}{(y'(t))^2 + (y(t))^2}\right| \leq \frac{4M_1^5(3M_2 + 18M_1^2)}{m_1 m_2^4}. \tag{15}$$

Thus there exists a constant $K = K(M_1, M_2, m_1, m_2)$, mentioned in (3), such that

$$\pi \leq \int_{t_1}^{t_2} \left|\frac{d}{dt} \arctan \frac{y(t)}{y'(t)}\right| dt \leq \int_{t_1}^{t_2} K \, dt. \tag{16}$$

The last inequality implies (2).

To prove Corollary 1 we consider an auxiliary function

$$f(t) = y(t) - \tilde{y}(t) - y(t_1) + \tilde{y}(t_1) + \frac{t - t_1}{t_2 - t_1}(y(t_1) - \tilde{y}(t_1)). \tag{17}$$

We have $f(t_1) = f(t_2) = 0$. Rewriting (1) in terms of f we come to following differential equation

$$f'(t) = A(t)f^2(t) + B(t)f(t) + C(t), \tag{18}$$

where

$$\left\{\begin{array}{c} A(t) := a(t), \\ B(t) := 2a(t)\left[\tilde{y}(t) + \left(y(t_1) - \tilde{y}(t_1)\right)\left(1 - \frac{t-t_1}{t_2-t_1}\right)\right] + b(t), \\ C(t) := a(t)\left[\tilde{y}(t) + \left(y(t_1) - \tilde{y}(t_1)\right)\left(1 - \frac{t-t_1}{t_2-t_1}\right)\right]^2 \\ +b(t)\left[\tilde{y}(t) + \left(y(t_1) - \tilde{y}(t_1)\right)\left(1 - \frac{t-t_1}{t_2-t_1}\right)\right] + c(t) - \tilde{y}' + \frac{y(t_1) - \tilde{y}(t_1)}{t_2 - t_1}. \end{array}\right. \tag{19}$$

Let us denote by $\overline{M}_1$, $\overline{M}_2$, $\overline{m}_1$ and $\overline{m}_2$ the corresponding values from (18), calculated for the coefficients $A(t)$, $B(t)$ and $C(t)$. For the test function $\tilde{y}$ we define

$$Y_1 := \sup_{t_1 < t < t_2} |\tilde{y}(t)|, \quad Y_2 := \sup_{t_1 < t < t_2} |\tilde{y}'(t)|, \quad Y_3 := \sup_{t_1 < t < t_2} |\tilde{y}''(t)|.$$

To short the formulas we make use also the following notations:

$$Y_0 := |y(t_1) - \tilde{y}(t_1)|, \quad T(t) := \frac{t - t_1}{t_2 - t_1}$$

and

$$L := |Y_0 T'(t)| = \left| \frac{y(t_1) - \tilde{y}(t_1)}{t_2 - t_1} \right|.$$

Then the coefficients in (19) can be estimated in terms of (4) in the following way (here we take into account that $T(t) \in (0,1)$):

$$|A(t)| = |a(t)| \leq M_1,$$

$$|B(t)| \leq M_1 \big[2(Y_1 + Y_0) + 1 \big] \leq M_1 \big[(Y_1 + Y_0)^2 + (Y_1 + Y_0) + 2 \big],$$

$$|C(t)| \leq M_1 \big[(Y_1 + Y_0)^2 + (Y_1 + Y_0) + 1 \big] + Y_2 + L,$$

and hence

$$\overline{M}_1 \leq M_1 \big[(Y_1 + Y_0)^2 + (Y_1 + Y_0) + 2 \big] + Y_2 + L. \tag{20}$$

Clearly

$$|A'(t)| = |a'(t)| \leq M_2,$$

$$|B'(t)| \leq 2M_2 \big[Y_1 + Y_0 \big] + 2M_1 \big[Y_2 + L \big] + M_2,$$

$$|C'(t)| \leq M_2 \big[Y_1 + Y_0 \big]^2 + 2M_1 \big[Y_1 + Y_0 \big] \big[Y_2 + L \big]$$

$$+ M_2 \big[Y_1 + Y_0 \big] + M_1 \big[Y_2 + L \big] + M_2 + Y_3$$

and thus

$$\overline{M}_2 \leq M_2 \Big[(Y_1 + Y_0)^2 + (Y_1 + Y_0) + 2 \Big]$$

$$+ 2M_1 \Big[Y_1 + Y_0 + 1 \Big] \Big[Y_2 + L \Big] + Y_3. \tag{21}$$

Applying now Theorem 1 to the equation (18), taking into account (20), (21) and observing that $\overline{m}_2 = m_2$ we obtain inequality (5) with a constant

$$K^* := 3(\overline{M}_2 + 6\overline{M}_1^2) \max \left\{ \frac{4}{\overline{m}_1^2}; \frac{|4\overline{M}_1|^5}{\overline{m}_1 \overline{m}_2^4}; \frac{4}{\overline{m}_2^2} \right\}$$

$$= K^*(t_1, t_2, M_1, M_2, m_1, \overline{m}_2, Y_1, Y_2, Y_3). \tag{22}$$

References

1. O. Akira, *Diffusion and Ecological Problems: Mathematical Models*, Springer-Verlag, Berlin-Heidelberg-New York (1980).
2. G. A. Barsegian, *Gamma-lines: on the Geometry of Real and Complex Functions*, Taylor and Francis, London-New York (2002).
3. G. A. Barsegian, K. Barseghyan, A Method for Studying Oscillations of Nonlinear Differential Equations. Applications to Some Equations in Biology and Economics, *Topics in Analysis and its Applications*, editors: G. A. Barsegian and H. G. W. Begehr, NATO Science series, **147** (2004), p. 123-148.
4. J. -E. Cohen, Population Growth and the Earth's Human Carrying Capacity, *Science*, **269** (1995), p. 341-346.
5. B. D. Coleman, Nonautonomous Logistic Equations as Models of the Adjustment of Populations to Environmental Change, *Mathematical Biosciences*, **45** (1978), p. 159-173
6. J. C. Frauenthal, *Mathematical Modelling in Epidemiology*, Springer-Verlag, Berlin-Heidelberg-New York (1980).
7. M. E. Gyrtin, R. C. MacCamy, Nonlinear Age-Dependent Population Dynamics, *Arch. Rat. Mech. Anal.*, **53** (1974), p. 281-300.
8. T. Haavelmo, *A study in the Theory of Economic Evolution*, North-Holland, Amsterdam (1964).
9. G. Jarne, J. Sánchez-Chóliz, F. Fatás-Villafranca, "S-shaped" Economic Dynamics. The Logistic and Gompertz curves generalized, *The Electronic Journal of Evolutionary and Economic Modelling and Economic Dynamics*, **1048** (2005), p. 1-37.
10. P. S. Meyer, J. H. Ausubel, Carrying Capacity: A Model with Logistically Varying Limits, *Technological Forecasting and Social Changes*, **61(3)** (1999), p. 209-214.
11. J. J. Shepherd, L. Stojkov, The Logistic Population Model with Slowly Varying Carrying Capacity, *ANZIAM J.*, **47** (2007), p. C492-C506.
12. W. -B. Zhang, Multiregional Dynamics Based on Creativity and Knowledge Diffusion, *The Annals of Regional Science*, **25** (1991), p. 179-191.
13. W. -B. Zhang, *Economic Growth theory: Capital, Knowledge and Economic Structures*, Ashgate Publishing Company, Burlington (2005).

A Heine-Stieltjes theorem for higher order differential operators

Petter Brändén

Department of Mathematics, Royal Institute of Technology,
SE-100 44 Stockholm, Sweden
and
Department of Mathematics, Stockholm University,
SE-106 91 Stockholm, Sweden
E-mail: pbranden@math.su.se

The Heine–Stieltjes Theorem describes the polynomial solutions, (v, f) such that $T(f) = vf$, to specific second order differential operators, T, with polynomial coefficients. We extend the theorem to concern all (non-degenerate) differential operators preserving the property of having only real zeros, thus solving a conjecture of B. Shapiro. The new methods developed are used to describe intricate interlacing relations between the zeros of different pairs of solutions. This extends recent results of Bourget, McMillen and Vargas for the Heun equation and answers their question on how to generalize their results to higher degrees. Many of the results are new even for the classical case. This is an extended abstract, and proofs will appear elsewhere.

Keywords: Heine–Stieltjes Theorem; Heine–Stieltjes polynomials; Van Vleck polynomials; hyperbolic polynomials; real zeros; interlacing zeros

1. An Extension of the Heine–Stieltjes Theorem

Special cases of the following general eigen-problem have been studied frequently since the 1830's.

Problem 1.1. *Consider a linear operator $T : \mathbb{C}[z] \to \mathbb{C}[z]$. Describe all pairs $(v, f) \in \mathbb{C}[z] \times \mathbb{C}[z]$, with $f \not\equiv 0$, such that $T(f) = vf$.*

Recently Problem 1.1 has received new interest from different perspectives, see e.g. Refs. 2,3,6,8.

We say that v above is a *(generalized) Van Vleck polynomial* and that f is a *(generalized) Stieltjes polynomial* for T.

The following theorem is known as the Heine–Stieltjes Theorem[11] and is due to Stieltjes,[10] except for the statement regarding the location of the zeros of the Van Vleck polynomials in (3), which is due to Van Vleck.[12]

Theorem 1.1. *Suppose that $T : \mathbb{C}[z] \to \mathbb{C}[z]$ is a differential operator of the form $T = Q_2(z)D^2 + Q_1(z)D$, where $D = d/dz$, $Q_2(z) = \prod_{j=1}^{d}(z - \alpha_j)$ and*

$$Q_1(z) = Q_2(z) \sum_{j=1}^{d} \frac{\gamma_j}{z - \alpha_j},$$

where $\alpha_1 < \cdots < \alpha_d$ are real, $d \geq 2$, and $\gamma_1, \ldots, \gamma_d$ are positive. Let $n \geq 2$ be an integer. Then

(1) There are exactly $\binom{n+d-2}{n}$ pairs $(v, f) \in \mathbb{C}[z] \times \mathbb{C}[z]$ such that $T(f) = vf$ and f is monic of degree n;
(2) For each Van Vleck polynomial there is a unique monic Stieltjes polynomial;
(3) The zeros of v and f in (1) are real, simple and belong to the interval (α_1, α_d).

Example 1.1. A much studied instance of the Heine–Stieltjes theorem is the Lamé equation:

$$Q(z)D^2 f(z) + \frac{1}{2}Q'(z)Df(z) = v(z)f(z),$$

where $Q(z) = (z - \alpha_1)(z - \alpha_2)(z - \alpha_3)$ and $\alpha_1 < \alpha_2 < \alpha_3$.

Other classical examples are the *Jacobi polynomials*, $P_n^{(\alpha,\beta)}(z)$, with $\alpha, \beta > -1$, which are the solutions to

$$(z^2 - 1)D^2 f(z) + ((\alpha + \beta + 2)z - (\beta - \alpha))\, Df(z) = n(n + \alpha + \beta + 1)f(z).$$

The problem we address here is how to generalize the Heine–Stieltjes Theorem to larger families of linear operators. Which properties of T in the Heine–Stieltjes Theorem are essential? We will see that the essential characteristic of T is that it preserves the property of having only real zeros.

A polynomial $f \in \mathbb{R}[z]$ is *hyperbolic* if all its zeros are real and a linear operator $T : \mathbb{R}[z] \to \mathbb{R}[z]$ *preserves hyperbolicity* if $T(f)$ is hyperbolic or identically zero whenever f is hyperbolic. Such operators were recently characterized in Ref. 1. Our extension of Theorem 1.1 is to differential operators, of finite but arbitrary order, that preserve hyperbolicity. Let

$$T = \sum_{k=M}^{N} Q_k(z)D^k, \qquad \text{where } Q_M(z)Q_N(z) \not\equiv 0, \tag{1}$$

be a differential operator with polynomial coefficients. The number

$$r = \max\{\deg Q_k - k : M \leq k \leq N\}$$

is called the *Fuchs index* of T, and T is *non-degenerate* if $\deg Q_N = N + r$.

For a positive integer m, let $[m] = \{1, \ldots, m\}$. An *ordered partition of* $[m]$ *into two parts* is a pair (A, B) where $A, B \subseteq [m]$, $A \cup B = [m]$ and $A \cap B = \emptyset$. Let f and g be two hyperbolic polynomials of degree r and n, respectively and let $\alpha_1 \le \alpha_2 \le \cdots \le \alpha_{n+r}$ be the zeros of fg. Let further (A, B) be an ordered partition of the set $[n + r]$, such that $|A| = r$ and $|B| = n$. The pair (f, g) is said to have *zero pattern* (A, B) if the zeros of f are $\{\alpha_i : i \in A\}$ and the zeros of g are $\{\alpha_i : i \in B\}$. Note that the zero pattern is unique if and only if f and g are co-prime.

It was conjectured by B. Shapiro[8,9] that a Heine–Stieltjes Theorem should hold for non-degenerate hyperbolicity preserving differential operators. The solution to this conjecture, Theorem 1.2, is our main result. For proofs, see Ref. 4. Theorem 1.1 is the special case of Theorem 1.2 when $N = 2$ and $Q_0(z) = 0$, see Ref. 4.

Theorem 1.2. *Let* $T = \sum_{k=0}^{N} Q_k(z) D^k$, *where* $Q_N(z) \not\equiv 0$, *be a non-degenerate hyperbolicity preserving differential operator with nonnegative Fuchs index* r, *and suppose that* $n \in \mathbb{N}$ *is such that there is no* $\theta \in \mathbb{R}$ *such that* $Q_0(\theta) = \cdots = Q_n(\theta) = 0$. *Then*

(1) If f_1, f_2 *are monic polynomial for which* (v, f_1) *and* (v, f_2) *are solutions to Problem 1.1, then* $f_1 = f_2$;

(2) For all $S \subseteq [n + r]$ *of cardinality* r *there is a unique pair* (v, f) *of hyperbolic polynomials with zero pattern* $(S, [n+r] \setminus S)$ *such that* $T(f) = vf$, *with* f *monic of degree* n. *These are all Stieltjes and Van Vleck polynomials. Hence there are exactly* $\binom{n+r}{r}$ *pairs of polynomials* (v, f) *such that* $T(f) = vf$, *with* f *monic of degree* n;

(3) v *and* f *are co-prime and their zeros lie in the convex hull of the zeros of* Q_e, *where* $e = \min(n, N)$;

(4) If there is no $\theta \in \mathbb{R}$ *such that* $Q_1(\theta) = \cdots = Q_n(\theta) = 0$ *then* v *and* f *have simple zeros which all lie in the interior of the convex hull of the zeros of* Q_e, *where* $e = \min(n, N)$.

That v and f are co-prime in the original Heine–Stieltjes theorem is due to Shah.[7]

The coefficients $Q_j(z)$, for $0 \le j \le N$, in Theorem 1.2 are hyperbolic, see Ref. 4. Hence the convex hull of the zeros of $Q_e(z)$ in Theorem 1.2 (3) and (4) is a closed interval.

Example 1.2. A natural class of differential operators satisfying the hypothesis of Theorem 1.2 consists of operators of the form

$$T(f)(z) = D^{N-M}(PD^M f)(z) = \sum_{k=M}^{N} \binom{N-M}{k-M} P^{(N-k)}(z) D^k f(z),$$

where P is a hyperbolic polynomial of degree $N+r$, with no zero of multiplicity larger than $N-M$.

The following theorem offers a way of (in theory) generating Stieltjes and Van Vleck polynomials. Note that there is no restriction on the common zeros of the Q_ks.

Theorem 1.3. *Let* $T = \sum_{k=0}^{N} Q_k(z)D^k$ *be a non-degenerate hyperbolicity preserving differential operator with nonnegative Fuchs index r. Suppose that $n \in \mathbb{N}$ is such that $T(z^n) \not\equiv 0$ and that $A \subseteq [n+r]$ is a set of cardinality r. Let a be the smallest zero of $Q_e(z)$, where $e = \min(n, N)$.*

Define two sequences of polynomial $\{f_i\}_{i=0}^{\infty}$ and $\{v_i\}_{i=1}^{\infty}$ recursively by $f_0(z) = (z-a)^n$ and

$$T(f_i) = v_{i+1}f_{i+1},$$

where $v_{i+1}f_{i+1}$ is the unique factorization of $T(f_i)$ such that f_{i+1} is monic and the pair (v_{i+1}, f_{i+1}) has zero pattern (A, B), where $B = [n+r] \setminus A$. Then

$$\lim_{i \to \infty} f_i = f \quad and \quad \lim_{i \to \infty} v_i = v,$$

where $T(f) = vf$ and (v, f) have zero pattern (A, B).

2. Interlacing Properties

In this section we will investigate interlacing properties of zeros of Stieltjes polynomial and of Van Vleck polynomials. The first two theorems concern Stieltjes polynomials of the same or consecutive degrees. These theorems generalize the theorems obtained by Bourget and McMillen[2] for the Heun equation. In Ref. 2 it was asked if and how their results could be extended to the case of Fuchs index greater than one. Theorems 2.1 and 2.2 answers this. Our theorems concerning Van Vleck polynomials generalize to arbitrary nonnegative Fuchs index the results of Bourget, McMillen and Vargas[3] regarding the Heun equation.

Let $\alpha_1 \leq \alpha_2 \leq \cdots \leq \alpha_n$ and $\beta_1 \leq \beta_2 \leq \cdots \leq \beta_m$ be the zeros (counted with multiplicities) of two hyperbolic polynomials f and g, with $\deg f = n$

and $\deg g = m$. We say that these zeros *interlace* if they can be ordered so that either $\alpha_1 \le \beta_1 \le \alpha_2 \le \beta_2 \le \cdots$ or $\beta_1 \le \alpha_1 \le \beta_2 \le \alpha_2 \le \cdots$, in which case $|n - m| \le 1$. Note that by our convention, the zeros of any two polynomials of degree 0 or 1 interlace. It is not difficult to show that if the zeros of f and g interlace then the *Wronskian* $W[f, g] := f'g - fg'$ is either nonnegative or nonpositive on the whole real axis $\mathbb{R}$. In the case that $W[f, g] \le 0$ we say that f and g are in *proper position*, denoted $f \ll g$. For technical reasons we also say that the zeros of the polynomial 0 interlace the zeros of any (nonzero) hyperbolic polynomial and write $0 \ll f$ and $f \ll 0$. Note that if f and g are (nonzero) hyperbolic polynomials such that $f \ll g$ and $g \ll f$ then f and g must be constant multiples of each other, that is, $W[f, g] \equiv 0$.

Definition 2.1. Let $\binom{[n+r]}{n}$ denote the set of all n element subsets of $[n+r]$, and let $T = \sum_{k=M}^{N} Q_k(z)D^k$, where $Q_M Q_N \not\equiv 0$, be a non-degenerate hyperbolicity preserver with nonnegative Fuchs index r. Let further $B \in \binom{[n+r]}{n}$ and $A = [n + r] \setminus B$, where $n \ge M$. We denote by $(v_{A,n}, f_{B,n})$ the pair of polynomials with zero-pattern (A, B) afforded by Theorem 1.3. That is, $f_{B,n}$ is monic of degree n and $T(f_{B,n}) = v_{A,n} f_{B,n}$.

Henceforth it will tacitly be assumed that $n \ge M$ as in Definition 2.1.

A pair $(\mathbf{x}, \mathbf{y})$ of real vectors $\mathbf{x} = (x_1, \ldots, x_k)$ and $\mathbf{y} = (y_1, \ldots, y_k)$ are said to be in *proper position*, written $\mathbf{x} \ll \mathbf{y}$, if

$$x_1 \le y_1 \le x_2 \le y_2 \le \cdots \le x_k \le y_k.$$

If A is a set of integers let $[A]$ denote the vector obtained by ordering its elements increasingly. Define a relation $\rightarrow$ on $\binom{[n+r]}{n}$ by: $A \rightarrow B$ if $2[A] \ll 2[B] + \mathbf{1}$, where $\mathbf{1} = (1, \ldots, 1)$.

The relation $\rightarrow$ may be described differently as follows. If $\mathbf{x} = (x_1, \ldots, x_{n+r}) \in \mathbb{R}^{n+r}$ and $B \in \binom{[n+r]}{n}$, let $\mathbf{x}_B$ be the vector obtained by deleting the coordinates labeled by $[n + r] \setminus B$. Then $B \rightarrow C$ if for all $\mathbf{x}, \mathbf{y} \in \mathbb{R}^{n+r}$ with $\mathbf{x} \ll \mathbf{y}$ we have $\mathbf{x}_B \ll \mathbf{y}_C$.

Theorem 2.1. *Let T be a non-degenerate hyperbolicity preserver with nonnegative Fuchs index r and let $B, C \in \binom{[n+r]}{n}$. If $B \rightarrow C$ then $f_{B,n} \ll f_{C,n}$.*

Moreover, if there is no $\theta \in \mathbb{R}$ for which $Q_1(\theta) = \cdots = Q_n(\theta) = 0$, then $f_{B,n}$ and $f_{C,n}$ are co-prime, i.e., their zeros strictly interlace.

Remark 2.1. Given $B = \{b_1 < \cdots < b_n\} \subseteq [n+r]$ there are exactly $b_1(b_2 - b_1) \cdots (b_n - b_{n-1})$ sets $A \in \binom{[n+r]}{n}$ for which $A \rightarrow B$. Indeed, to construct A we pick an integer from the interval $[b_k + 1, b_{k+1}]$, independently for each

k. By a counting argument there are exactly $\binom{2n+r}{r}$ relations (including the trivial ones) in $\binom{[n+r]}{n}$.

Similarly, a pair $(\mathbf{x}, \mathbf{y})$ of real vectors $\mathbf{x} = (x_1, \ldots, x_k)$ and $\mathbf{y} = (y_1, \ldots, y_{k+1})$ are said to be in *proper position*, written $\mathbf{x} \ll \mathbf{y}$, if

$$y_1 \leq x_1 \leq y_2 \leq x_2 \leq \cdots \leq x_k \leq y_{k+1}.$$

The relation $\to$ extends naturally to concern $\binom{[n+r]}{n} \times \binom{[n+1+r]}{n+1}$: $A \to B$ if $2[A] \ll 2[B] + 1$. Again, $A \to B$ if and only if for all $\mathbf{x} \in \mathbb{R}^{n+r}$ and $\mathbf{y} \in \mathbb{R}^{n+r+1}$ with $\mathbf{x} \ll \mathbf{y}$ we have $\mathbf{x}_A \ll \mathbf{y}_B$.

The next theorem describes interlacing relationships between Stieltjes polynomials of consecutive degrees.

Theorem 2.2. *Let T be a non-degenerate hyperbolicity preserver with non-negative Fuchs index r and let $B \in \binom{[n+r]}{n}$ and $C \in \binom{[n+1+r]}{n+1}$. If $B \to C$ then $f_{B,n} \ll f_{C,n+1}$.*

Moreover, if there is no $\theta \in \mathbb{R}$ for which $Q_1(\theta) = \cdots = Q_n(\theta) = 0$, then $f_{B,n}$ and $f_{C,n}$ are co-prime, i.e., their zeros strictly interlace.

We have the following theorem describing interlacing properties of Van Vleck polynomials that have Stieltjes polynomials of the same degree.

Theorem 2.3. *Let T be a non-degenerate hyperbolicity preserver with non-negative Fuchs index r and let $A \in \binom{[n+r-1]}{r}$. Then $v_{A,n} \ll v_{A+\mathbf{1},n}$, where $A + \mathbf{1} := \{a + 1, a \in A\}$.*

Moreover, if there is no $\theta \in \mathbb{R}$ for which $Q_1(\theta) = \cdots = Q_n(\theta) = 0$, then $v_{A,n}$ and $v_{A+\mathbf{1},n}$ are co-prime, i.e., their zeros strictly interlace.

The next theorem describes interlacing properties of Van Vleck polynomials that have Stieltjes polynomials of consecutive degrees.

Theorem 2.4. *Let T be a non-degenerate hyperbolicity preserver with nonnegative Fuchs index r and let $A \in \binom{[n+r]}{r}$. Then $v_{A,n+1} \ll v_{A,n} \ll v_{A+\mathbf{1},n+1}$.*

Moreover, if there is no $\theta \in \mathbb{R}$ for which $Q_1(\theta) = \cdots = Q_n(\theta) = 0$, then the interlacing of the zeros is strict.

Suppose that $T = \sum_{k=M}^{N} Q_k(z) D^k$ is a non-degenerate hyperbolicity preserver with Fuchs index $r = 1$, and that $Q_M(\theta) = \cdots = Q_n(\theta)$ for no $\theta \in \mathbb{R}$. Then the Van Vleck polynomials are of degree one and there are exactly $n + 1$ of them. Define the nth *spectral polynomial*, $p_n(z)$, to be the monic polynomial whose zeros are precisely the zeros of the $n+1$ Van Vleck polynomials.

Corollary 2.1. *Suppose that $T = \sum_{k=M}^{N} Q_k(z)D^k$ is a non-degenerate hyperbolicity preserver with Fuchs index $r = 1$, and that $Q_M(\theta) = \cdots = Q_n(\theta)$ for no $\theta \in \mathbb{R}$. Then the zeros of p_n and p_{n+1} interlace. Moreover p_n and p_{n+1} have no zeros in common.*

For the Heun equation Corollary 2.1 reduces to the main result, Theorem 2.1, of Ref. 3.

3. Further Directions

The Jacobi orthogonal polynomials arise as the Stieltjes polynomials corresponding to the case when $Q_2(z)$ is of degree 2 in the classical Heine–Stieltjes theorem, see Example 1.1. Given a hyperbolicity preserving differential operator as in Theorem 1.2 (4) of Fuchs index $r = 0$, it is natural to ask whether the Stieltjes polynomials form an orthogonal family. This was suggested by T. McMillen.

The linear operators considered in this paper are all finite order differential operators. One may try to extend our results to hyperbolicity preservers that are not finite order differential operators.

Acknowledgments

The author is partially supported by the Göran Gustafsson Foundation.

References

1. J. Borcea and P. Brändén, *Ann. of Math. (2)* **170** (2009).
2. A. Bourget and T. McMillen, *Preprint*, arXiv:0903.0644.
3. A. Bourget, T. McMillen, and A. Vargas, *Proc. Amer. Math. Soc.* **137** (2009).
4. P. Brändén, *Preprint*, arXiv:0907:0648.
5. E. Heine, *Handbuch der Kugelfunctionen. Theorie und Anwendungen. Band I, II*, (Physica-Verlag, Würzburg, 1961).
6. A. Martínez-Finkelshtein and E. B. Saff, *J. Approx. Theory* **118** (2002).
7. G. M. Shah, *Proc. Amer. Math. Soc.* **19** (1968).
8. B. Shapiro, *Preprint*, arXiv:0812.4193.
9. B. Shapiro, *Private communication*, 2005.
10. T. J. Stieltjes, *Acta Math.* **6** (1885).
11. G. Szegő, *Orthogonal polynomials* (American Mathematical Society, Providence, 1975).
12. E. B. Van Vleck, *Bull. Amer. Math. Soc.* **4** (1898).

Extended Laguerre inequalities and a criterion for real zeros

David A. Cardon

Department of Mathematics,
Brigham Young University,
Provo, Utah 84602, USA
E-mail: cardon@math.byu.edu
http://www.math.byu.edu/cardon

Let $f(z) = e^{-bz^2} f_1(z)$ where $b \geq 0$ and $f_1(z)$ is a real entire function of genus 0 or 1. We give a necessary and sufficient condition in terms of a sequence of inequalities for all of the zeros of $f(z)$ to be real. These inequalities are an extension of the classical Laguerre inequalities.

Keywords: Laguerre-Pólya class; real zeros; Laguerre inequalities.

1. Introduction

The Laguerre-Pólya class, denoted $\mathcal{LP}$, is the collection of real entire functions obtained as uniform limits on compact sets of polynomials with real coefficients having only real zeros. It is known that a function f is in $\mathcal{LP}$ if and only if it can be represented as

$$f(z) = e^{-bz^2} f_1(z) \tag{1}$$

where $b \geq 0$ and where $f_1(z)$ is a real entire function of genus 0 or 1 having only real zeros. The basic theory of $\mathcal{LP}$ can be found in Chap. 8 of Ref. 6 and Chap. 5.4 of Ref. 8.

In this paper, we extend a theorem of Csordas, Patrick, and Varga on a necessary and sufficient condition for certain real entire functions to belong to the Laguerre-Pólya class. They proved the following:

Theorem 1.1. *Let*

$$f(z) = e^{-bz^2} f_1(z), \qquad (b \geq 0,\ f(z) \not\equiv 0),$$

where $f_1(z)$ is a real entire function of genus 0 or 1. Set

$$L_n[f](x) = \sum_{k=0}^{2n} \frac{(-1)^{k+n}}{(2n)!} \binom{2n}{k} f^{(k)}(x) f^{(2n-k)}(x) \tag{2}$$

for $x \in \mathbb{R}$ and $n \geq 0$. Then $f(z) \in \mathcal{LP}$ if and only if

$$L_n[f](x) \geq 0 \tag{3}$$

for all $x \in \mathbb{R}$ and all $n \geq 0$.

Patrick proved the forward direction in Thm. 1 of Ref. 7. The reverse direction was proved by Csordas and Varga in Thm. 2.9 of Ref. 4. Theorem 1.1 is significant because it gives a nontrivial sequence of inequality conditions that hold for functions in the Laguerre-Pólya class. The case $n = 1$ reduces to the classical Laguerre inequality which says that if $f(z) \in \mathcal{LP}$, then

$$[f'(x)]^2 - f(x)f''(x) \geq 0$$

for $x \in \mathbb{R}$. Consequently, inequalities like those in Theorem 1.1 are sometimes called Laguerre-type inequalities. Csordas and Escassut discuss the inequalities $L_n[f](x) \geq 0$ and related Laguerre-type inequalities in Ref. 3. Other results on similar inequalities of Turán and Laguerre types can be found in Refs. 1 and 2.

2. An extension of Laguerre-type inequalities

In this section, we extend Theorem 1.1 and give new necessary and sufficient inequality conditions for a function to belong to the Laguerre-Pólya class.

First we generalize the operator L_n defined in Theorem 1.1. Let

$$g(z) = \sum_{\ell=0}^{M} c_\ell z^\ell = \prod_{j=1}^{M} (z + \alpha_j) \tag{4}$$

be a polynomial with complex roots. Define $\Phi(z,t)$ as the product

$$\Phi(z,t) = \prod_{j=1}^{M} f(z + \alpha_j t). \tag{5}$$

The coefficients of the Maclaurin series of $\Phi(z,t)$ with respect to t are functions of z, and we write:

$$\Phi(z,t) = \sum_{k=0}^{\infty} A_k(z) t^k, \tag{6}$$

where

$$A_k(z) = \frac{1}{k!} \left[\frac{d^k}{dt^k} \Phi(z,t) \right]_{t=0} = \frac{1}{k!} \left[\frac{d^k}{dt^k} \prod_{j=1}^{M} f(z + \alpha_j t) \right]_{t=0}. \tag{7}$$

Since $f(z)$ is entire, each $A_k(z)$ is entire. Another expression for $A_k(z)$ is given in (11). The choice $g(z) = 1 + z^2 = (z-i)(z+i)$ produces $A_{2k+1}(z) = 0$ and $A_{2k}(z) = L_k[f](z)$ as in (2) of Theorem 1.1. Thus, we may regard the sequence of functions $A_k(z)$ as a generalization of the sequence $L_k[f](z)$. We note that the zeros of $A_k(z)$ were studied by Dilcher and Stolarsky in Ref. 5. In §3, we give several examples of $A_k(z)$ for interesting choices of $g(z)$.

Theorem 2.1. *Let $f(z) = e^{-bz^2} f_1(z)$, where $f_1(z) \not\equiv 0$ is a real entire function of genus 0 or 1 and $b \geq 0$. Assume $g(z)$ in (4) is an even polynomial with non-negative real coefficients having at least one non-real root. Then $f \in \mathcal{LP}$ if and only if*

$$A_k(x) \geq 0 \tag{8}$$

for all $x \in \mathbb{R}$ and all $k \geq 0$.

Corollary 2.1. *The choice $g(z) = 1 + z^2$ in Theorem 2.1 gives $f(z) \in \mathcal{LP}$ if and only if $L_k[f](x) \geq 0$ for all $x \in \mathbb{R}$ and all $k \geq 0$, as stated in Theorem 1.1.*

Proof of Theorem 2.1. Since $g(z)$ is an even polynomial, it follows that α_j is a root if and only if $-\alpha_j$ is a root with the same multiplicity. So,

$$\Phi(z,t) = \prod_{j=1}^{M} f(z + \alpha_j t) = \prod_{j=1}^{M} f(z - \alpha_j t) = \Phi(z,-t).$$

Hence, $A_k(z) \equiv 0$ for all odd k and we may write

$$\Phi(z,t) = \sum_{k=0}^{\infty} A_{2k}(z) t^{2k}.$$

Now assume $A_{2k}(x) \geq 0$ for all $x \in \mathbb{R}$ and all $k \geq 0$. Let $f(z) = e^{-bz^2} f_1(z)$ where $b \geq 0$ and $f_1(z)$ is a real entire function of genus 0 or 1, and assume $f(z)$ is not identically zero. Suppose, by way of contradiction, that $f(z)$ has a non-real root, say z_0. Let α_s be any fixed non-real root of $g(z)$ and write

$$z_0 = x_0 + \alpha_s t_0,$$

where both x_0 and t_0 are real. Then $f(z_0) = f(x_0 + \alpha_s t_0) = 0$, and

$$0 = \Phi(x_0, t_0) = \prod_{j=1}^{M} f(x_0 + \alpha_j t_0) = \sum_{k=0}^{\infty} A_{2k}(x_0) t_0^{2k}.$$

Assume $t \neq 0$. Then the nonnegativity of $A_{2k}(x_0)$ implies $A_{2k}(x_0) = 0$ for all k. This in turn implies $\Phi(x_0, t)$ is identically zero for all complex t. But that is false since $f(z)$ is a nonzero entire function. Therefore, $t_0 = 0$. Then $z_0 = x_0 + \alpha_s t_0 = x_0$ is also real, contradicting the choice of z_0. Thus, all the roots of $f(z)$ are real and $f(z) \in \mathcal{LP}$.

Conversely, assuming $f(z) \in \mathcal{LP}$, we will show that $A_{2k}(x) \geq 0$ for all $x \in \mathbb{R}$ and all $k \geq 0$. We will show this when $f(z)$ is a polynomial and the result for arbitrary $f(z) \in \mathcal{LP}$ will follow by taking limits. Let

$$f(z) = \prod_{i=1}^{n}(z + r_i)$$

where $r_1, \ldots, r_n$ are real. Calculating $\Phi(z, t)$ gives

$$\Phi(z, t) = \prod_{j=1}^{M} f(z + \alpha_j t) = \prod_{j=1}^{M}\prod_{i=1}^{n}(z + \alpha_j t + r_i)$$

$$= \prod_{i=1}^{n}\prod_{j=1}^{M}\left((z + r_i) + \alpha_j t\right) = \prod_{i=1}^{n}\sum_{\ell=0}^{M} c_\ell (z + r_i)^\ell t^{M-\ell}, \qquad (9)$$

where $g(z) = \prod_{j=1}^{M}(z + \alpha_j) = \sum_{\ell=0}^{M} c_\ell z^\ell$. Since $g(z)$ is an even polynomial, $c_\ell = 0$ for odd ℓ and

$$\Phi(z, t) = \prod_{i=1}^{n}\sum_{\ell=0}^{M/2} c_{2\ell}(z + r_i)^{2\ell} t^{M-2\ell} = \sum_{k=0}^{nM/2} A_{2k}(z) t^{2k}. \qquad (10)$$

From (10), $A_{2k}(z)$ is the sum of products of terms of the form $c_{2\ell}(z + r_i)^{2\ell}$. Because $c_{2\ell} \geq 0$ and $(z + r_i)^{2\ell}$ is a square, it follows that $A_{2k}(x) \geq 0$ for real x.

Now let $f(z) \in \mathcal{LP}$ be an arbitrary function that is not a polynomial. Then there exist polynomials $f_n(z) \in \mathcal{LP}$ such that

$$\lim_{n \to \infty} f_n(z) = f(z)$$

uniformly on compact sets. The derivatives also satisfy

$$\lim_{n \to \infty} f_n^{(k)}(z) = f^{(k)}(z)$$

uniformly on compact sets. If we write

$$\Phi_n(z, t) = \prod_{j=1}^{M} f_n(z + \alpha_j t) = \sum_{k=0}^{\infty} A_{n,2k}(z) t^{2k},$$

we see from (7) that

$$\lim_{n\to\infty} A_{n,2k}(z) = A_{2k}(z)$$

uniformly on compact sets. Since $A_{n,2k}(x) \geq 0$ for real x, the limit also satisfies this inequality. Thus, for arbitrary $f(z) \in \mathcal{LP}$, $A_{2k}(x) \geq 0$ for $x \in \mathbb{R}$ and $k \geq 0$, completing the proof of the theorem. $\qquad\square$

3. Discussion and examples

The function $A_k(z)$ in (7) is described in terms of the kth derivative of a product of entire functions. Either by using the generalized product rule for derivatives or by expanding each $f(z + \alpha_j t)$ as a series and multiplying series, one obtains the following formula for $A_k(z)$:

$$A_k(z) = \sum_{\lambda \vdash k} \frac{m_\lambda(\alpha_1, \ldots, \alpha_M)}{\lambda_1! \cdots \lambda_r!} f(z)^{M-r} \prod_{j=1}^{r} f^{(\lambda_j)}(z), \qquad (11)$$

where $\lambda \vdash k$ means that the sum is over all unordered partitions λ of k,

$$\lambda = (\lambda_1, \ldots, \lambda_r) \qquad k = \lambda_1 + \cdots + \lambda_r,$$

where r is the length of the partition λ, and where $m_\lambda(\alpha_1, \ldots, \alpha_M)$ is the monomial symmetric function of M variables for the partition λ evaluated at the roots $\alpha_1, \ldots, \alpha_M$.

The coefficients c_ℓ of $g(z) = \sum_{\ell=0}^{M} c_\ell z^\ell = \prod_{j=1}^{M}(z + \alpha_j)$ are elementary symmetric function of $\alpha_1, \ldots, \alpha_M$. The monomial symmetric functions $m_\lambda(\alpha_1, \ldots, \alpha_M)$ appearing in (11) can therefore be calculated in terms of $c_0, \ldots, c_M$ without direct reference to $\alpha_1, \ldots, \alpha_M$. We see that if $c_0, \ldots, c_M$ are real, then $m_\lambda(\alpha_1, \ldots, \alpha_M)$ is also real. However, in general $m_\lambda(\alpha_1, \ldots, \alpha_M)$ is not necessarily positive even if all the c_ℓ are positive. So, in the setting of Theorem 2.1, the type of summation appearing in (11) will typically involve both addition and subtraction, and the nonnegativity of $A_k(x)$ for real x is not directly obvious from this representation.

Example 3.1. Let $g(z) = 4 + z^4$ and let $f(z) \in \mathcal{LP}$. Then $\Phi(z, t)$ is

$$f(z+(1+i)t)f(z+(1-i)t)f(z+(-1+i)t)f(z+(-1-i)t) = \sum_{k=0}^{\infty} A_{2k}(z)t^{2k}.$$

A small calculation shows that

$$\tfrac{3}{2}A_4(x) = -f(z)^3 f^{(4)}(z) + 3f(z)^2 f''(z)^2 + 6f'(z)^4$$
$$+ 4f(z)^2 f^{(3)}(z)f'(z) - 12f(z)f'(z)^2 f''(z).$$

According to Theorem 2.1 this expression is nonnegative for all $x \in \mathbb{R}$.

Example 3.2. Let $f(z) = \prod_{i=1}^{n}(z + r_i)$ where $r_1, \ldots, r_n \in \mathbb{R}$ and let

$$g(z) = z^{2m} + 1 = \prod_{j=1}^{2m}(z + \omega^{2j-1})$$

where $\omega = \exp(2\pi i / 4m)$ and $m \in \mathbb{N}$. Calculating as in (9) gives

$$\Phi(z,t) = \prod_{i=1}^{n}\left((z + r_i)^{2m} + t^{2m}\right)$$

$$= f(z)^{2m} \prod_{i=1}^{n}\left(1 + \frac{t^{2m}}{(z + r_i)^{2m}}\right)$$

$$= f(z)^{2m} \sum_{k=0}^{n} e_k\left(\frac{1}{(z+r_1)^{2m}}, \ldots, \frac{1}{(z+r_n)^{2m}}\right)(t^{2m})^k,$$

where e_k is the kth elementary symmetric function of n variables evaluated at $(z + r_1)^{-2m}, \ldots, (z + r_n)^{-2m}$. Thus, if $x \in \mathbb{R}$,

$$A_{2mk}(x) = f(x)^{2m} e_k\left(\frac{1}{(x+r_1)^{2m}}, \ldots, \frac{1}{(x+r_n)^{2m}}\right)$$

is expressed as a sum of squares of real numbers and is therefore nonnegative. Dilcher and Stolarsky studied the zeros of $A_{2mk}(x)$. (See Prop. 2.3 and §3 of Ref. 5).

Example 3.3. This example illustrates how certain modifications to Theorem 2.1 are possible. Let $f(z)$ be a polynomial with *negative* roots. Then $f(z) = \prod_{i=1}^{n}(z + r_i)$ where each $r_i > 0$. Let

$$g(z) = 1 + z + z^2 = (z + e^{\pi i/3})(z + e^{-\pi i/3}).$$

Although $g(z)$ is not even as in the hypothesis of the theorem, its coefficients are nonnegative. Then

$$\Phi(z,t) = f(z + te^{\pi i/3})f(z + te^{-\pi i/3})$$

$$= \underbrace{f(z)^2}_{A_0(z)} + \underbrace{f(z)f'(z)}_{A_1(z)}\, t + \underbrace{\tfrac{1}{2!}\left(2f'(z)^2 - f(z)f''(z)\right)}_{A_2(z)} t^2$$

$$+ \underbrace{\tfrac{1}{3!}\left(3f'(z)f''(z) - 2f(z)f'''(z)\right)}_{A_3(z)} t^3$$

$$+ \underbrace{\tfrac{1}{4!}\left(6f''(z)^2 - 4f'(z)f^{(3)}(z) - f(z)f^{(4)}(z)\right)}_{A_4(z)} t^4 + \cdots$$

On the other hand, calculating as in (9) gives

$$\Phi(z, t) = \prod_{i=1}^{n} \left((z + r_i)^2 + (z + r_i)t + t^2 \right)$$

$$= f(z)^2 \prod_{i=1}^{n} \left(1 + \frac{t}{z + r_i} + \frac{t^2}{(z + r_i)^2} \right)$$

$$= f(z)^2 \sum_{k=0}^{2n} \left(\sum_{\substack{\lambda \vdash k \\ \lambda_j \leq 2}} m_\lambda\left(\tfrac{1}{z+r_1}, \ldots, \tfrac{1}{z+r_n}\right) \right) t^k,$$

where the inner sum is over all unordered partitions λ of k whose parts satisfy $\lambda_j \leq 2$ and where m_λ is the monomial symmetric function in n variables for the partition λ evaluated at $(z + r_1)^{-1}, \ldots, (z + r_n)^{-1}$. From the last expression, we see that each

$$A_k(x) \geq 0$$

for all $x \geq 0$ and all $k \geq 0$.

References

1. David A. Cardon and Adam Rich, *Turán inequalities and subtraction-free expressions*, JIPAM. J. Inequal. Pure Appl. Math. **9** (2008), no. 4, Article 91, 11 pp. (electronic).
2. Thomas Craven and George Csordas, *Iterated Laguerre and Turán inequalities*, JIPAM. J. Inequal. Pure Appl. Math. **3** (2002), no. 3, Article 39, 14 pp. (electronic).
3. George Csordas and Alain Escassut, *The Laguerre inequality and the distribution of zeros of entire functions*, Ann. Math. Blaise Pascal **12** (2005), no. 2, 331–345.
4. George Csordas and Richard S. Varga, *Necessary and sufficient conditions and the Riemann hypothesis*, Adv. in Appl. Math. **11** (1990), no. 3, 328–357.
5. Karl Dilcher and Kenneth B. Stolarsky, *On a class of nonlinear differential operators acting on polynomials*, J. Math. Anal. Appl. **170** (1992), no. 2, 382–400.
6. Boris Ja. Levin, *Distribution of zeros of entire functions*, revised ed., Translations of Mathematical Monographs, vol. 5, American Mathematical Society, Providence, R.I., 1980.
7. Merrell L. Patrick, *Extensions of inequalities of the Laguerre and Turán type*, Pacific J. Math. **44** (1973), 675–682.
8. Qazi I. Rahman and Gerhard Schmeisser, *Analytic theory of polynomials*, London Mathematical Society Monographs. New Series, vol. 26, The Clarendon Press Oxford University Press, Oxford, 2002.

New classes of stable polynomials and polynomials with real negative roots

Marios Charalambides

General Department, Frederick University,
Nicosia, 1303, Cyprus
E-mail: bus.chm@fit.ac.cy

The author presents a general stability theorem. Polynomials whose coefficients are successive derivatives of a class of orthogonal polynomials evaluated at $x = c$, where c is a constant, are shown to fit in this general framework. Special reference is made to the ones related to the classical orthogonal polynomials. Related families of polynomials with real negative roots are also introduced.

Keywords: Stable polynomials; positive pair; orthogonal polynomials.

1. Introduction

In two recent papers [1,2], new properties of a class of Jacobi polynomials[6] and generalized Laguerre polynomials[6] were presented. As a result, new classes of stable polynomials and polynomials with real negative roots were introduced. The purpose of this paper is to provide a general theorem on stability and to show how the previous results fit in this general framework.

Stability analysis plays an important role in many areas of pure and applied mathematics such as the Routh-Hurwitz problem [11], matrix analysis [10], operators preserving stability [5] and numerical analysis [3,4,7,8].

A real polynomial, $p(z)$, is said to be *stable* (or a *Hurwitz polynomial*), if all the zeros of $p(z)$ lie in the open left half-plane, $\Re(z) < 0$. The characterization of stable polynomials that is most useful for the purpose of this paper is given by the *Hermite-Biehler Theorem*[8] in terms of positive pairs, defined below.

Definition 1.1. Two real polynomials $\Omega_1(z)$ and $\Omega_2(z)$ of degree n and $n-1$ (or n), respectively, form a *positive pair* if: (a) the zeros $z_1, \ldots, z_n$ of Ω_1 and $z'_1, \ldots, z'_{n-1}$ (or $z'_1, \ldots, z'_n$) of Ω_2 are real, negative and distinct; (b) the zeros strictly interlace (or alternate) as follows: $z_1 < z'_1 < \cdots < z'_{n-1} <$

$z_n < 0$ (or $z_1' < z_1 < \cdots < z_n' < z_n < 0$); and (c) the highest coefficients of $\Omega_1(z)$ and $\Omega_2(z)$ are of like sign.

Theorem 1.1 (Hermite-Biehler Theorem). *A real polynomial $p(z) = \Omega_1(z^2) + z\Omega_2(z^2)$ is stable, if and only if $\Omega_1(z)$ and $\Omega_2(z)$ form a positive pair.*

An outline of this paper is as follows: In Section 2, a general stability theorem is introduced (Theorem 2.1). Next, it is shown that the theorem is valid for a family of Jacobi and generalized Laguerre polynomials. In Section 3, related families of polynomials that have only real negative zeros are proved with the aid of the Hermite-Biehler Theorem (Theorem 3.1).

2. New Classes of Stable Polynomials

Below, a general stability theorem is presented which is the main result of this paper.

Theorem 2.1. *Let $(a, b) \subseteq \mathbb{R}$, z a complex number, $z \neq 0$, and $W : (a, b) \to \mathbb{R}_{\geq 0}$ a differentiable weight function. Also, let $f(x; z) : (a, b) \to \mathbb{C}$ where $|f|^2, \frac{d}{dx}|f|^2 \in L_W^1$ and $f(x; z)$ is a solution to*

$$f(x; z) - z\frac{df(x; z)}{dx} = q(x) \tag{1}$$

where $q : (a, b) \to \mathbb{R}$. Further, suppose that

$$\int_a^b \Re\left(zf(x; z)\right) q(x)W(x)dx = -\lambda\Re(z) \tag{2}$$

for some $\lambda \geq 0$.

(1) If $W(x)$ is non-decreasing and
$$\lim_{x \to a^+} \{|f(x; z)|^2 W(x)\} \geq \lim_{x \to b^-} \{|f(x; z)|^2 W(x)\}, \text{ then } \Re(z) < 0 \text{ while,}$$
(2) If $W(x)$ is non-increasing and
$$\lim_{x \to a^+} \{|f(x; z)|^2 W(x)\} \leq \lim_{x \to b^-} \{|f(x; z)|^2 W(x)\}, \text{ then } \Re(z) > 0.$$

Proof. For simplicity $\overline{f}$ denote the complex conjugate of f. Rearrange equation (1) in the form:

$$\frac{1}{z}(f(x; z) - q(x)) = \frac{df(x; z)}{dx} \tag{3}$$

Multiply equation (3) by $W(x)\overline{f(x; z)}$ and add to this its complex conjugate to obtain:

$$\left(\frac{1}{z} + \frac{1}{\bar{z}}\right)|f|^2 W(x) - \left(\frac{\bar{f}}{z} + \frac{f}{\bar{z}}\right) q(x)W(x) = \frac{d|f|^2}{dx}W(x). \qquad (4)$$

Next, integrate equation (4) over (a,b) to get:

$$\left(\frac{1}{z} + \frac{1}{\bar{z}}\right)\int_a^b |f|^2 W(x)dx - \int_a^b \left(\frac{\bar{f}}{z} + \frac{f}{\bar{z}}\right) q(x)W(x)dx$$

$$= \int_a^b \frac{d|f|^2}{dx}W(x)dx. \qquad (5)$$

The conditions of Theorem 2.1 along with an integration by parts argument on the right hand side of equation (5) yields:

$$\frac{2\Re(z)}{|z|^2}\left(\int_a^b |f|^2 W(x)dx + \lambda\right) = \lim_{x \to b^-}\{|f(x;z)|^2 W(x)\}$$

$$- \lim_{x \to a^+}\{|f(x;z)|^2 W(x)\} - \int_a^b |f|^2 \frac{dW}{dx}dx. \qquad (6)$$

Since $\lambda \geq 0$, the bracket on the left hand side of equation (6) is positive. Hence, hypothesis *(a)* of the theorem asserts that the right hand side of equation (6) is negative and thus $\Re(z) < 0$, while hypothesis *(b)* asserts that the right hand side is positive and thus $\Re(z) > 0$. $\qquad \square$

Remark 2.1. If $q(x)$ is a polynomial of degree n then $f(x;z) = \sum_{k=0}^n q^{(k)}(x)z^k$, where $q^{(k)}(x) := \frac{d^k q(x)}{dx^k}$. In addition, if $c \in (a,b)$ and hypothesis *(a)* of Theorem 2.1 holds, the polynomial $f(c;z) = \sum_{k=0}^n q^{(k)}(c)z^k$ is stable. If instead hypothesis *(b)* of Theorem 2.1 holds, the polynomial $f(c;-z) = \sum_{k=0}^n q^{(k)}(c)(-z)^k$ is stable. Further, if $f(x;z)$ is defined for an interval larger than (a,b) where equation (1) is satisfied, the above hold for all c in the new interval.

Recall that the Jacobi polynomials are defined by $P_n^{(\alpha,\beta)}(x) := \frac{1}{2^n}\sum_{k=0}^n \binom{n+\alpha}{k}\binom{n+\beta}{n-k}(x-1)^{n-k}(x+1)^k$, where $\alpha, \beta > -1$ and are orthogonal over the interval $(-1,1)$ with respect to the weight function $w(x) = (1-x)^\alpha(1+x)^\beta$. They are generalizations of several families of orthogonal polynomials such as the Chebyshev polynomials of first and second kind ($\alpha = \beta = -\frac{1}{2}$ and $\alpha = \beta = \frac{1}{2}$, respectively), the Legendre polynomials ($\alpha = \beta = 0$), and the Gegenbauer (ultraspherical) polynomials ($\alpha = \beta = \nu - \frac{1}{2}$). The next theorem, which gives stable polynomials generated by a family of Jacobi polynomials[6] was originally proved in [2]. These polynomials fit the general framework of Theorem 2.1 with an additional condition.

Theorem 2.2.[2] *Let $P_n^{(\alpha,\beta)}(x)$ denote the Jacobi polynomial of degree n, where $n \geq 2$. Then the polynomial*

$$\phi_n^{(\alpha,\beta)}(z) := \sum_{k=0}^{n} z^k D^k P_n^{(\alpha,\beta)}(1) \tag{7}$$

is stable if $-1 < \alpha \leq 1$ and $\beta > -1$.

Proof. Let the interval of Theorem 2.1 be $(a,b) = (-1,1)$ and $q(x)$ of equation (1) be $q(x) = P_n^{(\alpha,\beta)}(x)$. By Remark 2.1, $f(x;z) = \sum_{k=0}^{n} P_n^{(\alpha,\beta)}(x)z^k$, which satisfies equation (1) for $x \in \mathbb{R}$. Also, define $W(x) = \frac{1+x}{1-x}w(x) = (1-x)^{\alpha-1}(1+x)^{\beta+1}$, where $w(x)$ is the weight function of the Jacobi polynomial. If we further demand that $f(1;z) = 0$, we deduce from [2, eq. 22, 23] that

$$\int_{-1}^{1} \Re(zf(x;z))\, W(x)\, P_n^{(\alpha,\beta)}(x)\, dx = -\Re(z) \int_{-1}^{1} \left(P_n^{(\alpha,\beta)}(x)\right)^2 w(x)\, dx. \tag{8}$$

Thus, $\lambda = \int_{-1}^{1} \left(P_n^{(\alpha,\beta)}(x)\right)^2 w(x)\, dx > 0$. Also notice that

$$\lim_{x\to-1^+} \{|f(x;z)|^2 W(x)\} = \lim_{x\to-1^+} \{|f(x;z)|^2(1-x)^{\alpha-1}(1+x)^{\beta+1}\} = 0 \tag{9}$$

for $\beta > -1$. Since $f(1;z) = 0$, deduce that

$$\lim_{x\to1^-} \{|f(x;z)|^2 W(x)\} = \lim_{x\to1^-} \{|f(x;z)|^2(1-x)^{\alpha-1}(1+x)^{\beta+1}\} = 0 \tag{10}$$

for $\alpha > -1$. Finally, since

$$\frac{d}{dx}\{W(x)\} = (2 - \alpha + \beta - (\alpha+\beta)x)\,(1-x)^{\alpha-2}(1+x)^{\beta} \geq 0 \tag{11}$$

for $-1 < \alpha \leq 1$ and $\beta > -1$, $W(x)$ is non-decreasing and by hypothesis *(a)* of Theorem 2.1 the result follows. $\square$

Remark 2.2. Theorem 2.2 ensures that the polynomial (7) is stable for the special cases of the Chebyshev polynomials of first and second kind, the Legendre polynomials and the Gegenbauer polynomials for $-\frac{1}{2} < \nu \leq \frac{3}{2}$.

A remarkable connection between the stable polynomial (7) and the Bessel polynomials was identified in [2]. Recall that the Bessel polynomial [6, p. 181] is defined as $y_n(x) := {}_2F_0(-n, 1+n; -; -\frac{x}{2}) =$

$\sum_{k=0}^{n} \frac{(n+k)!}{(n-k)!k!} \left(\frac{x}{2}\right)^k$. It is known that the Bessel polynomials are stable with simple zeros [9, Theorem 1, p. 74 and Corollary 2, p. 82]. A short proof of the stability of the Bessel polynomials was given in [2] by showing that they can also be written as $y_n(x) \equiv \sum_{k=0}^{n} P_n^{(k)}(1)\, x^k$, where $P_n(x)$ is the Legendre polynomial. Thus, the result followed from Theorem 2.2.

Recall that the generalized Laguerre polynomials are defined by $L_n^{(\alpha)}(x) := \sum_{k=0}^{n}(-1)^k \binom{n+\alpha}{n-k}\frac{x^k}{k!}$, where $\alpha > -1$. They are orthogonal over the interval $(0, \infty)$ with respect to the weight function $w(x) = x^\alpha e^{-x}$ and are generalizations of the classical Laguerre polynomials ($\alpha = 0$). The next theorem, which gives stable polynomials generated by a family of generalized Laguerre polynomials,[6] was originally proved in [1]. These polynomials fit the general framework of Theorem 2.1 with an additional condition.

Theorem 2.3.[1] *For each $n \geq 1$, the polynomial*

$$p_n^{(\alpha)}(z) = \sum_{k=0}^{n} D^k L_n^{(\alpha)}(0)\,(-z)^k = \sum_{k=0}^{n} \binom{n+\alpha}{n-k} z^k, \tag{12}$$

is stable for $-1 < \alpha \leq 1$.

Proof. Let the interval of Theorem 2.1 be $(a, b) = (0, \infty)$ and $q(x)$ of equation (1) be $q(x) = L_n^{(\alpha)}(x)$. By Remark 2.1, $f(x; z) = \sum_{k=0}^{n} D^k L_n^{(\alpha)}(x) z^k$, which satisfies equation (1) for $x \in \mathbb{R}$. Also, define $W(x) = x^{-1} w(x) = x^{\alpha-1} e^{-x}$, where $w(x)$ is the weight function of the generalised Laguerre polynomial. If we further demand that $f(0; z) = 0$, deduce that

$$\int_0^\infty \Re(z f(x; z))\, L_n^{(\alpha)}(x) W(x)\, dx = \int_0^\infty \Re\left(z\frac{f(x; z)}{x}\right) L_n^{(\alpha)}(x) x^\alpha e^{-x}\, dx = 0 \tag{13}$$

since $L_n^{(\alpha)}(x)$ is orthogonal to all polynomials with degree less than n. This suggests that $\lambda = 0$. Moreover, given that $f(0; z) = 0$ then

$$\lim_{x \to 0^+} \{|f(x; z)|^2 W(x)\} = \lim_{x \to 0^+} \{|f(x; z)|^2 x^{\alpha-1} e^{-x}\} = 0 \tag{14}$$

for $\alpha > -1$ while,

$$\lim_{x \to \infty} \{|f(x; z)|^2 W(x)\} = \lim_{x \to \infty} \{|f(x; z)|^2 x^{\alpha-1} e^{-x}\} = 0. \tag{15}$$

Finally, since

$$\frac{d}{dx}\{W(x)\} = (\alpha - 1 - x)\, x^{\alpha-1} e^{-x} \leq 0 \tag{16}$$

for $-1 < \alpha \leq 1$, $W(x)$ is non-increasing, and by hypothesis *(b)* of Theorem 2.1 the result follows. $\square$

Remark 2.3.[1] A very nice connection between stable polynomial (12) and the binomial series was identified in [1]. The authors showed that the reverse polynomial of $p_n^{(\alpha)}(z)$ defined by $b_n^{(\alpha)}(z) := p_n^{(\alpha)}\left(\frac{1}{z}\right)$, $\quad z \neq 0$, is given by $b_n^{(\alpha)}(z) = \sum_{k=0}^{n} \binom{n+\alpha}{k} z^k$. Notice that this is the nth partial sum of the binomial series $(1+z)^{n+\alpha}$, $|z| < 1$. Since the reverse polynomial of a stable polynomial is stable [1, sec. 3], Theorem 2.3 suggests that $b_n^{(\alpha)}(z)$ is stable for $-1 < \alpha \leq 1$.

Theorem 2.1 is quite general since it does not demand $q(x)$ to be a polynomial. Even in the case that $q(x)$ is a polynomial, the only known examples of functions f satisfying Theorem 2.1 are the ones described by Theorems 2.2 and 2.3. These examples need extra conditions to satisfy the theorem that compromize the end result as the constant c is just one of the endpoints. These observations have led to the formulation of the following open problem:

Open Problem 2.1. Determine whether there exists a function $f(x; z)$ that fits in the general framework of Theorem 2.1 with no additional conditions.

3. Polynomials with Real Negative Roots

With the aid of Remark 2.1 and Theorems 2.1 and Hermite-Biehler, we obtain next a family of polynomials that has only real and negative zeros.

Theorem 3.1. *Let $f(x; z)$ and $q(x)$ as in Theorem 2.1 where $q(x)$ is now a polynomial of degree n. If $f(c; z)$ is stable, then the polynomials*

$$\Phi_1(z) = \sum_{k=0}^{\left[\frac{n}{2}\right]} q^{(2k)}(c)\, z^k \quad and \quad \Phi_2(z) = \sum_{k=0}^{\left[\frac{n}{2}\right]} q^{(2k+1)}(c)\, z^k \qquad (17)$$

form a positive pair. If instead $f(c; -z)$ is stable then the polynomials $\Phi_1(z)$ and $-\Phi_2(z)$ form a positive pair.

The theorem follows from a direct application of the Hermite Biehler Theorem on the stable polynomials $f(c; z)$ and $f(c; -z)$, respectively.

Remark 3.1. For the special case of the Jacobi polynomials, Theorems 3.1 and 2.2 imply that the polynomials $\sum_{k=0}^{n} z^k D^{2k} P_n^{(\alpha,\beta)}(1)$ and $\sum_{k=0}^{n} z^k D^{2k+1} P_n^{(\alpha,\beta)}(1)$ form positive pairs, that is they have real negative roots that interlace. This result was originally proved in [2]. For the Gegenbauer case, it was also shown in [3] that the above polynomials obey a three-term recurrence relation plus a constant term that vanishes for $\nu = \frac{1}{2}$ (Legendre polynomials) and for $\nu = \frac{3}{2}$. This and the classical theorem of Favard [6, p.21] suggest that the latter form orthogonal polynomial sequences.

Remark 3.2. For the special case of the generalized Laguerre polynomials, Theorems 3.1 and 2.3 imply that the polynomials $\sum_{k=0}^{n} z^k D^{2k} L_n^{(\alpha)}(0)$ and $-\sum_{k=0}^{n} z^k D^{2k+1} L_n^{(\alpha)}(0)$ form positive pairs, as well. This result was originally proved in [1]. It was also shown that these polynomials also obey a three-term recurrence relation plus a constant term that vanishes for $\alpha = 0$ (Laguerre polynomials) and for $\alpha = 1$. For these cases the polynomials also form orthogonal polynomial sequences.

Acknowledgments

I thank the referee for providing constructive comments and help in improving the contents of this paper.

References

1. M. Charalambides and G. Csordas, *A New Property of a Class of Generalized Laguerre Polynomials, Submitted to Proc. Amer. Math. Soc.* (June 2009).
2. M. Charalambides, G. Csordas and F. Waleffe, *A New Property of a Class of Jacobi Polynomials, Proc. Amer. Math. Soc.* **133**, 3551–3560 (2005).
3. M. Charalambides and F. Waleffe, *Spectrum of the Jacobi tau approximation for the second derivative operator, SIAM J. Numer. Anal.* **46**, 280–294 (2007).
4. M. Charalambides and F. Waleffe, *Gegenbauer tau methods with and without spurious eigenvalues, SIAM J. Numer. Anal.* **47**, 48–68 (2008).
5. T. Craven and G. Csordas, *The Fox-Wright functions and Laguerre multiplier sequences, J. Math. Anal. Appl.* **314**, (2006).
6. T. S. Chihara, *An Introduction to Orthogonal Polynomials*, (Gordon and Breach Sci. Pub., New York, 1978).
7. D. Gottlieb, *The stability of pseudospectral-Chebyshev methods, Math. Comp.* **36**, 107–118 (1981).
8. D. Gottlieb and L. Lustman, *The spectrum of the Chebyshev collocation operator for the heat equation, SIAM J. Numer. Anal.* **20**, 909–921 (1983).
9. E. Grosswald, *Bessel Polynomials*, (Lecture Notes in Mathematics, **698**, Springer, Berlin, 1978).
10. O. Holtz, *Hermite-Biehler, Routh-Hurwitz, and total positivity, Linear Algebra Appl.* **372**, 105–110 (2003).
11. M. Marden, *Geometry of Polynomials*, (Math. Surveys no. 3, Amer. Math. Soc. Providence, RI, 1966).
12. Q. I. Rahman, G. Schmeisser, *Analytic Theory of Polynomials*, (Oxford University Press, 2002).

On a general concept of order of a meromorphic function

A. Fernandez Arias* and J. Perez Alvares

Matemáticas Fundamentales, U.N.E.D,
Madrid, 28040, Spain
** E-mail: afernan@mat.uned.es*
www.uned.es

A general notion of order for a meromorphic function is introduced. Peter T. Yu. Chern [2] considered the notion of logarithmic order to analyze the value distribution of meromorphic functions of order zero. This turns out to be a particular case of this more general concept of order which was already suggested in the classical literature by E. Lindelöf, G. Valiron and others. In this way we extend some of the conclusions of P.T. Yu. Chern.

Keywords: Meromorphic function; order; exceptional value.

1. Introduction

In [2] Peter Tien Yu Chern introduced the notion of logarithmic order of a meromorphic function with the object of a finer analysis of the value distribution of meromorphic functions of classical order zero. To this aim he compares the growth of the characteristic function $T\,(r,f)$ with the powers $(\log r)^{\lambda}$, instead of the powers r^{λ} as in the classical notion of order.

R. Nevanlinna [4] mentions that E. Lindelöf already considered as comparison functions those of the more general form

$$r^{\lambda}\,(\log r)^{\mu_1}\,(\log\log)^{\mu_2}\,...\,(\log_l r)^{\mu_l}\,...$$

and in connection with this he refers to works of P. Boutroux , O. Blumenthal and G. Valiron.

Here we retake these considerations about the notion of growth of a meromorphic function and relate it to the work of Peter Chern.

---

*Partially supported by the Grant MTM-090960 of the Spanish Ministry of Education.

2. $(l, \lambda, \mu^{l)})$ — orders of growth

Given $l \in \mathbb{N}$, $\lambda \in \mathbb{R}^+$, $\mu^{l)} = (\mu_1, ..., \mu_l) \in \mathbb{R}_+^l$, we shall say that a positive increasing function $S(r)$ defined for $r > 0$, is of order $(l, \lambda, \mu^{l)})$ order if

$$\limsup_{r \to \infty} \frac{S(r)}{r^\lambda (\log r)^{\mu_1} (\log \log r) \ldots (\log_l r)^\mu} < \infty \, ,$$

for every $\mu > \mu_l$, and

$$\limsup_{r \to \infty} \frac{S(r)}{r^\lambda (\log r)^{\mu_1} (\log \log r) \ldots (\log_l r)^\mu} = \infty \, ,$$

for every $\mu < \mu_l$, where $\log_l r$ is the $l - th$ iterated logarithmic, that is $\log_l r = \log (\log_{l-1} r)$.

It might happen for a given $S(r)$, that there is no λ or μ_l for which this condition holds. In the first case $S(r)$ is of infinite order in the classical sense, in the second case $S(r)$ is of finite order λ, but not of order $(l, \lambda, \mu^{l)})$ for any $\mu^{l)}$.

The following result gives a useful criterion for a positive increasing function to be of order $(l, \lambda, \mu^{l)})$.

Theorem 2.1. *Let $S(r)$ be a positive increasing function defined for $r > 0$. Then $S(r)$ is of order $(l, \lambda, \mu^{l)})$ if and only if the integral*

$$\int^\infty \frac{S(r)\, dr}{r^{\lambda+1} (\log r)^{\mu_1+1} (\log_2 r)^{\mu_2+1} \ldots (\log)^{\mu+1}} \, ,$$

is convergent for $\mu > \mu_l$ and divergent for $\mu < \mu_l$.

3. $(l, \lambda, \mu^{l)})$ — exponent of convergence

Let $f(z)$ be a meromorphic function in the plane and $a \in \widehat{\mathbb{C}} = \mathbb{C} \cup \{\infty\}$, we denote by $\{z_j(a)\}_{j \in \mathbb{N}}$ the sequence of $a-$points of $f(z)$, that is, the roots of $f(z) = a$, where

$$|z_j(a)| = r_j(a) \le r_{j+1}(a) = |z_{j+1}(a)| \, .$$

We shall say that the exponent of convergence of these $a-$points is $(l, \lambda(a), \mu^{l)}(a))$, where $l \in \mathbb{N}$, $\mu^{l)}(a) \in \mathbb{R}_+^l$, if the series

$$\sum_{j=1}^\infty \frac{1}{r_j(a)^{\lambda(a)} (\log r_j(a))^{\mu_1(a)} \ldots (\log_l r_j(a))^\mu}$$

converges for every $\mu > \mu_l(a)$ and diverges for every $\mu < \mu_l(a)$.

It is well-known that if a meromorphic function $f(z)$ is of finite order in the classical sense, then the order of $n(r, f = a)$, the counting function of $a-$points, equals the exponent of convergence of the $a-$points. From the following theorem follows that the situation is analogous for the $\left(l, \lambda, \mu^{l)}\right)$ $-$order and the $\left(l, \lambda(a), \mu^{l)}(a)\right)$ exponent of convergence of a.

Theorem 3.1. *If $f(z)$ is a meromorphic function in $\mathbb{C}$, $a \in \widehat{\mathbb{C}}$ and the $r_j(a)$ are the moduli of the $a-$ points as above then for $\left(l, \lambda, \mu^{l)}\right)$, with $\lambda > 0$, the series*

$$\sum_{j=1}^{\infty} \frac{1}{r_j(a)^{\lambda} \left(\log r_j(a)\right)^{\mu_1} \ldots \left(\log_l r_j(a)\right)^{\mu_l}}$$

and the integrals

$$\int^{\infty} \frac{n(t, f = a)\, dt}{t^{\lambda+1} \left(\log t\right)^{\mu_1} \ldots \left(\log_l t\right)^{\mu_l}}$$

and

$$\int^{\infty} \frac{N(t, f = a)\, dt}{t^{\lambda+1} \left(\log t\right)^{\mu_1} \ldots \left(\log_l t\right)^{\mu_l}},$$

are either simultaneously convergent or simultaneously divergent.

The case $\lambda = 0$ must be treated separately, in fact, we obtain the following result, which for $l = 1$ becomes Lemma 4.1 in Peter Tien-Yu Chern [2]

Theorem 3.2. *If $f(z)$ is a meromorphic function in $\mathbb{C}$, $a \in \widehat{\mathbb{C}}$ and the $r_j(a)$ are the moduli of the $a-$ points as above, then for $\left(l, \lambda = 0, \mu^{l)}\right)$, the convergence of the series*

$$(I) = \sum_{j=1}^{\infty} \frac{1}{\left(\log r_j(a)\right)^{\mu_1} \ldots \left(\log_l r_j(a)\right)^{\mu_l}} \tag{1}$$

$$= \lim_{J \to \infty} \sum_{j=1}^{J} \frac{1}{\left(\log r_j(a)\right)^{\mu_1} \ldots \left(\log_l r_j(a)\right)^{\mu_l}}, \tag{2}$$

implies the convergence of the integral

$$(II) = \int^{\infty} \frac{n(t, f = a)\, dt}{t \left(\log t\right)^{\mu_1+1} \left(\log_2 t\right)^{\mu_2} \ldots \left(\log_l t\right)^{\mu_l}} \tag{3}$$

$$= \lim_{R \to \infty} \int^{R} \frac{n(t, f = a)\, dt}{t \left(\log t\right)^{\mu_1+1} \left(\log_2 t\right)^{\mu_2} \ldots \left(\log_l t\right)^{\mu_l}}, \tag{4}$$

and the convergence of the previous integral implies the convergence of

$$(III) = \int^{\infty} \frac{N\left(t, f = a\right) dt}{t \left(\log t\right)^{\mu_1 + 1} \left(\log_2 t\right)^{\mu_2} \dots \left(\log_l t\right)^{\mu_l}} \tag{5}$$

$$= \lim_{R \to \infty} \int^{R} \frac{N\left(t, f = a\right) dt}{t \left(\log t\right)^{\mu_1 + 1} \left(\log_2 t\right)^{\mu_2} \dots \left(\log_l t\right)^{\mu_l}}. \tag{6}$$

Conversely the convergence of the integral (III), *implies*

$$\int^{R} \frac{n\left(t, f = a\right) dt}{t \left(\log t\right)^{\mu_1 + 1} \left(\log_2 t\right)^{\mu_2} \dots \left(\log_l t\right)^{\mu_l}} = o\left(\log R\right),$$

and the convergence of (II) *implies*

$$\sum_{j=1}^{J} \frac{1}{\left(\log r_j\left(a\right)\right)^{\mu_1} \dots \left(\log_l r_j\left(a\right)\right)^{\mu_l}} = o\left(\log r_J\left(a\right)\right).$$

4. On Borel exceptional values of meromorphic functions of $\left(l, \lambda, \mu^{l)}\right)$ order

For a meromorphic function in the plane $f\left(z\right)$ o finite classical order λ and for any $a \in \widehat{\mathbb{C}}$ the counting functions $n\left(r, f = a\right)$ and $N\left(r, f = a\right)$, both have the same order. Next, we describe this behaviour for meromorphic functions of finite $\left(l, \lambda, \mu^{l)}\right)$ order.

First of all, we show the following theorem

Theorem 4.1. *Let* $f\left(z\right)$ *be a non-constant meromorphic function in the plane of finite classical order* $\lambda > 0$. *Then the functions* $n\left(r, f = a\right)$ *and* $N\left(r, f = a\right)$ *are of the same order* $\left(l, \lambda, \mu^{l)}\right)$.

Proof. By Theorem (3.1), the integrals

$$\int^{\infty} \frac{n\left(t, f = a\right) dt}{t^{\lambda + 1} \left(\log t\right)^{\mu_1} \dots \left(\log_l t\right)^{\mu_l}},$$

and

$$\int^{\infty} \frac{N\left(t, f = a\right) dt}{t^{\lambda + 1} \left(\log t\right)^{\mu_1} \dots \left(\log_l t\right)^{\mu_l}},$$

are simultaneously convergent or simultaneously divergent. Then the statement of the theorem follows from Theorem 2.1. q.e.d. $\qquad \square$

Next we consider the case $\lambda = 0$.

Theorem 4.2. *Let $f(z)$ be a meromorphic function in the plane of zero order in the classical sense. Then the orders $\left(l, 0, \mu_n^{l)}\right), \left(l, 0, \mu_N^{l)}\right)$ of the functions $n(t, f = a), N(t, f = a)$ satisfy the following relation*

$$\mu_N^1 \leq \mu_n^1 + 1.$$

Proof. This follows from Theorem 2.1 and Theorem 3.2. q.e.d. $\qquad\square$

A value $a \in \widehat{\mathbb{C}}$ is called a Borel exceptional value of $f(z)$ if the order of $n(r, f = a)$ is less than the order of $f(z)$, i.e. if the inequality

$$\limsup_{r \to \infty} \frac{\log n(r, f = a)}{\log r} < \lambda,$$

holds for a, where λ is the order of $f(z)$.

We shall now define a $\left(l, \lambda, \mu^{l)}\right) -$ Borel exceptional value of a meromorphic function $f(z)$ as a value $a \in \widehat{\mathbb{C}}$ for which the $\left(l, \lambda(a), \mu^{l)}(a)\right) -$order of $n(r, f = a)$ satisfies

$$\lambda(a) < \lambda$$

or

$$\lambda(a) = \lambda \ , \ \mu_1(a) = \mu_1, ..., \mu_j(a) < \mu_j \ ,$$

for some $j = 1, ..., l$, in the case $\lambda > 0$ and

$$\lambda(a) = \lambda \ , \ \mu_1(a) = \mu_1 - 1, ..., \mu_j(a) < \mu_j \ ,$$

for some $j = 1, ..., l$ if $\lambda = 0$, where $\left(\lambda, l, \mu^{l)}\right)$ is the order of $f(z)$.

Theorem 4.3 (The $\left(l, \lambda, \mu^{l)}\right) -$ Borel exceptional value Theorem). *If $f(z)$ is a non-constant meromorphic function of $\left(\lambda_f, l, \mu_f^l\right)$ finite order, then $f(z)$ has at most two $\left(\lambda, l, \mu^{l)}\right) -$ Borel exceptional values.*

Proof. If $f(z)$ is a non-constant rational function then $f(z)$ has order $\left(l, 0, \mu^{l)} \equiv (1, 0, ...0)\right)$, whereas the order of $n(r, f = a)$ is zero for every $a \in \widehat{\mathbb{C}}$, so that the conclusion follows immediately.

Now we shall assume $f(z)$ to be a transcendental meromorphic function of finite $\left(\lambda, l, \mu^{l)}\right) -$order with $\lambda > 0$.

Suppose that there are three distinct $\left(\lambda, l, \mu^{l)}\right)$ $-$Borel exceptional values of $f(z)$, say a_i , $i = 1, 2, 3$ so that

$$\limsup_{r \to \infty} \frac{n(r, a_i)}{r^\lambda (\log r)^{\mu_1} \dots (\log_l r)^{\mu_l - \epsilon}} = 0 \ ,$$

then by Theorem 4.1, we also have

$$\limsup_{r \to \infty} \frac{N(r, a_i)}{r^\lambda (\log r)^{\mu_1} \dots (\log_l r)^{\mu_l - \epsilon}} = 0 \ .$$

The Second Fundamental Theorem of Nevanlinna asserts

$$T(r, f) \leq \sum_{i=1}^{3} N(r, a_i) - N_1(r, f) + S(r, f) \ ,$$

where

$$N_1(r, f) \geq 0 \text{ and } S(r, f) = o(T(r, f)) \ ,$$

as $r \to \infty$, so that

$$T(r, f)(1 + o(1)) \leq \sum_{i=1}^{3} N(r, f = a_i) \tag{7}$$

$$\leq Cr^\lambda (\log r)^{\mu_1} \dots (\log_l r)^{\mu_l - \epsilon} \ , \tag{8}$$

what contradicts the fact that $f(z)$ is of $\left(\lambda, l, \mu^{l)}\right)$ order.

Finally, we shall consider $f(z)$ to be a transcendental meromorphic function of order $\left(l, \lambda = 0, \mu^{l)}\right)$, and again assume that there are three distinct $\left(l, \lambda = 0, \mu^{l)}\right)$ $-$Borel exceptional values a_i $i = 1, 2, 3$ of $f(z)$ so that

$$\limsup_{r \to \infty} \frac{n(r, a_i)}{(\log r)^{\mu_1 - 1} \dots (\log_l r)^{\mu_l - \epsilon}} = 0 \ , i = 1, 2, 3,$$

then recalling again Theorem 4.2, we get

$$\limsup_{r \to \infty} \frac{N(r, a_i)}{(\log r)^{\mu_1} \dots (\log_l r)^{\mu_l - \epsilon}} = 0 \ , i = 1, 2, 3, \dots$$

Again we make use of the Second Main Theorem of Nevanlinna and obtain

$$T(r, f)(1 + o(1)) \leq \sum_{i=1}^{3} N(r, f = a_i) \tag{9}$$

$$\leq Cr^\lambda (\log r)^{\mu_1} \dots (\log_l r)^{\mu_l - \epsilon} , \tag{10}$$

what again contradicts the fact that $f(z)$ is of $\left(\lambda, l, \mu^{l)}\right)$ order. q.e.d. $\square$

References

1. M. Cartwright, *Integral functions*, Cambridge University Press, 1956.
2. T. Y. Peter Chern, On meromorphic functions with finite logarithmic order, *Transactions of the American Math. Soc.* **358** (2005) 2, 473-489.
3. W. K. Hayman, *Meromorphic Functions*, Oxford Clarendon Press, 1975.
4. R. Nevanlinna, *Le Théoreme de Picard-Borel et la Theorie des Fonctions Meromorphes*, Chelsea Publishing Company, New York, 1974.

Séries universelles construites à l'aide de la fonction zeta de Riemann

A. Poirier

*Département de mathématiques et statistiques, Université de Montréal,
Montréal, Québec H3C 3J7, Canada
Courriel: apoirier@dms.umontreal.ca
www.umontreal.ca*

À partir de la fonction zeta, on construit une série de puissance dans $\mathbb{C}^n$, centrée en un point a, dont les sommes partielles approchent, sur tout compact polynômialement convexe ne contenant pas le point a, toute fonction holomorphe.

Mots-Clés: Séries universelles; fonction zeta de Riemann.

1. Théorème d'universalité de Voronin

Cet article se base sur le théorème de S. M. Voronin [4] suivant.

Théorème 1.1 (Voronin).

Soit la bande critique $\Omega = \{z \in \mathbb{C} : \frac{1}{2} < \Re(z) < 1\}$. Pour tout $a \in \Omega$ et pour tout $n \in \mathbb{N}$, la courbe

$$\left\{ \left(\zeta(a + \imath t), \zeta'(a + \imath t), ..., \zeta^{(n)}(a + \imath t) \right) : t \in \mathbb{R} \right\}$$

est dense dans $\mathbb{C}^{n+1}$.

Dans le livre de Karatsuba et Voronin [3], ce théorème est démontré comme corollaire du grand théorème d'universalité de Voronin [5]. Ce théorème très puissant affirme que pour tout compact $K \subset \Omega$, K^C connexe, et pour toute fonction holomorphe et n'ayant pas de zéros sur K, il existe une suite $\{\tau_j\}$ de nombres réels telle que

$$\zeta(z + \imath \tau_j) \longrightarrow g(z)$$

uniformément sur K. En effet, si on veut une démonstration plus simple du théorème 1.1, soit $(y_0, y_1, ..., y_n)$, $y_0 \neq 0$, et soit f le polynôme

$$f(z) = y_0 + y_1(z - a) + y_2 \frac{(z - a)^2}{2} + ... + y_n \frac{(z - a)^n}{n!}.$$

Soit K un disque fermé dans Ω centré en a sur lequel f n'a pas de zéros. Par le grand théorème de Voronin [5], il existe une suite de nombres réels $\{t_j\}$ telle que

$$\zeta(z + \imath t_j) \longrightarrow f(z)$$

uniformément sur K. Si on considère un disque fermé $K' \subset K^0$, K^0 étant l'intérieur de K, tel que $a \in K'$, on a la convergence des dérivées de ζ vers les dérivées de f. En particulier, on a que

$$\zeta^{(k)}(a + \imath t_j) \longrightarrow f^{(k)}(a)$$

pour $k = 0, 1, ..., n$. Mais, par la construction de f, on a que $f^{(k)}(a) = y_k$ et on a ainsi démontrer que le théorème 1.1 est en fait un corollaire du grand théorème de Voronin [5].

2. Approximation dans $\mathbb{C}$

À l'aide de ce théorème, on pourra améliorer le théorème de P.M. Gauthier et R. Clouâtre [2], qui dit que les translatés de polynômes de Taylor de la fonction ζ de Riemann peuvent approcher, sur tout compact K tel que son complément est connexe, toute fonction holomorphe sur ce dernier. Notons par $H(K)$ l'espace des fonctions holomorphes au voisinage de K.

Théorème 2.1.

Soit $\epsilon > 0$. Pour tout compact K tel que K^C est connexe, pour toute fonction $g \in H(K)$ pour tout $a \in \Omega$, il existe $\tau \in \mathbb{R}$ et $n \in \mathbb{N}$ tels que

$$\left| g(z) - \sum_{k=0}^{n} \frac{\zeta^{(k)}(a + \imath\tau)}{k!}(z - a)^k \right| < \epsilon, \quad \forall z \in K.$$

Notons d'abord, en remplaçant $z - a$ par $(z + \imath\tau) - (a + \imath\tau)$, que ce polynôme d'approximation est le polynôme de Taylor de ζ, centré en $a + \imath\tau$ et évalué en $z + \imath\tau$.

Preuve.

Fixons d'abord $\epsilon > 0$, K un compact du plan complexe et $a \in \Omega$. Puisque l'ensemble des polynômes est dense dans $H(K)$, il est suffisant de démontrer le théorème pour les polynômes. Soit p un polynôme. On peut alors trouver un $n \in \mathbb{N}$ tel que

$$p(z) = \sum_{k=0}^{n} \frac{p^{(k)}(a)}{k!}(z-a)^k.$$

Considérons maintenant le point $\big(p(a), p'(a), ..., p^{(n)}(a)\big)$. Puisqu'il s'agit d'un point dans $\mathbb{C}^{n+1}$, on peut appliquer le théorème 1.1 et en conclure qu'il existe une suite $\{\tau_j\}$, $\tau_j \in \mathbb{R}$ telle que

$$\zeta^{(k)}(a + \imath\tau_j) \longrightarrow p^{(k)}(a)$$

pour $k = 0, 1, ..., n$, et on en déduit la conclusion du théorème.

La famille des polynômes n'est pas dense dans $H(K)$ si K^C n'est pas connexe. Par contre, les fonctions rationnelles le sont et nous utiliserons cela afin de démontrer le théorème qui suit.

Théorème 2.2.

Soit $\epsilon > 0$. Pour tout compact $K \subset \mathbb{C}$, pour tout $a \in \Omega$ et pour toute fonction $g \in H(K)$, il existe $n,m \in \mathbb{N}$ $\tau,\gamma \in \mathbb{R}$ tels que

$$\left| g(z) - \left(\sum_{k=0}^{n} \frac{\zeta^{(k)}(a + \imath\tau)}{k!}(z-a)^k \Big/ \sum_{k=0}^{m} \frac{\zeta^{(k)}(a + \imath\gamma)}{k!}(z-a)^k \right) \right| < \epsilon$$

pour tout $z \in K$.

Preuve.

Par le théorème de Runge, il suffit d'approximer les fonctions rationnelles p/q, où q est sans zéros sur K. Puisque g n'a pas de zéros sur K, on peut supposer aussi que p n'a pas de zéros sur K. Le théorème découle alors du théorème précédent.

3. Approximation dans $\mathbb{C}^n$

Dans $\mathbb{C}^n$, nous allons maintenant démontrer qu'un polynôme construit à partir d'un produit de n fonctions ζ peut approcher toute fonction holomorphe sur n'importe quel compact $K \subset \mathbb{C}^n$ pour lequel les polynômes sont denses dans l'espace $H(K)$ des fonctions holomorphes sur K. Nous appelerons ces compacts des compacts de Runge. Définissons d'abord la fonction $Z : \mathbb{C}^n \longrightarrow \mathbb{C}$ de la manière suivante

$$Z(z) = \zeta(z_1)\zeta(z_2)...\zeta(z_n)$$

où $z = (z_1, z_2, ..., z_n) \in \mathbb{C}^n$. Soit le multi-indice $\alpha = (\alpha_1, \alpha_2, ..., \alpha_n)$, $\alpha_j \in \mathbb{N} \cup \{0\}$ pour $j = 1, 2, ..., n$. Il s'en suit que

$$Z^{(\alpha)} = \zeta^{(\alpha_1)}(z_1)\zeta^{(\alpha_2)}(z_2)...\zeta^{(\alpha_n)}(z_n).$$

Notons d'abord $\alpha! = \alpha_1!\alpha_2!...\alpha_n!$ et $z^\alpha = z_1^{\alpha_1} z_2^{\alpha_2}...z_n^{\alpha_n}$. Nous noterons aussi la longueur de α comme étant $|\alpha| = \alpha_1 + \alpha_2 + ... + \alpha_n$. Avant d'arriver au théorème, on introduit la notion de polynôme dispersé comme suit.

Définition 3.1.

Un polynôme dispersé de Z de degré m est de la forme

$$\sum_{|\alpha| \leq m} \frac{Z^{(\alpha)}(a + z_\alpha)}{\alpha!}(z - a)^\alpha$$

où $a \in \mathbb{C}^n$ et $z_\alpha \in \mathbb{C}^n$. On dit que le polynôme est dispersé sur un ensemble E si $\{a + z_\alpha : |\alpha| \leq m\} \subset E$.

Passons maintenant au théorème d'approximation par la fonction Z de fonctions à plusieurs variables. Soit $\Omega^n = \{z \in \mathbb{C}^n : \frac{1}{2} < \Re(z_k) < 1, k = 1, 2, ..., n\}$.

Théorème 3.1.

Pour tout $a \in \Omega^n$, les polynômes dispersés de Z sur le n-plan $\{z \in \mathbb{C}^n : \Re(z) = \Re(a)\}$ sont denses dans les polynômes sur tout compact de $\mathbb{C}^n$.

Preuve.

Fixons d'abord $\epsilon > 0$, $a \in \Omega^n$, K un compact de $\mathbb{C}^n$ et p un polynôme à n variables. Alors, il existe un m tel que

$$p(z) = \sum_{|\alpha| \leq m} \frac{p^{(\alpha)}(a)}{\alpha!} (z - a)^\alpha$$

pour tout $z \in \mathbb{C}^n$. Pour chaque α, on peut, par le théorème de Voronin [4], trouver une suite $\{\tau_{\alpha,j}\}_{j=1}^\infty$, $\tau_{\alpha,j} \in \mathbb{R}^n$, telle que

$$Z^{(\alpha)}(a + \imath\tau_{\alpha,j}) \longrightarrow p^{(\alpha)}(a). \tag{1}$$

Par exemple, puisque Z est le produit de n fonction ζ, on peut définir la première coordonnée de $\tau_{\alpha,j}$, notée τ_{α,j_1}, telle que

$$\zeta^{(\alpha_1)}(a_1 + \imath\tau_{\alpha,j_1}) \longrightarrow p^{(\alpha)}(a)$$

et ensuite définir les $n - 1$ autres coordonnées de manière à ce que

$$\zeta^{(\alpha_k)}(a_k + \imath\tau_{\alpha,j_k}) \longrightarrow 1$$

pour $k = 2, 3, ..., n$. Ainsi, on obtiendrait (1), ce qui mène à la conclusion du théorème.

Spécifiquement sur les compacts de Runge, nous avons le corollaire suivant:

Corollaire 3.1.

Pour tout compact de Runge $K \subset \mathbb{C}^n$, pour tout $a \in \Omega^n$, les polynômes dispersés de Z sur le n-plan $\{z \in \mathbb{C}^n : \Re(z) = \Re(a)\}$ sont denses dans $H(K)$.

Remarquons qu'il est possible, pour plusieurs $a \in \Omega^n$, d'approcher tout polynôme par des polynômes dispersés de Z sur une droite de Ω^n. Avant de passer au théorème, définissons d'abord le point $e^j \in \mathbb{C}^n$, où $e^j_k = 0$ si $k \neq j$ et $e^j_j = 1$. En effet, on a le théorème suivant pour tout $a \in \Omega^n$ tel que $Z^\alpha(a) \neq 0$ pour tout α.

Théorème 3.2.

Pour presque tout $a \in \Omega^n$, pour tout $j \in \{1, 2, ..., n\}$, les polynômes dispersés de Z sur la droite $\{z \in \mathbb{C}^n : \Re(z_j) = \Re(a_j), z_k = 0, k = 1, 2, ..., j-1, j+1, ..., n\}$ sont denses dans les polynômes sur tout compact de $\mathbb{C}^n$.

Similairement au théorème 3.1, on a le corollaire suivant:

Corollaire 3.2.

Pour presque tout $a \in \Omega^n$ et pour pour tout compact de Runge $K \subset \mathbb{C}^n$, les polynômes dispersés de Z sur la droite $\{z \in \mathbb{C}^n : \Re(z) = \Re(a_k),\ z_j = 0\ si\ j \neq k\}$ sont denses dans l'espace des fonctions holomorphes $H(K)$.

4. Séries universelles

Nous allons terminer cet article avec un théorème inspiré d'un théorème de Clouâtre [1]. Dans sa thèse, Clouâtre démontre le lemme suivant.

Lemme 4.1.

Soit $K \subset \mathbb{C}^n$ un sous-ensemble compact polynômialement convexe avec $0 \in K^C$. Alors, pour tout nombre naturel k, pour toute fonction $g \in H(K)$, il existe un polynôme $r(z)$ tel que son monôme de plus petit degré est de degré au moins k et tel que

$$sup_K |r(z) - g(z)| < 2^{-(k+2)}.$$

Par un simple changement de variable, nous avons le lemme suivant.

Lemme 4.2.

Soit $K \subset \mathbb{C}^n$ un sous-ensemble compact polynômialement convexe avec $a \in K^C$. Alors, pour tout nombre naturel k, pour toute fonction $g \in H(K)$, il existe une approximation polynômiale p de g de la forme suivante

$$sup_K \left| \sum_{k \leq |\alpha| \leq m} \frac{p^{(\alpha)}(a)}{\alpha!}(z - a)^\alpha - g(z) \right| < 2^{-(k+2)}.$$

En combinant ce lemme avec la preuve du théorème 3.2, on a le lemme suivant.

Lemme 4.3.

Soit $a \in \Omega^n$ comme dans le théorème 3.2 et soit $K \subset \mathbb{C}^n$ un sous-ensemble compact polynômialement convexe avec $a \in K^C$. Alors, pour tout nombre naturel k, pour toute fonction $g \in H(K)$, il existe une approximation polynômiale p de g de la forme suivante

$$\sup_K \left| \sum_{k \leq |\alpha| \leq m} \frac{Z^{(\alpha)}(a + \imath \tau_\alpha e^j)}{\alpha!} (z - a)^\alpha - g(z) \right| < 2^{-(k+2)}.$$

Clouâtre a montré l'existence de séries universelles dans $\mathbb{C}^n$. En suivant la preuve de Clouâtre avec l'aide des lemmes précedent, on en tire l'existence d'une série de Taylor dispersée de la fonction Z qui est universelle, dispersée sur un ensemble dénombrables de points sur une droite.

Théorème 4.1.

Pour presque tout $a \in \Omega^n$, pour tout $j = 1, 2, \cdots, n$, il existe des τ_α telles que, pour tout compact $K \subset \mathbb{C}^n$ polynômialement convexe tel que $a \notin K$, et pour toute fonction $f \in H(K)$, il existe une suite de nombres naturels $\{n_k\}$ telle que

$$\sum_{|\alpha| \leq n_k} \frac{Z^{(\alpha)}(a + \imath \tau_\alpha e^j)}{\alpha!} (z - a)^\alpha \longrightarrow f(z)$$

uniformément sur K lorsque $k \longrightarrow \infty$.

Pour $n = 1$, la fonction Z est en fait la fonction ζ de Riemann. Nous avons donc l'universalité de la fonction zeta de Riemann à la manière du théorème 4.1.

Remerciements

Recherche subventionnée par le CRSNG, sous la direction de P.M. Gauthier.

References

1. R. Clouâtre, *Universal Series in $\mathbb{C}^N$*, Canad. Math. Bull. (accepté).
2. P. M. Gauthier, R. Clouâtre, Approximation by translates of Taylor polynomials of the Riemann zeta function, *Comput. Methods Funct. Theory* **8** (2008), no. 1–2, 15–19.
3. A. A. Karatsuba and S. M. Voronin, *The Riemann Zeta-Function.*, Hawthorn, NY: de Gruyter, 1992.
4. S. M. Voronin, Ω-theorems of the theory of the Riemann zeta-function, (Russe) *Izv. Akad. Nauk SSSR Ser. Mat.* **52** (1988), no.2, 424–436, 448.
5. S. M. Voronin, A theorem on the "universality" of the Riemann zeta-function, (Russe) *Izv. Akad. Nauk SSSR Ser. Mat.* **39** (1975), no. 3, 475–486, 703.

II.1. Clifford and quaternion analysis

Organisers: I. Sabadini, F. Sommen

The session "Clifford and quaternion analysis" hosted contributions in the fields of theoretical quaternionic and Clifford analysis and, more in general, hypercomplex analysis. By hypercomplex analysis we intend the study of the function theory related to the Dirac operator and systems of partial differential operators taking values in a Clifford algebra. The talks presented in the session covered topics varying from the study of monogenic functions, its generalisations to higher spin operators such as the Rarita-Schwinger system, Clifford analysis on superspace, Clifford-Radon and Fourier transforms. Some talks also discussed the relationship between Dirac or Dirac type operators with representation theory and Physics. All together the contributions of the 36 participants to the session gave the most updated state-of-the-art of the researches in this field.

A higher order integral representation formula in isotonic Clifford analysis

Juan Bory Reyes

Department of Mathematics, Universidad de Oriente,
Santiago de Cuba 90500, Cuba
E-mail: jbory@rect.uo.edu.cu

Helmuth R. Malonek and Dixan Peña Peña

Department of Mathematics, Aveiro University,
3810-193 Aveiro, Portugal
E-mails: hrmalon@ua.pt,
dixanpena@ua.pt; dixanpena@gmail.com

Frank Sommen

Department of Mathematical Analysis, Ghent University,
9000 Gent, Belgium
E-mail: fs@cage.ugent.be

Let Δ be the Laplace operator in $\mathbb{R}^{2n}$. The aim of this paper is to present an integral representation formula for the solutions of the generalized isotonic system

$$\underline{\partial}_1 \Delta^k f + i\Delta^k \tilde{f}\underline{\partial}_2 = 0, \quad k \in \mathbb{N}_0,$$

where $\underline{\partial}_1$, $\underline{\partial}_2$ are Dirac type operators and where $\tilde{f}$ stands for the main involution in the complex Clifford algebra $\mathbb{C}_n$. Two special cases of this formula are also discussed, yielding generalizations of the Bochner-Martinelli formula for the holomorphic functions and for the biregular functions.

Keywords: Clifford algebras; Δ^k-isotonic functions; integral representation formula.

1. Polymonogenic functions

Let $\mathbb{C}_m$ be the complex Clifford algebra constructed over the orthonormal basis $(e_1, \ldots, e_m)$ of the Euclidean space $\mathbb{R}^m$ (see Ref. 7). The multiplication in $\mathbb{C}_m$ is determined by the relations $e_j e_k + e_k e_j = -2\delta_{jk}$, $j, k = 1, \ldots, m$, where δ_{jk} is the Kronecker delta. A general element of $\mathbb{C}_m$ is of the form $a = \sum_A a_A e_A$, $a_A \in \mathbb{C}$, where for $A = \{j_1, \ldots, j_k\} \subset \{1, \ldots, m\}$, $j_1 < \cdots <$

j_k, $e_A = e_{j_1} \ldots e_{j_k}$. For the empty set $\emptyset$, we put $e_\emptyset = 1$, the latter being the identity element. If the coefficients a_A are real rather than complex, then the respective algebra is denoted by $\mathbb{R}_{0,m}$.

Notice that any $a \in \mathbb{C}_m$ may also be written as $a = \sum_{k=0}^{m} [a]_k$ where $[a]_k$ is the projection of a on $\mathbb{C}_m^{(k)}$. Here $\mathbb{C}_m^{(k)}$ denotes the subspace of k-vectors defined by

$$\mathbb{C}_m^{(k)} = \left\{ a \in \mathbb{C}_m : a = \sum_{|A|=k} a_A e_A, \quad a_A \in \mathbb{C} \right\}.$$

In particular, $\mathbb{C}_m^{(0)}$ and $\mathbb{C}_m^{(1)}$ are called, respectively, the space of scalars and vectors in $\mathbb{C}_m$. The main involution $a \to \tilde{a}$ and the conjugation $a \to \bar{a}$ are respectively given by

$$\tilde{a} = \sum_A a_A \tilde{e}_A, \quad \tilde{e}_A = (-1)^{|A|} e_A,$$

$$\bar{a} = \sum_A \bar{a}_A \bar{e}_A, \quad \bar{e}_A = (-1)^{\frac{|A|(|A|+1)}{2}} e_A,$$

where $\bar{a}_A$ stands for the classical complex conjugation. One easily checks that $\widetilde{ab} = \tilde{a}\tilde{b}$ and $\overline{ab} = \bar{b}\bar{a}$, for any $a, b \in \mathbb{C}_m$. By means of the conjugation, a norm $|a|$ may be defined for each $a \in \mathbb{C}_m$ by putting $|a|^2 = [a\bar{a}]_0 = \sum_A |a_A|^2$.

The product of two Clifford vectors $\underline{x} = \sum_{j=1}^{m} x_j e_j$ and $\underline{y} = \sum_{j=1}^{m} y_j e_j$ splits into a scalar and a 2-vector

$$\underline{x}\,\underline{y} = \underline{x} \bullet \underline{y} + \underline{x} \wedge \underline{y},$$

where $\underline{x} \bullet \underline{y} = -\langle \underline{x}, \underline{y} \rangle = -\sum_{j=1}^{m} x_j y_j$ equals, up to a minus sign, the standard Euclidean inner product between $\underline{x}$ and $\underline{y}$, while $\underline{x} \wedge \underline{y} = \sum_{j=1}^{m} \sum_{k=j+1}^{m} e_j e_k (x_j y_k - x_k y_j)$ represents the standard outer (or wedge) product between them.

Let us denote by $\underline{\partial} = \sum_{j=1}^{m} e_j \partial_{x_j}$ the Dirac operator. It factorizes the Laplace operator Δ in $\mathbb{R}^m$, i.e.

$$\Delta = \sum_{j=1}^{m} \partial_{x_j}^2 = -\underline{\partial}^2. \tag{1}$$

The fundamental solution of $\underline{\partial}$ is the so-called Cauchy kernel

$$E(\underline{x}) = -\frac{1}{\omega_m} \frac{\underline{x}}{|\underline{x}|^m}, \quad \underline{x} \in \mathbb{R}^m \setminus \{0\},$$

where ω_m is the area of the unit sphere in $\mathbb{R}^m$.

A $\mathbb{C}_m$-valued function $f \in C^k(\Omega)$, $k \in \mathbb{N}$, is called (left) k-monogenic in the open set $\Omega \subset \mathbb{R}^m$ if and only if it fulfills in Ω the equation $\underline{\partial}^k f = 0$ (see Refs. 2,8,11,12,15,16). It is worth mentioning that for the particular case $k = 1$, we get the well-known monogenic functions which generalize to higher dimension the holomorphic functions in the complex plane (see e.g. Refs. 4,9).

In order to work with sets with very general boundaries, integration will be carried out with respect to the $(m-1)$-dimensional Hausdorff measure $\mathcal{H}^{m-1}$; and we will use the general exterior unit normal vector introduced by Federer (see e.g. Ref. 10). Throughout the paper, Ω will stand for a simply connected bounded and open set in $\mathbb{R}^m$ with boundary S such that $\mathcal{H}^{m-1}(S) < \infty$. The exterior unit normal vector on S at the point $\underline{y} \in S$ defined by Federer, will be denoted by $\underline{\nu}(\underline{y}) = \sum_{j=1}^{m} \nu_j(\underline{y})e_j$.

From now on, for simplicity (but without loss of generality) we will assume that $k < m$. Let us now consider the following finite sequence of functions $\{E_j(\underline{x})\}$, $j = 1, \ldots, k$, defined on $\mathbb{R}^m \setminus \{0\}$ and satisfying $\underline{\partial} E_j(\underline{x}) = E_{j-1}(\underline{x})$ with $E_1(\underline{x}) = E(\underline{x})$ (see e.g. Refs. 3,16). Specifically,

$$E_j(\underline{x}) = \begin{cases} c_j \dfrac{\underline{x}}{|\underline{x}|^{m-j+1}}, & \text{if } j \text{ odd,} \\[2ex] c_j \dfrac{1}{|\underline{x}|^{m-j}}, & \text{if } j \text{ even,} \end{cases}$$

where c_j are real constants given by

$$c_j = \begin{cases} -\dfrac{c_{j-1}}{j-1}, & \text{if } j \text{ odd,} \\[2ex] -\dfrac{c_{j-1}}{m-j}, & \text{if } j \text{ even,} \end{cases}$$

$j = 2, \ldots, k$. Clearly, $c_1 = -1/\omega_m$.

We end this section with the integral representation formula for the polymonogenic functions (see Refs. 1,2,8,16).

Theorem 1.1. *Suppose that $f \in C^{k-1}(\Omega \cup S)$. If f is k-monogenic in Ω, then*

$$f(\underline{x}) = \sum_{j=1}^{k}(-1)^{j-1} \int_S E_j(\underline{y} - \underline{x})\underline{\nu}(\underline{y})\underline{\partial}^{j-1} f(\underline{y})d\mathcal{H}^{m-1}(\underline{y}), \quad \underline{x} \in \Omega.$$

We notice that this theorem was stated in Ref. 16 assuming that S is a piecewise C^1 compact surface. But with the help of the general version of the Gauss-Green Theorem provided by Federer (see Ref. 10) and using the same proof given in Ref. 16, it follows that Theorem 1.1 is still valid for surfaces satisfying $\mathcal{H}^{m-1}(S) < \infty$.

2. Generalized isotonic functions

Let us now assume that the dimension of $\mathbb{R}^m$ is even. Let $m = 2n$ and denote by $\mathbb{C}_n$ the complex Clifford algebra generated by $(e_1, \ldots, e_n)$. Next, we introduce the primitive idempotent $I = \prod_{j=1}^{n} I_j$, with $I_j = \frac{1}{2}(1 + i e_j e_{n+j})$, $j = 1, \ldots, n$. It easily follows that the following conversion relations hold

$$e_{n+j} I = i e_j I, \quad j = 1, \ldots, n, \tag{2}$$

and for $a \in \mathbb{C}_n$ we also have that

$$aI = 0 \Leftrightarrow a = 0. \tag{3}$$

Below, we will need the following Clifford vectors and their corresponding Dirac operators:

$$\underline{x}_1 = \sum_{j=1}^{n} x_j e_j, \qquad \underline{\partial}_1 = \sum_{j=1}^{n} e_j \partial_{x_j}$$

$$\underline{x}_2 = \sum_{j=1}^{n} x_{n+j} e_j, \qquad \underline{\partial}_2 = \sum_{j=1}^{n} e_j \partial_{x_{n+j}}.$$

Now consider two real Clifford vectors $\underline{x}, \underline{y} \in \mathbb{R}^{2n}$, which may be written as $\underline{x} = \sum_{j=1}^{n} (x_j e_j + x_{n+j} e_{n+j})$, $\underline{y} = \sum_{j=1}^{n} (y_j e_j + y_{n+j} e_{n+j})$. For $a \in \mathbb{C}_n$ it follows that

$$\underline{x} a I = \underline{x}_1 a I + \sum_{j=1}^{n} x_{n+j} e_{n+j} a I = \underline{x}_1 a I + \tilde{a} \sum_{j=1}^{n} x_{n+j} e_{n+j} I,$$

whence application of (2) yields (see Ref. 5)

$$\underline{x} a I = (\underline{x}_1 a + i \tilde{a} \underline{x}_2) I. \tag{4}$$

From the above equality, we deduce that

$$\underline{x}\,\underline{y} a I = \big(\underline{x}_1(\underline{y}_1 a + i \tilde{a} \underline{y}_2) + (a \underline{y}_2 - i \underline{y}_1 \tilde{a}) \underline{x}_2\big) I. \tag{5}$$

If we now take a function $f : \Omega \subset \mathbb{R}^{2n} \to \mathbb{C}_n$, $f \in C^{2k+1}(\Omega)$, then we learn from (1) and (4) that

$$\underline{\partial}^{2k+1}(fI) = (-1)^k \underline{\partial}((\Delta^k f)I) = (-1)^k \big(\underline{\partial}_1 \Delta^k f + i \Delta^k \tilde{f} \underline{\partial}_2\big) I,$$

whence it follows from (3) that fI is $(2k+1)$-monogenic if and only if

$$\underline{\partial}_1 \Delta^k f + i \Delta^k \tilde{f} \underline{\partial}_2 = 0. \tag{6}$$

Definition 2.1. A $\mathbb{C}_n$-valued function $f \in C^{2k+1}(\Omega)$, $k \in \mathbb{N}_0$, which satisfies in Ω the equation (6), is said to be Δ^k-isotonic in Ω.

It should be noted that the case $k = 0$ corresponds to the so-called isotonic functions introduced in Ref. 18 (see also Refs. 5,14,17). As we mentioned at the beginning of the paper, we will be concerned with finding an integral representation formula for the Δ^k-isotonic functions. For that purpose, we put $\underline{\nu}_1(\underline{y}) = \sum_{j=1}^n \nu_j(\underline{y})e_j$ and $\underline{\nu}_2(\underline{y}) = \sum_{j=1}^n \nu_{n+j}(\underline{y})e_j$, $\underline{y} \in S$.

Let $f \in C^{2k}(\Omega \cup S)$ be a Δ^k-isotonic function in Ω. By Theorem 1.1, we see that for $\underline{x} \in \Omega$

$$f(\underline{x})I = \sum_{j=1}^{k+1}(-1)^{j-1}c_{2j-1}\int_S \frac{(\underline{y} - \underline{x})\underline{\nu}(\underline{y})\left(\Delta^{j-1}f(\underline{y})\right)I}{|\underline{y} - \underline{x}|^{2(n-j+1)}}\, d\mathcal{H}^{2n-1}(\underline{y})$$

$$+ \sum_{j=1}^{k}(-1)^{j}c_{2j}\int_S \frac{\underline{\nu}(\underline{y})\partial\left(\left(\Delta^{j-1}f(\underline{y})\right)I\right)}{|\underline{y} - \underline{x}|^{2(n-j)}}\, d\mathcal{H}^{2n-1}(\underline{y}).$$

Here we recall that for simplicity we assume $k < n$. Now applying (5) and (3), we get the following result.

Theorem 2.1. *Let* $f \in C^{2k}(\Omega \cup S)$ *be a* $\mathbb{C}_n$*-valued function. If* f *is* Δ^k*-isotonic in* Ω*, then for* $\underline{x} \in \Omega$ *we have that*

$$f(\underline{x}) =$$

$$\sum_{j=1}^{k+1}(-1)^{j-1}c_{2j-1}\int_S \left(\frac{(\underline{y}_1 - \underline{x}_1)\left(\underline{\nu}_1(\underline{y})\Delta^{j-1}f(\underline{y}) + i\Delta^{j-1}\tilde{f}(\underline{y})\underline{\nu}_2(\underline{y})\right)}{|\underline{y} - \underline{x}|^{2(n-j+1)}}\right.$$

$$\left.+ \frac{\left(\Delta^{j-1}f(\underline{y})\underline{\nu}_2(\underline{y}) - i\underline{\nu}_1(\underline{y})\Delta^{j-1}\tilde{f}(\underline{y})\right)(\underline{y}_2 - \underline{x}_2)}{|\underline{y} - \underline{x}|^{2(n-j+1)}}\right)\, d\mathcal{H}^{2n-1}(\underline{y})$$

$$+ \sum_{j=1}^{k}(-1)^{j}c_{2j}\int_S \left(\frac{\underline{\nu}_1(\underline{y})\left(\partial_1\Delta^{j-1}f(\underline{y}) + i\Delta^{j-1}\tilde{f}(\underline{y})\partial_2\right)}{|\underline{y} - \underline{x}|^{2(n-j)}}\right.$$

$$\left.+ \frac{\left(\Delta^{j-1}f(\underline{y})\partial_2 - i\partial_1\Delta^{j-1}\tilde{f}(\underline{y})\right)\underline{\nu}_2(\underline{y})}{|\underline{y} - \underline{x}|^{2(n-j)}}\right)\, d\mathcal{H}^{2n-1}(\underline{y}).$$

3. Two special cases

In this last section, we will briefly discuss two particular cases which arise when considering the equation (6).

Case 1: It is easily seen that if f takes values in the space of scalars $\mathbb{C}$, then f is Δ^k-isotonic if and only if

$$\left(\partial_{x_j} + i\partial_{x_{n+j}}\right)\Delta^k f = 0, \quad j = 1,\ldots,n. \tag{7}$$

Corollary 3.1. *Assume that $f \in C^{2k}(\Omega \cup S)$ is a $\mathbb{C}$-valued function. If f satisfies (7) in Ω, then for $\underline{x} \in \Omega$ we have that*

$$f(\underline{x}) =$$
$$\sum_{j=1}^{k+1} (-1)^j c_{2j-1} \int_S \frac{\langle (\underline{y}_1 - \underline{x}_1) - i(\underline{y}_2 - \underline{x}_2), \underline{\nu}_1(\underline{y}) + i\underline{\nu}_2(\underline{y}) \rangle \Delta^{j-1} f(\underline{y})}{|\underline{y} - \underline{x}|^{2(n-j+1)}} \, d\mathcal{H}^{2n-1}(\underline{y})$$
$$+ \sum_{j=1}^{k} (-1)^{j+1} c_{2j} \int_S \frac{\langle \underline{\nu}_1(\underline{y}) - i\underline{\nu}_2(\underline{y}), \partial_1 \Delta^{j-1} f(\underline{y}) + i\partial_2 \Delta^{j-1} f(\underline{y}) \rangle}{|\underline{y} - \underline{x}|^{2(n-j)}} \, d\mathcal{H}^{2n-1}(\underline{y}).$$

Case 2: If f takes values in the real Clifford algebra $\mathbb{R}_{0,n}$, then f is Δ^k-isotonic if and only if

$$\partial_1 \Delta^k f = \Delta^k f \partial_2 = 0. \tag{8}$$

Corollary 3.2. *Assume that $f \in C^{2k}(\Omega \cup S)$ is an $\mathbb{R}_{0,n}$-valued function. If f satisfies (8) in Ω, then for $\underline{x} \in \Omega$ we have that*

$$f(\underline{x}) = \sum_{j=1}^{k+1} (-1)^{j-1} c_{2j-1}$$
$$\times \int_S \frac{(\underline{y}_1 - \underline{x}_1)\underline{\nu}_1(\underline{y})\Delta^{j-1} f(\underline{y}) + \Delta^{j-1} f(\underline{y})\underline{\nu}_2(\underline{y})(\underline{y}_2 - \underline{x}_2)}{|\underline{y} - \underline{x}|^{2(n-j+1)}} \, d\mathcal{H}^{2n-1}(\underline{y})$$
$$+ \sum_{j=1}^{k} (-1)^j c_{2j} \int_S \frac{\underline{\nu}_1(\underline{y})\partial_1 \Delta^{j-1} f(\underline{y}) + \left(\Delta^{j-1} f(\underline{y})\partial_2 \right)\underline{\nu}_2(\underline{y})}{|\underline{y} - \underline{x}|^{2(n-j)}} \, d\mathcal{H}^{2n-1}(\underline{y}).$$

These results easily follow from the formula in Theorem 2.1. The solutions of the overdetermined system (7) (resp. (8)) may be seen as a "mixture" of polyharmonic functions and holomorphic functions (resp. biregular functions). It must be remarked that the integral representations obtained in Corollaries 3.1 and 3.2 are, respectively, generalizations of the classical Bochner-Martinelli formula for holomorphic functions and the Bochner-Martinelli formula for biregular functions (see Refs. 6,13).

Acknowledgment

J. Bory Reyes wishes to thank the Department of Mathematics, University of Aveiro, where the paper was written, for the invitation and hospitality. D. Peña Peña was supported by a Post-Doctoral Grant of *Fundação para a Ciência e a Tecnologia*, Portugal (grant number: SFRH/BPD/45260/2008).

References

1. H. Begehr, *Integral representations in complex, hypercomplex and Clifford analysis*, Integral Transforms Spec. Funct. 13 (2002), no. 3, 223–241.
2. F. Brackx, *On (k)-monogenic functions of a quaternion variable*, Funct. theor. Methods Differ. Equat. 22–44, Res. Notes in Math., no. 8, Pitman, London, 1976.
3. F. Brackx, B. De Knock, H. De Schepper, D. Eelbode, *A calculus scheme for Clifford distributions*, Tokyo J. Math. 29 (2006), no. 2, 495–513.
4. F. Brackx, R. Delanghe and F. Sommen, *Clifford analysis*, Research Notes in Mathematics, 76, Pitman (Advanced Publishing Program), Boston, MA, 1982.
5. F. Brackx, H. De Schepper and F. Sommen, *The Hermitian Clifford analysis toolbox*, Adv. Appl. Clifford Algebr. 18 (2008), no. 3-4, 451–487.
6. F. Brackx and W. Pincket, *A Bochner-Martinelli formula for the biregular functions of Clifford analysis*, Complex Variables Theory Appl. 4 (1984), no. 1, 39–48.
7. W. K. Clifford, *Applications of Grassmann's Extensive Algebra*, Amer. J. Math. 1 (1878), no. 4, 350–358.
8. R. Delanghe and F. Brackx, *Hypercomplex function theory and Hilbert modules with reproducing kernel*, Proc. London Math. Soc. (3) 37 (1978), no. 3, 545–576.
9. R. Delanghe, F. Sommen and V. Souček, *Clifford algebra and spinor-valued functions*, Mathematics and its Applications, 53, Kluwer Academic Publishers Group, Dordrecht, 1992.
10. H. Federer, *Geometric measure theory*, Die Grundlehren der mathematischen Wissenschaften, Band 153 Springer-Verlag New York Inc., New York 1969.
11. G. Laville and I. Ramadanoff, *Holomorphic Cliffordian functions*, Adv. Appl. Clifford Algebras 8 (1998), no. 2, 323–340.
12. H. R. Malonek and G. Ren, *Almansi-type theorems in Clifford analysis*, Math. Methods Appl. Sci. 25 (2002), no. 16-18, 1541–1552.
13. E. Martinelli, *Sulle estensioni della formula integrale di Cauchy alle funzioni analitiche di più variabili complesse*, Ann. Mat. Pura Appl. (4) 34, (1953), 277–347.
14. R. Rocha Chávez, M. Shapiro and F. Sommen, *Integral theorems for functions and differential forms in $\mathbb{C}^m$*, Research Notes in Mathematics, 428, Chapman & Hall/CRC, Boca Raton, FL, 2002.
15. J. Ryan, *Iterated Dirac operators in $\mathbb{C}^n$*, Z. Anal. Anwend. 9, No.5, 385–401 (1990).
16. J. Ryan, *Basic Clifford analysis*, Cubo Mat. Educ. 2 (2000), 226–256.
17. I. Sabadini and F. Sommen, *Hermitian Clifford analysis and resolutions*, Clifford analysis in applications, Math. Methods Appl. Sci. 25 (2002), no. 16-18, 1395–1413.
18. F. Sommen and D. Peña Peña, *Martinelli-Bochner formula using Clifford analysis*, Archiv der Mathematik, 88 (2007), no. 4, 358–363.

A version of Fueter's theorem in Dunkl-Clifford analysis

Minggang Fei

School of Mathematical Sciences, University of Electronic Science and Technology of China, Chengdu, 610054, P. R. China
and
Departamento de Matemática, Universidade de Aveiro, Aveiro, P-3810-193, Portugal
E-mail: fei@uestc.edu.cn

Shanshan Li

School of Computer Science and Technology, Southwest University for Nationalities, Chengdu, 610041, P. R. China
E-mail: shanshanli@swun.cn

In relation to the solutions of Vekua-type systems for axial Dunkl-monogenic functions, a version of Fueter's Theorem in the case of finite reflection groups is considered and studied.

Keywords: Dunkl-Dirac operator; Fueter's Theorem; Vekua-type system.

1. Introduction

We consider the differential-difference operators T_i, $i = 1, \cdots, d$, on $\mathbb{R}^d$, associated with a positive root system R_+, a nonnegative multiplicity function κ and index $\gamma_\kappa > 0$, introduced by C. F. Dunkl[3] and called Dunkl operators in the literature which are invariant under reflection groups. These operators are very important in pure mathematics and in physics. They provide a useful tool in the study of special functions with root systems, moreover these operators are pairwise commuting, so the commutative algebra generated by them has been used in the study of certain exactly solvable models of quantum mechanics, namely the Calogero-Moser-Sutherland models, which deal with system of identical particles in one dimensional space.[7]

Based on Dunkl operators P. Cerejeiras et al[1] recently introduced a Dirac operator, called Dunkl-Dirac operator, which is invariant under reflection groups and also factorizes the Dunkl Laplacian. Starting from the latter operator we will prove a version of Fueter's Theorem for Dunkl-monogenic functions, as given in the following:

Theorem 1.1. *Let W be a finite reflection group which leaves x_0-axis invariant and f be a holomorphic function in an open set B in the upper half complex plane given by*

$$f = u(s,t) + \mathbf{i}v(s,t), \qquad z = s + \mathbf{i}t.$$

Then in the set $\vec{B}$ the function

$$\Delta_h^{\gamma_\kappa + \frac{d-1}{2}}\left(u(x_0, |\underline{x}|) + \frac{\underline{x}}{|\underline{x}|}v(x_0, |\underline{x}|) \right)$$

is Dunkl-monogenic whenever $\gamma_\kappa + (d-1)/2$ is a positive integer. Hereby Δ_h denotes the Dunkl-Laplacian.

2. Preliminaries

We denote by $\mathbb{R}_{0,d}$ be the real Clifford algebra constructed over the orthonormal basis $(\mathbf{e}_1, \cdots, \mathbf{e}_d)$ of the Euclidean space $\mathbb{R}^d$.[2] An important subspace of the real Clifford algebra $\mathbb{R}_{0,d}$ is the so called space of paravector $\mathbb{R}_1^d = \mathbb{R}_1 \oplus \mathbb{R}^d$, being sums of scalars and vectors. A $\mathbb{R}_{0,d}$−valued function f over $\Omega \subset \mathbb{R}_1^d$ has a representation $f = \sum_A \mathbf{e}_A f_A$, with component $f_A : \Omega \to \mathbb{R}$.

The reflection $\sigma_\alpha x$ of a given vector $x \in \mathbb{R}_1^d$ on the hyperplane H_α orthogonal to $\alpha \neq 0$ is given by

$$\sigma_\alpha x := x - 2\frac{\langle \alpha, x \rangle}{|\alpha|^2}\alpha.$$

A finite set $R \subset \mathbb{R}_1^d \backslash \{0\}$ is called a root system if $R \cap \mathbb{R}_1^d \cdot \alpha = \{\alpha, -\alpha\}$ and $\sigma_\alpha R = R$ for all $\alpha \in R$. For a given root system R the reflections σ_α, $\alpha \in R$, generate a finite group $W \subset O(d)$, called the finite reflection group associated with R. For a given $\beta \in \mathbb{R}^d \backslash \bigcup_{\alpha \in R} H_\alpha$, we fix the positive subsystem $R_+ = \{\alpha \in R | \langle \alpha, \beta \rangle > 0\}$, i.e. for each $\alpha \in R$ either $\alpha \in R_+$ or $-\alpha \in R_+$. Sometimes we will only consider reflections which only act in $\mathbb{R}^d$. In this case we denote α or β by $\underline{\alpha}$ or $\underline{\beta}$.

A function $\kappa : R \to \mathbb{C}$ on a root system R is called a multiplicity function if it is invariant under the action of the associated reflection group W. For abbreviation, we introduce the index $\gamma_\kappa = \sum_{\alpha \in R_+} \kappa(\alpha)$. For each fixed positive subsystem R_+ and multiplicity function κ we have, as invariant operators, the differential-difference operators (also called Dunkl operators):

$$T_i f(x) = \frac{\partial}{\partial x_i} f(x) + \sum_{\alpha \in R_+} \kappa(\alpha) \frac{f(x) - f(\sigma_\alpha x)}{\langle \alpha, x \rangle} \alpha_i, \qquad i = 0, 1, \cdots, d, \quad (1)$$

for $f \in C^1(\mathbb{R}_1^d)$. In the case $\kappa = 0$, the T_i, $i = 0, 1, \cdots, d$, reduce to the corresponding partial derivatives. This also give us the justification to think of these differential-difference operators as the equivalent of partial derivatives in the case of finite reflection groups. In this paper, we will assume throughout that $\kappa \geq 0$ and $\gamma_\kappa > 0$. More importantly, these operators mutually commute; that is, $T_i T_j = T_j T_i$. This property allows us to define a Dunkl-Dirac operator in $\mathbb{R}^d$ for the corresponding reflection group W given by

$$\underline{D}_h f = \sum_{i=1}^d \mathbf{e}_i T_i f. \qquad (2)$$

The Dunkl Laplacian $\underline{\Delta}_h$ in $\mathbb{R}^d$ associated with the finite reflection group W and the multiplicity function κ is defined by

$$\underline{\Delta}_h f = -\underline{D}_h^2 f = \sum_{i=1}^d T_i^2 f$$

$$= \Delta_{\underline{x}} f + 2 \sum_{\underline{\alpha} \in R_+} \kappa(\underline{\alpha}) \frac{\langle \underline{\alpha}, \nabla_{\underline{x}} f(\underline{x}) \rangle}{\langle \underline{\alpha}, \underline{x} \rangle} - 2 \sum_{\underline{\alpha} \in R_+} \kappa(\underline{\alpha}) \frac{f(\underline{x}) - f(\sigma_{\underline{\alpha}} \underline{x})}{\langle \underline{\alpha}, \underline{x} \rangle^2} \quad (3)$$

for any $f \in C^2(\mathbb{R}^d)$, where $\Delta_{\underline{x}}$ and $\nabla_{\underline{x}}$ are usual Laplacian and gradient operator in $\mathbb{R}^d$.

We now introduce the Dunkl-Cauchy-Riemann operator in $\mathbb{R}_1^d$

$$D_h = T_0 + \underline{D}_h,$$

and Dunkl Laplacian in $\mathbb{R}_1^d$

$$\Delta_h = T_0^2 + \underline{\Delta}_h.$$

In this paper we will assume that our group W will leave the x_0-axis invariant. Since in this case we have $T_0 = \partial_{x_0}$ the Dunkl-Cauchy-Riemann operator and Dunkl-Laplacian in $\mathbb{R}_1^d$ can also be written by

$$D_h = \partial_{x_0} + \underline{D}_h, \qquad (4)$$

and

$$\Delta_h = \partial_{x_0}^2 + \underline{\Delta}_h. \qquad (5)$$

Functions belonging to the kernel of Dunkl-Dirac operator $\underline{D}_h$ or Dunkl-Cauchy-Riemann operator D_h will be called Dunkl-monogenic functions. As usual, functions belonging to be the kernel of Dunkl Laplacian will be called Dunkl-harmonic functions.

3. Proof of the Main Theorem

The classical Fueter's Theorem and its generalizations[5] provide us with Dunkl-monogenic functions of the form

$$A(x_0, r) + \underline{\omega} B(x_0, r) \tag{6}$$

whereby $\underline{x} \in \mathbb{R}^d$, $r = |\underline{x}|$, $\underline{x} = r\underline{\omega}$ and A and B are scalar-valued functions.

To study this system we need the representation of the Dunkl-Dirac operator in terms of spherical coordinates:[4]

Proposition 3.1. *In spherical coordinates the Dunkl-Dirac operator has the form:*

$$\underline{D}_h = \underline{\omega}\left(\partial_r + \frac{1}{r}\Gamma_{\underline{\omega}}\right),$$

where $\Gamma_{\underline{\omega}}$ is a first differential-difference operator which satisfies

$$\text{(i) } \Gamma_{\underline{\omega}} f(r) = 0,$$
$$\text{(ii) } \Gamma_{\underline{\omega}}(\underline{\omega} f(r)) = \Gamma_{\underline{\omega}}(\underline{\omega}) f(r),$$
$$\text{(iii) } \Gamma_{\underline{\omega}}(\underline{\omega}) = (2\gamma_\kappa + d - 1)\underline{\omega},$$

where $f(r) = f(|\underline{x}|)$ is a radial function defined in $\mathbb{R}^d$.

Using the obtained properties in Proposition 3.1 we find that the assumed Dunkl-monogenicity of (6) requires that our functions A and B satisfy the following Vekua-type system

$$\begin{cases} \partial_{x_0} A - \partial_r B = \frac{2\gamma_\kappa + d - 1}{r} B, \\ \\ \partial_{x_0} B + \partial_r A = 0. \end{cases} \tag{7}$$

Now let us outline our proof. First, we would like to remark that this version of Fueter's Theorem provides us with axial monogenic functions, i.e.

$$\Delta_h^{\gamma_\kappa + (d-1)/2}\left(u(x_0, |\underline{x}|) + \underline{\omega} v(x_0, |\underline{x}|)\right) = A(x_0, r) + \underline{\omega} B(x_0, r)$$

for some scalar-valued and continuously differentiable functions A and B. Hence, the proof consists in showing that A and B satisfy our Vekua-type system (7). To this end, we start with the following lemma which shows that the iterated Dunkl-Laplacian Δ_h^m, for any positive integer m, keeps functions of the form $A(x_0, r) + \underline{\omega}B(x_0, r)$ invariant whenever A and B are scalar-valued harmonic functions in $\mathbb{R}^2$.

Lemma 3.1. *Let $h(x_0, r)$ be a scalar-valued harmonic function in $\mathbb{R}^2$, i.e.*

$$\partial_{x_0}^2 h + \partial_r^2 h = 0. \tag{8}$$

And Let D_r and D^r be differential operators defined by $D_r(0)\{f\} = D^r(0)\{f\} = f$ and

$$D_r(m)\{f\} = \left(\frac{1}{r}\partial_r\right)^m \{f\},$$

$$D^r(m)\{f\} = \partial_r\left(\frac{D^r(m-1)\{f\}}{r}\right)$$

for $m \geq 1$. Then we have

$$\Delta_h^m h(x_0, r) = \prod_{i=1}^m (2\gamma_\kappa + d - (2i - 1))D_r(m)\{h(x_0, r)\}$$

and

$$\Delta_h^m(\underline{\omega}h(x_0, r)) = \underline{\omega}\prod_{i=1}^m (2\gamma_\kappa + d - (2i - 1))D^r(m)\{h(x_0, r)\},$$

with m being a positive integer.

Proof. We will prove this lemma by induction. When $m = 1$, we need to show that the following identities hold

$$\Delta_h h = (2\gamma_\kappa + d - 1)D_r(1)\{h\} \tag{9}$$

and

$$\Delta_h(\underline{\omega}h) = \underline{\omega}(2\gamma_\kappa + d - 1)D^r(1)\{h\}. \tag{10}$$

To prove (9), we start from

$$\Delta_h = \partial_{x_0}^2 - \underline{D}_h\underline{D}_h, \qquad \underline{D}_h = \underline{\omega}\left(\partial_r + \frac{1}{r}\Gamma_{\underline{\omega}}\right).$$

Then using Proposition 3.1 and (8) we get

$$
\begin{aligned}
\Delta_h h &= \partial_{x_0}^2 h - \underline{\omega}(\partial_r + \frac{1}{r}\Gamma_{\underline{\omega}})(\underline{\omega}\partial_r h) \\
&= \partial_{x_0}^2 h + \partial_r^2 h + \frac{2\gamma_\kappa + d - 1}{r}\partial_r h \\
&= \frac{2\gamma_\kappa + d - 1}{r}\partial_r h \\
&= (2\gamma_\kappa + d - 1)D_r(1)\{h\}.
\end{aligned}
$$

To prove (10), again applying Proposition 3.1 and (8) we obtain

$$
\begin{aligned}
\Delta_h(\underline{\omega}h) &= \underline{\omega}\partial_{x_0}^2 h - \underline{\omega}(\partial_r + \frac{1}{r}\Gamma_{\underline{\omega}})(-\partial_r h - \frac{2\gamma_\kappa + d - 1}{r}h) \\
&= \underline{\omega}\left(\partial_{x_0}^2 h + \partial_r^2 h + (2\gamma_\kappa + d - 1)\partial_r\{\frac{h}{r}\}\right) \\
&= \underline{\omega}(2\gamma_\kappa + d - 1)D^r(1)\{h\}.
\end{aligned}
$$

Summarizing we have that the lemma is true in the case $m = 1$. Assume that our formulae hold for a positive integer m, we have to show them for $m + 1$.

We thus get

$$
\begin{aligned}
\Delta_h^{m+1} h &= \prod_{i=1}^{m}(2\gamma_\kappa + d - (2i - 1))\Delta_h D_r(m)\{h\}, \\
&= \prod_{i=1}^{m}(2\gamma_\kappa + d - (2i - 1)) \\
&\quad \cdot(\partial_{x_0}^2 D_r(m)\{h\} + \partial_r^2 D_r(m)\{h\} + (2\gamma_\kappa + d - 1)D_r(m + 1)\{h\}) \\
&= \prod_{i=1}^{m}(2\gamma_\kappa + d - (2i - 1)) \\
&\quad \cdot(D_r(m)\{\partial_{x_0}^2 h + \partial_r^2 h\} + (2\gamma_\kappa + d - (2m + 1))D_r(m + 1)\{h\}) \\
&= \prod_{i=1}^{m+1}(2\gamma_\kappa + d - (2i - 1))D_r(m + 1)\{h\},
\end{aligned}
$$

which establishes the first formula. The other one may be proved in a similar way. $\quad\square$

We are now ready to present our proof of Theorem 1.1.

Proof. By Lemma 3.1, we get that

$$\Delta_h^{\gamma_\kappa+(d-1)/2}\left(\left(u(x_0,|\underline{x}|)+\frac{\underline{x}}{|\underline{x}|}v(x_0,|\underline{x}|)\right)\right)$$
$$=(2\gamma_\kappa+d-1)!!(A(x_0,r)+\underline{\omega}B(x_0,r)),$$

with

$$A=D_r\left(\gamma_\kappa+\frac{d-1}{2}\right)\{u\},$$

$$B=D^r\left(\gamma_\kappa+\frac{d-1}{2}\right)\{v\}.$$

The task is now to show that A and B satisfy the Vekua-type system (7). In order to do that, it will be necessary to use the assumptions on u and v and some well known results about differential operators D_r and D^r, for example, contained in Lemma 2 in Peña Peña et al's paper.[6]

Indeed, we obtain

$$\partial_{x_0}A-\partial_r B=D_r\left(\gamma_\kappa+\frac{d-1}{2}\right)\{\partial_{x_0}u\}-\partial_r D^r\left(\gamma_\kappa+\frac{d-1}{2}\right)\{v\}$$
$$=D_r\left(\gamma_\kappa+\frac{d-1}{2}\right)\{\partial_r v\}-\partial_r D^r\left(\gamma_\kappa+\frac{d-1}{2}\right)\{v\}$$
$$=\frac{2\gamma_\kappa+d-1}{r}D^r\left(\gamma_\kappa+\frac{d-1}{2}\right)\{v\}$$
$$=\frac{2\gamma_\kappa+d-1}{r}B$$

and

$$\partial_{x_0}B+\partial_r A=D^r\left(\gamma_\kappa+\frac{d-1}{2}\right)\{\partial_{x_0}v\}+\partial_r D_r\left(\gamma_\kappa+\frac{d-1}{2}\right)\{u\}$$
$$=D_r\left(\gamma_\kappa+\frac{d-1}{2}\right)\{\partial_{x_0}v\}+D^r\left(\gamma_\kappa+\frac{d-1}{2}\right)\{\partial_r u\}$$
$$=D_r\left(\gamma_\kappa+\frac{d-1}{2}\right)\{\partial_{x_0}v+\partial_r u\}$$
$$=0,$$

which completes the proof. $\square$

References

1. P. Cerejeiras, U. Kähler and G. Ren, *Clifford analysis for finite reflection groups*, Complex Var. Elliptic Equ. 51 (5-6) (2006), 487-495.

2. R. Delanghe, F. Sommen and V. Souček, *Clifford algebra and spinor valued functions*, A function theory for Dirac operator, Kluwer, Dordrecht, 1992.

3. C. F. Dunkl, *Differential-difference operators associated to reflection groups*, Trans. Amer. Math. Soc. 311 (1989), 167-183.

4. M. Fei, P. Cerejeiras and U. Kähler, *Fueter's Theorem and its Generalizations in Dunkl-Clifford Analysis*, J. Phys. A: Math. Theor. 42(2009), 395209(15pp).

5. K. I. Kou, T. Qian and F. Sommen, *Generalizations of Fueter's Theorem*, Meth. Appl. Anal. 9 (2) (2002), 273-290.

6. D. Peña Peña, T. Qian and F. Sommen, *An alternative proof of Fueter's theorem*, Complex Var. Elliptic Equ. 51 (8-11) (2006), 913-922.

7. M. Rösler, *Dunkl Operators: Theory and Applications*, Lecture Notes in Math. 1817 (2003), 93-135.

Boundary values of pluriholomorphic functions in $\mathbb{C}^2$

A. Perotti

Department of Mathematics, University of Trento,
Trento, I-38100, Italy
E-mail: perotti@science.unitn.it
www.science.unitn.it/˜perotti/

We consider the Dirichlet problem for pluriholomorphic functions of two complex variables. Pluriholomorphic functions are solutions of the system $\frac{\partial^2 g}{\partial \bar{z}_i \partial \bar{z}_j} = 0$ for every i, j. The key point is the relation between pluriholomorphic functions and pluriharmonic functions. The link is constituted by the Fueter-regular functions of one quaternionic variable. We apply previous results about the boundary values of pluriharmonic functions and new results on L^2 traces of regular functions to obtain a characterization of the traces of pluriholomorphic functions.

Keywords: Pluriholomorphic functions; pluriharmonic functions; quaternionic regular functions.

1. Introduction

We discuss a boundary value problem in two complex variables on a class of pseudoconvex domains containing the unit ball B. The class consists of domains Ω that satisfy a $L^2(\partial\Omega)$-estimate (cf. Sec. 3.2). We conjecture that the estimate holds on every strongly pseudoconvex domain in $\mathbb{C}^2$.

We relate the Dirichlet boundary value problems for pluriholomorphic functions and pluriharmonic functions by means of a class of quaternionic regular functions (cf. Sec. 3), a variant of Fueter-regular functions studied by many authors (see for instance Refs. 11,13,15). Pluriholomorphic functions are solutions of the system $\frac{\partial^2 g}{\partial \bar{z}_i \partial \bar{z}_j} = 0$ for $1 \leq i, j \leq 2$ (see e.g. Refs. 1–4,6–8). The Dirichlet problem for this system is not well posed and the homogeneous problem can have infinitely many independent solutions. As noted in Ref. 8, the Dirichlet problem for pluriharmonic functions has a different character, related to strong ellipticity.

The key point is that if $f = f_1 + f_2 j$ is regular, then f_1 is pluriholomorphic (and harmonic) if and only if f_2 is pluriharmonic.

We begin by giving an application of an existence principle in Functional Analysis proved by Fichera in the 50's. We obtain a result on the boundary values of class $L^2(\partial\Omega)$ of regular functions: every function f_1 which belongs to the class $L^2(\partial\Omega)$ together with its normal derivative $\overline{\partial}_n f_1$ is the first complex component of a regular function on Ω, of class $L^2(\partial\Omega)$. On the unit ball B, where computation of L^2-estimates can be more precise, the result is optimal. We show that the condition on the normal derivative cannot be relaxed and therefore the operation of regular conjugation is not bounded in the harmonic Hardy space $h^2(B)$.

In Sec. 4 we apply the results on the traces of pluriharmonic functions proved in Refs. 5,12 and obtain a characterization of the traces of pluriholomorphic functions. We generalize some results obtained by Detraz[6] and Dzhuraev[7] on the unit ball (cf. also Refs. 1–4,8). We show that if Ω satisfies the $L^2(\partial\Omega)$-estimate, a function $h \in L^2(\partial\Omega)$ with $\overline{\partial}_n h \in L^2(\partial\Omega)$ is the trace of a harmonic pluriholomorphic function on Ω if and only if it satisfies an orthogonality condition (see Theorem 4.2 for the precise statement).

See Ref. 14 for complete proofs of the results presented here.

2. Pluriholomorphic functions

Let Ω be a bounded domain in $\mathbb{C}^2$. A complex valued function $g \in C^2(\Omega)$ is *pluriholomorphic* on Ω if it satisfies the PDE system

$$\frac{\partial^2 g}{\partial \overline{z}_i \partial \overline{z}_j} = 0 \quad \text{on } \Omega \quad (1 \le i, j \le 2).$$

We refer to the works of Detraz,[6] Dzhuraev[7,8] and Begehr[1,2] for properties of pluriholomorphic functions of two or more variables. For the one-variable case there is a vast literature by Balk and his school (in their papers pluriholomorphic functions are called polyanalytic functions of order two). Every function g in the space $Phol(\Omega)$ of pluriholomorphic functions on Ω has a representation of the form: $g = g_0 + \sum_{i=1}^{2} \overline{z}_i g_i$, where g_0, g_1, g_2 are holomorphic on Ω. Note that the g_i's can be less regular than g on $\overline{\Omega}$.

2.1. *Dirichlet problem for pluriholomorphic functions*

Given a continuous function h on $\partial\Omega$, the following boundary problem

$$\begin{cases} g \in C^2(\Omega) \cap C(\overline{\Omega}) \\ \dfrac{\partial^2 g}{\partial \overline{z}_i \partial \overline{z}_j} = 0 \quad \text{on } \Omega \\ g_{|\partial\Omega} = h \quad \text{on } \partial\Omega \end{cases}$$

is not well posed (also for $n = 1$). Several facts about this problem can be found in the works of Begehr and Dzhuraev.[1-4] For instance, on the unit ball B the homogeneous problem with boundary datum $h = 0$ has infinitely many independent solutions $u_k(z) = (|z_1|^2 + |z_2|^2 - 1)z_1^{k_1} z_2^{k_2}$, $|k| = k_1 + k_2 \geq 0$. Detraz[6] proved that the nonhomogeneous problem imposes compatibility conditions on h: if $S = \partial B$ and $h \in C^2(S)$, then h must satisfy the tangential equation $LLh = 0$ on S, where $L = z_2 \frac{\partial}{\partial \bar{z}_1} - z_1 \frac{\partial}{\partial \bar{z}_2}$.

3. Fueter regular functions

We identify the space $\mathbb{C}^2$ with the set $\mathbb{H}$ of quaternions by means of the mapping that associates the pair $(z_1, z_2) = (x_0 + ix_1, x_2 + ix_3)$ with the quaternion $q = z_1 + z_2 j = x_0 + ix_1 + jx_2 + kx_3 \in \mathbb{H}$. A quaternionic function $f = f_1 + f_2 j \in C^1(\Omega)$ is *(left) regular* (or *hyperholomorphic*) on Ω if

$$\mathcal{D}f = 2\left(\frac{\partial}{\partial \bar{z}_1} + j\frac{\partial}{\partial \bar{z}_2}\right) = \frac{\partial f}{\partial x_0} + i\frac{\partial f}{\partial x_1} + j\frac{\partial f}{\partial x_2} - k\frac{\partial f}{\partial x_3} = 0 \quad \text{on } \Omega.$$

With respect to this definition of regularity, the space $\mathcal{R}(\Omega)$ of regular functions contains the identity mapping and every holomorphic mapping (f_1, f_2) on Ω (w.r.t. the standard complex structure) defines a regular function $f = f_1 + f_2 j$. We recall some properties of regular functions, for which we refer to the papers of Sudbery,[16] Shapiro and Vasilevski[15] and Nōno:[11]

(1) The complex components are both holomorphic or both non-holomorphic.
(2) Every regular function is harmonic.
(3) If Ω is pseudoconvex, every complex harmonic function is the complex component of a regular function on Ω.
(4) The space $\mathcal{R}(\Omega)$ of regular functions on Ω is a *right* $\mathbb{H}$-module with integral representation formulas.

3.1. *A link between pluriholomorphic functions and pluriharmonic functions*

In terms of its complex components f_1 and f_2 (called quaternionic *conjugate harmonic* functions), the regularity of $f = f_1 + f_2 j$ is equivalent to

$$\frac{\partial f_1}{\partial \bar{z}_1} = \frac{\partial \overline{f_2}}{\partial z_2}, \quad \frac{\partial f_1}{\partial \bar{z}_2} = -\frac{\partial \overline{f_2}}{\partial z_1}.$$

It follows easily that if $f = f_1 + f_2 j$ is regular, then f_1 is pluriholomorphic (and harmonic) if, and only if, f_2 is pluriharmonic. i.e. $\partial\bar{\partial}f_2 = 0$ on Ω.

The Dirichlet problem for pluriharmonic functions

$$\begin{cases} g \in C^2(\Omega) \cap C(\overline{\Omega}) \\ \dfrac{\partial^2 g}{\partial z_i \partial \bar{z}_j} = 0 \quad \text{on } \Omega \\ g_{|\partial\Omega} = h \quad \text{on } \partial\Omega \end{cases}$$

is characterized by strong ellipticity. The solution, if it exists, is unique, and the system can be splitted into equations for the real and imaginary parts of g.

3.2. *Quaternionic harmonic conjugation*

In the following we shall assume that the domain Ω satisfies the $L^2(\partial\Omega)$-estimate

$$|(f, Lg)| \le C\|\partial_n f\|\|\bar{\partial}_n g\| \tag{1}$$

for every complex harmonic functions f, g on Ω, of class C^1 on $\overline{\Omega}$, where L is the tangential Cauchy-Riemann operator

$$L = \frac{1}{|\partial\rho|} \left(\frac{\partial\rho}{\partial\bar{z}_2} \frac{\partial}{\partial\bar{z}_1} - \frac{\partial\rho}{\partial\bar{z}_1} \frac{\partial}{\partial\bar{z}_2} \right),$$

(ρ a defining function for Ω) $\bar{\partial}_n f$ is the normal part of $\bar{\partial}f$ on $\partial\Omega$:

$$\bar{\partial}_n f = \sum_k \frac{\partial f}{\partial\bar{z}_k} \frac{\partial\rho}{\partial z_k} \frac{1}{|\bar{\partial}\rho|} \quad \text{or} \quad \bar{\partial}_n f d\sigma = *\bar{\partial}f_{|\partial\Omega}$$

and $\partial_n f$ is the normal part of ∂f. Here $*$ is the Hodge operator on forms.

Theorem 3.1.

(i) On the unit ball B, the estimate (1) is satisfied with constant $C = 1$.
(ii) If Ω satisfies the estimate (1), then it is a domain of holomorphy.

We conjecture that the estimate holds for every (strongly) pseudoconvex domain in $\mathbb{C}^2$. We now prove a result about the existence of a quaternionic conjugate harmonic in the space $L^2(\partial\Omega)$. We use the following Hilbert subspace of complex valued functions in $L^2(\partial\Omega)$:

$$\overline{W}^1_n(\partial\Omega) = \{f \in L^2(\partial\Omega) \mid \bar{\partial}_n f \in L^2(\partial\Omega)\}$$

with product $(f, g)_{\overline{W}^1_n} = (f, g) + (\bar{\partial}_n f, \bar{\partial}_n g)$.

Theorem 3.2. *Assume $\partial\Omega$ connected. Given $f_1 \in \overline{W}_n^1(\partial\Omega)$, there exists $f_2 \in L^2(\partial\Omega)$ (unique up to a CR function) such that $f = f_1 + f_2 j$ is the trace of a regular function on Ω. Moreover, f_2 satisfies the estimate*

$$\inf_{f_0} \|f_2 + f_0\|_{L^2(\partial\Omega)} \le C\|f_1\|_{\overline{W}_n^1(\partial\Omega)},$$

where the infimum is taken among the CR functions $f_0 \in L^2(\partial\Omega)$.

On the unit ball B, a sharper estimate can be proved.

Theorem 3.3. *Given $f_1 \in \overline{W}_n^1(S)$, there exists $f_2 \in L^2(S)$ (unique up to a CR function) such that $f = f_1 + f_2 j$ is the trace of a regular function on B. Moreover, f_2 satisfies the estimate*

$$\inf_{f_0 \in CR(S)} \|f_2 + f_0\|_{L^2(S)} \le \|\overline{\partial}_n f_1\|_{L^2(S)}.$$

Remark 3.1. The condition on f_1 cannot be relaxed: $\exists f_1 \in L^2(S)$ for which does not exist any $L^2(S)$ function f_2 such that $f_1 + f_2 j$ is the trace of a regular function on B. This means that the operation of quaternionic regular conjugation is not bounded in the harmonic Hardy space $h^2(B)$. This is different from pluriharmonic conjugation (cf. Stout[17]) and in particular from the (complex) one-variable situation.

Example 3.1. A function $f_1 \in L^2(S)\backslash\overline{W}_n^1(S)$ with the required properties is $f_1 = z_2(1 - \bar{z}_1)^{-1}$.

4. Boundary values

We recall a characterization of the boundary values of pluriharmonic functions, proposed by Fichera in the 1980's and proved in Refs. 5 and 12. Let Ω have connected boundary of class C^1. Let

$$Harm_0^1(\Omega) = \{\phi \in C^1(\overline{\Omega}) \mid \phi \text{ is harmonic on } \Omega, \ \overline{\partial}_n\phi \text{ is real on } \partial\Omega\}.$$

This space can be characterized by means of the Bochner-Martinelli operator of the domain Ω. Cialdea[5] proved the following result for boundary values of class L^2 (and more generally of class L^p).

Theorem 4.1. *Let $g \in L^2(\partial\Omega)$ be complex valued. Then g is the trace of a pluriharmonic function on Ω if and only if the following orthogonality condition is satisfied:*

$$\int_{\partial\Omega} g * \overline{\partial}\phi = 0 \quad \forall\phi \in Harm_0^1(\Omega).$$

4.1. *Boundary values of pluriholomorphic functions*

If $f \in C^1(\overline{\Omega})$ is regular, on the boundary $\partial\Omega$ it satisfies the equations $\overline{\partial}_n f_1 = -\overline{L(f_2)}$, $\overline{\partial}_n f_2 = \overline{L(f_1)}$. More generally, if $f = f_1 + f_2 j : \partial\Omega \to \mathbb{H}$ is a function of class $L^2(\partial\Omega)$ and it is the trace of a regular function on Ω, then it satisfies the integral condition

$$\int_{\partial\Omega} f_1 \,\overline{\partial}\phi \wedge d\zeta = -2 \int_{\partial\Omega} \overline{f_2} *\overline{\partial}\phi \quad \forall\phi \in Harm^1(\Omega).$$

If $\partial\Omega$ is connected, it can be proved that also the converse is true. As a corollary, we get the following result.

Theorem 4.2. *Assume that Ω has connected boundary and satisfies the $L^2(\partial\Omega)$-estimate. Let $h \in \overline{W}_n^1(\partial\Omega)$. Then h is the trace of a harmonic pluriholomorphic function on Ω if and only if the following orthogonality condition is satisfied:*

$$\int_{\partial\Omega} h \,\overline{\partial}\phi \wedge d\zeta = 0 \quad \forall\phi \in Harm_0^1(\Omega). \tag{2}$$

On the unit ball B we can get a more precise result:

(1) If $h \in Phol(B) \cap C^1(\overline{B})$, then Lh is CR on S.
(2) If $h \in Phol(B) \cap C^2(\overline{B})$, then $LLh = 0$ on S.
(3) If $h \in C^1(S)$ and h is the trace on S of a pluriholomorphic function on $B \Rightarrow Lh \in CR(S) \Rightarrow$ condition (2) is satisfied.
(4) If h is of class $C^{1+\alpha}(S)$, with $\alpha > 0$, the conditions in (3) are all equivalent.

In particular, we get Detraz's result for C^2 functions: h is the trace on S of a pluriholomorphic function on B if and only if $LLh = 0$ on S.

Acknowledgments

This work was partially supported by MIUR (Project "Proprietà geometriche delle varietà reali e complesse") and GNSAGA of INdAM.

References

1. Begehr, H., Complex analytic methods for partial differential equations, *ZAMM* 76 (1996), Suppl. 2, 21–24.
2. Begehr, H., Boundary value problems in $\mathbb{C}$ and $\mathbb{C}^n$, *Acta Math. Vietnam.* 22 (1997), 407–425.
3. Begehr, H. and Dzhuraev, A., *An Introduction to Several Complex Variables and Partial Differential Equations*, Addison Wesley Longman, Harlow, 1997.

4. Begehr, H. and Dzhuraev, A., Overdetermined systems of second order elliptic equations in several complex variables. In: *Generalized analytic functions* (Graz, 1997), Int. Soc. Anal. Appl. Comput., 1, Kluwer Acad. Publ., Dordrecht, 1998, pp. 89–109.

5. Cialdea, A., On the Dirichlet and Neumann problems for pluriharmonic functions. In: *Homage to Gaetano Fichera*, Quad. Mat., 7, Dept. Math., Seconda Univ. Napoli, Caserta, 2000, pp. 31–78.

6. Detraz, J., Problème de Dirichlet pour le système $\partial^2 f/\partial \bar{z}_i \partial \bar{z}_j = 0$. (French), *Ark. Mat.* 26 (1988), no. 2, 173–184.

7. Dzhuraev, A., On linear boundary value problems in the unit ball of $\mathbb{C}^n$, *J. Math. Sci. Univ. Tokyo* 3 (1996), 271–295.

8. Dzhuraev, A., Some boundary value problems for second order overdetermined elliptic systems in the unit ball of $\mathbb{C}^n$. In: *Partial Differential and Integral Equations* (eds.: H. Begehr et al.), Int. Soc. Anal. Appl. Comput., 2, Kluwer Acad. Publ., Dordrecht, 1999, pp. 37–57.

9. Fichera, G., Alcuni recenti sviluppi della teoria dei problemi al contorno per le equazioni alle derivate parziali lineari. (Italian) In: *Convegno Internazionale sulle Equazioni Lineari alle Derivate Parziali*, Trieste, 1954, Edizioni Cremonese, Roma, 1955, pp. 174–227.

10. Fichera, G., *Linear elliptic differential systems and eigenvalue problems.* Lecture Notes in Mathematics, 8 Springer-Verlag, Berlin-New York, 1965.

11. Nōno, K., *α-hyperholomorphic function theory*, Bull. Fukuoka Univ. Ed. III 35 (1985), 11–17.

12. Perotti, A., Dirichlet Problem for pluriharmonic functions of several complex variables, *Communications in Partial Differential Equations*, 24, nn.3&4, (1999), 707-717.

13. Perotti, A., Quaternionic regular functions and the $\overline{\partial}$-Neumann problem in $\mathbb{C}^2$, *Complex Variables and Elliptic Equations*, Vol. 52, No. 5, 439–453 (2007).

14. Perotti, A., Dirichlet problem for pluriholomorphic functions of two complex variables, *J. Math. Anal. Appl.*, Vol. 337/1, 107–115 (2008).

15. Shapiro M.V. and Vasilevski, N.L., Quaternionic ψ-hyperholomorphic functions, singular integral operators and boundary value problems. I. ψ-hyperholomorphic function theory, *Complex Variables Theory Appl.* 27 no.1 (1995), 17–46.

16. Sudbery, T., Quaternionic analysis, *Mat. Proc. Camb. Phil. Soc.* 85 (1979), 199–225.

17. Stout, E. L., H^p-functions on strictly pseudoconvex domains, *Amer. J. Math.* 98 n.3 (1976), 821–852.

Some properties of k-biregular function space in real Clifford analysis

Yuying Qiao[1], Yonghong Xie[1,*] and Heju Yang[1,2]

[1] *College of Mathematics and Information,*
Hebei Normal University, Shijiazhuang 050016, P. R. China
E-mail: qiaoyuying@hebtu.edu.cn
[2] *College of Science, Hebei University of Science and Technology,*
Shijiazhuang, 050018, P. R. China
E-mail: yangheju@hebust.edu.cn

In this paper, we discuss some properties of k-biregular function space in real Clifford analysis.

Keywords: k-regular function; k-biregular function; Plemelj formula; Cauchy integral formula.

1. Introduction

In 1976, Brackx F first introduced k-regular function of the real quaternion and gave the Cauchy integral formula and Taylor expansion of regular function of the real quaternion[1]. K-regular function is a natural generalization of the regular function. It is a new kind of function including regular function. In 1977, Delanghe R and Brackx F studied k-regular function which is defined in R^{n+1} with values in $C(V_{n,0})$ and obtained its Cauchy integral formula[2]. In Ref[3], higher order Cauchy-Pompeiu formula for r times continuously differentiable function and Cauchy integral formula for the k-regular function in universal Clifford analysis were studied.

The k-biregular function is the extension of the biregular function. Although many k-biregular functions aren't biregular functions, most results of biregular function can be extended to k-biregular function. In this paper, we study Cauchy-Pompeiu formula and some space properties for k-biregular function in real Clifford analysis.

---

*Research supported by National Natural Science Foundation of China (No.10771049, 10671207) and National Natural Science Foundation of Hebei (No.A2007000225).

2. Preliminaries

2.1. *Clifford Algebra $C(V_{n,0})$*

Suppose $e_1, e_2, \cdots, e_n$ is a basis of the $n(\geq 2)$-dimensional real linear space $V_{n,0}$. And let $C(V_{n,0})$ be a 2^n-dimensional real linear space, whose basis is

$$\{e_A | A = (h_1, h_2, \cdots h_r) \in PN, 1 \leq h_1 < h_2 < \cdots < h_r \leq n\},$$

where $N = \{1, 2, \cdots, n\}$. PN is a set composed of all subsets of $\{1, 2, \cdots, n\}$ and the subsets are arranged according to natural sequence. Suppose when $e_A = e_0$, $A = \emptyset$ and

$$\begin{cases} e_i^2 = -1, & i = 1, 2, \cdots, n; \\ e_i e_j = -e_j e_i, & 1 \leq i < j \leq n; \\ e_{h1} e_{h2} \cdots e_{hr} = e_{h_1 h_2 \cdots h_r}, & 1 \leq h_1 < \cdots < h_n \leq n. \end{cases} \tag{1}$$

$$\begin{cases} \overline{e_i} = -e_i, & i = 1, 2, \cdots n; \\ \overline{\lambda \mu} = \overline{\mu}\,\overline{\lambda}, & \lambda \in C(V_{n,0}), \mu \in C(V_{n,0}). \end{cases} \tag{2}$$

Hence any element $a \in C(V_{n,0})$ has the type $a = \sum_A a_A e_A$, where $a_A(\in R)$ are real numbers. We call $C(V_{n,0})$ Clifford algebra. The norm for an element in $C(V_{n,0})$ is taken to be $|\lambda| = \left(\sum_A \lambda_A^2\right)^{\frac{1}{2}}$. It is easy to see $|x|^2 = -x^2$. when $x \in R^n$,

2.2. *Differential Operators*

Let

$$C^{(r)}(\Omega, C(V_{n,0})) = \left\{ f \,\middle|\, \begin{array}{l} f : \Omega \to C(V_{n,0}), f(x) = \sum_A f_A(x) e_A, \\ f_A(x) \in C^r(\Omega, R), x \in \Omega \subset R^n \end{array} \right\}.$$

And we introduce the Dirac operators in $C^{(r)}(\Omega, C(V_{n,0}))$ as follows:

$$Df = \sum_{i=1}^{n} e_i \frac{\partial f}{\partial x_i} = \sum_{i=1}^{n} \sum_A e_i e_A \frac{\partial f_A}{\partial x_i}, \quad fD = \sum_{i=1}^{n} \frac{\partial f}{\partial x_i} e_i = \sum_{i=1}^{n} \sum_A e_A e_i \frac{\partial f_A}{\partial x_i}.$$

Definition 1 Let $\Omega \subset R^n$ be a nonempty and open subset, $f \in C^{(r)}(\Omega, C(V_{n,0}))$, $(r \geq k$, $k < n)$. If $D^k f(x) = 0$ $(f(x)D^k = 0)$ for any $x \in \Omega$, then f is called a left $k-$regular (right $k-$regular) function on Ω. Usually left $k-$regular functiom is called $k-$regular function for short.

Definition 2 Suppose $\Omega = \Omega_1 \times \Omega_2$ is a nonempty, open and connected set in $R^m \times R^{m_0}$ and $f \in F_\Omega^{(r)}$. If for any $z = (x, y) \in \Omega$, we have
$$\begin{cases} D_x^k f(x, y) = 0 \\ f(x, y)D_y^k = 0 \end{cases},$$
f is called as a $k-$biregular function in Ω, where $r \geq k$ and $k < \min\{m, m_0\}$. We denote the set of all $k-$biregular function in Ω as $R^k(\Omega, C(V_{n,0}))$.

2.3. *Higher Order Kernel Function*

$$H_j^*(x) = \frac{A_j}{\omega_n} \frac{\overline{x}^j}{|x|^n}, \quad j < n, \tag{3}$$

$$A_j = \begin{cases} \dfrac{1}{2^{i-1}(i-1)! \displaystyle\prod_{r=1}^{i}(2r - n)}, & j = 2i, j < n, i = 1, 2, \cdots, \\[2em] \dfrac{1}{2^{i}i! \displaystyle\prod_{r=1}^{i}(2r - n)}, & j = 2i + 1, j < n, i = 0, 1, \cdots, \end{cases} \tag{4}$$

where $x \in R_0^n = R^n/\{0\}$, $\omega_n = \dfrac{2\pi^{\frac{n}{2}}}{\Gamma(\frac{n}{2})}$ is the area of the unit sphere in R^n. The function $H_j^*(x)$ defined in (3) is called a higher order kernel function. We can obtain
$$\begin{cases} H_1^*(x) = \dfrac{1}{\omega_n} \dfrac{\overline{x}}{|x|^n}, & x \in R_0^n, \\[1.5em] H_{2i+1}^*(x) = \dfrac{1}{2i} H_{2i}^*(x)\overline{x}, & x \in R_0^n, \ i = 0, 1, \cdots, 2i + 1 < n, \\[1.5em] H_{2i}^*(x) = \dfrac{1}{(2i - n)} H_{2i-1}^*(x)\overline{x}, & x \in R_0^n, \ i = 1, 2, \cdots, 2i < n. \end{cases} \tag{5}$$

$$\begin{cases} DH_1^*(x) = H_1^*(x)D = 0, & x \in R_0^n, \\[1em] DH_{j+1}^*(x) = H_{j+1}^*(x)D = H_j^*(x), & x \in R_0^n, \ 1 \leq j < n - 1. \end{cases} \tag{6}$$

Obviously, the higher order kernel function $H_k^*(x)$ is a k−regular function.

2.4. *Differential Form*

Let

$$d\widehat{x}_i = dx_1 \wedge \cdots \wedge dx_{i-1} \wedge dx_{i+1} \cdots \wedge dx_n, \quad i = 1, 2, \cdots, n,$$

$$d\sigma = \sum_{i=1}^{n}(-1)^{i-1}e_i d\widehat{x}_i, \ \overrightarrow{n} = \sum_{i=1}^{n}e_i n_i,$$

n_i is the ith component of the unit normal vector $\overrightarrow{n}$, $d\sigma = \overrightarrow{n}\,ds$. ds is the area differential, dx^n is the volume differential, i.e. $dx^n = dx_1 \wedge \cdots \wedge dx_n$.

3. Some Properties of k-Biregular Function in Clifford Analysis

Suppose $\Omega = \Omega_1 \times \Omega_2 \subset R^m \times R^{m_0}$ is a nonempty, connected and open set and $m \geq 2$, $m_0 \geq 2$. $\partial\Omega$ is a differentiable, oriented and compact Liapunov surface[3].

Lemma 1[3] Let $\Omega \subset R^n$ be a connected open set, $f \in C^{(r)}(\overline{\Omega}, C(V_{n,0}))$, $k \leq r, k < n$ and $H_j^*(x)$ is the function as equation (3). Then for any $z \in \Omega$, we can obtain

$$f(z)=\sum_{j=0}^{k-1}(-1)^j\int_{\partial\Omega} H_{j+1}^*(x-z)d\sigma_x[D^j f(x)]+(-1)^k\int_{\Omega} H_k^*(x-z)[D^k f(x)]dx^n$$

or

$$f(z)=\sum_{j=0}^{k-1}(-1)^j\int_{\partial\Omega} (f(x)D^j)d\sigma_x H_{j+1}^*(x-z)+(-1)^k\int_{\Omega} (f(x)D^k)H_k^*(x-z)dx^n.$$

Theorem 1 (Cauchy-Pompeiu formula) Let Ω and $\partial\Omega$ be as stated above, $\overline{\Omega} = \Omega \cup \partial\Omega, f \in F_\Omega^{(r)}, k \leq r, k < \min\{m, m_0\}$ and $H_j^*(x)$ is the above function. Then for any $z = (x, y) \in \Omega$, we have

$$f(x,y)$$

$$= \sum_{i=0,j=0}^{k-1}(-1)^{i+j}\int_{\partial\Omega_1 \times \partial\Omega_2} H_{i+1}^*(u-x)d\sigma_u \cdot [D_u^i f(u,v)D_v^j] \cdot d\sigma_v H_{j+1}^*(v-y)$$

$$+(-1)^k\sum_{i=0}^{k-1}(-1)^i\int_{\partial\Omega_1 \times \Omega_2} H_{i+1}^*(u-x)d\sigma_u \cdot [D_u^i f(u,v)D_v^k] \cdot H_k^*(v-y)dv^{m_0}$$

$$+(-1)^k \sum_{j=0}^{k-1}(-1)^j \int_{\Omega_1 \times \partial\Omega_2} H_k^*(u-x)\cdot [D_u^k f(u,v)D_v^j]\cdot d\sigma_v H_{j+1}^*(v-y)du^m$$

$$+\int_{\Omega_1 \times \Omega_2} H_k^*(u-x)\cdot [D_u^k f(u,v)D_v^k]\cdot H_k^*(v-y)du^m dv^{m_0}.$$

Proof: For any $z \in \Omega$, fixing the variable y, $f(x,y)$ is a $k-$regular function about variable x. By Lemma 1, we have

$$f(x,y)$$

$$=\sum_{i=0}^{k-1}(-1)^i \int_{\partial\Omega_1} H_{i+1}^*(u-x)d\sigma_u [D_u^i f(u,y)] + (-1)^k \int_{\Omega_1} H_k^*(u-x)[D_u^k f(u,y)]du^m.$$

Fixing the variable u, by Lemma 1 again we have

$$D_u^i f(u,y) = \sum_{j=0}^{k-1}(-1)^j \int_{\partial\Omega_2} (D_u^i f(u,v)D_v^j)d\sigma_v H_{j+1}^*(v-y)$$

$$+(-1)^k \int_{\Omega_2} (D_u^i f(u,v)D_v^k)H_k^*(v-y)dv^{m_0}.$$

Put the equation into the first equation and we can get the result.

Theorem 2 (Cauchy integral formula) Let $\Omega, \partial\Omega$ be as stated above and $f \in R^k(\Omega, C(V_{n,0})), k < \min\{m, m_0\}$. Then for any $(x,y) \in \Omega$, we have

$$f(x,y)$$

$$= \sum_{i=0,j=0}^{k-1}(-1)^{i+j} \int_{\partial\Omega_1 \times \partial\Omega_2} H_{i+1}^*(u-x)d\sigma_u \cdot [D_u^i f(u,v)D_v^j]\cdot d\sigma_v H_{j+1}^*(v-y).$$

Proof: It can be proved by Theorem 1.

Definition 3 Let Ω be as stated above and $F = \{f\}$ is a set of $k-$biregular function in Ω. If for any sequence $\{f_n\} \subset F$, there exists a subsequence which is uniformly convergent in any closed subset of Ω, F is called as a normal family in Ω.

Definition 4   Suppose Ω be as stated above and $F = \{f\}$ is a family of $k-$biregular functions in Ω. If for any compact set K, there exists a constant $M(K)$, such that $|D_x^i f(z) D_y^j| \leq M(K)$ when $z \in K$ and $f \in F$, $i, j = 0, 1, \cdots, k-1$, then F is called as a locally and uniformly bounded family in Ω.

Lemma 2   For any x, z and $z_0 \in R^n$, when $j \geq 0$ we have

$$\left| \frac{(x-z)^{j+1}}{|x-z|^n} - \frac{(x-z_0)^{j+1}}{|x-z_0|^n} \right|$$

$$\leq \left[\sum_{k=1}^{n-1} \frac{(|x|+|z_0|)^j}{|x-z|^k |x-z_0|^{n-k}} + \sum_{k=1}^{j} \frac{(|x|+|z_0|)^{j-k}}{|x-z|^{n-k}} \right] |z - z_0|.$$

We omit the proof here.

Theorem 3   Let Ω be as stated above, $F = \{f\}$ be a $k-$biregular function family in Ω and F be locally and uniformly bounded in Ω. Then F is a normal family in Ω.

Proof: For any $\{f_n\} \subset F$, we only need to prove that we can get a subsequence of $\{f_n\}$, which is uniformly convergent in any closed subset of Ω. We prove it by following three steps.

(1)   First we prove that for any given bounded and closed set $E \subset \Omega$, $\{f_n(z)\}$ satisfies for any $\varepsilon > 0$, there exists a number $\delta > 0$, such that for any $z_1, z_2 \in E$ and $|z_1 - z_2| < \delta$, we have $|f_n(z_1) - f_n(z_2)| < \varepsilon$, $n = 1, 2, \cdots$.

(2)   Next we prove that we can choose a subsequence $\{f_{n,n}(z)\}$ from $\{f_n(z)\}$, such that it is convergent at all rational points (whose coordinates are rational numbers) in Ω.

(3)   Lastly we prove $\{f_{n,n}(z)\}$ is uniformly convergent in any bounded and closed set $E \subset \Omega$.

Remark: Lemma 2 is important in this proof. We omit the details.

Theorem 4   Let Ω be as stated above, $\{f_n\}$ be a $k-$biregular function sequence in Ω and the sequence be locally and uniformly bounded in Ω. Then if $\{f_n\}$ is convergent in some connected open set $U \subset \Omega$, $\{f_n\}$ is uniformly convergent in any closed subset of Ω.

Proof: First we prove $\{f_n\}$ is convergent in Ω. We suppose there exists a point $a \in \Omega$, such that $\{f_n(a)\}$ isn't convergent on it. Then we choose two subsequences $\{f_{n,1}(a)\}$ and $\{f_{n,2}(a)\}$ from $\{f_n(a)\}$, which converge to Clifford numbers b_1, b_2, where $b_1 \neq b_2$. Then $\{f_{n,1}(z)\}$ and $\{f_{n,2}(z)\}$ are locally and uniformly bounded in Ω. By Theorem 3, we can choose two subsequences $\{f_{n,1}^*(z)\}$ and $\{f_{n,2}^*(z)\}$ from $\{f_{n,1}(z)\}$ and $\{f_{n,2}(z)\}$ respectively, which is uniformly convergent to $f_1^*(z)$ and $f_2^*(z)$ in any closed subset of Ω. Because

$$\lim_{n \to \infty} f_{n,1}^*(a) = f_1^*(a) = b_1, \quad \lim_{n \to \infty} f_{n,2}^*(a) = f_2^*(a) = b_2,$$

we know $f_1^*(a) \neq f_2^*(a)$. But $\{f_n(z)\}$ is convergent in some connected open set $U \subset \Omega$, i.e. for any $z \in U, \lim_{n \to \infty} f_n(z) = f(z)$. Because $\{f_{n,1}^*(z)\}, \{f_{n,2}^*(z)\}$ are subsequences of $f_n(z)$, we have

$$\lim_{n \to \infty} f_{n,1}^*(z) = f_1^*(z), \quad \lim_{n \to \infty} f_{n,2}^*(z) = f_2^*(z).$$

Hence for any $z \in U$, we know $f_1^*(z) = f_2^*(z)$. Then $f_1^*(a) = f_2^*(a)$. It contradicts the hypothesis $f_1^*(a) \neq f_2^*(a)$. Then $\{f_n(z)\}$ is convergent in Ω.

Similar to the part 3 of the proof of Theorem 3 we can prove $\{f_n\}$ is uniformly convergent in any closed subset of Ω. Then the result is true.

Remark: These results denote that $k-$biregular function has the basic properties of the holomorphic function with a complex variable.

References

1. F.Brackx. On (k)-monogenic functions of a quaternion variable. Function Theoretic Methods in Differential Equations, 1976, 8: 22-44
2. R.Delanghe, F.Brackx. Hypercomplex function theory and Hilbert modules with reproducing kernel. Proc.London Math.Soc., 1978, 37: 545-576
3. H.Begehr, J.Du, Z.Zhang. On higher order Cauchy-Pompeiu formula in Clifford analysis and its application. Gen.Math, 2003, 11: 5-26

II.2. Analytical, geometrical and numerical methods in Clifford- and Cayley-Dickson algebras

Organisers: K. Gürlebeck, V. Kisil, W. Sprößig

The session "Analytical, geometrical and numerical methods in Clifford- and Cayley-Dickson-algebras" took place on 13th and 14th July 2009 within the 7th ISAAC Congress in London. There were 15 talks (including a distinguished by T. Fokas) which broadly represent various areas of hypercomplex analysis and applications. Presented topics were boundary value problems (W. Sprößig), reproducing kernels (R. Krausshar), orthogonal polynomials (S. Bock), wavelets (S. Bernstein), integral transforms (V. Kisil), physical models (J. Tolksdorf, R. Farwell) and signal processing (G. Scheuermann, J. Hogan). The variety of approaches included both continuous and discrete analysis (U. Kähler, N. Faustino), commutative (D. Pinotsis) and non-commutative algebras, pure topics (S. Georgiev) and applied questions (N. Vieira, R. Leandre, J. Helmstetter). The further strength of the field was demonstrated on a sister session on Clifford and quaternion analysis.

Further results in discrete Clifford analysis

N. Faustino

Departamento de Matemática, Universidade de Aveiro,
Aveiro, P- 3810-193, Portugal
E-mail: nfaust@ua.pt
www.nelson-faustino.tk

We present a short and self-contained exposition of discrete Clifford analysis from the umbral calculus perspective.

Discrete Clifford analysis in its minimal form, deals with the canonical description of raising, lowering and degree-preserving Clifford-valued operators (i.e. the umbral counterparts of Dirac, vector variable operator and Euler operator, respectively) in terms of symmetries underlying the orthosymplectic Lie algebra $\mathrm{osp}(1|2)$ as a refinement of the Lie algebra $\mathrm{sl}_2(\mathbb{R})$.

A concrete example of canonical discretization of Dirac operators and vector-variable operators will be consider in connection with the quantum harmonic oscillator.

Keywords: Umbral Clifford analysis; orthosymplectic Lie algebra representation; discrete harmonic oscillator.

1. Umbral Calculus Revised

In this section we review some definitions regarding umbral calculus. The presented proofs and further results can be found in[1,10,11] or alternatively in,[6] Chapter 1.

For the ring of polynomials $\mathbb{R}[\underline{x}]$ over $\underline{x} = (x_1, x_2, \ldots, x_n) \in \mathbb{R}^n$ and for a vector of non-negative integers $\alpha = (\alpha_1, \alpha_2, \ldots \alpha_n)$, we set by $\underline{x}^\alpha = x_1^{\alpha_1} x_2^{\alpha_2} \ldots x_n^{\alpha_n}$ a monomial over $\underline{x}$, and a polynomial $p(\underline{x})$ as a (finite) linear combination of monomials $\underline{x}^\alpha$. The gradient operator will be denoted by the n−tuple $\partial_{\underline{x}} := (\partial_{x_1}, \partial_{x_2}, \ldots, \partial_{x_n})$ while the algebra of all linear operators acting on $\mathbb{R}[\underline{x}]$ will be denoted by $\mathrm{End}(\mathbb{R}[\underline{x}])$.

In addition, we will also take into account the compact notations $\partial_{\underline{x}}^\alpha := \partial_{x_1}^{\alpha_1} \partial_{x_2}^{\alpha_2} \ldots \partial_{x_n}^{\alpha_n}$, $\alpha! = \alpha_1! \alpha_2! \ldots \alpha_n!$, $\binom{\beta}{\alpha} = \frac{\beta!}{\alpha!(\beta-\alpha)!}$. An operator $Q \in \mathrm{End}(\mathbb{R}[\underline{x}])$ is shift-invariant if the equation

$$[Q, T_{\underline{y}}]P(\underline{x}) := Q(T_{\underline{y}}P(\underline{x})) - T_{\underline{y}}(Q\, P(\underline{x})) = 0$$

fulfils for any $P \in \mathbb{R}[\underline{x}]$ and $\underline{y} \in \mathbb{R}^n$. Hereby $[\mathbf{a}, \mathbf{b}] := \mathbf{ab} - \mathbf{ba}$ denotes the commuting bracket between $\mathbf{a}$ and $\mathbf{b}$.

Under the shift-invariance condition for Q, the first expansion theorem (c.f.[1]) states that any linear operator $Q : \mathbb{R}[\underline{x}] \rightarrow \mathbb{R}[\underline{x}]$ is shift-invariant if and only if it can be represented as the formal power series $Q = \sum_{|\alpha|=0}^{\infty} \frac{a_\alpha}{\alpha!} \partial_{\underline{x}}^\alpha$, with $a_\alpha = [Q\underline{x}^\alpha]_{\underline{x}=\underline{0}}$.

Now let $O_{\underline{x}} = (O_{x_1}, O_{x_2}, \ldots, O_{x_n})$ be a multivariate operator. We say that $O_{\underline{x}}$ is shift-invariant if its operator components $O_{x_1}, O_{x_2}, \ldots, O_{x_n}$ are shift-invariant. Moreover $O_{\underline{x}}$ is a multivariate delta operator if $O_{x_j}(x_k) = c\delta_{jk}$, holds for all $j, k = 1, 2, \ldots, n$, where c is a non-vanishing constant.

It can be shown that if $O_{\underline{x}}$ is a multivariate delta operator, any multivariate delta operator $O_{\underline{x}}$ uniquely determines a unique polynomial sequence of binomial type, (c.f.[6] Theorems 1.1.12 and 1.1.13)

$$V_\beta(\underline{x} + \underline{y}) = \sum_{|\beta|=0}^{|\alpha|} \binom{\beta}{\alpha} V_\alpha(\underline{x}) V_{\beta-\alpha}(\underline{y}) \tag{1}$$

satisfying $V_{\underline{0}}(\underline{x}) = 1$, $V_\alpha(\underline{0}) = \delta_{\alpha,0}$ and $O_{x_j} V_\alpha(\underline{x}) = \alpha_j V_{\alpha-\mathbf{v}_j}(\underline{x})$, where $\mathbf{v}_j$ stands the $j-$element of the canonical basis of $\mathbb{R}^n$.

The incherle derivative of each O_{x_j} with respect to the coordinate x_j is defined as the commutator acting on $f(\underline{x})$, that is

$$O'_{x_j} f(\underline{x}) := [O_{x_j}, x_j] f(\underline{x}) = O_{x_j}(x_j f(\underline{x})) - x_j(O_{x_j} f(\underline{x})), \tag{2}$$

The Pincherle derivative (2) plays an important role in the construction of basic polynomial sequences. In particular, notice that O'_{x_j} is shift-invariant whenever O_{x_j} is shift-invariant (c.f.,[6] Lemma 1.2.1). Since from the definition $O'_{x_j}(\mathbf{1}) = O_{x_j}(x_j)$ is a non-vanishing constant and thus $(O'_{x_j})^{-1}$ exists (c.f.,[6] Corollary 1.1.4).

Thus, the sequences $V_\alpha(\underline{x})$ are computed viz the action of the raising operator $(\underline{x}')^\alpha := \prod_{k=1}^n (x'_k)^{\alpha_k}$ on the polynomial $\Phi = \mathbf{1}$, i.e.

$$V_\alpha(\underline{x}) = (\underline{x}')^\alpha \mathbf{1}, \quad \text{where} \quad x'_k := x_k(O'_{x_k})^{-1}. \tag{3}$$

The properties of basic polynomial sequences are naturally handled within the the extension of the mapping $\Psi_{\underline{x}} : \underline{x}^\alpha \mapsto V_\alpha(\underline{x})$ (known in[10] as the Scheffer map) to $\mathbb{R}[\underline{x}]$. whose inverse $\Psi_{\underline{x}}^{-1}$ is given by the linear extension of $\Psi_{\underline{x}}^{-1} : V_\alpha(\underline{x}) \mapsto \underline{x}^\alpha$ to $\mathbb{R}[\underline{x}]$.

Furthermore, the following intertwining properties that fulfil on $\mathbb{R}[\underline{x}]$

$$O_{x_j}\Psi_{\underline{x}} = \Psi_{\underline{x}}\partial_{x_j}, \text{ and } x_j'\Psi_{\underline{x}} = \Psi_{\underline{x}}x_j$$

assure that x_j' and O_{x_j} generate the ose algebra, i.e. the free algebra generated from the $\Phi = \mathbf{1}$ (vacuum vector) and by $2n + 1$ elements $x_1', \ldots, x_n', O_{x_1}, \ldots, O_{x_n}, \mathbf{id}$ satisfying $O_{x_j}(\Phi) = 0$, for $j = 1, \ldots, n$ and the graded commutating relations:

$$[O_{x_j}, O_{x_k}] = 0 = [x_j', x_k'], [O_{x_j}, x_k'] = \delta_{jk}\mathbf{id}. \tag{4}$$

The algebra generated by the commuting relations (4) is also known in the literature as the Weyl-Heisenberg algebra while the Rodrigues formula (3) is nothing else than the basic lemma of quantum field theory.

In the quantum mechanical description of umbral calculus described above, the raising operators $x_j' : V_\alpha(\underline{x}) \mapsto V_{\alpha+\mathbf{v}_j}(\underline{x})$ are uniquely determined by assuming the commuting relations (4) as constraints. This in turn give us many degrees of freedom for the construction of x_j'. For example, in[4] it was consider $x_j' = \frac{1}{2}(x_j(O_{x_j}')^{-1} + (O_{x_j}')^{-1}x_j)$ as a special type of canonical discretization which sifts self-adjoint operators from *continuum* to equidistant lattices.

2. Orthosymplectic Lie Algebra Representation of Umbral Clifford Analysis

In what follows we will retain the notation introduced in Section 1. Additionally, by means of the anti-commuting bracket between $\mathbf{a}$ and $\mathbf{b}$, $\{\mathbf{a}, \mathbf{b}\} = \mathbf{ab} + \mathbf{ba}$, we define the Clifford algebra of signature $(0, n)$ as the algebra $Cl_{0,n}$ determined by the set of vectors $\mathbf{e}_1, \mathbf{e}_2, \ldots, \mathbf{e}_n$ satisfying $\{\mathbf{e}_j, \mathbf{e}_k\} = -2\delta_{jk}$.

The tensor product $\mathcal{P} = \mathbb{R}[\underline{x}] \otimes Cl_{0,n}$ denotes the algebra of Clifford-valued polynomials and $\text{End}(\mathcal{P})$ corresponds the algebra of all linear operators acting on $\mathcal{P}$.

The Heisenberg-Weyl character of the operators x_j' and O_{x_j} together with the radial character of the generators of the Clifford algebra $Cl_{0,n}$ (c.f.[12]) suggests to define umbral Clifford analysis as the study of the algebra of differential operators $\text{Alg}\{x_j', O_{x_j}, \mathbf{e}_j : j = 1, \ldots, n\}$.

By taking linear combinations of elements of the above algebra, we introduce the umbral of Dirac operator, vector variable and Euler operator as the following vector-field operators $D' = \sum_{j=1}^{n} \mathbf{e}_j O_{x_j}$, $x' = \sum_{j=1}^{n} \mathbf{e}_j x_j'$ and $E' = \sum_{j=1}^{n} x_j' O_{x_j}$, respectively.

Here, we would like to point out that D', x' and E' shall be understood as basic left endomorphisms acting on the algebra $\mathrm{End}(\mathcal{P})$, i.e.

$$x' : F(\underline{x}) \mapsto \sum_{j=1}^{n} \mathbf{e}_j x'_j(F(\underline{x})), \qquad D' : F(\underline{x}) \mapsto \sum_{j=1}^{n} \mathbf{e}_j O_{x_j}(F(\underline{x}))$$

$$\text{and} \quad E' : F(\underline{x}) \mapsto \sum_{j=1}^{n} x'_j O_{x_j}(F(\underline{x})).$$

The Heisenberg-Weyl character of the operators x'_j and O_{x_j} together with the anti-commutation rules $\{\mathbf{e}_j, \mathbf{e}_k\} = -2\delta_{jk}$ leads to the following lemma:

Lemma 2.1. *The operators $x', D', E' \in End(\mathcal{P})$ satisfy the following anti-commutation relations*

$$\{x', x'\} = -2\sum_{j=1}^{n}(x'_j)^2, \ \{D', D'\} = -2\sum_{j=1}^{n} O_{x_j}^2, \ \{x', D'\} = -2E' - n\mathbf{id}.$$

From Lemma 2.1, $(x')^2 = \frac{1}{2}\{x', x'\}$ and $-(D')^2 = -\frac{1}{2}\{D', D'\}$ are scalar-valued operators and then correspond to generalizations of the norm squared of a vector variable in the Euclidean space and the Laplacian operator, respectively. On the other hand, the third relation of Lemma 2.1 gives the action of the umbral Euler operator on the algebra $\mathrm{End}(\mathcal{P})$ as

$$E' = \sum_{j=1}^{n} x'_j O_{x_j} = -\frac{1}{2}\left(\{x', D'\} + n\mathbf{id}\right).$$

Moreover, the action of both sides of the above identity on the polynomial $\Phi = \mathbf{1}$ allows us to recast the dimension of the ambient space $\mathbb{R}^n$ as $n = -D'(x'\mathbf{1})$. From the border point of quantum mechanics, this corresponds to twice of the ground level energy associated to the ground level eigenstate $\Phi = \mathbf{1}$.

The subsequent lemma gives a canonical realization of a Lie superalgebra isomorphic to the orthosymplectic Lie algebra of type $\mathrm{osp}(1|2)$ in terms of the operators x', D' and $E' + \frac{n}{2}\mathbf{id}$

Lemma 2.2. *The operators x', D' and $E'_{n/2} := E' + \frac{n}{2}\mathbf{id}$ generate a finite-dimensional ie superalgebra in $End(\mathcal{P})$. The remaining commutation relations are*

$$[x', (x')^2] = 0, \qquad [x', -\Delta'] = -2D', \qquad [E'_{n/2}, x'] = x'$$

$$[D', (x')^2] = 2x', \qquad [D', -\Delta'] = 0, \qquad [E'_{n/2}, D'] = -D'$$

$$[(x')^2, -\Delta'] = 4E'_{n/2}, \ \left[E'_{n/2}, -(x')^2\right] = -2(x')^2, \ \left[E'_{n/2}, -\Delta'\right] = 2\Delta'$$

The proof of the above lemma follows the same order of ideas of the proof of Lemma 5.2.1 obtained in[6] (see Subsection 5.2 for further details).

Recall that the orthosymplectic Lie algebra of type $\mathrm{osp}(1|2)$ (c.f.[8]) is defined as a Lie superalgebra given by the direct sum of linear spaces

$$\mathrm{span}\left\{\mathbf{p}^-,\mathbf{p}^+,\mathbf{q}\right\} \oplus \mathrm{span}\left\{\mathbf{r}^-,\mathbf{r}^+\right\}$$

equipped with the standard graded commutator $[\cdot,\cdot]$ such that $\mathbf{p}^-,\mathbf{p}^+,\mathbf{q},\mathbf{r}^-$ and $\mathbf{r}^+$ satisfy the following standard commutation relations (see e.g.[8]):

$$\left[\mathbf{q},\mathbf{p}^\pm\right] = \pm\mathbf{p}^\pm, \quad \left[\mathbf{p}^+,\mathbf{p}^-\right] = 2\mathbf{q},$$

$$\left[\mathbf{q},\mathbf{r}^\pm\right] = \pm\frac{1}{2}\mathbf{p}^\pm, \quad \left[\mathbf{r}^+,\mathbf{r}^-\right] = \tfrac{1}{2}\mathbf{q},$$

$$\left[\mathbf{p}^\pm,\mathbf{r}^\mp\right] = -\mathbf{r}^\pm, \quad \left[\mathbf{r}^\pm,\mathbf{r}^\pm\right] = \pm\tfrac{1}{2}\mathbf{p}^\pm.$$

Here we would like to remark that the Lie algebra $\mathrm{sl}_2(\mathbb{R})$ appears as a refinement of $\mathrm{osp}(1|2)$, in the sense that the canonical generators $\mathbf{p}^-,\mathbf{p}^+,2\mathbf{q}$ itself generate $\mathrm{sl}_2(\mathbb{R})$. Furthermore, the normalization

$$\mathbf{p}^- = -\tfrac{1}{2}\Delta', \; \mathbf{p}^+ = -\tfrac{1}{2}(x')^2, \; \mathbf{q} = \tfrac{1}{2}\left(E' + \tfrac{n}{2}\mathbf{id}\right), \mathbf{r}^+ = \frac{1}{2\sqrt{2}}ix', \; \mathbf{r}^- = \frac{1}{2\sqrt{2}}iD'$$

lead to standard commutation relations for $\mathrm{osp}(1|2)$, as desired. Moreover, $\mathbf{p}^+ = \tfrac{1}{2}(x')^2$, $\mathbf{p}^- = \tfrac{1}{2}\Delta'$ and $2\mathbf{q} = E' + \tfrac{n}{2}\mathbf{id}$ are the canonical generators of the Lie algebra $\mathrm{sl}_2(\mathbb{R})$.

3. The Discrete Harmonic Oscillator

We now consider the discrete harmonic oscillator on an equidistant lattice of mesh-size $h > 0$ as the chain of n independent operators given by the Hamiltonian operator acting on $f(\underline{x})$:

$$\mathcal{H}'f(\underline{x}) = \sum_{j=1}^{n} \frac{2f(\underline{x}) - f(\underline{x}+h\mathbf{v}_j) - f(\underline{x}-h\mathbf{v}_j)}{2h^2} + \frac{1}{2}(x'_j)^2 f(\underline{x}).$$

Using the language of umbral Clifford analysis, the Hamiltonian operator can be rewritten as $\mathcal{H}' = \tfrac{1}{2}\left((D')^2 - (x')^2\right)$, where D' is a central difference operator on an equidistant lattice with mesh-width $\tfrac{h}{2}$, given by $D' = \sum_{j=1}^{n} \mathbf{e}_j \frac{1}{h}\left(T_{\frac{h}{2}\mathbf{v}_j} - T_{-\frac{h}{2}\mathbf{v}_j}\right)$.

Here we would like to notice that $-(D')^2$ corresponds to the star-Laplacian on an equidistant lattice of mesh-width $h > 0$, that is

$$-(D')^2 = \sum_{j=1}^{n} \frac{T_{h\mathbf{v}_j} - 2\mathbf{id} + T_{-h\mathbf{v}_j}}{h^2}.$$

Contrary to N. Faustino and U. Kähler where the usage of forward/backward difference Dirac operator were considered, the above splitting allows us to refine discrete harmonic analysis on an equidistant lattice with mesh-width h in terms of central difference Dirac operators on equidistant lattices with mesh-width $\frac{h}{2}$.

Therefore, the construction of the operator x' shall be take into account the formal series expansion for the Pincherle derivative given by formal series expansion $\frac{1}{2}\left(T_{\frac{h}{2}\mathbf{v}_j} + T_{-\frac{h}{2}\mathbf{v}_j}\right) = \cosh\left(\frac{h}{2}\partial_{x_j}\right)$. Here we would like to point out that according to,[4] inversion of $\frac{1}{2}\left(T_{\frac{h}{2}\mathbf{v}_j} + T_{-\frac{h}{2}\mathbf{v}_j}\right)$ shall be computed under periodic boundary conditions on the equidistant lattice $h\mathbb{Z}^n$.

Analogously to Subsection 2, it is also possible to obtain representations of the Lie superalgebra $\mathrm{osp}(1|2)$.

Indeed $\mathcal{H}'$ is splitted in terms of the ladder operators $D'_{\pm} = \frac{1}{\sqrt{2}}(x' \mp D')$, that is $\{D'_+, D'_-\} = -2\mathcal{H}'$ and moreover under straightforward computations, it follows that Lemma 2.2 also fulfils for the operators D'_+, D'_- and $\mathcal{H}'$, respectively (see,[6] Subsection 5.2 for further details).

4. Conclusion and Outlook

The umbral Clifford analysis approach appears as a mimetic transcription of classical Clifford analysis (c.f[3]) and classical harmonic analysis (c.f.[9]) by means representations of $\mathrm{osp}(1|2)$ and $\mathrm{sl}_2(\mathbb{R})$, respectively, establishing a contact with the celebrated Wigner Quantum systems introduced in[13] by E.P. Wigner. On the other hand, the intertwining properties at the level of the algebra $\mathrm{End}(\mathcal{P})$:

$$\Psi_{\underline{x}}D = D'\Psi_{\underline{x}}, \ \Psi_{\underline{x}}x = x'\Psi_{\underline{x}}, \ \Psi_{\underline{x}}E = E'\Psi_{\underline{x}}$$

allows us to interpret classical and discrete Clifford analysis as two quantal systems on which the Sheffer operator $\Psi_{\underline{x}}$ acts as a gauge transformation preserving the canonical relations between both systems. Hereby D and x stand the classical Dirac operator and vector derivative while $\Psi_{\underline{x}}$ denotes the Sheffer map introduced in Section 1.

In conclusion, this approach merge the radial algebra approach proposed by F. Sommen[12] for Clifford analysis with the quantum mechanical approach for umbral calculus described by Di Bucchianico, Loeb and Rota,[2] giving in this way a core of promising applications on both fields.

More concrete applications will appear in the forthcoming paper.[7]

References

1. A. Di Bucchianico and D.E. Loeb *Operator expansion in the derivative and multiplication by x*, Vol. 4 (Integral Transf. Spec. Func., 1996) pp. 49–68.
2. A. Di Bucchianico, D.E. Loeb and G.C. Rota, *Umbral calculus in Hilbert space* In: B. Sagan and R.P. Stanley (eds.), Mathematical Essays in Honor of Gian-Carlo Rota , pp. 213–238, Birkhäuser, Boston, 1998.
3. R. Delanghe, F. Sommen and V. Souček, *Clifford algebras and spinor-valued functions* (Kluwer Academic Publishers, 1992).
4. A. Dimakis, F. Mueller-Hoissen and T. Striker, *Umbral calculus, discretization, and quantum mechanics on a lattice* Vol. 29 (J. Phys. A, 1996) pp. 6861–6876.
5. N. Faustino, U. Kähler, *Fischer Decomposition for Difference Dirac Operators*, Vol. 17, no. 1 (Adv. Appl. Cliff. Alg., 2007) pp. 37–58.
6. N. Faustino, *Discrete Clifford Analysis*, PhD thesis, Universidade de Aveiro (Aveiro, Portugal, 2009), pp ix+130.
7. N. Faustino and G. Ren, *Almansi Theorems in Umbral Clifford Analysis and the Quantum Harmonic Oscillator*, in preparation.
8. L. Frappat, A. Sciarrino, P. Sorba, *Dictionary of Lie algebras and super algebras*, (Academic Press, New York, 2000).
9. R. Howe, E. Tan, *Nonabelian harmonic analysis: Applications of $SL(2, \mathbb{R})$* (Universitext. Springer-Verlag, New York, 1992).
10. S. Roman, *The Umbral Calculus* (Academic Press, San Diego, 1984).
11. S. Roman and G-C. Rota, *The umbral calculus* (Adv. Math. 27, 1978) pp. 95–188.
12. F. Sommen, *An Algebra of Abstract vector variables*, Vol. 54, no. 3 (Portugaliae Math., 1997) pp. 287–310.
13. E.P. Wigner, *Do the Equations of Motion Determine the Quantum Mechanical Commutation Relations?* Vol. 77 (Phys. Rev., 1950) pp. 711–712.

A note on the linear systems in quaternions

Svetlin G. Georgiev

University of Sofia,
Mathematics and Informatics, Department of Differential Equations,
Blvd "Tzar Osvoboditel" 15, Sofia 1000, Bulgaria
E-mail: sgg2000bg@yahoo.com

In this paper we investigate linear systems in quaternions. We give conditions under which they have solutions.

Keywords: Quaternion systems; existence; uniqueness.

1. Introduction

In this paper we will consider the linear system

$$(1) \qquad \sum_{l=1}^{r} \sum_{m=1}^{n} p_{lm}^{s} x_m q_{lm}^{s} = A_s, \quad s = 1, 2, \ldots, n,$$

where $n, r \geq 1$ are given constants, p_{lm}^{s}, q_{lm}^{s}, A_s, $l = 1, \ldots, r$, $m = 1, \ldots, n$, $s = 1, \ldots, n$, are given real quaternions, x_m, $m = 1, \ldots, n$, are unkown real quaternions.

Here we propose an algorithm for finding a solution to the system (1). Also, we give necessary and sufficient condition for solvability of the system (1) and some examples.

In [1] is investigated the equation

$$\sum_{j=1}^{\nu} a^{(j)} x b^{(j)} = e,$$

where $a^{(j)}$, $b^{(j)}$, $j = 1, 2, \cdots, \nu$, are given real quaternions and x is unknown real quaternion. The method in [1] is based on a fixed point formulation.

In [2] are proposed the solvability conditions and general solutions to the pairs of equations in quaternions

$$\overline{x} a y = b$$
$$\overline{y} b x = a,$$

a and b are given real quaternions, x and y are unknown real quternions.

Here we propose a new method for investigation of the system (1). This nice method gives new results.

2. Preliminaries

The concept of the quaternion was introduced by Hamilton in 1843. It is an extension of complex number to 4D algebra. The space of quaternion is denoted by $\mathbb{H}$. Every element of the space $\mathbb{H}$ is a linear combination of a real scalar and three orthogonal imaginary units, denoted by i, j, k with real cofficients

$$\mathbb{H} = \left\{ a = a_0 + ia_1 + ja_2 + ka_3 \quad : \quad a_0, a_1, a_2 \in \mathbb{R} \right\}.$$

The elements i, j and k obey Hamilton's rules

$$i^2 = j^2 = k^2 = -1,$$
$$ij = -ji = k, \quad ik = -ki = -1, \quad jk = -kj = i,$$
$$ijk = -1.$$

Every quaternion

$$a = a_0 + ia_1 + ja_2 + ka_3, \quad a_0, a_1, a_2, a_3 \in \mathbb{R},$$

has a quaternion conjugate

$$\overline{a} = a_0 - ia_1 - ja_2 - ka_3.$$

For the quaternion conjugation we have

$$\overline{a+b} = \overline{a} + \overline{b}, \quad \overline{\overline{a}} = a, \quad \overline{ab} = \overline{b}\,\overline{a}, \quad \forall a, b \in \mathbb{H}.$$

For

$$a = a_0 + ia_1 + ja_2 + ka_3, \quad a_0, a_1, a_2, a_3 \in \mathbb{R},$$
$$b = b_0 + ib_1 + jb_2 + kb_3, \quad b_0, b_1, b_2, b_3 \in \mathbb{R},$$
$$c = c_0 + ic_1 + jc_2 + kc_3, \quad c_0, c_1, c_2, c_3 \in \mathbb{R},$$

we have

$$ab = a_0 b_0 - a_1 b_1 - a_2 b_2 - a_3 b_3 +$$
$$+ i(a_0 b_1 + a_1 b_0 + a_2 b_3 - a_3 b_2) +$$
$$+ j(a_0 b_2 - a_1 b_3 + a_2 b_0 + a_3 b_1) +$$
$$+ k(a_0 b_3 + a_1 b_2 - a_2 b_1 + a_3 b_0),$$

and

$$abc = (a_0 b_0 c_0 - a_1 b_1 c_0 - a_2 b_2 c_0 - a_3 b_3 c_0 - a_0 b_1 c_1 - a_1 b_0 c_1 - a_2 b_3 c_1 + a_3 b_2 c_1 -$$
$$- a_0 b_2 c_2 + a_1 b_3 c_2 - a_2 b_0 c_2 - a_3 b_1 c_2 - a_0 b_3 c_3 - a_1 b_2 c_3 + a_2 b_1 c_3 - a_3 b_0 c_3) +$$
$$+ i(a_0 b_0 c_1 - a_1 b_1 c_1 - a_2 b_2 c_1 - a_3 b_3 c_1 + a_0 b_1 c_0 + a_1 b_0 c_0 + a_2 b_3 c_0 - a_3 b_2 c_0 +$$
$$+ a_0 b_2 c_3 - a_1 b_3 c_3 + a_2 b_0 c_3 + a_3 b_1 c_3 - a_0 b_3 c_2 - a_1 b_2 c_2 + a_2 b_1 c_2 - a_3 b_0 c_2) +$$
$$+ j(a_0 b_0 c_2 - a_1 b_1 c_2 - a_2 b_2 c_2 - a_3 b_3 c_2 - a_0 b_1 c_3 - a_1 b_0 c_3 - a_2 b_3 c_3 + a_3 b_2 c_3 +$$
$$+ a_0 b_2 c_0 - a_1 b_3 c_0 + a_2 b_0 c_0 + a_3 b_1 c_0 + a_0 b_3 c_1 + a_1 b_2 c_1 - a_2 b_1 c_1 - a_3 b_0 c_1) +$$
$$+ k(a_0 b_0 c_3 - a_1 b_1 c_3 - a_2 b_2 c_3 - a_3 b_3 c_3 + a_0 b_1 c_2 + a_1 b_0 c_2 + a_2 b_3 c_2 - a_3 b_2 c_2 -$$
$$- a_0 b_2 c_1 + a_1 b_3 c_1 - a_2 b_0 c_1 - a_3 b_1 c_1 - a_0 b_3 c_3 - a_1 b_2 c_3 + a_2 b_1 c_3 - a_3 b_0 c_3).$$

The scalar part is

$$Sc(a) = a_0,$$

the vector part of the quaternion a is

$$Vec(a) = a_1 i + a_2 j + a_3 k.$$

Also we have

$$Sc(abc) = Sc(bca),$$

i.e. we have a cyclic multiplication symmetry.

We write

$$|a| = \sqrt{a\bar{a}} = \sqrt{a_0^2 + a_1^2 + a_2^2 + a_3^2},$$

and we have

$$|ab| = |a||b|,$$
$$a^{-1} = \frac{\bar{a}}{|a|^2}.$$

3. Formulation and proof of main results

Let

$$
\begin{aligned}
(2)\qquad
B^s_{lm00} &= p^s_{lm0}q^s_{lm0} - p^s_{lm1}q^s_{lm1} - p^s_{lm2}q^s_{lm2} - p^s_{lm3}q^s_{lm3}, \\
B^s_{lm10} &= p^s_{lm1}q^s_{lm0} - p^s_{lm0}q^s_{lm1} - p^s_{lm3}q^s_{lm2} - p^s_{lm2}q^s_{lm3}, \\
B^s_{lm20} &= -p^s_{lm2}q^s_{lm0} + p^s_{lm3}q^s_{lm1} - p^s_{lm0}q^s_{lm2} - p^s_{lm1}q^s_{lm3}, \\
B^s_{lm30} &= -p^s_{lm1}q^s_{lm0} - p^s_{lm2}q^s_{lm1} + p^s_{lm1}q^s_{lm2} - p^s_{lm0}q^s_{lm3}, \\
B^s_{lm01} &= p^s_{lm0}q^s_{lm1} + p^s_{lm1}q^s_{lm0} + p^s_{lm2}q^s_{lm3} - p^s_{lm3}q^s_{lm2}, \\
B^s_{lm11} &= -p^s_{lm1}q^s_{lm1} + p^s_{lm0}q^s_{lm0} + p^s_{lm2}q^s_{lm2} + p^s_{lm3}q^s_{lm3}, \\
B^s_{lm21} &= -p^s_{lm2}q^s_{lm1} - p^s_{lm3}q^s_{lm0} + p^s_{lm0}q^s_{lm3} - p^s_{lm1}q^s_{lm2}, \\
B^s_{lm31} &= -p^s_{lm3}q^s_{lm1} + p^s_{lm2}q^s_{lm0} - p^s_{lm1}q^s_{lm3} - p^s_{lm0}q^s_{lm2}, \\
B^s_{lm02} &= p^s_{lm0}q^s_{lm2} - p^s_{lm1}q^s_{lm3} + p^s_{lm2}q^s_{lm0} - p^s_{lm3}q^s_{lm1}, \\
B^s_{lm12} &= -p^s_{lm1}q^s_{lm2} - p^s_{lm0}q^s_{lm3} + p^s_{lm3}q^s_{lm0} - p^s_{lm2}q^s_{lm1}, \\
B^s_{lm22} &= -p^s_{lm2}q^s_{lm2} + p^s_{lm3}q^s_{lm3} + p^s_{lm0}q^s_{lm0} + p^s_{lm1}q^s_{lm1}, \\
B^s_{lm32} &= -p^s_{lm3}q^s_{lm2} - p^s_{lm2}q^s_{lm3} - p^s_{lm1}q^s_{lm0} + p^s_{lm0}q^s_{lm1}, \\
B^s_{lm03} &= -p^s_{lm0}q^s_{lm3} + p^s_{lm1}q^s_{lm2} - p^s_{lm2}q^s_{lm1} - p^s_{lm3}q^s_{lm3}, \\
B^s_{lm13} &= -p^s_{lm1}q^s_{lm3} + p^s_{lm0}q^s_{lm2} - p^s_{lm3}q^s_{lm1} + p^s_{lm0}q^s_{lm3}, \\
B^s_{lm23} &= -p^s_{lm2}q^s_{lm3} - p^s_{lm3}q^s_{lm2} - p^s_{lm0}q^s_{lm1} - p^s_{lm1}q^s_{lm3}, \\
B^s_{lm33} &= -p^s_{lm3}q^s_{lm3} + p^s_{lm2}q^s_{lm2} + p^s_{lm1}q^s_{lm1} - p^s_{lm0}q^s_{lm0}.
\end{aligned}
$$

Then the system (1) obtains the following form

$$
\sum_{l=1}^{r}\sum_{m=1}^{n}\left[\left(B^s_{lm00}x_{m0} + B^s_{lm10}x_{m1} + B^s_{lm20}x_{m2} + B^s_{lm30}x_{m3}\right)+\right.
$$

$$
+i\left(B^s_{lm01}x_{m0} + B^s_{lm11}x_{m1} + B^s_{lm21}x_{m2} + B^s_{lm31}x_{m3}\right)+
$$

$$
+j\left(B^s_{lm02}x_{m0} + B^s_{lm12}x_{m2} + B^s_{lm22}x_{m2} + B^s_{lm32}x_{m3}\right)+
$$

$$
\left.+k\left(B^s_{lm03}x_{m0} + B^s_{lm13}x_{m1} + B^s_{lm23}x_{m2} + B^s_{lm33}x_{m3}\right)\right] =
$$

$$
A_{s0} + iA_{s1} + jA_{s2} + kA_{s3},
$$

from here

$$\sum_{l=1}^{r} \sum_{m=1}^{n} \left(B_{lm00}^{s} x_{m0} + B_{lm10}^{s} x_{m1} + B_{lm20}^{s} x_{m2} + B_{lm30}^{s} x_{m3} \right) = A_{s0}$$

$$\sum_{l=1}^{r} \sum_{m=1}^{n} \left(B_{lm01}^{s} x_{m0} + B_{lm11}^{s} x_{m1} + B_{lm21}^{s} x_{m2} + B_{lm31}^{s} x_{m3} \right) = A_{s1}$$

$$\sum_{l=1}^{r} \sum_{m=1}^{n} \left(B_{lm02}^{s} x_{m0} + B_{lm12}^{s} x_{m1} + B_{lm22}^{s} x_{m2} + B_{lm32}^{s} x_{m3} \right) = A_{s2}$$

$$\sum_{l=1}^{r} \sum_{m=1}^{n} \left(B_{lm03}^{s} x_{m0} + B_{lm13}^{s} x_{m1} + B_{lm23}^{s} x_{m2} + B_{lm33}^{s} x_{m3} \right) = A_{s3}.$$

Let

$$A = \begin{pmatrix} \sum_{l=1}^{r} B_{l100}^{1} & \sum_{l=1}^{r} B_{l200}^{1} & \cdots & \sum_{l=1}^{r} B_{ln30}^{1} \\ & \cdots & & \\ \sum_{l=1}^{r} B_{l103}^{1} & \sum_{l=1}^{r} B_{l203}^{1} & \cdots & \sum_{l=1}^{r} B_{ln33}^{1} \\ & \cdots & & \\ \sum_{l=1}^{r} B_{l100}^{n} & \sum_{l=1}^{r} B_{l200}^{n} & \cdots & \sum_{l=1}^{r} B_{ln30}^{n} \\ & \cdots & & \\ \sum_{l=1}^{r} B_{l103}^{n} & \sum_{l=1}^{r} B_{l203}^{n} & \cdots & \sum_{l=1}^{r} B_{ln33}^{n} \end{pmatrix},$$

where A is $(4n \times 4n)$ matrix. Let also Δ is its determinant.

Then the following theorems follow immediately.

Theorem 3.1. *The system* (1) *has exactly one solution if and only if*

$$\Delta \neq 0.$$

Let the conditions of the above theorem hold. With Δ_j we denote the determinant which is obtained from the determinant Δ replacing the j th column vector with the column vector A_s. Then

$$x_{10} = \tfrac{\Delta_1}{\Delta}, \quad x_{11} = \tfrac{\Delta_2}{\Delta}, \quad x_{12} = \tfrac{\Delta_3}{\Delta}, \quad x_{14} = \tfrac{\Delta_4}{\Delta},$$

$$x_{20} = \tfrac{\Delta_5}{\Delta}, \quad x_{21} = \tfrac{\Delta_6}{\Delta}, \quad x_{22} = \tfrac{\Delta_7}{\Delta}, \quad x_{23} = \Delta_8,$$

$$\cdots$$

$$x_{n0} = \tfrac{\Delta_{4n-3}}{\Delta}, \quad x_{n1} = \tfrac{\Delta_{4n-2}}{\Delta}, \quad x_{n2} = \tfrac{\Delta_{4n-1}}{\Delta}, \quad x_{n3} = \tfrac{\Delta_{4n}}{\Delta}.$$

From here

$$
\begin{aligned}
x_1 &= \tfrac{1}{\Delta}\left(\Delta_1 + i\Delta_2 + j\Delta_3 + k\Delta_4\right), \\
x_2 &= \tfrac{1}{\Delta}\left(\Delta_5 + i\Delta_6 + j\Delta_7 + k\Delta_8\right), \\
&\cdots \\
x_n &= \tfrac{1}{\Delta}\left(\Delta_{4n-3} + i\Delta_{4n-2} + j\Delta_{4n-1} + k\Delta_{4n}\right).
\end{aligned}
$$

(3)

The steps for finding the solution of the system (1) are as follows

1) *Finding the quantities* (2);

2) *Finding the quantities*

$$\sum_{l=1}^{r} B^s_{lmab}, \quad 0 \le m, a, b \le 3, \quad 1 \le s \le n;$$

3) *Finding* Δ;

4) *Finding*

$$\Delta_i, \quad 1 \le i \le 4n;$$

5) *Obtaining* (3).

Example 3.1. Let us consider the system

$$
\begin{cases}
ix + yj = 1 + i + j \\
xk + yi = 1 - i - j - k.
\end{cases}
$$

Using the above scheme for its solution we have

$$
\begin{cases}
x = -1 - i - \tfrac{1}{2}j - \tfrac{3}{2}k \\
y = -\tfrac{1}{2} + \tfrac{1}{2}i - 2k.
\end{cases}
$$

Theorem 3.2. *Let*

$$\Delta = 0$$

and $A_s \neq 0$ for some $1 \leq s \leq n$. Then the system (1) has not any solutions.

Theorem 3.3. *Let*

$$\Delta = 0, \quad A_s = 0 \quad \forall s.$$

Then the system (1) has infinitely many solutions.

References

1. D. JANOVSKA, G. OPFER. Linear equations in quaternions, Numerical Mathematics and Advanced Applications, Proceeding of ENUMATH 2005, A.B. Castro, D. Gomez, P. Quintita, P. Saldago(eds), Springer, Berlin, Heidelberg, New York, 2006, 945-953, ISBN 3-540-34287-7.
2. Y. TIAN. Solving two pairs of quadratic equations in quaternions, Adv. appl. Clifford alg., 2009, Online First, DOI 10.1007/s00006-003-000.

Wavelets beyond admissibility

Vladimir V. Kisil

School of Mathematics, University of Leeds, Leeds LS2 9JT, UK
E-mail: kisilv@maths.leeds.ac.uk

The purpose of this paper is to articulate an observation that many interesting types of *wavelets* (or *coherent states*) arise from group representations which *are not* square integrable or vacuum vectors which *are not* admissible.

Keywords: Wavelets; coherent states; group representations; Hardy space; functional calculus; Berezin calculus; Radon transform; Möbius map; maximal function; affine group; special linear group; numerical range.

1. Covariant Transform

A general group-theoretical construction[1–6] of *wavelets* (or *coherent states*) starts from an square integrable (*s.i.*) representation. However, such a setup is restrictive and is not necessary, in fact.

Definition 1.1. Let ρ be a representation of a group G in a space V and F be an operator from V to a space U. We define a *covariant transform* $\mathcal{W}$ from V to the space $L(G, U)$ of U-valued functions on G by the formula:

$$\mathcal{W} : v \mapsto \hat{v}(g) = F(\rho(g^{-1})v), \qquad v \in V,\ g \in G. \tag{1}$$

Remark 1.1. We do not require that operator F shall be linear.

Remark 1.2. Usefulness of the covariant transform is in the reverse proportion to the dimensionality of the space U. The covariant transform encodes properties of v in a function $\mathcal{W}v$ on G. For a low dimensional U this function can be ultimately investigated by means of harmonic analysis. Thus $\dim U = 1$ is the ideal case, however, it is unattainable sometimes, see Ex. 2.4 below.

Theorem 1.1. *The covariant transform $\mathcal{W}$ (1) intertwines ρ and the left regular representation Λ on $L(G, U)$:*

$$\Lambda(g) : f(h) \mapsto f(g^{-1}h). \tag{2}$$

Proof. We have a calculation similar to wavelet transform [3, Prop. 2.6]:

$$[\mathcal{W}(\rho(g)v)](h) = F(\rho(h^{-1})\rho(g)v) = [\mathcal{W}v](g^{-1}h) = \Lambda(g)[\mathcal{W}v](h). \qquad \square$$

Corollary 1.1. *The image space $\mathcal{W}(V)$ is invariant under the left shifts on G.*

2. Examples of Covariant Transform

Example 2.1. Let V be a Hilbert space with an inner product $\langle \cdot, \cdot \rangle$ and ρ be a unitary representation. Let $F : V \to \mathbb{C}$ be a functional $v \mapsto \langle v, v_0 \rangle$ defined by a vector $v_0 \in V$. Then the transformation (1) is the well-known expression for a *wavelet transform* [4, (7.48)] (or *representation coefficients*):

$$\mathcal{W} : v \mapsto \hat{v}(g) = \langle \rho(g^{-1})v, v_0 \rangle = \langle v, \rho(g)v_0 \rangle, \qquad v \in V, \ g \in G. \quad (3)$$

The family of vectors $v_g = \rho(g)v_0$ is called *wavelets* or *coherent states*. In this case we obtain scalar valued functions on G, thus the fundamental rôle of this example is explained in Rem. 1.2.

    This scheme is typically carried out for a s.i. representation ρ and v_0 being an admissible vector.[1,2,4–6] In this case the wavelet (covariant) transform is a map into the s.i. functions[7] with respect to the left Haar measure.

However s.i. representations and admissible vectors does not cover all interesting cases.

Example 2.2. Let G be the "$ax + b$" (or *affine*) group [4, § 8.2]: the set of points (a, b), $a \in \mathbb{R}_+$, $b \in \mathbb{R}$ in the upper half-plane with the group law:

$$(a, b) * (a', b') = (aa', ab' + b) \quad (4)$$

and left invariant measure $a^{-2}\, da\, db$. Its isometric representation on $V = L_p(\mathbb{R})$ is given by the formula:

$$[\rho_p(a, b)\, f](x) = a^{\frac{1}{p}} f\,(ax + b). \quad (5)$$

We consider the operators $F_\pm : L_2(\mathbb{R}) \to \mathbb{C}$ defined by:

$$F_\pm(f) = \frac{1}{2\pi i} \int_{\mathbb{R}} \frac{f(t)\, dt}{t \mp i}. \quad (6)$$

Then the covariant transform (1) is the Cauchy integral from $L_2(\mathbb{R})$ to the Hardy space in the upper/lower half-plane $H_2(\mathbb{R}_\pm^2)$. Although the representation (5) is s.i. for $p = 2$, the function $\frac{1}{t \pm i}$ is not an admissible vacuum vector. Thus the complex analysis become decoupled from the traditional wavelets theory. As a result the application of wavelet theory shall relay on an extraneous mother wavelets.[8]

However many important objects in complex analysis are generated by inadmissible mother wavelets like (6). For example, if $F : L_2(\mathbb{R}) \to \mathbb{C}$ is defined by $F : f \mapsto F_+f + F_-f$ then the covariant transform (1) is simply the *Poisson integral*. If $F : L_2(\mathbb{R}) \to \mathbb{C}^2$ is defined by $F : f \mapsto (F_+f, F_-f)$ then the covariant transform (1) represents a function on the real line as a jump between functions analytic in the upper and the lower half-planes. This makes a decomposition of $L_2(\mathbb{R})$ into irreducible components of the representation (5). Another interesting but non-admissible vector is the Gaussian e^{-x^2}.

Example 2.3. For the group $G = SL_2(\mathbb{R})$[14] let us consider the unitary representation ρ on the space of s.i. function $L_2(\mathbb{R}_+^2)$ on the upper half-plane through the Möbius transformations:

$$\rho(g) : f(z) \mapsto \frac{1}{(cz+d)^2} \, f\left(\frac{az+b}{cz+d}\right), \qquad g^{-1} = \begin{pmatrix} a & b \\ c & d \end{pmatrix}.$$

Let F_i be the functional $L_2(\mathbb{R}_+^2) \to \mathbb{C}$ of pairing with the lowest/highest i-weight vector in the corresponding irreducible component of the discrete series [14, Ch. VI]. Then we can build an operator F from various F_i similarly to the previous example, e.g. this generalises the representation of an s.i. function as a sum of analytic ones from different irreducible subspaces.

Covariant transform is also meaningful for principal and complementary series of representations of the group $SL_2(\mathbb{R})$,[9] which are not s.i.

Example 2.4. A straightforward generalisation of Ex.2.1 is obtained if V is a Banach space and $F : V \to \mathbb{C}$ is an element of V^*. Then the covariant transform coincides with the construction of wavelets in Banach spaces.[3]

The next stage of generalisation is achieved if V is a Banach space and $F : V \to \mathbb{C}^n$ be a linear operator. Then the corresponding covariant transform is a map $\mathcal{W} : V \to L(G, \mathbb{C}^n)$. This is closely related to M.G. Krein's works on *directing functionals*,[10] see also *multiresolution wavelet analysis*,[11] Clifford-valued Bargmann spaces[12] and [4, Thm. 7.3.1].

Example 2.5. A step in a different direction is a consideration of non-linear operators. Take again the "$ax+b$" group and its representation (5). We define F to be a homogeneous but non-linear functional $V \to \mathbb{R}_+$:

$$F(f) = \frac{1}{2} \int_{-1}^{1} |f(x)| \, dx.$$

The covariant transform (1) becomes:

$$[\mathcal{W}_p f](a,b) = \frac{1}{2} \int\limits_{-1}^{1} \left| a^{\frac{1}{p}} f\left(ax+b\right) \right| dx = a^{\frac{1}{p}} \frac{1}{2a} \int\limits_{b-a}^{b+a} |f(x)|\, dx.$$

Obviously $M_f(b) = \max_a [\mathcal{W}_\infty f](a,b)$ coincides with the Hardy *maximal function*, which contains important information on the original function f. However, the full covariant transform is even more detailed. For example, $\|f\| = \max_b [\mathcal{W}_\infty f](\frac{1}{2}, b)$ is the shift invariant norm.[13]

From the Cor. 1.1 we deduce that the operator $M : f \mapsto M_f$ intertwines ρ_p with itself $\rho_p M = M \rho_p$.

Example 2.6. Let $V = L_c(\mathbb{R}^2)$ be the space of compactly supported bounded functions on the plane. We take F be the linear operator $V \to \mathbb{C}$ of integration over the real line:

$$F : f(x,y) \mapsto F(f) = \int_{\mathbb{R}} f(x,0)\, dx.$$

Let G be the group of Euclidean motions of the plane represented by ρ on V by a change of variables. Then the wavelet transform $F(\rho(g)f)$ is the *Radon transform*.

Example 2.7. Let a representation ρ of a group G act on a space X. Then there is an associated representation ρ_B of G on a space $V = B(X,Y)$ of linear operators $X \to Y$ defined by the identity:

$$(\rho_B(g)A)x = A(\rho(g^{-1})x), \qquad x \in X,\ g \in G,\ A \in B(X,Y).$$

Following the Remark 1.2 we take F to be a functional $V \to \mathbb{C}$, for example F can be defined from a pair $x \in X$, $l \in Y^*$ by the expression $F : A \mapsto \langle Ax, l \rangle$. Then the covariant transform:

$$\mathcal{W} : A \mapsto \hat{A}(g) = F(\rho_B(g)A),$$

this is an example of *covariant calculus*.[3,15]

Example 2.8. A modification of the previous construction is obtained if we have two groups G_1 and G_2 represented by ρ_1 and ρ_2 on X and Y^* respectively. Then we have a covariant transform $B(X,Y) \to L(G_1 \times G_2, \mathbb{C})$ defined by the formula:

$$\mathcal{W} : A \mapsto \hat{A}(g_1, g_2) = \langle A\rho_1(g_1)x, \rho_2(g_2)l \rangle.$$

This generalises *Berezin functional calculi*.[3]

Example 2.9. Let us restrict the previous example to the case when $X = Y$ is a Hilbert space, $\rho_1 = \rho_2 = \rho$ and $x = l$ with $\|x\| = 1$. Than the range of the covariant transform:

$$\mathcal{W} : A \mapsto \hat{A}(g) = \langle A\rho(g)x, \rho(g)x \rangle$$

is a subset of the *numerical range* of the operator A.

Example 2.10. The group $SL_2(\mathbb{R})$ consists of 2×2 matrices of the form $\begin{pmatrix} \alpha & \beta \\ \bar{\beta} & \bar{\alpha} \end{pmatrix}$ with the unit determinant [14, § IX.1]. Let A be an operator with the spectral radius less than 1. Then the associated Möbius transformation

$$g : A \mapsto g \cdot A = \frac{\alpha A + \beta I}{\bar{\beta}A + \bar{\alpha}I}, \qquad \text{where} \quad g^{-1} = \begin{pmatrix} \alpha & \beta \\ \bar{\beta} & \bar{\alpha} \end{pmatrix} \in SL_2(\mathbb{R}),$$

produces a well-defined operator with the spectral radius less than 1 as well. Thus we have a representation of $SL_2(\mathbb{R})$. A choise of an operator F will define the corresponding covariant transform. In this way we obtain generalisations of *Riesz–Dunford functional calculus*.[15]

3. Inverse Covariant Transform

An object invariant under the left action Λ (2) is called *left invariant*. For example, let L and L' be two left invariant spaces of functions on G. We say that a pairing $\langle \cdot, \cdot \rangle : L \times L' \to \mathbb{C}$ is *left invariant* if

$$\langle \Lambda(g)f, \Lambda(g)f' \rangle = \langle f, f' \rangle, \quad \text{for all} \quad f \in L, \ f' \in L'. \tag{7}$$

Remark 3.1.

(1) We do not require the pairing to be linear in general.
(2) If the pairing is invariant on space $L \times L'$ it is not necessarily invariant (or even defined) on the whole $C(G) \times C(G)$.
(3) In a more general setting we shall study an invariant pairing on a homogeneous spaces instead of the group. However due to length constraints we cannot consider it here beyond the Example 3.2.
(4) An invariant pairing on G can be obtained from an invariant functional l by the formula $\langle f_1, f_2 \rangle = l(f_1 \bar{f}_2)$.

For a representation ρ of G in V and $v_0 \in V$ we fix a function $w(g) = \rho(g)v_0$. We assume that the pairing can be extended in its second component to this V-valued functions, say, in the weak sense.

Definition 3.1. Let $\langle \cdot, \cdot \rangle$ be a left invariant pairing on $L \times L'$ as above, let ρ be a representation of G in a space V, we define the function $w(g) = \rho(g)v_0$ for $v_0 \in V$. The *inverse covariant transform* $\mathcal{M}$ is a map $L \to V$ defined by the pairing:

$$\mathcal{M} : f \mapsto \langle f, w \rangle, \qquad \text{where } f \in L. \tag{8}$$

Example 3.1. Let G be a group with a unitary s.i. representation ρ. An invariant pairing of two s.i. functions is obviously done by the integration over the Haar measure:

$$\langle f_1, f_2 \rangle = \int_G f_1(g) \bar{f}_2(g) \, dg.$$

For an admissible vector v_0,[7] [4, Chap. 8] the inverse covariant transform is known in this setup as *reconstruction formula*.

Example 3.2. Let ρ be a s.i. representation of G modulo a subgroup $H \subset G$ and let $X = G/H$ be the corresponding homogeneous space with a quasi-invariant measure dx. Then integration over dx with an appropriate weight produces an invariant pairing. The inverse covariant transform is a more general version [4, (7.52)] of the *reconstruction formula* mentioned in the previous example.

Let ρ be not a s.i. representation (even modulo a subgroup) or let v_0 be inadmissible vector of a s.i. representation ρ. An invariant pairing in this case is not associated with an integration over any non singular invariant measure on G. In this case we have a *Hardy pairing*. The following example explains the name.

Example 3.3. Let G be the "$ax + b$" group and its representation ρ (5) from Ex. 2.2. An invariant pairing on G, which is not generated by the Haar measure $a^{-2}da\,db$, is:

$$\langle f_1, f_2 \rangle = \lim_{a \to 0} \int_{-\infty}^{\infty} f_1(a,b)\,\bar{f}_2(a,b)\,db. \tag{9}$$

For this pairing we can consider functions $\frac{1}{2\pi i(x+i)}$ or e^{-x^2}, which are not admissible vectors in the sense of s.i. representations. Then the inverse covariant transform provides an *integral resolutions* of the identity.

Similar pairings can be defined for other semi-direct products of two groups. We can also extend a Hardy pairing to a group, which has a subgroup with such a pairing.

Example 3.4. Let G be the group $SL_2(\mathbb{R})$ from the Ex. 2.3. Then the "$ax + b$" group is a subgroup of $SL_2(\mathbb{R})$, moreover we can parametrise $SL_2(\mathbb{R})$ by triples (a, b, θ), $\theta \in (-\pi, \pi]$ with the respective Haar measure [14, III.1(3)]. Then the Hardy pairing

$$\langle f_1, f_2 \rangle = \lim_{a \to 0} \int_{-\infty}^{\infty} f_1(a, b, \theta)\, \bar{f}_2(a, b, \theta)\, db\, d\theta. \tag{10}$$

is invariant on $SL_2(\mathbb{R})$ as well. The corresponding inverse covariant transform provides even a finer resolution of the identity which is invariant under conformal mappings of the Lobachevsky half-plane.

A further study of covariant transform and its inverse shall be continued elsewhere.

References

1. A. Perelomov, *Generalized coherent states and their applications* (Springer-Verlag, Berlin, 1986).
2. Feichtinger, Hans G. and Groechenig, K.H., *J. Funct. Anal.* **86**, 307 (1989).
3. V. V. Kisil, *Acta Appl. Math.* **59**, 79 (1999), E-print: arXiv:math/9807141 .
4. S. T. Ali, J.-P. Antoine and J.-P. Gazeau, *Coherent States, Wavelets and Their Generalizations* (Springer-Verlag, New York, 2000).
5. H. Führ, *Abstract Harmonic Analysis of Continuous Wavelet Transforms*, Lecture Notes in Mathematics, Vol. 1863 (Springer-Verlag, Berlin, 2005).
6. J. G. Christensen and G. Ólafsson, *Acta Appl. Math.* **107**, 25 (2009).
7. M. Duflo and C. C. Moore, *J. Functional Analysis* **21**, 209 (1976).
8. O. Hutník, *Integral Equations Operator Theory* **63**, 29 (2009).
9. V. V. Kisil, *Complex Variables Theory Appl.* **40**, 93 (1999), E-print: arXiv:funct-an/9712003.
10. M. G. Kreĭn, *Akad. Nauk Ukrain. RSR. Zbirnik Prac' Inst. Mat.* **1948**, 83 (1948), MR#14:56c, reprinted in.[16]
11. O. Bratteli and P. E. T. Jorgensen, *Integral Equations Operator Theory* **28**, 382 (1997), E-print: arXiv:funct-an/9612003.
12. J. Cnops and V. V. Kisil, *Math. Methods Appl. Sci.* **22**, 353 (1999), E-print: arXiv:math/9806150. Zbl 1005.22003.
13. A. Johansson, *Systems Control Lett.* **57**, 105 (2008).
14. S. Lang, SL₂(**R**) (Springer-Verlag, New York, 1985).
15. V. V. Kisil, Spectrum as the support of functional calculus, in *Functional analysis and its applications*, North-Holland Math. Stud. Vol. 197, pp. 133–141, (Elsevier, Amsterdam, 2004). E-print: arXiv:math.FA/0208249.
16. M. G. Kreĭn, *Izbrannye Trudy. II* (Akad. Nauk Ukrainy Inst. Mat., Kiev, 1997). MR#96m:01030.

226

Itô formula for an integro-differential operator without an associated stochastic process

R. Léandre

*Institut de Mathématiques de Bourgogne, Université de Bourgogne,
Dijon, 21000, France
E-mail: Remi.leandre@u-bourgogne.fr*

We give an Itô formula for a (non-Markovian) semi-group associated to an integro-differential operator L. There is until now no stochastic process of the jump type associated to L.

Keywords: Itô formula; integro differential generator.

1. Introduction

Let us recall what is the Itô formula for a purely discontinuous martingales $t \to M_t$ with values in $\mathbb{R}$ [2]. Let f be a C^2 function on $\mathbb{R}$. We have

$$f(M_t) = f(M_0)$$
$$+ \int_0^t f'(M_{s-})\delta M_s + \sum_{s \leq t} f(M_s) - f(M_{s-}) - f'(M_{s-})\Delta M_s \quad (1)$$

It is the generalization of the celebrated Itô formula for the Brownian motion $t \to B_t$ on $\mathbb{R}$ [2]

$$f(B_t) = f(B_0) + \int_0^t f'(B_s)\delta B_s + 1/2 \int_0^t f''(B_s)ds \quad (2)$$

A lot of of stochastic analysis tools for diffusions were translated by Léandre in semi-group theory in [4], [5], [6], [8], [9], [12], [13], [14], [16]. Some basical tools of stochastic analysis for the study of jump processes were translated by Léandre in semi-group theory in [10], [11], [18]. For review on that, we refer to the review of Léandre [7], [15].

Léandre has extended the Itô formula for the Brownian motion to the case of some classical partial differential equations in [17], [19], [20], [21]. In such a case, there is until now no convenient measure on a convenient path space associated to this partial differential equation.

Our main in this work is to extend the Itô formula for jump process for an integro-differential generator when there is until now no stochastic process associated.

Let L be the generator acting on smooth function f on $\mathbb{R}$ with bounded derivatives at each order:

$$Lf(x) = \int_{\mathbb{R}} (f(x+z) - f(x) - z^2/2 f''(x)) \frac{1}{|z|^{7/2}} dz \tag{3}$$

Theorem 1.1. *L is a negative symmetric densely defined operator on $L^2(\mathbb{R})$. It admits therefore a self-adjoint extension still denoted by L. To this self-adjoint extension is associated a semi-group P_t.*

Let us suppose that the function f has a compact support. Let us consider a smooth function h on $\mathbb{R}^2$ with bounded derivatives at each order. We consider the operator L^f acting on h:

$$L^f h(x,y) = \int_{\mathbb{R}} \frac{1}{|z|^{7/2}} \left(h(x+z, y + f(x+z) - f(x)) - h(x,y) \right.$$
$$- z^2/2 \left(\frac{\partial^2}{\partial x^2} h(x,y) + 2 \frac{\partial^2}{\partial x \partial y} h(x,y) f'(x) + \frac{\partial}{\partial y} h(x,y) f''(x) \right.$$
$$\left. \left. + \frac{\partial^2}{\partial y^2} h(x,y) (f'(x))^2 \right) \right) dz \tag{4}$$

Theorem 1.2. *L^f is negative symmetric on $L^2(\mathbb{R}^2)$. It admits therefore a self-adjoint extension still denoted by L^f. To this self-adjoint extension is associated a semi-group P_t^f.*

P_t and P_t^f are related by the following formula, which is an interpretation of (1) in this context:

Theorem 1.3. *(Itô) Let $\hat{h}$ be the L^2 function on $\mathbb{R}$*

$$\hat{h}(x) = h(x, f(x)) \tag{5}$$

Then

$$P_t[\hat{h}](x) = P_t^f[h](x, f(x)) \tag{6}$$

if f and g have compact supports.

2. Proof of Theorem 1.1 and Theorem 1.2

We begin by the proof of Theorem 1.1.

Proof. Let Q_t be the semi-group associated to $\Delta = 1/2\frac{\partial^2}{\partial x^2}$. We have the relation

$$Lf(x) = C \int_0^\infty t^{-5/2}(Q_t f(x) - f(x) - t/2 f"(x))dt \tag{7}$$

Let us show this relation.

$$Q_t f(x) = \frac{1}{\sqrt{2\pi t}} \int_\mathbb{R} \exp[-\frac{z^2}{2\pi t}]f(x+z)dz \tag{8}$$

Moreover if $z \neq 0$

$$\int_0^\infty t^{-5/2}\frac{1}{\sqrt{2\pi t}}\exp[-\frac{z^2}{2t}]dt = \int_0^\infty t^{-5/2}\frac{1}{\sqrt{2\pi t}}\exp[-\frac{1}{2t}]dt\frac{1}{|z|^{7/2}} \tag{9}$$

On the other hand

$$\frac{1}{\sqrt{2\pi t}} \int_\mathbb{R} z^2/2\exp[-\frac{z^2}{2t}]dz = t/2 \tag{10}$$

Formula (7) shows clearly that L is symmetric. If we consider the spectral resolution of $(-\Delta)$, the associated spectrum is positive. On the other hand, if we consider $\lambda \geq 0$

$$\int_0^\infty t^{-5/2}(\exp[-\lambda t] - 1 + \lambda t)dt = C\lambda^{3/2} \geq 0 \tag{11}$$

by a simple change of variable. $\qquad \square$

Remark: Following the terminology of [22], we can say that $L = -C(-\Delta)^{3/2}$

We give now the Proof of Theorem 1.2.:

Proof. We can perform the change of variable $(x, y) \to (x, y + f(x))$. We put $\overline{h}(x, y) = h(x, y + f(x))$. The main observation is then:

$$L\overline{h}(x,y) = \int_\mathbb{R} \Big(h(x+z, y+f(x+z)) - h(x, y+f(x))$$
$$- z^2/2\frac{\partial^2}{\partial x^2}\overline{h}(x,y) \Big) \frac{1}{|z|^{7/2}}dz$$
$$= [L^f h](x, y+f(x)) \tag{12}$$

Therefore

$$\int_{\mathbb{R}^2} L^f g(x,y)g(x,y)dxdy = \int_{\mathbb{R}^2} [L^f g](x, y+f(x))g(x, y+f(x))dxdy$$
$$= \int_{\mathbb{R}^2} [L\overline{h}](x,y)\overline{h}(x,y)dxdy \tag{13}$$

which is negative by the previous theorem.

Let us show the symmetry by an analog argument. We have

$$\int_{\mathbb{R}^2} L^f g^1(x,y) g^2(x,y) dxdy = \int_{\mathbb{R}^2} [L^f g^1](x, y+f(x)) g^2(x, y+f(x)) dxdy$$
$$= \int_{\mathbb{R}^2} [L\bar{g}^1](x,y)\bar{g}^2(x,y) dxdy \tag{14}$$

By the previous theorem, this last quantity is equal to

$$\int_{\mathbb{R}^2} [L\bar{g}^2](x,y)\bar{g}^1(x,y) dxdy \tag{15}$$

But

$$\int_{\mathbb{R}^2} [L\bar{g}^2](x,y)\bar{g}^1(x,y) dxdy = \int_{\mathbb{R}^2} [L^f g^2](x, y+f(x)) g^1(x, y+f(x)) dxdy$$
$$= \int_{\mathbb{R}^2} L^f g^2(x,y) g^1(x,y) dxdy \tag{16}$$

This shows the theorem. $\square$

3. Proof of the Itô formula

Let us show by induction on n that

$$(L^f)^n[h])(x, y+f(x)) = L^n\bar{h}(x,y) \tag{17}$$

For $n = 1$ it is proved in (12). Let us suppose it is proved for n. Let us show it is still proved for $n+1$. We have

$$(L^f)^{n+1}[h](x, y+f(x)) = [(L^f)^n(L^f h)](x, y+f(x)) = L^n(\overline{L^f h})(x,y) \tag{18}$$

by induction. But $\overline{L^f h} = L\bar{h}$. Therefore the result. Let us suppose that f is a finite sum of $\cos[ay]\exp[-by^2]$ and of $\sin[ay]\exp[-by^2]$ and that h is a finite sum of product of the previous expressions, but depending of x and y. We get

$$P_t[\hat{h}](x) = \sum \frac{t^n}{n!} L^n\hat{h}(x) = \sum \frac{t^n}{n!} L^n\bar{h}(x,0)$$
$$= \sum \frac{t^n}{n!} [(L^f)^n h](x, f(x)) = [P_t^f h](x, f(x)) \tag{19}$$

This formula can be extended by continuity to the case where h is smooth with compact support. It remains to extend it by continuity to any function f smooth with compact support. It is the purpose of the sequel.

Lemma 3.1. *If h is smooth with compact support and if f is smooth with derivatives bounded and in L^2, $(x,y) \rightarrow [P_t^f h](x,y)$ is smooth with bounded* *derivatives in L^2. Morever the uniform norms of its derivatives as well as their L^2 norms can be estimated in terms of the unifom norms of the derivatives of f and h and their L^2 norms.*

Proof. Let us introduce the operator L^y:

$$L^y h(x,y) = \int_{\mathbb{R}} (h(x,y+z) - h(x,y) - z^2/2 \frac{\partial^2}{\partial y^2} h(x,y)) \frac{dz}{|z|^{7/2}} \qquad (20)$$

L^y and L^f commute. Let n a positive integer. $\phi_t^n = ((L^y)^n + (L^f)^n) P_t^f h(x,y)$ satisfy to the problem:

$$\phi_0^n = ((L^y)^n + (L^f)^n) h(x,y) \,; \, d/dt \phi_t^n = L^f \phi_t^n \qquad (21)$$

Therefore

$$\phi_t^n = P_t^f [((L^y)^n + (L^f)^n) h](x,y) \qquad (22)$$

We would like to estimate the uniform norm as well as the L^2 norm of a function $(x,y) \rightarrow h(x,y)$ in terms of the L^2 norms of $((L^y)^n + (L^f)^n)) h$. It is possible. It is enough to estimate the uniform norms and the L^2 norms of the derivatives of $(x,y) \rightarrow h(x,y+f(x)) = \overline{h}(x,y)$. But

$$[((L^y)^n + (L^f)^n))(h)](x,y+f(x)) = (L^n + (L^y)^n)\overline{h}(x,y) =$$
$$C^n [(-\frac{\partial^2}{\partial x^2})^{3n/2} + (-\frac{\partial^2}{\partial y^2})^{3n/2}]\overline{h}(x,y) \qquad (23)$$

The result goes by the classical Sobolev imbedding theorem. $\qquad \square$

We can give now the proof of the Itô formula:

Proof. Let f^n of the type studied before such that f^n as well as all its derivatives tends uniformly and in L^2 to f. The result will come from the fact that $P_t^{f^n} h \rightarrow P_t^f h$ uniformly. Let us denote

$$\phi_t^n = P_t^f h - P_t^{f^n} h \qquad (24)$$

ϕ_t^n is solution of the parabolic equation with second member starting from 0:

$$d/dt \phi_t^n = L^f \phi_t^n + (L^f - L^{f^n}) P_t^{f^n} h \qquad (25)$$

This equation can be solved by the method of variation of constant. The result comes from the previous lemma. $\qquad \square$

References

1. P. Auscher, P. Tchamitchian, *Square root problems for divergence operators and related topics* (Asterisque 249 S.M.F., Paris, 1998).
2. C. Dellacherie, P.A. Meyer, *Probabilités et potentiel. Théorie des martingales*, (Hermann, Paris, 1980).
3. J. Dieudonné, *Eléments d'analyse. VIII.* (Gauthier-Villars, Paris, 1978)
4. R. Léandre, in *Festchrift in honour of K. Sinha. Proc. Indian. Acad. Sci (Math. Sci)* **116**, 507 (2006) *arXiv:0707.2143v1[math.PR]*
5. R. Léandre, in *Mathematical methods in engineerings* (Springer, Heidelberg, 2007).
6. R. Léandre, *Mathematische Zeitschrift* **258**, 893 (2008).
7. R. Léandre, in *Simulation, Modelling and Optimization* (C.D. WSEAS, Athens, 2006), *WSEAS transactions on mathematics* **5**, 1205, (2006).
8. R. Léandre, in *Applied mathematics* (W.S.E.A.S. press, Athens, 2007).
9. R. Léandre, *WSEAS Transactions on mathematics* **6**, 755 (2007).
10. R. Léandre, in *Num. Ana. Applied. Mathematics* (A.I.P. Proceedings 936, A.I.P., Melville, 2007).
11. R. Léandre, in *Fractional order systems. Jour.Euro.Syst.Automatises* **42**, 715, (2008).
12. R. Léandre, in *Control, Automation, Robotics and Vision* (C.D., I.E.E.E. Catalog Number, CFP08532-CDR, 2008)
13. R. Léandre, in *Applied computing conference*, (W.S.E.A.S. press, Athens, 2008).
14. R. Léandre, *WSEAS transactions on mathematics* **7**, 244 (2008).
15. R. Léandre, *Far East Journal of Mathematical Sciences* **30**, 1 (2008).
16. R. Léandre, in *Num. Ana. Applied. Mathematics* (A.I.P. Proceedings 1048, A.I.P., Melville, 2008).
17. R. Léandre, in *Non-Euclidean geometry and its applications. Acta Physica Debrecina* **42**, 133 (2008).
18. R. Léandre, Regularity of a degenerated convolution semi-group without to use the Poisson process. To appear in *Nonlinear Science and Complexity*
19. R. Léandre, in *Fractional differentiation and its applications* (C.D., Cankaya University, Ankara, 2008)
20. R. Léandre, Itô-Stratonovitch formula for the Schroedinger equation associated to a big order operator on a torus. To appear in *Workshop on fractional differentiation. Physica Scripta.*
21. R. Léandre, Itô-Stratonovitch formula for the wave equation on a torus. To appear in *Computation of stochastic systems.*
22. K. Yosida, *Functional analysis* (Springer, Heidelberg, 1977).

Integral theorems in a commutative three-dimensional harmonic algebra

S.A. Plaksa and V.S. Shpakivskyi

Institute of Mathematics of the National Academy of Sciences of Ukraine,
3 Tereshchenkivska Street,
Kiev, Ukraine
E-mails: shpakivskyi@mail.ru,
plaksa@imath.kiev.ua

For monogenic functions taking values in a three-dimensional commutative harmonic algebra with the unit and two-dimensional radical, we have proved analogs of classical integral theorems of the theory of analytic functions of the complex variable: the Cauchy integral theorems for surface integral and curvilinear integral, the Morera theorem and the Cauchy integral formula.

Keywords: Laplace equation; harmonic commutative Banach algebra; monogenic function; integral Cauchy theorem; integral Cauchy formula.

1. Introduction

Let $\mathbb{A}_3$ be a three-dimensional commutative associative Banach algebra with the unit 1 over the field of complex numbers $\mathbb{C}$. Let $\{1, \rho_1, \rho_2\}$ be a basis of the algebra $\mathbb{A}_3$ with the multiplication table $\rho_1 \rho_2 = \rho_2^2 = 0$, $\rho_1^2 = \rho_2$.

The algebra $\mathbb{A}_3$ is *harmonic* (see[1,2]) because there exist *harmonic* bases $\{e_1 = 1, e_2, e_3\}$ in $\mathbb{A}_3$ satisfying the following condition

$$e_1^2 + e_2^2 + e_3^2 = 0. \tag{1}$$

Consider the linear envelope $E_3 := \{\zeta = x + ye_2 + ze_3 \ : \ x, y, z \in \mathbb{R}\}$ generated by the vectors $1, e_2, e_3$ over the field of real numbers $\mathbb{R}$. For a set $S \subset \mathbb{R}^3$ consider the set $S_\zeta := \{\zeta = x + ye_2 + ze_3 : (x, y, z) \in S\} \subset E_3$ congruent to S. In what follows, $\zeta = x + ye_2 + ze_3$ and $x, y, z \in \mathbb{R}$.

A continuous function $\Phi : \Omega_\zeta \to \mathbb{A}_3$ is *monogenic* in a domain $\Omega_\zeta \subset E_3$ if Φ is differentiable in the sense of Gateaux in every point of Ω_ζ, i.e. if for every $\zeta \in \Omega_\zeta$ there exists an element $\Phi'(\zeta) \in \mathbb{A}_3$ such that

$$\lim_{\varepsilon \to 0+0} \left(\Phi(\zeta + \varepsilon h) - \Phi(\zeta) \right) \varepsilon^{-1} = h \Phi'(\zeta) \quad \forall h \in E_3.$$

It follows from the equality (1) and the equality

$$\triangle_3 \Phi := \frac{\partial^2 \Phi}{\partial x^2} + \frac{\partial^2 \Phi}{\partial y^2} + \frac{\partial^2 \Phi}{\partial z^2} = \Phi''(\zeta)(e_1^2 + e_2^2 + e_3^2)$$

that every twice monogenic function $\Phi : \Omega_\zeta \to \mathbb{A}_3$ satisfies the three-dimensional Laplace equation $\triangle_3 \Phi = 0$.

In the paper[3] for functions differentiable in the sense of Lorch in an arbitrary convex domain of commutative associative Banach algebra, some properties similar to properties of holomorphic functions of complex variable (in particular, the integral Cauchy theorem and the integral Cauchy formula, the Taylor expansion and the Morera theorem) are established. In the paper[4] the convexity of the domain is withdrawn in the mentioned results from.[3]

In this paper we establish similar results for monogenic functions $\Phi : \Omega_\zeta \to \mathbb{A}_3$ given only in a domain Ω_ζ of the linear envelope E_3 instead of domain of whole algebra $\mathbb{A}_3$. Let us note that *a priori* the differentiability of the function Φ in the sense of Gateaux is a restriction weaker than the differentiability of this function in the sense of Lorch. Moreover, note that the integral Cauchy formula established in the papers[3,4] is not applicable for a monogenic function $\Phi : \Omega_\zeta \to \mathbb{A}_3$ because it deals with an integration along a curve on which the function Φ is not given, generally speaking.

Note that as well as in,[3,4] some hypercomplex analogues of integral Cauchy theorem for curvilinear integral are established in the papers.[5,6] In the papers[5,7–9] similar theorems are established for surface integral.

2. Cauchy integral theorem for a surface integral

A function $\Phi(\zeta)$ of the variable $\zeta \in \Omega_\zeta$ is monogenic if and only if the following Cauchy – Riemann conditions are satisfied (see Theorem 1.3[2]):

$$\frac{\partial \Phi}{\partial y} = \frac{\partial \Phi}{\partial x} e_2, \qquad \frac{\partial \Phi}{\partial z} = \frac{\partial \Phi}{\partial x} e_3. \tag{2}$$

Along with monogenic functions, consider a function $\Psi : \Omega_\zeta \to \mathbb{A}_3$ having continuous partial derivatives of the first order in a domain Ω_ζ and satisfying the equation

$$\frac{\partial \Psi}{\partial x} + \frac{\partial \Psi}{\partial y} e_2 + \frac{\partial \Psi}{\partial z} e_3 = 0 \tag{3}$$

in every point of this domain.

In the scientific literature the different denominations are used for functions satisfying equations of the form (3). For example, in the papers[5,6,10] they are called regular functions, and in the papers[7,8,11] they are called monogenic functions. As well as in the papers,[9,12,13] we call a *hyperholomorphic* function if it satisfies the equation (3).

It is well known that in the quaternion analysis the classes of functions determined by means conditions of the form (2) and (3) do not coincide (see[5,14]).

Note that in the algebra $\mathbb{A}_3$ the set of monogenic functions is a subset of the set of hyperholomorphic functions because every monogenic function $\Phi : \Omega_\zeta \to \mathbb{A}_3$ satisfies the equality (3) owing to the conditions (1), (2). But, there exist hyperholomorphic functions which are not monogenic. For example, the function $\Psi(x + ye_2 + ze_3) = ze_2 - ye_3$ satisfies the condition (3), but it does not satisfy the equalities of the form (2).

Let Ω be a bounded closed set in $\mathbb{R}^3$. For a continuous function $\Psi : \Omega_\zeta \to \mathbb{A}_3$ of the form

$$\Psi(x + ye_2 + ze_3) = \sum_{k=1}^{3} U_k(x, y, z)e_k + i \sum_{k=1}^{3} V_k(x, y, z)e_k, \qquad (4)$$

where $(x, y, z) \in \Omega$, we define a volume integral by the equality

$$\int_{\Omega_\zeta} \Psi(\zeta)dxdydz := \sum_{k=1}^{3} e_k \int_{\Omega} U_k(x, y, z)dxdydz + i \sum_{k=1}^{3} e_k \int_{\Omega} V_k(x, y, z)dxdydz.$$

Let Σ be a quadrable surface in $\mathbb{R}^3$ with quadrable projections on the coordinate planes. For a continuous function $\Psi : \Sigma_\zeta \to \mathbb{A}_3$ of the form (4), where $(x, y, z) \in \Sigma$, we define a surface integral on Σ_ζ with the differential form $\sigma_{\alpha_1,\alpha_2,\alpha_3} := \alpha_1 dydz + \alpha_2 dzdxe_2 + \alpha_3 dxdye_3$, where $\alpha_1, \alpha_2, \alpha_3 \in \mathbb{R}$, by the equality

$$\int_{\Sigma_\zeta} \Psi(\zeta)\sigma_{\alpha_1,\alpha_2,\alpha_3} := \sum_{k=1}^{3} e_k \int_{\Sigma} \alpha_1 U_k(x, y, z)dydz + \sum_{k=1}^{3} e_2 e_k \int_{\Sigma} \alpha_2 U_k(x, y, z)dzdx +$$

$$+ \sum_{k=1}^{3} e_3 e_k \int_{\Sigma} \alpha_3 U_k(x, y, z)dxdy + i \sum_{k=1}^{3} e_k \int_{\Sigma} \alpha_1 V_k(x, y, z)dydz +$$

$$+ i \sum_{k=1}^{3} e_2 e_k \int_{\Sigma} \alpha_2 V_k(x, y, z)dzdx + i \sum_{k=1}^{3} e_3 e_k \int_{\Sigma} \alpha_3 V_k(x, y, z)dxdy.$$

A connected homeomorphic image of a square in $\mathbb{R}^3$ is called a *simple surface*. A surface is *locally-simple* if it is simple in a certain neighborhood of every point.

If a simply connected domain $\Omega \subset \mathbb{R}^3$ have a closed locally-simple piece-smooth boundary $\partial\Omega$ and a function $\Psi : \Omega_\zeta \to \mathbb{A}_3$ is continuous together with partial derivatives of the first order up to the boundary $\partial\Omega_\zeta$, then the following analogue of the Gauss – Ostrogradsky formula is true:

$$\int_{\partial\Omega_\zeta} \Psi(\zeta)\sigma = \int_{\Omega_\zeta} \left(\frac{\partial\Psi}{\partial x} + \frac{\partial\Psi}{\partial y}e_2 + \frac{\partial\Psi}{\partial z}e_3 \right) dxdydz, \tag{5}$$

where $\sigma := \sigma_{1,1,1} \equiv dydz + dzdxe_2 + dxdye_3$. Now, the next theorem is a result of the formula (5) and the equality (3).

Theorem 2.1. *Suppose that Ω is a simply connected domain with a closed locally-simple piece-smooth boundary $\partial\Omega$. Suppose also that the function $\Psi : \overline{\Omega_\zeta} \to \mathbb{A}_3$ is continuous in the closure $\overline{\Omega_\zeta}$ of domain Ω_ζ and is hyper-holomorphic in Ω_ζ. Then*

$$\int_{\partial\Omega_\zeta} \Psi(\zeta)\sigma = 0.$$

3. Cauchy integral theorem for a curvilinear integral

Let γ be a Jordan rectifiable curve in $\mathbb{R}^3$. For a continuous function $\Psi : \gamma_\zeta \to \mathbb{A}_3$ of the form (4), where $(x, y, z) \in \gamma$, we define an integral along the curve γ_ζ by the equality

$$\int_{\gamma_\zeta} \Psi(\zeta)d\zeta := \sum_{k=1}^{3} e_k \int_\gamma U_k(x,y,z)dx + \sum_{k=1}^{3} e_2 e_k \int_\gamma U_k(x,y,z)dy +$$

$$+ \sum_{k=1}^{3} e_3 e_k \int_\gamma U_k(x,y,z)dz + i \sum_{k=1}^{3} e_k \int_\gamma V_k(x,y,z)dx +$$

$$+ i \sum_{k=1}^{3} e_2 e_k \int_\gamma V_k(x,y,z)dy + i \sum_{k=1}^{3} e_3 e_k \int_\gamma V_k(x,y,z)dz,$$

where $d\zeta := dx + e_2 dy + e_3 dz$.

If a function $\Phi : \Omega_\zeta \to \mathbb{A}_3$ is continuous together with partial derivatives of the first order in a domain Ω_ζ, and Σ is a piece-smooth surface in Ω, and the edge γ of surface Σ is a rectifiable Jordan curve, then the following analogue of the Stokes formula is true:

$$\int\limits_{\gamma_\zeta} \Phi(\zeta)d\zeta = \int\limits_{\Sigma_\zeta} \left(\frac{\partial\Phi}{\partial x}e_2 - \frac{\partial\Phi}{\partial y}\right)dxdy + \left(\frac{\partial\Phi}{\partial y}e_3 - \frac{\partial\Phi}{\partial z}e_2\right)dydz+$$

$$+ \left(\frac{\partial\Phi}{\partial z} - \frac{\partial\Phi}{\partial x}e_3\right)dzdx. \tag{6}$$

Now, the next theorem is a result of the formula (6) and the equalities (2).

Theorem 3.1. *Suppose that $\Phi : \Omega_\zeta \to \mathbb{A}_3$ is a monogenic function in a domain Ω_ζ, and Σ is a piece-smooth surface in Ω, and the edge γ of surface Σ is a rectifiable Jordan curve. Then*

$$\int\limits_{\gamma_\zeta} \Phi(\zeta)d\zeta = 0. \tag{7}$$

Now, similarly to the proof of Theorem 3.2[4] we can prove the following

Theorem 3.2. *Let $\Phi : \Omega_\zeta \to \mathbb{A}_3$ be a monogenic function in a domain Ω_ζ. Then for every closed Jordan rectifiable curve γ homotopic to a point in Ω, the equality (7) is true.*

For functions taking values in the algebra $\mathbb{A}_3$, the following Morera theorem can be established in the usual way.

Theorem 3.3. *If a function $\Phi : \Omega_\zeta \to \mathbb{A}_3$ is continuous in a domain Ω_ζ and $\int_{\partial\triangle_\zeta} \Phi(\zeta)d\zeta = 0$ for every triangle $\triangle_\zeta$ for which $\overline{\triangle_\zeta} \subset \Omega_\zeta$, then the function Φ is monogenic in the domain Ω_ζ.*

4. Cauchy integral formula

In what follows, we consider a harmonic basis $\{e_1, e_2, e_3\}$ with the following decomposition with respect to the basis $\{1, \rho_1, \rho_2\}$:

$$e_1 = 1, \qquad e_2 = i + \frac{1}{2}i\rho_2, \qquad e_3 = -\rho_1 - \frac{\sqrt{3}}{2}i\rho_2.$$

It follows from Lemma 1.1[2] that

$$\zeta^{-1} = \frac{1}{x+iy} + \frac{z}{(x+iy)^2}\rho_1 + \left(\frac{i}{2}\frac{\sqrt{3}z-y}{(x+iy)^2} + \frac{z^2}{(x+iy)^3}\right)\rho_2 \tag{8}$$

for all $\zeta = x + ye_2 + ze_3 \in E_3 \setminus \{ze_3 : z \in \mathbb{R}\}$. Thus, it is obvious that the straight line $\{ze_3 : z \in \mathbb{R}\}$ is contained in the radical of the algebra $\mathbb{A}_3$.

Using the equality (8), it is easy to calculate that

$$\int_{\tilde{\gamma}_\zeta} \tau^{-1} d\tau = 2\pi i, \tag{9}$$

where $\tilde{\gamma}_\zeta := \{\tau = x + y e_2 : x^2 + y^2 = R^2\}$.

Theorem 4.1. *Let Ω be a domain convex in the direction of the axis Oz and $\Phi : \Omega_\zeta \to \mathbb{A}_3$ be a monogenic function in the domain Ω_ζ. Then for every point $\zeta_0 \in \Omega_\zeta$ the following equality is true:*

$$\Phi(\zeta_0) = \frac{1}{2\pi i} \int_{\gamma_\zeta} \Phi(\zeta) \, (\zeta - \zeta_0)^{-1} \, d\zeta, \tag{10}$$

where γ_ζ is an arbitrary closed Jordan rectifiable curve in Ω_ζ, which around once the straight line $\{\zeta_0 + z e_3 : z \in \mathbb{R}\}$.

Proof. We represent the integral from the right-hand side of equality (10) as the sum of the following two integrals:

$$\int_{\gamma_\zeta} \Phi(\zeta) \, (\zeta - \zeta_0)^{-1} \, d\zeta = \int_{\gamma_\zeta} (\Phi(\zeta) - \Phi(\zeta_0)) \, (\zeta - \zeta_0)^{-1} \, d\zeta +$$

$$+ \Phi(\zeta_0) \int_{\gamma_\zeta} (\zeta - \zeta_0)^{-1} \, d\zeta =: I_1 + I_2.$$

Inasmuch as the domain Ω is convex in the direction of the axis Oz and the curve γ_ζ around once the straight line $\{\zeta_0 + z e_3 : z \in \mathbb{R}\}$, γ is homotopic to the circle $K(R) := \{(x - x_0)^2 + (y - y_0)^2 = R^2, z = z_0\}$, where $\zeta_0 = x_0 + y_0 e_2 + z_0 e_3$. Then using the equality (9), we have $I_2 = 2\pi i \Phi(\zeta_0)$.

Let us prove that $I_1 = 0$. First, we choose on the curve γ two points A and B in which there are tangents to γ, and we choose also two points A_1, B_1 on the circle $K(\varepsilon)$ which is completely contained in the domain Ω. Let γ^1, γ^2 be connected components of the set $\gamma \setminus \{A, B\}$. By K^1 and K^2 we denote connected components of the set $K(\varepsilon) \setminus \{A_1, B_1\}$ in such a way that after a choice of smooth arcs Γ^1, Γ^2 each of the closed curves $\gamma^1 \cup \Gamma^2 \cup K^1 \cup \Gamma^1$ and $\gamma^2 \cup \Gamma^1 \cup K^2 \cup \Gamma^2$ will homotopic to a point of the domain $\Omega \setminus \{(x_0, y_0, z) : z \in \mathbb{R}\}$.

Then it follows from Theorem 3.2 that

$$\int_{\gamma_\zeta^1 \cup \Gamma_\zeta^2 \cup K_\zeta^1 \cup \Gamma_\zeta^1} (\Phi(\zeta) - \Phi(\zeta_0)) \, (\zeta - \zeta_0)^{-1} \, d\zeta = 0, \tag{11}$$

$$\int\limits_{\gamma_\zeta^2 \cup \Gamma_\zeta^1 \cup K_\zeta^2 \cup \Gamma_\zeta^2} \left(\Phi(\zeta) - \Phi(\zeta_0)\right)\left(\zeta - \zeta_0\right)^{-1} d\zeta = 0. \qquad (12)$$

Inasmuch as each of the curves Γ_ζ^1, Γ_ζ^2 has different orientations in the equalities (11), (12), after addition of the mentioned equalities we obtain

$$\int\limits_{\gamma_\zeta} \left(\Phi(\zeta) - \Phi(\zeta_0)\right)\left(\zeta - \zeta_0\right)^{-1} d\zeta = \int\limits_{K_\zeta(\varepsilon)} \left(\Phi(\zeta) - \Phi(\zeta_0)\right)\left(\zeta - \zeta_0\right)^{-1} d\zeta, \qquad (13)$$

where the curves $K_\zeta(\varepsilon)$, γ_ζ have the same orientation.

The integrand in the right-hand side of the equality (13) is bounded by a constant which does not depend on ε. Therefore, passing to the limit in the equality (13) as $\varepsilon \to 0$, we obtain $I_1 = 0$ and the theorem is proved. $\square$

Using the formula (10), we obtain the Taylor expansion of monogenic function in the usual way. Thus, as in the complex plane, one can give different equivalent definitions of monogenic function $\Phi : \Omega_\zeta \to \mathbb{A}_3$.

References

1. I. P. Mel'nichenko, Algebras of functionally-invariant solutions of the three-dimensional Laplace equation, *Ukr. Math. J.*, **55** (2003), no. 9, 1551–1557.
2. I. P. Mel'nichenko and S. A. Plaksa, *Commutative algebras and spatial potential fields*, Kiev: Inst. Math. NAS Ukraine, 2008. [in Russian]
3. E. R. Lorch, The theory of analytic function in normed abelin vector rings, *Trans. Amer. Math. Soc.*, **54** (1943), 414–425.
4. E. K. Blum, A theory of analytic functions in banach algebras, *Trans. Amer. Math. Soc.*, **78** (1955), 343–370.
5. A. Sudbery, Quaternionic analysis, *Math. Proc. Camb. Phil. Soc.*, **85** (1979), 199–225.
6. F. Colombo, I. Sabadini and D. Struppa, *Slice monogenic functions*, arXiv:0708.3595v2 [math.CV] 25 Jan 2008.
7. F. Brackx and R. Delanghe, Duality in Hypercomplex Functions Theory, *J. Funct. Anal.*, **37** (1980), no. 2, 164–181.
8. S. Bernstein, Factorization of the nonlinear Schrödinger equation and applications, *Complex Variables and Elliptic Equations*, **51** (2006), no. 5–6, 429–452.
9. V. V. Kravchenko and M. V. Shapiro, *Integral representations for spatial models of mathematical physics*, Pitman Research Notes in Mathematics, Addison Wesley Longman Inc., 1996.
10. W. Sprößig, Eigenvalue problems in the framework of Clifford analysis, *Advances in Applied Clifford Algebras*, **11** (2001), 301–316.
11. J. Ryan, Dirac operators, conformal transformations and aspects of classical harmonic analysis, *J. of Lie Theory*, **8** (1998), 67–82.

12. W. Sprössig, Quaternionic analysis and Maxwell's equations, *CUBO A Math. J.*, **7** (2005), no. 2, 57–67.

13. B. Schneider and E. Karapinar, A note on biquaternionic MIT bag model, *Int. J. Contemp. Math. Sci.*, **1** (2006), no. 10, 449–461.

14. A. S. Meylekhzon, On monogenity of quaternions, *Doklady Acad. Nauk SSSR*, **59** (1948), no. 3, 431–434. [in Russian]

Segre quaternions, spectral analysis and a four-dimensional Laplace equation

D.A. Pinotsis

*Wellcome Trust Centre for Neuroimaging, University College London,
United Kingdom
E-mail: d.pinotsis@ucl.ac.uk*

We introduce some novel formulae in the theory of Segre quaternions, namely
a Dbar (or Pompeiu-Borel) formula as well as certain variations of the fundamental theorem of calculus. The latter enable us to obtain the solution to
a boundary value problem of a four dimensional Laplace equation by using
spectral analysis.

Keywords: Segre quaternions; Dbar formula; Laplace equation; boundary value
problems; Fundamental Theorem of Calculus.

1. Introduction

Segre or commutative quaternions appeared in a paper by Segre[13] in 1892.
Segre's construction involves commutative quaternion units, namely the
units e_j, $j = 1, 2, 3$ which satisfy

$$e_1 e_2 = e_2 e_1 = e_3, \qquad e_1^2 = e_2^2 = -1, \tag{1}$$

as opposed to the widely known Hamilton quaternions where the units
anticommute.[5,6,9] The commutativity is obtained at the expense of the
existence of zero divisors in the relevant theory, in particular

$$\frac{1 + e_3}{2}, \frac{1 - e_3}{2}. \tag{2}$$

Since their appearence, several authors have used Segre quaternions in various applications; for some recent results see eg.[1,2,7,8] . Here, we first introduce some novel formulae in the theory of Segre quaternions. In particular,
we present a generalisation of an important formula of complex analysis,
the so-called Dbar (or Pompeiu-Borel) formula as well as some variations of
the Fundamental Theorem of Calculus (different from the ones presented
in[11]). Also, we discuss a boundary value problem for a four dimensional

generalisation of the Laplace equation and we assert that this problem can be solved through spectral analysis[3] by reducing it to two copies of a boundary value problem for the usual two-dimensional Laplace equation. The relevant proofs are included in[10] .

2. Segre quaternions

A quaternionic variable can be represented as

$$z = x_0 + e_1 x_1 + e_2 x_2 + e_3 x_3, \quad x_0, x_j, \text{real} \quad j = 1, 2, 3, \tag{3}$$

where the units e_j satisfy the relations (1). Equivalently, we can write the quaternionic variable z in the following two forms, namely

$$z = z_1 + e_2 z_2, \tag{4}$$

where $z_1 = x_0 + e_1 x_1$, $z_2 = x_2 + e_1 x_3$, as well as the so-called idempotent representation

$$z = z_+ \frac{1 - e_3}{2} + z_- \frac{1 + e_3}{2}, \tag{5}$$

where $z_+ = z_1 + e_1 z_2$ and $z_- = z_1 - e_1 z_2$. It is convenient to define the spaces $Q^\pm$ which are the spaces spanned by the variables z_+ and z_- respectively. Namely, given the complex variables $z_1, z_2 \in C^1$, the spaces $Q^\pm$ are defined as

$$Q^\pm = \{z_\pm \in Q^\pm, z_\pm = z_1 \pm e_1 z_2\}. \tag{6}$$

Also, the following definitions are quite useful.

Given four real-valued functions $x_0(t)$, $x_j(t)$, $j = 1, ..., 3$, we define a curve $\mathcal{C} = \mathcal{C}(t)$ in R^4 and the complex curves $\mathcal{C}(t)^\pm$ in the spaces $Q^\pm$ by

$$\mathcal{C}(t): \quad z_1(t) + e_2 z_2(t) = x_0(t) + e_1 x_1(t) + e_2 x_2(t) + e_3 x_3(t), \quad a \le t \le b, \tag{7}$$

where $a, b \in R$, and

$$\mathcal{C}^\pm(t): \quad z_1(t) \pm e_1 z_2(t) = x_0(t) \mp x_3(t) + e_1(x_1(t) \pm x_2(t)). \tag{8}$$

We can write formally

$$\mathcal{C}(t) = \mathcal{C}^+(t) \frac{1 - e_3}{2} + \mathcal{C}^-(t) \frac{1 + e_3}{2}. \tag{9}$$

A curve $\mathcal{C}(t)$ induces the curves $\mathcal{C}^\pm(t)$ and conversely, a pair of curves $\mathcal{C}^\pm(t)$ defines a curve $\mathcal{C}(t)$.

Let ω be a domain in R^4, and let $f : \omega \to R^4$, $(x_0, x_1, x_2, x_3) \to f(x_0, x_1, x_2, x_3)$ be a Segre quaternion–valued function defined in ω. Then

f is given in terms of the four real–valued functions f_0, f_j $j = 1, 2, 3$ by the expression

$$f(x_0, x_1, x_2, x_3) = f_0(x_0, x_1, x_2, x_3) + \sum_{j=1}^{3} e_j f_j(x_0, x_1, x_2, x_3). \tag{10}$$

Also, letting $u = f_0 + e_1 f_1$, $v = f_2 + e_1 f_3$, we find

$$f(x_0, x_1, x_2, x_3) = u(x_0, x_1, x_2, x_3) + e_2 v(x_0, x_1, x_2, x_3). \tag{11}$$

Demanding now that u, v are complex functions of z_1 and z_2, we can define a function $f(z_1, z_2)$ of two complex variables. In particular, $f(z_1, z_2)$ is defined by

$$f(z_1, z_2) = u(z_1, z_2) + e_2 v(z_1, z_2). \tag{12}$$

3. A novel Dbar or Pompeiu-Borel formula for functions of two complex variables

In this and the following sections, we introduce some novel formulae for functions of two complex variables. Consider the subclass of functions defined by (12) where $u(z_1, z_2)$ and $v(z_1, z_2)$ are analytic functions w.r.t. the variables z_1 and z_2, namely they satisfy

$$u_{\overline{z_1}} = u_{\overline{z_2}} = v_{\overline{z_1}} = v_{\overline{z_2}} = 0. \tag{13}$$

A function $f(z_1, z_2)$ belonging to the above class admits an integral representation given by a generalisation of the well-known Dbar or Pompeiu-Borel formula of complex analysis. This generalisation is included below and its validity rests upon the assumption that the domain $D \subset C^2$ where $f(z_1, z_2)$ is valid has a so-called type-I boundary.

Consider now a point P in C^2 and let $P^{\pm}$ be its projections to the spaces $Q^{\pm}$. Then, a curve $\mathcal{C}$ in $C^2 \cong R^4$ is of type-I, if the induced curves $\mathcal{C}_{\pm}$ satisfy the following two conditions:

(C-i) The curves $\mathcal{C}_{\pm}$ are positively oriented ;

(C-ii) The points $\mathcal{P}_{\pm}$ are inside the curves $\mathcal{C}_{\pm}$ respectively.

Let u, v be complex analytic functions of z_1 and z_2, D be a domain in C^2, and let $f : D \to C^2$, $(z_1, z_2) \to f(z_1, z_2)$ be a function defined in D. Then the function $f(z_1, z_2)$ admits the integral representation

$$f(z_1, z_2) = \frac{1}{2\pi e_1} \int_{\mathcal{C}} \frac{f(\zeta_1, \zeta_2)}{\zeta - z} (d\zeta_1 + e_2 d\zeta_2)$$

$$- \frac{e_2}{2\pi e_1} \int_{D} \frac{(\partial_{\zeta_1} + e_2 \partial_{\zeta_2}) f(\zeta_1, \zeta_2)}{\zeta - z} d\zeta_1 \wedge d\zeta_2, \tag{14}$$

where $D \subset R^4$ with a smooth type-I boundary $\mathcal{C}$.

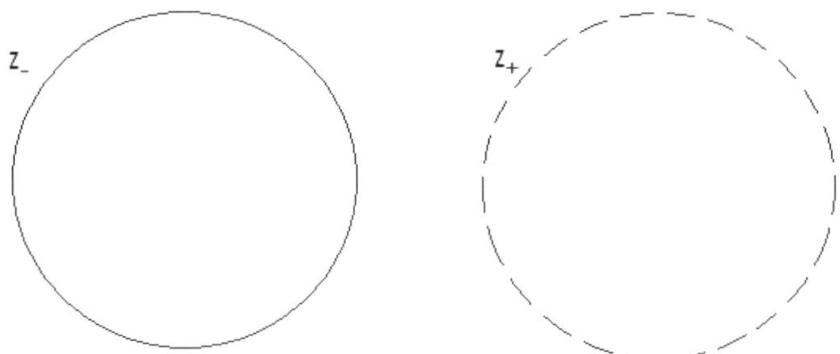

Fig. 1. Let z_+ and z_- belong to the above circles which are assumed positively oriented. Then equation (9) and the condition *(C-ii)* define a type-I curve.

4. Some new variations of the fundamental theorem of calculus

The function $f(z_1, z_2) = u(z_1, z_2) + e_2 v(z_1, z_2)$ is holomorphic iff $u(z_1, z_2)$ and $v(z_1, z_2)$ are holomorphic w.r.t z_1 and z_2 and

$$\frac{\partial u}{\partial z_1} = \frac{\partial v}{\partial z_2}, \qquad \frac{\partial u}{\partial z_2} = -\frac{\partial v}{\partial z_1}. \tag{15}$$

A holomorphic quaternion-valued function can be represented as[11]

$$f(z_1, z_2) = f_+(z_1 + e_1 z_2)\frac{1 - e_3}{2} + f_-(z_1 - e_1 z_2)\frac{1 + e_3}{2}, \tag{16}$$

where $f_+(z_1 + e_1 z_2)$ and $f_-(z_1 - e_1 z_2)$ are complex holomorphic functions given by
$f_+(z_1 + e_1 z_2) = u(z_1, z_2) + e_1 v(z_1, z_2)$ and $f_-(z_1 - e_1 z_2) = u(z_1, z_2) - e_1 v(z_1, z_2)$.
Also, it can be shown that $f(z_1, z_2)$ can be written as a series involving $z = z_1 + e_2 z_2$ only, i.e

$$f(z) = u(z_1, z_2) + e_2 v(z_1, z_2), \tag{17}$$

and $(\partial_{\zeta_1} + e_2 \partial_{\zeta_2}) f(\zeta) = 0$ therefore we obtain the analogue of the Cauchy formula

$$f(z) = \frac{1}{2\pi e_1} \int_C \frac{f(\zeta)}{\zeta - z} d\zeta. \tag{18}$$

For holomorphic functions the following theorems are valid:
Let $f(z) : D \to \mathbf{C}^2$ be a holomorphic function and let C be a curve $z(t) = z_1(t) + e_2 z_2(t), t \in [a, b]$, with continuous derivative and trace in D. Then

$$\frac{1}{2} \int_C (\partial_{z_1} - e_2 \partial_{z_2}) f(z) dz = f[z(b)] - f[z(a)]. \tag{19}$$

Also, let $f(z) : D \to \mathbb{C}^2$ be a holomorphic function and let $\mathcal{C}$ be a curve $\tilde{z}(t) = z_1(t) - e_2 z_2(t), t \in [a, b]$, with continuous derivative and trace in D. Then

$$\frac{1}{2} \int_{\mathcal{C}} (\partial_{z_1} + e_2 \partial_{z_2}) f(\tilde{z}) d\tilde{z} = f[\tilde{z}(b)] - f[\tilde{z}(a)]. \tag{20}$$

Next, we define k to be the following quaternionic variable:

$$k = k_1 + e_2 k_2, \quad k_1 = \lambda_0 + e_1 \lambda_1, \quad k_2 = \lambda_2 + e_1 \lambda_3.$$

We call this variable the spectral variable and an equation which contains this variable a spectral equation.

Using the first of the above theorems for $f = \nu e^{-e_1 k z}$, we obtain the following result:

Let the quaternion-valued functions $\nu(z, k)$ and $F(z)$ be holomorphic functions with respect to z which satisfy the equation

$$\frac{1}{2}(\partial_{z_1} - e_2 \partial_{z_2})\nu(z, k) - e_1 k \nu(z, k) = F(z), \tag{21}$$

in a domain D. Then, the function $\nu(z, k)$ is given by the expression

$$\nu(z, k) = \int_{\mathcal{C}} F(\zeta) e^{e_1 k(z - \zeta)} d\zeta, \tag{22}$$

where $\mathcal{C}$ is a curve with continuous derivative, joining $z(a)$ with $z(b) = z$, and a is s.t. $\nu(z(a)) e^{-e_1 k z(a)} = 0$.

5. Boundary value problems for a four-dimensional Laplace equation

Let the function $\nu(z, k)$ satisfy the system

$$\frac{1}{2}(\partial_{z_1} - e_2 \partial_{z_1})\nu(z, k) - e_1 k \nu(z, k) = \frac{1}{2}(\partial_{z_1} - e_2 \partial_{z_2})Q(z_1, z_2), \tag{23}$$

$$(\partial_{z_1} + e_2 \partial_{z_2})\nu(z, k) = 0. \tag{24}$$

Then $Q(z, k)$ satisfies

$$Q_{z_1 z_1}(z_1, z_2) + Q_{z_2 z_2}(z_1, z_2) = 0, \tag{25}$$

therefore (23)–(24) is a *Lax pair* of (25). We claim that we can solve boundary value problems for equation (25) by mapping them to a pair of identical boundary value problems for the usual Laplace equation

$$q_{w\bar{w}} = 0, \qquad w = x + iy. \tag{26}$$

For example, assume that equation (26) is valid in the quarter space with Dirichlet boundary conditions $g_1(x)$ and $g_2(y)$. The solution can be obtained using the approach of,[3] see also[4] . This solution is given in terms of certain known functions called spectral functions, which we denote here by $\nu_{\pm}$. The function $\nu(z,k)$ is given in terms of these functions by the expression

$$\nu(z,k) = \nu_+(z_+,k)\frac{1-e_3}{2} + \nu_-(z_-,k)\frac{1+e_3}{2}. \tag{27}$$

Therefore, assuming that the solution of a boundary value problem for (26) is known, this equation and (23) yield the solution of (25). Consider the

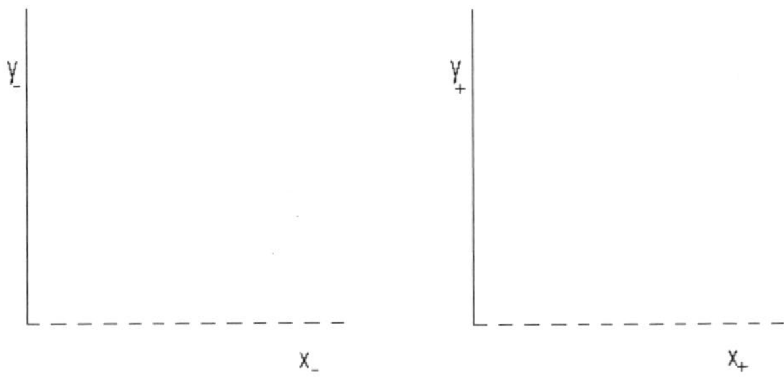

Fig. 2.

curve obtained by first joining the solid and dashed curves of figure 2 and then combining the resulting four dimensional curves together, namely

$$C = C_1 \cup C_2, \tag{28}$$

where

$$C_1 = \{y_+ = 0\}\frac{1-e_3}{2} + \{y_- = 0\}\frac{1+e_3}{2}, \tag{29}$$

and similarly for C_2. We then have,

$$C_1 = x_0 + e_3 x_3, \qquad C_2 = e_1 x_1 + e_2 x_2. \tag{30}$$

Furthermore, if z_+ is in the quarter space, i.e $x_+, y_+ \geq 0$ and similarly for z_-, then the quaternion z is in the domain Ω, where

$$\Omega = \{z = x_0 + x_j e_j \in R^4,\ |x_1| \geq x_2,\ |x_0| \geq x_3\}. \tag{31}$$

In summary, the boundary value problem for equation (25) valid in the domain Ω defined by (31) with

$$\frac{1}{2}(\partial_{z_1} - e_2\partial_{z_2})Q(z) = F(z), \quad z \in \Omega \cap \mathcal{C}_1 \tag{32}$$

and a similar boundary condition for $\mathcal{C}_2$, can be solved by mapping it to two copies of the Dirichlet problem for the Laplace equation in the quarter plane.

Acknowledgement

The author wishes to thank the anonymous referee for useful suggestions. Also, he expresses his gratitude to Professor A S Fokas for most valuable discussions.

References

1. F. Catoni, R. Cannata and P. Zampetti, An introduction to commutative quaternions, *Adv. Appl. Cliff. Alg.* **16**, 1 (2006)
2. A. Vajiac, *Singularities of Functions of One and Several Bicomplex Variables* (presentation at the 7th ISAAC conference, London 2009)
3. A.S. Fokas and A.A. Kapaev, On a Transform Method for the Laplace Equation in a Polygon, IMA J. Appl. Math., **68**, 355 (2003)
4. A.S. Fokas and D.A. Pinotsis, The Dbar formalism for certain non homogeneous linear elliptic equations in two dimensions, *Eur. J. Appl. Math.*, **17**, 3, 323–346 (2006)
5. A.S. Fokas and D.A. Pinotsis, Quaternions, evaluation of integrals and boundary value problems, *Comput. Meth. & Func. Th.* **7**, 2, 443 (2007)
6. K. Gürlebeck and W. Sprössig, *Quaternionic and Clifford Calculus for Physicists and Engineers*, Wiley (1997)
7. V.V. Kravchenko, On the relationship between p-analytic functions and the Schrödinger equation *Z. Anal. Anwendungen* **24** 487–496 (2005)
8. D. Rochon and M. Shapiro, On algebraic properties of bicomplex and hyperbolic numbers, *Anal. Univ. Oradea*, fasc. math., vol. 11 (2004).
9. D.A. Pinotsis, Quaternionic analysis, elliptic problems and a physical application of the Dbar formalism, *Adv. Appl. Cliff. Alg.* (to appear)
10. D.A. Pinotsis, Commutative Quaternions, Spectral Analysis and Boundary Value Problems (preprint)
11. G.B. Price, *An introduction to multicomplex spaces and functions*, Dekker, New York (1991)
12. J. Ryan, Complexified Clifford Analysis, *Complex Variables Theory and Appl.* **1** , 119–149 (1982)
13. C. Segre, Le Rappresentazioni Reali delle Forme Complesse e Gli Enti Iperalgebrici, *Math. Ann.*, **40**, 413 (1892)

III.2. Reproducing kernels and related topics

Organisers: A. Berlinet, S. Saitoh

Since the first works laying its foundations as a subfield of complex analysis, the theory of reproducing kernels has proved to be a powerful tool in many fields of pure and applied mathematics. The aim of this session was to gather researchers interested in theoretical as well as applied modern problems involving this theory.

The speakers of the session were
B. Abdous: A general theory for kernel estimation of smooth functionals
J.R. Higgins: Paley-Wiener spaces and their reproducing formulae
D.M. Onchis: Irregular sampling in multiple-window spline-type spaces
K. Fujita: Integral formulas on the boundary of some ball
K. Takemura: Free boundary value problem for $(-1)^M(d/dx)^{2M}$ and the best constant of Sobolev inequality
S.D. Sharma: Weighted composition operators on some spaces of analytic functions

The consequent papers gathered in this volume are by B. Abdous and A. Berlinet, K. Fujita, J.R. Higgins, H.G. Feichtinger and D.M. Onchis, A.K. Sharma and S.D. Sharma. H. Fujiwara gave a distinguished talk common to III.2 and III.3.

Reproducing kernel Hilbert spaces and local polynomial estimation of smooth functionals

B. Abdous

Université Laval, Québec, Canada
Médecine Sociale et Préventive,
Québec, Qc, Canada, G1K 7P4
E-mail: belkacem.abdous@msp.ulaval.ca
www.uresp.ulaval.ca/babdous/

A. Berlinet

Université Montpellier 2,
Institut de Mathématiques et de Modélisation de Montpellier,
Place Eugène Bataillon 34095 Montpellier Cedex France
E-mail: berlinet@iutmontp.univ-montp2.fr

We outline a general method that estimates smooth functionals of a probability distribution from a sample of observations, restricting the framework to local polynomial fitting. The construction of the estimators is based on a weighted least squares criterion and reproducing kernel Hilbert spaces theory. We briefly discuss their asymptotic properties and review applications to classical bivariate risk measures estimation.

Keywords: Kernel smoothing; local estimation; reproducing kernel Hilbert spaces; bivariate risk measures.

1. Introduction

Let $X_1, \ldots, X_n$ be independent observations from an unknown distribution function F defined on a domain $\Omega = \mathbb{R}, \mathbb{R}^k, [a, b], [a, \infty), \ldots$ and denote by $\widehat{F}_n(x)$ the associated empirical distribution function or any consistent estimate (e.g. the Kaplan-Meier estimate for censored data). Let $\Phi(x, F)$ be an arbitrary and given functional. Our aim is to construct nonparametric estimates of $\Phi(x, F)$ and possibly its derivatives (or partial derivatives). Statistical literature abounds in examples of functionals estimation problems. Examples and references will be provided hereafter. For the sake of presentation clarity, we will restrict ourselves to univariate domains. But, as illustrated in the last section, multivariate problems might be handled in a similar manner.

The distribution function F being unknown, a pilot estimator of $\Phi(x, F)$ might be defined by

$$\Phi_n(x) = \Phi(x, \widehat{F}_n) \qquad \text{with} \qquad \widehat{F}_n(t) = \frac{1}{n} \sum_{j=1}^{n} \mathbf{I}_{(X_j \leq t)}, \quad t \in \mathbb{R},$$

when F is estimated by the empirical distribution function. Furthermore, under some regularity assumptions (e.g. Hadamard differentiability), asymptotic properties of $(\Phi(x, \widehat{F}_n) - \Phi(x, F))$ can be easily deduced from that of $(\widehat{F}_n(x) - F(x))$, see for example.[1,2] That said, $\Phi(\cdot, \widehat{F}_n)$ is in general a rough estimator and it needs to be smoothed.

Let $r \geq 0$ be a fixed integer and assume that $\Phi(x, F)$ is $(r + 1)$ times continuously differentiable at x. Then, for any y in a neighborhood of x, the function $\Phi(y, F)$ can be locally approximated by a polynomial of order r. Indeed, Taylor's expansion ensures that

$$\Phi(y, F) = \sum_{i=0}^{r} \frac{(y - x)^i}{i!} \Phi^{(i)}(x, F) + \frac{(y - x)^{r+1}}{(r + 1)!} \Phi^{(r+1)}(\xi, F) \qquad (1)$$
$$\simeq P_r(y - x)$$

for some $\xi \in (x, y)$. Thus, one might use least squares method to fit a polynomial of order r to the empirical estimate $\Phi(x, \widehat{F}_n)$. This could be achieved by minimizing with respect to the coefficients a_i's the following criterion

$$J(a_0, \ldots, a_r) = \int_{\Omega} K\left(\frac{y - x}{h}\right) \left[\Phi_n(y) - \sum_{k=0}^{r} a_k (y - x)^k\right]^2 dy, \qquad (2)$$

where the smoothing parameter h controls the size of the neighborhood of x, while K denotes a weight function (or kernel). Based on Taylor's expansion above, the obtained solution $(\hat{a}_0(x), \ldots, r! \hat{a}_r(x))$ should approximate $(\Phi(x, F), \ldots, \Phi^r(x, F))$. A slightly different version of this approach, also known as "local polynomial fitting", is widely used in regression estimation problems. More generally, this estimation procedure is quite close in spirit to local likelihood fitting techniques. See[3-5] for more details.

The rest of the paper is organized as follows: In Section 2, we will use reproducing kernel Hilbert spaces to provide an explicit expression of the solution $(\hat{a}_0(x), \ldots, r! \hat{a}_r(x))$. Some of its properties will be summarized as well. The last section will present several recent applications. Space is missing to evoke more applications, for instance in Extreme Value Theory.

2. Representation and asymptotic behavior of the proposed estimators

2.1. *Construction of the estimators*

To ease our presentation, let us assume that Φ's domain is equal to $\mathbb{R}$. Other cases will be considered later on. Without loss of generality, assume that K is a probability density supported on the interval $[-1, 1]$. Let $\mathcal{P}_r$ be the space of polynomials of degree less than or equal to r and denote by $Q_0(z), Q_1(z), \ldots, Q_r(z)$ an orthonormal basis of $\mathcal{P}_r$ considered as a subspace of $L_2(K\,\lambda)$. Rewrite the weighted least squares criterion in (2) as follows

$$J(a_0, \ldots, a_r) = \int_{-1}^{1} \left[\Phi_n(x + uh) - \sum_{k=0}^{r} a_k h^k u^k \right]^2 K(u)du.$$

Then, the polynomial $\widehat{P}_r(u) = \sum_{k=0}^{r} \widehat{a}_k h^k u^k$ that minimizes J might be viewed as the $L_2(K\lambda)$ projection of $\Phi_n(x + hu)$ onto $\mathcal{P}_r$. Therefore, the function $\mathcal{K}(u, v) = \sum_{k=0}^{r} Q_k(u)Q_k(v)$ being a reproducing kernel of $\mathcal{P}_r$ we can write

$$\widehat{P}_r(u) = \int_{-1}^{1} \mathcal{K}(u, v)\, \Phi_n(x + hv)K(v)\, du.$$

Next, upon noting that for any $m = 0, \ldots, r$, the derivatives of $\widehat{P}_r$ satisfy $\widehat{P}_r^{(m)}(0) = m! h^m \widehat{a}_m$, we see that

$$m! h^m \widehat{a}_m = \int_{-1}^{1} \sum_{k=0}^{r} Q_k^{(m)}(0)Q_k(v)\Phi_n(x + hv)K(v)dv.$$

In virtue of Taylor expansion (1), the quantity

$$\Phi_{n,m}(x) := m! \widehat{a}_m = \frac{1}{h^m} \int_{-\infty}^{\infty} \Phi_n(z) \frac{1}{h} K^{[m,r]}\left(\frac{z - x}{h} \right) dz,$$

could be used to estimate $\Phi^{(m)}(x, F)$ for $m = 0, \ldots, r$, where

$$K^{[m,r]}(u) = \sum_{k=0}^{r} Q_k^{(m)}(0)Q_k(u)K(u). \tag{3}$$

Note that the kernels $K^{[m,r]}(\cdot)$ do not depend on the chosen orthonormal basis for $\mathcal{P}_r = \text{span}\{1, u, u^2, \ldots, u^r\}$. Their construction could be based on the classical Christoffel-Darboux formula, see[6] and the references therein for more details on orthogonal polynomials. Besides, these kernels possess many interesting properties, see.[7,8] For instance, they belong to the hierarchies of higher order kernels generated by $K(\cdot)$ and satisfy for $k \leq r$ the identity

$$\int_{-1}^{1} u^k K^{[m,r]}(u)du = \left\{ \begin{array}{ll} m! & k = m \\ 0 & k \neq m \end{array} \right. .$$

This property is very useful when dealing with asymptotic behaviors of the the estimators $\Phi_{n,m}(x)$. In addition, if the support Ω has one or two finite ends, then the kernels $K^{[m,r]}(\cdot)$ avoid boundary effects by taking different forms that depend on the region where the estimation is performed. For instance, if $\Omega = [a,b]$, with $-\infty < a < b < \infty$, then we have

$$\Phi_{n,m}(x) = \int_{a}^{b} \frac{1}{h^{m+1}} K^{[m,r]}\left(\frac{z-x}{h}\right) \Phi_n(z)dz$$

$$= \left\{ \begin{array}{ll} \dfrac{1}{h^m} \displaystyle\int_{-\alpha}^{1} K_L^{[m,r]}(z,\alpha)\Phi_n(x+hz)dz, & \text{if } x = a + \alpha h \text{ with } \alpha \in (0,1]; \\[2ex] \dfrac{1}{h^m} \displaystyle\int_{-1}^{1} K_I^{[m,r]}(z)\Phi_n(x+hz)dz, & \text{if } x \in (a+h, b-h); \\[2ex] \dfrac{1}{h^m} \displaystyle\int_{-1}^{\alpha} K_R^{[m,r]}(z,\alpha)\Phi_n(x+hz)dz, & \text{if } x = b - \alpha h \text{ with } \alpha \in (0,1]. \end{array} \right.$$

where the interior region kernel $K_I^{[m,r]}$ is given by (3) and has a unique form for all points x in the interior region $I = (a-h, b+h)$, while the boundary kernels $K_R^{[m,r]}$ and $K_L^{[m,r]}$ vary with x and can be obtained via (3) as well.

2.2. *Asymptotic behavior*

Under standard regularity conditions, it can be shown that the estimators $\Phi_{n,m}(x)$ converge in various modes to the true function $\Phi^{(m)}(x,F)$. Hereafter, we summarize some of these results, but details are omitted and can be found in.[9]

Theorem 2.1. *Suppose that the first m derivatives of $\Phi(x,F)$ exist and they belong to $L^p(\mathbb{R})$ for some $1 \leq p \leq \infty$. Assume that the point x is a Lebesgue point of $\Phi^{(m)}(x,F)$ and there exists an open neighborhood $\mathcal{N}_x$ of x such that for any sequence of distribution functions $\{F_n\}_{n=1}^{\infty}$ converging to F in the sup-norm,*

$$\lim_{n \to \infty} \sup_{y \in \mathcal{N}_x} |\Phi(y,F_n) - \Phi(y,F)| = 0$$

almost surely. Then $\Phi_{n,m}(x)$ converges almost surely to $\Phi^{(m)}(x,F)$ whenever the sequence of estimated distribution functions $\{F_n\}_{n=1}^{\infty}$ converges almost surely to F and the bandwidth converges to zero slowly enough.

Similarly, asymptotic normality follows from standard arguments and some additional regularity assumptions such us Hadamard differentiability. Here again we refer to[9] for details.

Next, in practice one has to select an "optimal" smoothing parameter h. A plethora of bandwidth selection methods have been proposed in density and regression estimation problems. Quite often, one relies on asymptotic expansions of a given error criterion such as the L_1-error, or the pointwise Mean Square Error (MSE) or its integrated version (MISE). However, the functional $\Phi(\cdot, F)$ being too general, useful and explicit asymptotic expansions of these loss functions are hard to come by. That said, if one is willing to specify $\Phi(\cdot, F)$, (e.g., mean residual life function,[10] extremes dependence function,[11] equipercentile equating,[12] ...), then, explicit asymptotic expansions of MSE might be exhibited.

3. Applications

The above approach has been applied in several univariate estimation problems, see.[9] In the sequel, we will focus on some classical bivariate risk measures estimation problems only. Indeed, let (X, Y) be a bivariate random variable with density function $f(x, y)$, cumulative distribution function $F(x, y)$, survival function $S(x, y) = \mathbb{P}(X > x, Y > y)$, marginal densities f_X, f_Y and survival marginals $S_X(x) = \mathbb{P}(X > x)$ and $S_Y(y) = \mathbb{P}(Y > y)$. Denote the conditional survival functions of Y and X by $S_Y(y|x) = \mathbb{P}(Y > y|X = x)$ and $S_X(x|y) = \mathbb{P}(X > x|Y = y)$ respectively. Denote the partial derivatives of $S(\cdot, \cdot)$ by $S_{ij}(x, y) = \partial^{i+j}S(x, y)/\partial x^i \partial y^j$, for $0 \leq i + j \leq 2$. Several risk or association measures are expressed in terms of the survival function $S(\cdot, \cdot)$ and its derivatives:

- Bivariate hazard rate

$$\lambda(x; y) = \frac{f(x, y)}{S(x, y)} = \frac{S_{11}(x, y)}{S(x, y)}, \text{ for } (x, y) : S(x, y) \neq 0$$

- Conditional hazard rate of Y given $X = x$

$$\lambda(y|x) = \lim_{\Delta y \to 0} \frac{\mathbf{P}(Y \leq y + \Delta y|Y > y, X = x)}{\Delta y}$$

$$= \frac{f_Y(y|x)}{S_Y(y|x)} = -\frac{S_{11}(x, y)}{S_{10}(x, y)}, \text{ for } (x, y) : S_{10}(x, y) \neq 0$$

- Conditional hazard rate of Y given $X > x$

$$\lambda(y|X > x) = \lim_{\Delta y \to 0} \frac{\mathbf{P}(Y \le y + \Delta y | Y > y, X > x)}{\Delta y}$$

$$= \frac{f_Y(y|X > x)}{S_Y(y|X > x)} = -\frac{S_{01}(x,y)}{S(x,y)}, \ \text{for } (x,y) : S(x,y) \ne 0$$

- The cross-ratio function (a local measure of association)

$$\theta(x,y) = \frac{\lambda(y|X = x)}{\lambda(y|X > x)} = \frac{S_{11}(x,y)S(x,y)}{S_{01}(x,y)S_{10}(x,y)},$$

provided that $S_{01}(x,y)S_{10}(x,y) \ne 0$.

- The global odds ratio function (a measure of association as well)

$$\mathrm{OR}(x,y) = \frac{\mathbb{P}(X > x|Y > y)\mathbb{P}(X \le x|Y \le y)}{\mathbb{P}(X \le x|Y > y)\mathbb{P}(X > x|Y \le y)}$$

$$= \frac{S(x,y)\big[1 - S(x,0) - S(0,y) + S(x,y)\big]}{\big[S(x,0) - S(x,y)\big]\big[S(0,y) - S(x,y)\big]}$$

- The conditional residual life function of Y given $X > x$

$$e(y|X > x) = E\big[Y - y|Y > y, X > x\big] = \frac{1}{S(x,y)} \int_y^\infty S(x,v)dv$$

if $S(x,y) \ne 0$.

Clearly, plug-in estimates of these functions are obtained upon estimating the bivariate survival function and its derivatives. To this end, an empirical estimate of $S(x,y)$ is needed first. In practice, the survival times X and Y are often subject to right censoring. Denote the associated censoring variables by C^X and C^Y respectively and set $\overline{G}$ for the survival functions of the couple (C^X, C^Y). Under this standard setting, the observed variables are $(\tilde{X}, \tilde{Y}, \delta^X, \delta^Y)$ where $\tilde{X} = X \wedge C^X$, $\tilde{Y} = Y \wedge C^Y$, $\delta^X = \mathbb{I}_{\{X \le C^X\}}$ and $\delta^Y = \mathbb{I}_{\{Y \le C^Y\}}$, where $\mathbb{I}(\cdot)$ stands for the indicator function and $x \wedge y = \min(x,y)$. The cases of unique censoring: $C^X \equiv C^Y \equiv C$ and univariate censoring: $C^X \equiv \infty$ (or $C^Y \equiv \infty$) are particular cases of this formulation. Next, it is easy to see that when (X,Y) and (C^X, C^Y) are independent, the survival function $\overline{H}$ of $(\tilde{X}, \tilde{Y})$ satisfies: $\overline{H}(x,y) = S(x,y)\overline{G}(x,y)$.

Thus, given a sample of independent and identically distributed random variables $(\tilde{X}_i, \tilde{Y}_i, \delta_i^X, \delta_i^Y)$, $i = 1, \ldots, n$ having the same distribution as $(\tilde{X}, \tilde{Y}, \delta^X, \delta^Y)$, a nonparametric estimate of the targeted survival function $S(x,y)$ is provided by

$$S_n(x,y) = \overline{H}_n(x,y)/\overline{G}_n(x,y)$$

where $\overline{H}_n(x,y) = \frac{1}{n}\sum_{i=1}^{n} 1_{\{\tilde{X}_i > x, \tilde{Y}_i > y\}}$ is the classical empirical estimate based on a complete sample, while $\overline{G}_n(x,y)$ is an estimate of $\overline{G}(x,y)$ constructed from either univariate or bivariate Kaplan-Meier estimates of the censoring times. We refer to[13-15] for more details.

Now, suppose that the survival function S is smooth enough and consider its Taylor expansion of total order r

$$S(\mathbf{z}) = \sum_{j=0}^{r} \left[(\mathbf{z} - \mathbf{z}_0)^T \nabla\right]^j S(\mathbf{z}_0) + o\left(\|\mathbf{z} - \mathbf{z}_0\|^r\right)$$

where

$$\left[(\mathbf{z} - \mathbf{z}_0)^T \nabla\right]^j S(\mathbf{z}_0) = \sum_{i=0}^{j} \frac{(x - x_0)^i}{i!} \frac{(y - y_0)^{j-i}}{(j-i)!} S_{ij}(x_0, y_0)$$

and $S_{ij}(x,y) = \partial^{i+j} S(x,y)/\partial x^i \partial y^j$. Then, a bivariate polynomial local fitting of S and its partial derivative can be obtained by minimizing the criterion

$$\int_{\mathbb{R}^{+2}} K_H(\mathbf{u} - \mathbf{z}) \left[S_n(\mathbf{u}) - P_r(\mathbf{u} - \mathbf{z})\right]^2 d\mathbf{u}.$$

where $P_r(\mathbf{z}) = \sum_{k=0}^{r}\sum_{i=0}^{k} c_{i,k-i} x^i y^{k-i}$ and $K : \mathbb{R}^2 \mapsto \mathbb{R}^+$ is an arbitrary kernel with $K_H(\mathbf{z}) = |H|^{-1} K(H^{-1}\mathbf{z})$, and H is a non-singular 2×2 smoothing matrix. Bivariate orthogonal polynomials together with reproducing kernel Hilbert spaces can be used to provide explicit representations of the estimators. Details can be found in.[16]

References

1. D. D. Boos and R. J. Serfling, *Ann. Statist.* **8**, 618 (1980).
2. L. T. Fernholz, *von Mises calculus for statistical functionals*, Lecture Notes in Statistics, Vol. 19 (Springer-Verlag, New York, 1983).
3. J. Fan and I. Gijbels, *Local polynomial modelling and its applications*, Monographs on Statistics and Applied Probability, Vol. 66 (Chapman & Hall, London, 1996).
4. C. Loader, *Local regression and likelihood*, Statistics and Computing (Springer-Verlag, New York, 1999).
5. J. Jiang and K. Doksum, *Sci. China Ser. A* **47**, 114 (2004).
6. A. Berlinet and C. Thomas-Agnan, *Reproducing kernel Hilbert spaces in probability and statistics* (Kluwer Academic Publishers, Boston, MA, 2004).
7. A. Berlinet, Reproducing kernels and finite order kernels, in *Nonparametric functional estimation and related topics (Spetses, 1990)*, , NATO Adv. Sci.

Inst. Ser. C Math. Phys. Sci. Vol. 335 (Kluwer Acad. Publ., Dordrecht, 1991) pp. 3–18.

8. A. Berlinet, *Probab. Theory Related Fields* **94**, 489 (1993).
9. B. Abdous, A. Berlinet and N. Hengartner, *Rev. Roumaine Math. Pures Appl.* **48**, 217 (2003).
10. B. Abdous and A. Berred, *J. Statist. Plann. Inference* **132**, 3 (2005).
11. B. Abdous and K. Ghoudi, *J. Nonparametr. Stat.* **17**, 915 (2005).
12. B. Abdous and K. ElFassi, *Local polynomial fitting of an equipercentile equating function : Asymptotic mean square error*, technical report, Université Laval (Québec, Canada, 2009).
13. W. Wang and M. T. Wells, *Biometrika* **84**, 863 (1997).
14. D. Y. Lin and Z. Ying, *Biometrika* **80**, 573 (1993).
15. D. M. Dabrowska, *Ann. Statist.* **16**, 1475 (1988).
16. B. Abdous and E. Bensaid, *J. Nonparametr. Statist.* **13**, 77 (2001).

Constructive reconstruction from irregular sampling in multi-window spline-type spaces

H.G. Feichtinger and D.M. Onchiş

NuHAG, Faculty of Mathematics,
Vienna, Austria
E-mails: darian.onchis@univie.ac.at, hans.feichtinger@univie.ac.at
www.nuhag.eu

The constructive recovery of functions in multi-window spline-type spaces from irregular samples are treated, using the concepts of frames and reproducing kernels. It is well known that one can describe iterative methods, which guarantee stable reconstruction, convergent in a variety of functions spaces, at least if the sampling set is dense enough. The corresponding theoretical statements either work in the context of continuous variables or discrete signals, but rarely consider the effect of errors that unavoidably occur if the continuous problem is implemented in a discrete (finite) setting. The analysis of the situation, using function spaces methods (in particular Wiener amalgams) and the description of implementable variants of these iterative algorithms with guaranteed rates of convergence are treated in the necessary detail here. It relies on the results of an earlier paper by the authors, demonstrating the fact that one can guarantee a good approximation of the biorthogonal family for the set of generators of the space. It is important to note, that the imperfections due to discretization do not spoil the chance of perfect reconstruction (in the limit), but only lead to a mild degradation of the rate of convergence, compared to the ideal case.

Keywords: Iterative algorithms; multi-window spline-type spaces; reproducing kernels; sampling.

1. Introduction

Existing iterative algorithms for the irregular sampling problem [1-6,8,13] presume that the "realization of the projection" of a given quasi-interpolation (such as nearest neighborhood interpolation, piecewise linear interpolation or more general interpolation procedures) can be easily obtained, because the (orthogonal) projection on the spline-type space is "given explicitly by a formula" and also known to be continuous with respect to a family of weighted L^p-norms. We will show how one can realize the reconstruction of a function in multi-window spline-spaces from irregular sampling sets in a more realistic (not purely functional analytic) way, using only FFTs and finite vectors, while still measuring the errors in the continuous context.

The structure of this contribution is as follows: In the first sections we introduce the necessary results concerning spline-type spaces, followed by a description of the irregular sampling problem and its connections to frame theory. The main section goes on treats a constructive iterative realization of the reconstruction of functions in multi-window spline-type spaces from irregular samples.

2. Multi-window spline-type spaces

Although the theory is valid over LCA groups, we work with general lattices $\Lambda \lhd G = \mathbb{R}^d$, which are of the form $\Lambda = A\mathbb{Z}^d$, for some non-singular $d \times d$ matrix A.

Let us first give the definition of *spline-type spaces* (sometimes also called shift invariant spaces, or principal shift-invariant spaces,[10] when they are generated by a *single* function and its translates). Recall that a Banach space of sequences $\boldsymbol{B}_d$ (indexed by a set Λ) is called *solid* if for any sequence $\mathbf{c} \in \boldsymbol{B}_d$ and $\mathbf{d}$ with $|\mathbf{d}(\lambda)| \leq |\mathbf{c}(\lambda)|$ for all $\lambda \in \Lambda$ (i.e. pointwise domination), implies $\mathbf{d} \in \boldsymbol{B}_d$ and $\|\mathbf{d}\|_{\boldsymbol{B}_d} \leq \|\mathbf{c}\|_{\boldsymbol{B}_d}$. We write T_x for the *translation operators*, given by $T_x\varphi(z) = \varphi(z - x)$, $\quad z, x \in \mathbb{R}^d$.

Definition 2.1. Let Λ be a discrete subgroup of $\mathbb{R}^d$, and φ be any element in some translation-invariant Banach space $(\boldsymbol{B}\,\|\cdot\|_{\boldsymbol{B}})$ of functions or distributions on $\mathbb{R}^d$. Then the closed linear span of the family $(T_\lambda g)_{\lambda \in \Lambda}$ is a closed, Λ-invariant subspace of $(\boldsymbol{B}, \|\cdot\|_{\boldsymbol{B}})$. We denote this space by $\boldsymbol{V}_{\phi,\Lambda}^{\boldsymbol{B}}$, and call it a **spline-type space** (in $\boldsymbol{B}$) if the family $(T_\lambda g)_{\lambda \in \Lambda}$ is a *Riesz projection basis* for $\boldsymbol{V}_{\phi,\Lambda}^{\boldsymbol{B}}$, i.e., if the following two properties are valid:

(1) There is some naturally associated solid Banach space $(\boldsymbol{B}_d\,\|\cdot\|_{\boldsymbol{B}_d})$ of sequences on Λ such that the synthesis mapping (representation operator R) described by $R : \mathbf{c} \mapsto \sum_{\lambda \in \Lambda} c_\lambda T_\lambda \varphi$ is well-defined and defines a continuous bijection between $\boldsymbol{B}_d$ and $\boldsymbol{V}_{\phi,\Lambda}^{\boldsymbol{B}}$.

(2) There exists a bounded linear mapping C (coefficient mapping) defined on all of $\boldsymbol{B}$ such that $C \circ R = Id_{\boldsymbol{B}_d}$ (and $C \circ R = Id_{\boldsymbol{V}}$).

For convenience we write $\boldsymbol{V}_{\phi,\Lambda}$ for the case $\boldsymbol{B} = \boldsymbol{L}^2(\mathbb{R}^d)$ and $\boldsymbol{V}_{\varphi,\Lambda}^p$ if $\boldsymbol{B} = \boldsymbol{L}^p(\mathbb{R}^d)$ in the sequel.

Remark 2.1. The existence of the left inverse C to the given synthesis operator R is in fact equivalent to the assumption that there exists a bounded and linear projection operator P from $\boldsymbol{B}$ onto $\boldsymbol{V}_{\phi,\Lambda}^{\boldsymbol{B}}$.

Remark 2.2. In the case of Hilbert spaces, every closed linear subspace allows for a corresponding orthogonal projection and therefore it suffices to rephrase condition (1) in the usual way by assuming that there exist constants $C, D > 0$ such that for all $\mathbf{c} \in \ell^2(\Lambda)$ one has:

$$C\|\mathbf{c}\|_{\ell^2} \leq \|\sum_{\lambda \in \Lambda} c_\lambda T_\lambda \varphi\|_{L^2} \leq D\|\mathbf{c}\|_{\ell^2}.$$

In other words: a family is a Riesz basis for a $\boldsymbol{V}_{\phi,\Lambda}$ ($p = 2$) if and only if it is an $\ell^2(\Lambda)$-Riesz projection basis. It is also an $\ell^1(\Lambda)$-Riesz projection basis for $\boldsymbol{V}_{\phi,\Lambda} \cap \boldsymbol{L}^1(\mathbb{R}^d)$ (with the L^1-norm) if the corresponding synthesis and analysis windows φ and $\tilde{\varphi}$ are in $\boldsymbol{W}(\boldsymbol{C}_0, \boldsymbol{\ell}^1)(\mathbb{R}^d)$, i.e. if they are absolutely Riemann integrable. A practical sufficient condition for this to be true is $|\varphi(z)| \leq C(1 + |z|)^{-d-1}, z \in \mathbb{R}^d$ and the same goes for the dual atom $\tilde{\varphi}$. Here $\tilde{\varphi}$ is the generator for the biorthogonal Riesz basis $(T_\lambda \tilde{\varphi})_{\lambda \in \Lambda}$. Following standard terminology we call $\tilde{\varphi}$ the *dual atom* to φ (with respect to Λ). The stated condition ensures among others that $f * \tilde{\varphi}$ is in $\boldsymbol{W}(\boldsymbol{C}_0, \boldsymbol{\ell}^1)$ for all $f \in \boldsymbol{L}^1(\mathbb{R}^d)$, its samples are in $\ell^1(\Lambda)$, and this is one of the reasons why we have chosen this particular function space setting.

Remark 2.3. For the applications and numerical realization we will use only $\Lambda = \mathbb{Z}^d \lhd \mathbb{R}^d$, because by a simple transformation (automorphism of $\mathbb{R}^d$) of the lattice and the corresponding generator system one can reduce the general problem to this setting, at the cost of replacing the set of generators by their transformed version. Since all the function spaces used are invariant under affine transformations $\alpha^*(f) : z \mapsto f(\alpha(z))$ for arbitrary automorphism of the underlying group this is no restriction of generality.

Corollary 2.1. *The orthogonal projection from* $\boldsymbol{L}^2(\mathbb{R}^d)$ *onto* $\boldsymbol{V}_{\phi,\Lambda}$ *is given by*

$$f \mapsto P_{\varphi,\Lambda}(f) = \sum_{\lambda \in \Lambda} [f * \varphi^*](\lambda) \cdot T_\lambda \tilde{\varphi} = \sum_{\lambda \in \Lambda} [f * \tilde{\varphi}^*](\lambda) \cdot T_\lambda \varphi. \qquad (1)$$

For $\varphi \in \boldsymbol{W}(\boldsymbol{C}_0, \boldsymbol{\ell}^1)(\mathbb{R}^d)$, *this mapping is (uniformly) bounded with respect to all* $\boldsymbol{L}^p$-*norms, with* $p \in [1, \infty]$.

Remark 2.4. As a consequence of the above corollary we can explicitly write the dual atom as linear combinations of the original atom with the coefficients coming from inverting the Gramian matrix, whose entries are the values of the autocorrelation function $\varphi * \varphi^*$ for φ, sampled over Λ.

$$\tilde{\varphi} = \sum_{\lambda \in \Lambda} b_\lambda T_\lambda \varphi. \qquad (2)$$

For the multi-window spline-type case or finitely generated multi-spline type spaces, we consider a set of functions in $L^2(\mathbb{R}^d)$, namely $\Phi = \{\varphi_1, \varphi_2, ..., \varphi_k\}$. The functions $\varphi_1, \varphi_2, ..., \varphi_k$ are called *set of generators* for the space $V = V_{\Phi,\Lambda}$.

It is a well known fact that a (bounded), indexed set of vectors $(g_\rho)_{\rho \in J}$ in a Hilbert-space (here in $\mathcal{H} = L^2(\mathbb{R}^d)$) is a Riesz basis (for its closed linear span) if and only if its Gramian, given by

$$G = (g_{\rho,\rho'}) = ((\langle g_{\rho'}, g_\rho \rangle))_{\rho,\rho' \in J}$$

is boundedly invertible on $\ell^2(\Lambda)$. In fact, the inverse Gramian matrix allows to find the biorthogonal family $(\tilde{g}_\rho)_{\rho \in J}$ in the closed linear span of the family $(g_\rho)_{\rho \in J}$. We have $\tilde{g}_{\rho'} = \sum_\rho d_\rho^{\rho'} g_\rho$, where the coefficients are taken from the ρ'−th row (or column) of G^{-1}. Conversely, the existence of a biorthogonal family generating a bounded mapping from $L^2(\mathbb{R}^d)$ into $\ell^2(J)$ implies that $(g_\rho)_{\rho \in J}$ is a Riesz basis.

Definition 2.2. Given $\Phi = (\varphi_1, ..., \varphi_k)$ and $\mathbf{c} = (c_1, ..., c_k)$ we write the multi-window spline-type space - in analogy to ordinary spline-type spaces - in the form

$$V^p_{\Phi,\Lambda} = \{f = \sum_{\lambda \in \Lambda} \mathbf{c}_\lambda T_\lambda \Phi := \sum_{i=1}^k \sum_{\lambda \in \Lambda} c_\lambda^i T_\lambda \varphi^i \,|\, \mathbf{c}^i = (\mathbf{c}_\lambda^i) \in \ell^p(\Lambda),$$

$$\text{for } i = 1, \ldots, k\} \quad (3)$$

3. Sampling in spline-type spaces

The classical sampling theorem applies to band-limited functions f in

$$\mathcal{B} = \{f \in L^2(R) : supp\hat{f} \subseteq [-1/2, 1/2]\} \quad (4)$$

The Shannon-Whittaker-Kotelnikov sampling theorem provides exact reconstruction of $f \in \mathcal{B}$ from its samples at the integer lattice $\mathbb{Z}$:

$$f(t) = \sum_{j=-\infty}^{\infty} f(j) \frac{sin\, \pi(t-j)}{\pi(t-j)} \quad (5)$$

This representation of f is not optimal,[2] because the SINC-kernel has poor decay at infinity and therefore many samples $f(j)$ are needed to approximate $f(t)$ even only locally. This motivates the use of spline-type spaces. One replaces the SINC kernel by a function φ satisfying:

1. φ is stable, i.e. $f = \sum_{\lambda \in \Lambda} c_\lambda T_\lambda \varphi$ satisfies: $\|f\|_2 \asymp \|c\|_2 \quad \forall f \in V_{\varphi, \mathbb{Z}^d}.$

2. Point evaluations $f \mapsto f(x)$ must make sense.

This can be guaranteed if φ is a continuous function in $\boldsymbol{W}(\boldsymbol{C}_0, \boldsymbol{\ell}^1)$:

$$\|\varphi\|_{\boldsymbol{W}(\boldsymbol{C}_0, \boldsymbol{\ell}^1)} = \sum_{k \in \mathbb{Z}^d} sup_{t \in [0,1]^d} |\varphi(t+k)| < \infty$$

General properties of frames and RKHS (Reproducing kernel Hilbert spaces) imply:

Lemma 3.1. *(a) If φ is stable, there exists a dual generator $\tilde{\varphi} \in V_{\varphi, \mathbb{Z}^d}$,*

with $\langle \varphi, \tilde{\varphi}(\cdot - k) \rangle = \delta_{k0}, \quad \forall k \in \mathbb{Z}^d \quad$ *and* $\quad f = \sum_{k \in \mathbb{Z}^d} \langle f, \tilde{\varphi}(\cdot - k) \rangle \varphi(\cdot - k).$

(b) If $\varphi \in \boldsymbol{W}(\boldsymbol{C}_0, \boldsymbol{\ell}^1)$, then $V_{\varphi, \mathbb{Z}^d}$ is a RKHS with kernel K such that $f(x) = \langle f, K_x \rangle$, for

$$K_x(t) = \sum_{k \in \mathbb{Z}^d} \varphi(x - k) \overline{\tilde{\varphi}(t - k)} \tag{6}$$

Lemma 3.2. *The following conditions are equivalent to the fact that $\mathcal{X} = \{x_j : j \in J\}$, is a set of sampling for $V_{\varphi, \mathbb{Z}^d}$:*
(a) Sampling inequality. There exists $A, B > 0$, such that

$$A\|f\|_2^2 \le \sum_{j \in J} |f(x_j)|^2 \le B\|f\|_2^2, \quad \forall f \in \mathbb{Z}^d.$$

(b) Frame condition. There exists $A, B > 0$, such that

$$A\|f\|_2^2 \le \sum_{j \in J} |\langle f, K_{x_j} \rangle|^2 \le B\|f\|_2^2, \quad \forall f \in \mathbb{Z}^d.$$

(c) Frame operator. $Sf = \sum_{j \in J} f(x_j) K_{x_j}$ is invertible on $V_{\varphi, \mathbb{Z}^d}$.

(d) The set of reproducing kernels $\{K_{x_j} : x_j \in \mathcal{X}\}$ forms a frame for $V_{\varphi, \mathbb{Z}^d}$.

(e) Frame reconstruction. There exists a dual frame $\widetilde{K_{x_j}}$, such that

$$f = \sum_{j \in J} \langle f, K_{x_j} \rangle \widetilde{K_{x_j}} = \sum_{j \in J} f(x_j) \widetilde{K_{x_j}} \ for \ f \in V_{\varphi, \mathbb{Z}^d}.$$

(f) Projection operator. The orthogonal projection P from L^2 onto $V_{\varphi, \mathbb{Z}^d}$ is

$$Pf = \sum_{j \in J} \langle f, K_{x_j} \rangle \widetilde{K_{x_j}}.$$

4. Realizable iterative reconstruction algorithms

Our paper[12] describes a constructive approach to the numerical realization of the dual generators for a multi-window spline-type family. Given a finite family of functions $\Phi = (\varphi_1, ..., \varphi_k)$ from the Wiener algebra $\boldsymbol{W}(\boldsymbol{C}_0, \boldsymbol{\ell}^1)(\mathbb{R}^d)$ and some $\varepsilon > 0$ there is a way to calculate the biorthogonal family $\tilde{\Phi}$ up to an error less than ε in the $\boldsymbol{W}(\boldsymbol{C}_0, \boldsymbol{\ell}^1)$-norm, using only a finite number of matrix inversions. In fact, what is found is a family of coefficients, describing these approximation of the individual elements $\tilde{\varphi}_i^n, i \in I$ as finite linear combinations of the original elements. The approach guarantees convergence of these (more and more refined approximations) to $\tilde{\varphi}_i$:

$$\|\tilde{\varphi}_i - \tilde{\varphi}_i^n\|_{\boldsymbol{W}(\boldsymbol{C}_0, \boldsymbol{\ell}^1)} \to 0 \quad \text{for} \quad n \to \infty. \tag{7}$$

While the standard methods described in the literature[1–6,8,13] pretend that perfect reconstruction can be carried out by an iterative procedure, they progress as follows: First of all the so-called *quasi-interpolators* Q_X (linear combinations of the elements of a sufficient fine BUPU = bounded and uniformly small partition of unity) are applied, and then the projection P on the spline-type space has to be performed. Although this approach can be viewed as constructive from an analysis point of view we want to shed some light on the practical realization of these steps. Although the ideas presented here cover mostly the finite discrete context they are even more relevant for the general (continuous and even multi-dimensional) setting.

We will discuss two cases: in the first case the "usual frame point of view" will be taken:

Lemma 4.1. *Assume that the family $\left(K_{x_j}\right)_{j \in J}$ is a frame for the spline-type space $V_{\varphi, \mathbb{Z}^d}$. Let us write K^n for the computable approximate kernel, obtained by using the approximate dual $\tilde{\varphi}_i^n$ instead of $\tilde{\varphi}_i$ in (6). Then for any sufficiently large n the family $(K_{x_j}^n)_{j \in J}$ is a frame as well, close to the original one, so that frame iterations still work and the resulting limits of the iterations of the frame algorithm gives a good approximate reconstruction.*

More precisely: In the above situation one can find for any given $\delta > 0$ some $N > 0$ and K such that one can guarantee: the reconstructions $f_{n,k}$ obtained by doing $k \geq K$ steps with any of the approximate kernels $K_{x_j}^n$, for $n \geq N$ will satisfy

$$\|f - f_{n,k}\|_2 \leq \delta \|f\|_2 \quad \text{for all} \quad f \in V_{\varphi, \mathbb{Z}^d}. \tag{8}$$

The proof, based on a perturbation argument, is left to the reader. It is similar to the situation when one has jitter error (i.e. unknown positions), cf. for example[7,9,11] . Note however, that unfortunately in this situation one cannot expect to obtain "perfect reconstruction" by doing a large number of iterations, for fixed (maybe large) n. Also nothing is said about general p for this situation.

The standard iterative algorithm works under the assumption that an ideal operator of the form

$$f \mapsto Af := PQ_Xf \tag{9}$$

where P is the orthogonal projection onto $V_{\varphi,\mathbb{Z}^d}$ and Q_X is a quasi-interpolant using only the sampling values $(f(x_j)) = f|_X$. In reality however one will not have (except for the finite/discrete setting) a perfect projection, but only the approximate one, using the approximate dual elements. Let us call the corresponding operators P_n and the resulting approximation operator $A_n := P_nQ_X$ (we avoid to discuss the problem of approximate BUPUs, for reasons of space, although that argument would be similar). Using this we obtain:

Theorem 4.1. *Assume that reconstruction under prefect conditions is possible. Then one has for fixed BUPU Ψ and corresponding quasi-interpolation operator Q_X and $\varepsilon' > 0$ and n large enough:*

$$\|Af - A_nf\|_p \leq \varepsilon' \cdot \|f\|_p \quad for\ all \quad f \in V^p_{\Phi,\Lambda}, \quad for\ any \quad p \in [1,\infty]. \tag{10}$$

Consequently the iterative algorithm using A_n instead of A provides perfect recovery of any $f \in V^p_{\varphi,\mathbb{Z}^d}$, with convergence in $\boldsymbol{L}^p$.

The proof relies on the estimate (7) which guarantees (using convolution relations for Wiener amalgam spaces) the claimed small error in the operator norm, uniformly for all $\boldsymbol{L}^p$-norms.

Note that in the same way the robustness of the approximate reconstructions can be verified. For example, the performance of the realistic iterative reconstruction method described above with respect to jitter error can be garanteed, very much as in the standard descriptions.

5. Conclusions

We have pointed out that the procedures described in the papers [1–6,8,13–15] are constructive from an analytic point of view, but not *realizable* on any computer. Hence we suggest to replace them by (numerically realizable) constructive iterative versions, making use of approximately dual generators for those spaces. The main theorem claims that one has still perfect reconstruction, and only a mild deterioration in terms of improvement per iteration.

Computing the iterations as above, i.e. with guaranteed small error in the Wiener algebra norm (or even a weighted version of it) one can guarantee convergence of the iterative algorithm to the correct set of coefficients, if precise knowledge about data is given, and if the samples of the original building blocks at the sampling points are given precisely. The same robustness claims as for the standard case (jitter errors, approximate knowledge about the generators, small local averages, etc.) apply in this situation, because all the errors can be controlled using operator norms.

Acknowledgement

The research was supported by the project EUCETIFA (The Marie-Curie Excellence Grant "European Center of Time-Frequency Analysis", 2006-2009) and the Compututational Science priority project NAHA of the University of Vienna.

References

1. A. Aldroubi. Non-uniform weighted average sampling and reconstruction in shift-invariant and wavelet spaces. *Appl. Comput. Harmon. Anal.*, 13(2):151–161, 2002.
2. A. Aldroubi and H. G. Feichtinger. Complete iterative reconstruction algorithms for irregularly sampled data in spline-like spaces. pages 1857–1860, April 1997.
3. A. Aldroubi and H. G. Feichtinger. Exact iterative reconstruction algorithm for multivariate irregularly sampled functions in spline-like spaces: The L^p-Theory. *Proc. Amer. Math. Soc.*, 126(9):2677–2686, 1998.
4. A. Aldroubi and K. Gröchenig. Beurling-Landau-type theorems for non-uniform sampling in shift invariant spline spaces. *J. Fourier Anal. Appl.*, 6(1):93–103, 2000.
5. A. Aldroubi and K. Gröchenig. Nonuniform sampling and reconstruction in shift-invariant spaces. *SIAM Rev.*, 43(4):585–620, 2001.
6. A. Aldroubi, Q. Sun, and W.-S. Tang. Nonuniform average sampling and reconstruction in multiply generated shift-invariant spaces. *Constructive Approximation*, 20(2):173–189, 2004.
7. P. G. Casazza, C. Zhao, and P. Zhao. Perturbation of regular sampling in shift-invariant spaces for frames. *IEEE Trans. Inform. Theory*, 52(10):4643–4648, 2006.
8. W. Chen, S. Itoh, and J. Shiki. On sampling in shift invariant spaces. *IEEE Trans. Inf. Theory*, 48(10):2802–2810, 2002.
9. O. Christensen, H. O. Kim, R. Y. Kim, and J. K. Lim. Perturbation of frame sequences in shift-invariant spaces. *J. Geom. Anal.*, 15(2):181–192, 2005.

10. C. de Boor, R. A. DeVore, and A. Ron. The structure of finitely generated shift-invariant spaces in $L^2(\mathbb{R}^d)$. *J. Funct. Anal.*, 119(1):37–78, 1994.

11. H. G. Feichtinger, U. M. Molter, and J. L. Romero. Perturbation techniques in irregular spline-type spaces. *Int. J. Wavelets Multiresolut. Inf. Process.*, 6(2):249-277, 2008.

12. H. G. Feichtinger and D. Onchis. Constructive realization of dual systems for generators of multi-window spline-type spaces. *J. Comput. Appl. Math.*, submitted, 2009.

13. A. G. García, M.A. Hernández-Medina, and G. Perez Villalón. Generalized sampling in shift-invariant spaces with multiple stable generators. *J. Math. Anal. Appl.*, 337(1):69–84, 2008.

14. W. Sun and X. Zhou. Sampling theorem for wavelet subspaces: error estimate and irregular sampling. *Signal Process.*, 48(1):223–226, 2002.

15. X. Zhou and W. Sun. On the sampling theorem for wavelet subspaces. *J. Fourier Anal. Appl.*, 5(4):347–354, 1999.

Integral formulas on the boundary of some ball

K. Fujita

Faculty of Culture and Education, Saga University,
Saga, 840-8502, Japan
E-mail: keiko@cc.saga-u.ac.jp

We have studied integral representations for holomorphic functions on the "N_p-ball" which is a generalization of the Lie ball. In our previous papers, we gave an integral formula whose integral was taken over the boundary of the N_p-ball. In this paper, we will consider the integral formula represented by an iterated integral and consider the 2-dimensional case in detail.

Keywords: Integral formula; Szegö kernel; Cauchy-Hua kernel; Lie ball; Lie sphere.

1. Introduction

For $z = (z_1, \cdots, z_{n+1}), w = (w_1, \cdots, w_{n+1}) \in \mathbf{C}^{n+1}$, we denote by

$$z \cdot w = z_1 w_1 + \cdots + z_{n+1} w_{n+1}, \quad z^2 = z \cdot z, \quad \|z\|^2 = z \cdot \overline{z}.$$

The open Lie ball is defined by

$$\tilde{B}(r) = \left\{ z \in \mathbf{C}^{n+1}; \ L(z) \equiv \sqrt{\|z\|^2 + \sqrt{\|z\|^4 - |z^2|^2}} < r \right\}$$

and the closed Lie ball is defined by $\tilde{B}[r] = \{ z \in \mathbf{C}^{n+1}; \ L(z) \leq r \}$. The Shilov boundary of the Lie ball, which we call the *Lie sphere*, is given by

$$\Sigma_r = \left\{ e^{i\theta} x \in \mathbf{C}^{n+1}, \theta \in \mathbf{R}, \ x \in S_r \right\},$$

where $S_r = \{ x \in \mathbf{R}^{n+1}; x^2 = r^2 \}$ is the real sphere. We denote by $\mathcal{O}(\tilde{B}(r))$ the space of holomorphic functions on $\tilde{B}(r)$ and by $C(\tilde{B}[r])$ the space of continuous functions on $\tilde{B}[r]$. For holomorphic functions on $\tilde{B}(r)$ we have the following Cauchy-Hua integral formula.

Theorem 1.1. *(Cauchy-Hua integral formula, [4])*
For $f \in \mathcal{O}(\tilde{B}(r)) \cap C(\tilde{B}[r])$, we have the following integral representation:

$$f(z) = \int_{\Sigma_r} f(w) H_r(z, w) d\Sigma_r(w), \quad z \in \tilde{B}(r),$$

where $d\Sigma_r$ is the normalized invariant measure on Σ_r and

$$H_r(z, w) = \frac{r^{2(n+1)}}{(r^4 - 2r^2 z \cdot \overline{w} + z^2 \overline{w}^2)^{(n+1)/2}}$$

is the Cauchy-Hua kernel.

In 1999, Morimoto [5] gave a generalized Cauchy-Hua integral formula for holomorphic functions on the following subspace of the Lie ball;

$$\tilde{B}(\rho, r) = \{z \in \tilde{B}(r); |z^2| < |\rho|^2\}, \quad \tilde{B}[\rho, r] = \{z \in \tilde{B}[r]; |z^2| \le |\rho|^2\},$$

where $\rho \in \mathbf{C}$. For the generalized Cauchy-Hua integral formula for $f \in \mathcal{O}(\tilde{B}(\rho, r))$, the integration is taken over the generalized Lie Sphere:

$$\Sigma_{\rho, r} = \{e^{i\theta} z \in \mathbf{C}^{n+1}; \theta \in \mathbf{R}, z \in \tilde{S}_{\rho, r}\},$$

where $\tilde{S}_{\rho, r} = \{z \in \tilde{S}_\rho; L(z) = r\}$ and $\tilde{S}_\rho = \{z \in \mathbf{C}^{n+1}; z^2 = \rho^2\}$. When $\rho = r$, $\tilde{B}(\rho, r) = \tilde{B}(r)$, $\tilde{B}[\rho, r] = \tilde{B}[r]$, $\tilde{S}_{\rho, r} = S_r$ and $\Sigma_{\rho, r} = \Sigma_r$. Note that

$$\partial \tilde{B}(\rho, r) = \{z \in \mathbf{C}^{n+1}; |z^2| = |\rho|^2, L(z) = r\} = \Sigma_{\rho, r}.$$

On the other hand, we introduced an N_p-ball which is a generalization of the Lie ball. By the fact that the N_p-ball is represented by a union of the subspaces of the Lie ball, we considered an integral formula represented by an iterated integral in [2] and [3] (see section 2). The purpose of this paper is to revise the measures introduced in [2] and [3] , and to consider the 2-dimensional case in detail.

2. Integral formula over the boundary of the N_p-ball

2.1. N_p-ball (A generalization of the Lie ball)

Consider the function

$$N_p(z) = \left(\left((\|z\|^2 + \sqrt{\|z\|^4 - |z^2|^2})^{p/2} + (\|z\|^2 - \sqrt{\|z\|^4 - |z^2|^2})^{p/2}\right) / 2\right)^{1/p}.$$

For $p \ge 1$, $N_p(z)$ is a norm on $\mathbf{C}^{n+1}$ and we define the N_p-balls by

$$\tilde{B}_p(r) = \{z \in \mathbf{C}^{n+1}; N_p(z) < r\}, \quad \tilde{B}_p[r] = \{z \in \mathbf{C}^{n+1}; N_p(z) \le r\}.$$

Since $L(z) = \lim_{p \to \infty} N_p(z)$, we take $\tilde{B}_\infty(r)$ for $\tilde{B}(r)$. Note that $N_2(z) = \|z\|$ is the complex Euclidean norm, $2^{-1/p} L(z) \le N_p(z) \le L(z)$ and $\tilde{B}(r) \subset \tilde{B}_p(r) \subset \tilde{B}(2^{1/p}r)$. Further N_p-balls are represented by

$$\tilde{B}_p(r) = \bigcup\{\tilde{B}[\lambda, R]; 0 < \lambda \le R, R^p + \lambda^{2p}/R^p < 2r^p\},$$

$$\tilde{B}_p[r] = \bigcup\{\tilde{B}[\lambda, R]; 0 \le \lambda \le R, R^p + \lambda^{2p}/R^p \le 2r^p\}.$$

2.2. *The boundary of the N_p-ball*

For the Lie ball $\tilde{B}(r)$, its Shilov boundary is the Lie sphere Σ_r which is not equal to the topological boundary $\partial\tilde{B}(r) = \{z \in \mathbf{C}^{n+1}; L(z) = r\}$. When $p \neq \infty$, the boundary of the N_p-ball

$$\partial\tilde{B}_p(r) = \{z \in \mathbf{C}^{n+1}; N_p(z) = r\}$$

may be the same as the Shilov boundary of $\tilde{B}_p(r)$. For fixed r, put

$$R_\lambda = (r^p + \sqrt{r^{2p} - \lambda^{2p}})^{1/p}.$$

For $0 \leq \lambda \leq r$, R_λ is monotone decreasing in λ and $\tilde{B}(\lambda, R_\lambda) \neq \emptyset$. Further

$$\partial\tilde{B}_p(r)\bigcap\partial\tilde{B}(R_\lambda) = \{z; L(z) = R_\lambda, N_p(z) = r\}$$

$$= \{z; L(z) = R_\lambda, N_p(z) = ((R_\lambda^p + (|z^2|/R_\lambda)^p)/2)^{1/p} = r\}$$

$$= \{z; L(z) = R_\lambda, |z^2| = \lambda^2\} = \partial\tilde{B}(\lambda, R_\lambda) = \Sigma_{\lambda, R_\lambda},$$

and $\bigcup_{0\leq\lambda\leq r}\left(\partial\tilde{B}_p(r)\bigcap\partial\tilde{B}(R_\lambda)\right) = \partial\tilde{B}_p(r)\bigcap\left(\bigcup_{0\leq\lambda\leq r}\partial\tilde{B}(R_\lambda)\right)$. Because $\bigcup_{0\leq\lambda\leq r}\partial\tilde{B}(R_\lambda) = \tilde{B}[2^{1/p}r] \setminus \tilde{B}(r)$ and $\tilde{B}(r) \subset \tilde{B}_p(r) \subset \tilde{B}(2^{1/p}r)$ we have

$$\partial\tilde{B}_p(r) = \bigcup_{0\leq\lambda\leq r} \Sigma_{\lambda, R_\lambda}. \tag{1}$$

Moreover,

$$\Sigma_{\lambda, R_\lambda} \bigcap \Sigma_{\lambda', R_{\lambda'}} = \emptyset \quad \text{if} \quad \lambda \neq \lambda'. \tag{2}$$

Since we have (1) and (2), for $f \in \mathcal{O}(\tilde{B}_p(r)) \cap C(\tilde{B}_p[r])$, there is a measure $d\mu(\lambda)$ such that

$$\int_{\partial\tilde{B}_p(r)} f(z)dV_{p,r}'(z) = \int_0^r \int_{\Sigma_{\lambda, R_\lambda}} f(z)d\Sigma_{\lambda, R_\lambda}'(z)d\mu(\lambda),$$

where $dV_{p,r}'(z)$ is the $O(n+1)$-invariant measure on $\partial\tilde{B}_p(r)$ and $d\Sigma_{\lambda, R}'$ is the $O(n+1)$-invariant measure on $\Sigma_{\lambda, R_\lambda}$. Put

$$\mathrm{Vol}(\Sigma_{\lambda, R}) = \int_{\Sigma_{\lambda, R}} d\Sigma_{\lambda, R}', \quad \mathrm{Vol}(\tilde{S}_{\lambda, R}) = \int_{\tilde{S}_{\lambda, R}} d\tilde{S}_{\lambda, R}',$$

where $d\tilde{S}_{\lambda, R}'$ is the $O(n+1)$-invariant measure on $\tilde{S}_{\lambda, R}$. For $z \in \Sigma_{\lambda, R}$, write $z = e^{i\theta}z'$, $z' \in \tilde{S}_{\lambda, R}$. Then for $f \in \mathcal{O}(\tilde{B}(r)) \cap C(\tilde{B}[r])$, we have

$$\int_{\Sigma_{\lambda, R}} f(z)\frac{d\Sigma_{\lambda, R}'(z)}{\mathrm{Vol}(\Sigma_{\lambda, R})} = \frac{1}{2\pi}\int_0^{2\pi} \int_{\tilde{S}_{\lambda, R}} f(e^{i\theta}z')\frac{d\tilde{S}_{\lambda, R}'(z')}{\mathrm{Vol}(\tilde{S}_{\lambda, R})}d\theta.$$

Thus we have

$$\int_{\partial\tilde{B}_p(r)} f(z)dV_{p,r}'(z) = \int_0^r \int_0^{2\pi} \int_{\tilde{S}_{\lambda, R_\lambda}} f(z)d\tilde{S}_{\lambda, R_\lambda}'d\theta d\mu(\lambda).$$

2.3. *Integral formula represented by an iterated integral*

Let $P_{k,n}(t)$ be the Legendre polynomial of degree k in $\mathbf{C}^{n+1}$;

$$P_{k,n}(t) = \frac{k!\Gamma(n-1)}{\Gamma(k+n-1)} \sum_{l=0}^{[k/2]} (-1)^l \frac{\Gamma(k+(n-1)/2-l)}{l!(k-2l)!\Gamma((n-1)/2)} (2t)^{k-2l}, \quad n \geq 2,$$

$$P_{k,1}(t) = T_k(t) = \left\{ (t+i\sqrt{1-t^2})^k + (t-i\sqrt{1-t^2})^k \right\}/2.$$

We define the homogeneous harmonic extended Legendre polynomial by

$$\tilde{P}_{k,n}(z,w) = (\sqrt{z^2})^k (\sqrt{w^2})^k P_{k,n}\big(\frac{z}{\sqrt{z^2}} \cdot \frac{w}{\sqrt{w^2}}\big).$$

Note that $\tilde{P}_{k,n}(z,w) = \tilde{P}_{k,n}(w,z)$ and $(\frac{\partial^2}{\partial z_1^2} + \cdots + \frac{\partial^2}{\partial z_n^2})\tilde{P}_{k,n}(z,w) = 0$. The dimension $N(k,n)$ of the space of homogeneous harmonic polynomials of degree k in $\mathbf{C}^{n+1}$ is known as

$$N(0,1) = 1, \ N(k,1) = 2, \ k = 1, 2, \cdots,$$
$$N(k,n) = (2k+n-1)(k+n-2)!/(k!(n-1)!), \quad n \geq 2.$$

For $\lambda > 0$ put

$$L_{k,\lambda,r} = \lambda^{2k} P_{k,n}\left(\frac{1}{2}\left(\frac{\lambda^2}{r^2} + \frac{r^2}{\lambda^2}\right)\right), \quad L_{k,0,r} = \frac{\Gamma(k+(n+1)/2)r^{2k}}{N(k,n)\Gamma((n+1)/2)k!}.$$

For $R_\lambda = (r^p + \sqrt{r^{2p} - \lambda^{2p}})^{1/p}$, put

$$c_{k,l,r}^p = \frac{1}{N(k-2l,n)} \int_0^r \lambda^{4l} L_{k-2l,\lambda,R_\lambda} \operatorname{Vol}(\Sigma_{\lambda,R_\lambda}) d\mu(\lambda),$$

$$H_{p,r}(z,w) = \sum_{k=0}^{\infty} \sum_{l=0}^{[k/2]} (c_{k,l,r}^p)^{-1} (z^2)^l (\overline{w}^2)^l \tilde{P}_{k-2l,n}(z,\overline{w}).$$

Then we can prove the following theorem just as in the proof of Theorem 3.3 in [3] . The difference is that we define $dV'_{p,r}$ by the invariant measure on $\partial\tilde{B}_p(r)$ but in [3] we just set $dV'_{p,r}$ by $d\Sigma'_{\lambda,R_\lambda} d\lambda$.

Theorem 2.1. *Let $f \in \mathcal{O}(\tilde{B}_p(r)) \cap C(\tilde{B}_p[r])$. Then we have*

$$f(z) = \int_{\partial\tilde{B}_p(r)} f(w)H_{p,r}(z,w)dV'_{p,r}(w), \quad z \in \tilde{B}_p(r).$$

In [1] , we denote by $L^2(\partial\tilde{B}_p(r), dS_{p,r})$ the space of square integrable function on $\partial\tilde{B}_p(r)$ with respect to the *normalized* invariant measure $dS_{p,r}$ on $\partial\tilde{B}_p(r)$ and by $H\mathcal{O}(\partial\tilde{B}_p(r))$ be the space of the closure in

$L^2(\partial\tilde{B}_p(r), dS_{p,r})$ with the topology of the restrictions to $\partial\tilde{B}_p(r)$ of the elements of $\mathcal{O}(\tilde{B}_p(r)) \cap C(\tilde{B}_p[r])$. Then we proved the following theorem:

Theorem 2.2. *(Theorem 4.1 in[1]) The Szegö kernel of $H\mathcal{O}(\partial\tilde{B}_p(r))$ is given by*

$$S_{p,r}(z,w) = \sum_{k=0}^{\infty} \sum_{l=0}^{[k/2]} (\alpha_{k,l,r}^p)^{-1}(z^2)^l(\overline{w}^2)^l \tilde{P}_{k-2l,n}(z,\overline{w}),$$

where

$$\alpha_{k,l,r}^p = \int_{\partial\tilde{B}_p(r)} |(\zeta^2)^l \tilde{P}_{k-2l,n}(\zeta,\omega)|^2 dS_{p,r}(\zeta), \quad \omega \in S_1.$$

By the definition of $dV'_{p,r}$ and $dS_{p,r}$, putting $\mathrm{Vol}(\partial\tilde{B}_p(r)) = \int_{\partial\tilde{B}_p(r)} dV'_{p,r}$,

$$H_{p,r}(z,w) = S_{p,r}(z,w) \times \mathrm{Vol}(\partial\tilde{B}_p(r)). \tag{3}$$

In the case of $n = 1$, $\alpha_{k,l,r}^{1,p}$ can be calculated as follows:

$$\alpha_{k,l,r}^{1,p} = \frac{\Gamma(4/p+1)\Gamma((2k-2l+2)/p)\Gamma((2l+2)/p)2^{2k/p}r^{2k}}{pN(k-2l,1)\Gamma(2/p+1)^2\Gamma((2k+4)/p)}. \tag{4}$$

Further, in $\mathbf{C}^2$, it is easy to calculate that

$$\mathrm{Vol}(\partial\tilde{B}_p(r)) = \pi^2 2^{4/p}r^3\Gamma(2/p+1)^2/\Gamma(4/p+1). \tag{5}$$

By (3), (4) and (5), the 2-dimensional $c_{k,l,r}^p$ and $H_{p,r}(z,w)$ are given concretely. We will look for a 2-dimensional $d\mu(\lambda)$ in the following section.

3. 2-dimensional case

For $z = (z_1, z_2) \in \mathbf{C}^2$ we have

$$N_p(z) = \left(\frac{|z_1 + iz_2|^p + |z_1 - iz_2|^p}{2}\right)^{1/p}, \quad L(z) = \max|z_1 \pm iz_2|.$$

Thus for p with $1 \le p < \infty$, the N_p-norm is equivalent to the L_p-norm and the Lie norm is equivalent to the supremum norm in $\mathbf{C}^2$.

Put $\zeta_1 = z_1 + iz_2$, $\zeta_2 = z_1 - iz_2$. Then $z^2 = z_1^2 + z_2^2 = \zeta_1\zeta_2$ and

$$N_p(z) = ((|\zeta_1|^p + |\zeta_2|^p)/2)^{1/p}, \quad L(z) = \max\{|\zeta_1|, |\zeta_2|\}.$$

Thus

$$\tilde{B}_p(r) \simeq \{(\zeta_1, \zeta_2) \in \mathbf{C}^2; |\zeta_1|^p + |\zeta_2|^p < 2r^p\},$$
$$\tilde{B}(r) \simeq \{(\zeta_1, \zeta_2) \in \mathbf{C}^2; |\zeta_1| < r, |\zeta_2| < r\}.$$

On the other hand,

$$\Sigma_{\lambda,R} \simeq \{(\zeta_1,\zeta_2) \in \mathbf{C}^2; |\zeta_1\zeta_2| = \lambda^2, \max\{|\zeta_1|, |\zeta_2|\} = R\} \equiv \tilde{\Sigma}_{\lambda,R}$$
$$= \{(\zeta_1,\zeta_2) \in \mathbf{C}^2; |\zeta_1| = R, |\zeta_2| = \lambda^2/R\}$$
$$\bigcup \{(\zeta_1,\zeta_2) \in \mathbf{C}^2; |\zeta_1| = \lambda^2/R, |\zeta_2| = R\}.$$

Thus $\mathrm{Vol}(\tilde{\Sigma}_{\lambda,R}) = 2(2\pi R)(2\pi\lambda^2/R) = 8\pi^2\lambda^2$ and $\mathrm{Vol}(\Sigma_{\lambda,R}) = 2\pi^2\lambda^2$. When $R_\lambda = (r^p + \sqrt{r^{2p} - \lambda^{2p}})^{1/p}$, $\lambda^2/R_\lambda^2 = (r^p - \sqrt{r^{2p} - \lambda^{2p}})^{2/p}/\lambda^2$ and

$$L_{k,\lambda,r} = \lambda^{2k} P_{k,1}\left((\lambda^2/R_\lambda^2 + R_\lambda^2/\lambda^2)/2\right)$$
$$= \left(\left(r^p + \sqrt{r^{2p} - \lambda^{2p}}\right)^{2k/p} + \left(r^p - \sqrt{r^{2p} - \lambda^{2p}}\right)^{2k/p}\right)/2.$$

If we put $\mathrm{Vol}(\Sigma_{\lambda,R_\lambda})d\mu(\lambda) = \frac{4\pi^2 r^{p-1}\lambda^3}{\sqrt{r^{2p}-\lambda^{2p}}}d\lambda$ and $\sqrt{r^{2p} - \lambda^{2p}} = r^p s$, then

$$c_{k,l,r}^p = \frac{1}{N(k-2l,1)} \int_0^r \lambda^{4l} L_{k-2l,\lambda,R_\lambda} \mathrm{Vol}(\Sigma_{\lambda,R_\lambda})d\mu(\lambda)$$
$$= \frac{2\pi^2 r^{2k+3}}{pN(k-2l,1)} \int_{-1}^1 (1-s)^{(2l+2)/p-1}(1+s)^{(2k-2l+2)/p-1}ds$$
$$= \frac{\pi^2\Gamma((2k-2l+2)/p)\Gamma((2l+2)/p)2^{2k/p+4/p}r^{2k+3}}{pN(k-2l,1)\Gamma((2k+4)/p)}.$$

Thus when we take $4\pi^2 r^{p-1}\lambda^3 d\lambda/\sqrt{r^{2p} - \lambda^{2p}}$ for $\mathrm{Vol}(\Sigma_{\lambda,R_\lambda})d\mu(\lambda)$, the relation $c_{k,l,r}^p = \mathrm{Vol}(\partial\tilde{B}_p(r))\alpha_{k,l,r}^{1,p}$ is satisfied. Therefore, in the 2-dimensional case, if we put

$$d\mu(\lambda) = \frac{4\pi^2 r^{p-1}\lambda^3}{\sqrt{r^{2p} - \lambda^{2p}}}d\lambda/\mathrm{Vol}(\Sigma_{\lambda,R_\lambda}) = \frac{2r^{p-1}\lambda}{\sqrt{r^{2p} - \lambda^{2p}}}d\lambda,$$

then for $f \in \mathcal{O}(\tilde{B}_p(r)) \cap C(\tilde{B}_p[r])$ we have

$$\int_{\partial\tilde{B}_p(r)} f(z)dV_{p,r}'(z) = \int_0^r \int_{\Sigma_{\lambda,R_\lambda}} f(z)d\Sigma_{\lambda,R_\lambda}'(z)\frac{2r^{p-1}\lambda}{\sqrt{r^{2p} - \lambda^{2p}}}d\lambda$$
$$= \int_0^r \int_0^{2\pi} \int_{\tilde{S}_{\lambda,R_\lambda}} f(z)d\tilde{S}_{\lambda,R_\lambda}' d\theta \frac{2r^{p-1}\lambda}{\sqrt{r^{2p} - \lambda^{2p}}}d\lambda,$$

and for $z \in \tilde{B}_p(r)$ we have

$$f(z) = \int_{\partial\tilde{B}_p(r)} f(w)H_{p,r}(z,w)dV_{p,r}'(w)$$
$$= \int_0^r \int_0^{2\pi} \int_{\tilde{S}_{\lambda,R_\lambda}} f(w)H_{p,r}(z,w)d\tilde{S}_{\lambda,R_\lambda}' d\theta \frac{2r^{p-1}\lambda}{\sqrt{r^{2p} - \lambda^{2p}}}d\lambda,$$

where

$$H_{p,r}(z,w) = \sum_{k=0}^{\infty} \sum_{l=0}^{[k/2]} \frac{pN(k-2l,1)\Gamma((2k+4)/p)(z^2)^l(\overline{w}^2)^l \tilde{P}_{k-2l,n}(z,\overline{w})}{\pi^2\Gamma((2k-2l+2)/p)\Gamma((2l+2)/p)2^{2k/p+4/p}r^{2k+3}}.$$

At last, we will check that

$$\int_0^r \int_{\Sigma_{\lambda,R_\lambda}} d\Sigma'_{\lambda,R_\lambda}(z) \frac{2r^{p-1}\lambda}{\sqrt{r^{2p}-\lambda^{2p}}} d\lambda = \int_0^r 2\pi^2\lambda^2 \frac{2r^{p-1}\lambda}{\sqrt{r^{2p}-\lambda^{2p}}} d\lambda$$

$$= \frac{4\pi^2 r^3}{p} \int_0^1 (1+s)^{(2-p)/p}(1-s)^{(2-p)/p} ds = \frac{4\pi^2 r^3}{p} \frac{\Gamma(2/p)^2}{2\Gamma(4/p)} 2^{4/p-1}$$

$$= \mathrm{Vol}(\partial \tilde{B}_p(r)) = \int_{\partial \tilde{B}_p(r)} dV'_{p,r}.$$

Acknowledgment

The author would like to thank Professor Mitsuo Morimoto for his useful discussion.

References

1. K. Fujita, Reproducing kernels for holomorphic functions on some balls related to the Lie ball, Ann. Polo. Math., 91.2-3, 2007, 219–234.
2. K. Fujita, Integral representations on some balls related to the Lie ball, Proceedings of international symposium on new development of geometric function theory and its applications, School of mathematical sciences faculty of science & technology university Kebangsaan Malaysia, 2008, 110–117.
3. K. Fujita, Integral representations for holomorphic functions on some balls, Proceedings of the 16th International conference on finite or infinite dimensional complex analysis and applications, 2009, 93–98.
4. L. K. Hua, Harmonic Analysis of Functions of Several Complex Variables in Classical Domain, Moskow 1959, (in Russian); Translations of Math. Monographs vol. 6, Amer. Math. Soc., Providence, Rhode Island, 1979.
5. M. Morimoto, A generalization of the Cauchy-Hua integral formula on the Lie ball, Tokyo J. Math, Vol 22, No. 1, 1999, 177–192.

Paley–Wiener spaces and their reproducing formulae

J.R. Higgins

4, rue du Bary, 11250 Montclar, France

A dictionary is presented listing concepts from sampling theory in reproducing kernel spaces and their counterparts in harmonic analysis. These concepts help us to discuss concrete and discrete reproducing formulae in the setting of operators on Paley–Wiener spaces. The Riesz transforms provide an example.

Keywords: Reproducing kernel; sampling series.

1. Introduction

The classical Paley–Wiener space of functions f which are L^2 and continuous over $\mathbb{R}$ and whose Fourier transforms are supported on $[-\pi, \pi]$ has a concrete reproducing equation (r.e. for short):

$$f(s) = \int_{\mathbb{R}} f(t) \operatorname{sinc}(s - t)dt, \quad (s \in \mathbb{R}), \quad \operatorname{sinc} s := (\sin \pi s)/\pi s,$$

and a discrete reproducing equation, or sampling series:

$$f(s) = \sum_{n \in \mathbb{Z}} f(n) \operatorname{sinc}(s - n), \quad (s \in \mathbb{R}).$$

There is clearly a striking comparison between the two. In the present article we show that there are 'operator' forms of this concrete–discrete comparison.

2. The Dictionary

A mathematical dictionary can stimulate the search for results in one language which are only known in the other, and indicate possible methods. An example is the approximate sampling theorem (see Beaty et al.[1]), known in harmonic analysis but not in the reproducing kernel theory. Please see the remarks following the Dictionary for notations, definitions etc.; I will denote a suitable indexing set.

Table 1.   DICTIONARY

	Reproducing kernel theory	**Harmonic analysis**
1.	E, an abstract set	G, an LCA group with dual Γ
2.	H, a separable Hilbert space	$L^2(\Omega),\ \Omega \subset \Gamma$
3.	$K_t \in H,\ (t \in E)$	$(t,\cdot)$, a character of Γ
4.	$k(s,t) = \langle K_t, K_s \rangle_H$	$\int_\Omega (t,\gamma)(s,\gamma)\,d\gamma$ $= \int_\Omega (t-s,\gamma)\,d\gamma,\ = \chi_\Omega^\vee(t-s)$
5.	$\mathcal{K}g := \langle g, K_t \rangle_H = f(t)$	$\mathcal{F}^{-1}$
6.	R realized as $\{f = \mathcal{K}g, g \in H\}$ $\ni \ \|f\|_R = \|g\|_H$	$PW_\Omega(G) =$ $\{f : f = g^\vee,\ g \in L^2(\Gamma),\ g$ null outside $\Omega\}$
7.	For $f \in R$ the r.e. is $f(t) = \langle f, k(\cdot, t) \rangle_R$	For $f \in PW_\Omega$ the r.e. is $f(t) = \int_G f(\tau)\chi_\Omega^\vee(t-\tau)\,d\tau$
8.	Let there exist $\ \{s_n\}_{n \in I} \subset E \ni$ $\{K_{s_n}\}$ is an ON basis for H $f(t) = \sum_{n \in I} f(s_n) k(s_n, t)$	$f(t) = \sum_{h \in \mathfrak{H}} f(h)\chi_\Omega^\vee(t-h)$
	$\vdots$	$\vdots$

The left-hand column 'Reproducing kernel theory' is a *précis* of Saitoh's approach to the theory[2] of reproducing kernels insofar as it leads up to the sampling theorem (item 8) in that setting. With its rather few assumptions it is a foundational approach to sampling theory. In this column R denotes the Hilbert space having $k(s,t)$ as its reproducing kernel. By taking $E = \mathbb{R}$, $H = L^2(-\pi, \pi)$, $K_t = e^{it\cdot}$ etc., the classical case is obtained.

The right-hand column 'Harmonic analysis' requires some explanation. All necessary background, and much more, is to be found in Dodson.[3] Let G be a locally compact abelian (LCA) group (written additively). Let (t, γ) be a character of G, that is, a continuous homomorphism of G into the circle group. Let $G^\wedge = \Gamma$ denote the dual group of continuous characters on G. We assume that Γ has a countable discrete subgroup Λ.

Haar measures on groups are normalised in the standard way,[3] and this means in particular that there is a measurable transversal (a complete set of coset representatives) $\Omega \subset \Gamma$ of Γ/Λ, of finite Haar measure.

Now

$$\mathfrak{H} = \Lambda^\perp := \{h \in G : (h, \lambda) = 1, (\lambda \in \Lambda)\}$$

is a subgroup of G and is called the 'annihilator' of Λ. We assume that $\mathfrak{H}$ is discrete; it follows that the quotient group Γ/Λ is compact.

The Fourier transform on $L^2(G)$ is defined in the usual way:

$$f^\wedge(\gamma) = (\mathcal{F}f)(\gamma) := \int_G f(t)(t, \gamma)\,dt,$$

in the L^2 sense. Integrations are understood to be with respect to the appropriate Haar measure. The inverse Fourier transform of f will be denoted by $f^\vee$ or by $\mathcal{F}^{-1}f$. The characteristic function of Ω is denoted by χ_Ω.

A discussion of the sampling series in item 8, due to I. Kluvánek, will be found in Dodson.[3]

3. Operator kernels and reproducing formulae

For general discussions of operators on reproducing kernel Hilbert spaces see: Pedrick,[4] Burbea and Masani,[5] Alpay,[6] Rochberg[7] and Meschkowski.[8] More details of the calculations in this section can be found in Higgins.[9]

Let $\mathcal{B}$ be a bijection on R, and let $\mathcal{B}^*$ denote the adjoint operator.

Definition 3.1. The kernel

$$h(s,t) := \big(\mathcal{B}^* k(\cdot, t)\big)(s), \qquad s, t \in E \tag{1}$$

will be called the *operator kernel* of $\mathcal{B}$.

By using $\mathcal{B}^{*-1}$ in Definition 3.1 to obtain $k(\cdot, t)$, and substituting it in the r.e. (Dictionary, item 7) we obtain the *operator reproducing equation*

$$f(t) = \langle (\mathcal{B}^{-1}f)(\cdot), h(\cdot, t)\rangle. \tag{2}$$

Definition 3.2.

$$h^*(s,t) := \overline{h(t,s)} \qquad ((t,s) \in E \times E)$$

will be called the *adjoint operator kernel* of $\mathcal{B}$.

It can then be proved that we have the *adjoint operator reproducing equation*

$$f(t) = \langle (\mathcal{B}^*)^{-1}f, h^*(\cdot, t)\rangle. \tag{3}$$

We assume the existence of a sequence $\{s_n\} \subset E$, $n \in I$, such that $\{\overline{h(s_n, t)}\}$ is an orthogonal basis for R with normalising factors ν_n, so that $\{\nu_n \overline{h(s_n, t)}\}$ is orthonormal. At the expense of more technicalities we could have assumed that $\{h(s_n, t)\}$ is just a basis for R, or a frame. It will be noticed that this is the first assumption about E that we have made beyond those adopted in the Dictionary, items 1, 2 and 3.

Let $f \in R$. Its expansion in our assumed orthonormal basis is

$$f(t) = \sum_{n \in I} c_n \nu_n \overline{h(s_n, t)}, \tag{4}$$

$$c_n = \langle f, \nu_n \overline{h(s_n, \cdot)} \rangle = \overline{\nu_n} \langle f, h^*(\cdot, s_n) \rangle = \overline{\nu_n} (\mathcal{B}^* f)(s_n),$$

using the adjoint operator r.e. (3). So (4) is

$$f(t) = \sum_{n \in I} |\nu_n|^2 (\mathcal{B}^* f)(s_n)\, \overline{h(s_n, t)}, \tag{5}$$

the *operator sampling series*.

It will be seen that the comparison between (2) and (5) does not have quite the exactness found in the classical case (§1), neither does (8) compare exactly with (9). Nevertheless there do exist reasonable concrete–discrete comparisons of reproducing formulae in the operator setting.

4. Multiplier operators on PW_Ω

From now on we work in the LCA group setting, i.e., the right-hand side of the Dictionary. Let $\mu(\gamma)$ be a non-null complex valued function on Γ such that, for some constants α and β,

$$\begin{cases} 0 < \alpha \le |\mu(\gamma)| \le \beta < \infty, & \text{(Haar) a.a.}\, \gamma \in \Omega; \\[2mm] \mu(\gamma) = 0, & \gamma \notin \Omega. \end{cases} \tag{6}$$

Let $\mathcal{M}$ denote the operator of multiplication by $\mu(\gamma)$, acting on $L^2(\Omega)$.

Definition 4.1. Let $f \in PW_\Omega$. The operator $\mathcal{T}$ is defined by

$$(\mathcal{T} f)(s) := (\mathcal{F}^{-1} \mathcal{M} \mathcal{F} f)(s).$$

Some facts about $\mathcal{M}$ are collected in the following three Lemmas. Proofs are straightforward and only brief sketches will be given.

Lemma 4.1. *Let $\varphi \in L^2(\Omega)$. Then*
i. $\mathcal{M}^* \varphi$ consists in multiplication by $\overline{\mu(x)}$;
ii. $\mathcal{M}^{-1} \varphi$ consists in multiplication by $[\mu(x)]^{-1}$;
iii. $\mathcal{M}$ is an isomorphism (linear bijection) on $L^2(\Omega)$.

Proof. Proofs by direct verification. $\qquad\qquad\qquad\qquad\qquad\qquad$ □

Lemma 4.2. *Let $\{\varphi_n\}$, $n \in I$, be an orthonormal basis for $L^2(\Omega)$. Then*
i. $\{(\mathcal{M}\varphi_n)(\lambda)\} = \{\mu(\lambda)\varphi_n(\lambda)\}$ is a Riesz basis for $L^2(\Omega)$;

ii. $\{\mu(\lambda)\varphi_n(\lambda)\}$ *is an orthonormal basis if* $|\mu(\lambda)| = 1$, *a.a.* $\lambda \in \Omega$;

iii. *The basis biorthonormal to* $\{\mu(\lambda)\varphi_n(\lambda)\}$ *is*

$$\{((\mathcal{M}^*)^{-1}\varphi_n)(\lambda)\} = \left\{ \frac{1}{\overline{\mu(\lambda)}}\,\varphi_n(\lambda) \right\}, \qquad n \in I.$$

Proof. Direct verification, using basic properties of Riesz bases (e.g., Young [10, p. 31]). $\qquad\qquad\square$

Lemma 4.3. *The set* $\{\mathcal{F}^{-1}\mathcal{M}\varphi_n\}$ *is a Riesz (respectively orthonormal) basis for* PW_Ω *whenever* $\{\mathcal{M}\varphi_n\}$ *is a Riesz (respectively orthonormal) basis for* $L^2(\Omega)$.

Proof. This result follows from the unitary character of $\mathcal{F}$ on $\mathbb{R}$. $\qquad\square$

It is straightforward to prove that the operator $\mathcal{T}$ is an isomorphism on PW_Ω. The adjoint operator $\mathcal{T}^*$ is given by $\mathcal{T}^* f = \mathcal{F}^{-1}\mathcal{M}^*\mathcal{F}f$.

To find the operator kernel for $\mathcal{T}$ we have, from (1) and the Dictionary, item 4,

$$h(s,t) = \left(\mathcal{T}^* \chi_\Omega^\vee(\cdot - t)\right)(s), \quad (s,t \in G).$$

Therefore by direct calculation, using the 'shift' property of $\mathcal{F}$,

$$h(s,t) = \overline{\mu(\cdot)}^\vee (s - t).$$

From (2) the operator r.e. becomes

$$f(t) = \int_G (\mathcal{T}^{-1}f)(s)\overline{\mu}^\vee(s - t)\,ds, \tag{8}$$

and since $\overline{h(s,t)} = \overline{\mu(\cdot)}^\vee(s - t) = \mu^\vee(t - s)$, the operator sampling series (5) becomes:

$$f(t) = \sum_{n \in I} |\nu_n|^2 (\mathcal{T}^*f)(s_n)\,\mu^\vee(t - s_n) = \sum_{n \in I} |\nu_n|^2(\mathcal{T}^*f)(s_n)\,\overline{\mu^\vee(s_n - t)}.$$
$$\tag{9}$$

5. Examples

Example 5.1. The Hilbert transform $\mathcal{H}$ on $PW_{[-\pi,\pi]}$ is obtained by taking $G = \mathbb{R}$, $\Omega = [-\pi, \pi]$, $\mathcal{T} = \mathcal{H} := \mathcal{F}^{-1}\mathcal{M}\mathcal{F}$ where $\mathcal{M}$ denotes multiplication by $-i\,\mathrm{sgn}(y)$. We have $\mathcal{H}^{-1} = -\mathcal{H} = \mathcal{H}^*$.

$$\overline{\mu}^{\vee}(s-t) = \frac{1}{\sqrt{2\pi}}\,\frac{i}{\sqrt{2\pi}}\int_{-\pi}^{\pi}\mathrm{sgn}(y)e^{i(s-t)y}\,dy$$

$$= -\operatorname{sinc}\tfrac{1}{2}(s-t)\sin\tfrac{\pi}{2}(s-t). \tag{10}$$

Note that $|\mu| = 1$ and $\{-i\,\mathrm{sgn}(y)e^{iyn}/\sqrt{2\pi}\}$ is an orthonormal basis for $L^2(-\pi,\pi)$, as in $ii.$ of Lemma 4.2. With these specialisations the formulae (8) and (9) give a textbook example.[11]

Example 5.2. The Riesz transforms.

Take G to be $\mathbb{R}^d$, $(d \in \mathbb{N})$, and $\Omega = [-\pi,\pi]^d$. Let $\boldsymbol{t} = (t_1,\ldots,t_d)$ and let $\boldsymbol{y} = (y_1,\ldots,y_d)$ etc.

Definition 5.1. Let $f \in L^2(\mathbb{R}^d)$, and define

$$\mathcal{R}_j f := \mathcal{F}^{-1}\mathcal{M}_j\mathcal{F}f, \quad j = 1,\ldots,d,$$

$\mathcal{M}_j$ denoting multiplication by $-iy_j/|\boldsymbol{y}|\;\chi_{[-\pi,\pi]^d}(\boldsymbol{y})$. The operator $\mathcal{R}_j$ is called the jth Riesz transform. The case $d = 1$ gives the Hilbert transform of the previous Example 5.1.

Since $\mathcal{M}_j$ is not of the type (6) the Riesz transforms are not easy to handle in the present context. However, we can form an operator involving all d Reisz transforms on $\mathbb{R}^d$ whose multiplier is of the right type. Indeed, let $\mathcal{I}$ denote the identity operator and put $\mathcal{R} = \mathcal{I} + \mathcal{R}_1 + \cdots + \mathcal{R}_d$. Now

$$|\mu(\boldsymbol{y})|^2 = \left|1 - i\frac{\sum_{j=1}^d y_j}{\sqrt{\sum_{k=1}^d y_k^2}}\right|^2 = 1 + \frac{\left\{\sum_{j=1}^d y_j\right\}^2}{\sum_{k=1}^d y_k^2} \geq 1. \tag{11}$$

Also, from the second equality in (11),

$$|\mu(\boldsymbol{y})|^2 \leq 1 + 1 + 2\frac{\sum_{i,j=1;\,i\neq j}^d |y_i y_j|}{\sum_{k=1}^d y_k^2} = 2\left(1 + \sum_{i,j=1;\,i\neq j}^d U_i U_j\right) \tag{12}$$

say, where

$$U_i := \frac{|y_i|}{\sqrt{\sum_{k=1}^d y_k^2}} \geq 0,$$

and similarly for U_j, so that $\sum_{i=1}^d U_i^2 = 1$, hence $U_i \leq 1$, $(i = 1\ldots,d)$. We find from (12) that $|\mu|$ is bounded above, and by (11) it is bounded below.

In view of the fact that $\{e^{-i\,\boldsymbol{y}\cdot\boldsymbol{k}}/(2\pi)^{d/2}\}$, $(\boldsymbol{k}\in\mathbb{Z}^d,\,\boldsymbol{y}\in[-\pi,\pi]^d)$ is an orthonormal basis for $L^2([-\pi,\pi]^d)$, the multiplier μ given in (11) ensures that $\{\mu(\boldsymbol{y})e^{-i\boldsymbol{y}\cdot\boldsymbol{k}}/(2\pi)^{d/2}\}$ is a Riesz basis for $L^2([-\pi,\pi]^d)$. Then by Lemma 4.3 the inverse fourier transforms of these functions form a Riesz basis for PW_Ω.

These are the basic facts for this example but we shall not pursue them any further here. They ensure that there are reproducing formulae of the type (8) and (9) for the operator $\mathcal{R}$ associated with the Riesz transforms.

Acknowledgements

This article stems from a talk given in the special session *Reproducing Kernels and Related Topics* at the 7th ISAAC Congress, London, July 2009, chaired by Prof. A. Berlinet and Prof. S. Saitoh. The author is very grateful to the chairmen for their kind assistance, especially with the bibliography.

References

1. M. G. Beaty, M. M. Dodson, S. P. Eveson and J. R. Higgins, On the approximate form of Kluvánek's theorem, *J. Approx. Th.*, To appear.
2. S. Saitoh, *Integral transforms, reproducing kernels and their applications* (Longman, Harlow, 1997).
3. M. M. Dodson, Groups and the sampling theorem, *Samp. Theory Signal Image Process.* **6**, 1 (2007).
4. G. B. Pedrick, *Theory of reproducing kernels for Hilbert spaces of vector valued functions*, Tech. rep., University of Kansas (July 1957).
5. J. Burbea and P. Masani, Hilbert spaces of Hilbert space valued functions, in *Probability theory on vector spaces II. Lecture notes in mathematics*, (Springer, Berlin, 1980) pp. 1–20.
6. D. Alpay, The Schur algorithm, reproducing kernel Hilbert spaces and system theory, in *SMF/AMS Texts and Monographs, vol. 5,*, (Amer. Math. Soc., Société Mathématique de France, 2001)
7. R. Rochberg, Toeplitz and Hankel operators on the Paley–Wiener space, *Integral Equations Operator Theory* **10** (1987) pp. 187–235.
8. H. Meschkowski, *Hilbertsche Räume mit Kernfunktion* (Springer–Verlag, Berlin, 1962).
9. J. R. Higgins, Sampling in reproducing kernel theory and harmonic analysis – a comparison. In preparation, (2009).
10. R. Young, *An introduction to nonharmonic Fourier series* (Academic Press, New York, 1980).
11. J. R. Higgins, *Sampling theory in Fourier and signal analysis: foundations* (Clarendon Press, Oxford, 1996).

Weighted composition operators between some spaces of analytic functions

S.D. Sharma

Department of Mathematics, University of Jammu,
Jammu-180006, J&K, India
E-mail: somdatt_jammu@yahoo.co.in

Ajay K. Sharma

School of Mathematics, Shri Mata Vaishno Devi University,
Kakryal, Katra-182320, J&K, India
E-mail: aksju_76@yahoo.com

In this paper, we consider weighted composition operators between Hardy, weighted Bergman, α-Bloch and $\mathcal{A}^{-\alpha}$-spaces.

Keywords: Weighted composition operator; Hardy space; weighted Bergman space; α-Bloch space.

1. Introduction

In what follows, we denote by $\mathbb{D}$ the open unit disk, $\partial\mathbb{D}$ the unit circle, $H(\mathbb{D})$ the space of holomorphic functions on $\mathbb{D}$, Ψ the set of holomorphic-self maps of $\mathbb{D}$, that is, $\Psi = \{\varphi : \varphi \in H(\mathbb{D}) \text{ and } \varphi(\mathbb{D}) \subset \mathbb{D}\}$, $dA(z) = dxdy/\pi$ and $dm = d\zeta/2\pi$ the normalized Lebesgue measures on $\mathbb{D}$ and $\partial\mathbb{D}$, respectively. X and Y will always be used to denote Banach spaces consisting of analytic functions on $\mathbb{D}$. For norm of X, we write $||\cdot||_X$. The closed unit ball of the Banach space X is the set $B_X = \{x \in X : ||x||_X \leq 1\}$. Given two Banach spaces X and Y, $\mathcal{L}(X,Y)$ denotes the set of all bounded linear operators from X to Y. Recall that an operator $T : X \to Y$ is said to be compact if for every sequence $\{x_n\}$ in B_X, $\{Tx_n\}$ admits a norm convergent subsequence in Y.

Let $\varphi \in \Psi$ and $\psi \in H(\mathbb{D})$. Then we can define a linear operator

$$W_{\psi,\varphi}f(z) = \psi(z)f(\varphi(z))$$

for $f \in H(\mathbb{D})$ and $z \in \mathbb{D}$, called a weighted composition operator. In fact,

$W_{\psi,\varphi} = M_\psi C_\varphi$, where $M_\psi f(z) = \psi(z) f(z)$ is the multiplication operator induced by ψ and $C_\varphi f(z) = f(\varphi(z))$ is the composition operator induced by φ. Our aim is to find:

For what φ and ψ is $W_{\psi,\varphi} : X \to Y$ bounded? A compact operator? What is the essential norm of $W_{\psi,\varphi}$? Recall that the essential norm $||T||_e$ of a bounded linear operator on a Banach space X is given by

$$||T||_e = \inf\{||T - K|| : K \text{ is compact on } X\}.$$

Clearly, T is compact if and only if $||T||_e = 0$. Recently, several authors have studied weighted composition operators on different spaces of analytic functions (see, for example [1],[3],[5,6],[7],[8],[10] and [11]).

2. Preliminaries

In this section we give the definitions and basic properties of some spaces of analytic functions. We also collect some essential facts that will be needed later.

For $0 < p < \infty$, we denote by H^p the Hardy space of all functions analytic in the unit disk $\mathbb{D}$ for which

$$||f||_{H^p} = \sup_{0 < r < 1} \left(\int_0^{2\pi} |f(re^{i\theta})|^p \frac{d\theta}{2\pi} \right)^{1/p} < \infty.$$

For $p \geq 1$, H^p is a Banach space and is a Hilbert space for $p = 2$. For $p = \infty$ the Banach space H^∞ consists of bounded analytic functions with norm $||f||_\infty = \sup_{z \in \mathbb{D}} |f(z)|$.

Let $\alpha \in (-1, \infty)$ and $d\nu_\alpha(z) = (\alpha + 1)(1 - |z|^2)^\alpha dA(z)$, $z \in \mathbb{D}$. Then $d\nu_\alpha$ is a probability measure on $\mathbb{D}$. For $0 < p < \infty$ the weighted Bergman space A_α^p is the set of all holomorphic functions f on $\mathbb{D}$ with norm

$$||f||_{A_\alpha^p} = \left(\int_\mathbb{D} |f(z)|^p d\nu_\alpha(z) \right)^{1/p} < \infty.$$

Let $\alpha > 0$. A function f holomorphic in $\mathbb{D}$ is said to belong to the α-Bloch space $\mathcal{B}^\alpha$ if $\sup_{z \in \mathbb{D}}(1 - |z|^2)^\alpha |f'(z)| < \infty$ and to the little α-Bloch space $\mathcal{B}^\alpha$ if $\lim_{|z| \to 1}(1 - |z|^2)^\alpha |f'(z)| = 0$. For $f \in \mathcal{B}^\alpha$ define

$$||f||_{\mathcal{B}^\alpha} = |f(0)| + \sup_{z \in \mathbb{D}}(1 - |z|^2)^\alpha |f'(z)|.$$

With this norm $\mathcal{B}^\alpha$ is a Banach space and the little α-Bloch space $\mathcal{B}_0^\alpha$ is a closed subspace of the α-Bloch Space. Note that $\mathcal{B}^1 = \mathcal{B}$, the usual Bloch space.

For any $\alpha > 0$, the space $\mathcal{A}^{-\alpha}$ and $\mathcal{A}_0^{-\alpha}$, consists of analytic functions f

in $\mathbb{D}$ such that $||f||_{\mathcal{A}^{-\alpha}} = \sup_{z\in\mathbb{D}}(1-|z|^2)^\alpha|f(z)| < \infty$ and $\lim_{|z|\to 1^-}(1-|z|^2)^\alpha|f(z)| = 0$, respectively. If $\alpha > 1$, it is known that $f \in \mathcal{B}^\alpha$ if and only if $f \in \mathcal{A}^{-(\alpha-1)}$ or the antiderivative of f is in $\mathcal{B}^{\alpha-1}$.

The growth of functions in these spaces is essential in our study. To this end, the following estimates will be useful.

For every z in $\mathbb{D}$, we have

$$|f(z)| \leq \frac{||f||_X}{(1-|z|^2)^\eta}, \tag{1}$$

where $\eta = 0, \alpha, 1/p$ or $(2+\beta)/p$, if $X = H^\infty$, $\mathcal{A}^{-\alpha}$ H^p or $\mathcal{A}^p_\beta$, respectively, and $f \in X$. For the cases of H^∞ and $\mathcal{A}^{-\alpha}$, (1) is trivial. Also the remainder cases appear in [12].

For general background on weighted Bergman spaces $\mathcal{A}^p_\beta$, Bloch spaces $\mathcal{B}^\alpha$ and $\mathcal{B}^\alpha_0$ and growth spaces $\mathcal{A}^{-\alpha}$ and $\mathcal{A}^{-\alpha}_0$, one may consult [4] and [12] and the references therein.

3. Boundedness

In this section we present the main theorems that characterize those ψ and φ for which $W_{\psi,\varphi}$ acts boundedly between Hardy spaces, weighted Bergman spaces, α-Bloch spaces and $\mathcal{A}^{-\alpha}$-spaces. The proofs usually consist of considering some carefully chosen test functions and precise growth estimates for the functions belonging to the space in question.

Let γ and λ be two real numbers. For holomorphic maps ψ and φ of $\mathbb{D}$ such that $\varphi(\mathbb{D}) \subset \mathbb{D}$, define $\Lambda(\psi, \varphi, \gamma, \lambda)$ as

$$\Lambda(\psi, \varphi, \gamma, \lambda) = \sup_{z\in\mathbb{D}} \frac{(1-|z|^2)^\gamma}{(1-|\varphi(z)|^2)^\lambda}|\psi(z)|.$$

We consider $W_{\psi,\varphi} : X \to H^\infty$, when $X = H^\infty, H^p, \mathcal{A}^{-\alpha}$, or $\mathcal{A}^p_\beta$ first.

Theorem 3.1. *Let* $1 \leq p < \infty$, $\alpha > 0$, $\beta > -1$, $\psi \in H(\mathbb{D})$, $\varphi \in \Psi$ *and* $W_{\psi,\varphi} : X \to H^\infty$, *where* $X = H^\infty, H^p, \mathcal{A}^{-\alpha}$, *or* $\mathcal{A}^p_\beta$. *Then* $W_{\psi,\varphi} \in \mathcal{L}(X, H^\infty)$ *if and only if* $\Lambda(\psi, \varphi, 0, \eta) < \infty$, *where* $\eta = 0, \alpha, 1/p$ *or* $(2+\beta)/p$, *respectively for* $X = H^\infty, \mathcal{A}^{-\alpha}, H^p$ *or* $\mathcal{A}^p_\beta$.

Though we could not find suitable references for all the parts involved in Theorem 3.1, it seems that they are well known. We prove it for the convenince of the readers.

Proof of Theorem 3.1. First suppose that $W_{\psi,\varphi} \in \mathcal{L}(X, H^\infty)$. For $a \in \mathbb{D}$, consider the function $f_{a,\eta} = ((1-|\varphi(a)|^2)/(1-\overline{\varphi(a)}z)^2)^\eta$, where $\eta =$

$0, \alpha, 1/p$ or $(2+\beta)/p$ respectively, depending upon $X = H^\infty, \mathcal{A}^{-\alpha}, H^p$ or $\mathcal{A}_\beta^p$. Then $||f_{a,\eta}||_X \leq C$ and so $||W_{\psi,\varphi} f_{a,\eta}||_\infty \leq C||W_{\psi,\varphi}||$. Moreover,

$$||W_{\psi,\varphi} f_{a,\eta}||_\infty = \sup_{z \in \mathbb{D}} |\psi(z) f_{a,\eta}(\varphi(z))| \geq |\psi(a)||f_{a,\eta}(\varphi(a))| = \frac{|\psi(a)|}{(1 - |\varphi(a)|^2)^\eta}.$$

Since $a \in \mathbb{D}$ is arbiyrary, we have $\Lambda(\psi, \varphi, 0, \eta) < \infty$.

Conversely, suppose that $\Lambda(\psi, \varphi, 0, \eta) < \infty$. Using (1), we have

$$||W_{\psi,\varphi}|| = \sup_{||f||_X = 1} ||W_{\psi,\varphi} f||_\infty = \sup_{||f||_X = 1} \sup_{z \in \mathbb{D}} |\psi(z)||f(\varphi(z))|$$

$$\leq \sup_{||f||_X = 1} \sup_{z \in \mathbb{D}} \frac{|\psi(z)|}{(1 - |\varphi(z)|^2)^\eta} ||f||_X = \sup_{z \in \mathbb{D}} \frac{|\psi(z)|}{(1 - |\varphi(z)|^2)^\eta}.$$

where $\eta = 0, \alpha, 1/p$ or $(2+\beta)/p$ respectively, depending upon $X = H^\infty, \mathcal{A}^{-\alpha}, H^p$ or $\mathcal{A}_\beta^p$. Hence $W_{\psi,\varphi} \in \mathcal{L}(X, H^\infty)$. $\qquad\square$

In the next result, we consider $W_{\psi,\varphi} : X \to \mathcal{A}^{-\alpha}$, when $X = H^p, \mathcal{A}^{-\gamma}$ or $\mathcal{A}_\beta^p$.

Theorem 3.2. *Let* $1 \leq p < \infty$, $\alpha > 0$, $\gamma > 0$, $\beta > -1$, $\psi \in H(\mathbb{D})$, $\varphi \in \Psi$ *and* $W_{\psi,\varphi} : X \to \mathcal{A}^{-\alpha}$, *where* $X = H^p, \mathcal{A}^{-\gamma}$ *or* $\mathcal{A}_\beta^p$. *Then* $W_{\psi,\varphi} \in \mathcal{L}(X, \mathcal{A}^{-\alpha})$ *if and only if* $\Lambda(\psi, \varphi, \alpha, \eta) < \infty$, *where* $\eta = \gamma, 1/p$ *or* $(2+\beta)/p$, *respectively for* $X = \mathcal{A}^{-\gamma}$, H^p *or* A_β^p.

The proof follows on similar lines as the proof of Theorem 3.1. In case $X = \mathcal{A}_\beta^p$, the proof can be found in [10] whereas the boundedness of $W_{\psi,\varphi} : \mathcal{A}^{-\gamma} \to \mathcal{A}^{-\alpha}$ is an easy corollory of a more general result of Montes-Rodriguez [8].

The next result characterizes bounbedness of $W_{\psi,\varphi} : X \to \mathcal{B}^\beta$, when $X = H^p, \mathcal{B}^\alpha$, or $\mathcal{A}_\gamma^p$.

Theorem 3.3. *Let* $1 \leq p < \infty$, $\alpha > 0$, $\beta > 0$, $\gamma > -1$, $\psi \in H(\mathbb{D})$, $\varphi \in \Psi$ *and* $W_{\psi,\varphi} : X \to \mathcal{B}^\beta$, *where* $X = \mathcal{B}^\alpha$, H^p *or* $\mathcal{A}_\gamma^p$.

(i) *If* $0 < \alpha < 1$, *then* $W_{\psi,\varphi} \in \mathcal{L}(\mathcal{B}^\alpha, \mathcal{B}^\beta)$ *if and only if* $\Lambda(\psi', \varphi, \beta, \alpha) < \infty$.

(ii) $W_{\psi,\varphi} \in \mathcal{L}(\mathcal{B}, \mathcal{B}^\beta)$ *if and only if* $\Lambda(\psi\varphi', \varphi, \beta, 1) < \infty$ *and*

$$\sup_{z \in \mathbb{D}} |\psi'(z)|(1 - |z|^2)^\beta \log \frac{1}{1 - |\varphi(z)|^2} < \infty.$$

(iii) *If* $X = \mathcal{B}^\alpha(\alpha > 1)$, H^p *or* $\mathcal{A}_\gamma^p$, *then* $W_{\psi,\varphi} \in \mathcal{L}(X, \mathcal{B}^\beta)$ *if and only if* $\Lambda(\psi', \varphi, \beta, \lambda) < \infty$ *and* $\Lambda(\psi\varphi', \varphi, \beta, 1 + \lambda) < \infty$, *where* $\lambda = \alpha - 1$ *if* $X = \mathcal{B}^\alpha(\alpha > 1)$, $\lambda = 1/p$ *if* $X = H^p$ *and* $\lambda = (2 + \gamma)/p$ *if* $X = \mathcal{A}_\gamma^p$.

The boundedness of weighted composition operators between Bloch spaces was characterized in [6] and [7], whereas the boundedness of $W_{\psi,\varphi}$ from H^p or $\mathcal{A}^p_\gamma$ spaces to β-Bloch spaces can be obtained by modifying the arguments in [11]. Similarly, by modifying the arguments in [11], we have the following result.

Theorem 3.4. *Let* $1 \le p < \infty$, $\alpha > 0$, $\beta > 0$, $\gamma > -1$, $\psi \in H(\mathbb{D})$, $\varphi \in \Psi$ *and* $W_{\psi,\varphi} : X \to \mathcal{B}^\beta_0$, *where* $X = \mathcal{B}^\alpha$, H^p, *or* $\mathcal{A}^p_\gamma$. *Then* $W_{\psi,\varphi} \in \mathcal{L}(X, \mathcal{B}^\beta_0)$ *if and only if (i)* $W_{\psi,\varphi} \in \mathcal{L}(X, \mathcal{B}^\beta)$, *(ii)* $\psi \in \mathcal{B}^\beta_0$ *and (iii)* $\lim_{|z|\to 1}(1 - |z|^2)^\beta |\psi(z)\varphi'(z)| = 0$.

4. Compactness and essential norms of weighted composition operators

Let γ and λ be two real numbers. For holomorphic maps ψ and φ of $\mathbb{D}$ such that $\varphi(\mathbb{D}) \subset \mathbb{D}$, define $\Gamma(\psi, \varphi, \gamma, \lambda)$ as

$$\Gamma(\psi, \varphi, \gamma, \lambda) = \lim_{\delta \to 1} \sup_{|\varphi(z)| > \delta} \frac{(1 - |z|^2)^\gamma}{(1 - |\varphi(z)|^2)^\lambda} |\psi(z)|.$$

We next compute the essential norm of $W_{\psi,\varphi} : X \to H^\infty$, when $X = H^p, \mathcal{A}^{-\alpha}$ or $\mathcal{A}^p_\beta$.

Theorem 4.1. *Let* $1 < p < \infty$, $\alpha > 0$, $\beta > -1$, $\psi \in H(\mathbb{D})$, $\varphi \in \Psi$, $W_{\psi,\varphi} \in \mathcal{L}(X, H^\infty)$, *where* $X = H^p, \mathcal{A}^{-\alpha}$ *or* $\mathcal{A}^p_\beta$. *Then*

$$\Gamma(\psi, \varphi, 0, \eta) \le \|W_{\psi,\varphi}\|_e \le 2\Gamma(\psi, \varphi, 0, \eta),$$

where $\eta = \alpha, 1/p$ *or* $(2 + \beta)/p$, *respectively for* $X = \mathcal{A}^{-\alpha}, H^p$ *or* $\mathcal{A}^p_\beta$.

Proof. We consider the upper estimate first.

For any fixed $0 < r < 1$, it is easy to check that $W_{\psi,r\varphi} : X \to H^\infty$ is compact. Thus

$$\|W_{\psi,\varphi}\|_e \le \|W_{\psi,\varphi} - W_{\psi,r\varphi}\|.$$

Now for any $0 < \delta < 1$, we have

$$\|W_{\psi,\varphi} - W_{\psi,r\varphi}\| = \sup_{\|f\|_X = 1} \|W_{\psi,\varphi}f - W_{\psi,r\varphi}f\|_\infty$$

$$= \sup_{\|f\|_X = 1} \sup_{z \in \mathbb{D}} |\psi(z)||f(\varphi(z)) - f(r\varphi(z))|$$

$$\le \|\psi\|_\infty \sup_{\|f\|_X = 1} \sup_{|\varphi(z)| \le \delta} |f(\varphi(z)) - f(r\varphi(z))|$$

$$+ \sup_{\|f\|_X = 1} \sup_{|\varphi(z)| > \delta} |\psi(z)||f(\varphi(z)) - f(r\varphi(z))|.$$

We can choose r sufficiently close to 1 such that the first term on the right hand side is less than any given $\epsilon > 0$. Again the second term on the right hand side is dominated by

$$\sup_{||f||_X=1} \sup_{|\varphi(z)|>\delta} |\psi(z)|(|f(\varphi(z))| + |f(r\varphi(z))|),$$

which is dominated by

$$\sup_{||f||_X=1} \sup_{|\varphi(z)|>\delta} \left(\frac{||f||_X}{(1 - |\varphi(z)|^2)^\eta} + \frac{||f||_X}{(1 - r^2|\varphi(z)|^2)^\eta} \right)$$

$$\leq \sup_{|\varphi(z)|>\delta} \frac{|\psi(z)|}{(1 - |\varphi(z)|^2)^\eta} + \sup_{|\varphi(z)|>\delta} \frac{|\psi(z)|}{(1 - r^2|\varphi(z)|^2)^\eta}.$$

Now letting $\delta \to 1$ first and then $r \to 1$, we get the desired upper bound. We now turn to the lower estimate.

Let $K : X \to H^\infty$ be any compact operator. For $a \in \mathbb{D}$, consider the function $f_{a,\eta} = (1 - |a|^2)/(1 - \bar{a}z)^2)^\eta$, where $\eta = \alpha, 1/p$ or $(2 + \beta)/p$ respectively, depending upon $X = \mathcal{A}^{-\alpha}, H^p$ or $\mathcal{A}^p_\beta$. Then $||f_{a,\eta}||_X \leq 1$ and $f_{a,\eta}$ converges weakly to 0 as $|a| \to 1$, so $||Kf_{a,\eta}|| \to 0$ as $|a| \to 1$. Thus

$$\begin{aligned}
||W_{\psi,\varphi} - K|| &\geq \limsup_{|a|\to 1} ||(W_{\psi,\varphi} - K)f_{a,\eta}||_\infty \\
&\geq \limsup_{|a|\to 1} ||(W_{\psi,\varphi}f_{a,\eta}||_\infty - ||Kf_{a,\eta}||_\infty) \\
&= \limsup_{|a|\to 1} \sup_{z\in\mathbb{D}} |\psi(z)||f_{a,\eta}(\varphi(z))| \\
&\geq \limsup_{|a|\to 1} \sup_{|\varphi(z)|>\delta} |\psi(z)||f_{a,\eta}(\varphi(z))|,
\end{aligned}$$

Letting $\delta \to 1$, we have $|\varphi(z)| \to 1$. Thus setting $a = \varphi(z)$, we obtain the desird lower bound. $\qquad \square$

In the next result, we consider $W_{\psi,\varphi} : X \to \mathcal{A}^{-\alpha}$, when $X = H^p$, $\mathcal{A}^{-\gamma}$ or $\mathcal{A}^p_\beta$.

Theorem 4.2. *Let* $1 \leq p < \infty$, $\alpha > 0$, $\gamma > 0$, $\beta > -1$, $\psi \in H(\mathbb{D})$, $\varphi \in \Psi$, $W_{\psi,\varphi} : X \to \mathcal{A}^{-\alpha}$, *where* $X = H^p$, $\mathcal{A}^{-\gamma}$ *or* $\mathcal{A}^p_\beta$ *and* $W_{\psi,\varphi} \in \mathcal{L}(X, \mathcal{A}^{-\alpha})$. *Then* $\Gamma(\psi, \varphi, \alpha, \eta) \leq ||W_{\psi,\varphi}||_e \leq 2\Gamma(\psi, \varphi, \alpha, \eta)$, *where* $\eta = \gamma$, $1/p$ *or* $(2 + \beta)/p$, *respectively for* $X = \mathcal{A}^{-\gamma}$, H^p *or* $\mathcal{A}^p_\beta$.

The proof follows exactly on same lines as the proof of Theorem 4.1. So we omit the details.

Acknowledgments

The first author would like to thank Department of Science and Techenology, Government of India for providing travel support and University of Jammu for providing partial support to participate and present this paper in the 7th ISAAC held at London.

The second author is thankful to NBHM/DAE, India for partial support (Grant No. 48/4/2009/R&D-II/426).

References

1. M. D. Contreras and S. Diaz-Madrigal, *Compact-type operators defind on* H^∞, *Contem. Math.*, **232**, 111-118, (1999).

2. C. C. Cowen and B. D. MacCluer, *Composition operators on spaces of analytic functions,* (CRC Press Boca Raton, New York, 1995).

3. Z. Cuckovic and R. Zhao, *Weighted composition operators on the Bergman space J. London Math. Soc.,* **70**, 499-511, (2004).

4. H. Hedenmalm, B. Korenblum and K. Zhu, *Theory of Bergman spaces,* (Springer, New York, Berlin, etc. 2000).

5. B. D. MacCluer and R. Zhao, *Essential norms of weighted composition operators between Bloch-type spaces Rocky Mountain J. of Math.* **33**, 1437-1458, (2003).

6. S. Ohno, *Weighted composition operators on the Bloch space, Bull. Austral. Math. Soc.,* **63**, 177-185, (2001).

7. S. Ohno, K. Stroethoff and R. Zhao, *Weighted composition operators between Bloch-type spaces, Rocky Mountain J. Math.,* **33**, 191-215, (2003).

8. A. Montes-Rodriguez, *Weighted composition operators on weighted Banach spaces of analytic functions, J. London. Math. Soc.,* **61**, 872-884, (2000).

9. J. H. Shapiro, *Composition operators and classical function theory,* (Springer-Verlag, New York. 1993).

10. Ajay K. Sharma and S. D. Sharma, *Weighted composition operators between Bergman-type spaces, Communications of the Korean Mathematical society,* **21**, 465-474, (2006).

11. Ajay K. Sharma and Rekha Kumari, *Weighted composition operators between Bergman and Bloch spaces, Communications of the Korean Mathematical society,* **22**, 373-382, (2007).

12. K. Zhu, *Operator theory in function spaces,* (Marcel Dekker, New York, 1990).

III.3. Modern aspects of the theory of integral transforms

Organisers: A. Kilbas, S. Saitoh

This session was a continuation of sessions organized on 3rd, 5th and 6th ISAAC Congresses where different problems of one- and multi-dimensional integral transforms and their applications were discussed. A subject of most reports in session III.3 of 7th ISAAC Congress was basically connected with theoretical problems of classical and new integral transforms and their applications to differential and integral equations. The session comprised 11 talks on variety of topics which include the theoretical aspects such as new classes of polynomials related to the Kontorovich-Lebedev transform and some aspects of modified Kontorovich-Lebedev transform, Feynman integrals over functional and abstract Wiener spaces, integral transforms with the Mittag-Leffler type functions, fractional calculus of variations, Blaschke product with several parameters, integral transforms related to generalized convolutions and their applications to the solutions of integral equations, applications of classical Fourier, Laplace and Mellin integral transforms to fractional differential equations and applications of fractional calculus transforms to equations with symmetrized and distributed order fractional derivatives.

The following is the list of speakers in session III.3: Liubov Britvina, Qiuhui Chen, Dong Hyun Cho, Diana Dolicanin, Anatoly Kilbas, Bong Jin Kim, Sanja Konjik, Anna Koroleva, Ljubica Oparnica, Juri Rapoport, Semyon Yakubovich.

Hiroshi Fujiwara gave a distinguished talk jointly in Sessions III.2 and III.3 in which he presented a numerical real inversion formula and many successful numerical experiments concerning the inversion of the Laplace transform on the basis of some reproducing kernel Hilbert space and a new inversion algorithm employing the Tikhonov regularization and the concept of infinite precision algorithm.

Numerical real inversion of the Laplace transform by reproducing kernel and multiple-precision arithmetic

H. Fujiwara

Graduate School of Informatics, Kyoto University,
Kyoto, 606-8501, Japan
E-mail: fujiwara@acs.i.kyoto-u.ac.jp

The aim of the paper is to show numerical real inversion of the Laplace transform by multiple-precision arithmetic. Proposed real inversion formula is based on Tikhonov regularization on a reproducing kernel Hilbert space. A small regularization parameter is involved for accurate approximation in regularization, which causes numerical instability of computational processes. We design and implement a fast multiple-precision arithmetic environment for large scale scientific numerical computations to overcome the numerical instability, and we realize accurate numerical inversions.

Keywords: Multiple-precision arithmetic, numerical instability, numerical real inversion of the Laplace transform, reproducing kernel Hilbert space.

1. Introduction

In the present paper we shall introduce a multiple-precision arithmetic environment "exflib" and discuss its effectiveness in numerically unstable problems. In particular, we shall realize numerical real inversion of the Laplace transform by multiple-precision arithmetic. Numerically unstable problems arise in most inverse problems in engineering, geophysics, or medicine, and their reliable numerical treatments are important. But numerical instability yields the rapid growth of computational errors and computation fails.

The main interest of theoretical numerical analysis for partial differential equations or integral equations is consistency, stability and convergence of discretization schemes. If a problem is well-posed in the sense of Hadamard, we expect stability of its discretization scheme in an appropriate norm. Here a problem is said to be well-posed if and only if there exists a unique solution and it depends continuously on the data. Ill-posedness is the opposite concept of well-posedness, and instability is serious in numerical treatments on digital computers. We give a remark that the stability of a discretization scheme does not always imply the stability of its numerical process. In other words, growth of computational errors possibly makes computations meaningless though the original problem is mathematically well-posed and its discretization scheme is stable in some norm. This is because the stability of a discretization scheme is a property of its exact solution, which cannot be realized on digital computers due to rounding errors.

To overcome numerical instability we propose the use of multiple-precision arithmetic. Multiple-precision arithmetic enables us to approximate a real value with arbitrary accuracy on digital computers. If we take enough precision in each numerical process, we can hide the influence of rounding errors and realize numerical computation without rounding errors *virtually*. Because multiple-precision arithmetic is not implemented in the standard computational environments, we design and implement a new software environment "exflib".

In numerical computations of partial differential equations or integral equations discretization errors should be also considered. In recent researches, the use of multiple-precision arithmetic and a high-accurate discretization enables us to realize direct numerical simulations of ill-posed problems.[1,2] But it requires analyticity of functions, and is not applicable to problems involving singularities.

We show another approach to numerical computation of ill-posed problems in the present paper: application of multiple-precision arithmetic to Tikhonov regularization. To achieve an accurate approximation by regularization a small regularization parameter is required, which implies numerical instability. We use multiple-precision arithmetic to realize reliable numerical computations with a small regularization parameter. Based on this idea we establish a numerical real inversion of the Laplace transform.

2. Floating-Point Arithmetic and Numerical Instability

In digital arithmetic, a real number is treated in a fixed-point format or a floating-point format, and the latter is mainly used in scientific computations. A floating-point format consists of three components: a sign part, an exponent part, and a fractional part. Two kinds of floating-point arithmetic are defined as the standard in IEEE754.[3] One is single precision, and the other is double precision. The latter is used in the most scientific numerical computations, and it has approximately 16.0 decimal digits accuracy.

Since a floating-point format has a finite precision in its fractional part, the floating-point values exist on the real axis discretely. Thus real values and results of floating-point arithmetic are approximated with floating-point values. The approximation is called rounding, and error in approximation is called a rounding error. Though each rounding error is relatively small, accumulations of rounding errors sometimes cause serious influence, in particular, in a numerically unstable process.

We show an example of numerical instability in numerical processes of $\{a_n\}$ defined by

$$a_{n+2} = \frac{34}{11}a_{n+1} - \frac{3}{11}a_n, \quad a_0 = 1, \quad a_1 = \frac{1}{11}, \tag{1}$$

whose exact solution is $a_n = (1/11)^n$. Fig. 1 shows numerical results of straightforward computation of Eq. (1) on two computer environments. In the figure, $+$ and $\times$ shows results by Opteron and UltraSPARC-III respectively. Difference is remarkable and both do not approximate the exact solution. We remark that characteristic values of Eq. (1) are 3 and 1/11, and the former which is greater than one causes the numerical instability.

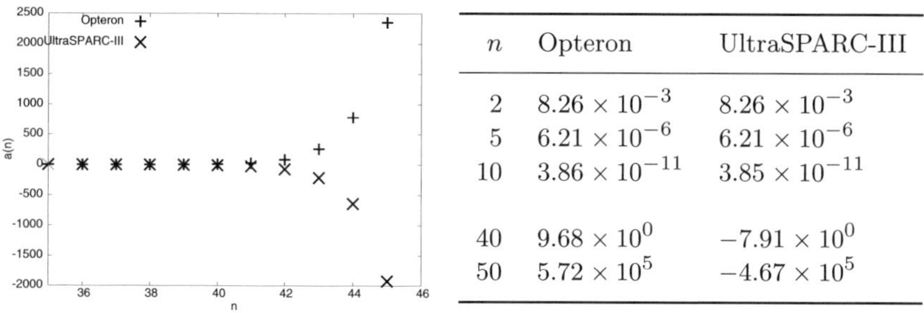

n	Opteron	UltraSPARC-III
2	8.26×10^{-3}	8.26×10^{-3}
5	6.21×10^{-6}	6.21×10^{-6}
10	3.86×10^{-11}	3.85×10^{-11}
40	9.68×10^{0}	-7.91×10^{0}
50	5.72×10^{5}	-4.67×10^{5}

Fig. 1. Numerical instability of the scheme Eq. (1)

3. Multiple-Precision Arithmetic Environment

We introduce a new multiple-precision arithmetic environment "exflib" (extended precision floating-point arithmetic library). Floating-point multiple-precision arithmetic allocates a lot of bits to its fractional part for accurate approximations of real numbers on digital computers. The proposed environment is designed for scientific numerical computation with one hundred up to several thousand decimal digits, and advantages of the proposed environment lie in fast computation, small memory requirement, and useful interface.

We use 64-bit integers and ALU (Arithmetic Logical Unit) operations for a data structure and arithmetic algorithms in the proposed multiple-

Table 1.   Computation time for basic arithmetic (unit: micro-sec.)

| digits | operation | | Bi-Opteron246 2GHz | | | Athlon64 2.2GHz |
		Maple	Mathematica	MPFR[5]	Pari	proposed method
100	mul	22	1.5	0.24	0.29	0.13
	div	26	5.9	0.52	1.2	0.30
1000	mul	100	17	9.2	15	3.4
	div	93	50	16	23	8.0
10000	mul	3400	580	360	1400	257
	div	3000	1800	730	1500	606

Note: digits means the precision in decimal, mul: multiplication, div: division

precision arithmetic. Each real number is approximated and normalized in the form $(-1)^s \times 2^e \times 1.F$, where a sign part s has one bit, an exponent part e has 63 bits, and a fractional part $F = f_1 f_2 \cdots f_n$ has $64 \times n$ bits (each f_i has 64 bits). We store it in a memory as a 64-bit unsigned integer array with $(n+1)$ elements. The precision of the type is thus $\log_{10} 2^{64n+1} \approx 19.26 \times n$ decimal digits, where the parameter n is defined at a compile time.

The proposed environment provides basic arithmetic operations with rounding, comparisons, output and assignment, and built-in functions. We use the elementary algorithms for addition and multiplication, Karatsuba's method for multiplication with large digits, and improved Ozawa's method for division.[4] Basic arithmetic rules of multiple-precision numbers, and multiplication and division of a multiple-precision number by an integer are written in an assembly language for fast computation.

In scientific computations FORTRAN and the programming language C are standard, so that we design the environment as a library for the programming languages. The proposed environment is equipped with polymorphic interface for four basic arithmetic rules, comparisons, assignments, output, and built-in functions. Polymorphic interface realizes high usability and portability from existing user programs written in the language C or FORTRAN.

We show arithmetic speed in Table 1. The proposed environment realizes faster computation than existing software environments.

The proposed environment is distributed via the Internet.[6] We implement it on AMD64, EM64T, SPARC V9, and IA32 architectures. It is executable on UNIX-like operating system (Solaris, Linux, FreeBSD, Mac OSX) and Windows operating system with commercial and noncommercial C++ and FORTRAN90 compilers. The proposed environment is applicable to parallel computations with MPI or OpenMP.

4. Numerical Real Inversion of the Laplace Transform

In this section, we apply the proposed multiple-precision arithmetic environment to real inversion of the Laplace transform. The Laplace transform,

$$\mathcal{L}f(p) = F(p) = \int_0^\infty e^{-pt} f(t)\, dt,$$

is fundamental in mathematical science and engineering. To find $f(t)$ from a given image function $F(p), p \geq 0$, is called the *real* inversion of the Laplace transform. Useful numerical methods are little known.[7,8]

Let w and u be positive and continuous functions defined on $[0, \infty)$. We introduce the reproducing kernel Hilbert space H_w[9] which consists of an absolutely continuous function $f(t)$ on $t \geq 0$ and $f(0) = 0$ with a finite norm

$$\|f\|_{H_w}^2 := \int_0^\infty |f'(t)|^2 w(t) dt < \infty.$$

The reproducing kernel of H_w is given by $K_t(s) = \int_0^{\min(s,t)} w(\xi)^{-1} d\xi$. Put

$$L_u^2 := \left\{ g \,;\, \|g\|_{L_u^2}^2 := \int_0^\infty |g(p)|^2 u(p) dp < \infty \right\}.$$

We consider the linear operator L given by

$$Lf(p) := p\mathcal{L}f(p), \quad f \in H_w.$$

If weight functions w and u satisfy

$$\int_0^\infty \int_0^\infty e^{-2pt} w(t)^{-1} u(p)\, dt dp < \infty,$$

then the operator $L : H_w \to L_u^2$ is compact.[10] Therefore the operator L does not have a bounded inverse, and the equation $Lf = G, f \in H_w, G \in L_u^2$, is ill-posed in the sense of Hadamard. We consider the Tikhonov functional as $J_\alpha(f_\alpha) = \|Lf - G\|_{L_u^2}^2 + \alpha\|f\|_{H_w}^2$, where the positive number α is called a regularization parameter. For any fixed α, there exists a unique minimizer $f_\alpha \in H_w$ of J_α, and the regularized solution f_α is given by[11,12]

$$f_\alpha(t) = \int_0^\infty pF(p) H_{\alpha,t}(p) u(p)\, dp, \tag{2}$$

where $H_{\alpha,t}$ is a unique solution of the integral equation of the second kind

$$(\alpha I + LL^*)H_{\alpha,t} = LK_t. \tag{3}$$

If a Laplace image $F \in \mathcal{L}(H_w)$, then f_α converges to the exact original function $f = \mathcal{L}^{-1}F$ in H_w as α tends to zero. If $F \notin \mathcal{L}(H_w)$, then f_α gives an approximation of $\mathcal{L}^{-1}F$ in the sense of minimizing J_α.

We show numerical examples of real inversion by Eq. (2) and Eq. (3) with multiple-precision arithmetic. We use $w(t) = (t+1)^{-2}$ and $u(p) = \exp(-p - 1/p)$, and set $\alpha = 10^{-100}$ or $\alpha = 10^{-400}$. Eq. (3) is discretized by the Nyström method with the double exponential rule,[13] and the discretized linear equation is solved by the LU decomposition without preconditioners. For the case $\alpha = 10^{-400}$, the computation is done with 600 decimal digits precision with discretization parameter 6000, and the condition number in 2-norm is approximately 10^{400}. The computational time of LU decomposition is about 100 minutes with 24 process parallel computation on Opteron 2214 (2.2GHz).

Example 4.1. The Laplace transform of a square function is

$$\mathcal{L}f(p) = \frac{e^{-p}}{p(1 + e^{-p})}.$$

Fig. 2 shows numerical results of $f_\alpha(t)$ for some α. The exact original function has discontinuities at integers. The regularized solution $f_\alpha(t)$ with a small regularization parameter gives a good approximation.

Example 4.2. Consider the Laplace transform arising in a circuit:[14]

$$\mathcal{L}f(p) = \frac{1}{p(p+1)} \left(\frac{1}{2p} - \frac{e^{-2p}}{1 - e^{-2p}} \right)$$

Numerical results of $f_\alpha(t)$ are shown in Fig. 3. The exact solution oscillates, and our method with small regularization parameter results in a good approximation.

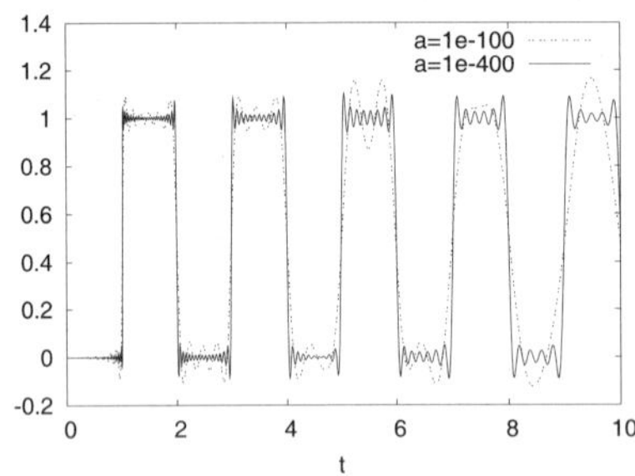

Fig. 2.   Numerical f_α for Example 4.1, dotted curve : $\alpha = 10^{-100}$ with 200 digits computation, solid curve : $\alpha = 10^{-400}$ with 600 digits computation.

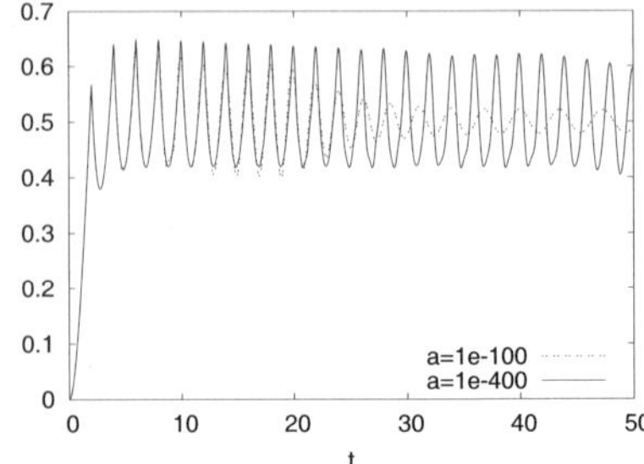

Fig. 3.   Numerical f_α for Example 4.2, dotted curve : $\alpha = 10^{-100}$ with 200 digits computation, solid curve : $\alpha = 10^{-400}$ with 600 digits computation.

5. Concluding Remarks

We design and implement a fast multiple-precision arithmetic environment, exflib, for large scale scientific numerical computations. It realizes fast and reliable numerical computations for small regularization parameters. It is applicable to accurate real inversion of the Laplace transform involving singularities by Tikhonov regularization on the reproducing kernel Hilbert space.

Acknowledgements

The author wishes to express his gratitude to Professor Yuusuke Iso, Professor Saburou Saitoh, and Professor Luís Castro for their many suggestions throughout the research. This work is supported in part by the Grant-in-Aid for Young Scientists (B) (No.20740057) and Grand-in-Aid for Scientific Research (B) (No.19340022).

References

1. H. Imai and T. Takeuchi, *GAKUTO Internat. Ser. Math. Sci. Appl.* **17**, 323 (2002).
2. H. Fujiwara, H. Imai, T. Takeuchi, and Y. Iso, *Trans. Japan Soc. Ind. Appl. Math.* **15**, 419 (2005).
3. IEEE standard for binary floating-point arithmetic (1985).
4. D. E. Knuth, *The art of computer programming, vol. 2, semi numerical algorithms*, 3rd edn. (Addison-Wesley, 1998).
5. `http://www.medicis.polytechnique.fr/~pphd/mpfr/timings.html`.
6. `http://www-an.acs.i.kyoto-u.ac.jp/~fujiwara`.
7. A. M. Cohen, *Numerical methods for Laplace transform inversion* (Springer, 2007).
8. J. P. Kaipio and E. Somersalo, *Statistical and computational inverse problems* (Springer, 2004).
9. S. Saitoh, *Far East J. Math. Sci.* **11**, 53 (2003).
10. H. Fujiwara, T. Matsuura, S. Saitoh and Y. Sawano, *J. Anal. Appl.* **6**, 55 (2008).
11. T. Matsuura, A. Al-Shuaibi, H. Fujiwara and S. Saitoh, *J. Anal. Appl.* **5**, 123 (2007).
12. H. Fujiwara, *RIMS Kokyuroku* **1618**, 192 (2008).
13. H. Takahashi and M. Mori, *Publ. Res. Inst. Math. Sci.* **9**, 721 (1974).
14. N. W. McLachlan, *Complex variable theory and transform calculus with technical applications*, 2nd edn. (Cambridge University Press, 1953).

Some aspects of modified Kontorovitch-Lebedev integral transforms

Juri M. Rappoport*

*Department of Mathematical Sciences,
Russian Academy of Sciences,
Vlasov street, Building 27, Apt.8,
Moscow 117335, Russia
E-mail:jmrap@landau.ac.ru*

A proof of inversion formulas of the modified Kontorovitch-Lebedev integral transforms is developed. The Parseval equations for modified Kontorovitch-Lebedev integral transforms are proved and sufficient conditions for them are found. Some new representations of these transforms by means of Fourier and Laplace transforms are justified. The inequalities which give estimations for their kernels - the real and imaginary parts of the modified Bessel functions of the second kind $ReK_{\frac{1}{2}+i\beta}(x)$ and $ImK_{\frac{1}{2}+i\beta}(x)$ for all values of the variables x and β are obtained. The applications of Kontorovitch-Lebedev transforms to the solution of some mixed boundary value problems in the wedge domains are accomplished. The solution of the appropriate dual and singular integral equations is considered. The numerical aspects of using of these transforms are elaborated in detail.

Keywords: Kontorovitch–Lebedev integral transform; modified Bessel function; Laplace transform; Fourier transform.

1. Some properties of the functions $ReK_{\frac{1}{2}+i\beta}(x)$ and $ImK_{\frac{1}{2}+i\beta}(x)$

It is possible to write the kernels of the modified Kontorovitch–Lebedev integral transforms in the form $ReK_{\frac{1}{2}+i\beta}(x) = \dfrac{K_{\frac{1}{2}+i\beta}(x)+K_{\frac{1}{2}-i\beta}(x)}{2}$ and $ImK_{\frac{1}{2}+i\beta}(x) = \dfrac{K_{\frac{1}{2}+i\beta}(x)-K_{\frac{1}{2}-i\beta}(x)}{2i}$, where $K_\nu(x)$ is the modified Bessel function of the second kind (also called Macdonald function).

*Work was partially supported by the U S Civilian Research and Development Foundation, Grant RM1-361 and by the Program of the Fundamental Research "Mathematical Modeling" from the Presidium of the Russian Academy of Sciences, Grant 1035.

The functions $ReK_{\frac{1}{2}+i\beta}(x)$ and $ImK_{\frac{1}{2}+i\beta}(x)$ have integral representations [1,2]

$$ReK_{\frac{1}{2}+i\beta}(x) = \int_0^\infty e^{-x\cosh t}\cosh\frac{t}{2}cos(\beta t)dt, \tag{1}$$

$$ImK_{\frac{1}{2}+i\beta}(x) = \int_0^\infty e^{-x\cosh t}\sinh\frac{t}{2}sin(\beta t)dt. \tag{2}$$

It follows from (1)-(2) that it is possible to write $ReK_{\frac{1}{2}+i\beta}(x)$ in the form of the Fourier cosinus-transform

$$ReK_{\frac{1}{2}+i\beta}(x) = \left(\frac{\pi}{2}\right)^{\frac{1}{2}}F_C[e^{-x\cosh t}\cosh\frac{t}{2};t\to\beta], \tag{3}$$

and $ImK_{\frac{1}{2}+i\beta}(x)$ in the form of the Fourier sinus-transform

$$ImK_{\frac{1}{2}+i\beta}(x) = (\frac{\pi}{2})^{\frac{1}{2}}F_S[e^{-x\cosh t}\sinh\frac{t}{2};t\to\beta]. \tag{4}$$

The inversion formulas have the respective forms

$$F_C[ReK_{\frac{1}{2}+i\beta}(x);\beta\to t] = (\frac{\pi}{2})^{\frac{1}{2}}e^{-x\cosh t}\cosh\frac{t}{2},$$

$$F_S[ImK_{\frac{1}{2}+i\beta}(x);\beta\to t] = (\frac{\pi}{2})^{\frac{1}{2}}e^{-x\cosh t}\sinh\frac{t}{2} \tag{5}$$

or, in integral form,

$$\int_0^\infty ReK_{\frac{1}{2}+i\beta}(x)\cos(t\beta)d\beta = \frac{\pi}{2}e^{-x\cosh t}\cosh\frac{t}{2}, \tag{6}$$

$$\int_0^\infty ImK_{\frac{1}{2}+i\beta}(x)\sin(t\beta)d\beta = \frac{\pi}{2}e^{-x\cosh t}\sinh\frac{t}{2}. \tag{7}$$

For the computation of certain integrals of the functions $ReK_{\frac{1}{2}+i\beta}(x)$ and $ImK_{\frac{1}{2}+i\beta}(x)$ integral identities are useful which reduce this problem to the computation of some other integrals over elementary functions.

Proposition 1.1.

If f is absolutely integrable on $[0,\infty)$, then the following identities hold,

$$\int_0^\infty ReK_{\frac{1}{2}+i\beta}(x)f(\beta)d\beta = (\frac{\pi}{2})^{\frac{1}{2}}\int_0^\infty e^{-x\cosh t}\cosh\frac{t}{2}F_C(t)dt, \tag{8}$$

$$\int_0^\infty ImK_{\frac{1}{2}+i\beta}(x)f(\beta)d\beta = (\frac{\pi}{2})^{\frac{1}{2}}\int_0^\infty e^{-x\cosh t}\sinh\frac{t}{2}F_S(t)dt, \tag{9}$$

where $F_C(t)$ is the Fourier cosinus-transform of $f(\beta)$, and $F_S(t)$ the Fourier sinus-transform of $f(\beta)$.

Proof. Multiplying both sides of the equalities (3) and (4) by $f(\beta)$, integrating with respect to β from 0 to ∞ , and applying Fubini's theorem for singular integrals with parameter, we obtain (8) and (9). $\square$

Proposition 1.2.

   If f is absolutely integrable on $[0, \infty)$, then the following identities hold

$$\int_0^\infty ReK_{\frac{1}{2}+i\beta}(x)F_C(\beta)d\beta = (\frac{\pi}{2})^{\frac{1}{2}} \int_0^\infty e^{-x\cosh t} \cosh \frac{t}{2} f(t)dt, \qquad (10)$$

$$\int_0^\infty ImK_{\frac{1}{2}+i\beta}(x)F_S(\beta)d\beta = (\frac{\pi}{2})^{\frac{1}{2}} \int_0^\infty e^{-x\cosh t} \sinh \frac{t}{2} f(t)dt. \qquad (11)$$

Proof. This follows from (6)-(7) and from Fubini's theorem. $\square$

The equations (8)-(11) are useful for the simplification and the calculation of different integrals containing $ReK_{1/2+i\beta}(x)$ and $ImK_{1/2+i\beta}(x)$.

We use the representation (1) for the evaluation of the LAPLACE transformation of $ReK_{\frac{1}{2}+i\beta}(x)$. We have

$$L[ReK_{\frac{1}{2}+i\beta}(x); \beta] = \int_0^\infty \cos(\beta t) \cosh \frac{t}{2} \int_0^\infty e^{-(p+\cosh t)x} dxdt =$$

$$= \int_0^\infty \frac{\cos(\beta t) \cosh \frac{t}{2}}{\cosh t + \cosh \alpha} dt \ (p = \cosh \alpha) \ =$$

$$= \sqrt{\frac{\pi}{2}} F_C(\frac{\cosh \frac{t}{2}}{\cosh t + \cosh \alpha}) = \frac{\pi}{2} \frac{\cos(\alpha\beta)}{\cosh \frac{\alpha}{2} \cosh(\pi\beta)}.$$

Equivalently, we can write

$$L^{-1}[\frac{\cos(\beta \cosh^{-1} p)}{\sqrt{\frac{p+1}{2}}}] = (\frac{\pi}{2})^{-1} \cosh(\pi\beta) ReK_{\frac{1}{2}+i\beta}(x). \qquad (12)$$

For the evaluation of the Laplace transform of $ImK_{\frac{1}{2}+i\beta}(x)$ we utilize the representation (2). We have

$$L[ImK_{\frac{1}{2}+i\beta}(x); p] = \sqrt{\frac{\pi}{2}} F_S(\frac{\sinh \frac{t}{2}}{\cosh t + \cosh \alpha}) = \frac{\pi}{2} \frac{\sin(\alpha\beta)}{\cosh(\pi\beta) \sinh \frac{\alpha}{2}},$$

or, equivalently,

$$L^{-1}[\frac{\sin(\beta \cosh^{-1} p)}{\sqrt{\frac{p-1}{2}}}] = \sqrt{\frac{\pi}{2}} \cosh(\pi\beta) ImK_{\frac{1}{2}+i\beta}(x).$$

We note that these equations can also be obtained directly from the formula for the Laplace transforms [3] of $K_\nu(x)$ by separating real and imaginary parts.

It follows from (1) that for all $\beta \in [0, \infty)$

$$|ReK_{\frac{1}{2}+i\beta}(x)| \leq K_{\frac{1}{2}}(x) = (\frac{\pi}{2x})^{\frac{1}{2}}e^{-x}, \tag{13}$$

and it follows from (2) that for all $\beta \in [0, \infty)$

$$|ImK_{\frac{1}{2}+i\beta}(x)| \leq \int_0^\infty e^{-x\cosh t}\sinh\frac{t}{2}dt = (\frac{\pi}{2x})^{\frac{1}{2}}e^x[1 - \phi((2x)^{\frac{1}{2}})] \leq B\frac{e^{-x}}{x}, \tag{14}$$

where B is some positive constant.

2. The modified Kontorovitch–Lebedev integral transforms

Integral transforms containing integration with respect to the index of the Bessel function play an important role for the solution of some classes of the problems in mathematical physics [3-5]. In particular, for the solution of mixed boundary value problems for the Helmholtz equation in wedge-shaped and conic domains, the modified Kontorovitch–Lebedev transforms [1, 2] are used,

$$F_+(\beta) = \int_0^\infty f(x)\frac{K_{\frac{1}{2}+i\beta}(x) + K_{\frac{1}{2}-i\beta}(x)}{2}dx, \ 0 \leq \beta < \infty, \tag{15}$$

$$F_-(\beta) = \int_0^\infty f(x)\frac{K_{\frac{1}{2}+i\beta}(x) - K_{\frac{1}{2}-i\beta}(x)}{2i}dx, \ 0 \leq \beta < \infty. \tag{16}$$

The inversion formulas have the form (17) and (18), respectively,

$$f(x) = \frac{4}{\pi^2}\int_0^\infty \cosh(\pi\beta)F_+(\beta)\frac{K_{\frac{1}{2}+i\beta}(x) + K_{\frac{1}{2}-i\beta}(x)}{2}d\beta, \ 0 < x < \infty, \tag{17}$$

$$f(x) = \frac{4}{\pi^2}\int_0^\infty \cosh(\pi\beta)F_-(\beta)\frac{K_{\frac{1}{2}+i\beta}(x) - K_{\frac{1}{2}-i\beta}(x)}{2i}d\beta, \ 0 < x < \infty. \tag{18}$$

Thus, the modified Kontorovitch–Lebedev integral transforms $F_+(\beta)$ and $F_-(\beta)$ of the function $f(x)$ defined on the positive real semiaxis are defined by the formulas

$$REK[f(x); \beta] = \int_0^\infty f(x)ReK_{\frac{1}{2}+i\beta}(x)dx, \tag{19}$$

$$IMK[f(x); \beta] = \int_0^\infty f(x)ImK_{\frac{1}{2}+i\beta}(x)dx, \tag{20}$$

and are written in the form

$$F_+(\beta) = REK[f(x); \beta], \ F_-(\beta) = IMK[f(x); \beta].$$

It follows from the inequalities (13)-(14) that for all $\beta \in (0, \infty)$

$$|F_+(\beta)| \leq \int_0^\infty |f(x)| K_{\frac{1}{2}}(x) dx \qquad (21)$$

and

$$|F_-(\beta)| \leq B \int_0^\infty |f(x)| \frac{e^{-x}}{x} dx. \qquad (22)$$

Proposition 2.1.

If the following conditions for function $f(x)$ are valid

1. $\frac{f(x)}{x} \in L(0, \frac{1}{2})$,

2. $f(x) e^{-x} x^{-\frac{1}{2}} \in L(\frac{1}{2}, \infty)$,

then the modified Kontorovitch–Lebedev integral transforms $F_+(\beta)$ and $F_-(\beta)$ are well defined, the integrals (19)-(20) converge uniformly in β and determine a continuous functions of β bounded for $\beta \in [0, \infty)$.

Proof. The proposition follows from the inequalities (13)-(14) and from criteria of uniform convergence of integrals with a parameter. $\qquad \square$

The integrals (19)-(20) are thus well defined and the integrals (21)-(22) converge uniformly in β and determine a continuous function of β.

The proofs of the inversion formulas and Parseval equalities for these transforms are investigated in [5-11].

3. The representation of the modified Kontorovitch–Lebedev integral transforms in the form of a composition of simple transformations

The problem of the evaluation of the Kontorovitch– Lebedev integral transforms is simplified by means of their decompositions in the form of compositions of simpler integral transformations, in particular, Fourier and Laplace transforms.

The representation of the Kontorovitch–Lebedev integral transform in the form of a composition of Laplace and Fourier transforms, derived earlier, has the form

$$K[f(x); \beta] = \sqrt{\frac{\pi}{2}} F_C[L(x^{-1} f(x); \cosh u); u \to \beta].$$

Proposition 3.1.

If the following condition is satisfied

$$f(x) x^{-\frac{1}{2}} e^{-x} \in L(0, \infty),$$

then the following representations of the modified KONTOROVITCH–
LEBEDEV *integral transforms hold,*

$$REK[f(x);\beta] = \sqrt{\frac{\pi}{2}}F_C[\cosh\frac{u}{2}L(f(x);\cosh u);u \to \beta], \qquad (23)$$

$$IMK[f(x);\beta] = \sqrt{\frac{\pi}{2}}F_S[\sinh\frac{u}{2}L(f(x);\cosh u);u \to \beta]. \qquad (24)$$

Proof. The representation (23) follows from the sequence of the equalities

$$REK[f(x);\beta] = \int_0^\infty f(x)ReK_{\frac{1}{2}+i\beta}(x)dx =$$

$$= \int_0^\infty f(x)\int_0^\infty e^{-x\cosh t}\cos(\beta t)\cosh\frac{t}{2}dtdx =$$

$$= \int_0^\infty \cos(\beta t)\cosh\frac{t}{2}\int_0^\infty f(x)e^{-x\cosh t}dxdt$$

and Fubini's theorem.

The equality (24) is proved analogously. $\qquad\square$

Proposition 3.2.

If the following conditions are satisfied
1. $f(x)x^{-\frac{1}{2}}\ln^{2k}x \in L(0,\frac{1}{2})$,
2. $f(x)x^{-\frac{1}{2}}e^{-x} \in L(\frac{1}{2},\infty)$,
then the following representation holds,

$$F_+^{(2k)}(\beta) = (-1)^k\sqrt{\frac{\pi}{2}}F_C[\cosh(\frac{t}{2})t^{2k}L(f(x);\cosh t);t \to \beta]. \qquad (25)$$

Proof. It follows from Proposition 3.1 that the following equality holds,

$$F_+(\beta) = \int_0^\infty \cos(\beta t)\cosh\frac{t}{2}\int_0^\infty f(x)e^{-x\cosh t}dxdt. \qquad (26)$$

We note that on the basis of Fubini's theorem

$$|\int_0^\infty t^{2k}\cos(\beta t)\cosh\frac{t}{2}\int_0^\infty f(x)e^{-x\cosh t}dxdt| \leq$$

$$\leq \int_0^\infty t^{2k}\cosh\frac{t}{2}\int_0^\infty |f(x)|e^{-x\cosh t}dxdt = \qquad (27)$$

$$= \int_0^\infty |f(x)|\int_0^\infty t^{2k}\cosh\frac{t}{2}e^{-x\cosh t}dtdx.$$

The last integral exists by virtue of the conditions 1,2 of the proposition
and asymptotics for the integral $\int_0^\infty t^n\cosh\frac{t}{2}e^{-x\cosh t}dt \sim x^{-\frac{n+1}{2}}e^{-x}$ for
$x \to \infty$, and $\int_0^\infty t^n\cosh\frac{t}{2}e^{-x\cosh t}dt \sim x^{-\frac{1}{2}}\ln^n x$ for $x \to 0$.

We obtain from the theorem on differentiation of an improper integral depending on a parameter, in view of a convergence of the integral (27),

$$F_+^{(2k)}(\beta) = (-1)^k \int_0^\infty t^{2k} \cosh \frac{t}{2} \cos(\beta t) \int_0^\infty f(x)e^{-x\cosh t} dx dt.$$

i.e. the proposition (25) is valid.   □

Proposition 3.3.

If the following conditions are satisfied
1. $f(x)x^{-\frac{1}{2}} \ln^{2k+1} x \in L(0, \frac{1}{2})$,
2. $f(x)x^{-\frac{1}{2}} e^{-x} \in L(\frac{1}{2}, \infty)$, then the following representation holds,

$$F_+^{2k+1}(\beta) = (-1)^k \sqrt{\frac{\pi}{2}} F_S[\cosh \frac{t}{2} L(f(x); \cosh t); t \to \beta]. \qquad (29)$$

The proof of Proposition 3.3 is analogous to the proof of Proposition 3.2.

Similar results are valid for the transformation (16).

Thus, knowledge of the Fourier and Laplace transforms is sufficient for the evaluation of the Kontorovitch–Lebedev integral transforms.

The Kontorovitch–Lebedev integral transforms may be expressed in terms of general Meyer integral transforms of special index and argument.

The applications for the solution of mixed boundary value problems in wedge domains, dual integral equations, numerical algorithms, approximation and computation of the kernels of the modified Kontorovitch–Lebedev integral transforms are described in [12-16].

References

1. N. N. Lebedev and I. P. Skalskaya, *Some integral transforms related to Kontorovitch– Lebedev transforms*, The questions of the mathematical physics, Leningrad, Nauka, (1976), 68–79 [in Russian].
2. N. N. Lebedev and I. P. Skalskaya, *The dual integral equations connected with the* KONTOROVITCH–LEBEDEV *transform*, Prikl. Matem. and Mechan., **38** (1974), N 6, 1090–1097 [in Russian].
3. S. Saitoh, *Integral Transforms, Reproducing Kernels and their Applications*, Pitman Res. Notes in Math. Series (369), Addison Wesley Longman Ltd., UK, 1997.
4. A. A. Kilbas and M. Saigo, *H-Transform. Theory and Applications*, Chapman and Hall/CRC, Boca Raton-London-New York- Washington D.C., 2004.
5. S. B. Yakubovich, *Index Transforms*, World Scientific Publishing, Singapore, New Jersey, London and Hongkong, 1996.

6. V. B. Poruchikov and J.M. Rappoport, *Inversion formulas for modified Kontorovitch–Lebedev transforms*, Diff. Uravn., **20** (1984), N 3, 542–546 [in Russian].

7. J. M. Rapppoport, *Some properties of modified Kontorovitch-Lebedev integral transforms*, Diff. Uravn., **21** (1985), N 4, 724–727 [in Russian].

8. J. M. Rappoport, *Some results for modified Kontorovitch–Lebedev integral transforms*, Proceedings of the 7th International Colloquium on Finite or Infinite Dimensional Complex Analysis, Marcel Dekker Inc., (2000), 473–477.

9. J. M. Rappoport, *The properties, inequalities and numerical approximation of modified Bessel function*, Electronic Transactions on Numerical Analysis, **25** (2006), 454–466.

10. J. M. Rappoport, *About modified Kontorovitch–Lebedev integral transforms and their kernels*, Imperial College of Science, Technology and Medicine, Department of Mathematics, London, 2009, preprint 09P/001, 31p.

11. J. M. Rappoport, *Some integral equations with modified Bessel function*, Proceedings of the 5th ISAAC Congress. More Progresses in Analysis. World Scientific Publishing, (2009), 269–278.

12. J. M. Rappoport, *The canonical vector-polynomials at computation of the Bessel functions of the complex order*, Comput. Math. Appl., **41** (2001), N 3/4, 399–406.

13. J. M. Rappoport, *Some numerical quadrature algorithms for the computation of the Macdonald function*, Proceedings of the Third ISAAC Congress. Progress in Analysis. Volume 2. World Scientific Publishing, (2003), 1223–1230.

14. B. R. Fabijonas, D. L. Lozier and J. M. Rappoport, *Algorithms and codes for the Macdonald function: Recent progress and comparisons*, Journ. Comput. Appl. Math., **161** (2003), N 1, 179–192.

15. J. M. Rappoport, *Dual integral equations for some mixed boundary value problems*, Proceedings of the 4th ISAAC Congress. Advances in Analysis. World Scientific Publishing, (2005), 167–176.

16. J. M. Rappoport, *Chebyshev polynomial approximations for some hypergeometric systems*, Isaac Newton Institute of Mathematical Sciences, Cambridge, 2009, preprint NI09059-DIS, 8p.

III.4. Spaces of differentiable functions of several real variables and applications

Organisers: V. Burenkov, S. Samko

Image normalization of WHOs in diffraction problems

A. Moura Santos

*Department of Mathematics, I.S.T., Technical University of Lisbon,
1049-001 Lisbon, Portugal
E-mail: amoura@math.ist.utl.pt*

We give a survey of the image normalization technique for certain classes of
Wiener-Hopf operators (WHOs) associated to ill-posed boundary-transmission
value problems. We briefly describe the method of normalization for the positive half-plane and apply it to the diffraction problem with generalized Dirichlet
boundary conditions. Then, we consider a WHO equivalent to a convolution
type operator associated to a boundary value problem on a strip. Finally, we
give an example of a generalization of the method for a junction of two infinite
half-planes.

Keywords: Diffraction boundary value problems; ill-posed problems; image
normalization; Wiener-Hopf operators.

1. Motivation and some notation

The normalization problem for bounded linear operators acting between
Banach spaces[1] can be solved by various methods. Here we describe how
to apply an image normalization method to convert not normally solvable
Wiener-Hopf operators (WHOs) into operators with closed image.[2]

In order to study these operators, we begin with the formulation of the
following boundary-transmission value problem for the diffraction of a wave
by different plane structures Ω in the natural setting of locally finite energy
norm.

Problem $\mathcal{P}$. Find $\varphi \in L^2(\mathbb{R}^2)$ with $\varphi_{|\mathbb{R}\times\mathbb{R}_\pm} = \varphi^\pm \in H^1(\mathbb{R} \times \mathbb{R}_\pm)$ such
that

$$\left(\Delta + k_0^2\right)\varphi^\pm \;=\; 0 \text{ in } \quad \mathbb{R} \times \mathbb{R}_\pm\,,$$

$$\begin{cases} \varphi_0^+ - \varphi_0^- = 0 \\ \varphi_1^+ - \varphi_1^- = 0 \end{cases} \text{ on } \quad \mathbb{R} \setminus \overline{\Omega}\,, \tag{1}$$

$$B_j\varphi(x) = \sum_{\sigma_1+\sigma_2\leq m_j} b^+_{\sigma,j}(D^\sigma\varphi^+)(x,0) + b^-_{\sigma,j}(D^\sigma\varphi^-)(x,0) = g_j(x) \text{ on } \Omega, \quad (2)$$

where the boundary Ω can represent either the half-line, $\Omega = \mathbb{R}_+$, either a strip, $\Omega =]0,a[$, $a > 0$, or even a finite union of strips [a]. Moreover, $\sigma = (\sigma_1,\sigma_2)$, $\sigma_j \in \mathbb{N}_0$, and $m_j \in \mathbb{N}_0$ represent the order of the boundary operator B_j, with $j = 1,2$ corresponding to the banks of Ω. For instance, for $\sigma_1 + \sigma_2 \leq 1$, B_j consists of a linear combination of Dirichlet, Neumann, and oblique derivative data. We can also consider linear combinations of higher derivatives both normal and tangential, with coefficients $b^\pm_{\sigma,j} \in \mathbb{C}$ depending on the materials of the boundary Ω. In the Helmholtz equation k_0 stands for the complex wave number such that $\operatorname{Re} k_0 > 0$ and $\operatorname{Im} k_0 > 0$.

Following the standard operator procedure of the classical survey of Meister and Speck,[3] we describe Problem $\mathcal{P}$ by a unique equation

$$\mathcal{P}\phi = g, \quad (3)$$

where $\mathcal{P}$ is a linear operator *associated* to Problem $\mathcal{P}$ which acts like

$$\mathcal{P} : D(\mathcal{P}) \to H^{1/2-m_1}(\Omega) \times H^{1/2-m_2}(\Omega).$$

The domain $D(\mathcal{P})$ is given by the subspace of $H^1(\mathbb{R} \times \mathbb{R}_\pm)$ whose elements fulfill the Helmholtz equation and all the homogeneous transmission conditions in (1) and the image space is characterized by $g_j \in H^{1/2-m_1}(\Omega) \times H^{1/2-m_2}(\Omega)$ according to the trace theorem and the representation formula applied to (2).[3]

Based on operator theory results, we write the equivalence $\mathcal{P} = EWF$, where E and F are bounded invertible linear operators and W is the convolution type operator given by

$$W : \tilde{H}^{1/2}(\Omega) \times \tilde{H}^{-1/2}(\Omega) \to H^{1/2-m_1}(\Omega) \times H^{1/2-m_2}(\Omega) \quad (4)$$

acting between subspaces $\tilde{H}^s(\Omega)$ of H^s-distributions supported on $\overline{\Omega}$ and $H^s(\Omega)$ restrictions of H^s distributions on Ω. For $\Omega = \mathbb{R}_+$, we write $\tilde{H}^s(\Omega) = H^s_+$ and (4) defines a WHO. In the vector case for $s = (s_1, ..., s_j, ..., s_n) \in \mathbb{R}^n$ we define the product spaces $H^s = \bigotimes_{j=1}^n H^{s_j}$, $H^s_+ = \bigotimes_{j=1}^n H^{s_j}_+, ...$ and each Bessel potential space of order s is given by

$$H^s = \{\phi \in \mathcal{S}' : \mathcal{F}^{-1}\lambda^s \cdot \mathcal{F}\phi \in L^2\}$$

---

[a]As a consequence of the physics of the wave diffraction the boundary Ω can be identified with subsets of the real line.

with $\lambda(\xi) = \sqrt{\xi^2 + 1}$, $\xi \in \mathbb{R}$, and $\mathcal{F}$ the Fourier transformation [b].

There are classes of WHOs in (4) (e.g. operators associated with certain Dirichlet and Neumann boundary value problems,[3] impedance,[4] oblique derivative[5]) that are not normally solvable, i.e. the image of the associated WHO are not closed. We say that the corresponding Problem $\mathcal{P}$ is ill-posed.

Before we write down the fundamental result of *image normalization* for the class of not normally solvable WHOs, let us describe in more detail the structure of these operators. For the next considerations, let be the boundary $\Omega = \mathbb{R}_+$ in (4), i.e. the half-plane, and write

$$W = r_+ A \left.\right|_{H_+^r} : H_+^r \to H^s(\mathbb{R}_+), \tag{5}$$

where $A = \mathcal{F}^{-1}\Phi \cdot \mathcal{F}$ is a translation invariant homeomorphism with a matrix *Fourier symbol* $\Phi \in L_{\mathrm{loc}}^\infty$. Lifting the WHO W into L^2, we obtain the equivalent *lifted* WHO

$$W_0 = r_+ A_0 \left.\right|_{[L_+^2]^n} : [L_+^2]^n \to [L^2(\mathbb{R}_+)]^n, \tag{6}$$

where $A_0 = \mathcal{F}^{-1}\Phi_0 \cdot \mathcal{F}$, $\Phi_0 \in L^\infty(\mathbb{R})^{n \times n}$ and moreover $\Phi_0 \in \mathcal{G}C^\nu(\ddot{\mathbb{R}})^{n \times n}$, i.e. the lifted Fourier symbol belongs to the invertible algebra of Hölder continuous, on $\ddot{\mathbb{R}} = [-\infty, +\infty]$, $n \times n$ matrix functions. This implies that we should always have $\det \Phi_0(\xi) \neq 0$, $\xi \in \ddot{\mathbb{R}}$. [c]

For the lifted WHO in (6) the following Fredholm criterium is well-known.[6] The operator W_0 is normally solvable iff

$$\det(\mu\Phi_0(-\infty) + (1 - \mu)\Phi_0(+\infty)) \neq 0, \quad \mu \in]0, 1[. \tag{7}$$

Finally, we shall also use the zero extension operator $\ell^{(0)}$ and the following Bessel potential operators[2] for $w \in \mathbb{C}$, $k_0 \in \mathbb{C}$, $\mathrm{Im}\, k_0 > 0$

$$\Lambda_\pm^w = \mathcal{F}^{-1}\lambda_\pm^w \cdot \mathcal{F} : H^s \to H^{s-\mathrm{Re}w},$$

with scalar Fourier symbols $\lambda_\pm^w(\xi) = (\xi \pm k_0)^w$.

---

[b]In the context of Problem $\mathcal{P}$, it is also used equivalently $\lambda(\xi) = \sqrt{\xi^2 - k_0^2}$ with branch cuts along $\pm k_0 \pm i\epsilon$, $\epsilon \geq 0$
[c]In this paper we always consider coefficients in (2) such that $\det\Phi_0 \neq 0$.

2. Image normalization in scalar and matrix cases

Our approach[2] to the normalization of the WHOs is based on two central ideas: the first is to respect the physics behind the operator, and the second is a mathematical argument of a *minimal* change in the spaces. Thus, we want the domain of the operator to be a space of locally finite energy, and we want to change the image space in a *minimal* way.

The following scalar result[2] helps to understand the method.

Theorem 2.1. *Let us consider the scalar WHO, which acts symmetrically, i.e. $r = s$ in (4)*

$$W_s = W_s(\Phi) = r_+ A \mid_{H_+^s} : H_+^s \to H^s(\mathbb{R}_+).$$

Then for the critical orders[7] $s + \sigma + 1/2 \in \mathbb{Z}$, where $\sigma = \frac{1}{2\pi i} \int_{\mathbb{R}} d \arg \Phi$, the operator W_s is not normally solvable. With $w = \sigma + i\tau$, $\tau = \frac{1}{2\pi} \ln |\Phi(-\infty)/\Phi(+\infty)|$, we define the image normalized operator $\check{W}_s$ by

$$\check{W}_s = \mathrm{Rst} W_s : H_+^s \to \check{H}^{s-i\tau}(\mathbb{R}_+)$$

where $\check{H}^{s-i\tau}(\mathbb{R}_+) = r_+ \Lambda_-^{-s+i\tau-1/2} H_+^{-1/2} \subset H^{\mathrm{Rew}}(\mathbb{R}_+)$. The image space of $\check{W}_s$ solves the normalization problem for $\{W_s = W_s(\Phi) : \Phi \in \mathcal{G}C^\nu(\ddot{\mathbb{R}}), \nu \in \ ]0, 1[, \mathrm{im} W_s \neq \overline{\mathrm{im} W_s}\}$.

The normalization in the $n \times n$ matrix case is based on the same idea of using the jump at infinity of the lifted Fourier symbol to change the image space in a *minimal* way. In the 2×2 matrix case, the statement reads.

Theorem 2.2. *Let W be the WHO in (5) with $r = (r_1, r_2)$, $s = (s_1, s_2)$. Suppose that the Fourier symbol of the corresponding lifted WHO Φ_0 has a jump at infinity. Let*

$$\Phi_0^{-1}(-\infty)\Phi_0(+\infty) = T^{-1} \mathrm{diag}(\lambda_1, \lambda_2) T, \tag{8}$$

where $\lambda_1 = e^{2i\pi w_1}, \lambda_2 = e^{2i\pi w_2}$ are the eigenvalues of Φ_0 and e.g. $\mathrm{Re} w_1 = -1/2$ and $\mathrm{Re} w_2 \neq -1/2$. Then W is not normally solvable for the given s_1 and $\sigma_1 = -1/2$, but we can define the image normalized operator $\check{W}$ by

$$\check{W} : H_+^r \to r_+ \Lambda_-^{-s} \ell^{(0)} \{ \check{H}^0(\mathbb{R}_+) \times L^2(\mathbb{R}_+) \}$$

and $\check{H}^0(\mathbb{R}_+) = r_+ \Lambda_-^{-1/2} H_+^{-1/2}$ is a proper dense subspace of $L^2(\mathbb{R}_+)$. The image space of $\check{W}$ solves the normalization problem for W.

3. An ill-posed problem on the half-plane

We aim now to apply Theorem 2.2 to particular cases of ill-posed bounded value problems. Beginning with the half-plane, i.e. $\Omega = \mathbb{R}_+$, we consider Problem $\mathcal{P}$ with the following generalized Dirichlet boundary condition.

$$\begin{cases} a_0^+ \varphi_0^+ + a_0^- \varphi_0^- = g_1 \\ b_0^+ \varphi_0^+ + b_0^- \varphi_0^- = g_2 \end{cases} \quad \text{on } \mathbb{R}_+ \tag{9}$$

and $g_j \in H^{1/2}(\mathbb{R}_+)$. In this case, the Fourier symbol of the lifted WHO reads

$$\Phi_0 = -\frac{1}{2} \begin{bmatrix} -\rho(a_0^+ - a_0^-) & a_0^+ + a_0^- \\ -\rho(b_0^+ - b_0^-) & b_0^+ + b_0^- \end{bmatrix} \tag{10}$$

where $\rho = (\lambda_-/\lambda_+)^{1/2}$ which doesn't fulfill the Fredholm criterium (7) for $\mu = 1/2$. The proof of the following statement is similar to the discussion that can be found in[2] for the matrix impedance case with an obvious shift by one of the order of the image space [d]

Theorem 3.1 (generalized Dirichlet). *Let $\det\Phi_0 \neq 0$, i.e. $a_0^+ b_0^- \neq a_0^- b_0^+$, then the WHO W associated with Problem $\mathcal{P}$ with the boundary condition (9), a particular case of (2), which acts like*

$$W : H_+^{1/2} \times H_+^{-1/2} \to [H^{1/2}(\mathbb{R}_+)]^2.$$

is not-normally solvable due to the jump at infinity of the lifted Fourier symbol. In this case we consider the corresponding image normalized operator $\check{W}$ defined by

$$\check{W} = \mathrm{Rst}\,W : H_+^{1/2} \times H_+^{-1/2} \to r_+ \Lambda_-^{-s} \ell^{(0)} \{\check{H}^0(\mathbb{R}_+) \times L^2(\mathbb{R}_+)\},$$

where $s = (1/2, 1/2)$, and $\check{H}^0(\mathbb{R}_+) = r_+ \Lambda_-^{-1/2} H_+^{-1/2}$ is a proper dense subspace of $L^2(\mathbb{R}_+)$. The image space of $\check{W}$ solves the normalization problem for the WHO.

4. Image normalization for strips

For structures like strips [e], we must prove first the equivalence of the operator $\mathcal{P}$ in (3) to a Toeplitz operator W defined e.g. on the interval $\Omega =]0, a[$.

---

[d]In the cited paper we used a slightly different notation for $\check{H}^0$ but it is the same space.
[e]Here we can include finite unions of strips.

From this convolution type operator, via operator matrix identities, we arrive at a WHO algebraically equivalent to W up to invertible operators E and F, not necessarily bounded. But, for both the operators, Toeplitz and WH with closed images, we obtain an equivalence after extension with invertible operators E and F.[8] We give an example of a boundary condition of type (2) on a strip, namely the oblique derivative boundary condition.[9]

$$
\begin{cases}
a_0^+ \varphi_{\tan}^+ + b_0^+ \varphi_1^+ = g_1 \\
a_0^- \varphi_{\tan}^+ - b_0^- \varphi_1^- = g_2
\end{cases}
\quad \text{on } \Omega
\tag{11}
$$

and $g_j \in H^{-1/2}(\Omega)$, $\Omega =\,]0, a[$, $a > 0$.

Theorem 4.1 (strip with oblique derivatives). *The convolution type operator W on the finite interval associated with Problem $\mathcal{P}$ with boundary conditions (11) which acts as*

$$
W : \tilde{H}^{s-1/2}(\Omega) \times \tilde{H}^{s-3/2}(\Omega) \to [H^{s-3/2}(\Omega)]^2,
$$

where $s - 1/2 \notin \mathbb{N}$ [f] is not-normally solvable for $s = \frac{2n+1}{4}, n \in \mathbb{N}$, since the 4×4 Fourier symbol of the equivalent lifted WHO W_0 has a jump at infinity. The corresponding image normalized operators are given by

$$
\check{W}_{\mathrm{o}} = \mathrm{Rst} W_0 : [L_+^2(\mathbb{R}_+)]^4 \to [L^2(\mathbb{R}_+)]^2 \times \check{H}^0(\mathbb{R}_+) \times L^2(\mathbb{R}_+)
$$

for n odd and

$$
\check{W}_{\mathrm{e}} = \mathrm{Rst} W_0 : [L_+^2(\mathbb{R}_+)]^4 \to [L^2(\mathbb{R}_+)]^3 \times \check{H}^0(\mathbb{R}_+)
$$

for n even, respectively. The image spaces of $\check{W}_{\mathrm{o}}$, and $\check{W}_{\mathrm{e}}$ solve the normalization problem for the WHO.

5. Image normalization for the junction of two half-planes

Finally, we can generalize our Problem $\mathcal{P}$ to include another class of boundary value problems which in general claims for normalization, e.g. whenever the boundary conditions given at the junction of two half-planes have both even order. This class consists of boundary value problems with different data on the two half-planes [g], i.e. we formulate a Problem $\mathcal{P}'$ which together

---

[f]These space orders are a consequence of considering regularity orders which include the finite energy norm space.
[g]This class of problems were studied in the Master Thesis of N. Bernardino (2004), in Portuguese, to be published soon.

with the Helmholtz equation, and the boundary condition (2) on $\Omega = \mathbb{R}_+$, includes a similar condition on the complementary half-plane $\mathbb{R} \setminus \overline{\Omega} = \mathbb{R}_-$, instead of the transmission condition (1). Similarly to Section 1, we can define the operator $\mathcal{P}'$ associated to the problem and work with the equivalent WHO. Let us consider e.g. the following boundary conditions of even orders m_j, $j = 1, 2$.

$$\begin{cases} \varphi_{m_1}^+ = h_1 \\ \varphi_{m_1}^- = h_2 \end{cases} \text{ on } \quad \mathbb{R} \setminus \overline{\Omega}, \tag{12}$$

$$\begin{cases} a_0^+ \varphi_{m_2}^+ + a_0^- \varphi_{m_2}^- = g_1 \\ b_0^+ \varphi_{m_2}^+ + b_0^- \varphi_{m_2}^- = g_2 \end{cases} \text{ on } \Omega \tag{13}$$

and $h_j \in H^{1/2 - m_1}(\mathbb{R}_-)$, $g_j \in H^{1/2 - m_2}(\mathbb{R}_+)$. The Fourier symbol of the lifted WHO is given by

$$\Phi_0 = \rho^{1 - m_1 - m_2} \begin{bmatrix} a_0^+ & a_0^- \\ b_0^+ & b_0^- \end{bmatrix}$$

where $\rho = (\lambda_- / \lambda_+)^{1/2}$, and doesn't satisfy the Fredholm criterium (7) for $\mu = 1/2$. The following result can be proved based on Theorem 2.2.

Theorem 5.1 (higher orders derivatives on 2 half-planes). *Let* $\det \Phi_0 \neq 0$, *i.e.* $a_0^+ b_0^- \neq a_0^- b_0^+$, *then the WHO W associated with Problem $\mathcal{P}'$ with boundary conditions (12) and (13), respectively on $\mathbb{R}_-$ and $\mathbb{R}_+$, where m_1 and m_2 are even numbers, and which acts as*

$$W : [H_+^{1/2 - m_1}]^2 \to [H^{1/2 - m_1 - m_2}(\mathbb{R}_+)]^2.$$

is not-normally solvable due to the jump at infinity of the lifted Fourier symbol. In this case we consider the corresponding image normalized operator $\check{W}$ defined by

$$\check{W} = \mathrm{Rst} W : [H_+^{1/2 - m_1}]^2 \to r_+ \Lambda_-^{-s} \ell^{(0)} \{ \check{H}^0(\mathbb{R}_+) \times \check{H}^0(\mathbb{R}_+) \},$$

and $s = (1/2 - m_2, 1/2 - m_2)$. The image space of $\check{W}$ solves the normalization problem for the WHO.

References

1. V.G. Kravchenko, *On normalization of singular integral operators,* Soviet Math. Dokl. **32**, 880-883 (1985).

2. A. Moura Santos, F.-O. Speck, F.S. Teixeira, *Minimal normalization of Wiener-Hopf operators in spaces of Bessel potentials*, J. Math. Anal. Appl. **225**, 501-531 (1998).
3. E. Meister, F.-O. Speck, *Modern Wiener-Hopf methods in diffraction theory*, Pitman Res. Notes Math. Ser. **216**, 130-171 (1989).
4. E. Meister, F.-O. Speck, *Diffraction problems with impedance conditions*, Appl. Anal. **22**, 193-211 (1986).
5. A. Moura Santos, F.-O. Speck, *Sommerfeld diffraction problems with oblique derivatives*, Math. Meth. Appl. Sci. **20**, 635-652 (1997).
6. S.G. Mikhlin, S. Prössdorf, *Singular Integral Operators* (Springer, Berlin 1986).
7. R. Duduchava, *Integral Equations with Fixed Singularities* (Teubner Texte zur Mathematik, Teubner, Leipzig 1979).
8. L.P. Castro, F.-O. Speck, *Relations between convolution type operators on intervals and on the half-line*, Integr. Equ. Oper. Theory **37**, 169-207 (2000).
9. L.P. Castro, A. Moura Santos, *An operator approach for an oblique derivative boundary-transmission problem*, Math. Meth. Appl. Sci. **27**, 1469-1491 (2004).

Weighted estimates for the averaging integral operator and reverse Hölder inequalities

Bohumír Opic

Institute of Mathematics, AS CR,
Žitná 25, 11567 Praha 1, Czech Republic
E-mail: opic@math.cas.cz

Let $1 < p < +\infty$ and let v be a weight on $(0, +\infty)$ such that $v(x)x^\rho$ is equivalent to a non-decreasing function on $(0, +\infty)$ for some $\rho \geq 0$. Let A be the averaging operator given by $(Af)(x) := \frac{1}{x}\int_0^x f(t)\,dt$, $x \in (0, +\infty)$, and let $L^p(v)$ denote the weighted Lebesgue space of all measurable functions f on $(0, +\infty)$ for which $\left(\int_0^{+\infty} |f(x)|^p v(x)\,dx\right)^{1/p} < +\infty$. Then the following statements are equivalent:

(i) A is bounded on $L^p(v)$;

(ii) A is bounded on $L^{p-\varepsilon}(v)$ for some $\varepsilon \in (0, p-1)$;

(iii) A is bounded on $L^p(v^{1+\varepsilon})$ for some $\varepsilon > 0$;

(iv) A is bounded on $L^p(v(x)x^\varepsilon)$ for some $\varepsilon > 0$.

Moreover, if A is bounded on $L^p(v)$, then A is bounded on $L^q(v)$ for all $q \in [p, +\infty)$.

We also show that the boundedness of the averaging operator A on the space $L^p(v)$ implies that, for all $r > 0$, the weight $v^{1-p'}$ satisfies the reverse Hölder inequality over the interval $(0, r)$ with respect to the measure dt, while the weight v satisfies the reverse Hölder inequality over the interval $(r, +\infty)$ with respect to the measure $t^{-p}\,dt$.

Assume moreover that $p \leq q < +\infty$ and that w is a weight on $(0, +\infty)$ such that
$$[w(x)x]^{1/q} \approx [v(x)x]^{1/p} \quad \text{for all } x \in (0, +\infty).$$
Then the operator A is bounded on $L^p(v)$ if and only if the operator $A : L^p(v) \to L^q(w)$ is bounded. This enables us to transfer our results on the boundedness of the operator A mentioned above to the case when the operator A acts between two weighted Lebesgue spaces $L^p(v)$ and $L^q(w)$.

Keywords: Averaging integral operator; weighted Lebesgue spaces; weights; Hardy-type inequalities; reverse Hölder inequalities.

1. Introduction and survey of results

Let $1 < p < +\infty$ and let v be a weight on $(0, +\infty)$, i.e., a measurable function which is positive a.e. on $(0, +\infty)$. By $L^p(v) \equiv L^p((0, +\infty); v)$ we denote the weighted Lebesgue space of all measurable functions f on $(0, +\infty)$ for which the norm

$$\|f\|_{p,v} = \left(\int_0^{+\infty} |f(x)|^p v(x)\, dx \right)^{1/p}$$

is finite.

We shall consider one of the basic operators in the mathematical analysis, the averaging operator A defined by

$$(Af)(x) := \frac{1}{x} \int_0^x f(t)\, dt, \quad x \in (0, +\infty).$$

It is well known (see [7] or, e.g., [10], [6]) that if $1 < p < +\infty$ and v is a weight on $(0, +\infty)$, then the averaging operator

$$A : L^p(v) \to L^p(v) \quad \text{is bounded} \tag{1}$$

if and only if $v \in M_p$, where M_p is the class of all weights v on $(0, +\infty)$ which satisfy

$$\sup_{r>0} \left(\int_r^{+\infty} v(t) t^{-p}\, dt \right)^{1/p} \left(\int_0^r v(t)^{1-p'}\, dt \right)^{1/p'} < +\infty, \tag{2}$$

where $p' = p/(p-1)$.

Throughout the paper we use the following convention: For two non-negative expressions (i.e. functions or functionals) F and G the symbol $F \lesssim G$ (or $F \gtrsim G$) means that $F \leq cG$ (or $cF \geq G$), where c is a positive constant independent of appropriate quantities involved in F and G. We shall write $F \approx G$ (and say that F and G are equivalent) if both relations $F \lesssim G$ and $F \gtrsim G$ hold.

Our first result reads as follows.

Theorem 1.1. *Let $1 < p < +\infty$ and v be a weight on $(0, +\infty)$ such that*

$$v(x)x^\rho \text{ is equivalent to a non-decreasing function on } (0, +\infty) \tag{3}$$

for some $\rho \geq 0$.

Then the following statements are equivalent:
(i) A is bounded on $L^p(v)$;
(ii) A is bounded on $L^{p-\varepsilon}(v)$ for some $\varepsilon \in (0, p-1)$;
(iii) A is bounded on $L^p(v^{1+\varepsilon})$ for some $\varepsilon > 0$;
(iv) A is bounded on $L^p(v(x)x^\varepsilon)$ for some $\varepsilon > 0$.

Moreover, if A is bounded on $L^p(v)$, then A is bounded on $L^q(v)$ for all $q \in [p, +\infty)$.

Proof. Theorem 1.1 follows from [11, Theorems 1-3]. $\square$

Remark 1.1. Let A_p, $1 < p < +\infty$, be the A_p-class of B. Muckenhoupt of those weights v on $(0, +\infty)$ for which the Hardy-Littlewood maximal operator associated with the interval $(0, +\infty)$ is bounded on the space $L^p(v)$. Since this Hardy-Littlewood maximal operator dominates the averaging operator A, we see that $A_p \subset M_p$.

If (3) holds with $\rho = 0$, then v is equivalent to a non-decreasing function on $(0, +\infty)$. It is known (cf. [4, Theorem 6.1] or [3, Proposition 2.3]) that a non-decreasing weight v satisfies $v \in M_p$ if and only if it belongs to the A_p-class. Moreover, it can be shown that a non-decreasing weight v from the class M_p also belongs to the C_p-class of Calderón introduced in [1] as the family of those weights v on $(0, +\infty)$ for which both the operator A and its adjoint A' are bounded on the space $L^p(v)$. Since

$$v \in A_p \Longrightarrow v \in A_{p-\varepsilon} \qquad \text{for some } \varepsilon \in (0, p-1),$$
$$v \in A_p \Longrightarrow v^{1+\varepsilon} \in A_p \qquad \text{for some } \varepsilon > 0,$$
$$v \in A_p \Longrightarrow v \in A_q \qquad \text{for all } q \in [p, +\infty),$$
$$v \in C_p \Longrightarrow v(x)x^\varepsilon \in M_p \quad \text{for some } \varepsilon > 0$$

(cf. [8] or [5] for the first three implications, and [1, Proposition 2.4] for the last one), Theorem 1.1 with $\rho = 0$ also follows from properties of weights $v \in A_p \cap C_p$.

On the other hand, there are weights in the M_p-class which satisfy (3) but which do not belong to $A_p \cap C_p$. A simple example is $v(t) = t^\beta$, $t > 0$, with $\beta \leq -1$. (Note that the weight $v(t) = t^\beta$, $t > 0$, with $\beta \in \mathbb{R}$, belongs to the A_p-class or the C_p-class if and only if $-1 < \beta < p-1$. However, v belongs to the M_p-class if and only if $\beta < p-1$. Moreover, it is clear that v satisfies (3) for any $\beta \in \mathbb{R}$.)

Remark 1.2. Denote by D_p, $1 < p < +\infty$, the subset of the M_p-class consisting of those weights v on $(0, +\infty)$ which satisfy condition (3). In particular, our results imply that the D_p-class possesses similar properties to those of the A_p-class. Namely,

$$v \in D_p \Longrightarrow v \in D_{p-\varepsilon} \qquad \text{for some } \varepsilon \in (0, p-1),$$
$$v \in D_p \Longrightarrow v^{1+\varepsilon} \in D_p \qquad \text{for some } \varepsilon > 0, \tag{4}$$
$$v \in D_p \Longrightarrow v \in D_q \qquad \text{for all } q \in [p, +\infty).$$

Moreover,

$$v \in D_p \Longrightarrow v(x)x^\varepsilon \in D_p \qquad \text{for some } \varepsilon > 0.$$

It is well-known that a weight $v \in A_p$ possesses a better integrability than that mentioned in the A_p-condition and that such a weight v satisfies a reverse Hölder inequality. Implication (4) shows that also a weight $v \in D_p$ possesses better integrability properties than those mentioned in the definition of the D_p-class (cf. (2)).

We have already mentioned that the weight $v \in M_p$ satisfying (3) with $\rho = 0$ belongs to the A_p-class. Since $v \in A_p$ implies that $v^{1-p'} \in A_{p'}$, the following two reverse Hölder inequalities hold for such a weight (cf. [8] or [5]):

$$\left(\frac{1}{r} \int_0^r [v(t)^{1-p'}]^{1+\delta} \, dt \right)^{1/(1+\delta)} \lesssim \frac{1}{r} \int_0^r v(t)^{1-p'} \, dt,$$

$$\left(\frac{1}{r} \int_0^r v(t)^{1+\delta} \, dt \right)^{1/(1+\delta)} \lesssim \frac{1}{r} \int_0^r v(t) \, dt,$$

for all $r > 0$ and some $\delta > 0$.

The next theorem shows that the former inequality remains true even when $\rho \geq 0$ in (3) while the latter inequality is then replaced by the reverse Hölder inequality for the weight v, the interval $(r, +\infty)$ and the measure $t^{-p} \, dt$.

Theorem 1.2 (see [9, Theorem 2]). *Let $1 < p < +\infty$ and let v be a weight on $(0, +\infty)$ such that (3) holds. Assume that the averaging operator*

$$A : L^p(v) \to L^p(v) \quad \text{is bounded.}$$

Then there is $\delta_0 > 0$ such that

$$\left(\frac{1}{r} \int_0^r [v(t)^{1-p'}]^{1+\delta} \, dt \right)^{1/(1+\delta)} \lesssim \frac{1}{r} \int_0^r v(t)^{1-p'} \, dt$$

and

$$\left(\frac{1}{r^{1-p}} \int_r^{+\infty} v(t)^{1+\delta} t^{-p} \, dt \right)^{1/(1+\delta)} \lesssim \frac{1}{r^{1-p}} \int_r^{+\infty} v(t) t^{-p} \, dt$$

for all $r > 0$ and $\delta \in [0, \delta_0)$.

Now we turn our attention to the case when the averaging operator A acts between two weighted Lebesgue spaces $L^p(v)$ and $L^q(w)$. By [2] (see also, e.g., [10] or [6]), if $1 < p \le q < +\infty$ and w, v are weights on $(0, +\infty)$, then the averaging operator $A : L^p(v) \to L^q(w)$ is bounded if and only if

$$\sup_{r>0} \left(\int_r^{+\infty} w(t)t^{-q}\, dt \right)^{1/q} \left(\int_0^r v(t)^{1-p'}\, dt \right)^{1/p'} < +\infty. \tag{5}$$

There is a natural question whether one can transfer results of Theorem 1.1 to such a case. If the weights v and w are tied up in a convenient way, then one can use the following theorem to this end.

Theorem 1.3 (see [9, Theorem 1]). *Let $1 < p \le q < +\infty$ and let v, w be weights on $(0, +\infty)$ such that (3) holds and*

$$[w(x)x]^{1/q} \approx [v(x)x]^{1/p} \quad \text{for all } x \in (0, +\infty). \tag{6}$$

Then the averaging operator

$$A : L^p(v) \to L^p(v) \quad \text{is bounded} \tag{7}$$

if and only if the operator

$$A : L^p(v) \to L^q(w) \quad \text{is bounded.}$$

Remark 1.3. Together with (7), the assumptions of Theorem 1.3 ensure that

$$\left(\int_r^{+\infty} w(t)t^{-q}\, dt \right)^{1/q} \left(\int_0^r v(t)^{1-p'}\, dt \right)^{1/p'} \approx 1 \quad \text{for all } r > 0,$$

which means that (w, v) is the optimal couple of weights for which (5) holds.
Note also that assumption (6) is satisfied when $w = v$ and $q = p$.

An analogue of Theorem 1.1 involving two weights reads as follows.

Theorem 1.4. *Let $1 < p \le q < +\infty$ and let v, w be weights on $(0, +\infty)$ such that (3) and (6) hold. Then the following statements are equivalent:*
(i) $A : L^p(v) \to L^q(w)$ is bounded;
(ii) $A : L^{p-\varepsilon}(v) \to L^{q-\varepsilon q/p}(w)$ is bounded for some $\varepsilon \in (0, p-1)$;
(iii) $A : L^p(v^{1+\varepsilon}(x)) \to L^q(w^{1+\varepsilon}(x)x^{\varepsilon(1-q/p)})$ is bounded for some $\varepsilon > 0$;
(iv) $A : L^p(v(x)x^\varepsilon) \to L^q(w(x)x^{\varepsilon q/p})$ is bounded for some $\varepsilon > 0$.
Moreover, if $A : L^p(v) \to L^q(w)$ is bounded, then $A : L^P(v) \to L^Q(w)$ is bounded, where $P \in [p, +\infty)$ and $Q = Pq/p$.

Proof. Theorem 1.4 is a consequence of Theorems 1.1 and 1.3. (Theorem 1.4 also follows from [11, Theorems 1-3].) $\qquad\qquad\square$

2. Notes on proofs

We conclude our contribution with notes on proofs of Theorems 1.1-1.3. One of the main ingredients of these proofs is the following lemma.

Lemma 2.1 (see [11, Lemma B]). *Let* $1 < p \le q < +\infty$ *and let* v, w *be weights on* $(0, +\infty)$ *such that* (3) *and* (6) *hold. Assume that the averaging operator* $A : L^p(v) \to L^q(w)$ *is bounded. Then there exists a positive constant* α_0 *such that*

$$\int_0^r [v(t)t^\alpha]^{1-p'} \, dt \approx [v(r)r^\alpha]^{1-p'} r$$

and

$$\int_r^{+\infty} w(t)t^{\alpha-q} \, dt \approx [w(r)r^{\alpha-q}] \, r$$

for all $r > 0$ *and* $\alpha \in [0, \alpha_0)$.

Note also that the essential role in the proof of Lemma 2.1 is played by the next two assertions, Lemma 2.2 and its dual version Lemma 2.3.

Lemma 2.2 (see [12, Lemma 2]). *Let* $\varphi : (0, +\infty) \to (0, +\infty)$. *If there is a constant* $c_0 > 0$ *such that*

$$\int_r^{+\infty} \varphi(t) \frac{dt}{t} \le c_0 \varphi(r) \quad \text{for all } r > 0, \tag{8}$$

then there exist positive constants α_1 *and* c *such that*

$$\int_r^{+\infty} \varphi(t)t^\alpha \frac{dt}{t} \le c\varphi(r)r^\alpha \quad \text{for all } r > 0 \text{ and } \alpha \in [0, \alpha_1).$$

Remark 2.1. In fact, it is proved in [12] that the last inequality holds for all $r > 0$ and some $\alpha > 0$. However, checking the proof of Lemma 2 in [12], one can see that Lemma 2.2 holds, e.g., with $\alpha_1 = (2c_0)^{-1}$ (and then one can put $c = 2c_0$), where c_0 is the constant in (8).

Lemma 2.3. *Let* $\varphi : (0, +\infty) \to (0, +\infty)$. *If there is a constant* $c_0 > 0$ *such that*

$$\int_0^r \varphi(t) \frac{dt}{t} \le c_0 \varphi(r) \quad \text{for all } r > 0,$$

then there exist positive constants β_1 *and* c *such that*

$$\int_0^r \varphi(t)t^{-\beta} \frac{dt}{t} \le c\varphi(r)r^{-\beta} \quad \text{for all } r > 0 \text{ and } \beta \in [0, \beta_1).$$

Proof. Lemma 2.3 can be obtained from Lemma 2.2 by the change of variables $t \mapsto t^{-1}$. $\qquad\square$

Acknowledgments

The research was supported by the grant no. 201/08/0383 of the Czech Science Foundation and by the Institutional Research Plan no. AV0 Z10190503 of the Academy of Sciences of the Czech Republic.

References

1. J. Bastero, M. Milman and F. J. Ruiz, *On the connection between weighted norm inequalities, commutators and real interpolation.* Mem. Amer. Math. Soc. **731**, Providence, 2001.
2. J. S. Bradley, *Hardy inequalities with mixed norms.* Canad. Math. Bull. **21** (1978), 405–408.
3. J. Cerdà and J. Martín, *Weighted Hardy inequalities and Hardy transforms of weights.* Studia Math. **139** (2000), 189–196.
4. D. Cruz-Uribe, SFO, *Piecewise monotonic doubling measures.* Rocky Mountain J. Math. **26** (1996), 545–583.
5. J. García-Cuerva and J. L. Rubio de Francia, *Weighted norm inequalities and related topics.* North-Holland Mathematics Studies **116**, North-Holland, Amsterdam, 1985.
6. A. Kufner and L.-E. Persson, *Weighted inequalities of Hardy type.* World Scientific Publishing Co., Inc., River Edge, NJ, 2003.
7. B. Muckenhoupt, *Hardy's inequality with weights.* Studia Math. **44** (1972), 31–38.
8. B. Muckenhoupt, *Weighted norm inequalities for the Hardy maximal function.* Trans. Amer. Math. Soc. **165** (1972), 207–226.
9. B. Opic, *The averaging integral operator between weighted Lebesgue spaces and reverse Hölder inequalities.* Preprint **187**, IM ASCR, Prague, 2009.(Accepted for publication in Complex Variables and Elliptic Equations.)
10. B. Opic and A. Kufner, *Hardy-type inequalities.* Pitman Research Notes in Mathematics Series **219**, Longman Scientific & Technical, Harlow, 1990.
11. B. Opic and J. Rákosník, *On weighted estimates for the averaging integral operator.* Preprint **186**, IM AS CR, Prague, 2009.
12. E. Nakai, *Hardy-Littlewood maximal operator, singular integral operators and the Riesz potentials on generalized Morrey spaces.* Math. Nachr. **166** (1994), 95–103.

IV.2. Dispersive equations

Organisers: F. Hirosawa, M. Reissig

The topic of this session was chosen with expectation towards the development of new approaches to dispersive equations and related topics, combining ideas of colleagues from different specialisations and countries. Indeed, 27 lectures were given in this session; 15 of them were 45 minutes talks and the other 12 lectures were 30 minutes. The lectures covered the theory of linear and nonlinear hyperbolic equations, Schrödingr equations and parabolic equations, in particular. Many lectures presented new remarkable results on equations of dispersive type. By joining this session, some colleagues found new ideas to develop their own research, some colleagues discovered a relation between their present research topics and another one, and some colleagues could start joint research projects with other participants. For this reason, we think that the session "Dispersive equations" was very successful.

The following colleagues presented lectures in Session IV.2: Q-H. Choi (Incheon), F. Colombini (Pisa), M. D'Abbicco (Bari), D. Del Santo (Trieste), M.R. Ebert (Sao Paulo), D. Fang (Hangzhou), A. Galstyan (Edinburg), V. Georgiev (Pisa), M. Ghisi (Pisa), M. Gobbino (Pisa), T. Herrmann (Freiberg), F. Hirosawa (Yamaguchi), T. Jung (Kunsan), L. Karp (Karmiel), H. Kubo (Tohoku), T. Matsuyama (Hiratsuka), K. Mochizuki (Tokyo), H. Nakazawa (Chiba), T. Nishitani (Osaka), R. Picard (Dresden), M. Pivetta (Trieste), M. Reissig (Freiberg), J. Saito (Tokyo), R. Suzuki (London, Tokyo), H. Uesaka (Tokyo), K. Yagdjian (Edinburg), B. Yordanov (Tennessee-Knoxville).

Some $L^p - L^q$ estimates for hyperbolic systems

M. D'Abbico and S. Lucente

Department of Mathematics, University of Bari,
Bari, 70125, Italy
E-mail: dabbicco@dm.uniba.it, lucente@dm.uniba.it
www.uniba.it

G. Taglialatela

Dipartimento di Scienze Economiche e Metodi Matematici, University of Bari,
Bari, 70124, Italy
E-mail: taglia@dse.uniba.it
www.uniba.it

The results here stated have been presented by the first author during the 7th ISAAC Congress. Details are given in Ref. 2 by the three authors.

We establish $L^p - L^q$ estimates for the solution of the strictly hyperbolic first order linear systems with bounded time dependent coefficients. In the equation setting, Reissig and others obtained such estimates by using WKB representation of the solutions. Here the crucial point is to find minimal assumptions on the coefficients of the system so that Reissig's approach still works for systems.

Keywords: Regularly hyperbolic systems; $L^p - L^q$ estimates.

1. Introduction

A great deal of work has been devoted to establish $L^p - L^q$ estimates for hyperbolic equations. This theory goes back to the Strichartz' works in the seventies (see Ref. 14), and for the time dependent coefficients case it develops up to the recent works by Reissig and others (See Ref. 7–12).

We consider the Cauchy Problem for an $M \times M$ system:

$$D_t U = \sum_{j=1}^{n} A_j(t)\, D_{x_j} U + B(t) U\,, \qquad U(0, x) = U_0(x)\,, \qquad (1)$$

with the standard notation $D_t = -i\partial_t$ and $D_{x_j} = -i\partial_{x_j}$. Here and in the following, $t \geq 0$. We assume that $A_j(t)$ and $B(t)$ are complex-valued matrices and $t \mapsto A_j(t)$ and $t \mapsto B(t)$ are bounded, smooth functions.

Denoting by $V(t,\xi) = \widehat{U}(t,\xi)$ the Fourier transform of U with respect to the x-variable, the Cauchy Problem (1) can be rewritten as

$$D_t V = A(t,\xi)\,V + B(t)V\,, \qquad V(0,\xi) = \widehat{U}_0(\xi)\,, \tag{2}$$

being $A(t,\xi) = \sum_{j=1}^{n} A_j(t)\xi_j$.

Assumption 1.1. *We assume that $A(t,\xi)$ is regularly hyperbolic, i.e. its eigenvalues $\tau_j(t,\xi)$ are real and there exists $c_0 > 0$ such that for any $(t,\xi) \in \mathbb{R}_+ \times \mathbb{R}^n$ and for any $j \neq k$, one has*

$$|\tau_k(t,\xi) - \tau_j(t,\xi)| \geq c_0\,|\xi|\,.$$

Following Reissig as in Ref. 9 we introduce two functions

$$g_\rho : [0,+\infty[\ \to [3^\rho,+\infty[\ ,\ \text{such that}\ g_\rho(t) = (\log(t+e^3))^\rho\,,$$
$$f_\rho : [0,+\infty[\ \to [e^3\,3^{-\rho},+\infty[\ ,\ \text{such that}\ f_\rho(t) = (t+e^3)\big(g_\rho(t)\big)^{-1}\,,$$

with $\rho \in \mathbb{R}$. We remark that g_ρ is strictly increasing provided that $\rho > 0$, whereas f_ρ is strictly increasing for any $\rho \leq 3$. By means of f_ρ one classifies the oscillating behavior of the coefficients.

Definition 1.1. Let $\rho \leq 3$ and $m \in \mathbb{R}$. We define:

$$T_\rho\{m\} = \left\{ a \in \mathcal{C}^\infty(\mathbb{R}_+) \ \middle|\ |D_t^k a(t)| \leq C_k\,(f_\rho(t))^{m-k}\ ,\ k \in \mathbb{N} \right\}.$$

We say that a matrix belongs to $T_\rho\{m\}$ if all its entries belong to $T_\rho\{m\}$.

It holds $T_\rho\{m_1\} \subset T_\rho\{m_2\}$ with $m_1 \leq m_2$. From now on, we fix $\gamma \in [0,1]$.

Assumption 1.2. *We assume that $A_j \in T_\gamma\{0\}$ for any $j = 1,\dots,n$.*

Assumption 1.3. *We assume that $B \in T_\gamma\{-1\}$ and there exists $c_1 > 0$ such that for any $t \in \mathbb{R}_+$, one has*

$$\left\| \int_0^t B(r)\,dr \right\| \leq c_1 g_\gamma(t)\,. \tag{3}$$

By using Proposition 6.4 in Ref. 6 one can see that Assumption 1.1 implies the existence of a diagonalizing matrix $N(t,\xi)$, i.e. for any $(t,\xi) \in \mathbb{R}_+ \times \mathbb{R}^n$ the matrix $N(t,\xi)\,A(t,\xi)\,N^{-1}(t,\xi)$ is diagonal.

Proposition 1.1. *One can take $N(t,\xi)$ such that*
(1) each entry is a smooth function in t, homogeneous of order zero in ξ,
(2) for any $\xi \in S_\xi^{n-1}$ one has $N(\cdot,\xi) \in T_\gamma\{0\}$,
(3) $c_2 := \inf_{(t,\xi)\in\mathbb{R}_+\times\mathbb{R}n}|\det N(t,\xi)| > 0$.

We need other technical assumptions on $N(t,\xi)$. The next one corresponds to a control of the interaction between the coefficients $A_j(t)$ and $B(t)$.

Assumption 1.4. *For any $(t,\xi) \in \mathbb{R}_+ \times \mathbb{R}^n$, we put*

$$\zeta(t,\xi) = \big(D_t N(t,\xi) + N(t,\xi) B(t)\big) N^{-1}(t,\xi). \tag{4}$$

We assume that there exists $c_3 > 0$ such that for any $(t,\xi) \in \mathbb{R}_+ \times \mathbb{R}^n$ and any $j = 1, \ldots, M$, one has

$$\left| \int_0^t \Im\big(\zeta_{jj}(r,\xi)\big) \, dr \right| \leq c_3. \tag{5}$$

We will see that $\zeta(\cdot,\xi) \in T_\gamma\{-1\}$ for any $\xi \in \mathbb{R}^n$. In particular, condition (5) holds true whenever $\Im\big(\zeta_{jj}(\cdot,\xi)\big) \in L^1(\mathbb{R}_+)$, for any $\xi \in \mathbb{R}^n$.

Assumption 1.5. *There exists $t_0 > 0$ such that*

$$c_4 := \inf \left\{ |\det A(t,\xi)| \, : \, (t,\xi) \in [t_0, +\infty) \times S^{n-1}_\xi \right\} \neq 0, \tag{6}$$

and the Gaussian curvature of

$$\Sigma_q(t) := \left\{ \xi \in \mathbb{R}^n \, : \, \frac{1}{t} \int_0^t \tau_q(s,\xi) \, ds = 1 \right\}, \qquad q = 1, \ldots, M, \tag{7}$$

is bounded away from zero uniformly with respect to $t \in [t_0, \infty)$ for any $q = 1, \ldots, M$.

Main result of this paper is the following.

Theorem 1.1. *Consider (1) a regularly hyperbolic Cauchy Problem, i.e. Assumption 1.1 is verified. Let Assumptions 1.2 and 1.3 be satisfied in correspondence of a fixed $\gamma \in [0,1]$. Suppose in addiction that Assumptions 1.4 and 1.5 hold true.*
Then there exists a constant $C = C(n,p) > 0$ such that for the solution $U \in C^\infty\big(\mathbb{R}_+ \times \mathbb{R}^n, \mathbb{C}^M\big)$ of (1), the following $L^p - L^q$ estimate holds:

$$\|U(t,\cdot)\|_{L^q} \leq C\,(1+t)^{-\frac{n-1}{2}\left(\frac{1}{p}-\frac{1}{q}\right)+s_0}\|U_0\|_{H^{N_p,p}}, \tag{8}$$

where $1/p + 1/q = 1$, $1 < p \leq 2$, $N_p \geq n(1/p - 1/q)$ and s_0 depends on γ:

- *if $\gamma = 0$, then $s_0 = 0$;*
- *if $\gamma \in (0,1)$, then for any $\epsilon > 0$ one can take $s_0 = \epsilon$ in (8); moreover, in such a case, $C = C(n,p,\epsilon)$;*
- *if $\gamma = 1$, then there exists a finite $s_0 > 0$ such that (8) holds.*

We remark that, in facts, if $\gamma = 1$ then (8) is a decay estimate if and only if $s_0 < \frac{n-1}{2}\left(\frac{1}{p} - \frac{1}{q}\right)$.

We can read the previous result as an analysis of the influence of the coefficients oscillations on the decay rate. Following Reissig, we see that the previous theorem gives the $L^p - L^q$ estimates for very slow ($\gamma = 0$), slow ($\gamma \in]0,1[$) and fast ($\gamma = 1$) oscillations of $A_j(t)$ and $B(t)$. Instead, in the system framework, it seems difficult to consider very fast oscillations. Roughly speaking this means that it is not clear how to apply the Floquet theory to show that Assumption 1.2 can not be removed in Theorem 1.1. Having in mind Remark 4.5 in Ref. 9, being (8) an extension of the decay estimates of the wave equations with bounded speed of propagation, we see that our result is related to the $L^p - L^q$ decay estimates for the wave-type equations with increasing time-coefficients. On the other hand, the H^s well posedness for such equations corresponds to the weakly hyperbolic theory and to a strictly hyperbolic theory with non Lipschitz-continuous coefficients. It arises the problem to get similar estimates for weakly hyperbolic systems or strictly hyperbolic systems with non-smooth coefficients.

Example 1.1. If A_j are constant matrices, then:

- Assumption 1.1 means that $A(\xi)$ is strictly hyperbolic;
- Assumption 1.2 is trivially satisfied for any γ, in particular for $\gamma = 0$;
- Assumptions 1.3 and 1.4 hold true for $\gamma = 0$ whenever $B(t) \in T_0\{-1\} \cap L^1(\mathbb{R}_+)$. Indeed, in this case $\zeta(t,\xi) = N(\xi)\,B(t)\,N^{-1}(\xi) \in L^1(\mathbb{R}_+)$.
- In Assumption 1.5, condition (6) becomes $\det A(\xi) \neq 0$ for any $\xi \in S^{n-1}$, whereas $\Sigma_q = \{\xi \in \mathbb{R}^n : \tau_q(\xi) = 1\}$ in (7) For a comprehensive analysis of the constant coefficients equations the reader can see Ref. 12.

Example 1.2. If A is a diagonal matrix, then Assumption 1.4 reduces to

$$\left| \int_0^t \left| \Im\big(b_{jj}(r)\big) \, dr \right| \leq C \,, \qquad j = 1, \ldots, M,$$

being $\zeta = B(t)$ in (4). In particular, if $B(t) \in L^1(\mathbb{R}_+) \cap T_0\{-1\}$, then Assumption 1.3 holds with $\gamma = 0$ and Assumption 1.4 is satisfied.

Example 1.3. A regularly hyperbolic 2×2 system where $A(t,\xi)$ has identically vanishing trace, satisfies Assumption 1.5; indeed, Lemma 3.1 in Ref. 8 is applicable since the eigenvalues $\tau_\pm$ are exactly given as square roots of a quadratic form in ξ, namely $\tau_\pm = \pm\sqrt{-\det A(t,\xi)}$.

We can apply Theorem 1.1 to some scalar regularly hyperbolic equations, establishing the following corollary; in particular, it covers some results obtained by Reissig for second order differential equations.

Corollary 1.1. *Let h be a **scalar** regularly hyperbolic operator of order m having the form*

$$h(t, D_t, D_x)u = D_t^m u - \sum_{j=1}^{m} a_j(t)\, D_t^{m-j}\, D_x^j u - \sum_{j+k=m-1} b_{j,k}(t)\, D_t^j\, D_x^k u\,.$$

The principal symbol of h can be written as:

$$P(t, \tau, \xi) = \tau^m - \sum_{j=1}^{m} a_j(t)\,\tau^{m-j}\,\xi^j = \prod_{j=1}^{m} \left(\tau - \tau_j(t)\xi\right).$$

For fixed $\gamma \in [0,1]$, assume that

- *$a_j \in T_\gamma\{0\}$ for any $j = 1, \ldots, m$;*
- *$b_{j,k} \in T_\gamma\{-1\} \cap L^1$ for any $j + k = m - 1$;*
- *$\inf_{t \geq t_0} |a_m(t)| \neq 0$ for some $t_0 > 0$;*
- *there exists $C > 0$ such that for any $t \geq 0$,*

$$\left| \int_0^t \partial_s \tau_j(s) \sum_{l \neq j} \frac{1}{\tau_l(s) - \tau_j(s)}\, d\,s \right| \leq C \quad j = 1, \ldots, m. \tag{9}$$

Let $u : \mathbb{R}_+ \times \mathbb{R}^n \to \mathbb{R}$ be the solution of the Cauchy Problem

$$\begin{cases} h(t, D_t, D_x)u = 0\,, & (t, x) \in \mathbb{R}_+ \times \mathbb{R}\,, \\ D_t^j u(0, x) = u_j(x)\,, & x \in \mathbb{R}\,,\ j = 0, \ldots, m - 1\,. \end{cases}$$

Denoting by $U_0(x) := \left(D_x^{m-1} u_0\,, D_x^{m-2} u_1\,, \ldots, D_x u_{m-2}\,, u_{m-1} \right)^T$ one has

$$\|U(t, \cdot)\|_{L^2} \leq C\,(1 + t)^{s_0}\|U_0\|_{L^2}\,,$$

with $s_0 = s_0(\gamma) \geq 0$ given by Theorem 1.1 and

$$U = \left(D_x^{m-1} u\,, D_t D_x^{m-2} u\,, \ldots, D_t^{m-2} D_x u\,, D_t^{m-1} u \right)^T.$$

For proving Theorem 1.1 one divides the phase space $[0, +\infty[\, \times \mathbb{R}^n$ in two zones respectively named *pseudo-differential zone* and *hyperbolic zone*: for a given $N > 0$ and $\rho \in [0, 3]$ one defines

$$Z_{pd}(N, \rho) = \left\{ (t, \xi) \in [0, +\infty[\, \times \mathbb{R}^n \ \Big|\ |\xi|\, f_\rho(t) \leq N \right\},$$

$$Z_{hyp}(N, \rho) = \left\{ (t, \xi) \in [0, +\infty[\, \times \mathbb{R}^n \ \Big|\ |\xi|\, f_\rho(t) > N \right\}.$$

In both zones we derive some properties for the fundamental solution of (2). One can use the boundedness of $A_j(t)$ and the Assumption 1.3, to directly derive an estimate for the fundamental solution of (2) in the *pseudo-differential zone.*

In the *hyperbolic zone,* as usual for strictly hyperbolic systems, one can reduce the principal part of the system in diagonal form. Thanks to Assumption 1.4 on the diagonalizer, the lower order term of this system has sufficiently good decay properties as $t \to \infty$, and one can perform successive diagonalizations. In order to derive an estimate for the fundamental solution of (2) in the *hyperbolic zone,* we divide it into the *oscillating subzone* and the *regular subzone,*

$$Z_{osc}(N,\gamma) = \left\{ (t,\xi) \in [0,+\infty[\times \mathbb{R}^n \ \middle| \ N < |\xi| \, f_\gamma(t), \ |\xi| \, f_{2\gamma}(t) \leq 2\,N \right\},$$

$$Z_{reg}(N,\gamma) = \left\{ (t,\xi) \in [0,+\infty[\times \mathbb{R}^n \ \middle| \ |\xi| \, f_{2\gamma}(t) > 2\,N \right\};$$

then we obtain estimates for the fundamental solution in both these subzones, and we use these to achieve a WKB representation.

We prove that if $V(t,\xi) = (v_1(t,\xi), \ldots, v_M(t,\xi))$ is the solution of (2) with initial data $U_0(x) = (\varphi_1(x), \ldots, \varphi_M(x))$, then there exist $b_{k,\ell}^{(j)} : \mathbb{R}_+ \times \mathbb{R}^n \to \mathbb{C}$ with $j, k, \ell = 1, \ldots, M$, such that

$$v_j(t,\xi) = \sum_{k,\ell=1}^{M} b_{k\ell}^{(j)}(t,\xi) \exp\left(i \int_0^t \tau_k(s,\xi) d s \right) \widehat{\varphi_\ell}(\xi), \qquad (10)$$

and the coefficients satisfy the following estimates:

$$|b_{k\ell}^{(j)}(t,\xi)| \leq C_{N,\gamma} \exp\left(C_{N,\gamma} g_\gamma(t) \right) \qquad\qquad (t,\xi) \in Z_{pd} \cup Z_{osc},$$

$$|\partial_\xi^\alpha b_{k\ell}^{(j)}(t,\xi)| \leq C_{N,\gamma} |\xi|^{-|\alpha|} g_{2|\alpha|\gamma}(t) \exp\left(C_{N,\gamma} g_\gamma(t) \right) \qquad (t,\xi) \in Z_{reg}, |\alpha| \leq L,$$

for sufficiently large $L \in \mathbb{N}$.

Thanks to this representation in the *pseudo-differential zone* and in the *oscillating subzone,* Theorem 1.11 in Ref. 3 comes directly into play and gives the desired estimate. The same approach works in the *regular subzone* for small time. In the *regular subzone* for large time we need a Paley Littlewood partition of unity and we can apply stationary phase methods. In particular we separately derive $L^1 - L^\infty$ and $L^2 - L^2$ estimates, obtaining $L^p - L^q$ ones by interpolation arguments. The $L^2 - L^2$ estimates easily follow from Lemma 3 in Ref. 1, whereas to get $L^1 - L^\infty$ estimates, we use a Littman-type lemma (see Ref. 5). Indeed, we can apply Theorem 4.1 in Ref. 10 for deriving an $L^1 - L^\infty$ estimate combining Assumption 1.5, the boundedness of $A_j(t)$ together with the above estimates for the coefficients $b_{k\ell}$.

Gluing together the obtained estimates, the conclusion of the proof of Theorem 1.1 follows from

$$g_{2L\gamma}(t) \exp\left(Cg_\gamma(t)\right) \lesssim (1+t)^{s_0(\gamma)},$$

with $s_0(0) = 0$, $s_0(\gamma)$ arbitrary small for $\gamma \in]0,1[$ and $s_0(1) > 0$.

Remark 1.1. After the paper in Ref. 2 was accepted, we became aware that M. Ruzhansky and J. Wirth gain some results on the special case $\gamma = 0$ of Theorem 1.1, in a paper appeared in Ref. 13.

References

1. Brenner P., *On $L_p - L_q$ estimates for the wave equation.*, Math. Z. **145** (1975) 251–254.
2. M. D'Abbicco, S. Lucente and G. Taglialatela, *$L^p - L^q$ Estimates for Regularly Hyperbolic Systems*, in *Advances in Diff. Eq.* **14**, 2009, pp. 801–834.
3. L. Hormander, *Estimates for translation invariant operators in L_p spaces*, in *Acta Math.* **104**, 1960, pp. 93–140.
4. E. Jannelli, *On the symmetrization of the principal symbol of hyperbolic equations*, in *Comm. Partial Differential Equations* **14**, 1989, pp. 1617–1634.
5. W. Littman, *Fourier transforms of surface carried measures and differentiability of surface averages*, in *Bull. Amer. Math. Soc.* **69**, 1963, pp. 766–770.
6. S. Mizohata, *The Theory of Partial Differential Equations*, (Cambridge Univ. Press, London, 1973).
7. M. Reissig, On $L^p - L^q$ estimates for solutions of a special weakly hyperbolic equation, in *Proc. Conference on Nonlinear Evolution Equations and Infinite-dimensional Dynamical Systems* (Shanghai, 1995).
8. M. Reissig, *$L^p - L^q$ decay estimates for wave equation with bounded time dependent coefficients*, in *Dispersive nonlinear problems in mathematical physics, Quad. Mat., 15, Dept. Math., Seconda Univ. Napoli* (Caserta, 2004)
9. M. Reissig and J. Smith, *$L^p - L^q$ estimate for wave equation with bounded time dependent coefficient*, in *Hokkaido. Math. J.* **34**, 2005, pp. 541–586.
10. M. Reissig and K. Yagdjian, *$L^p - L^q$ decay estimates for the solutions of strictly hyperbolic equations of second order with increasing in time coefficients*, in *Math. Nachr.* **214**, 2000, pp. 71–104.
11. M. Reissig and K. Yagdjian, *About the influence of oscillations on Strichartz-type decay estimates*, in *Partial differential operators, Rend. Sem. Mat. Univ. Politec. Torino 58* (Torino, 2000).
12. M. Ruzhansky and J. Smith, *Dispersive and Strichartz estimates for hyperbolic equations with constant coefficients, MSJ Memoirs* **22**, (Mathematical Society of Japan, Tokyo, 2010).
13. M. Ruzhansky and J. Wirth, *Dispersive estimates for t-dependent hyperbolic systems*, in *Rend.Sem.Mat.Univ.Pol.Torino* **64**, 2008, pp. 339–349.
14. R. S. Strichartz, *A priori estimates for the wave equations and some applications*, in *J. Functional Analysis* **5**, 1970, pp. 218–235.

The wave equation in the Einstein and de Sitter spacetime

A. Galstyan

Department of Mathematics, University of Texas-Pan American,
Edinburg, TX, 78539, USA
E-mail: agalstyan@utpa.edu
http://www.math.utpa.edu/agalstyan

T. Kinoshita

Institute of Mathematics, University of Tsukuba,
Tsukuba Ibaraki 305-8571, Japan
E-mail: kinosita@math.tsukuba.ac.jp

K. Yagdjian

Department of Mathematics, University of Texas-Pan American,
Edinburg, TX, 78539, USA
E-mail: yagdjian@utpa.edu

We consider the wave propagating in the the Einstein & de Sitter spacetime. The covariant d'Alambert's operator in the Einstein & de Sitter spacetime belongs to the family of the non-Fuchsian partial differential operators. We introduce the initial value problem for this equation and give the explicit representation formulas for the solutions.

Keywords: Wave equation; Einstein & de Sitter model; non-Fuchsian equations.

1. Introduction

The current note is concerned with the wave propagating in the universe modeled by the cosmological models with expansion. We are motivated by the significant importance of the solutions of the partial differential equations arising in the cosmological problems for our understanding of the universe. While there exists extensive literature on the hyperbolic equations, the question of initial value problems for the wave equation in the curved spaces with singularities, and, in particular, in the Einstein & de Sitter spacetime, which are well posed and preserve many features of the classical waves, remains unresolved.

The Einstein & de Sitter model of the universe is the simplest non-empty expanding model with the line-element

$$ds^2 = -dt^2 + a_0^2 t^{4/3} \left(dx^2 + dy^2 + dz^2 \right)$$

in comoving coordinates.[6] It was first proposed jointly by Einstein & de Sitter (the EdeS model).[5] The observations of the microwave radiation fit in with this model.[4] The result of this case also correctly describes the early epoch, even in a universe with curvature different from zero.[1] Recently it was used in Ref. 11 to study cosmological black holes. The covariant d'Alambert's operator in the Einstein & de Sitter spacetime is

$$\Box_g \psi = -\left(\frac{\partial}{\partial t} \right)^2 \psi + t^{-4/3} \sum_{i=1,2,3} \left(\frac{\partial}{\partial x^i} \right)^2 \psi - \frac{2}{t} \frac{\partial}{\partial t} \psi \,.$$

Consequently, the covariant wave equation with the source term f written in the coordinates is

$$\left(\frac{\partial}{\partial t} \right)^2 \psi - t^{-4/3} \sum_{i=1,2,3} \left(\frac{\partial}{\partial x^i} \right)^2 \psi + \frac{2}{t} \frac{\partial}{\partial t} \psi = f \,. \tag{1}$$

The last equation belongs to the family of the non-Fuchsian partial differential equations. There is very advanced theory of such equations (see, e.g., Ref. 8 and references therein). In this note we investigate the initial value problem for this equation and give the representation formulas for the solutions with any dimension $n \in \mathbb{N}$ of the spatial variable $x \in \mathbb{R}^n$.

The equation (1) is strictly hyperbolic in the domain with $t > 0$. On the hypersurface $t = 0$ its coefficients have singularities that make the study of the initial value problem difficult. Then, the speed of propagation is equal to $t^{-\frac{2}{3}}$ for every $t \in \mathbb{R} \setminus \{0\}$. The equation (1) is not Lorentz invariant, which brings additional difficulties.

The classical works on the Tricomi and Gellerstedt equations (see, e.g, Ref. 2) appeal to the singular Cauchy problem for the Euler-Poisson-Darboux equation,

$$\Delta u = u_{tt} + \frac{c}{t} u_t \,, \qquad c \in \mathbb{C} \,, \tag{2}$$

and to the Asgeirsson mean value theorem when handling a high-dimensional case. Here Δ is the Laplace operator on the flat metric, $\Delta := \sum_{i=1}^n \frac{\partial^2}{\partial x_i^2}$.

We use the approach suggested in Ref. 12 and reduce the problem for equation (1) to the Cauchy problem for the free wave equation in Minkowski spacetime. More precisely, in the present note we utilize the solution $v = v(x, t; b)$ to the Cauchy problem

$$\begin{cases} v_{tt} - \triangle v = 0, & t > 0, \ x \in \mathbb{R}^n, \\ v(x, 0) = \varphi(x, b), & v_t(x, 0) = 0, \qquad x \in \mathbb{R}^n, \end{cases} \tag{3}$$

with the parameter $b \in B \subseteq \mathbb{R}$. We denote that solution by $v_\varphi = v_\varphi(x, t; b)$. In the case of function φ independent of parameter, we skip b and simply write $v_\varphi = v_\varphi(x, t)$. There are well-known explicit representation formulas for the solution of the last problem.

In particular, if f is independent of t, then $v_f(x, t; b)$ does not depend on b and we briefly write $v_f(x, t)$.

The straightforward application of the formulas obtained in Ref. 12 to the Cauchy problem for equation (1) decidedly does not work, but it reveals a surprising link to the Einstein & de Sitter spacetime. To demostrate that link we note that the "principal part" of equation (1) belongs to the family of the Tricomi-type equations (in the case of odd l it is Gellerstedt equation): $u_{tt} - t^l \triangle u = 0$, where $l \in \mathbb{N}$. According to Ref. 12 the solution to the Cauchy problem

$$u_{tt} - t^l \triangle u = f(x, t), \quad u(x, 0) = \varphi_0(x), \quad u_t(x, 0) = \varphi_1(x),$$

with the smooth functions f, φ_0, and φ_1, can be represented as follows:

$$\begin{aligned} u(x, t) = {}& 2^{2-2\gamma} \frac{\Gamma(2\gamma)}{\Gamma^2(\gamma)} \int_0^1 (1 - s^2)^{\gamma-1} v_{\varphi_0}(x, \phi(t)s)ds \tag{4} \\ & + t 2^{2\gamma} \frac{\Gamma(2 - 2\gamma)}{\Gamma^2(1 - \gamma)} \int_0^1 (1 - s^2)^{-\gamma} v_{\varphi_1}(x, \phi(t)s)ds \\ & + 2c_k \int_0^t db \int_0^{\phi(t) - \phi(b)} dr\, E(r, t; 0, b) v_f(x, r; b), \quad x \in \mathbb{R}^n, \ t > 0, \end{aligned}$$

with the kernel

$$E(r, t; 0, b) := ((\phi(b) + \phi(t))^2 - r^2)^{-\gamma} F\left(\gamma, \gamma; 1; \frac{(\phi(t) - \phi(b))^2 - r^2}{(\phi(t) + \phi(b))^2 - r^2}\right). \tag{5}$$

Here $F(\gamma, \gamma; 1; \zeta)$ is the hypergeometric function, while $k := l/2$, $\phi(t) := \frac{t^{k+1}}{k+1}$, $\gamma := \frac{k}{2k+2}$, and $c_k = (k+1)^{-\frac{k}{k+1}} 2^{-\frac{1}{k+1}}$.

Suppose now that we are looking for the simplest possible kernel $E(r, t; 0, b)$ (5) of the last integral transform. In the hierarchy of the hypergeometric functions the simplest one, that is different from the constant, is a linear function. That simplest function $F(a, b; 1; \zeta)$ has the parameters

$a = b = -1$ and coincides with $1 + \zeta$. The parameter l leading to such function $F(-1, -1; 1; \zeta) = 1 + \zeta$ is exactly the exponent $l = -4/3$ of the wave equation (and of the metric tensor) in the Einstein & de Sitter spacetime.

It is evident that the first term of the representation (4), as it is written, is meaningless if $\gamma = -1$. This indicates the fact that the Cauchy problem is not well-posed for the equation with $l = -4/3$. The main result of this paper is the following theorem, which shows how the "lower order term" of the equation (1) affects the Cauchy problem.

Theorem 1.1. *Assume that $\varphi_0, \varphi_1 \in C^\infty(\mathbb{R}^n)$, $f(x, t) \in C^\infty(\mathbb{R}^n \times (0, \infty))$, and that with some $\varepsilon > 0$ one has*

$$|\partial_x^\alpha f(x, t)| + |t \partial_t \partial_x^\alpha f(x, t)| \le C_\alpha t^{\varepsilon - 2} \text{ for all } x \in \mathbb{R}^n, \text{ and for all small } t > 0,$$

and for every α, $|\alpha| \le [(n + 1)/2]$. Then the solution $\psi = \psi(x, t)$ to

$$\begin{cases} \psi_{tt} - t^{-4/3} \triangle \psi + 2t^{-1} \psi_t = f(x, t), & t > 0, \ x \in \mathbb{R}^n, \\ \lim_{t \to 0} t\psi(x, t) = \varphi_0(x), \\ \lim_{t \to 0} \left(t\psi_t(x, t) + \psi(x, t) + 3t^{-1/3} \triangle \varphi_0(x) \right) = \varphi_1(x), \ x \in \mathbb{R}^n, \end{cases} \tag{6}$$

is given by

$$\psi(x, t) = \frac{3}{2} t^2 \int_0^1 db \int_0^{1 - b^{1/3}} ds\, bv_f(x, 3t^{1/3}s; tb)\left(1 + b^{2/3} - s^2\right) \tag{7}$$
$$+ t^{-1} v_{\varphi_0}(x, 3t^{1/3}) - 3t^{-2/3}\left(\partial_t v_{\varphi_0}\right)(x, 3t^{1/3})$$
$$+ \frac{3}{2} \int_0^1 v_{\varphi_1}(x, 3t^{1/3}s)(1 - s^2)ds\,.$$

The theorem gives a structure of singularity of the solution at the point $t = 0$, which hints at the proper initial conditions which have to be prescribed for the solution. The initial conditions prescribed in the previous theorem are the Cauchy conditions modified to the so-called *weighted initial conditions* in order to adjust them to the equation. For the Euler-Poisson-Darboux equation (2) one can find such weighted initial conditions, for instance, in book Ref. 10. The existence and uniqueness of the solutions for the initial value problem with the weighted initial conditions for the equation (2) with the time-dependent c are proved in Ref. 3 by application of the Fourier transform in x-variable.

The EdeS model recently became a focus of interest for an increasing number of authors. (See, e.g., Ref. 6, Ref. 7, Ref. 9, Ref. 11 and references therein.) We believe that the initial value problem and the explicit representation formulas obtained in our paper fill the gap in the existing literature on the wave equation in the EdeS spacetime.

2. Outline of the proof of Theorem 1.1

If we denote $\mathcal{L} := \partial_t^2 - t^{-\frac{4}{3}} \triangle + 2t^{-1}\partial_t$, $\mathcal{S} := \partial_t^2 - t^{-\frac{4}{3}}\triangle$, then we can easily check for $t \neq 0$ the following operator identity $t^{-1} \circ \mathcal{S} \circ t = \mathcal{L}$. The last equation suggests a change of unknown function ψ with u such that $\psi = t^{-1}u$. Then the problem for u is as follows:

$$\begin{cases} u_{tt} - t^{-4/3} \triangle u = g(x,t), & t > 0, \ x \in \mathbb{R}^n, \\ \lim_{t \to 0} u(x,t) = \varphi_0(x), \ \lim_{t \to 0}\left(u_t(x,t) + 3t^{-1/3} \triangle \varphi_0(x)\right) = \varphi_1(x), \ x \in \mathbb{R}^n, \end{cases} \tag{8}$$

where $g(x,t) = tf(x,t)$. Thus, it is enough to find a representation of the solution of the last problem. We discuss it in three cases of: $(\mathbf{f})$ with $\varphi_0 = \varphi_1 = 0$; $(\boldsymbol{\varphi_0})$ with $f = 0$ and $\varphi_1 = 0$; $(\boldsymbol{\varphi_1})$ with $f = 0$ and $\varphi_0 = 0$.

The operator $\mathcal{S}$ belongs to the family of the Tricomi-type operators $\mathcal{T} := \partial_t^2 - t^l\triangle$, where $l \in \mathbb{R}$. The Cauchy problem for such operators with positive l is well investigated (see, e.g., Ref. 13 and references therein). The fundamental solutions of the operator and the representation formulas for the solutions of the Cauchy problem are given in Ref. 12. The results of Ref. 12 allow us to write an *ansatz* for the solutions of the equation of (8). Here we use this *ansatz* to consider the weighted initial value problem (8) with data on the plane $t = 0$ where coefficients and source term are singular. As we already emphasized, it is interesting that the case of $l = -4/3$, that is the case of EdeS spacetime, is an exceptional case in the sense that it simplifies the Gauss' hypergeometric function $F(\gamma, \gamma; 1; z)$ appearing in the fundamental solutions constructed in Ref. 12, to the linear function $F(-1, -1; 1; z) = 1 + z$.

The case of (f). Assume that f satisfies assumptions of the Theorem 1.1 . We have to prove that the solution to

$$\mathcal{L}\psi = f, \ t > 0, \ x \in \mathbb{R}^n \ ; \ \lim_{t \to 0} \psi(x,t) = 0, \ \lim_{t \to 0} \psi_t(x,t) = 0, \ x \in \mathbb{R}^n,$$

is given by

$$\psi(x,t) = \frac{3}{2}t^2 \int_0^1 db \int_0^{1-b^{1/3}} ds \, bv_f(x, 3t^{1/3}s; tb)\left(1 + b^{2/3} - s^2\right). \tag{9}$$

Here the function $v_f(x, r; t)$ is given by well-known explicit representation formulas for the solution of the problem (3) with $\varphi(x, b) = f(x, b)$.

It is sufficient to check the properties of the function $u = u(x,t) = t\psi(x,t)$, which solves the equation $\mathcal{S}u = g$ with $g(x,t) = tf(x,t)$. Hence, we can restrict ourselves to the representation

$$u(x,t) = \frac{1}{18} \int_0^t dl \int_0^{\phi(t)-\phi(l)} dr\, v_g(x,r;l)\left(\phi^2(t) + \phi^2(l) - r^2\right) \quad (10)$$

and take into account the identity $v_g(x,r;b) = bv_f(x,r;b)$. Here $\phi(t) := 3t^{1/3}$. First we prove that the integral is convergent and that it represents the $C^2(\mathbb{R}^n \times (0,\infty))$-function. It is evident that $v(x,r;b) \in C^2(\mathbb{R}^n \times (0,\infty))$ and $|v_g(x,\phi(t)s;tb)| \le Ct^{-1+\varepsilon}b^{-1+\varepsilon}$. Then we use the last inequality and representation (10) to check initial conditions $\lim_{t\to 0} u(x,t) = 0$, $\lim_{t\to 0} u_t(x,t) = 0$ and verify the equation. Thus, for this case the theorem is proven.

The case of (φ_0). One can find in the literature different approaches for the construction of the null-solutions of the Fuchsian and non-Fuchsian partial differential equations. (See, e.g., Ref 8.) In the following lemma we construct a null-solution of the equation of (6) with the prescribed initial conditions and with $f = 0$. We believe that this result is new and it does not follow directly from the mentioned above.

Lemma 2.1. *The function* $u(x,t) = v_{\varphi_0}(x, 3t^{1/3}) - 3t^{1/3}(\partial_r v_{\varphi_0})(x, 3t^{1/3})$ *solves the problem*

$$\begin{cases} \mathcal{S}u = 0, \quad x \in \mathbb{R}^n, \quad t > 0, \\ \lim_{t\to 0} u(x,t) = \varphi_0(x), \quad \lim_{t\to 0}\left(u_t(x,t) + 3t^{-1/3}\triangle\varphi_0(x)\right) = 0, \quad x \in \mathbb{R}^n. \end{cases}$$

Here $v_\varphi(x, 3t^{1/3})$ *is a value of the solution* $v(x,r)$ *to the Cauchy problem for the wave equation,* $v_{rr} - \triangle v = 0$, $v(x,0) = \varphi(x)$, $v_t(x,0) = 0$, *taken at the point* $(x,r) = (x, 3t^{1/3})$.

We prove lemma by straightforward calculations.

Corollary 2.1. *The function* $\psi = t^{-1}u(x,t)$ *solves the problem (6) with* $\varphi_1 = 0$ *and without source term* $f = 0$, *that is*

$$\begin{cases} \psi_{tt}(x,t) - t^{-4/3}\triangle\psi(x,t) + 2t^{-1}\psi_t(x,t) = 0, \\ \lim_{t\to 0} t\psi(x,t) = \varphi_0, \quad \lim_{t\to 0}\left(t\psi_t(x,t) + \psi(x,t) + 3t^{-\frac{1}{3}}\triangle\varphi_0(x)\right) = 0, \quad x \in \mathbb{R}^n. \end{cases}$$

In particular, the corollary shows that for the given dimension $n \in \mathbb{N}$ Huygens's Principle is valid for some particular waves propagating in the Einstein & de Sitter model of the universe if and only if it is valid for the waves propagating in Minkowski spacetime (cf. with Ref. 12).

The case of (φ_1). In this case $f = 0$ and $\varphi_0 = 0$.

Lemma 2.2. *For $\varphi_1 \in C_0^\infty(\mathbb{R}^n)$ the function*

$$u(x,t) = t\frac{3}{2} \int_0^1 v_{\varphi_1}(x, \phi(t)s)(1 - s^2)ds, \quad x \in \mathbb{R}^n, \ t > 0, \qquad (11)$$

solves the problem

$$\mathcal{S}u = 0, \ x \in \mathbb{R}^n, \ t > 0 \, ; \lim_{t \to 0} u(x,t) = 0, \ \lim_{t \to 0} u_t(x,t) = \varphi_1(x), \ x \in \mathbb{R}^n.$$

Here $v_\varphi(x, \phi(t)s)$ is a value of the solution $v(x,r)$ to the Cauchy problem for the wave equation, $v_{rr} - \triangle v = 0$, $v(x,0) = \varphi(x)$, $v_t(x,0) = 0$, taken at the point $(x,r) = (x, \phi(t)s)$, while $\phi(t) = 3t^{1/3}$.

We prove lemma by straightforward calculations.

Corollary 2.2. *The function $\psi = t^{-1}u(x,t)$ solves the problem (6) with $\varphi_0 = 0$ and without source term $f = 0$, that is*

$$\begin{cases} \psi_{tt} - t^{-4/3} \triangle \psi + 2t^{-1}\psi_t = f(x,t), & t > 0, \ x \in \mathbb{R}^n, \\ \lim_{t \to 0} t\psi(x,t) = 0, & \lim_{t \to 0} \left(t\psi_t(x,t) + \psi(x,t)\right) = \varphi_1(x), \ x \in \mathbb{R}^n, \end{cases}$$

The last corollary completes the proof of Theorem 1.1. $\square$

Corollary 2.2 shows that for all $n \in \mathbb{N}$ Huygens's Principle is not valid for the waves propagating in the Einstein & de Sitter model of the universe.

References

1. T.-P. Cheng, *Relativity, Gravitation And Cosmology: A Basic Introduction* (Oxford University Press, Oxford, New York, 2005) .
2. S. Delache, J. Leray, *Bull. Soc. Math. France* **99**, 313 (1971).
3. D. Del Santo, T. Kinoshita, M. Reissig, *Rend. Istit. Mat. Univ. Trieste* **39**, 141 (2007) .
4. P. A. M. Dirac, *Proc. Roy. Soc. London Ser.* A **365** 19 (1979) .
5. A. Einstein, W. de Sitter, *Proc. Natn. Acad.Sci. U.S.A.* **18**, 213 (1932) .
6. G. Ellis, H. van Elst, *Cargèse Lectures 1998: Cosmological Models.*
7. O. Gron, S. Hervik, *Einstein's general theory of relativity: with modern applications in cosmology* (Springer-Verlag, New York, LLC, 2007) .
8. T. Mandai, *J. Math. Soc. Japan* **45**, 511 (1993) .
9. A. D. Rendall, *Partial differential equations in general relativity.* Oxford Graduate Texts in Mathematics, **16** (Oxford University Press, Oxford, 2008) .
10. M. M. Smirnov, *Equations of mixed type*, (AMS, Providence, R.I., 1978) .
11. J. Sultana, C. C. Dyer, *General Relativity and Gravitation* **37**, 1347 (2005) .
12. K. Yagdjian, *J. Differential Equations* **206**, 227 (2004) .
13. K. Yagdjian, *The Cauchy Problem for Hyperbolic Operators. Multiple Characteristics. Micro-Local Approach* (Akademie Verlag, Berlin, 1997) .

P-evolution operators with characteristics of variable multiplicity

T. Herrmann* and M. Reissig

*Fakultät für Mathematik und Informatik, TU Bergakademie Freiberg,
Freiberg, 09599, Germany
*E-mail: torsten.herrmann@math.tu-freiberg.de
www.mathe.tu-freiberg.de/ana/mitarbeiter/torsten-herrmann*

We are interested in H^∞ well-posedness of the Cauchy problem in 1-D for p-evolution equations with characteristics of variable multiplicity. Here of interest is the precise loss of derivatives of the solution and it's first derivative in time. We shall use among other things the theory of special functions and Floquet theory to prove optimality of our results.

Keywords: P-evolution equation; degenerate Cauchy problem; loss of regularity.

1. Introduction

A well-known result for the Cauchy problem for the wave equation is the L^2 well-posedness, that is, if $u_0 \in H^s$ and $u_1 \in H^{s-1}$, then the unique solution to

$$\partial_t^2 u - \partial_x^2 u = 0, \quad u(0,x) = u_0(x), \quad u_t(0,x) = u_1(x)$$

satisfies $u \in C([0,T], H^s) \cap C^1([0,T], H^{s-1})$. In general one cannot expect this L^2 well-posedness if we consider

$$\partial_t^2 u - \lambda^2(t)\partial_x^2 u = 0, \quad u(0,x) = u_0(x), \quad u_t(0,x) = u_1(x),$$

where the non-negative shape function λ may have zeros. This was shown in Taniguchi-Tozaki[1] for $\lambda(t) = t^l$ and in Aleksandryan[2] for $\lambda(t) = 1/t^2 \exp(-1/t)$. The present note is devoted to the following Cauchy problem of p-evolution type:

$$D_t^2 u - \lambda^2(t)b^2(t)D_x^{2p} - a_{p+k}(t)D_x^{p+k}u = 0,$$
$$u(0,x) = u_0(x), \quad u_t(0,x) = u_1(x),$$

with $1 \leq k < p$ and $p \in \mathbb{N}$. Here $\lambda \in C^2[0, T]$ is a shape function describing the degeneration of the Cauchy problem only in $t = 0$. The function $b \in C^2(0, T]$ describes the oscillating behavior of the coefficient. We are interested in results about the H^∞ well-posedness of this Cauchy problem with derivation of the sharp loss of regularity.

2. Model cases

In this section we consider generalizations of two model cases which are introduced in Taniguchi-Tozaki[1] and Aleksandryan.[2] The first one is of Taniguchi/Tozaki - type,

$$D_t^2 u - t^{2l} D_x^{2p} u = 0, \quad u(0, x) = u_0(x), \quad u_t(0, x) = u_1(x). \tag{1}$$

The second one is of Alexandrian - type,

$$D_t^2 u - \frac{\exp\left(-\frac{2}{t}\right)}{t^4} D_x^{2p} u = 0, \quad u(0, x) = u_0(x), \quad u_t(0, x) = u_1(x). \tag{2}$$

The special structure of the coefficients allows to derive explicit representations of the solutions by using theory of special functions. In Herrmann[3] the following theorems are proved:

Theorem 2.1. *If we consider the Cauchy problem (1) under the assumptions $u_0 \in H^s$ and $u_1 \in H^{s - \frac{p}{l+1}}$, we can assure that there exists a unique solution $u \in C([0, T], H^s) \cap C^1\left([0, T], H^{s - \frac{p(l+2)}{2(l+1)}}\right)$.*

Theorem 2.2. *We consider the Cauchy problem (2). If $u_0 \in H^s$ and $u_1 \in \log\langle D_x\rangle H^s$, we can assure that there exists a unique solution $u \in C([0, T], H^s) \cap C^1\left([0, T], \log\langle D_x\rangle H^{s - \frac{p}{2}}\right)$.*

Remark 2.1.

- From $u \in C([0, T], H^s)$ and $u_0 \in H^s$ we conclude that we do not have any loss of regularity for the solution of our Cauchy problem, but from the regularity of u and u_1 we obtain a loss of regularity for $D_t u$.
- To obtain no loss of regularity for the solution it is important to have a certain difference of regularity between the data u_0 and u_1, that is $\langle D_x\rangle^{\frac{p}{l+1}}$ in Theorem 2.1 and $\log\langle D_x\rangle$ in Theorem 2.2.

The results are sharp because we use in the proof the exact representation of the solution.

3. A model with very slow oscillations

Now we want to consider a Cauchy problem with very slow oscillations in the sense of Reissig/Yagdjian.[4] Let us consider the Cauchy problem

$$D_t^2 u - \lambda^2(t) D_x^{2p} u - a_{p+k}(t) D_x^{p+k} u = 0,$$
$$u(0, x) = u_0(x), \quad u_t(0, x) = u_1(x), \tag{3}$$

with $1 \le k < p$. In the whole paper $\Lambda(t) = \int_0^t \lambda(\tau) d\tau$ denotes a primitive of $\lambda(t)$. The conditions of the shape function and the term of lower order are given by the following assumptions:

$$\lambda(0) = 0, \quad \lambda'(t) > 0 \text{ for } t > 0,$$
$$c_1 \frac{\lambda(t)^2}{\Lambda(t)} \le \lambda'(t) \le c_2 \frac{\lambda(t)^2}{\Lambda(t)}, \quad \tfrac{1}{2} \le c_1 \le c_2, \tag{4}$$

$$|D_t^l a_{p+k}(t)| \le C \Big(\frac{\lambda(t)}{\Lambda(t)} \Big)^l \frac{\lambda^2(t)}{\Lambda(t)^{\frac{p-k}{p}}}, \quad l = 0, 1, 2, \quad k = 1, ..., p-1. \tag{5}$$

Theorem 3.1. *Let us consider the Cauchy problem (3). If the conditions (4) and (5) are fulfilled and if we choose the data*

$$u_0 \in H^s, \quad u_1 \in \Big(\Lambda^{-1} \Big(\frac{N}{\langle D_x \rangle^p} \Big) \Big)^{-1} H^s,$$

for a sufficiently large positive N, then there exists a unique solution

$$u(t, \cdot) \in H^s \text{ with } D_t u(t, \cdot) \in \sqrt{ \lambda \Big(\Lambda^{-1} \Big(\frac{N}{\langle D_x \rangle^p} \Big) \Big) } H^{s-p}.$$

Remark 3.1.

- In Theorem 3.1 we present the exact difference of regularity between the data to guarantee H^∞ well-posedness with no loss of regularity for the solution. But there is a loss of regularity for the first derivative in time of the solution.
- From Section 2 we know that this loss really appears in the cases of our examples of Taniguchi-Tozaki or Alexandrian-type. In this sense our result is sharp.
- This Cauchy problem allows to treat only very slow oscillations, this means, oscillations which have no essential influence on the regularity properties of the solution.
- It is no problem to extend the ideas of the proof to the Cauchy problem

$$D_t^2 u - \lambda^2(t) D_x^{2p} u - \sum_{k=1}^{p-1} a_{p+k}(t) D_x^{p+k} u = 0,$$
$$u(0, x) = u_0(x), \quad u_t(0, x) = u_1(x).$$

Sketch of the proof. For the proof we use WKB analysis in the extended phase space. We split the extended phase space into two zones. In the first zone, the so-called pseudo-differential zone, we use an appropriate micro-energy to estimate the fundamental solution in a very effective way. In the second zone, the so-called evolution zone, we define a suitable micro-energy, we use two steps of a diagonalization procedure in scales of symbols and estimate the remainder in an effective way. Finally, we estimate the fundamental solution by using the theory of Volterra integral equations of second kind. $\qquad\square$

4. A finite degenerate model with fast oscillations

4.1. *An at most loss of derivatives*

In this section we consider the Cauchy problem

$$D_t^2 u - \lambda^2(t)b^2(t)D_x^{2p}u - a_{p+k}(t)D_x^{p+k}u = 0,$$
$$u(0,x) = u_0(x), \quad D_t u(0,x) = u_1(x), \tag{6}$$

for $1 \le k < p$. For the shape function $\lambda \in C^2$ we need the assumptions (4) and the condition for finite degeneracy

$$Ct^l \le \lambda(t) \le C't^{l'}. \tag{7}$$

Furthermore we need the following assumptions for $b(t)$ and $a_{p+k}(t)$:

$$d_1 \le b(t) \le d_2 \qquad \text{with constants } 0 < d_1 \le d_2,$$

$$|D_t^i b(t)| \lesssim \left(\tfrac{\lambda(t)}{\Lambda(t)}\nu(t)\right)^i, \qquad i = 1, 2, \tag{8}$$

$$|D_t^i a_{p+k}(t)| \lesssim \frac{\lambda^2(t)}{\Lambda(t)^{\frac{p-k}{p}}}\nu(t)^{\frac{p-k}{p}} \left(\tfrac{\lambda(t)}{\Lambda(t)}\nu(t)\right)^i, \qquad i = 0, 1, 2,$$

with $\nu(t) = \left(\log \tfrac{1}{\Lambda(t)}\right)^\gamma$ and $\gamma \in [0,1]$.

Theorem 4.1. *We consider the Cauchy problem (6) under the assumptions (4), (7) and (8). For initial data $u_0 \in H^s$ and $u_1 \in \lambda^{-1}\left(\left(\tfrac{\Lambda}{\nu}\right)^{-1}\left(\tfrac{N}{\langle D_x\rangle^p}\right)\right) H^{s-p}$, with a fixed positive constant N, there exists a unique solution $u(t,\cdot) \in \exp\left(C_N \left(\log\langle D_x\rangle\right)^\gamma\right) H^s$ with $D_t u \in \sqrt{\lambda\left(\left(\tfrac{\Lambda}{\nu}\right)^{-1}\left(\tfrac{N}{\langle D_x\rangle^p}\right)\right)} \exp\left(C_N \left(\log\langle D_x\rangle\right)^\gamma\right) H^{s-p}.$*

Remark 4.1.

- We see that the problem is H^∞ well-posed with an *at most* loss of derivatives for the solution $\exp\left(C_N \left(\log\langle D_x\rangle\right)^\gamma\right)$. If $\gamma = 1$, then the upper bound for the loss is finite, if $\gamma \in (0,1)$ it is arbitrary small.

- To prove Theorem 4.1 we need the precise difference of regularity between the data.
- We get a stronger loss of derivatives for the first derivative in time of the solution. But this (possible) loss is still finite.

Sketch of the proof. Once again we use WKB analysis in the extended phase space. This time the splitting of the phase space also depends on the function $\nu = \nu(t)$. The separating line is given by the set of (t_ξ, ξ) satisfying $\Lambda(t_\xi)|\xi|^p = N\nu(t_\xi)$ for a sufficiently large constant N. With this, $a_{p+k}(t)\xi^{p+k}$ is dominant in the pseudo-differential zone and $\lambda^2(t)b^2(t)\xi^{2p}$ is dominant in the evolution zone. In the pseudo-differential zone it is sufficient to estimate the micro-energy using the Matrizant representation. In the evolution zone we have to apply two steps of diagonalization procedure. After the first step of diagonalization we have a decoupled system of first order disturbed by a remainder. In the second step of diagonalization we diagonalize a part of the remainder and keep the structure of the other parts. After that we can estimate the decoupled system and deal with the remainder as if it is just an error term. Combining the estimates of both zones completes the proof. $\qquad\square$

4.2. *Optimality of the result*

Let us consider the finite degenerate Cauchy problem

$$D_t^2 u - \lambda^2(t)D_x^{2p}u - a_{p+k}(t)D_x^{p+k}u = 0,$$
$$u(0,x) = u_0(x), \quad D_t u(0,x) = u_1(x), \tag{9}$$

for $1 \le k < p$. For simplicity we set $b(t) \equiv 1$. We choose $a_{p+k}(t)$ in a special way:

$$a_{p+k}(t) = \frac{\lambda^2(t)}{\Lambda(t)^{\frac{p-k}{p}}}\left(\log\frac{1}{\Lambda(t)}\right)^{\gamma\frac{p-k}{p}}\omega\left(\left(\log\frac{1}{\Lambda(t)}\right)^{\gamma+1}\right), \tag{10}$$

where ω is a positive, smooth, 1-periodic and non-constant function. Applying partial Fourier transformation with respect to x yields for $v(t,\xi) = F_{x\to\xi}(u(t,x))$ the following Cauchy problem

$$D_t^2 v - \lambda^2(t)\xi^{2p}v - a_{p+k}(t)\xi^{p+k}v = 0,$$
$$v(0,\xi) = v_0(\xi), \quad D_t v(0,\xi) = v_1(\xi). \tag{11}$$

We introduce the function $\widetilde{\lambda} = \widetilde{\lambda}(t,\xi)$ which is connected to the separating line defined in the proof of Theorem 4.1:

$$\widetilde{\lambda}(t,\xi) = \begin{cases} \lambda(t_\xi) & \text{for } t \le t_\xi, \\ \lambda(t) & \text{for } t \ge t_\xi. \end{cases}$$

Theorem 4.2. *Consider the Cauchy problem (11) with the special choice of (10). There exists a positive, smooth, non-constant and 1-periodic function $\omega = \omega(t) \in C^\infty[0,T]$ such that the inequality*

$$\|(\widetilde{\lambda}(t,\xi)\langle\xi\rangle^p v, D_t v)(t,\cdot)\|_{L^{2,s}}$$
$$\lesssim \|\exp(C(\log\langle\xi\rangle)^{\gamma_0})(\widetilde{\lambda}(0,\xi)\langle\xi\rangle^p v, D_t v)(0,\cdot)\|_{L^{2,s}} \tag{12}$$

does not hold for $\gamma_0 < \gamma$, where C is an arbitrary positive constant which is independent of $t \in [0,T]$.

Remark 4.2. In Section 4.1 we have shown that the inequality

$$\|(\widetilde{\lambda}(t,\xi)\langle\xi\rangle^p v, D_t v)(t,\cdot)\|_{L^{2,s}}$$
$$\lesssim \|\exp(C(\log\langle\xi\rangle)^{\gamma})(\widetilde{\lambda}(0,\xi)\langle\xi\rangle^p v, D_t v)(0,\cdot)\|_{L^{2,s}} \tag{13}$$

holds for a fixed $\gamma \in [0,1]$. We understand optimality in the sense that from Theorem 4.2 there exists no $\gamma_0 < \gamma$ satisfying this inequality.

We choose the special coefficient (10) with a function ω which gives the relation to Floquet theory, the main tool of the proof. After some transformations we arrive at the following basic Cauchy problem:

$$\frac{d}{d\tau} X(s_0 + \tau; s_0) = \begin{pmatrix} 0 & 1 \\ -\lambda_0 \widetilde{\omega}(s_0 + \tau) & 0 \end{pmatrix} X(s_0 + \tau; s_0),$$
$$X(s_0; s_0) = \begin{pmatrix} 1 & 0 \\ 0 & 1 \end{pmatrix}, \tag{14}$$

where λ_0 is chosen as in the following lemma.

Lemma 4.1 (Floquet Lemma). [5,6] *Let $\widetilde{\omega}(s)$ be a positive, smooth, 1-periodic and non-constant function, and let $X(s_0 + \tau; s_0)$ be the solution to the first order system (14). Then there exists a positive real number λ_0 such that $X(s_0 + 1; s_0)$ has the eigenvalues μ and μ^{-1} satisfying $|\mu| > 1$.*

Sketch of the proof to Theorem 4.2. First, we transform the Cauchy problem to a Hill's equation. Then we want to apply the Floquet Lemma 4.1. A propagator $X(s-1, s, \xi)$ can transport the solution $(w(s,\xi), w_t(s,\xi))^T$ of the Hill's equation to $(w(s-1,\xi), w_t(s-1,\xi))^T$. We can show, that this propagator $X(s-1, s, \xi)$ and also the propagator $X(s-j, s-j+1, \xi)$ for $j \le n$, n large, but not to large, satisfies the conditions of the Floquet Lemma 4.1. This means we can apply n propagators to $(w(s,\xi), w_t(s,\xi))^T$ to transport the solution to $(w(s-n,\xi), w_t(s-n,\xi))^T$. In every step the propagator has an eigenvalue $\mu_j > 1$. This leads to an estimate from below for the solution of the Hill's equation. Backward transformation implies the statement. $\qquad\square$

5. Concluding remarks and summary

In Section 3 we have shown that for the p-evolution problem with slow oscillations and a given difference of regularity for the data the Cauchy problem is H^∞ well-posed with no loss of regularity for the solution. We have also shown, that for the first derivative in time of the solution we have to expect a loss of derivatives. In Section 2 we have presented two examples for p-evolution problems with very slow oscillations. In both the finite and the infinite degenerate Cauchy problem we observed the sharpness of the statements from Theorem 3.1. In Section 4 we allowed also fast oscillations for the Cauchy problem. But we had to constrain ourselves to a finite degenerate Cauchy problem. We could show in Section 4.1 that the problem is H^∞ well-posed with an at most finite loss of derivatives for the solution. Once again it is important to have a certain difference of regularity for the data and once again we have a loss of derivatives for $D_t u(t, \cdot)$ which is stronger than the one for $u(t, \cdot)$.

Still open is the question if the statements from Section 4 are valid for Cauchy problems with a degeneracy of infinite order and with fast oscillations.

References

1. K. Taniguchi and Y. Tozaki, *Math. Jap.* **25**, 279 (1980).
2. G. Aleksandryan, *Contemp. Math. Anal., Arm. Acad. Sci.* **19**, 33 (1984).
3. T. Herrmann, *Diploma Thesis* (2008).
4. M. Reissig and K. Yagdjian, *Osaka J. Math.* **36**, 437 (1999).
5. W. Magnus and S. Winkler, *Hill's equation* (Wiley & Sons New York et.al., 1966).
6. M. Eastham, *The spectral theory of periodic differential equations.* (Academic Press., 1973).

Backward uniqueness for the system of thermoelastic waves with non-Lipschitz continuous coefficients

Marco Pivetta

Dipartimento di matematica e informatica,
Universita di Trieste
Via Valerio 12/1, Trieste, 34127 Italy

Using the Carleman estimates developed by Koch and Lasiecka [KL] together with an approximation technique in the phase space, a uniqueness result for the backward Cauchy problem is proved for the system of themoelastic waves having coefficients which are in a class of log-Lipschitz continuous functions.

We deal with the following system:

$$\mathcal{P}u + \alpha(t,x)\partial_x\theta(t,x) + \beta(t,x)\theta(t,x) = f(t,x)$$

$$\mathcal{L}\theta + \delta(t,x)\partial_x\partial_t u(t,x) + \rho(t,x)\partial_t u(t,x) + \psi(t,x)u(t,x) = g(t,x) \qquad (1)$$

$$u(0,x) = u_0(x), \ \partial_t u(0,x) = u_1(x), \ \theta(0,x) = \theta_0(x) \qquad \text{on } \mathbb{R}$$

on $]0,T[\times\mathbb{R}$ with

$$\mathcal{P}u = \partial_t^2 u - a(t)\partial_x^2 u, \qquad \mathcal{L}\theta = \partial_t\theta + b(t)\partial_x^2\theta.$$

The system is obtained by an inversion of time for the model of a vibrating string with a heat flow and it is composed by two coupled equations, one hyperbolic, one parabolic (see [Bi], [CS] and [DI]). The others are coupling terms; the most important one among them is the second order term $\partial_x\partial_t u(t,x)$ in fact the coefficient δ will be more regular than the others. We underline that without this second order term we cannot talk at all about a thermoelastic system.

Our purpose is to prove a uniqueness result for the Cauchy problem, imposing low regularity on the coefficients $a(t)$ and $b(t)$ appearing on the principal parts of the two operators.

The proof follows some ideas implemented by H. Koch and I. Lasiecka [KL]; they deal with the 3D-case of thermoelastic waves with time and space variable coefficients, assuming some regularity on the principal coefficients.

Reduced to 1D-case, with only time dependence, their assumptions consider $a(t)$ and $b(t)$ as Lipschitz continuous functions. In their work they show the uniqueness result by proving two Carleman estimates (see [C] and [N]), one involving the hyperbolic operator, the other the parabolic one and finally combining the two estimates with two inequalities derived directly from the system.

In our work we want to improve the result by assuming non-Lipschitz continuity for the two functions a and b. We will follow the same scheme in [KL] readapting the estimates and the proofs to fit our hypothesis and to conclude with the same argument. We will analyse the outline of the proof showing directly how it works in our case.

The fundamental tools used to overcome the low regularity of the coefficients are standard regularization with mollifiers and the Littlewood-Paley decomposition, that gives a good characterization of Sobolev and Hölder spaces, allowing us to deduce some useful inequalities (see [AG]).

1. Definitions and main theorem

In order to state the main theorem about uniqueness, we need some preliminary definitions:

Definition 1.1. (modulus of continuity)

A function $\mu : [0,1] \to [0,1]$, continuous, strictly increasing and concave, with $\mu(0) = 0$ is called modulus of continuity.

Definition 1.2. A function $f : \mathbb{R} \to \mathbb{R}$ is called μ-continuous (or $\mathcal{C}^\mu$) if $f \in L^\infty$ and there exist a constant $K > 0$ such that

$$\sup_{|x-y|<\frac{1}{2}} |f(x) - f(y)| \leq K\mu(|x-y|)$$

Remark 1.1. One can easily see that if $\mu(t) = t$ then we obtain the class of Lipschitz functions, while taking $\mu(t) = t\,|\ln t|$ we obtain the so called log-Lipschitz functions.

We want to characterize the functions by the properties of their modulus of continuity, so we give the following definitions.

Definition 1.3. We say that a modulus of continuity μ satisfies Osgood condition (or simply is Osgood) if

$$\int_0^1 \frac{1}{\mu(s)}\,ds = +\infty.$$

Consider now a modulus of continuity satisfying Osgood condition. We can define the function $\phi : [1; +\infty] \to [0; +\infty]$ as follows

$$\phi(t) = \int_{\frac{1}{t}}^{1} \frac{1}{\mu(s)} ds.$$

A brief computation reveals that ϕ is invertible and so we can define another function Φ as follows

$$\Phi(\tau) = \int_{0}^{\tau} \phi^{-1}(t) dt.$$

This function will enter in the definition of the weight function for the Carleman estimate. We are ready to define an important property of the modulus of continuity:

Definition 1.4. We say that an Osgood modulus of continuity satisfies $\star$ condition if there exist $0 < a < 1$ such that

$$\lim_{\tau \to +\infty} \frac{\Phi'(\tau)}{(\Phi'(a\tau))^2} = 0.$$

Remark 1.2. In the following pages we will refer to a space of continuous functions with an $\star$-modulus of continuity writing $\mathcal{C}^{\mu_\star}$. The relation between $\star$-condition and Lipschitz continuity can be found in the Appendix A.

The linearity of the system allows us to state our main result as follows:

Theorem 1.1. *Let* $a(t) \in log - Lip([0,T])$ *and* $b(t) \in \mathcal{C}^{\mu_\star}([0,T])$ *and suppose that* $\alpha, \beta, \rho, \psi \in L^{\infty}([0,T], \mathcal{C}^{\sigma})$ *and* $\delta \in L^{\infty}([0,T], \mathcal{C}^{1+\sigma})$, *with* $\sigma \in]0,1[$ *then if*

$$u \in \mathcal{H} = H^2([0,T], L^2(\mathbb{R}_x)) \cap H^1([0,T], H^1(\mathbb{R}_x)) \cap L^2([0,T], H^2(\mathbb{R}_x))$$

$$\theta \in \mathcal{K} = H^1([0,T], L^2(\mathbb{R}_x)) \cap L^2([0,T], H^2(\mathbb{R}_x))$$

are solutions of the homogeneous system associated to (1) then

$$u \equiv \theta \equiv 0 \qquad on \ [0,T].$$

The proof is composed essentialy by the three steps: two Carleman estimates and a final combination of them.
In the next paragraph we focus on the two estimates.

2. Carleman estimates

In order to deal with the hyperbolic operator we first define a localized approximate energy:

$$e_\nu(t) = \int_{\mathbb{R}} a_{2^{-\nu}}(t) \left| \partial_x u_\nu(t,x) \right|^2 + \left| \partial_t u_\nu(t,x) \right|^2 + \left| u_\nu(t,x) \right|^2 dx,$$

With *localized* we mean that we take the ν-th components of the Littlewood-Paley decomposition of u, while *approximate* refers to the mollification of the function $a(t)$, namely

$$a_{2^{-\nu}}(t) = (\eta_{2^{-\nu}} * a)(t),$$

with η_ϵ standard family of mollifiers. This last tool is used essentially because we have to differentiate and the regularity of a is not enough to do this. Moreover we underline that the parameter of the approximation $2^{-\nu}$ is related to the component of the function u.

What we have in mind with this definition is to build a total energy summing up this components, as follows

$$E(t) = \sum_{\nu=0}^{+\infty} e^{-2\beta(\nu+1)t} e_{\nu(t)},$$

with β to be fixed in the next theorem.

From Littlewood-Paley theory we can deduce that there exists $C > 0$ such that

$$CE(t) \leq \|\partial_x u\|^2_{H^{-\beta * t}} + \|\partial_t u\|^2_{H^{-\beta * t}} + \|u\|^2_{H^{-\beta * t}} \leq \frac{1}{C} E(t) \qquad \beta = \beta^* \ln 2.$$

For this energy we prove the following theorem

Theorem 2.1. *There exist C_0, β and $\gamma_0 > 0$ such that for every $\gamma \geq \gamma_0$*

$$\int_0^{\frac{T}{2}} e^{\frac{2}{\gamma}\Phi(\gamma(T-t))} E(t)dt \leq \frac{C_0}{(\Phi'(\frac{\gamma T}{2}))^2} \int_0^{\frac{T}{2}} e^{\frac{2}{\gamma}\Phi(\gamma(T-t))} \|\mathcal{P}u\|^2_{H^{-\beta * t}} dt,$$

for every $u \in \mathcal{C}_0^\infty([0,T] \times \mathbb{R})$ with supp $u \subseteq [0, \frac{T}{2}] \times \mathbb{R}$, with Φ defined by the modulus of continuity of $b(t)$.

The proof is based on some ideas in [CL]. We have introduced a weight function; the same will be used for the parabolic operator, as the next theorem shows.

Theorem 2.2. *Let β be the same as in the previous theorem, then there exist K and $\gamma_0 > 0$ such that for every $\gamma > \gamma_0$ we have, for every $\theta \in C_0^\infty([0,T] \times \mathbb{R})$ with $\operatorname{supp} \theta \subseteq [0, \frac{T}{2}] \times \mathbb{R}$:*

$$\int_0^{\frac{T}{2}} e^{\frac{2}{\gamma}\Phi(\gamma(T-t))} \|\mathcal{L}\theta\|_{H^{-1-\beta*t}}^2 \, dt \geq K \int_0^{\frac{T}{2}} e^{\frac{2}{\gamma}\Phi(\gamma(T-t))}$$

$$\times \left(\gamma\|\theta\|_{H^{-1-\beta*t}}^2 + \sqrt{\gamma}\|\theta\|_{H^{-\beta*t}}^2 + \frac{1}{(\Phi'(\gamma T))}\|\theta\|_{H^{1-\beta*t}}^2 \right) dt.$$

The proof is based on some ideas used in [DSP].

A standard density argument allows us to state that the two estimates hold also for $u \in \mathcal{H}$ and $\theta \in \mathcal{K}$ with $u(0,x) = \partial_t(0,x) = \theta(0,x) = 0$, and $u \equiv \theta \equiv 0$ for $t \in \left[\frac{T}{2}, T\right]$. In the next paragraph we show the combination of this two estimates.

3. Uniqueness

First we obtain two other inequalities: we take two solutions v and ζ of the homogeneous system and multiply them by the following cut-off function

$$\chi(t) \in C^\infty(\mathbb{R}) \qquad \text{with} \qquad \chi(t) = \begin{cases} 1 & \text{for} \quad t \leq \dfrac{T}{4} \\ 0 & \text{for} \quad t \geq \dfrac{T}{3}, \end{cases}$$

thus obtaining two functions u and θ for which the two previous estimates hold. Directly from the system we obtain:

$$\|\mathcal{P}u\|_{H^{-\beta*t}}^2 \leq C \left(\|\theta\|_{H^{1-\beta*t}}^2 + \|\theta\|_{H^{-\beta*t}}^2 + \|\partial_t^2\chi(t)v + 2\partial_t\chi(t)\partial_t v\|_{L^2}^2 \right)$$

$$\|\mathcal{L}\theta\|_{H^{-1-\beta*t}}^2 \leq C'(\|\partial_t u\|_{H^{-\beta*t}}^2 + \|u\|_{H^{-\beta*t}}^2$$
$$+ \|\partial_t\chi(t)\partial_x v + \partial_t\chi(t)\zeta + \partial_t\chi(t)v\|_{L^2}^2)$$

after having restricted the interval $[0,T]$ to get $\beta*T < \delta$ to use the following:

Proposition 3.1. *Let $a \in C^\sigma(\mathbb{R}^n)$ and $s \in \mathbb{R}$, $0 \leq |s| < \sigma$. Then the map $u \mapsto au$ is continuous from H^s to H^s and*

$$\|au\|_{H^s} \leq C_n C_s \|a\|_{C^\sigma} \|u\|_{H^s}.$$

this last proposition can be proved using Bony's paraproduct ([B]).

Now we can resume in a symbolic way all these results as follows:

$$E(t) \lesssim\gtrsim \|\mathcal{P}u\| \lesssim\gtrsim \|\theta\| \lesssim\gtrsim \|\mathcal{L}\theta\| \lesssim\gtrsim E(t),$$

where we did not indicate the space of the various norm, we forgot the integral sign, the weight function and we got rid of some other terms because they play a less important role: the reason is that we want to focus only on the fact that combining our inequalities we reach an estimate which leads to our goal, in fact we can write:

$$\int_0^{\frac{T}{2}} e^{\frac{2}{\gamma}\Phi(\gamma(T-t))} E(t)dt \leq C \frac{\Phi'(\gamma T)}{\left(\Phi'(\frac{\gamma T}{2})\right)^2} \int_0^{\frac{T}{2}} e^{\frac{2}{\gamma}\Phi(\gamma(T-t))} E(t)dt$$

$$+ \frac{C}{2\left(\Phi'(\frac{\gamma T}{2})\right)^2} \int_{\frac{T}{4}}^{\frac{T}{2}} e^{\frac{2}{\gamma}\Phi(\gamma(T-t))} \|\partial_t\chi(t)\cdot\partial_t v + \partial_t^2\chi(t)\cdot v\|_{L^2}^2 dt$$

$$+ C\frac{\Phi'(\gamma T)}{\left(\Phi'(\frac{\gamma T}{2})\right)^2} \int_{\frac{T}{4}}^{\frac{T}{2}} e^{\frac{2}{\gamma}\Phi(\gamma(T-t))} (\|\partial_t\chi(t)\zeta\|_{L^2}^2 + \|\partial_t\chi(t)\partial_x v\|_{L^2}^2$$

$$+ \|\partial_t\chi(t)\cdot v\|_{L^2}^2)\, dt.$$

Now one can use the $\star$-condition with $a = \frac{1}{2}$, say that the first and third term on the right hand side are multiplied by a factor that goes to 0 as $\gamma \to \infty$ and conclude the proof. We show the last part of the proof in the Appendix A, together with an explanation about how one can extend the result when $a \neq \frac{1}{2}$.

Appendix A. The $\star$-condition

Briefly we make some comment about the $\star$-condition, asking if there exist non-Lipschitz functions satisfying this condition. For Lipschitz functions we have $\mu(s) = s$ which gives $\Phi'(\tau) = e^\tau$: simple computations show that our condition holds in this case, while for the log-Lipschitz case we get $\Phi'(\tau) = e^{e^\tau}$ and we cannot hope for the $\star$-condition to hold in this case. Between the two classes mentioned above we introduce the so called $\log_K$-Lipschitz one, whose modulus of continuity is $\mu(s) = s\,|\log s|^K$, with $K \in\,]0, 1[$, which leads to $\Phi'(\tau) = e^{((1-K)\tau)^{\frac{1}{1-K}}}$ and in this case the condition holds for $a > \left(\frac{1}{2}\right)^{1-K}$. We see that as we approach the log-Lipschitz class ($K \to 1$) the coefficient a gets closer and closer to 1. In order to conclude the proof in this general case we restrict our interval to $[0, hT]$, with $h = 1 - a$, obtaining

$$\frac{1}{2}\int_0^{\frac{hT}{8}} E(t)dt \leq$$

$$\frac{\widetilde{C}}{(\Phi'(a\gamma T))^2} e^{\frac{2}{\gamma}\left(\Phi\left(\frac{3}{4}\gamma hT\right)-\Phi\left(\frac{7}{8}\gamma hT\right)\right)} \int_{\frac{hT}{4}}^{\frac{hT}{2}} \|\partial_t\chi(t)\partial_t v + \partial_t^2\chi(t)v\|_{L^2}^2 dt$$

$$+ \frac{\Phi'(\gamma T)}{(\Phi'(a\gamma T))^2} e^{\frac{2}{\gamma}\left(\Phi\left(\frac{3}{4}\gamma hT\right)-\Phi\left(\frac{7}{8}\gamma hT\right)\right)} \int_{\frac{hT}{4}}^{\frac{hT}{2}} \|\partial_t\chi(t)\zeta\|_{L^2}^2 + \|\partial_t\chi(t)v\|_{L^2}^2$$

$$+ \|\partial_t\chi(t)\partial_x v(t,x)\|_{L^2}^2 dt,$$

where we have used the $\star$-condition with a arbitrary close to 1. Moreover thanks to the properties of Φ:

$$\lim_{\gamma\to+\infty} \frac{2}{\gamma}\left(\Phi\left(\frac{3}{4}\gamma hT\right) - \Phi\left(\frac{7}{8}\gamma hT\right)\right) = -\infty,$$

so $E(t) \equiv 0$ in $[0, \frac{hT}{8}]$ and the same for u and thus for θ. Reproducing the same argument we can cover all $[0, T]$, this concludes the proof.

References

AG.   S. Alinhac, P. Gérard, Pseudo-differential operators and the Nash-Moser theorem. Translated from the 1991 French original by Stephen S. Wilson. Graduate Studies in Mathematics, 82. American Mathematical Society, Providence, RI, 2007

B.   J. M. Bony, Calcul symbolique et propagation des singularités pour les équations aux dérivées partielles non linéaires, *Annales Scientifiques de l' Ecole Normale Supérieure* **14** (1981), 209–246.

Bi.   M. A. Biot, Thermoelasicity and irreversible thermodynamics, *J. Appl. Phys.* **27** (1956), 240-253.

C.   T. Carleman, Sur un problème d'unicité pur les systèmes d'équations aux dérivées partielles a deux variables indépendantes, *Ark. Mat, Astr. Fys* **26B** (1939), 1–9.

CS.   P. Chadwick, I. N. Sneddon, Plane Waves in an elastic solid conducting heat, *J. Mech. and Phys. of Solids* **6** (1958), 223-230.

CL.   F. Colombini, N. Lerner, Hyperbolic operators with non-Lipschitz coefficients, *Duke Math. J.* **77** (1995), 657–698.

DI.   M. Daimaruya, H. Ishikawa, On the Propagation of Thermoelastic Waves According to Coupled Thermoelastic Theory, *Bulletin of the JSME* Vol. **17** No. **110** (1974), 991-999.

DSP.   D. Del Santo, M. Prizzi, Backward uniqueness for parabolic operators whose coefficients are non-Lipschitz continuous in time, *J. Math. Purés Appl.* **84** (2005), 471–491.

KL.   H. Koch, I. Lasiecka, Backward uniqueness in linear thermoelasticity with time and space variable coefficients, *Fuctional Analysis and Evolution Equations. The Günter Lumer volume* (2007), 389–403.

N.   L. Nirenberg, *Lectures on linear partial differential equations*, Regional Conference Series in Mathematics **17**,
American Mathematical Society, Providence, Rhode Island, 1973.

IV.3. Control and optimisation of nonlinear evolutionary systems

Organisers: F. Bucci, I. Lasiecka

The session "Control and optimisation of nonlinear evolutionary systems" of the 7th ISAAC Congress at Imperial College London focused on new developments in the area of well-posedness, optimisation, and control of systems described by evolutionary Partial Differential Equations (PDEs) such as nonlinear wave and plate equations, Navier-Stokes and Euler equations, thermoelastic systems, models for viscoelasticity and electromagnetism. Of particular interest have been composite systems of PDEs which combine distinct dynamics in two (or more) separate regions, with a coupling on an interface between these regions. Examples of such systems are models for acoustic-structure, fluid-structure and magneto-structure interactions; these have a wide range of applications that include engineering (noise reduction in an acoustic cavity, control of turbulence), medicine (diagnostic imaging such as MRI, ultrasound), geophysics (reconstruction of seismic data) and others. Recent years have witnessed a rapid development of new mathematical tools in both analysis and geometry that allow to obtain various PDE estimates of inverse type. These enable to establish properties such as, for instance, controllability, reconstruction of the data, stabilisation.

The session aimed to provide a forum for discussion of significant questions arising in the study of the aforementioned topics, as well as to promote the scientific exchange between younger and established researchers, and between specialists belonging to different communities. While control problems may constitute the prime motivation for a deep PDE analysis, the PDE analysis necessitated by diverse applications may benefit from the ongoing and emerging interactions with control theory. The session took in both classical and more recent themes in control theory, such as modeling, stabilization, controllability, Carleman estimates, optimal control; control methods and inverse problems, control methods in the study of the asymptotic behaviour of infinite dimensional dynamical systems.

The session comprised talks on the following specific subjects: global well-posedness and long-time behavior of solutions to wave equations, optimal controls for parabolic variational inequalities, controllability of fluid-structure interaction problems, stability and uniform decay rate estimates for wave equations, qualitative aspects of the damped Korteweg-de Vries and Airy type equations, optimal control of waves in anisotropic media, global existence for the one-dimensional semilinear Tricomi-type equation, optimal control of a thermoelastic structural acoustic model, the Balayage Method for boundary control of a thermo-elastic plate, Hopf-Lax type formulas and Hamilton-Jacobi equations, on-line inversion techniques for the analysis of boundary control problems, dissipation in contact problems, heat equations with memory, observability problems for PDEs, invariant manifolds for parabolic problems with dynamical boundary conditions, evolution equations with memory terms, regularity properties of optimal control and Lagrange multipliers, stabilization of structure-acoustics interactions by localized nonlinear boundary feedbacks, exponential stability of the wave equation with boundary time varying delay, state estimation for parabolic systems, Euler flow and morphing shape metric. The twentythree speakers, representing academic institutions in Italy, France, Germany, Tunisia, Russia, U.S.A., Brasil and Japan, are listed (in alphabetical order) below: L. Bociu, M. Boukrouche, M. Boulakia, M. Cavalcanti, M. Daoulatli, V. Domingos Cavalcanti, M. Eller, G. Fragnelli, A. Galstyan, C. Lebiedzik, W. Littman, P. Loreti, V. Maksimov, M.G. Naso, L. Pandolfi, M. Renardy, R. Schnaubelt, D. Sforza, Y. Shvartsman, D. Toundykov, J. Valein, M. Yamamoto, J.-P. Zolesio.

Null boundary controllability of the Schrödinger equation with a potential

O. Arena

Dipartimento di Matematica e Applicazioni per l'Architettura,
Università degli Studi di Firenze,
piazza Ghiberti, 27 - 50122 Firenze, Italy
E-mail: arena@unifi.it

W. Littman

School of Mathematics,
University of Minnesota,
206 Church Street S.E., Minneapolis, MN 55455, USA
E-mail: littman@math.umn.edu

We discuss the null boundary controllability of the Schrödinger equation with a potential in a "*D*" shaped domain in $\mathbb{R}^2$. The controls are applied to the curved part of the boundary.

Keywords: Null boundary control; Schrödinger equation; Ovcyannikov theorem.

1. Introduction

We are interested in the study of a problem of null boundary controllability to the system:

$$i\frac{\partial}{\partial t}u(x,t) = \frac{1}{2}\triangle u(x,t) + V(x)u(x,t) \quad \text{in} \quad \Omega \times (0,+\infty) \tag{1}$$

$$u(x,0) = u_0(x), \quad x \in \Omega \tag{2}$$

$$u(x,0) = 0, \quad x \in \Gamma_1 \tag{3}$$

$$\mathcal{B}u(x,t) = g(x,t), \quad (x,t) \in (\partial\Omega\backslash\Gamma_1) \times (0,+\infty) \tag{4}$$

where Ω is a bounded domain of $\mathbb{R}^2$, Γ_1 is a segment of a straight line, part of the boundary $\partial\Omega$, $\mathcal{B}$ an appropriate boundary operator, u_0 initial data and g a boundary control function on $(\partial\Omega\setminus\Gamma_1) \times (0,+\infty)$, $\partial\Omega\setminus\Gamma_1$ a C^2 curve.

We study this problem in the case Ω has the following geometry in the $x \equiv (x_1, x_2)$-plane.

Assume, for simplicity, Ω be a semi-disc, for instance:

$$\Omega = \{(x_1, x_2) \in \mathbb{R}^2 : x_1^2 + x_2^2 < r^2, \quad x_1 > 0\}.$$

The null boundary controllability problem is the following: given u_0 in the appropriate space and $T > 0$, find g in the appropriate space such that the solution of (1), (2), (3), (4) vanishes for $t \geq T$. The boundary control g is defined on $(\partial\Omega\setminus\{x_1 = 0\}) \times (0, +\infty)$.

To solve such a problem we are going to apply the approach introduced by W. Littman, Ref. 2.

Many authors have studied the Schrödinger operator in the framework of null boundary control problems and different methods have been applied. We refer to the recent papers by W. Littman and S. Taylor (Refs. 4 and 5) where "direct methods" are employed. With regard to these, we choose the use of the so-called "balayage" method.

Furthermore, the problem we are discussing here exhibits a new feature for this method: the control function, is prescribed only in one part of the boundary.

We state the main result:

Theorem 1.1.   *Suppose the potential V to be a real analytic function in $\mathbb{R}^2$, even in the x_1 variable. Assume that the control is of the form:*

$$u(x,t) = g(x,t), \quad (x,t) \in (\partial\Omega\setminus\Gamma_1) \times (0,+\infty).$$

Let $T > 0$. Then, given $u_0 \in L^2(\Omega)$, there exists a boundary control function g such that $u(x,T) = 0$. Moreover, for every $\varepsilon > 0 : g \in C([0,T]; H^{\frac{1}{2}-\varepsilon}(\partial\Omega\setminus\Gamma_1))$.

Note that the assumptions on evenness and oddness will result, in section 3, in the oddness of $\omega(x_1, x_2, t)$ in x_1, hence $\omega(0, x_2, t) = 0$.

In section 2, we start by solving the pure initial value problem for (1) and then we multiply its solution $u = u(x, t)$ by an appropriate Gevrey function $\psi(t)$ (Ref. 1; pag. 146). At this stage, in section 3, we solve a Cauchy problem for a nonhomogeneous version of (1), for which a space variable plays the role of a time variable. There, the smoothness of the solution u of the initial value problem is crucial. In section 4, the boundary control function g is found.

2. The Cauchy problem for (1)

To study the Cauchy problem for the equation (1), we start by reflecting the domain Ω about the x_2-axis and extending u_0 to the entire resulting disk B as an odd function of x_1.

We define u_0 to be zero outside of B and extend the potential V as an even function about $x_1 = 0$. Solve, then, equation (1) in all x space to get the solution $u(x,t)$, which must be odd in x_1, and hence must vanish when $x_1 = 0$.

We also know from Refs 3 and 7 that the solution $u(x,t)$ is analytic in the space variables and of Gevrey class 2 in t. Indeed, the following result holds:

Theorem 2.1. *Consider the initial value problem for (1) in all $\mathbb{R}^2$. If the potential V belongs to a Gevrey class γ^δ and the initial value u_0 is in L^2 and has compact support, then the solution $u(x,t)$ belongs to a Gevrey class $\gamma^{2\delta}$, as a function of t, in compact sets of $\mathbb{R}^2 \times \{t > 0\}$.*

Recall that a function $F(x,t)$ is said to belong to the space of Gevrey δ class, γ^δ, with respect to the positive t variable, uniformly for (x,t) in the compact set $\mathcal{K}$ if, for every $(x,t) \in \mathcal{K}$ and for every $\theta > 0$:

$$\left| \frac{\partial^n}{\partial t^n} F(x,t) \right| \leq C_{\mathcal{K},\theta} \cdot \theta^n (n!)^\delta$$

for all $n = 1, 2, \ldots$, with $C_{\mathcal{K},\theta}$ a positive constant.

Let, now, $\psi(t)$ be a cut-off function defined as follows:

$$\psi(t) = \begin{cases} 1 \text{ for } t \leq \dfrac{T}{2} \\ 0 \text{ for } t \geq T \end{cases}$$

belonging to the Gevrey class $\gamma^{\frac{3}{2}}$.

Such a function can be constructed explicitly (see L. Hörmander, Ref. 1, pag. 146).

Multiply $u(x,t)$ by $\psi(t)$ and set:

$$f(x,t) = \left(-\triangle -2V(x) + 2\,i\,\frac{\partial}{\partial t} \right)(u\psi) = 2\,i\,u(x,t)\psi'(t). \qquad (5)$$

Note that $f(x,t)$ vanishes outside $\left[\frac{T}{2}, T\right]$ and the function $\overline{u}(x,t) = u(x,t)\psi(t)$ vanishes for $t \geq T$. We are interested, thus, in finding solutions $\omega = \omega(x,t)$ of the equation:

$$i\,\frac{\partial \omega}{\partial t} = \frac{1}{2}\,\triangle\,\omega + V(x)\omega + \frac{1}{2}f(x,t) \qquad (6)$$

that vanish outside $\left[\frac{T}{2}, T\right]$.

3. The nonhomogeneous equation (6)

We can solve the nonhomogeneous equation (6) for ω, as Cauchy problem with the variable x_2 as the time variable.

We solve (6) in the set $K \times [0, \eta] \times \{t > 0\}$, where K is a compact interval of the x_2-axis containing $[-r, r]$ and $\eta > r$. Note that $f(x, t)$, given by (5), is real analytic in x and Gevrey, as function of t, because of the choice of ψ. First, we transform the problem to a system. Set:

$$\frac{\partial \omega^{(1)}}{\partial x_2} = \omega^{(2)}, \quad \triangle_1 = \frac{\partial^2}{\partial x_1^2}, \quad \widetilde{w} = \begin{bmatrix} \omega^{(1)} \\ \omega^{(2)} \end{bmatrix}$$

and write (6) as:

$$\frac{d\widetilde{w}}{dx_2} = A(x_2)\widetilde{w} + \widetilde{F}(x_2) \tag{7}$$

where

$$A = \begin{bmatrix} 0 & 1 \\ -2V - \frac{\partial^2}{\partial x_1^2} + 2i\frac{\partial}{\partial t} & 0 \end{bmatrix}, \quad \widetilde{F} = \begin{bmatrix} 0 \\ f \end{bmatrix}. \tag{8}$$

We solve (7) subject to the initial condition $\widetilde{\omega}(0) = 0$ and prove the existence-uniqueness of the solution by using a generalization of the Cauchy-Kovalevski theorem, due to Ovcyannikov (see F.Trèves, Ref. 8). To this end, let us recall the following

Definition 3.1. $\{E_s\}$, $s \in [0, 1]$ is a scale of Banach spaces if, for any $s \in [0, 1]$, E_s is a linear subspace of E_0 and if $s' \leq s$, then $E_s \subset E_{s'}$ and the natural injection of E_s into $E_{s'}$, has norm no greater than one.

Now, we state the following version of the Ovcyannikov theorem.

Theorem 3.1.   *Let $\{E_s, ||\cdot||_s\}_{s \in [0,1]}$ be a scale of Banach spaces such that:*

(1) *For $s' < s$ and each $x_2 \in [0, \eta]$, let $A(x_2)$ be a bounded linear mapping from E_s into $E_{s'}$, with operator norm no greater than $M(s - s')^{-1} + P$, where $M > 0$ and $P \geq 0$ are constants independent of x_2, s and s'. Further, $A(x_2)$ is a continuous function of x_2 in the uniform operator topology.*

(2) *$y_0 \in E_1$ and $\widetilde{F} \in C([0, \eta]; E_1)$.*

Then the problem

$$\frac{d\widetilde{w}}{dx_2} = A(x_2)\widetilde{\omega} + \widetilde{F}(x_2), \quad \widetilde{\omega}(0) = y_0,$$

has a unique solution belonging to $C^1([0, \delta_0(1-s)]; E_s)$ for each $s \in [0,1]$, where $\delta_0 = min(\eta, (Me)^{-1})$.

We omit the proof of this theorem which is not difficult to obtain modifying the proof of the classical theorem given in Ref. 8.

To apply this theorem to our problem, we need to assume $y_0 = 0$. For details we refer to Ref. 4.

4. The control function g

We are now in a position to construct the boundary control function. Set:

$$U(x,t) = u(x,t)\psi(t) - \omega^{(1)}(x,t).$$

$U(x,t)$ satisfies equation (1), (2) in a bounded domain $\Omega^* \supset \overline{\Omega}$ and vanishes for $t \geq T$. $U(x,t)$ belongs to C^∞ for $t > 0$ and the boundary control function $g(x,t) = \mathcal{B}(x,t)u(x,t)$, for $x \in (\partial\Omega \backslash \Gamma_1)$, is as smooth as the boundary operator $\mathcal{B}$ and the boundary $\partial\Omega \backslash \Gamma_1$ allow, for $t > 0$. Now, for $t \geq 0$, let $\Phi(t) : L^2(\Omega) \to L^2(\Omega^*)$ denote the solution operator $u_0 \to U(\cdot, t)$. We point out that $\Phi(t)$ is C^∞ in the uniform operator topology for $t > 0$, but at $t = 0$ we only know that $\Phi(t)$ is strongly continuous, since $\Phi(t)$ turns out to be a C_0 semigroup for $0 \leq t \leq \frac{T}{2}$.

So, to improve regularity of the boundary control function as $t \to 0$, we proceed as in Ref. 4. We explain this procedure for the case of Dirichlet control, i.e. by assuming that:

$$u(x,t) = g(x,t), \quad x \in \partial\Omega \smallsetminus \Gamma_1.$$

i Solve (1) and (2) with zero Dirichlet data $u(x,t) = 0$, $x \in \partial\Omega$.
   Denote $u_1(x,t)$ the solution of this problem.
ii Let $u_2(x,t) = u_1(x,t)\psi(t)$, where $\psi(t)$ is the same cut-off function of section 2. The function u_2 satisfies the initial condition, has zero Dirichlet data and vanishes for $t \geq T$. The function u_2 is solution to the nonhomogeneous equation (6), where $f(x,t) = u_1(x,t)\psi'(t)$.
iii Using Duhamel's principle, one may express the solution of the same nonhomogeneous equation as:

$$W(x,t) = \int_0^t \Phi(t-\tau)f(x,\tau)d\tau.$$

Since $f(x,t)$ vanishes outside $[\frac{T}{2}, T]$, the function

$$\widetilde{U}(x,t) = u_1(x,t)\psi(t) - W(x,t)$$

satisfies (1), (2) and vanishes for $t \geq T$.

The boundary control is then given by

$$g(x,t) = \widetilde{U}(x,t) \quad \text{for } x \in \partial\Omega \smallsetminus \Gamma_1, \quad \text{for Dirichlet control.}$$

Note that, from Theorem 1.5 of Ref. 7, the solution operator $\Phi(t)$: $L^2(\Omega) \rightarrow H^1(\Omega^*)$ has norm bounded by C/t for some constant C. Then, by interpolation, $\Phi(t)$: $L^2(\Omega) \rightarrow H^s(\Omega^*)$ has operator norm bounded by C/t^s, $s \in [0,1]$. Moreover, for $0 \leq s < 1$, one has:

$$||W(\cdot,t)||_{H^s} \leq C||u_0||_{L^2} max(\psi') \int_0^t (t-\tau)^{-s} d\tau \leq \widetilde{C}||u_0||_{L^2}.$$

Assuming the Trace Theorem holds for $\partial\Omega \smallsetminus \Gamma_1$, we find that g belongs to $L^\infty((0,\infty); H^{\frac{1}{2}-\varepsilon}(\partial\Omega \smallsetminus \Gamma_1))$, $\varepsilon > 0$. Indeed, the strong continuity of Φ implies that $g \in C([0,\infty); H^{\frac{1}{2}-\varepsilon}(\partial\Omega \smallsetminus \Gamma_1))$, $\varepsilon > 0$.

Acknowledgments

Orazio Arena would like to thank the School of Mathematics of the University of Minnesota in Minneapolis for the warm hospitality, while he was visiting the School (April-July, 2009) and this research was started. His research has been supported by italian MIUR.

References

1. L. Hörmander, *Linear Partial Differential Operators*, Academic Press, New York, 1963.
2. W. Littman, *Boundary control Theory for Beams and Plates*, Proceedings, 24th Conference on Decision and Controll, Ft. Lauderdale, FL, 2007-2009, December 1985.
3. W. Littman and S. Taylor, *Smoothing Evolution Equations and Boundary Controllability*, Journal d'Analyse. Mathématique, *59*, 117-131, 1992.
4. W. Littman and S. Taylor, *The Heat and Schrödinger Equation Boundary Control whit One Shot*, Control Methods in PDE-Dynamical Systems, Contemporary Math. *426*, $\mathcal{AMS}$, Providence, RI, 299-305, 2007.
5. W. Littman and S. Taylor, *The Balayage Method: Boundary Control of a Thermo-Elastic Plate*, Applicationes Math., *35*, (4), 467-479, 2008.
6. A. Pazy, *Semigroups of Linear Operators and Applications to Partial Differential Equation*, Appl. Math. Sci. *44*, Springer, New York, 1983.
7. S. Taylor, *Gevrey Smoothing Properties of the Schrödinger Evolution Group in Weighted Sobodev Spaces*, Journal of Math. Anal. and Appl. *194*, 14-38, 1985.
8. F. Trèves, *Ovcyannikov Theorem and Hyperdifferentia Operators*, Notas Mat., no. 46, 1968.

V.1. Inverse problems

Organisers: Y. Kurylev, M. Yamamoto

Optimal combination of data modes in inverse problems: Maximum compatibility estimate

Mikko Kaasalainen

Department of Mathematics,
Tampere University of Technology,
P.O. Box 553, 33101 Tampere,
Finland

We present an optimal strategy for the relative weighting of different data modes in inverse problems, and derive the maximum compatibility estimate (MCE) that corresponds to the maximum likelihood or maximum a posteriori estimates in the case of a single data mode. MCE is not explicitly dependent on the noise levels, scale factors or numbers of data points of the complementary data modes, and can be determined without the mode weight parameters. As a case study, we consider the problem of reconstructing the shape of a body in $\mathbb{R}^3$ from the boundary curves (profiles) and volumes (brightness values) of its generalized projections.

Keywords: Inverse problems; computational geometry; three-dimensional polytopes.

1. Introduction

In many inverse problems, various complementary data modes are available. For example, constructing the shape model of a body in $\mathbb{R}^3$ is typically based on projectionlike data at various viewing geometries. In this paper, we consider the case where images $\mathcal{I}(\omega, \omega_0)$ (generalized projections) obtained at viewing and illumination directions $\omega, \omega_0 \in S^2$ are available, but the reliable infomation in these images is only contained in the boundary curves $\partial\mathcal{I}$ between the dark background or a shadow and the illuminated portion of the target surface. This is a typical case in adaptive optics data in astrophysics, where the coverage of viewing geometries is also seldom wide enough to enable a full reconstruction of the model from images alone.[2] Thus we include the possibility of augmenting the image dataset with a set of measured brightnesses (volumes of the generalized projections) $L(\omega, \omega_0)$ of the target at various observing geometries.

2. Case study: Generalized projections

We consider the inverse problem of determining the shape of a body $\mathcal{B} \in \mathbb{R}^3$ from some measured profiles of generalized projections $\partial \mathcal{I}(\omega_i, \omega_{0i})$, $i = 1, \ldots, n$ and their volumes $L(\omega_{0i}, \omega_i)$, $i = 1, \ldots, m$.[3,4]

Our goal is to construct a total goodness-of-fit measure χ^2_{tot}

$$\chi^2_{\mathrm{tot}} = \chi^2_L + \lambda_\partial \chi^2_\partial + \lambda_R g(P), \tag{1}$$

where L denotes brightness data, ∂ generalized profiles, and R regularizing functions $g(P)$ (see Ref. 4 for discussion of these), where $P \in \mathbb{R}^p$ is the vector of model parameters. Determining an optimal value for λ_∂ (and λ_R) is part of the inverse problem.

The volumes of generalized projections are also called total or disk-integrated brightnesses:[3]

$$L(\omega_0, \omega) = \int_{\mathcal{A}_+} R(x; \omega_0, \omega)\langle \omega, \nu(x)\rangle \, d\sigma(x), \tag{2}$$

where $\mathcal{A}_+$ is the set of visible and illuminated points $x \in \mathcal{B}$,[3] $\nu(x) \in S^2$ and $d\sigma(x)$ are, respectively, the outward surface normal and surface patch of $\mathcal{B}$, and $R(x; \omega_0, \omega) \in \mathbb{R}$ describes the intensity of scattered light at the point x on the surface. In its basic form,

$$\chi^2_L = \sum_i [L^{(\mathrm{obs})}(\omega_{0i}, \omega_i) - L^{(\mathrm{mod})}(\omega_{0i}, \omega_i)]^2 \tag{3}$$

(assuming a constant noise level; see Ref. 3 and references therein for modifications and variations of this). L-data on $S^2 \times S^2$ uniquely determine a convex body and the solution is stable,[3] but L-data do not carry information on nonconvexities in most realistically available $S^2 \times S^2$ geometries in practice.

For many typical adaptive optics targets in our solar system, the generalized profiles are starlike due to the proximity of ω and ω_0 and some regularity of the target shape at global scale.[2] Then we can write χ^2_∂ by considering, for each profile i, their observed and modelled maximal radii (from some point $\varkappa_0 \in \mathbb{R}^2$ within the profile) on the projection plane at direction angles α_{ij} (starting from a chosen coordinate direction):

$$\chi^2_\partial = \sum_{ij} [r^{(\mathrm{obs})}_{\max}(\alpha_{ij}) - r^{(\mathrm{mod})}_{\max}(\alpha_{ij})]^2. \tag{4}$$

We now represent the body $\mathcal{B}$ as a polytope. Let two vertices a and b of a facet have projection points $\varkappa_a$, $\varkappa_b$. The intersection point $\varkappa$ of the radius line at α and the projection of the facet edge ab is readily determined. The model $r_{\max}(\alpha)$ can now be determined by going through all eligible facet edges and their intersection points $\varkappa_{ab}(\alpha)$:

$$r_{\max}^{(\mathrm{mod})}(\alpha) = \max\left\{\|\varkappa_{ab}(\alpha) - \varkappa_0\|\,\big|\,a, b \in \mathcal{V}_+\right\}, \tag{5}$$

where $\mathcal{V}_+$ is the set of vertices of the set of facets $\tilde{\mathcal{A}}_+$ approximating $\mathcal{A}_+$. The set $\tilde{\mathcal{A}}_+$ is determined by ray-tracing.[3] In general, facet edge circuits $\partial\tilde{\mathcal{A}}_+$ approximating $\partial\mathcal{A}_+$ (and corresponding forms of χ_∂^2) can be automatically derived for non-starlike shape models or profiles as well.[4]

3. Maximum compatibility estimate

Let us choose as goodness-of-fit measures (from which probability distributions can be constructed) the χ^2-functions of n data modes. Our task is to construct a joint χ_{tot} with well-defined weighting for each data mode:

$$\chi_{\mathrm{tot}}^2(P, D) = \chi_1^2(P, D_1) + \sum_{i=2}^{n} \lambda_{i-1}\chi_i^2(P, D_i) \quad D = \{D_i, i = 1, \ldots, n\} \tag{6}$$

(to which regularization functions $g(P)$ can be added), where D_i denotes the data from the source i, and $P \in \mathbb{R}^p$ is the set of model parameter values. We assume the χ_i^2-space to be nondegenerate, i.e.,

$$\arg\min \chi_i^2(P) \neq \arg\min \chi_j^2(P), \quad i \neq j.$$

In two dimensions, denote

$$x(\lambda) := \{\chi_1^2|\min \chi_{\mathrm{tot}}^2; \lambda\}, \tag{7}$$
$$y(\lambda) := \{\chi_2^2|\min \chi_{\mathrm{tot}}^2; \lambda\}.$$

The curve

$$\mathcal{S}(\lambda) := [\log x(\lambda), \log y(\lambda)] \tag{8}$$

resembles the well-known "L-curve" related to, e.g., Tikhonov regularization.[1,6] However, here we make no assumptions on the shape of $\mathcal{S}$. The curve $\mathcal{S}$ is a part of the boundary $\partial\mathcal{R}$ of the region $\mathcal{R} \in \mathbb{R}^2$ formed by the mapping $\chi : \mathbb{R}^p \to \mathbb{R}^2$ from the parameter space $\mathbb{P}$ into χ_i^2-space:

$$\chi = \{\mathbb{P} \to (\log \chi_1^2, \log \chi_2^2)\}, \quad \mathcal{R} = \chi(\mathcal{P})$$

where the set $\mathcal{P}$ includes all the possible values of model parameters (assuming that χ is continuous and well-behaved such that a connected $\mathcal{R}$ and $\partial\mathcal{R}$ exist). If the possible values of χ_i^2 are not bounded, the remaining part $\partial\mathcal{R} \setminus \mathcal{S}$ stretches droplet-like towards (∞, ∞). The parameter λ describes the position on the interesting part $\mathcal{S} \subset \partial\mathcal{R}$, and it is up to us to define a criterion for choosing the optimal value of λ.

The logarithm ensures that the shape of $\mathcal{S}(\lambda)$ is invariant under unit or scale transforms in the χ_i^2 as they merely translate $\mathcal{S}$ in the $(\log \chi_1^2, \log \chi_2^2)$-plane. It also provides a meaningful metric for the $\log \chi_i^2$-space: distances depict the relative difference in χ^2-sense, removing the problem of comparing the absolute values of quite different types of χ_i^2. The endpoints of $\mathcal{S}(\lambda)$ are at $\lambda = 0$ and $\lambda = \infty$, i.e., at the values of χ_i^2 that result from using only one of the data modes in inversion. We can translate the origin of the $(\log \chi_1^2, \log \chi_2^2)$-plane to a more natural position by choosing the new coordinate axes to pass through these endpoints. Denote

$$\hat{x}_0 = \log x(\lambda)|_{\lambda=0} = \log \min \chi_1^2 \tag{9}$$
$$\hat{y}_0 = \log y(\lambda)|_{\lambda\to\infty} = \log \min \chi_2^2.$$

Then the "ideal point" $(\hat{x}_0, \hat{y}_0)$ is the new origin in the $(\log x, \log y)$-plane. A natural choice for an optimal location on $\mathcal{S}$ is the point closest to $(\hat{x}_0, \hat{y}_0)$, i.e., the parameter values $P_0 \in \mathbb{P}$ such that

$$P_0 = \arg\min \left([\log \chi_1^2(P) - \hat{x}_0]^2 + [\log \chi_2^2(P) - \hat{y}_0]^2 \right), \tag{10}$$

so we have, with λ as argument,

$$\lambda_0 = \arg\min \left([\log x(\lambda) - \hat{x}_0]^2 + [\log y(\lambda) - \hat{y}_0]^2 \right). \tag{11}$$

In this approach, neither the numbers of data points in each χ_i^2 nor the noise levels as such affect the solution for the optimal P_0 as their scaling effects cancel out in each quadratic term. P_0 is thus a pure compatibility estimate describing the best model compromise explaining the datasets of different modes simultaneously.

We call the point P_0 the *maximum compatibility estimate* (MCE), and λ_0 the *maximum compatibility weight* (MCW). This corresponds to the maximum likelihood estimate in the case of one data mode, or to the maximum a posteriori estimate as well since we can include regularization functions here. If regularizing is used, the weights for the functions are either determined in a similar manner (see below), or they can be fixed and the regularization terms are absorbed in χ_1^2 (otherwise $\mathcal{S} \subset \partial\mathcal{R}$ does not hold).

Another choice, frequently used in the L-curve approach, is to find the λ at which $\mathcal{S}$ attains its maximum curvature,[1,6] but evaluating this point is less robust than finding λ_0, and (11) is a more natural prescription, requiring no assumptions on the shape of $\mathcal{S}$. We make two implicit assumptions here:

(1) The solutions $P_{\partial\mathcal{R}}$ corresponding to points on $\partial\mathcal{R}$ should be continuous (and one-to-one) in $\mathbb{P}$-space along $\partial\mathcal{R}$ at least in the vicinity of the solution corresponding to λ_0. If this is not true (in practice, if $P_\lambda = \arg\min \chi^2_{\text{tot}}(P)$ makes large jumps in $\mathbb{P}$ for various λ around λ_0), one should be cautious about the uniqueness and stability of the chosen solution P_0, and restrict the regions of $\mathbb{P}$ included in the analysis.

(2) The optimal point λ_0 on $\mathcal{S}$ should be feasible: if we have upper limits ϵ_i to acceptable χ_i^2, the feasible region $\mathcal{F}$ is the rectangle $\bigcap_i \{\log \chi_i^2 \leq \log \epsilon_i\}$. If $[\log \chi_1^2(P_0), \log \chi_2^2(P_0)] \notin \mathcal{F}$ and $\mathcal{F} \cap \mathcal{R} \neq \emptyset$, we choose the point on the portion $\mathcal{S} \subset \mathcal{R}$ closest to the one corresponding to λ_0 (i.e., $\log \chi_i^2 = \log \epsilon_i$ for one i). If $\mathcal{F} \cap \mathcal{R} = \emptyset$, the data modes do not allow a compatible joint model, so either the model is incorrect for one or both data modes, or one or both ϵ_i have been estimated too low (e.g., systematic errors have not been taken into account). Note that model insufficiency should be taken into account in the estimation of ϵ_i.

Note that, in the interpretation $\mathcal{R} = \chi(\mathcal{P})$, λ, χ^2_{tot} and $\partial\mathcal{R}$ are all in fact superfluous quantities, and we can locate the point estimate MCE P_0 entirely without them with standard optimization procedures (and with no extra computational cost). However, it is useful (though computationally somewhat noisier) to approximate $\mathcal{S}$ via the minimization of χ^2_{tot} with sample values of λ (see Fig. 1a), as in addition to obtaining the MCW λ_0 (and hence MCE as well) we can plot $\mathcal{S}$ to examine the mutual behaviour of the complementary data sources (including the position of the feasibility region $\mathcal{F}$ w.r.t. $\mathcal{S}$). The solution for λ_0 is also needed for constructing distributions based on χ^2_{tot}. Another possibility to examine $\mathcal{R}$ and $\partial\mathcal{R}$ is direct adaptive Monte Carlo sampling, but this is computationally slow.

This approach straightforwardly generalizes to n χ^2-functions and $n-1$ parameters λ_i describing the position on the $n-1$-dimensional boundary surface $\partial\mathcal{R}$ of an n-dimensional domain $\mathcal{R}$: the MCE is

$$P_0 = \arg\min \sum_{i=1}^{n} \left[\log \frac{\chi_i^2(P)}{\chi_{i0}^2} \right]^2, \quad \chi_{i0}^2 := \min \chi_i^2(P), \qquad (12)$$

and the MCW is, for $\lambda \in \mathbb{R}^{n-1}$,

$$\lambda_0 = \arg\min \sum_{i=1}^{n} \left[\log \frac{\hat{\chi}_{i,\text{tot}}^2(\lambda)}{\chi_{i0}^2} \right]^2, \quad \hat{\chi}_{i,\text{tot}}^2(\lambda) := \left\{ \chi_i^2 \,\middle|\, \min \chi_{\text{tot}}^2; \lambda \right\}. \quad (13)$$

Another scale-invariant version of MCE can be constructed by plotting

χ_i^2 in units of χ_i^2/χ_{i0}^2 and shifting the new origin to $\chi_i^2/\chi_{i0}^2 = 1$:

$$P_0 = \arg\min \sum_{i=1}^{n} \left[\frac{\chi_i^2(P)}{\chi_{i0}^2} - 1 \right]^2, \quad \lambda_0 = \arg\min \sum_{i=1}^{n} \left[\frac{\hat{\chi}_{i,\text{tot}}^2(\lambda)}{\chi_{i0}^2} - 1 \right]^2. \quad (14)$$

This, however, is exactly the first-order approximation of (12) and (13) in $\delta \ll 1$ when $\chi_i^2/\chi_{i0}^2 = 1 + \delta$, giving virtually the same result as (12) and (13) as usually $\chi_i^2(P_0)/\chi_{i0}^2 - 1 \ll 1$ in the region around $\chi_i^2(P_0)$, and any larger ratios of χ_i^2/χ_{i0}^2 are not eligible for the optimal solution (see Fig. 1a).

It is possible to use this approach for general regularizing functions $g(P)$ as well (change $\chi_i^2 \to g(P)$ for some i), but in such cases the shape of $\mathcal{S}$ must be taken into account. If it is possible to have a solution $g(P') = 0$ for a regularizing function g (or an almost vanishing $g(P')$ such that $\log g(P') \to -\infty$), one should, e.g., set a lower practical limit to $g(P)$ by looking at the shape of S, and choose the λ_0 within the restricted part of S.

4. Numerical implementation

As examples of the optimal combining of brightness values and profile contours, we show some results for asteroid data.

Fig. 1a depicts a typical evaluation of the curve $\mathcal{S}$ for 2 Pallas at various choices of λ (or rather, this plot portrays the cross-section of the 2-surface $\partial\mathcal{R}$ in $\mathbb{R}^3$ with smoothness regularization weight fixed at its final optimal value). The values for χ_i^2 are normalized to be the rms deviations of model fits $d_i = \sqrt{\chi_i^2/N_i}$, as in logarithmic scale this corresponds only to a shift of origin and a uniform linear change of plot scaling. The plotted points outline the curve $\mathcal{S}(\lambda)$ that is rather an oblique line than an L-shape, and the ideal point region, i.e., the point closest to the lower left-hand corner, can directly be found. The endpoints $\lambda = 0$ and $\lambda = \infty$ stop at saturation regions rather than continue to large distances in the $\log \chi^2$-space. As can be seen from Fig. 1a, computational noise in the estimated points at $\lambda = 0$ and $\lambda = \infty$, corresponding to a small change of the position of the new origin w.r.t. $\mathcal{S}$, does not affect the estimated location of the optimal point on S significantly.

A sample observed vs. modelled profiles for 41 Daphne is shown in Fig. 1b. The starlike surface model was described by the exponential Laplace (spherical harmonics) series for the surface radius r (for explicit positivity):

$$r(\theta, \varphi) = \exp\left[\sum_{lm} c_{lm} Y_l^m(\theta, \varphi) \right], \quad (\theta, \varphi) \in S^2, \quad (15)$$

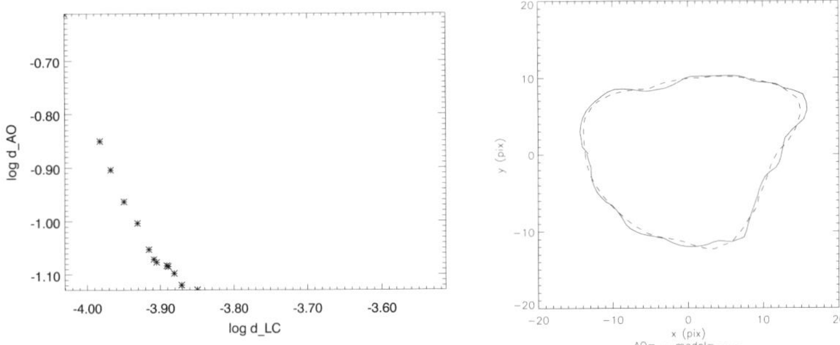

Fig. 1. (a) $\mathcal{S}$ curve plotted for 2 Pallas with various weights λ (LC for brightness data, AO for adaptive optics profiles). (b) Sample observed (solid line) vs. modelled (dashed line) AO profile contour for 41 Daphne. Coordinates are in pixel units.

truncated at $l = 8, m = 6$, with c_{lm} as the shape parameters to be solved for. Other model parameters are the profile offset $\varkappa_0$ for each image and the physical spin parameters describing the rotational transformations yielding the correct viewing and illumination directions (ω, ω_0) on the body.[3–5]

5. Discussion

The concept of the maximum compatibility estimate is directly applicable to any inverse problems with complementary data modes. The invariance properties of the MCE make it more generally usable than heuristic strategies for choosing the weights, especially when they use assumptions on the shape of $\partial\mathcal{R}$ or other case-specific characteristics. In our case study, the use of profiles is practical as it removes two sources of systematic errors inherent to using full images (brightness distributions $\mathcal{I}$ on the image plane): the errors in $\mathcal{I}$ from adaptive optics deconvolution and the model $\mathcal{I}$ errors due to the insufficently modellable light-scattering properties of the surface of the target body.

References

1. M. Belge, M. Kilmer, and E. Miller, *Inverse Problems* **18**, 1161 (2002).
2. B. Carry, C. Dumas, M. Kaasalainen, and 9 colleagues, *Icarus*, in press (2009).
3. M. Kaasalainen and L. Lamberg, *Inverse Problems* **22**, 749 (2006).
4. M. Kaasalainen, *Inverse Problems and Imaging*, submitted (2009).
5. M. Kaasalainen, J. Ďurech, B. Warner, Y. Krugly, and N. Gaftonyuk, *Nature* **446**, 420 (2007).
6. J. Kaipio and E. Somersalo, *Statistical and computational inverse problems* (Springer, New York 2005).

Explicit and direct representations of the solutions of nonlinear simultaneous equations

M. Yamada

University Education Center, Gunma University,
4-2, Aramaki-machi, Maebashi City, Gunma, 371-8510, Japan
E-mail: yamadama@cap.ocn.ne.jp

S. Saitoh

Department of Mathematics, University of Aveiro,
Aveiro, 3810-193, Portugal
E-mail: saburou.saitoh@gmail.com

In this paper we shall give explicit representations of inverse mappings of n-dimensional mappings (of the solutions of n-nonlinear simultaneous equations). We derive those concrete formulas from very general ideas for the representation of the inverse functions.

Keywords: Inverse function; integral representation; Green–Stokes formula; nonlinear equation.

1. Introduction

The 2nd author of this paper considered, for any mapping ϕ from an arbitrary abstract set into an arbitrary set, the representation of the inversion ϕ^{-1} in terms of the direct mapping ϕ and obtained some simple concrete formulas from some general ideas in ([1]). Here, we shall give furthermore some general methods and ideas for the inversion formulas for some general nonlinear mappings. We shall first state the principles for our methods for the representations of inverses of nonlinear mappings based on ([1]):

We shall consider some representation of the inversion ϕ^{-1} in terms of some integral form — at this moment, we shall need a natural assumption for the mapping ϕ . Then, we shall transform the integral representation by the mapping ϕ to the original space that is the defined domain of the mapping ϕ. Then, we will be able to obtain the representation of the inverse ϕ^{-1} in terms of the direct mapping ϕ. In [1], we considered the representation of the inverse ϕ^{-1} in some reproducing kernel Hilbert spaces, and in [4], we considered the representations of the inverse ϕ^{-1} for a very concrete situation and we gave a very fundamental representation of the inverse for some general functions on 1 dimensional spaces.

Indeed, note that

$$K(y_1, y_2) = \frac{1}{2} e^{-|y_1 - y_2|} \qquad y_1, \ y_2 \in [A, B]$$

is the reproducing kernel for the Sobolev Hilbert space H_K whose members are real-valued and absolutely continuous functions on $[A, B]$ and whose inner product is given by

$$(f_1, f_2)_{H_K} = \int_A^B \left(f_1'(y) f_2'(y) + f_1(y) f_2(y) \right) dy + f_1(A) f_2(A) + f_1(B) f_2(B)$$

([2]).

For a function $y = f(x)$ such that f is of C^1 class and strictly increasing, and $f'(x)$ is not vanishing on $[a, b]$ ($f(a) = A$, $f(b) = B$), the inverse function $f^{-1}(y)$ is single-valued and belongs to the space H_K and from the reproducing property, we obtain the representation, for any $y_0 \in [f(a), f(b)]$

$$\begin{aligned}
f^{-1}(y_0) &= \left(f^{-1}(\cdot), K(\cdot, y_0) \right)_{H_K} \\
&= \int_{f(a)}^{f(b)} \left((f^{-1})'(y) K_y(y, y_0) + f^{-1}(y) K(y, y_0) \right) dy \\
&\quad + a K(f(a), y_0) + b K(f(b), y_0).
\end{aligned} \tag{1}$$

From this representation, we obtained in ([4]) the very simple representation

$$f^{-1}(y_0) = \frac{a + b}{2} + \frac{1}{2} \int_a^b \operatorname{sign}(y_0 - f(x)) \, dx, \tag{2}$$

through involved calculations. Furthermore, by using the several reproducing kernel Hilbert spaces from [2] as in (1), we calculated similarly with the related assumptions. However, surprisingly enough, we obtain the same formula (2). For formula (2), we directly note that we do not need any smoothness assumption for the function $f(x)$, indeed, we need only the strictly increasing assumption.

Now, we would like to obtain some multi-dimensional versions. At this moment, it seems that we can not find some simple representations as in (1) by some concretely known reproducing kernels for some general domains.

In order to consider some general integral representations for some general domains and functions, we shall recall the fundamental fact:

We can represent a function f in terms of the delta function δ in the form

$$f(q) = \int_D f(p)\delta(p - q)dp \tag{3}$$

in some domain, symbolically. Meanwhile, a fundamental solution $G(p - q)$ for some linear differential operator L is given by the equation, symbolically

$$LG(p - q) = \delta(p - q).$$

So, from (3) we obtain the representation

$$f(q) = \int_D f(p)LG(p - q)dp.$$

Then, we can obtain the representation symbolically, by using the Green–Stokes formula, for some adjoint operator L^* for L,

$$f(q) = \int_D L^* f(p)G(p - q)dp + \text{some} \quad \text{boundary} \quad \text{integrals.} \tag{4}$$

We shall use this type representation. In this approach, we will meet the singular integral representation in the first term of (4), however, if $G(p - q)$ is integrable, then by a simple regularization for $G(p - q)$ we will be able to realize the representation in numerical treatments. In the separate paper [3] we discussed the natural regularization in the form, for example, for the singularity

$$\frac{1}{(|x - y|)^\alpha},$$

we consider the regularization

$$\frac{1}{(|x - y| + \delta)^\alpha}$$

for a small δ and we considered their error estimates.

We considered very concrete cases in the 2 dimensional spaces. It will seem that the results will depend on dimensions, domains and functions spaces dealing with.

Let $D \subset \mathbf{R}^2$ be a bounded domain with a finite number of piecewise C^1 class boundary components. Let f be a one-to-one C^1 class mapping from $\bar{D}$ into $\mathbf{R}^2$ and we assume that its Jacobian $J(x)$ is positive on D. We shall represent f as follows:

$$y_1 = f_1(x) = f_1(x_1, x_2)$$
$$y_2 = f_2(x) = f_2(x_1, x_2) \tag{5}$$

and the inverse mapping f^{-1} of f as follows:

$$x_1 = (f^{-1})_1(y) = (f^{-1})_1(y_1, y_2)$$
$$x_2 = (f^{-1})_2(y) = (f^{-1})_2(y_1, y_2). \tag{6}$$

Then, we represented

$$\begin{pmatrix} (f^{-1})_1(y^*) \\ (f^{-1})_2(y^*) \end{pmatrix} \tag{7}$$

in terms of the direct mapping (5).

By using the generalized Cauchy integral formula for complex-valued functions of C^1 class, we obtained

Proposition 1.1. *([5]) For the mappings (5) and (6) with (7), we obtain the representation, for any $y^* = (y_1^*, y_2^*) \in f(D)$,*

$$\begin{pmatrix} (f^{-1})_1(y^*) \\ (f^{-1})_2(y^*) \end{pmatrix} = \frac{1}{2\pi} \oint_{\partial D} \begin{pmatrix} x_1 \\ x_2 \end{pmatrix} d\mathrm{Arctan} \frac{f_2(x) - y_2^*}{f_1(x) - y_1^*}$$
$$- \frac{1}{2\pi} \int \int_D \frac{1}{|f(x) - y^*|^2} \mathrm{adj}\, J(x) \begin{pmatrix} f_1(x) - y_1^* \\ f_2(x) - y_2^* \end{pmatrix} dx_1 dx_2.$$

We can give the natural version for the 3 dimensional case by using the well-known Poisson integral formula, however, the representation is quite involved. Surprisingly enough we can give some unified and natural inversion formulas for the general dimensions that are better than the formula derived from the Poisson integral formula.

2. n-dimensional formulas

We shall give the very beautiful representation

Theorem 2.1. *Let D be a bounded domain in $\mathbf{R}^n$ with a finite number ∂D of C^1 class boundary components. Let f be a C^1 class real-valued function on $\bar{D}$. For any $\hat{x} \in D$ and for any $n \in \mathbf{N}$ we have the representation*

$$f(\hat{x}) = -c_n \big(df(x), dG_n(x - \hat{x}) \big) + c_n \int_{\partial D} f(x) * dG_n(x - \hat{x}). \tag{8}$$

Here, for $n \le 2, c_n = 1$ and for $n \ge 3, c_n = n - 2$. $*$ is the Hodge star operator, G_n the fundamental solution of the Laplacian $\Delta_n = \sum_{i=1}^{n} \frac{\partial^2}{\partial x_i^2}$, and $(\,,\,)$ the inner product of the vector space $A^k(D)$ comprising of the k order differential forms over D with finite L^2 norms that is

$$(\omega, \eta) = \int_D \omega \wedge *\eta = \int_D \eta \wedge *\omega \quad (\omega, \eta \in A^k(D)).$$

Lemma 2.1. *Let $U_\epsilon(0)$ be an ϵ neighbourhood with centre 0, then*

$$\int_{\partial U_\epsilon(0)} *dG_n(x) = \frac{1}{c_n}.$$

Proof.   Let A_n be $A_n = \frac{2\pi^{\frac{n}{2}}}{\Gamma(\frac{n}{2})}$, the surface measure of the n dimensional unit disk. Then,

$$G_n(x) = \frac{1}{c_n A_n} \begin{cases} |x| & (n = 1) \\ \log|x| & (n = 2) \quad \text{(logarithmic kernel)} \\ -\frac{1}{|x|^{n-2}} & (n \ge 3) \quad \text{(Newton kernel)}. \end{cases}$$

Hence, on $\mathbf{R}^n \setminus U_\epsilon(0)$ we have $dG_n(x) = \frac{\sum_{i=1}^{n} x_i dx_i}{c_n A_n |x|^n}$ $(\forall n \in \mathbf{N})$. Then, for $x = (x_1, \cdots, x_n)$

$$*dG_n(x) = \frac{\sum_{i=1}^{n} (-1)^{i-1} x_i dx_1 \wedge \cdots dx_{i-1} \wedge dx_{i+1} \wedge \cdots \wedge dx_n}{c_n A_n |x|^n} \quad (\forall n \in \mathbf{N}).$$

For a local coordinate $\phi : U_\epsilon(0) \to \mathbf{R}^n$, we denote the pull back $\phi^* * dG_n(x)$ of $*dG_n(x)$ by the polar coordinate, by using $x = \phi(\theta) = \epsilon\tilde{\phi}(\theta), \theta = (\theta_1, \cdots, \theta_{n-1}) \in [0, \pi] \times \cdots \times [0, \pi] \times [0, 2\pi]$,

$$\tilde{\phi}(\theta) = \left(\cos\theta_1, \sin\theta_1 \cos\theta_2, \cdots, \sin\theta_1 \cdots \sin\theta_{n-2} \sin\theta_{n-1}\right),$$

we have $\phi^* * dG_n(x) = \frac{\sin^{n-2}\theta_1 \cdots \sin\theta_{n-2}}{c_n A_n} d\theta_1 \wedge \cdots \wedge d\theta_{n-1}$. Hence,

$$\int_{\partial U_\epsilon(0)} *dG_n(x) = \int_{[0,\pi] \times \cdots \times [0,\pi] \times [0,2\pi]} \phi^* * dG_n(x) = \frac{1}{c_n}. \qquad \square$$

Proof of Theorem 2.1. Let $U_\epsilon(\hat{x})$ be a neighbourhood contained in D. Then, for $G_n(x - \hat{x}) \in C^\infty(D \setminus U_\epsilon(\hat{x}))$ and for $f \in C^1(D)$, $f(x) * dG_n(x - \hat{x}) \in A^1(D \setminus U_\epsilon(\hat{x}))$ that is a C^1 class differential. Hence, on $D \setminus U_\epsilon(\hat{x})$, for $f(x) * dG_n(x - \hat{x})$ we apply the Green–Stokes formula and we have

$$\int_{D\setminus U_\epsilon(\hat{x})} d\{f(x) * dG_n(x - \hat{x})\}$$

$$= \int_{\partial D} f(x) * dG_n(x - \hat{x}) - \int_{\partial U_\epsilon(\hat{x})} f(x) * dG_n(x - \hat{x}).$$

Let δ be the Dirac distribution and ω_v be $dx_1 \wedge \cdots \wedge dx_n$. Then, by $d * dG_n(x - \hat{x}) = \Delta_n G_n(x - \hat{x})\omega_v = \delta(x - \hat{x})\omega_v$, we have

$$d\{f(x) * dG_n(x - \hat{x})\} = df(x) \wedge *dG_n(x - \hat{x}) + (-1)^0 f(x) d * dG_n(x - \hat{x})$$

$$= df(x) \wedge *dG_n(x - \hat{x}) + \delta(x - \hat{x})f(x)\omega_v.$$

Hence,

$$\lim_{\epsilon \to 0} \int_{D\setminus U_\epsilon(\hat{x})} d\{f(x) * dG_n(x - \hat{x})\} = \big(df(x), dG_n(x - \hat{x})\big).$$

As in the proof of Lemma 2.1, from the polar coordinate representation $\phi^*(f(x)*dG_n(x-\hat{x})) = \phi^* f(x)\phi^* * dG_n(x-\hat{x}) = f(\hat{x}+\epsilon\tilde{\phi}(\theta))\phi^* * dG_n(x-\hat{x})$ and from Lemma 2.1,

$$\int_{\partial U_\epsilon(\hat{x})} f(x) * dG_n(x - \hat{x}) = \int_{[0,\pi]\times\cdots\times[0,\pi]\times[0,2\pi]} f(\hat{x} + \epsilon\tilde{\phi}(\theta))\phi^* dG_n(x - \hat{x})$$

$$= \frac{f(\hat{x})}{c_n} \quad (\epsilon \to 0).$$

We thus obtain the desired representation. $\qquad\qquad\qquad\qquad\square$

Theorem 2.2. *In the situation of Theorem 2.1 and we assume furthermore that f is a sense preserving C^1 class function on $\bar{D}$ in $\mathbf{R}^n$ with a single-valued inverse. Then, for $\hat{y} \in f(D)$, we obtain the representation*

$$f_i^{-1}(y_0) = -\int_D dx_i \wedge f^* * dG_n(y - y_0) + \int_{\partial D} x_i f^* * dG_n(y - y_0).$$

Here, f_i^{-1} denotes the i component of f^{-1}.

Proof. For the function f^{-1} on $f(\bar{D})$, we use the representation in Theorem 2.1 and we use the transform of the representation by f. Then, by using the formulas $f^* df_i^{-1}(y) = dx_i$, and $f^*\big(\omega, \eta\big) = \big(f^*\omega, f^*\eta\big)$, we obtain the desired representation. $\qquad\square$

In particular, for $n = 1$, we obtain (2), directly.

For $n = 2$, we obtain (13) and this formula may be represented as follows, from our general formula:

For any $\hat{y} \in f(D)$, we have

$$f_i^{-1}(\hat{y}) = \frac{1}{2\pi}\left(\int_{\partial D} x_i d\theta_i - \int_D dx_i \wedge d\theta_i\right)$$
$$i = 1, 2.$$

Here, $\theta_1 = \text{Arctan}\,\frac{f_1(x)-\hat{y}_1}{f_2(x)-\hat{y}_2}$, $\theta_2 = -\text{Arctan}\,\frac{f_2(x)-\hat{y}_2}{f_1(x)-\hat{y}_1}$.

The fundamental applications of Theorem 2.2 are the identification of the solution space; because for the out side of the solutions, the representations in the right hand side in Theorem 2.2 are zero and applications in the implicit function theory in various situations, because we can represent the implicit functions explicitly.

Acknowledgements

S. Saitoh was supported in part by Research Unit Mathematics and Applications, University of Aveiro, Portugal, through FCT - Portuguese Foundation for Science and Technology. He was also supported in part by the Grant-in-Aid for the Scientific Research (C)(2)(No. 21540111) from the Japan Society for the Promotion Science.

References

1. S. Saitoh, *Proc. Amer. Math. Soc.* **125**, 3633 (1997).
2. S. Saitoh, *Integral Transforms, Reproducing Kernels and Their Applications*, Pitman Res. Notes in Math. Series Vol. **369** (Addison Wesley Longman Ltd., UK, 1997).
3. Y. Sawano, M. Yamada and S. Saitoh, *Singular integrals and natural regularizations, Math. Inequal. Appl.* (to appear).
4. M. Yamada, T. Matsuura and S. Saitoh, *Fract. Calc. Appl. Anal.* **10**, 161 (2007).
5. M. Yamada and S. Saitoh, *Appl. Anal.* **88**, 151 (2009).

V.2. Stochastic analysis

Organisers: D. Crisan, T. Lyons

Stochastic Analysis aims to provide mathematical tools needed to describe and model high dimensional random systems, such tools and models include Stochastic Differential Equations and Stochastic Partial Differential Equations, Infinite Dimensional Stochastic Geometry, Random Media and Interacting Particle Systems, Super-processes, Stochastic Filtering, Martingales, etc. From its beginnings in the 20th century, it has emerged as a core area of Mathematics and its Applications and is currently undergoing quite rapid scientific development.

The Stochastic Analysis section organized during the 7th ISAAC Congress in the period 15–17 July 2008 provided a forum for researchers working on the different aspects of stochastic analysis to present their findings, and to interact with people working in the wider area of Analysis. The session comprised 25 talks on a variety of topics which include: cylindrical Levy processes in Banach spaces, integration by parts for locally smooth laws and applications to jump type diffusions, stochastic differential equations driven by non-linear Levy noise, Poisson equations with fractional noise, evolution equations for communities, a rough path approach to nonlinear stochastic PDEs, solving backward SDEs using cubature methods, unbiased perturbations of the Navier Stokes equation, equivalence of stochastic equations and martingale problems, uniqueness problems for SDEs, accelerated numerical schemes for nonlinear filtering, information and asset pricing, discrete-time models for the interest-rate term structure, risk-sensitive portfolio optimisation with jump diffusion asset prices. The following is the list of speakers to the session: D. Applebaum, V. Bally, M. Caruana, V. Kolokoltsov, I. Gyongy, M. Pistorius, M. Kelbert, A.B. Cruzeiro, A. Davie, M. Davis,

A. Papavasiliou, S. Jacka, L. Hughston, B. Rozovsky, M. Hairer, A. Mijatovic, D. Brody, X.-M. Li, T. Lyons, T. Kurtz, D. Crisan, M. Sanz-Solè, R. Tunaru, M. Tretyakov, E. Usoltseva. In addition, P. Malliavin gave a plenary talk on Thursday 16 July.

Information and asset pricing

Dorje C. Brody

Department of Mathematics, Imperial College London, London SW7 2AZ, UK
E-mail: dorje@imperial.ac.uk

The idea of information-based asset pricing theory, whereby one constructs from the outset the market filtration to deduce price processes of a range of assets to price and hedge financial instruments, is described.

Keywords: Information process; BHM theory; bridge processes.

1. Information-driven asset-price dynamics

The purpose of this paper is to survey briefly the information-based asset pricing framework.[1] Further details can be found in the literature.[2-8] The conventional paradigm for derivatives pricing and risk management relies on the direct modelling of price process of the underlying asset. Movements of prices are, however, induced primarily by market information. Thus, we model the flow of information about impending debt obligation and equity dividend payments, from which we derive the price processes.

We model the financial markets with the specification of a probability space $(\Omega, \mathcal{F}, \mathbb{Q})$ with filtration $\{\mathcal{F}_t\}$. The probability measure $\mathbb{Q}$ is the risk-neutral measure, and the filtration $\{\mathcal{F}_t\}$ is the 'market filtration'. We assume the absence of arbitrage and the existence of a pricing kernel so that the existence of a unique risk-neutral measure is ensured. The default-free discount-bond system $\{P_{tT}\}_{0 \leq t \leq T}$ is assumed deterministic. It follows that if the random variable H_T represents a cash-flow occurring at time T, then its value S_t at any earlier time t is given by $S_t = P_{tT}\mathbb{E}[H_T|\mathcal{F}_t]$. We shall model explicitly the market filtration $\mathcal{F}_t$, and 'derive' the price process.

Evidently, *some* partial information regarding the value of the impending repayment H_T is available at earlier times. This information will in general be imperfect. We shall assume that the following $\{\mathcal{F}_t\}$-adapted market information process is accessible to market participants:

$$\xi_t = \sigma H_T t + \beta_{tT}. \tag{1}$$

The parameter σ represents the rate at which the true value of H_T is 'revealed' as time progresses. Here the process $\{\beta_{tT}\}_{0 \leq t \leq T}$ is a standard Brownian bridge on $[0, T]$. We assume that $\{\beta_{tT}\}$ is independent of H_T, and thus represents pure noise.

2. Determining the price process and its dynamics

Consider the situation where $H_T = h_i$ ($i = 0, 1, \ldots, n$) with *a priori* probability $\mathbb{Q}[H_T = h_i] = p_i$. In the case of a defaultable bond, we think of $H_T = h_n$ as the case of no default, and other cases as various degrees of partial recovery. The conditional probability $\pi_{it} = \mathbb{Q}(H_T = h_i | \xi_t)$ is then:

$$\pi_{it} = \frac{p_i \exp\left[\frac{T}{T-t}(\sigma h_i \xi_t - \frac{1}{2}\sigma^2 h_i^2 t)\right]}{\sum_i p_i \exp\left[\frac{T}{T-t}(\sigma h_i \xi_t - \frac{1}{2}\sigma^2 h_i^2 t)\right]}. \tag{2}$$

Therefore, the conditional expectation $H_{tT} = \mathbb{E}[H_T | \xi_t]$ of H_T is $\sum_i h_i \pi_{it}$. The asset price process $\{S_t\}$ is then given by $S_t = P_{tT} H_{tT}$.

It is of interest to analyse the dynamics of $\{S_t\}$. The key relation we need for determining the dynamics of $\{S_t\}$ is that the conditional probability $\{\pi_{it}\}$ satisfies a diffusion equation of the form

$$\mathrm{d}\pi_{it} = \frac{\sigma T}{T - t}(h_i - H_{tT})\pi_{it}\mathrm{d}W_t. \tag{3}$$

The 'innovations' process $\{W_t\}_{0 \leq t < T}$ appearing here, defined by

$$W_t = \xi_t + \int_0^t \frac{1}{T - s} \xi_s \, \mathrm{d}s - \sigma T \int_0^t \frac{1}{T - s} H_{sT} \, \mathrm{d}s, \tag{4}$$

is an $\{\mathcal{F}_t\}$-Brownian motion. The Brownian motion $\{W_t\}$ arising in this way can thus be regarded as part of the information accessible to market participants. Unlike $\{\beta_{tT}\}$, the value of W_t contains 'real' information. It follows that for the asset price dynamics we have $\mathrm{d}S_t = r_t S_t \, \mathrm{d}t + \Sigma_{tT} \, \mathrm{d}W_t$. Here r_t is the deterministic short rate. The absolute bond volatility Σ_{tT} is given by $\Sigma_{tT} = \sigma T P_{tT} V_{tT}/(T - t)$. The process $\{V_{tT}\}$ appearing here is defined by the relation $V_{tT} = \sum_i (h_i - H_{tT})^2 \pi_{it}$.

The present framework allows for a simple and natural simulation methodology for the dynamics of defaultable bonds and related structure. For each 'run' of the simulation we choose at random a value for H_T, and a sample path for the Brownian bridge, i.e. for each $\omega \in \Omega$ we simulate $\xi_t(\omega) = \sigma t H_T(\omega) + \beta_{tT}(\omega)$ for $t \in [0, T]$. One way of simulating $\{\beta_{tT}\}$ is to write $\beta_{tT} = B_t - \frac{t}{T}B_T$, where $\{B_t\}$ is a standard Brownian motion.

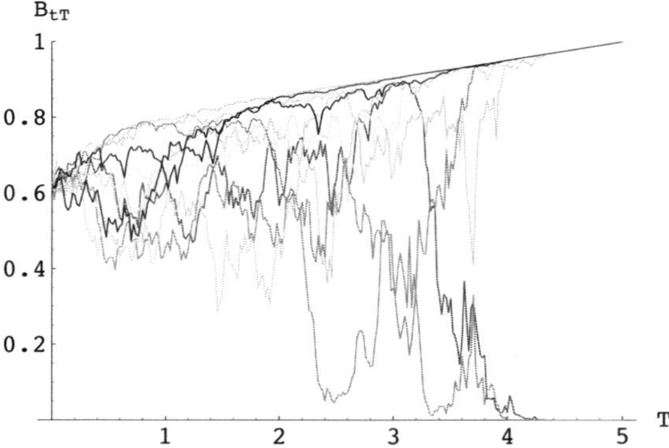

Fig. 1.   Examples of sample paths for $\sigma = 1$.

3. Defaultable bond option pricing

For a European call option on a credit-risky bond with strike K, option maturity t, and bond maturity T, the call price $C_0 = P_{0t}\mathbb{E}[(S_t - K)^+]$ is

$$C_0 = P_{0t}\,\mathbb{E}\left[\frac{1}{\Phi_t}\left(\sum_{i=0}^{n}\left(P_{tT}h_i - K\right)p_{it}\right)^+\right], \tag{5}$$

where $p_{it} = \pi_i \exp\left[\frac{T}{T-t}\left(\sigma h_i \xi_t - \frac{1}{2}\sigma^2 h_i^2 t\right)\right]$ are the unnormalised conditional probabilities, and we also have $\Phi_t = \sum_i p_{it}$. The martingale Φ_t^{-1} can be used as change of measure $(\mathbb{Q} \Rightarrow \mathbb{B})$ such that $\{\xi_t\}$ is a $\mathbb{B}$-Brownian bridge.

In the case of a binary bond $(H_T = h_0, h_1)$, we obtain:

$$C_0 = P_{0t}\left[\pi_1(P_{tT}h_1 - K)N(d_+) - \pi_0(K - P_{tT}h_0)N(d_-)\right], \tag{6}$$

with $\tau = tT/(T - t)$, and

$$d_\pm = \frac{\ln\left[\frac{\pi_1(P_{tT}h_1 - K)}{\pi_0(K - P_{tT}h_0)}\right] \pm \frac{1}{2}\sigma^2\tau(h_1 - h_0)^2}{\sigma\sqrt{\tau}(h_1 - h_0)}. \tag{7}$$

Option delta is specified by $\Delta = \partial C_0/\partial S_0$. A short calculation then gives

$$\Delta = \frac{(P_{tT}h_1 - K)N(d^+) + (K - P_{tT}h_0)N(d^-)}{P_{tT}(h_1 - h_0)}, \tag{8}$$

in the case of a binary option. It is interesting that an explicit formula for a consistent hedging strategy can be derived in the present framework, in contrast to other models employed in the market for defaultable bonds.

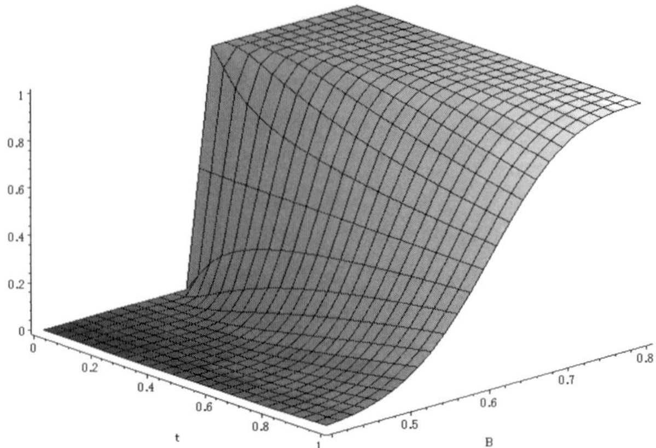

Fig. 2. Delta hedge for a defaultable digital bond option with $r = 5\%$, $\sigma = 25\%$, $T = 2$ year, and $K = 0.6$.

4. Geometric Brownian motion model

More generally, the 'signal' component of the information process is associated with some market factor: $\xi_t = \sigma t X_T + \beta_{tT}$, where the Brownian bridge $\{\beta_{tT}\}$ is independent of X_T. The simplest application of this technique arises in the case of geometric Brownian motion models. Consider a limited-liability company that makes a cash distribution S_T at time T. We assume that S_T has a log-normal distribution under $\mathbb{Q}$:

$$S_T = S_0 \exp\left(rT + \nu\sqrt{T}X_T - \tfrac{1}{2}\nu^2 T\right), \tag{9}$$

where the market factor X_T is normally distributed with mean zero and variance one, and $r > 0$ and $\nu > 0$ are constants. If we set $\sigma = 1/\sqrt{T}$, then a short calculation shows that

$$S_t = \mathrm{e}^{-r(T-t)}\mathbb{E}[S_T|\mathcal{F}_t] = S_0 \exp\left(rt + \nu W_t - \tfrac{1}{2}\nu^2 t\right), \tag{10}$$

which is just the familiar geometric Brownian motion model. An important point to note here is that the Brownian bridge process $\{\beta_{tT}\}$ appears quite naturally in this context. If we start with S_t then we can use the orthogonal decomposition: $W_t = \frac{t}{T}W_T + \left(W_t - \frac{t}{T}W_T\right)$. Thus by writing $X_T = W_T/\sqrt{T}$ and $\sigma = 1/\sqrt{T}$ we find that the right side of this expression is indeed the market information. In other words, *formulated in the information-based framework, the standard Black-Scholes-Merton theory can be expressed in terms of a normally distributed X-factor and an independent Brownian bridge noise process.*

5. Cumulative gains with gamma information

We consider now the situation where a random variable X_T represents the final result of a series of gains made over the interval $[0, T]$. There are many financial problems that are of this type. For example X_T might be proportional to the total number of accidents of a certain type that have occurred, representing the totality of claims payable for the given period. In this case ξ_t represents the claims payable or paid so far. Or X_T might represent the total loss on a credit portfolio. Our goal is to model the value process $\{S_t\}_{0 \le t \le T}$ of the claim that pays out X_T at time T. The market information in these cases are driven by a gamma bridge.

We begin with a general discussion on properties of gamma processes and associated bridges. We fix a probability space $(\Omega, \mathcal{F}, \mathbb{Q})$. By a *standard gamma process* $\{\gamma_t\}_{0 \le t \le \infty}$ with rate m we mean a process with independent increments such that $\gamma_0 = 0$ and such that the random variable $\gamma_u - \gamma_t$ for $u \ge t \ge 0$ has a gamma distribution with parameter $m(u - t)$:

$$g(x) = x^{m(u-t)-1} \mathrm{e}^{-x} / \Gamma[m(u - t)]. \tag{11}$$

It follows from $\Gamma[a + 1]/\Gamma[a] = a$ that $\mathbb{E}[\gamma_t] = mt$, which justifies the interpretation of the parameter m as the *mean growth rate* of the process. A straightforward calculation shows that $\mathbb{E}\left[\mathrm{e}^{\mathrm{i}\lambda\gamma_t}\right] = (1 - \mathrm{i}\lambda)^{-mt}$ for $t \ge 0$ and all $\lambda \in \mathbb{C}$ such that $\mathrm{Im}(\lambda) > -1$. It follows that $\mathrm{Var}[\gamma_t] = mt$. From the independent increments property we deduce that $\{(1 + \alpha)^{mt}\mathrm{e}^{-\alpha\gamma_t}\}$ is a geometric gamma martingale. In general, we let $\{L_n^{(k)}(z)\}$ denotes the associated Laguerre polynomials:

$$(1 + \alpha)^{mt}\mathrm{e}^{-\alpha z} = \sum_{n=0}^{\infty} L_n^{(mt-n)}(z)\,\alpha^n. \tag{12}$$

Then $\{L_n^{(mt-n)}(\gamma_t)\}_{n=0,1,\ldots,\infty}$ are martingales.

Now suppose that $\{\gamma_t\}_{0 \le t \le \infty}$ is a standard gamma process with rate m. For fixed T define the process $\{\gamma_{tT}\}_{0 \le t \le T}$ by $\gamma_{tT} = \gamma_t/\gamma_T$. Then clearly we have $\gamma_{0T} = 0$ and $\gamma_{TT} = 1$. We refer to $\{\gamma_{tT}\}$ as the standard *gamma bridge* over the interval $[0, T]$ associated with the gamma process $\{\gamma_t\}$. The density function of the random variable γ_{tT} is given by

$$f(y) = y^{mt-1}(1 - y)^{m(T-t)-1} / \mathrm{B}[mt, m(T - t)], \tag{13}$$

where $\mathrm{B}[a, b] = \Gamma[a]\Gamma[b]/\Gamma[a + b]$. The gamma bridge has the property that for all $T \ge t \ge 0$ the random variables γ_t/γ_T and γ_T are independent.[9]

6. Gamma information and the valuation of claims

Our objective now is to calculate the value at time t of the claim that pays X_T at time T, where X_T is a positive random variable. The market filtration is generated by an information process $\{\xi_t\}_{0\leq t\leq T}$ of the form

$$\xi_t = X_T\gamma_{tT}, \tag{14}$$

where $\{\gamma_{tT}\}$ is a $\mathbb{Q}$-gamma bridge with parameter m. The gamma bridge $\{\gamma_{tT}\}$ is independent of the random variable X_T, and represents the noise.

Since the gamma information process $\{\xi_t\}_{0\leq t\leq T}$ is Markovian, the value S_t of the claim at time t is thus given by $S_t = 1_{\{t<T\}}P_{tT}\mathbb{E}\left[X_T\,|\,\xi_t\right]$. By use of the Bayes formula, the conditional probability density for X_T, $\pi_t(x)\mathrm{d}x = \mathrm{d}\mathbb{Q}[X_T \leq x\,|\,\xi_t]$, can be obtained, from which we deduce that

$$S_t = P_{tT}\frac{\int_{\xi_t}^{\infty} p(x)x^{2-mT}\left(x - \xi_t\right)^{m(T-t)-1}\,\mathrm{d}x}{\int_{\xi_t}^{\infty} p(x)x^{1-mT}\left(x - \xi_t\right)^{m(T-t)-1}\,\mathrm{d}x}. \tag{15}$$

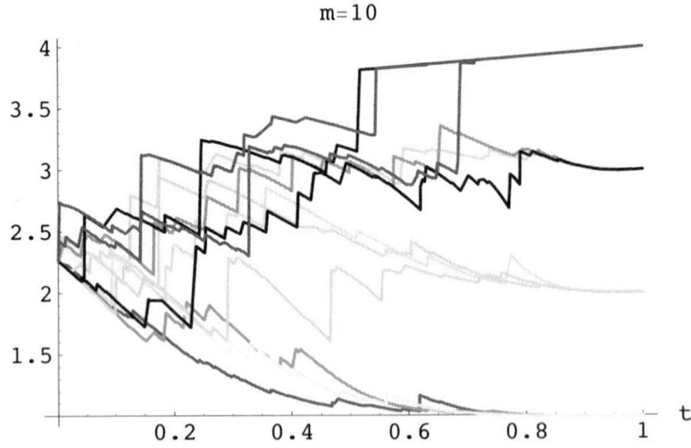

Fig. 3.   Sample paths for the price process. The rate parameters are $m = 10$ and $r = 5\%$.

We now consider the problem of evaluating derivatives whose payoffs are determined by the value ξ_τ of the market information at the derivative maturity date τ. We call such a derivative an *information derivative*. Our goal is to derive an expression for the price of an option that pays $(S_\tau - K)^+$ at some time $\tau \in [0, T]$. It is convenient, however, to consider first a more primitive type of derivative, a type of Arrow-Debreu security defined by the payoff $\delta(\xi_\tau - y)$ at time τ, where y is a parameter. We write $\mathfrak{A}_{s\tau}(y) = P_{s\tau}\mathbb{E}_s[\delta(\xi_\tau - y)]$ for the value at time s of such an elementary information derivative. The price of the Arrow-Debreu security then determines the risk-neutral probability density for the random variable ξ_τ. That is, if $H_\tau = H(\xi_\tau)$ is the payoff at time τ of a more general information derivative, then we have the following expression for its present value: $H_s = \int_0^\infty \mathfrak{A}_{s\tau}(y)H(y)\mathrm{d}y$. Evidently, from (ifgammaAsset) we see that the value S_τ of the asset at time τ can be expressed as a function of ξ_τ. Therefore, a derivative on the asset can be regarded as an example of an exotic information derivative. After some calculation we find:

$$\mathfrak{A}_{s\tau}(y) = P_{s\tau}\frac{y^{m\tau-1}}{\mathrm{B}[m\tau, m(T-\tau)]}\int_y^\infty \pi_s(x)\,x^{1-mT}(x-y)^{m(T-\tau)-1}\mathrm{d}x. \tag{16}$$

Next, consider the valuation of call options on the asset. The option maturity is set at time $\tau < T$, and the strike at K. The initial value $P_{0\tau}\mathbb{E}[(S_\tau - K)^+]$ of the call option is therefore given by

$$C_0 = \int_0^\infty \mathfrak{A}_{0\tau}(y)[S(\tau, y) - K]^+\mathrm{d}y, \tag{17}$$

where the function $S(\tau, y)$ is defined by

$$S(\tau, y) = P_{\tau T}\frac{\int_y^\infty p(x)x^{2-mT}(x-y)^{m(T-\tau)-1}\mathrm{d}x}{\int_y^\infty p(x)x^{1-mT}(x-y)^{m(T-\tau)-1}\mathrm{d}x}. \tag{18}$$

If y^* is the value that satisfies $S(\tau, y^*) = 0$, we find, after some algebra,

$$C_0 = P_{0\tau}\int_{y^*}^\infty p(x)\,(x\,P_{\tau T} - K)\,B(y^*/x)\,\mathrm{d}x, \tag{19}$$

where $B(u)$ is the right-cumulative Beta distribution function.

When X_T takes discrete values x_i with probability p_i the call option is:

$$C_0 = P_{0\tau}\sum_{i=0}^n p_i\,(x_i P_{\tau T} - K)\,B(y^*/x_i)\mathbf{1}_{\{x_i>y^*\}}. \tag{20}$$

In particular, if X_T is a binary variable, this reduces to

$$C_0 = P_{0\tau}\left[p_0\,(x_0 P_{\tau T} - K)\,B\left(\frac{y^*}{x_0}\right) + p_1\,(x_1 P_{\tau T} - K)\,B\left(\frac{y^*}{x_1}\right)\right], \tag{21}$$

where $y^* = (\theta x_1 - x_0)/(\theta - 1)$ and

$$\theta = \left[\frac{p_1(K - P_{\tau T}x_1)}{p_0(P_{\tau T}x_0 - K)}(\frac{x_1}{x_0})^{1-mT}\right]^{\frac{1}{m(T-\tau)-1}}. \tag{22}$$

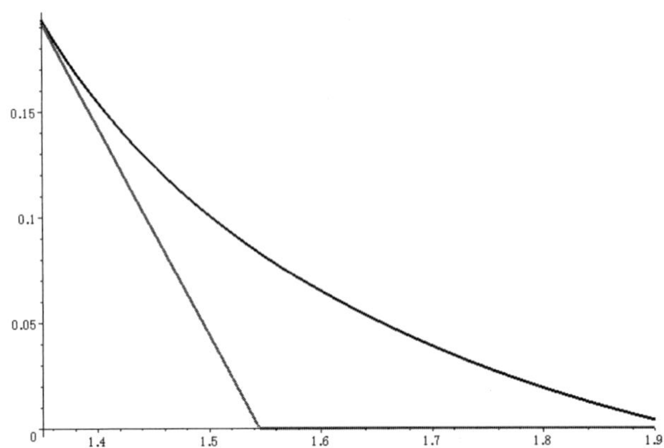

Fig. 4.   Call price as a function of the strike K. The parameters are: $r = 5\%$, $m = 4.5$, $T = 1$ year, $\tau = 0.3$ year, and $S_0 = 1.52196$.

In figure 4 we plot a typical call price as a function of the option strike K.

References

1. D. C. Brody, L. P. Hughston, & A. Macrina, in *Advances in Mathematical Finance, Festschrift volume in honour of D. Madan* (Birkhäuser, Basel, 2007).
2. D. C. Brody, L. P. Hughston, & A. Macrina, *Int. J. Theo. App. Finance* **11** 107 (2008).
3. D. C. Brody, L. P. Hughston, & A. Macrina, *Proc. R. Soc. London* A**464** 1801 (2008).
4. D. C. Brody, M. H. A. Davis, R. L. Friedman, & L. P. Hughston, *Proc. R. Soc. London* A**465** 1103 (2009).
5. D. C. Brody, & R. L. Friedman, *Risk Magazine* **December** 105 (2009)
6. L. P. Hughston, & A. Macrina, In *Advances in mathematics of finance* (ed. L. Stettner). Banach Centre Publications **83** 117 (Warsaw, Poland: Polish Academy of Sciences 2007).
7. A. Macrina, An information-based framework for asset pricing: X-factor theory and its applications. PhD thesis, King's College London (2006).
8. M. Rutkowski, & N. Yu, *Int. J. Theo. App. Finance* **10** 557 (2007).
9. M. Yor, in *Advances in Mathematical Finance, Festschrift volume in honour of D. Madan* (Birkhäuser, Basel, 2007).

Solving backward stochastic differential equations using the cubature method.
Application to nonlinear pricing

D. Crisan and K. Manolarakis

Department of Mathematics,
Imperial College London,
180 Queen's Gate
London, SW7 2AZ, UK

We propose a new algorithm for the numerical solution of backward stochastic differential equations (BSDEs) with the terminal condition being a function of X_T, where $X = \{X_t, t \in [0, T]\}$ is the solution to a standard stochastic differential equation. Using the property that the solution of a BSDE can be written as an integral of a certain functional against the law of the underlying diffusion, this new algorithm combines the Bouchard–Touzi[1]–Zhang[2] discretization of BSDEs with a weak approximation method known as cubature on Wiener space, constructed by Lyons and Victoir.[3] The main results concerning the propagation of the error are reported and a numerical example is included.

1. Introduction

On a complete probability space $(\Omega, \mathcal{F}, \mathbb{P})$ endowed with a filtration $\{\mathcal{F}_t\}_{t \geq 0}$ satisfying the usual conditions, we consider an $\mathcal{F}_t$-adapted d–dimensional Brownian motion W driving the following forward–backward stochastic differential equation:

$$
\begin{cases}
dX_t = V_0(X_t)dt + \sum_{j=1}^d V_j(X_t) \circ dW_t^j \\
-dY_t = f(t, X_t, Y_t, Z_t)dt - Z_t dW_t \\
X_0 = x, \quad Y_T = \Phi(X_T)
\end{cases}
\tag{1}
$$

In (1), $T > 0$ is the time horizon, $V_i : \mathbb{R}^d \to \mathbb{R}^d$, $i = 0, \ldots, d$ are appropriate vector fields, $\circ$ denotes stochastic integration in the Stratonovitch sense, $f : [0, T] \times \mathbb{R}^d \times \mathbb{R} \times \mathbb{R}^d \to \mathbb{R}$ is a real valued function called the driver and $\Phi : \mathbb{R}^d \to \mathbb{R}$ is the terminal condition.

A solution to (1) is a triplet of $\mathcal{F}_t$-*adapted* processes (X, Y, Z). The existence of the process X can be established by appealing to classical results concerning the existence and uniqueness of solutions to stochastic differential equations. On the other hand, it is due to the work of Pardoux and Peng,[4] that one can justify the existence of solutions (Y, Z) for the backward part of (1) when the driver f is Lipschitz in its spatial variables (see also the excellent survey article[5]).

One of the celebrated results in the theory of BSDEs, is the so-called Feynman-Kac formula that connects the stochastic flow process $(X^{t,x}, Y^{t,x}, Z^{t,x})$, $t \in [0, T]$, $x \in \mathbb{R}^d$ associated to the system (1):

$$\begin{cases} dX_s^{t,x} = V_0(X_s^{t,x})ds + \sum_{j=1}^d V_j(X_s^{t,x}) \circ dW_s^j, \\ -dY_s^{t,x} = f(s, X_s^{t,x}, Y_s^{t,x}, Z_s^{t,x})ds - Z_s^{t,x}dW_s, \quad s \in [t, T] \\ X_t^{t,x} = x, \qquad Y_T^{t,x} = \Phi(X_T^{t,x}) \end{cases} \quad (2)$$

with the solution of the semi–linear PDE (see Pardoux and Peng[6]).

$$\frac{du}{dt} = -\tilde{V}(x) \cdot \nabla u - \frac{1}{2} Tr \left[V(x) V^*(x) D^2 u \right] - f(t, x, u, \nabla u V(x)), \quad (3)$$

where

$$\tilde{V}(x) = V_0(x) - \frac{1}{2} \sum_{j=1}^d \nabla V_j(x) V_j(x). \quad (4)$$

In (3), V is the matrix valued function with columns $V_i(x)$, $i = 1, \ldots, d$, $V^*(x)$ is the transpose of $V(x)$ and u has final condition $u(T, x) = \Phi(x)$. If $u(t, x)$ is the (viscosity) solution of (3) which is once continuously differentiable with respect to x, then

$$Y_s^{t,x} = u(s, X_s^{t,x}), \qquad Z_s^{t,x} = \nabla u(s, X_s^{t,x}) V(X_s^{t,x}). \quad (5)$$

In effect, any algorithm of the numerical solution of a BSDE provides an approximation method for semilinear PDEs.

The paper is organized as follows: In Section 2, we introduce the Bouchard-Touzi-Zhang discretization of the solution of the BSDE followed by the cubature method presented in Section 3. In Section 4 we introduce the algorithm and the state the main convergence result. We conclude with a numerical result in Section 5. The interested reader can find proofs and further details in Crisan and Manolarakis.[7]

2. Discretization and simulation of BSDEs

The form of the system (1), or equivalently (2), together with the non linear Feynman Kac formula (5), makes it intuitively clear that the analysis of the behaviour of the processes Y and Z will depend, to a large extent, on the law of the process X. By writing the backward part of (1) in integral form and integrating, we obtain

$$Y_t^{t,x} = \mathbb{E}\left[\Phi(X_T^{t,x}) + \int_t^T f(s, X_s^{t,x}, Y_s^{t,x}, Z_s^{t,x}) ds\right]$$

or equivalently

$$Y_t^{t,x} = \mathbb{E}\left[\Phi(X_T^{t,x}) + \int_t^T f(s, X_s^{t,x}, u(s, X_s^{t,x}), u(s, X_s^{t,x})V(X_s^{t,x})) ds\right],$$

where u solves equation (3). In other words, there exists an implicitly defined functional $\Lambda_t \;:\; C\left([t,T]:\mathbb{R}^d\right) \;\to\; \mathbb{R}$, where $C\left([t,T]:\mathbb{R}^d\right) := \left\{\alpha : [t,T] \to \mathbb{R}^d \,\middle|\, \alpha \text{ is continuous}\right\}$ such that,

$$Y_t^{t,x} = \mathbb{E}\left[\Lambda_t(X_{\cdot}^{t,x})\right]$$

Hence, to approximate $Y_t^{t,x}$, one needs to approximate the functional Λ_t and integrate this approximation against the law of $X^{t,x}$ (or an approximation of this law). This first step is performed in the following manner: Given a partition $\pi = \{t = t_0 < \ldots < t_{n-1} < t_n = T\}$ of $[t,T]$ we define the family of one–step backward operators $\{R_i\}_{i=0}^{n-1}$ as

$$R_i g\,(x) = \mathbb{E}\left[g\left(X_{t_{i+1}}^{t_i,x}\right)\right]$$
$$+ h_{i+1} f\left(t_i, x, R_i g\,(x), \frac{1}{h_{i+1}}\mathbb{E}\left[g\left(X_{t_{i+1}}^{t_i,x}\right)\left(W_{t_{i+1}} - W_{t_i}\right)\right]\right) \tag{6}$$

and $h_j = t_j - t_{j-1}$. Observe that this definition corresponds to an Euler type discretization of the backward part of (2). For example, at time t_{n-1} an approximation of $Y_{t_{n-1}}^{t_{n-1},x}$ would be $R_{n-1}\Phi(x)$. More generally, we define

$$Y_{t_i}^{t_i,x,\pi} := R_i \circ R_{i+1} \ldots \circ R_{n-1}\Phi(x), \quad i = 0,\ldots,n-1, \quad x \in \mathbb{R}^d. \tag{7}$$

Obviously, we now have a discrete process Y_t^{π} (this may be defined as a step process changing values at the partition points according to (7)), but there still remains the issue of computing the involved (conditional) expectations. In general the law of $X_{\cdot}$ is not explicitly known and one needs to introduce an approximation of this law. Let the operator $\hat{\mathbb{E}}[\cdot]$ denote integration with respect to this approximation. We then define

$$\hat{R}_i g\,(x) = \hat{\mathbb{E}}[g(X_{t_{i+1}}^{t_i,x})]$$
$$+ h_{i+1} f(t_i, x, \hat{R}_i g\,(x), \frac{1}{h_{i+1}}\hat{\mathbb{E}}[g(X_{t_{i+1}}^{t_i,x})(W_{t_{i+1}} - W_{t_i})]) \tag{8}$$

and

$$\hat{Y}_{t_i}^{t_i,x,\pi} := \hat{R}_i \circ \hat{R}_{i+1} \ldots \circ \hat{R}_{n-1}\Phi(x), \quad i = 0,\ldots,n-1, \quad x \in \mathbb{R}^d. \tag{9}$$

A computable approximation of $Y_{t_i} = u(t_i, X_{t_i})$ is then $\hat{Y}_{t_i}^{t_i,X_{t_i},\pi} := \hat{R}_i \circ \hat{R}_{i+1} \ldots \circ \hat{R}_{n-1}\Phi(X_{t_i})$. The composition of $\hat{R}_{0,n-1} := \hat{R}_0 \ldots \hat{R}_{n-1}$ provides an approximation for $\mathbb{E}[\Lambda_t(X_\cdot^{t,x})]$ and, implicitly, for the backward component. The error is then measured in terms of the difference

$$\sup_x \left| \mathbb{E}[\Lambda_t(X_\cdot^{t,x})] - \hat{R}_{0,n-1}\Phi(x) \right| = \sup_x \left| Y_t^{t,x} - \hat{R}_{0,n-1}\Phi(x) \right|$$

or, alternatively, in terms of

$$\mathbb{E}\left[\left| Y_t - \hat{R}_{0,n-1}\Phi(X_t) \right| \right] = \int_{\mathbb{R}^d} \left| Y_t^{t,x} - \hat{R}_{0,n-1}\Phi(x) \right| P_{X_t}(dx).$$

In this algorithm the approximation to the law of $X_\cdot$ is provided by the cubature on Wiener space method.

3. Cubature

From this point on we use the standard identification of a (smooth) vector field $V : \mathbb{R}^d \to \mathbb{R}^d$ with the corresponding first order operator:

$$V_i \equiv \sum_{k=1}^{d} V_i^k \frac{\partial}{\partial x_k},$$

and for any multi idex $\alpha = (\alpha_1, \ldots, \alpha_k)$ we write $V_\alpha f := V_{\alpha_1} \ldots V_{\alpha_k} f$ for the iterated application of these vector fields. On the set of multi–indices, we consider the norm $\|\beta\| := |\beta| + card\{j : \beta_j = 0, \ 1 \leq j \leq |\beta|\}$ and the hierarchical sets

$$\mathcal{A}_m = \{\beta : \ \|\beta\| \leq m\}.$$

Given any sufficiently smooth function $f : \mathbb{R}^d \to \mathbb{R}$, the Stratonovich–Taylor expansion tells us that

$$f(X_t^{0,x}) = \sum_{\alpha \in \mathcal{A}_m} V_\alpha f(x) J_\alpha[1]_{0,t} + R_m(f,t,x),$$

where $J_\alpha[\cdot]_{s,t}$, $s < t$ stands for the iterated Stratonovich integral (see chapter 5 of Kloeden and Platen[8] for details). The term $R_m(f,t,x)$ is called the remainder and is generally small (for $t \ll 1$). Hence, one may approximate $\mathbb{E}[g(X_t^{0,x})]$ by computing the expectation of the iterated integrals $J_\alpha[1]_{0,t}$ up to a certain length, depending on the error estimate we wish to achieve.

Definition 3.1. We will say that the positive weights $\lambda_1, \ldots, \lambda_N$ and the paths of bounded variation $\omega_1, \ldots \omega_N : [0, T] \to \mathbb{R}^d$ define a cubature formula of degree m at time t, if and only if for any multi index $(i_1, \ldots, i_k) \in \mathcal{A}_m$

$$\mathbb{E}\left[\int_{0<t_1<\ldots<t_k<t} \circ dW_{t_1}^{i_1} \ldots \circ dW_{t_k}^{i_k}\right]$$
$$= \sum_{j=1}^{N} \lambda_j \int_{0<t_1<\ldots<t_k<t} d\omega_1^{i_1}(t_1) \ldots d\omega_N^{i_k}(t_k), \quad (10)$$

where we use the convention that $W^0(t) = t$ and similarly for any ω_j, $j = 1, \ldots, N$, $\omega^0(t) = t$.

The measure $\mathbb{Q}_t := \sum_{j=1}^{N} \lambda_j \delta_{\omega_j}$ is called the cubature measure at time t. The following result is due to Lyons and Victoir.[3]

Theorem 3.1. *There exist positive weights* $\lambda_1, \ldots, \lambda_N$ *and paths of bounded variation* $\omega_{T,1}, \ldots, \omega_{T,N}$ *with* $N \leq card\mathcal{A}_m$ *that define a cubature formula of degree* m *at time* T.

For our algorithm, the approximation to the law of the forward diffusion $X^{0,x}_{\cdot}$ is defined by means of the cubature measure. In the notation introduced in (8), we have

$$\hat{\mathbb{E}}\left[\cdot | X_{t_i} = x\right] = \mathbb{E}_{\mathbb{Q}_{h_{i+1}}}\left[\cdot | X_{t_i} = x\right].$$

The following are the details of the method:

4. The Algorithm

For any path ω among the ones defining the cubature formula we form the ODE corresponding to the forward differential equation of (1) by substituting the Brownian motion by ω. Let $\Xi_{T,x}(\omega)$ be the solution at time T of the ODE

$$\begin{cases} dy_{t,x} = \sum_{i=0}^{d} V_i(y_{t,x}) d\omega^i(t) \\ y_{0,x} = x \end{cases}$$

driven by ω, where we understand that $\omega^0(t) = t$. Then,

$$\mathbb{E}_{\mathbb{Q}_{h_{i+1}}}\left[g\left(X_{t_{i+1}}^{t_i,x}\right)\right] = \sum_{j=1}^{N} \lambda_j \, g\left(\Xi_{h_{i+1},x}(\omega_{h_{i+1},j})\right)$$

$$\mathbb{E}_{\mathbb{Q}_{h_{i+1}}}\left[g\left(X_{t_{i+1}}^{t_i,x}\right)\Delta W_{t_{i+1}}\right] = \sum_{j=1}^{N} \lambda_j \, g\left(\Xi_{h_{i+1},x}(\omega_{h_{i+1},j})\right)$$

$$\times \left(\omega_{h_{i+1},j}(t_{i+1}) - \omega_{h_{i+1},j}(t_i)\right)$$

Example 4.1. Consider a cubature formula defined with two paths ω_1, ω_2 and with two time steps $0 = t_0 < t_1 < t_2 = 1$. Then the forward diffusion will assume the values on the tree

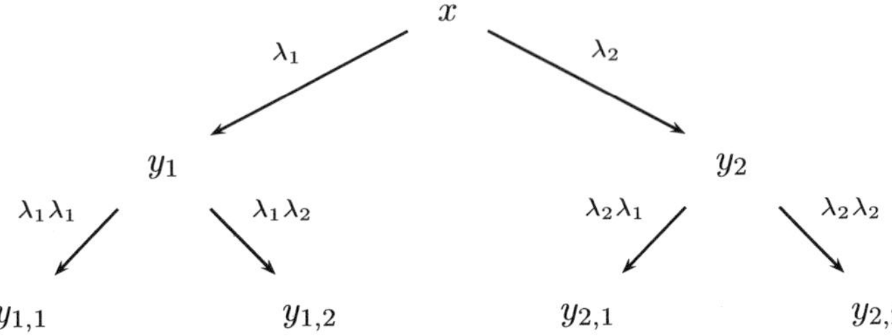

where

$$y_i = \Xi_{h_1,x}(\omega_{h_1,i}), \quad y_{i,j} = \Xi_{h_2,\Xi_{h_1,x}(\omega_{h_1,i})}(\omega_{h_2,j}), \quad i,j = 1,2,$$

given that $X_0 = x$. Hence, for $j = 1, 2$

$$\hat{R}_1 \Phi(y_j) =$$
$$\lambda_1 \Phi(y_{j,1}) + \lambda_2 \Phi(y_{j,2})$$
$$+ h_2 f\left(t_1, y_j, \hat{R}_1 \Phi(y_j), \frac{1}{h_2}\left(\lambda_1 \Phi(y_{j,1})\omega_{h_2,1}(t_2) + \lambda_2 \Phi(y_{j,2})\omega_{h_2,2}(t_2)\right)\right)$$

and

$$\hat{R}_0 \hat{R}_1 \Phi(x)$$
$$= \lambda_1 \hat{R}_1 \Phi(y_1) + \lambda_2 \hat{R}_1 \Phi(y_2)$$
$$+ h_1 f\left(0, x, \hat{R}_0 \hat{R}_1 \Phi(x), \frac{1}{h_1}\left(\lambda_1 \hat{R}_1 \Phi(y_1)\omega_{h_1,1}(t_1) + \lambda_2 \hat{R}_1 \Phi(y_2)\omega_{h_1,2}(t_1)\right)\right).$$

The analysis of the error depends heavily on the smoothness of the terminal condition $\Phi(\cdot)$. Moreover, the choice of the form of the partition depends on the same property. Below we use the standard notation

$C_b^m(\mathbb{R}^d;\mathbb{R}^d)$, $C_b^m(\mathbb{R}^d)$ to denote the space of $\mathbb{R}^d$ (resp. real)– valued functions that are m times continuously differentiable with bounded derivatives.

Theorem 4.1. *Let $m \geq 3$ be an integer and consider the family of operators $\{\hat{R}_i\}_{i=1}^{n-1}$ constructed according to (8) with a cubature formula of order m. Assume also that the vector fields V_i, $i = 0, \ldots, d \in C_b^\infty(\mathbb{R}^d;\mathbb{R}^d)$ and the driver $f \in C_b^{1,m}([0,T] \times \mathbb{R}^d \times \mathbb{R} \times \mathbb{R}^d)$. Then, if $\hat{Y}_0 = \hat{R}_0 \circ \ldots \circ \hat{R}_{n-1}\Phi(x)$, $X_0 = x$, there exists a constant C independent of the partition such that:*

(i) If $\Phi \in C_b^{m+2}(\mathbb{R}^d)$ then

$$|Y_0 - \hat{Y}_0^\pi| \leq C \sum_{i=0}^{n-1} \left(h_{i+1}^2(1 + \|u\|_{1,3}) + \sum_{j=3}^{4} h_{i+1}^{(j+1)/2} \sup_{\|I\|=j} \|V_I u\|_\infty \right. \quad (11)$$

$$\left. + \sum_{j=m+1}^{m+2} h_{i+1}^{j/2} \sup_{\|I\|=j,j-1} \|V_I u\|_\infty \right).$$

In particular, using an equidistant partition, i.e $t_i = Ti/n$, $\forall i = 0, \ldots, n$, gives us the estimate $|Y_0 - \hat{Y}_0^\pi| = O(1/n)$.

(ii) If Φ is only Lipschitz continuous, consider a partition of $[0,T]$, $\pi = \{0 = t_0 < t_1 < \ldots < t_n = T\}$ such that $h_i < h_n < 1$, $\forall i = 1, \ldots, n-1$. Then, the error satisfies

$$|Y_0 - \hat{Y}_0^\pi| \leq \begin{cases} C \left(h_n^{1/2} + \sum_{i=1}^{n-1} h_i^2 h_n^{-3/2} \right) & m = 3 \\ C \left(h_n^{1/2} + \sum_{i=1}^{n-1} \left(h_i^2 h_n^{-1} + h_i^{(m+1)/2} h_n^{-m/2} \right) \right) & m \geq 5 \end{cases}.$$

$$(12)$$

In particular , if one considers a partition of the form $t_i = T(1 - 1/n)i/n^\alpha$, $t_{n+1} = T$, i.e. a partition with $N = n^\alpha + 1$ points where $\alpha = 2\,\mathbf{1}_{m=3} + \frac{3}{2}\,\mathbf{1}_{m=5}$, then one obtains the error estimate

$$|Y_0 - \hat{Y}_0^\pi| = O\left(1/N^{1/4}\right)\mathbf{1}_{m=3} + O\left(1/N^{1/3}\right)\mathbf{1}_{m=5}.$$

5. A numerical example

To validate our algorithm we present an application on a forward backward system with a non linear driver. We consider an economy where the underlying stock, the dynamics of which are described by the forward part of (1), is assumed to evolve as a geometric Brownian motion. That is,

$$dX_t = (\mu - \frac{\sigma^2}{2})X_t dt + \sigma X_t \circ dW_t$$

We then wish to address the problem of pricing a call option in this market under the restriction that one is able to invest money in the money account at an interest rate r but borrows at an interest rate R with $R > r$. It is shown by El Karoui et. al.[5] that a self-financing trading strategy of portfolio Z and wealth process Y solves a BSDE with driver

$$f(t, x, y, z) = - (ry + z\theta - (R - r)(y - z/\sigma)_-)$$

where $(x)_-$ denotes the negative part of x and $\theta = (\mu - r)/\sigma$. The problem of pricing a call option corresponds to a terminal condition of the form $\Phi(x) = (x - K)_+$, where K is the strike.

Though this driver is not smooth, a standard molification argument coupled with Theorem 4.1 justifies the convergence of the algorithm in this case too (though no rates of convergence can be deduced). We test the algorithm with parameters

μ	r	R	σ	X_0	K
0.03	0.06	0.08	0.2	10	10

As explained by Gobet et. al.,[9] in such an economy the issuer of the call option keeps borrowing money to hedge the call option so that the price of the option is the Black-Scholes with interest rate R. Therefore the value Y_0 can be computed exactly, maxing the problem a suitable benchmark for testing the algorithm. In Figure 1 we plot the logarithm of the relevant error.

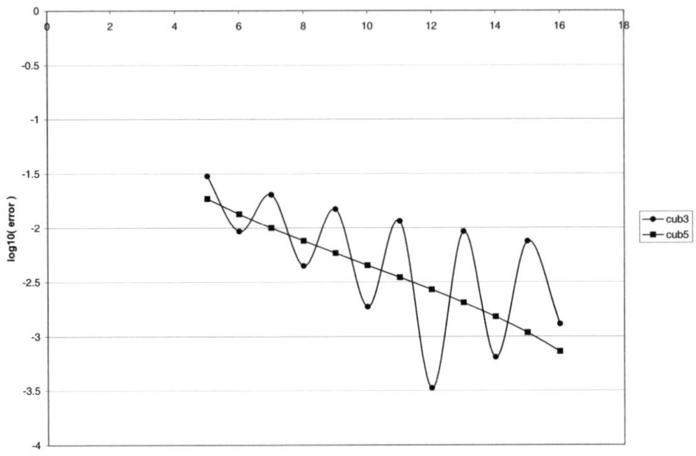

Fig. 1.   Call option in Black Scholes economy.

References

1. B. Bouchard and N. Touzi, *Stochastic processes and their applications* **111**, 175 (2004).
2. J. Zhang, *Annals of Applied Probability* , 459 (2004).
3. T. Lyons and N. Victoir, *Proc. Royal Soc. London* **468**, 169 (2004).
4. E. Pardoux and S. Peng, *Systems and control letters* **14**, 55 (1990).
5. N. El Karoui, S. Peng and M. Quenez, *Mathematical finance* **7**, 1 (1997).
6. E. Pardoux and S. Peng, *Lecture notes in control and information science* **176**, 200 (1992).
7. D. Crisan and K. Manolarakis, *submitted* .
8. P. Kloeden and E. Platen, *Numerical solutions of Stochastic Differential Equations* (Springer, 1999).
9. E. Gobet, J. Lemor and X. Warin, *Annals of Applied Probability* **15**, 2172 (2005).

A spectral gap for the Brownian bridge measure on hyperbolic spaces

X. Chen, X.-M. Li and B. Wu

Mathematics Institute, University of Warwick, Coventry CV4 7AL, UK

1. Introduction

Let N be a finite or infinite dimensional manifold, μ a probability measure, d a differential operator with domain a subspace of the the L^2 space of functions which restricted to differentiable functions is the usual differentiation operator. Let d^* be its L^2 adjoint with respect to this measure. We ask whether the operator $L = -d^*d$ has a spectral gap. If ∇ is the gradient operator associated to d through Riesz representation theorem, in the case that we have a Hilbert space structure, this is equivalent to a Poincaré inequality $\int_N (f - \bar{f})^2 \mu(dx) \leqslant \frac{1}{\lambda_1} \int_N |\nabla f|^2 \mu(dx)$, where f ranges through an admissible set of real valued functions on a space N and $\bar{f} = \int_N f \mu(dx)$. If N is a compact closed Riemannian manifold, dx the volume measure and ∇ the Riemannian gradient operator, the best constant in the Poincaré inequality is given by taking infimum of the Raleigh quotient $\frac{\int_N |df|^2 dx}{\int_N f^2 dx}$ over the set of non-constant smooth functions of zero mean and is the spectral gap for the Laplacian operator, its first non-trivial eigenvalue. The operator concerned is given by the gradient operator and depend on the measure μ. Poincaré inequality does not hold for $\mathbf{R}^n$ with Lebesgue measure, it does hold for the Gaussian measure. For the standard normalised Gaussian measure and ∇ the gradient operator in Malliavin calculus, the Poincaré constant is 1 and the corresponding eigenfunction of the Laplacian is the Hermitian polynomial $x/2$. If h is a smooth function μ a measure which is absolutely continuous with respect to the Lebesgue measure with density e^{-2h}, for any f in the domain of d,

$$\int_N |df|^2(x)\mu(dx) = -\int_N \langle f, \Delta f\rangle(x)\mu(dx) + 2\int_N f\langle df, dh\rangle\mu(dx).$$

The corresponding Poincaré inequality is then related to the Bismut-Witten Laplacian $\Delta^h := \Delta - 2L_{\nabla h}$ on $L^2(M, e^{-2h}dx)$, which is unitarily equivalent to $\square^h = \Delta - |dh|^2 + \Delta h$ on $L^2(M, dx)$. Here we take the convention that the Laplacian is a negative operator. The spectral property of Δ^h, hence the validity of the Poincaré inequality for μ is determined by the spectral property of the Schrödinger operator $\square^h$ on $L^2(M; dx)$.

Let M be a smooth finite dimensional Riemannian manifold which is stochastically complete. Fix a number $T > 0$. The path space on M based at x_0 is:

$$\mathcal{C}_{x_0} M = \{\sigma : [0, T] \to M, \sigma(0) = x_0 |\ \sigma \text{ is continuous}\}.$$

For $y_0 \in M$, we define the subspaces

$$\mathcal{C}_{x_0, y_0} M = \{\sigma \in \mathcal{C}_{x_0} M \mid \sigma(T) = y_0\}, L_{x_0} M = \{\sigma \in \mathcal{C}_{x_0} M \mid \sigma(T) = x_0\}.$$

In this article our state space would be the loop space $L_{x_0} M$ endowed with the Brownian Bridge measure. In the case of the Wiener space, the Brownian Bridge measure $\mu_{0,0}$ is the law of the Brownian bridge starting and ending at 0, one of whose realisation is $B_t - \frac{t}{T} B_T$. It can also be realised as solution to the time-inhomogeneous stochastic differential equation $dx_t = dB_t - \frac{x_t}{T-t} dt$. The Brownian bridge measure is a Radon Gaussian measure and Gaussian measure theory applies to give the required the Poincaré inequality. In fact the stronger Logarithmic Sobolev inequality holds:

$$\int f^2 \log \frac{f^2}{\mathbf{E}|f|^2} \mu(dx) \leqslant 2 \int |\nabla f|^2 \mu(dx).$$

This however may not hold in general. In fact as noted by L. Gross[14] Poincaré inequalities do not hold on the Lie group S^1 due to the lack of connectedness of the loop space. Following that A. Eberle[11] gave an example of a compact simply connected Riemannian manifold on which the Poincaré inequality does not hold .

There are two standard arguments to prove the spectral gap theorem. The first argument applies to a compact state manifold N where $\inf_{f \in H^1, |f|_{L^2}=1, \int f=0} \int_M |\nabla f|^2 dx$ is attained, by a non-constant function, due to the Rellich-Kondrachov compact embedding theorem of $H^{1,q}$ into L^p. The other approach is the dynamic one which applies to Gaussian measures, due to the commutation relation. Namely we consider a Markov process with semigroup P_t, a finite invariant measure $e^{-2h}dx$ (finiteness holds if $\inf_{|v|=1}\{Ric_x(v, v) + 2Hess_x(h)(v, v)\}$ is strongly stochastically positive[18,19]), and generator $\frac{1}{2}\Delta - L_{\nabla h}$. Suppose that $|dP_t f| \leq \frac{1}{\rho}|df|$ for some constant ρ, then

$$\int_M (f - \bar{f})^2 d\mu = \int_M (f^2 - \bar{f}^2) \, d\mu = \lim_{t \to \infty} \int_M (f^2 - (P_t f)^2)(x) d\mu(x)$$

$$= -\lim_{t \to \infty} \int_M \int_0^t \frac{\partial}{\partial s}(P_s f)^2 ds \, d\mu = \lim_{t \to \infty} \int_0^t \int_M (dP_s f)^2 \, d\mu \, ds$$

$$= \int_0^\infty \int_M (dP_s f)^2 \, d\mu \, ds \leq \frac{1}{\rho} \int_M |df|^2(x) \, d\mu.$$

In particular we have Poincaré inequality with respect to the measure e^{-2h} if the Bakry-Emery condition $\inf_{|v|=1}\{Ric_x(v,v) + 2Hess_x(h)(v,v)\} > \rho > 0$ holds. The dynamic argument can be modified leading to the beautiful Clark-Ocone formula approach.[6,13] However none of these approaches seems to work well for the Brownian bridge measure on a general path space. The main problem comes down to estimates on the heat kernel. However we do have one positive result for non-flat spaces: if M is the hyperbolic space we have indeed a spectral gap.[9] But it remains an open question whether the Logarithmic Sobolev inequality holds.

2. The Spectral Gap Theorem

A. Denote by Cyl the set of smooth cylindrical functions on $C_{x_0}M$,

$$Cyl_t = \{F | F(\sigma) = f(\sigma_{s_1}, \ldots, \sigma_{s_k}), f \in C_K^\infty(M^k), 0 < s_1 < \cdots < s_k \leq t < T\}.$$

The Brownian bridge measure μ_{x_0,y_0} is defined through integration of $F \in Cyl$,

$$p_T(x_0, y_0) \int_{C_{x_0} M} f(\sigma_{s_1}, \ldots, \sigma_{s_n}) d\mu_{x_0,y_0}(\sigma)$$

$$= \int_{M^n} f(x_1, \ldots, x_n) p_{s_1}(x_0, x_1) \ldots p_{s_n - s_{n-1}}(x_{n-1}, x_n) p_{T-s_n}(x_n, y_0) \Pi_{i=1}^n dx_i.$$

This cylindrical measure extends to a real measure if for some constants $\beta > 0$ and $\delta > 0$,

$$\int\int d(y,z)^\beta \frac{p_s(x_0, y) p_{t-s}(y, z) p_{T-t}(z, y_0)}{p_T(x_0, y_0)} dy dz \leq C|t - s|^{1+\delta}. \tag{1}$$

Throughout this article we shall assume the heat kernel satisfies the above inequality[10] and a number of assumptions, all of which hold true on the hyperbolic space.[2] See[9] for detail.

We now define the gradient operator. Take the Levi-Civita connection ∇, whose parallel translation along a path σ is denoted by $/\!/$. Define the tangent sub-space $H_\sigma = \{/\!/_s k_s : k \in L_0^{2,1}(T_{x_0} M)\}$, to $T_\sigma C_{x_0} M$ which we call the Bismut tangent space with Hilbert space norm induced from the Cameron Martin space. We identify $T_{x_0} M$ with $\mathbf{R}^n$. Let μ be a probability measure on $C_{x_0} M$ (including measures which concentrates on a subspace e.g. the loop space). The differential operator d, which restricts to differential functions is the usual exterior derivative from the space of L^2 functions to the space of L^2 H-valued 1-forms, is closable whenever Driver's integration by parts formula holds.[10] Assuming that this integration by parts formula holds, we define $\mathbb{D}^{1,2} \equiv \mathbb{D}^{1,2}(C_{x_0} M)$ to be the closure of smooth cylindrical function Cyl_t, $t < T$ under the following norm:

$$\sqrt{\int_{C_{x_0} M} |\nabla f|_{H_\sigma}^2 (\sigma) \mu(d\sigma) + \int f^2(\sigma) d\mu(\sigma)}$$

and this is the domain of the corresponding gradient defined by: $df(V) = \langle \nabla f, V \rangle_H$. **B.** Aida[2] showed that for M the standard hyperbolic space, of constant negative curvature -1, one has

$$\int_{C_{x_0} H^n} f^2 \log \frac{f^2}{|f|_{L^2(\mu_{x_0,y_0})}^2} d\mu_{x_0,y_0}(\gamma) \leqslant \int_{C_{x_0} H^n} C(\gamma) |\nabla f|^2 d\mu_{x_0,y_0}(\gamma) \quad (2)$$

for $C(\gamma) = C_1(n) + C_2(n) \sup_{0 \leqslant t \leqslant 1} d^2(\gamma_t, y_0)$. To obtain thisPoincaré inequality with modified Dirichlet form, he first proved an integration by parts formula from which a Clark-Ocone formula of the form:

$$\mathbf{E}^{\mu_{x_0,y_0}} \{F | \mathcal{G}_t\} = \mathbf{E}^{\mu_{x_0,y_0}} F + \int_0^t \langle H_s(\gamma), dW_s \rangle,$$

where W_t is the martingale part of the anti-development of the Brownian bridge and

$$H(s, \gamma) = \mathbf{E}^{\mu_{x_0,y_0}} \{L(\gamma) \frac{d}{ds} \nabla F(\gamma)(s) | \mathcal{G}_s\}$$

almost surely with respect to the product measure $dt \otimes \mu_{x_0,y_0}$. Here $\mathcal{G}_t$ is the filtration generated by $\mathcal{F}_t$ and the end point of the Brownian bridge. Unlike the case for Gaussian measures or for the Brownian motion measures, the function L in the Clark-Ocone formula is not a deterministic function, which underlines why the Clark-Ocone approach itself is not good enough to give the required inequality.

Theorem 2.1. *Let $M = H^n$, the hyperbolic space of constant curvature -1. Then Poincaré inequality holds for the Brownian bridge measure μ_{x_0,x_0}.*

402   X. Chen, X.-M. Li and B. Wu

The proof of the theorem is based on Lemmas 2.1 and 2.2, (2) and that

$$\int_{C_{x_0}M} e^{Cd^2(\sigma, y_0)} d\mu_{x_0, y_0}(\sigma) < \infty.$$

C. The Laplace Beltrami operator on a complete Riemannian manifold may not have a spectral gap. But it has always a local spectral gap, by restriction to an exhausting relatively compact open sets U_n. The constant λ_1 may blow up as n goes to infinity. Once a blowing up rate for local Poincaré inequalities are obtained, we arrived at the so called 'weak Poincaré inequalities' and in the case of Entropy the 'weak Logarithmic Sobolev inequalities':

$$\mathbf{Var}(f) \leqslant \alpha(s) \int |\nabla f|^2 d\mu + s|f|_\infty^2,$$

$$\mathbf{Ent}(f^2) \leqslant \beta(s) \int |\nabla f|^2 d\mu + s|f|_\infty^2.$$

Here $\mathbf{Var} f$ denotes the variance $\mathbf{E}(f - \mathbf{E}f)^2$ of a function f and $\mathbf{Ent}(f)$ its entropy $\mathbf{E}f \log \frac{f}{\mathbf{E}f}$, α and β are non-decreasing functions from $(0, \infty)$ to $\mathbf{R}_+$. We first state the following estimate:

Lemma 2.1. *Let μ be any probability measure on $C_{x_0}M$ with the property that there exists a positive function $u \in \mathbb{D}^{1,2}$ such that Aida's type inequality holds:*

$$\mathbf{Ent}(f^2) \leqslant \int u^2 |\nabla f|^2 d\mu, \qquad \forall f \in \mathbb{D}^{1,2} \cap L_\infty \tag{3}$$

Assume furthermore that $|\nabla u| \leqslant a$ and $\int e^{Cu^2} d\mu < \infty$ for some $C, a > 0$. Then for all functions f in $\mathbb{D}^{1,2} \cap L_\infty$

$$\mathbf{Ent}(f^2) \leqslant C|\log s| \int |\nabla f|^2 d\mu + s|f|_\infty^2, \qquad \forall f \in \mathbb{D}^{1,2}, s \in (0, r_0] \tag{4}$$

for some constants C, r_0.

D. Functional inequalities describe how the L^2 norms of a function is controlled by the homogeneous H^1 norm and possibly the L_∞ norm in the case of weak type inequalities. They describe the concentration phenomenon of the measure. Functional inequalities describe the concentration phenomenon of the measure. On the other hand concentration inequalities are related with isoperimetric inequalities. For example let $h = \inf_A \frac{\mu(\partial A)}{\min\{\mu(A), \mu(N/A)\}}$ where the infimum is taken over all open subsets of N. Then $h^2 \leqslant 4\lambda_1$ by Cheeger.[8] On the other hand if K is the lower bound of the Ricci curvature, $\lambda_1 \leqslant C(\sqrt{K}h + h^2)$,[5] See also[15,16] and[17] .

For finite dimensional spaces it was shown in[7] and[4] that one can pass from capacity type inequalities to weak Logarithmic Sobolev inequalities and vice versa with great precision. Although they did not phrase the theorem in terms of Malliavin calculus, most of their results work in infinite dimensional spaces and in particular the following lemma will transforms our estimates on the blowing up rate of the logarithmic Sobolev inequality into a spectral gap result.

Lemma 2.2. *If for all bounded functions in* $\mathbb{D}^{1,2}(C_{x_0}M)$, *the weak logarithmic Sobolev inequality holds for* $0 < s < r_0$, *some given* $r_0 > 0$,

$$\mathbf{Ent}(f^2) \leqslant C \log \frac{1}{s} \int |\nabla f|^2 d\mu + s|f|_\infty^2,$$

where $C > 0$ *is a constant, Poincaré inequality*

$$\mathbf{Var}(f) \leqslant \alpha \int |\nabla f|^2 d\mu, \qquad f \in \mathbb{D}^{1,2}(C_{x_0}M).$$

holds for some constant $\alpha > 0$.

Acknowledgement

This research is supported by the EPSRC (EP/E058124/1). We would like to thank Martin Hairer for stimulating discussions and for drawing our attention to some references.

References

1. Aida, S., Uniform positivity improving property, Sobolev inequalities, and spectral gaps. *J. Funct. Anal.*, 158(1):152–185, 1998.
2. Aida, S., Logarithmic derivatives of heat kernels and logarithmic Sobolev inequalities with unbounded diffusion coefficients on loop spaces. *J. Funct. Anal.*, 174(2):430–477, 2000.
3. Aida, S. and Masuda, T. and Shigekawa I., Logarithmic Sobolev inequalities and exponential integrability. *J. Funct. Anal.*, 126(1):83–101, 1994.
4. Barthe, F. and Cattiaux, P. and Roberto, C., Concentration for independent random variables with heavy tails. *AMRX Appl. Math. Res. Express*, (2):39–60, 2005.
5. Buser, P., A note on the isoperimetric constant. *Ann. Sci. École Norm. Sup. (4)*, 15(2):213–230, 1982.
6. Capitaine, M. and Hsu, E. P. and Ledoux, M., Martingale representation and a simple proof of logarithmic Sobolev inequalities on path spaces. *Electron. Comm. Probab.* 2 (1997), 71–81
7. Cattiaux, P. and Gentil, I. and and Guillin, A. Weak logarithmic sobolev inequalities and entropic convergence. *Prob. The.Rel. Fields*, 139:563–603, 2007.

8. Cheeger, J., A lower bound for the smallest eigenvalue of the Laplacian. In *Problems in analysis (Papers dedicated to Salomon Bochner, 1969)*, pages 195–199. Princeton Univ. Press, Princeton, N. J., 1970.

9. Chen, X. and Li, X.-M. and Wu, B., A Poincaré Inequality on Loop Spaces. 2009. Preprint.

10. Driver, B.K., Integration by parts and quasi-invariance for heat kernel measures on loop groups. *J. Funct. Anal.* 149, no. 2, 470–547, 1997.

11. Eberle, A., Absence of spectral gaps on a class of loop spaces. *J. Math. Pures Appl. (9)*, 81(10):915–955, 2002.

12. Elworthy, K. D. and Li, X.-M., Itô maps and analysis on path spaces. *Math. Z.* 257 (2007), no. 3, 643–706.

13. Fang S. Z., Inégalité du type de Poincar esur l'espace des chemins riemanniens, *C. R. Acad. Sci. Paris Sér. I Math.* 318 (1994) 257-260.

14. Gross, L., Logarithmic Sobolev inequalities on loop groups. *J. Funct. Anal.*, 102(2):268–313, 1991.

15. Ledoux, M., A simple analytic proof of an inequality by P. Buser. *Proc. Amer. Math. Soc.*, 121(3):951–959, 1994.

16. Ledoux, M., Isoperimetry and gaussian analysis. In *Ecole d'été de Probabiliti¿½s de St-Flour 1994. Lecture Notes in Math. 1648*, pages 165–294, 1996.

17. Ledoux, M.; and Talagrand, M., *Probability in Banach spaces: isoperimetry and processes*. Springer, 1991.

18. Li, X.-M Stochastic differential equations on noncompact manifolds: moment stability and its topological consequences. *Probab. Theory Related Fields*, 100(4):417–428, 1994.

19. Li, X.-M., On extensions of Myers' theorem. *Bull. London Math. Soc.*, 27(4):392–396, 1995.

20. Mathieu, P., Quand l'inegalite log-sobolev implique l'inegalite de trou spectral. In *Séminaire de Probabiliteés, Vol. XXXII,Lecture Notes in Math. , Vol. 1686*, pages 30–35. Springer-Verlag, Berlin, 1998.

21. Röckner, M. and Wang, F.-Y., Weak Poincaré inequalities and L^2-convergence rates of Markov semigroups. *J. Funct. Anal.*, 185(2):564–603, 2001.

Individual path uniqueness of solutions of SDE

A.M. Davie

School of Mathematics, University of Edinburgh,
King's Buildings, Mayfield Road,
Edinburgh EH9 3JZ, UK
E-mail: a.davie@ed.ac.uk

We consider the stochastic differential equation $dx(t) = f(t, x(t))dt + b(t, x(t))dW(t)$, $x(0) = x_0$ for $t \geq 0$, where $x(t) \in \mathbb{R}^n$, W is a standard d-dimensional Brownian motion, f is a bounded Borel function from $[0, \infty) \times \mathbb{R}^d$ to $\mathbb{R}^d$, and b is a matrix-valued function having rank n satisfying some regularity conditions. We show that, for almost all Brownian paths $W(t)$, there is a unique $x(t)$ satisfying this equation.

1. Introduction

In this paper we consider the stochastic differential equation

$$dx(t) = f(t, x(t))dt + b(t, x(t))dW(t), \qquad x(0) = x_0 \qquad (1)$$

for $t \geq 0$, where $x(t) \in \mathbb{R}^d$, W is a standard d-dimensional Brownian motion, f is a bounded Borel function from $[0, \infty) \times \mathbb{R}^d \to \mathbb{R}^d$ to $\mathbb{R}^d$, and b is an invertible $d \times d$ matrix valued function on $[0, \infty \times \mathbb{R}^d$ satisfying a suitable regularity condition. If b satisfies a Lipschitz condition in x, then it follows from a theorem of Veretennikov [4] that (2) has a unique strong solution, i.e. there is a unique process $x(t)$, adapted to the filtration of the Brownian motion, satisfying (1).

Here we consider a different question, posed by N. V. Krylov [2]: we choose a Brownian path W and ask whether (2) has a unique solution for that particular path. The first problem with this question is to interpret it, since the stochastic integral implied by the equation is not well-defined for individual paths. One case for which there is a simple interpretation is when $b(t, x)$ is the identity matrix for all t, x, since in that case we can write the equation as

$$x(t) = W(t) + x_0 + \int_0^t f(s, x(s))ds, \qquad t \geq 0 \qquad (2)$$

and the existence of a unique solution to (2), for almost every path W, was proved in [1].

In this paper we use an ideas from rough path theory to give an interpretation of (1) for b satisfying a slightly stronger regularity condition, and then prove that for almost all Brownian paths W there is a unique solution. The proof is similar to that in [1] but requires estimates for solutions of $dx = bdW$ similar to those for Brownian motion in Section 2 of [1], and different arguments including a suitable $T(1)$ theorem are needed for this.

2. Rough path interpretation

We now describe our interpretation of (1) precisely. We assume that $b_{ij}(t, x)$ is differentiable with respect to x and that b_{ij} and $\partial b_{ij}/\partial x_k$ satisfy locally a Hölder condition of order α for some $\alpha > 0$ in (t, x). We write (1) in components as

$$dx_i(t) = f_i(t, x(t))dt + \sum_{j=1}^{d} b_{ij}(t, x(t))dW_j(t)$$

for $i = 1, \cdots, n$. For $s < t$ we write $A_{rj}(s, t) = \int_s^t (W_r(\tau) - W_r(s))dW_j(\tau)$ and $\Psi_i(s, t, x) = \sum_{j=1}^{d} \int_s^t b_{ij}(\tau, x)dW_j(\tau)$ for $x \in \mathbb{R}^d$. On a suitable set of probability 1 these quantities can be defined simultaneously for all $0 \leq s < t$ and $x \in \mathbb{R}^d$, depending continuously on (s, t, x), and we assume that such definitions have been fixed. Then we define

$$R_i(s, t) = x_i(t) - x_i(s) - \int_s^t f_i(\tau, x(\tau))d\tau - \Psi_i(s, t, x(s)) - \sum_{j,r} g_{ijr}(x(s))A_{rj}(s, t)$$

where $g_{ijr} = \sum_{k=1}^{n} \frac{\partial b_{ij}}{\partial x_k} b_{kr}$.

We then say that $x(t)$ is a solution of (1) on an interval $0 \leq t < T$ where $0 < t \leq \infty$ if, for any T' with $0 < T' < T$ and any $\epsilon > 0$ we can find $\delta > 0$ such that if $0 < t_1 < t_2 < \cdots < t_N = T'$ and $t_i - t_{i-1} < \delta$ for $i = 1, \cdots, N$ then $\sum_{i=1}^{N} |R(t_{i-1}, t_i| < \epsilon$.

It is not hard to show that the strong solution of (1) is, for almost all Brownian paths, a solution in this sense.

We can now formulate the main result.

Theorem 2.1. *For almost every Brownian path W, there is a unique T with $0 < T \leq \infty$ and a unique continuous $x : [0, T) \to \mathbb{R}^d$ satisfying (1) in the above sense, and such that in case $T < \infty$ we have $|x(t)| \to \infty$ as $t \to T$.*

The proof of the theorem follows the same lines as [1]. First we can use a localisation argument to reduce the problem to proving uniqueness on $0 \le t \le 1$ when b is defined on $[0,1] \times \mathbb{R}^d$ and is constant outside a compact set. Then we need an estimate analogous to Proposition 2.1 of [1], where instead of $W(t)$ we have a solution $y(t)$ of the equation

$$dy(t) = b(t, y(t))dW(t) \tag{3}$$

The required estimate is the following:

Proposition 2.1. *Let g be a Borel function on $[0,1] \times \mathbb{R}^d$ with $|g(s,z)| \le 1$ everywhere, let $I \subseteq [0,1]$ be an interval, and let y satisfy the SDE (3). Then for any even positive integer p and $x \in \mathbb{R}^d$, we have*

$$\mathbb{E}\left(\int_I \{g(t, y(t) + x) - g(t, y(t))\}dt\right)^p \le C^p (p/2)! |x|^p |I|^{p/2}$$

where C is an absolute constant and $|x|$ denotes the usual Euclidean norm.

The proof of this is similar to that of Proposition 2.1 of [1] with some differences. One is that the reduction to the one-dimensional case no longer works, so the entire argument has to be carried out in $\mathbb{R}^d$ - this does not in fact change much. A more substantial difference is that we have to replace the Gaussian transition densities $E(t - s, z - y)$ by transition densities $p(s, t, y, z)$ for the diffusion defined by (3) and the second derivatives $D(t - s, z - y) = E''(t - s, z - y)$ by $q_{ij}(s, t, y, z) = \frac{\partial^2 p}{\partial y_i \partial z_j} p(s, t, y, z)$. Then the proof of Lemma 2.3 of [1], which uses translation invariance, no longer works. Instead we use the following:

Lemma 2.1. *The operators T_{ij} defined by*

$$T_{ij}h(s, y) = \int_0^1 \int_{\mathbb{R}^d} q_{ij}(s, t, y, z)dzdt$$

are bounded on $L^2([0,1] \times \mathbb{R}^d)$.

This is proved by applying a '$T(1)$ theorem' for an operator T of Calderon-Zygmund type, which asserts that given some mild bounds and regularity conditions on the kernel, such an operator is bounded on L^2 provided $T(1)$ and $T^*(1)$ are in BMO. Our operators T_{ij} are of the required type, provided we equip $\mathbb{R} \times \mathbb{R}^d$ with the parabolic letric $d(s, y; t, z) = (x - y| + |s - t|^{1/2}$, and we have in fact $T_{ij}(1) = 0$, and $T_{ij}^*(1) = 0$. The bounds and regularity conditions follow from Schauder estimates for $p(s, t, y, z)$.

The case $p = 2$ of Proposition 2.1 follows fairly easily from Lemma 2.1. The general case is proved by an inductive argument, similar to the proof of Proposition 2.1 of [1].

3. Outline of proof of theorem

For the proof of the theorem, we let $z(t)$ denote the strong solution of (1), so we have to show that for almost all paths W, z is the only solution of (1) in our "rough path' sense. We write $x(t) = z(t) + u(t)$ and then we have to show that the only solution u of

$$u(t) = \int_0^t \{f(s, z(s)+u(s)) - f(s, z(s))\} + \int_0^t \{b(s, z(s)+u(s)) - b(s, z(s))\} dW(s)$$

(4)

satisfying $u(0) = 0$ is $u(t) = 0$.

The basic idea is to approximate u by a sequence of step functions u_n, such that u_n is constant on each interval $I_{nk} = [k2^{-n}, (k+1)2^{-n}]$, $k = 0, 1, 2, \cdots, 2^n - 1$ and then use

$$\int_{I_{nk}} \{f(t, z(t) + u(t)) - f(t, z(t))\} dt$$

$$= \lim_{l \to \infty} \int_{I_{nk}} \{f(t, z(t) + u_l(t)) - f(t, z(t))\} dt$$

(5)

$$= \int_{I_{nk}} \{f(t, z(t) + u_n(t)) - f(t, z(t))\} dt$$

$$+ \sum_{l=n}^{\infty} \int_{I_{nk}} \{f(t, z(t) + u_{l+1}(t)) - f(t, z(t) + u_l(t))\} dt$$

with a similar expansion for $\int_{I_{nk}} \{b(t, z(t) + u(t)) - b(t, z(t))\} dW(t)$.

We introduce the notation

$$\sigma_{nk}(x) = \int_{I_{nk}} \{f(t, z(t)+x) - f(t, z(t))\} dt + \int_{I_{nk}} \{b(t, z(t)+x) - b(t, z(t))\} dW(t)$$

and

$$\rho_{nk}(x, y) = \sigma_{nk}(x) - \sigma_{nk}(y)$$

Using this notation, from (5), and the similar expansion for the dW, (4) implies

$$u((k+1)2^{-n}) - u(k2^{-n}) = \sigma_{nk}(u(k2^{-n}))$$

$$+ \sum_{l=n}^{\infty} \sum_{r=k2^{l-n}}^{(k+1)2^{l-n}-1} \rho_{l+1,2r+1}(u(2^{-l-1}(2r+1)), u(2^{-l}r))$$

(6)

The first stage of the proof is to show that with probability 1, for every cube $Q \subseteq \mathbb{R}^d$ there is a constant $C > 0$ such that the following bounds hold:

$$|\rho_{nk}(x,y)| \leq C\left\{n^{1/2} + \left(\log^+ \frac{1}{|x-y|}\right)^{1/2}\right\} 2^{-n/2}|x-y|$$

for all dyadic $x, y \in Q$ and all choices of integers n, k with $n > 0$ and $0 \leq k \leq 2^n - 1$;

for all $n \in \mathbb{N}$, $k \in \{0, 1, \cdots, 2^n - 1\}$ and dyadic $x \in Q$ we have

$$|\sigma_{nk}(x)| \leq Cn^{1/2}2^{-n/2}(|x| + 2^{-2^n})$$

together with some more technical bounds similar to those in [1].

To prove these bounds, the dt integrals are estimated in a similar manner to [1], using Proosition 2.1 and the fact that the process $z(t)$ has law mutuallly absolutely continuous with that of $y(t)$ in Proposition 2.1, by Girsanov's theorem. The dW integrals are easier to bound, using standard SDE methods, in view of the regularity assumed for b.

The next step is to show, that when the above bounds hold, the following is true:

Lemma 3.1. *There are positive constants K and m_0 such that, for all integers $m > m_0$, if u is a solution of (1) and for some $j \in \{0, 1, \cdots, 2^m - 1\}$ and some β with $2^{-2^{3m/4}} \leq \beta \leq 2^{-2^{2m/3}}$ we have $|u(j2^{-m})| \leq \beta$, then*

$$|u((j+1)2^{-m})| \leq \beta\{1 + K2^{-m}\log(1/\beta)\}$$

The proof of this Lemma uses (6) and the above bounds and largely follows the proof of Lemma 3.7 of [1]. Some modifications are required as in the present case u will not necessarily satisfy a Lipschitz condition as it does in [1].

Finally the theorem then follows from Lemma 3.1 exactly as in [1].

References

1. A. M. Davie, Uniqueness of solutions of stochastic differential equations, *Int. Maths. Res. Not.* **2007**, no. 24, Art. ID rnm124.
2. I Gyöngy, personal communication.
3. N. Ikeda and S. Watanabe, *Stochastic Differential Equations and Diffusion Processes*, North Holland, 1989.
4. A. Yu. Veretennikov, On strong solutions and explicit formulas for solutions of stochastic integral equations (Russian), *Mat. Sbornik (N.S.)* **111(153)** (1980), 434-452, 480. English transl. in *Math. USSR Sb.* **39** (1981), 387-403.

Periodic homogenization with an interface

Martin Hairer and Charles Manson

Courant Institute, NYU and University of Warwick

We consider a diffusion process with coefficients that are periodic outside of an 'interface region' of finite thickness. The question investigated in the articles[1,2] is the limiting long time / large scale behaviour of such a process under diffusive rescaling. It is clear that outside of the interface, the limiting process must behave like Brownian motion, with diffusion matrices given by the standard theory of homogenization. The interesting behaviour therefore occurs on the interface. Our main result is that the limiting process is a semimartingale whose bounded variation part is proportional to the local time spent on the interface. We also exhibit an explicit way of identifying its parameters in terms of the coefficients of the original diffusion.

Our method of proof relies on the framework provided by Freidlin and Wentzell[3] for diffusion processes on a graph in order to identify the generator of the limiting process.

1. Introduction

In this note, we report on recently obtained results[1,2] on the long-time large-scale behaviour of diffusions of the form

$$dX = b(X)\, ds + dB(s)\,, \qquad X(0) = x \in \mathbb{R}^d\,, \tag{1}$$

where B is a d-dimensional standard Wiener process. The drift b is assumed to be smooth and such that $b(x + e_i) = b(x)$ for the unit vectors e_i with $i = 2, \ldots, d$ (but not for $i = 1$). Furthermore, we assume that there exist smooth vector fields $b_\pm$ with unit period in *every* direction and $\eta > 0$ such that

$$b(x) = b_+(x)\,, \quad x_1 > \eta\,, \qquad b(x) = b_-(x)\,, \quad x_1 < -\eta\,. \tag{2}$$

Setting $X^\varepsilon(t) = \varepsilon X(t/\varepsilon^2)$, our aim is to characterise the limiting process $\bar{X} = \lim_{\varepsilon \to 0} X^\varepsilon$, if it exists. In the sequel, we denote by $\mathcal{L}$ the generator of X and by $\mathcal{L}_\pm$ the generators of the diffusion processes $X_\pm$ given by Eq. (1) with b replaced by $b_\pm$. The processes $X_\pm$ will be viewed as processes on the torus $\mathbb{T}^d$, and we denote by $\mu_\pm$ the corresponding invariant probability measures. In order to obtain a diffusive behaviour for X at large scales, we impose the centering condition $\int_{\mathbb{T}^d} b_\pm(x)\,\mu_\pm(x) = 0$.

Before stating the main result, first define the various quantities involved and their relevance.

We define the 'interface' of width η by $\mathcal{I}_\eta = \{x \in \mathbb{R}^d : x_1 \in [-\eta, \eta]\}$. In view of standard results from periodic homogenization,[4] any limiting process for X^ε should behave like Brownian motion on either side of the interface $\mathcal{I}_0 = \{x_1 = 0\}$, with effective diffusion tensors given by

$$D_{ij}^\pm = \int_{\mathbb{T}^d} (\delta_{ik} + \partial_k g_i^\pm)(\delta_{kj} + \partial_k g_j^\pm)\, d\mu_\pm \ . \tag{3}$$

(Summation of k is implied.) Here, the corrector functions $g_\pm : \mathbb{T}^d \to \mathbb{R}^d$ are the unique solutions to $\mathcal{L}_\pm g_\pm = -b_\pm$, centered with respect to $\mu_\pm$. Since $b_\pm$ are centered with respect to $\mu_\pm$, such functions do indeed exist.

This justifies the introduction of a differential operator $\bar{\mathcal{L}}$ on $\mathbb{R}^d$ defined in two parts by $\bar{\mathcal{L}}_+$ on $I_+ = \{x_1 > 0\}$ and $\bar{\mathcal{L}}_-$ on $I_- = \{x_1 < 0\}$ with

$$\bar{\mathcal{L}}_\pm = \frac{D_{ij}^\pm}{2}\partial_i \partial_j \ , \tag{4}$$

then one would expect any limiting process to solve a martingale problem associated to $\bar{\mathcal{L}}$. However, the above definition of $\bar{\mathcal{L}}$ is not complete, since we did not specify any boundary condition at the interface $\mathcal{I}_0$.

In the one dimensional case[1] the analysis is considerably simplified since

- The interface is zero dimensional in the limit and hence cannot exhibit any more complicated behavior than preferential exit behavior.
- The non-rescaled process is time-reversible and therefore admits an invariant measure for which one has an explicit expression.

Both of these clues allow us to make a reasonable guess that in one dimension the limiting process will be some (possibly different on each side of zero) rescaling of skew Brownian motion. Since the diffusion coefficients on either side of the interface are already determined by the theory of periodic homogenisation, the only parameter that remains to be determined is the relative probability of excursions to either side of the interface. This can be read off the invariant measure by using the fact that the rescaled invariant measure should converge to that of the limiting process.

One of the main ingredients in the analysis of the behavior of the limiting process at the interface is the invariant measure μ for the (original, not rescaled) process X. If we identify points that differ by integer multiples of e_j for $j = 2, \ldots, d$, we can interpret X as a process with state space

$\mathbb{R} \times \mathbb{T}^{d-1}$. It then follows from the results in[5] that this process admits a σ-finite invariant measure μ on $\mathbb{R} \times \mathbb{T}^{d-1}$.

Note that the invariant measure μ is *not* finite and can therefore not be normalised in a canonical way. However, if we define the 'unit cells' $C_j^{\pm}$ by

$$C_j^+ = [j, j+1] \times \mathbb{T}^{d-1} \tag{5}$$
$$C_j^- = [-j-1, -j] \times \mathbb{T}^{d-1} \tag{6}$$

then it is possible to make sense of the quantity $q_{\pm} = \lim_{j\to\infty} \mu(C_j^{\pm})$.

Let now $p_{\pm}$ be given by

$$p_{\pm} = \frac{q_{\pm} D_{11}^{\pm}}{q_+ D_{11}^+ + q_- D_{11}^-} \, , \tag{7}$$

Unlike in the one-dimensional case, these quantities are not sufficient to characterise the limiting process since it is possible that it picks up a non-trivial drift along the interface. It turns out that this drift can be described by drift coefficients α_j for $j = 2, \ldots, d$ given by

$$\alpha_j = 2 \left(\frac{p_+}{D_{11}^+} + \frac{p_-}{D_{11}^-} \right) \int_{\mathbb{R} \times \mathbb{T}^{d-1}} b_j(x) \, \mu(dx) \, , \tag{8}$$

where μ is normalised in such a way that $q_+ + q_- = 1$.

Given all of these ingredients, we construct an operator $\bar{\mathcal{L}}$ as follows. The domain $\mathcal{D}(\bar{\mathcal{L}})$ of $\bar{\mathcal{L}}$ consists of functions $f \colon \mathbb{R}^d \to \mathbb{R}$ such that

- f is continuous and its restrictions to I_+, I_-, and $\mathcal{I}_0$ are smooth.
- The partial derivatives $\partial_i f$ are continuous for $i \geq 2$.
- The partial derivative $\partial_1 f(x)$ has right and left limits $\partial_1 f|_{\pm}$ as $x \to \mathcal{I}_0$ and these limits satisfy the gluing condition

$$p_+ \partial_1 f|_+ - p_- \partial_1 f|_- + \sum_{j=2}^d \alpha_j \partial_j f = 0 \, . \tag{9}$$

For any $f \in \mathcal{D}(\bar{\mathcal{L}})$, we then set $\bar{\mathcal{L}} f(x) = \mathcal{L}_{\pm} f(x)$ for $x \in I_{\pm}$. With these definitions at hand, we can state the main result of the article:

Theorem 1.1. *The family of processes X^ε converges in law to the unique solution $\bar{X}$ to the martingale problem given by the operator $\bar{\mathcal{L}}$. Furthermore, there exist matrices $M_{\pm}$ and a vector $K \in \mathbb{R}^d$ such that this solution solves the SDE*

$$d\bar{X}(t) = \mathbf{1}_{\{\bar{X}_1 \leq 0\}} M_- dW(t) + \mathbf{1}_{\{\bar{X}_1 > 0\}} M_+ dW(t) + K \, dL(t) \, . \tag{10}$$

where L denotes the symmetric local time of $\bar{X}_1$ at the origin and W is a standard d-dimensional Wiener process. The matrices $M_\pm$ and the vector K satisfy

$$M_\pm M_\pm^T = D^\pm , \quad K_1 = p_+ - p_- , \quad K_j = \alpha_j , \tag{11}$$

for $j = \{2, \ldots, d\}$.

In Figure 1, we show an example of a numerical simulation of the process studied in this article. The figure on the left shows the small-scale structure (the periodic structure of the drift is drawn as a grid). One can clearly see the periodic structure of the sample path, especially to the left of the interface. One can also see that the effective diffusivity is not necessarily proportional to the identity. In this case, to the left of the interface, the process diffuses much more easily horizontally than vertically.

The picture to the right shows a simulation of the process at a much larger scale. We used a slightly different vector field for the drift in order to obtain a simulation that shows clearly the strong drift experienced by the process when it hits the interface. The remainder of this note is devoted to a short discussion of the proof of Theorem 1.1.

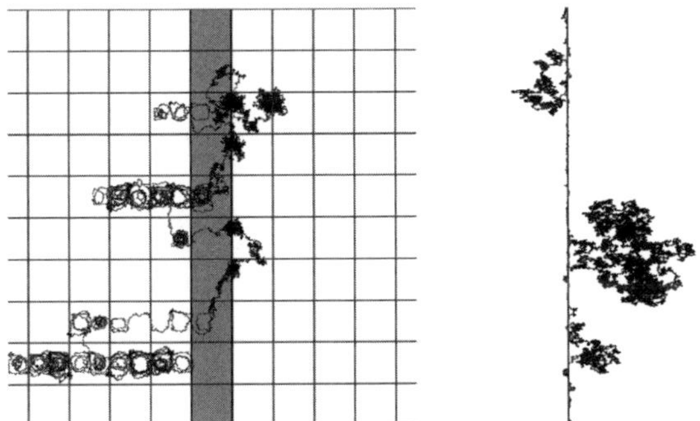

Fig. 1.   Sample paths at small (left) and large (right) scales.

2. Idea of proof

As is common in the theory of homogenization, the pattern of the proof is as follows: one first verifies tightness, then shows that any limit point satisfies the martingale problem associated to $\bar{\mathcal{L}}$, and then finally identifies solutions to this martingale problem as the unique solution to Eq. (10).

2.1. *Tightness of the rescaled processes*

We want to show that the modulus of continuity of X^ε is well-behaved uniformly in ε. The only barrier to this holding can easily be shown to be the drift picked up by the process in the interface. In order to bound this, we thus need to show that the process does not spend too much time there.

We decompose the trajectory for the process X^ε into excursions away from the interface, separated by pieces of trajectory inside the interface. We first show that if the process starts inside the interface, then the expected time spent in the interface before making a new excursion is of order ε^2. Then, we show that each excursion has a probability at least $\varepsilon/\sqrt{\delta}$ of being of length δ or more. This shows that in the time interval δ of interest, the process will perform at most of the order of $\sqrt{\delta}/\varepsilon$ excursions, so that the total time spent in the interface is indeed of the order $\varepsilon\sqrt{\delta}$. Since the drift of the rescaled process is of order $1/\varepsilon$, we conclude that the modulus of continuity is of order $\sqrt{\delta}$ everywhere.

2.2. *Identification of the limiting martingale problem*

In order to identify the martingale problem solved by the limiting process, it is possible to adapt a result obtained by Freidlin and Wentzell in the context of diffusions on graphs.[3] The main ingredients are the following. For $\delta = \varepsilon^\alpha$ with $\alpha \in (\frac{1}{2}, 1)$, denote by τ^δ the first hitting time of $\partial \mathcal{I}_\delta$ by X^ε. We then show that for $p_\pm$ and α_j as in Eq. (7) and Eq. (8), the convergences

$$\mathbb{P}^\varepsilon_x[X^\varepsilon(\tau^\delta) \in I_\pm] \to p_\pm , \qquad \frac{1}{\delta}\,\mathbb{E}^\varepsilon_x\Big[X^\varepsilon_j(\tau^\delta)\Big] \to \alpha_j , \qquad (12)$$

take place uniformly over $x \in \mathcal{I}_{\varepsilon\eta}$.

In order to show the first identity in (12), let τ_k be the first hitting time of $\partial \mathcal{I}_k$ by X, set $p^{x,k}_+ = \mathbb{P}_x(X(\tau_k) > 0)$, and consider

$$\bar{p}^k_+ = \sup_{x \in \mathcal{I}_\eta} p^{x,k}_+ , \qquad \underline{p}^k_+ = \inf_{x \in \mathcal{I}_\eta} p^{x,k}_+ . \qquad (13)$$

One can then show that $\lim_{k \to \infty} |\bar{p}^k_+ - \underline{p}^k_+| = 0$ using the fact that the process returns to any small neighborhood in $\mathcal{I}_\eta$ before τ_k with probability tending to 1 as $k \to \infty$, allowing the process to forget about its initial conditions through a coupling argument. The values $p_\pm$ can then be computed in a way similar to the one-dimensional case.

The main ingredient in this calculation is the fact that the invariant measure μ for the process X (which we can view as a recurrent process on

$\mathbb{R} \times \mathbb{T}^{d-1}$) gets closer and closer to multiples of $\mu_\pm$ away from the interface. This can be formalised as:

Proposition 2.1. *Let A denote a bounded measurable set and denote by μ the (unique up to scaling) invariant σ-finite measure of the process X. Denote furthermore by $\mu_\pm$ the invariant measure of the relevant periodic process, normalised in such a way that $\mu_\pm([k, k+1] \times \mathbb{T}^{d-1}) = 1$ for every $k \in \mathbb{Z}$. Then there exist normalisation constants $q_\pm$ such that,*

$$\lim_{k \to \infty} \left(|\mu(A + k) - q_+ \mu_+(A)| + |\mu(A - k) - q_- \mu_-(A)| \right) \to 0 . \tag{14}$$

(Here k is an integer.) Furthermore, this convergence is exponential, and uniform over the set A if we restrict its diameter.

In order to obtain an expression for the limiting values $p_\pm$, one can now argue as follows. Considering the first component of the limiting process, it is reasonable to expect that it converges to a rescaling Y of skew Brownian motion. This is characterised by three quantities: its diffusivity coefficients on either side of the interface (we already know that they are given by $D_{11}^\pm$) and a parameter p_+ such that, setting $p_- = 1 - p_+$,

$$\mathbb{P}_0^\varepsilon[Y(\tau^\delta) \in I_\pm] = p_\pm . \tag{15}$$

The invariant measure for Y is known to be proportional to Lebesgue measure on either side of the interface, with proportionality constants $q_\pm = \frac{p_\pm}{D_{11}^\pm}$. We can then simply solve this for $p_\pm$.

The second part of (12) is shown in two steps. With τ_k as before, we have the identity

$$\alpha_j = \lim_{k \to \infty} \frac{1}{k} \mathbb{E}_x \int_0^{\tau_k} b_j(X_s) \, ds , \tag{16}$$

for any fixed starting point x in the interface. If k is large, then the process X has had plenty of time to "equilibrate", so that it is natural to expect that α_j is proportional to $\int b_j(x) \, \mu(dx)$. The only question is: what should be the correct proportionality constant?

In order to answer this question, let us assume for the sake of the argument that $b_j = N^{-1} \mathbf{1}_{[-N,N]}$ for some fixed but large value of N. (Note that the fact that the function b_j appearing in Eq. (16) is given by the drift of the original diffusion is irrelevant to the argument, we could ask about the value of this limit for any function b that is localised around the interface.) We then have $\int b_j(x) \, \mu(dx) \approx 1$, thanks to the normalisation $q_+ + q_- = 1$. On the other hand, we know that the first component of the rescaled process converges to skew Brownian motion described by the parameters $p_\pm$ and $D_{11}^\pm$. Time-changing the process by a factor $D_{11}^\pm$ on either side of the origin, we can reduce ourselves to the case of standard skew-Brownian motion with parameters $p_\pm$. Since this consists of standard Brownian motion excursions biased to go to either side of the origin with respective probabilities $p_\pm$, this yields in this particular example

$$\alpha_j = \left(\frac{p_+}{D_{11}^+} + \frac{p_-}{D_{11}^-} \right) \lim_{k\to\infty} \frac{1}{k} \mathbb{E}_0 \int_0^{\tau_k} \mathbf{1}_{[-1,1]}(B(s))\, ds \;, \qquad (17)$$

where B is a standard Brownian motion. A simple calculation then shows that the term under the expectation is asymptotic to $2k$, so that we do indeed recover the proportionality constant from Eq. (8).

2.3. *Uniqueness of the martingale problem*

Finally, in order to show uniqueness of the martingale problem, we use Theorem 4.1 from[6] in conjunction with the Hille-Yosida theorem to ensure that the domain of the generator to our martingale problem is large enough. It is then possible to explicitly construct solutions to the system of SDEs given in (10) and to show that they solve the same martingale problem, thus concluding the proof.

References

1. M. Hairer and C. Manson, Periodic homogenization with an interface: the one-dimensional case, Preprint, (2009).
2. M. Hairer and C. Manson, Periodic homogenization with an interface: the multi-dimensional case, Preprint, (2009).
3. M. I. Freidlin and A. D. Wentzell, *Ann. Probab.* **21**, 2215 (1993).
4. A. Bensoussan, J. Lions and G. Papanicolaou, *Asymptotic analysis of periodic structures* (North-Holland, Amsterdam, 1978).
5. R. Z. Has'minskiĭ, *Teor. Verojatnost. i Primenen.* **5**, 196 (1960).
6. S. N. Ethier and T. G. Kurtz, *Markov processes: Characterization and convergence* (John Wiley & Sons Inc., New York, 1986).

Discrete-time interest rate modelling

Lane P. Hughston

Department of Mathematics, Imperial College London, London SW7 2AZ, UK
E-mail: l.hughston@imperial.ac.uk

Andrea Macrina

Department of Mathematics, King's College London, London WC2R 2LS, UK
Kyoto Institute of Economic Research, Kyoto University, Kyoto 606-8501, Japan
E-mail: andrea.macrina@kcl.ac.uk

This paper presents an axiomatic scheme for interest rate models in discrete time. We take a pricing kernel approach, which builds in the arbitrage-free property and provides a link to equilibrium economics. We require that the pricing kernel be consistent with a pair of axioms, one giving the inter-temporal relations for dividend-paying assets, and the other ensuring the existence of a money-market asset. We show that the existence of a positive-return asset implies the existence of a previsible money-market account. A general expression for the price process of a limited-liability asset is derived. This expression includes two terms, one being the discounted risk-adjusted value of the dividend stream, the other characterising retained earnings. The vanishing of the latter is given by a transversality condition. We show (under the assumed axioms) that, in the case of a limited-liability asset with no permanently-retained earnings, the price process is given by the ratio of a pair of potentials. Explicit examples of discrete-time models are provided.

Keywords: Interest rates models; pricing kernels; financial time series; Flesaker-Hughston models; transversality condition; financial bubbles.

1. Discrete-time asset pricing

Although discrete-time interest rate models are often introduced for computational purposes as a convenient approximation to the continuous-time situation, it is important to recognize that the theory can be developed in discrete time in an entirely satisfactory way in its own right, without reference to continuous time. Let $\{t_i\}_{i=0,1,2,\dots}$ denote a time sequence where t_0 is the present and $t_{i+1} > t_i$ for all $i \in \mathbb{N}_0$. We assume that $\{t_i\}$ is unbounded in the sense that for any T there exists a value of i such that $t_i > T$. The economy is represented by a probability space $(\Omega, \mathcal{F}, \mathbb{P})$ with a filtration $\{\mathcal{F}_i\}_{i \geq 0}$ which we call the "market filtration". Each asset is characterised by a pair $\{S_{t_i}\}_{i \geq 0}$ and $\{D_{t_i}\}_{i \geq 0}$ which we call the "value process" and the "dividend process". We interpret D_{t_i} as the random cash flow out paid by the asset at time t_i. Then S_{t_i} denotes the "ex-dividend" value of the asset at t_i. For simplicity, we often write $S_i = S_{t_i}$ and $D_i = D_{t_i}$. To ensure the absence of arbitrage, we assume the existence of a positive pricing kernel $\{\pi_i\}_{i \geq 0}$, and make the following assumptions:

Axiom A. *For any asset with value process $\{S_i\}_{i \geq \infty}$ and dividend process $\{D_i\}_{i \geq 0}$, the process $\{M_i\}_{i \geq 0}$ defined by*

$$M_i = \pi_i S_i + \sum_{n=0}^{i} \pi_n D_n \tag{1}$$

is a martingale.

Axiom B. *There exists a positive non-dividend-paying asset with value process $\{\bar{B}_i\}_{i \geq 0}$ such that $\bar{B}_{i+1} > \bar{B}_i$ for all $i \in \mathbb{N}_0$, and that for any $b \in \mathbb{R}$ there exists a time t_i such that $\bar{B}_i > b$.*

The notation $\{\bar{B}_i\}$ distinguishes the positive return asset from the previsible money-market account $\{B_i\}$ introduced later. Since $\{\bar{B}_i\}$ is non-dividend paying, Axiom A implies that $\{\pi_i \bar{B}_i\}$ is a martingale. Writing $\bar{\rho}_i = \pi_i \bar{B}_i$, we have $\pi_i = \bar{\rho}_i / \bar{B}_i$. Since $\{\bar{B}_i\}$ is increasing, $\{\pi_i\}$ is a supermartingale, and it follows from Axiom B that

$$\lim_{i \to \infty} \mathbb{E}[\pi_i] = 0. \tag{2}$$

We obtain the following result concerning limited-liability assets.

Proposition 1. *Let $S_i \geq 0$ and $D_i \geq 0$ for all $i \in \mathbb{N}$. We have*

$$S_i = \frac{m_i}{\pi_i} + \frac{1}{\pi_i} \mathbb{E}_i \left[\sum_{n=i+1}^{\infty} \pi_n D_n \right], \tag{3}$$

where $\{m_i\}$ is a non-negative martingale that vanishes if and only if:

$$\lim_{j \to \infty} \mathbb{E}[\pi_j S_j] = 0. \tag{4}$$

Thus $\{m_i\}$ represents that part of the value of the asset that is "never paid out". An idealised money-market account is of this nature, and so is a "permanent bubble" (cf. Tirole 1982). In the case of an asset for which the "transversality" condition is satisfied, the price is directly related to the future dividend flow:

$$S_i = \frac{1}{\pi_i} \mathbb{E}_i \left[\sum_{n=i+1}^{\infty} \pi_n D_n \right]. \tag{5}$$

This is the so-called "fundamental equation" often used as a basis for asset pricing (cf. Cochrane 2005). Alternatively we can write

$$S_i = \frac{1}{\pi_i}(\mathbb{E}_i[F_\infty] - F_i), \tag{6}$$

where $F_i = \sum_{n=0}^{i} \pi_n D_n$, and $F_\infty = \lim_{i \to \infty} F_i$. Hence the price of a pure dividend-paying asset can be expressed as a ratio of potentials, giving us a discrete-time analogue of a result of Rogers 1997.

2. Positive-return asset and pricing kernel

Let us introduce the notation

$$\bar{r}_i = \frac{\bar{B}_i - \bar{B}_{i-1}}{\bar{B}_{i-1}} \tag{7}$$

for the rate of return of the positive-return asset realised at time t_i on an investment made at t_{i-1}. The notation $\bar{r}_i$ is used to distinguish the rate of return on $\{\bar{B}_i\}$ from the rate of return r_i on the money market account $\{B_i\}$ introduced later.

Proposition 2. *There exists an asset with constant value $S_i = 1$ for all $i \in \mathbb{N}_0$, for which the associated cash flows are given by $\{\bar{r}_i\}_{i \geq 1}$.*

Proposition 3. *Let $\{\bar{B}_i\}$ be a positive-return asset satisfying Axioms A and B, and let $\{\bar{r}_i\}$ be its rate-of-return process. Then the pricing kernel can be expressed in the form $\pi_i = \mathbb{E}_i[G_\infty] - G_i$, where $G_i = \sum_{n=1}^{i} \pi_n \bar{r}_n$ and $G_\infty = \lim_{i \to \infty} G_i$.*

There is a converse to this result that allows one to construct a system satisfying Axioms A and B from a strictly-increasing non-negative adapted process that converges and satisfies an integrability condition.

Proposition 4. *Let $\{G_i\}_{i \geq 0}$ be a strictly-increasing adapted process with $G_0 = 0$, and $\mathbb{E}[G_\infty] < \infty$, where $G_\infty = \lim_{i \to \infty} G_i$. Let $\{\pi_i\}$, $\{\bar{r}_i\}$, and $\{\bar{B}_i\}$ be defined by $\pi_i = \mathbb{E}_i[G_\infty] - G_i$ for $i \geq 0$, $\bar{r}_i = (G_i - G_{i-1})/\pi_i$ for $i \geq 1$, and $\bar{B}_i = \prod_{n=1}^{i}(1 + \bar{r}_n)$ for $i \geq 1$, with $\bar{B}_0 = 1$. Let $\{\bar{\rho}_i\}$ be defined by $\bar{\rho}_i = \pi_i \bar{B}_i$ for $i \geq 0$. Then $\{\bar{\rho}_i\}$ is a martingale, and $\lim_{j \to \infty} \bar{B}_j = \infty$, from which it follows that $\{\pi_i\}$ and $\{\bar{B}_i\}$ satisfy Axioms A and B.*

3. Discrete-time discount bond systems

The price P_{ij} at t_i $(i < j)$ of a discount bond that matures at t_j is

$$P_{ij} = \frac{1}{\pi_i}\mathbb{E}_i[\pi_j]. \tag{8}$$

Since $\pi_i > 0$ for $i \in \mathbb{N}$, and $\mathbb{E}_i[\pi_j] < \pi_i$ for $i < j$, it follows that $0 < P_{ij} < 1$ for $i < j$. We observe that the "per-period" interest rate R_{ij} defined by $P_{ij} = 1/(1 + R_{ij})$ is positive. Since $\{\pi_i\}$ is given, there is no need to model the volatility structure of the bonds. Thus, our scheme differs from the discrete-time models discussed in Heath *et al.* 1990, and Filipović & Zabczyk 2002. As an example of a class of discrete-time models set $\pi_i = \alpha_i + \beta_i N_i$, where $\{\alpha_i\}$ and $\{\beta_i\}$ are positive, strictly-decreasing deterministic processes with $\lim_{i \to \infty} \alpha_i = 0$ and $\lim_{i \to \infty} \beta_i = 0$, and where $\{N_i\}$ is a positive martingale. Then we have

$$P_{ij} = \frac{\alpha_j + \beta_j N_i}{\alpha_i + \beta_i N_i},\tag{9}$$

giving a family of "rational" interest rate models. In a discrete-time setting we can produce models that do not necessarily have analogues in continuous time—for example, we can let $\{N_i\}$ be the martingale associated with a branching process.

Any discount bond system consistent with our scheme admits a representation of the Flesaker-Hughston type (Rutkowski 1997, Jin & Glasserman 2001, Cairns 2004, Musiela & Rutkowski 2005, Björk 2009). More precisely, we have:

Proposition 5. *Let $\{\pi_i\}$, $\{\bar{B}_i\}$, $\{P_{ij}\}$ satisfy Axioms A and B. Then there exists a family of positive martingales $\{m_{in}\}_{0 \le i \le n}$, $n \in \mathbb{N}$, such that*

$$P_{ij} = \frac{\sum_{n=j+1}^{\infty} m_{in}}{\sum_{n=i+1}^{\infty} m_{in}}.\tag{10}$$

4. Construction of the money-market asset

Let us look now at the situation where the positive-return asset is previsible. Thus we assume that B_i is $\mathcal{F}_{i-1}$-measurable and we drop the "bar" over B_i. In that case we have

$$P_{i-1,i} = \frac{1}{\pi_{i-1}} \mathbb{E}_{i-1}[\pi_i] = \frac{B_{i-1}}{\rho_{i-1}} \mathbb{E}_{i-1}\left[\frac{\rho_i}{B_i}\right] = \frac{B_{i-1}}{B_i}.\tag{11}$$

Hence, writing $P_{i-1,i} = 1/(1 + r_i)$ where $r_i = R_{i-1,i}$ we see that the rate of return on the money-market account is previsible, and is given by the one-period discount factor associated with the bond that matures at t_i.

Reverting to the general situation, it follows that if we are given a pricing kernel $\{\pi_i\}$ on $(\Omega, \mathcal{F}, \mathbb{P}, \{\mathcal{F}_i\})$, and a system of assets satisfying Axioms A and B, then we can construct a *candidate* for a previsible money market account by setting $B_0 = 1$ and

$$B_i = (1 + r_i)(1 + r_{i-1}) \cdots (1 + r_1), \tag{12}$$

for $i \geq 1$, where r_i is defined by

$$r_i = \frac{\pi_{i-1}}{\mathbb{E}_{i-1}[\pi_i]} - 1. \tag{13}$$

We refer to $\{B_i\}$ as the "natural" money-market account associated with $\{\pi_i\}$. To justify this terminology, we verify that $\{B_i\}$, so constructed, satisfies Axioms A and B. To this end, we note the following multiplicative decomposition. Let $\{\pi_i\}$ be a positive supermartingale satisfying $\mathbb{E}_i[\pi_j] < \pi_i$ for $i < j$ and $\lim_{j \to \infty}[\pi_j] = 0$. Then we can write $\pi_i = \rho_i / B_i$, where

$$\rho_i = \frac{\pi_i}{\mathbb{E}_{i-1}[\pi_i]} \frac{\pi_{i-1}}{\mathbb{E}_{i-2}[\pi_{i-1}]} \cdots \frac{\pi_1}{\mathbb{E}_0[\pi_1]} \pi_0 \tag{14}$$

for $i \geq 0$, and

$$B_i = \frac{\pi_{i-1}}{\mathbb{E}_{i-1}[\pi_i]} \frac{\pi_{i-2}}{\mathbb{E}_{i-2}[\pi_{i-1}]} \cdots \frac{\pi_1}{\mathbb{E}_1[\pi_2]} \frac{\pi_0}{\mathbb{E}_0[\pi_1]} \tag{15}$$

for $i \geq 1$, with $B_0 = 1$. In this scheme we have

$$\rho_i = \frac{\pi_i}{\mathbb{E}_{i-1}[\pi_i]} \rho_{i-1}, \tag{16}$$

with $\rho_0 = \pi_0$; and

$$B_i = \frac{\pi_{i-1}}{\mathbb{E}_{i-1}[\pi_i]} B_{i-1}, \tag{17}$$

with $B_0 = 1$. It is evident that $\{\rho_i\}$ is $\{\mathcal{F}_i\}$-adapted, and that $\{B_i\}$ is previsible and increasing. We establish the following:

Proposition 6. *Let $\{\pi_i\}$ be a non-negative supermartingale such that $\mathbb{E}_i[\pi_j] < \pi_i$ for all $i < j \in \mathbb{N}_0$, and $\lim_{i \to \infty} \mathbb{E}[\pi_i] = 0$. Let $\{B_i\}$ be defined by $B_0 = 1$ and $B_i = \prod_{n=1}^{i}(1 + r_n)$ for $i \geq 1$, where $1 + r_i = \pi_{i-1}/\mathbb{E}_{i-1}[\pi_i]$, and set $\rho_i = \pi_i B_i$ for $i \geq 0$. Then $\{\rho_i\}$ is a martingale, and the interest rate system defined by $\{\pi_i\}$, $\{B_i\}$, $\{P_{ij}\}$ satisfies Axioms A and B.*

A significant feature of Proposition 6 is that no integrability condition is required on $\{\rho_i\}$: the natural money market account defined above "automatically" satisfies Axiom A. Thus in place of Axiom B we can assume:

Axiom B*. *There exists a positive non-dividend paying asset, the money-market account $\{B_i\}_{i \geq 0}$, having the properties that $B_{i+1} > B_i$ for $i \in \mathbb{N}_0$, that B_i is $\mathcal{F}_{i-1}$-measurable for $i \in \mathbb{N}$, and that for any $b \in \mathbb{R}$ there exists a t_i such that $B_i > b$.*

The content of Proposition 6 is that Axioms A and B together are equivalent to Axioms A and B* together. Let us establish that the class of interest rate models satisfying Axioms A and B* is non-vacuous. In particular, consider the "rational" model defined for some choice of $\{N_i\}$. It is an exercise to see that the previsible money market account is given for $i = 0$ by $B_0 = 1$ and for $i \geq 1$ by

$$B_i = \prod_{n=1}^{i} \frac{\alpha_{n-1} + \beta_{n-1} N_{n-1}}{\alpha_n + \beta_n N_{n-1}}, \tag{18}$$

and that for $\{\rho_i\}$ we have

$$\rho_i = \rho_0 \prod_{n=1}^{i} \frac{\alpha_n + \beta_n N_n}{\alpha_n + \beta_n N_{n-1}}, \tag{19}$$

where $\rho_0 = \alpha_0 + \beta_0 N_0$. One can check for each $i \geq 0$ that ρ_i is bounded; therefore $\{\rho_i\}$ is a martingale, and $\{B_i\}$ satisfies Axioms A and B*.

5. Doob decomposition

Consider now the Doob decomposition given by $\pi_i = \mathbb{E}_i[A_\infty] - A_i$, with

$$A_i = \sum_{n=0}^{i-1} \left(\pi_n - \mathbb{E}_n[\pi_{n+1}] \right) \tag{20}$$

as discussed, e.g., in Meyer 1966. It follows that

$$A_i = \sum_{n=0}^{i-1} \pi_n \left(1 - \frac{\mathbb{E}_n[\pi_{n+1}]}{\pi_n} \right) = \sum_{n=0}^{i-1} \pi_n \left(1 - P_{n,n+1} \right) = \sum_{n=0}^{i-1} \pi_n r_{n+1} P_{n,n+1}, \tag{21}$$

where $\{r_i\}$ is the previsible short rate process. The pricing kernel can therefore be put in the form

$$\pi_i = \mathbb{E}_i \left[\sum_{n=i}^{\infty} \pi_n r_{n+1} P_{n,n+1} \right]. \tag{22}$$

Comparing (22) with the decomposition $\pi_i = \mathbb{E}_i[G_\infty] - G_i$, $G_i = \sum_{n=1}^{i} \pi_n \bar{r}_n$, given in Proposition 3, we see that by setting

$$\bar{r}_i = \frac{r_i \pi_{i-1} P_{i-1,i}}{\pi_i} \tag{23}$$

we obtain a positive-return asset based on the Doob decomposition.

6. Foreign exchange processes

An extension of the material presented here to models for foreign exchange and inflation is pursued in Hughston & Macrina (2008). In particular, since the money-market account is a positive-return asset, by Proposition 3 we can write:

$$\pi_i = \mathbb{E}_i \left[\sum_{n=i+1}^{\infty} \pi_n r_n \right]. \tag{24}$$

As a consequence, we see that the price process of a pure dividend-paying asset can be written in the following symmetrical form:

$$S_i = \frac{\mathbb{E}_i \left[\sum_{n=i+1}^{\infty} \pi_n D_n \right]}{\mathbb{E}_i \left[\sum_{n=i+1}^{\infty} \pi_n r_n \right]}. \tag{25}$$

In the case where $\{S_i\}$ represents a foreign currency, the dividend process is the foreign interest rate, and both $\{D_i\}$ and $\{r_i\}$ are previsible.

References

1. T. Björk (2009) *Arbitrage Pricing in Continuous Time*, Oxford Finance.
2. A. J. G. Cairns (2004) *Interest Rate Models: An Introduction*, Princeton University Press.
3. J. H. Cochrane (2005) *Asset Pricing*, Princeton University Press.
4. D. Filipovic & J. Zabczyk (2002) Markovian term structure models in discrete time, *Annals of Applied Probability* **12**, 710-729.
5. D. Heath, R. Jarrow & A. Morton (1990) Bond pricing and the term structure of interest rates: a discrete time approximation, *Journal of Financial and Quantitative Analysis* **25**, 419-440.
6. L. P. Hughston, & A. Macrina (2008) Information, inflation, and interest. In *Advances in Mathematics of finance* (ed. L. Stettner). Banach Centre Publications **83** 117 (Warsaw, Poland: Polish Academy of Sciences 2008).
7. Y. Jin & P. Glasserman (2001) Equilibrium positive interest rates: a unified view, *Review of Financial Studies* **14**, 187-214.
8. P. A. Meyer (1966) *Probability and Potentials*, Blaisdell.
9. M. Musiela & M. Rutkowski (2005) *Martingale Methods in Financial Modelling*, Springer.
10. L. C. G. Rogers (1997) The potential approach to the term structure of interest rate and foreign exchange rates, *Mathematical Finance* **7**, 157-176.
11. M. Rutkowski (1997) A note on the Flesaker-Hughston model of the term structure of interest rates, *Applied Mathematical Finance* **4**, 151-163.
12. J. M. Tirole (1982) On the probability of speculation under rational expectation, *Econometrica* **50**, 1163-1181.

Probabilistic representation for solutions of higher-order elliptic equations

Mark Kelbert

Mathematics Department, Swansea University,
Swansea, Wales, UK
E-mail: m.kelbert@swansea.ac.uk

This article presents a probabilistic representation for solutions of Lauricella problem and exploits it to obtain the upper bounds on the growth of these solutions when the domain increases.

Keywords: Lauricella problem; Brownian motion; torsion function; Bessel process.

1. Stochastic representation for Lauricella problem

The boundary value problem for biharmonic equation $\Delta^2 u = 0$ in a domain $D \subset \mathbf{R}^2$ with boundary conditions $\frac{\partial}{\partial x} u \big|_{\partial D} = f_1, \frac{\partial}{\partial y} u \big|_{\partial D} = f_2$ was subject of a prize offered by Paris Academy and obtained by J. Hadamard and G. Lauricella. They solved it by reducing to the solution of a Fredholm's integral equation. The problem with the boundary conditions $\Delta u \big|_{\partial D} = -g_1, u \big|_{\partial D} = g_0$ also appears naturally in elasticity theory. In this note we summarize some known results and use them to obtain new bounds on the growth of solutions. So, consider the Lauricella problem with an integer $m \geq 1$ and potential V

$$
\begin{aligned}
(\tfrac{1}{2}\Delta + V)^m u &= 0, & x &\in D, \\
(\tfrac{1}{2}\Delta + V)^k u &= (-1)^k g_k, & x &\in \partial D
\end{aligned}
\tag{1.1}
$$

where $k = 0, \ldots, m-1$, $(\tfrac{1}{2}\Delta + V)^0 := id$, $D \subset \mathbf{R}^d$ is a bounded domain with a regular boundary ∂D, $V \in C(\overline{D})$ and $g_k \in C(\partial D)$. Note that main results could be immediately extended to any generator L of Markov process.

Let $\lambda_1 = \lambda_1(D, V)$ denote the principle Dirichlet eigenvalue of $-\tfrac{1}{2}\Delta - V$ in D. Define $M_k := \max_{\partial D} |g_k|$, for $k = 0, \ldots, m-1$, and $\nu := \max_D V^+$.

Let $0 < \Psi \in C^\infty(D) \cap C(\overline{D})$ be the unique solution of the problem

$$-\frac{1}{2}\Delta\Psi - V\Psi = 1, \quad x \in D; \qquad \Psi = 0, \quad x \in \partial D$$

called *the torsion function* of $-\frac{1}{2}\Delta - V$ in D. Denote

$$\gamma(D, V) := \max_D \Psi. \tag{1.2}$$

It is known (see [4], Corollary 1.5) that $\lambda_1 > 0$ implies

$$\gamma(D, V) \leq Cm^{2/d}(D)$$

where $m(D)$ is the Lebesgue measure of D and $C = C(d, D, V) < \infty$.

Theorem 1.1. *Assume $\lambda_1 > 0$. Let $(B_0(t),\ t > 0)$ be the 'original' Brownian motion (BM) starting at $x \in D$, and τ_0 be its hitting time of the boundary ∂D. Next, $(B_1(t),\ t > 0)$ be the BM starting from the point $B_0(s_0)$, $s_0 \in (0, \tau_0)$, with increments $B_1(t) - B_0(s_0)$ independent of the BM $(B_0(t),\ t > 0)$, and τ_1 be the time of hitting ∂D for $(B_1(t),\ t > 0)$, and so on: $(B_{m-1}(t),\ t > 0)$ be the BM starting from the point $B_{m-2}(s_{m-2})$, $s_{m-2} \in (0, \tau_{m-2})$, with increments $B_{m-1}(t) - B_{m-2}(s_{m-2})$ independent of the BMs $(B_j(t),\ t > 0)$, $j = 0, \dots, m - 2$, and τ_{m-1} be the time of hitting ∂D for $(B_{m-1}(t),\ t > 0)$.*

The unique solution of the problem (1.1) is presented by

$$
\begin{aligned}
u(x) = \ &\mathbf{E}_x\left(g_0(B_0(\tau_0))\, \exp\left[\int_0^{\tau_0} V(B_0(u_0))\mathrm{d}u_0\right] \right) \\
&+ \mathbf{E}_x\left(\int_0^{\tau_0} \mathrm{d}s_0 \exp\left[\int_0^{s_0} V(B_0(u_0))\mathrm{d}u_0\right] \right. \\
&\qquad \times \left. \mathbf{E}_{B_0(s_0)}\left(g_1(B_1(\tau_1)) \exp\left[\int_0^{\tau_1} V(B_1(u_1))\mathrm{d}u_1\right] \right) \right) \\
&+ \cdots + \mathbf{E}_x\left(\int_0^{\tau_0} \mathrm{d}s_0 \exp\left[\int_0^{s_0} V(B_0(u_0))\mathrm{d}u_0\right] \right. \\
&\qquad \times \mathbf{E}_{B_0(s_0)}\left(\int_0^{\tau_1} \mathrm{d}s_1 \exp\left[\int_0^{s_1} V(B_1(u_1))\mathrm{d}u_1\right] \right. \\
&\qquad \times \cdots \times \mathbf{E}_{B_{m-3}(s_{m-3})}\left(\int_0^{\tau_{m-2}} \mathrm{d}s_{m-2} \right. \\
&\qquad \times \exp\left[\int_0^{s_{m-2}} V(B_{m-2}(u_{m-2}))\mathrm{d}u_{m-2}\right] \\
&\qquad \times \mathbf{E}_{B_{m-2}(s_{m-2})}\left(g_{m-1}(B_{m-1}(\tau_{m-1})) \right. \\
&\qquad \times \left.\left.\left. \exp\left[\int_0^{\tau_{m-1}} V(B_{m-1}(u_{m-1}))\mathrm{d}u_{m-1}\right] \right) \cdots \right) \right),
\end{aligned}
\tag{1.3}
$$

and admits the upper bound

$$\sup_{x \in D} |u(x)| \leq \left(M_{m-1}\gamma^{m-1} + M_{m-2}\gamma^{m-2} + \cdots + M_0 \right)\left(1 + \gamma\nu\right). \tag{1.4}$$

Remark 1.1. Observe that if the signs of the boundary values alternate, i.e. $g_l \geq 0$, $l = 0, 1, \ldots, m-1$, then the solution of problem (1.1) $u(x) \geq 0$.

Corollary 1.1. *Suppose the potential $V \equiv 0$. Then the unique solution of problem (1.1) admits the representation*

$$u(x) = \mathbf{E}_x\Big[\tfrac{1}{(m-1)!}\tau^{m-1}g_{m-1}(B(\tau))+$$
$$+ \tfrac{1}{(m-2)!}\tau^{m-2}g_{m-2}(B(\tau)) + \ldots + g_0(B(\tau))\Big], \ x \in D. \tag{1.5}$$

Here, $(B(t))$ is the standard BM starting at $x \in D$, and τ is hitting time of the boundary ∂D.

Remark 1.2. (a) For the first time the equality (1.5) appeared in [8], see [6] and [9] for more details. (b) Set $M := \max[M_0, \ldots, M_{m-1}]$. Then (1.4) could be replaced by

$$\sup_{x \in D} u(x) \leq M\left(\frac{\gamma^m - 1}{\gamma - 1}\right)(1 + \gamma\nu).$$

When $V \equiv 0$, after selecting $M_k \equiv 1$ and $M_i \equiv 0$ if $i \neq k$ estimate (1.4) leads to the bound

$$\sup_{x \in D} \mathbf{E}_x[\tau_D^k] \leq k! \sup_{x \in D}(\mathbf{E}_x[\tau_D])^k. \tag{1.6}$$

This is a version of Khas'minskii's lemma.

Theorem 1.2. *Suppose that $V \leq 0$. Then the following bound holds*

$$\sup_{x \in D} \mathbf{E}_x[\tau_D^k] \leq A_{d,k} m(D)^{2k/d}. \tag{1.7}$$

with

$$A_{d,k} = k!\left(\frac{d+2}{2\pi d}\left(\frac{d+2}{2}\right)^{2/d}\right)^k.$$

Remark 1.3. The constant $A_{d,k}$ obtained in Theorem 1.2 for $k > 2$ is not optimal. However, it is close to optimal for $k = 1$. The minimal constant in (1.7) for $k = 1$ and a given $m(D)$ is attained when D is a ball, see [1]. In the case $V \equiv 0$ various estimations of the torsion functions are available in the PDEs literature, in particular (e.g., see [2], Theorem 2.8), it is known that

$$\gamma(D) \leq a_d m(D)^{2/d},$$

where $a_d = d^{-1}(\omega_d)^{-2/d}$ and $\omega_d = \pi^{d/2}/\Gamma(d/2+1)$ denotes the volume of the unit ball in $\mathbf{R}^d$. The estimate is sharp when D is a ball. Clearly, for all $d = 1, 2, \ldots$

$$a_d < A_{d,1}.$$

Further, $A_{d,1}$ and a_d are monotone decreasing, and

$$\lim_{d \to \infty} A_{d,1} = \frac{1}{2\pi}, \ \lim_{d \to \infty} a_d = \frac{1}{4\pi e}.$$

Consider a class $\mathcal{G}(M)$ of vector-functions $G = (g_{m-1}, g_{m-2}, \ldots, g_1, g_0)$ on ∂D such that $|g_{m-1}| \le M$, $|g_{m-2}| \le M, \ldots, |g_0| \le M$, $g_i \in C(\partial D)$, $i = 0, \ldots, m-1$. We say that the sequence of domains (D_n) extends regularly to $\mathbf{R}^d$ if for any compact $K \subset D$ (i) $R_n(K) = \operatorname{dist}(K, \partial D_n) \to \infty$ as $n \to \infty$ and (ii) $\bar{R}_n(K)/R_n(K) = 1 + o(1)$ as $n \to \infty$ where $\bar{R}_n(K) = \sup_{y \in \partial D_n} \big[\operatorname{dist}(y, K)\big]$. For any $n = 0, 1, \ldots$ consider the classes of boundary values $\mathcal{G}^{(n)}(M)$ in domains $D_n, D_0 = D$. We assume that the constant M does not depend on n and omit the index n thereafter.

Theorem 1.3. *Let $\mathcal{U} = \mathcal{U}(M)$ be the set of solutions of the boundary value problem (1.1) in domains D_n, $D \subset D_n$, with $V \le 0$ and the boundary values from the set $\mathcal{G}(M)$. Then for any compact $K \subset D$ there exists $\bar{C} = \bar{C}(K, M, d, m)$ such that for $u \subset \mathcal{U}$*

$$\sup_{x \in K} \big|u(x)\big| \le \bar{C}(1 + \bar{R}_n^{2m-2}) \tag{1.8}$$

where $\bar{R}_n = \bar{R}_n(K) = \sup_{y \in \partial D_n} \big[\operatorname{dist}(y, K)\big]$.

2. Proofs

Lemma 2.1. Assume $\lambda_1(D, V) > 0$. Then for every $f \in C(D)$ and $g \in C(\partial D)$ the problem

$$\left(-\frac{1}{2}\Delta - V\right)u = f, \ x \in D, \quad u\big|_{x \in \partial D} = g \tag{2.1}$$

has a unique solution $u \in C^2(D) \cap C(\bar{D})$. Moreover,

$$u \le (K + \nu M)\gamma + M,$$

where $K := \max_D f$, $M := \max_{\partial D} g$ and $\nu := \max_D V^+$.

Lemma 2.2. Assume $\lambda_1(D, V) > 0$, and denote $M_k := \max_{\partial D} |g_k|$ for $k = 0, 1, \ldots, m-1$. Then the unique solution of problem (1.1) satisfies the bounds

$$u \le (M_{m-1}\gamma^{m-1} + M_{m-2}\gamma^{m-2} + \ldots + M_0)(1 + \gamma\nu). \tag{2.2}$$

Proof of Lemma 2.2. By Lemma 2.1 the problem

$$\left(-\frac{1}{2}\Delta - V\right)w_m = 0, x \in D, \quad w_m = g_{m-1}, x \in \partial D, \qquad (2.3)$$

admits the unique solution $w_m \in C(D)$. Next, consider the system

$$\left(-\frac{1}{2}\Delta - V\right)w_l = w_{l+1}, x \in D, \quad w_l = g_{l-1} = g, x \in \partial D, l = 1, \ldots, m-1. \qquad (2.4)$$

Lemma 2.1 guarantees that the l-th equation of system (2.4) admits the unique solution w_l. But then clearly, $u := w_1$ is a required solution of problem (1.1). To derive the bound (2.2), denote $\omega_m := \max_D w_m$. By (2.1) we obtain

$$\omega_m \le M_{m-1}(1 + \gamma\nu).$$

Further, for $l = m - 1, \ldots, 1$ from (2.1) we derive the recursive bound

$$\omega_l \le (\omega_{l+1} + M_{l-1}\nu)\gamma + M_{l-1}.$$

Solving the recursive inequality, we conclude that

$$\omega_1 \le (M_{m-1}\gamma^{m-1} + M_{m-2}\gamma^{m-2} + \cdots + M_0)(1 + \gamma\nu),$$

so the assertion follows. $\qquad\qquad\square$

Lemma 2.3. For any $g \in C(\partial D)$ define functions v_k for any $k = 1, \ldots, m$ as follows: for $k = 1$ the function v_1 is the solution of the problem

$$\left(\frac{1}{2}\Delta + V\right)v_1 = 0, \quad x \in D, \quad v_1|_{\partial D} = (-1)^{m-1}g \qquad (2.5)$$

whereas the functions $v_k, k = 2, \ldots, m$ are defined inductively as solutions of the following problem in D

$$\left(\frac{1}{2}\Delta + V\right)v_k = -v_{k-1}, x \in D, v_k|_{\partial D} = 0. \qquad (2.6)$$

Then

$$\begin{aligned}
v_m(x) = \mathbf{E}_x\Bigg(&\int_0^{\tau_0} ds_0 \exp\left[\int_0^{s_0} V(B_0(u_0))du_0\right] \\
&\times \mathbf{E}_{B_0(s_0)}\Bigg(\int_0^{\tau_1} ds_1 \exp\left[\int_0^{s_1} V(B_1(u_1))du_1\right] \\
&\times \cdots \times \mathbf{E}_{B_{m-3}(s_{m-3})}\Bigg(\int_0^{\tau_{m-2}} ds_{m-2} \\
&\times \exp\left[\int_0^{s_{m-2}} V(B_{m-2}(u_{m-2}))du_{m-2}\right] \\
&\times \mathbf{E}_{B_{m-2}(s_{m-2})}\Bigg(g_{m-1}(B_{m-1}(\tau_{m-1})) \\
&\times \exp\left[\int_0^{\tau_{m-1}} V(B_{m-1}(u_{m-1}))du_{m-1}\right]\Bigg)\cdots\Bigg)\Bigg),
\end{aligned} \qquad (2.7)$$

Theorem 1.1 follows from Lemmas 2.1-2.3. Corollary 1.1. could also be proved directly using Dynkin's identity (see [5]).

Proof of Theorem 1.2. Clearly,

$$\mathbf{E}_x[\tau^k] = \int_0^\infty k u^{k-1} \mathbf{P}_x(\tau > u)\mathrm{d}u. \tag{2.8}$$

Fix a value of u such that

$$\mathbf{P}_x(\tau > u) \le \mathbf{P}_x(B(u) \in D) \le \frac{m(D)}{(2\pi u)^{d/2}} = \theta < 1,$$

and estimate the integral in (2.8) as follows

$$\mathbf{E}_x[\tau^k] \le k u^k \sum_{n=0}^\infty (n+1)^{k-1} \mathbf{P}_x(\tau > nu).$$

By the Markov property, $\mathbf{P}_x(\tau > nu) \le \theta^n$. Hence,

$$\mathbf{E}_x[\tau^k] \le k u^k \sum_{n=0}^\infty (n+k-1)(n+k-2)\dots(n+1)\theta^k = k u^k \frac{(k-1)!}{(1-\theta)^k}. \tag{2.9}$$

The minimum of RHS is attained for $u_0 = (C(d+2)/2)^{2/d}$ where $C = \frac{m(D)}{(2\pi)^{d/2}}$. This argument establishes the inequality (1.7) and completes the proof. $\qquad\square$

Proof of Theorem 1.3. The radial part of Brownian motion $|B(t)|$ is the Bessel process of the order $\nu = \frac{d}{2} - 1$. Let $\tau_R = \inf[t : |B(t)| = R]$. In view of Eqn. 2.0.1, § 2.4, [3], for the Laplace transform $\psi_R(s, x) = \mathbf{E}_x\left[e^{-s\tau_R}\right]$

$$\psi_R(s, x) = \frac{x^{-\nu}\mathrm{I}_\nu(x\sqrt{2s})}{R^{-\nu}\mathrm{I}_\nu(R\sqrt{2s})}, x \le R, \tag{2.10}$$

where I_ν is the modified Bessel function $\mathrm{I}_\nu(x) = i^{-\nu}J_\nu(ix)$. Using the well-known asymptotic expansion

$$\mathrm{I}_\nu(x) = \sum_{k=0}^\infty \frac{(x/2)^{\nu+2k}}{k!\Gamma(\nu+k+1)}, \quad x \ll 1,$$

one obtains the moments $\mathbf{E}_x[\tau_R^l]$ for $x \le R$, e.g. the first three moments look as follows

$$\mathbf{E}_x[\tau_R] = \frac{(R^2 - x^2)}{2(\nu+1)},$$

$$\mathbf{E}_x[\tau_R^2] = \frac{(R^2 - x^2)(R^2\nu + 3R^2 - x^2\nu - x^2)}{8(\nu + 2)(\nu + 1)^2},$$

$$\mathbf{E}_x[\tau_R^3] = \frac{(R^2 - x^2)B(R,x)}{48(\nu + 3)(\nu + 2)(\nu + 1)^3}$$

where

$$B(R,x) = R^4\nu^2 + 8R^4\nu + 19R^4 - 10R^2x^2\nu - 2R^2x^2\nu^2 - 8R^2x^2 + x^4\nu^2 + 2x^4\nu + x^4.$$

Here, as before, $\nu = \frac{d}{2} - 1$. In is easy to check that $\sup_{x \in K} \mathbf{E}_x[\tau_R^l] \sim l!b(K,l,\nu)R^{2l}$ as $R \to \infty$ and $x \in K$. Now the result follows directly from (1.5) because moments $\mathbf{E}_x[\tau^l] \leq \mathbf{E}_x[\tau_{\bar{R}}^l]$. $\qquad\qquad\square$

Acknowledgement

The author is grateful to Vitaly Moroz and anonymous referee for helpful remarks.

References

1. Aizenman M., Simon B., Brownian motion and Harnack's inequality for Schrödinger operator, *Commun. Pure Appl. Math.*, **35**, 1982, 209–273
2. Bandle C., *Isoperimetric Inequalities and Applications*, Monograph Studies in Mathematics, Vol. 7, Boston: Pitman, 1980
3. Borodin A.N., Salminen P. *Handbook of Brownian Motion*, Basel: Birkhäuser, 1996
4. Cabré X., On the Alexandroff–Bakelman–Pucci estimate and the reversed Hölder inequality for solutions of elliptic and parabolic equations. *Comm. Pure Appl. Math.*, **48**, 1995, 539–570
5. Doleans-Dade C. and Mayer P.A., Intégrale stochastic per rapport aux martingale locales. Seminaire de Probabilities IV. *Lecture Notes in Maths 124*, Berlin: Springer, 77–107
6. Helms L.L., Biharmonic functions and Brownian motion. *J. Appl. Prob.*, **4**, 1967, 130–136
7. Kelbert M., Probabilistic representation of polyharmonic functions. *Probab. Theory Appl.*, **54**, 2009, 354–359 (in Russian)
8. Khas'minskii R.Z., Probabilistic representation of solutions of some differential equations. In the book: *Trydi VI Vses. Sovechaniya on Th. Prob. and Math. Stat.*, Vilnius, 1960, 177–183 (in Russian)
9. Mil'shtejn G.N., On the probability-theoretic solution of linear systems of elliptic and parabolic equations, *Probab. Theory Appl.*, **23**, 1978, n.4, 820-824

On additive time-changes of Feller processes

Aleksandar Mijatović and Martijn Pistorius*

*Department of Mathematics, Imperial College London,
180 Queen's Gate, South Kensington, London SW7 2AZ, UK
E-mails: a.mijatovic@imperial.ac.uk,
m.pistorius@imperial.ac.uk*

In this note we generalise the Phillips theorem [1] on the subordination of Feller processes by Lévy subordinators to the class of additive subordinators (i.e. subordinators with independent but possibly nonstationary increments). In the case where the original Feller process is Lévy we also express the time-dependent characteristics of the subordinated process in terms of the characteristics of the Lévy process and the additive subordinator.

Keywords: Subordination; semigroups; generators; time-dependent Markov processes.

1. Introduction

One of the established devices for building statistically relevant market models is that of the stochastic change of time-scale (e.g. Carr et al. [2]). Such a time change may be modelled as an independent additive subordinator $Z = \{Z_t\}_{t\geq 0}$, i.e. an increasing stochastic process with independent possibly nonstationary increments. If we subordinate a time-homogeneous Markov process $X = \{X_t\}_{t\geq 0}$ by Z, the resulting process $Y = \{X_{Z_t}\}_{t\geq 0}$ is a Markov process that will in general be time-inhomogeneous. The main result of this note shows that if X is a Feller process and Z satisfies some regularity assumptions, then Y is a time-inhomogeneous Feller process. The generator of Y is expressed in terms of the generator of X and the characteristics of Z. In the special case where X is a Lévy process it is shown that Y is an additive process with characteristics that are given explicitly in terms of the characteristics of X and of the additive subordinator Z. The explicit knowledge of the generator of Y is desirable from the viewpoint of pricing theory because contingent claims in the time-inhomogeneous market model

---

*The authors acknowledge support by EPSRC grant EP/D039053.

Y can be evaluated using algorithms that are based on the explicit form of the generator of the underlying process (see for example [3]).

2. Time-changed Feller processes

Throughout the paper we assume that $X = \{X_t\}_{t \geq 0}$ is a càdlàg Feller process with the state-space $\mathbb{R}^n$ for some $n \in \mathbb{N}$ and the infinitesimal generator $\mathcal{L}$ defined on a dense subspace $\mathcal{D}(\mathcal{L})$ in the Banach space of all continuous functions $C_0(\mathbb{R}^n)$ that vanish at infinity with norm $\|f\|_\infty := \sup_{x \in \mathbb{R}^n} |f(x)|$. The corresponding semigroup $(P_t)_{t \geq 0}$ is given by $P_t f(x) = \mathbf{E}^x[f(X_t)]$ for any $f \in C_0(\mathbb{R}^n)$, where the expectation is taken with respect to the law of X started at $X_0 = x$ (see Ethier and Kurtz [5] for the definition and properties of Feller semigroups).

Let $Z = \{Z_t\}_{t \geq 0}$ be an additive process, independent of X, with the Laplace exponent $\overline{\psi}_t(u) = \log \mathbf{E}[e^{-uZ_t}]$ given by $\overline{\psi}_t(u) := \int_0^t \psi_s(u)\mathrm{d}s$, where $\beta : \mathbb{R}_+ \to \mathbb{R}_+$, $g : \mathbb{R}_+ \times \mathbb{R}_+ \to \mathbb{R}_+$ are continuous and for all $s \in \mathbb{R}_+$, $u \in \mathbb{C}$ we have $\int_{(0,\infty)} (1 \wedge r) g(s,r)\mathrm{d}r < \infty$ and

$$\psi_s(u) = -u\beta(s) + \int_{(0,\infty)} (e^{-ur} - 1) g(s,r)\mathrm{d}r \quad \text{if} \quad \Re(u) \geq 0. \qquad (1)$$

In other words Z is a càdlàg process with nondecreasing paths such that the random variable $Z_t - Z_s$ is independent of Z_u for all $0 \leq u \leq s < t$ (see Jacod and Shiryaev [4, Ch. II, Sec. 4c], for a systematic treatment of additive processes).

In this paper we are interested in the process $(D, Y) = \{(D_t, Y_t)\}_{t \geq 0}$ defined by $D_t := D_0 + t$ and $Y_t := X_{Z_{D_t}}$ for some $D_0 \in \mathbb{R}_+$.

Theorem 2.1. *The process* (D, Y) *is Feller with the state-space* $\mathbb{R}_+ \times \mathbb{R}^n$ *and infinitesimal generator* $\mathcal{L}'$, *defined on a dense subspace of the Banach space* $C_0(\mathbb{R}_+ \times \mathbb{R}^n)$ *of continuous functions that vanish at infinity, given by*

$$\mathcal{L}'f(s,x) = \frac{\partial f}{\partial s}(s,x) + \beta(s)\mathcal{L}f_s(x) + \int_{(0,\infty)} [P_r f_s(x) - f(s,x)]\, g(s,r)\, \mathrm{d}r,$$

where $f \in C_0(\mathbb{R}_+ \times \mathbb{R}^n)$ *such that* $f_s(\cdot) := f(s, \cdot) \in \mathcal{D}(\mathcal{L})$ $\forall s \in \mathbb{R}_+$ *and the functions* $(s,x) \mapsto \mathcal{L}f_s(x)$ *and* $(s,x) \mapsto \frac{\partial f}{\partial s}(s,x)$ *are continuous and vanish at infinity.*

If Z is a Lévy subordinator, Theorem 2.1 reduces to the well-known Philips [1] theorem. If X is a Lévy process, then the time-changed process is an additive process with characteristics determined by those of Z and X.

Proposition 2.1. *Let X be a Lévy process with $X_0 = 0$ and characteristic triplet (c, Q, ν), where $c \in \mathbb{R}^n$, $Q \in \mathbb{R}^{n \times n}$ a nonnegative symmetric matrix and ν a measure on $\mathbb{R}^n \setminus \{0\}$ such that $\int_{\mathbb{R}^n \setminus \{0\}} (|x|^2 \wedge 1) \nu(dx)$. The process Y defined above (with $D_0 = 0$) is additive with càdlàg paths, jump measure*

$$\widetilde{\nu}_s(dx) = \beta(s)\nu(dx) + \int_{(0,\infty)} \mathbf{P}(X_r \in dx) g(s, r)\, dr,$$

nonnegative symmetric matrix $\widetilde{Q}_s = \beta(s)Q$, drift

$$\widetilde{c}_s = \beta(s)c + \int_{(0,\infty)} \mathbf{E}[X_r \mathbf{I}_{\{|X_r| \leq 1\}}] g(s, r)\, dr$$

and characteristic exponent $\overline{\Psi}_t(u) = \int_0^t \Psi_s(u)\, ds$ (recall that $\mathbf{E}[\mathrm{e}^{\mathrm{i}u \cdot Y_t}] = \mathrm{e}^{\overline{\Psi}_t(u)}$ for all $u \in \mathbb{R}^n$) where

$$\Psi_s(u) = \mathrm{i}u \cdot \widetilde{c}_s - \frac{1}{2} u \cdot \widetilde{Q}_s u + \int_{\mathbb{R}^n \setminus \{0\}} \left[\mathrm{e}^{\mathrm{i}u \cdot x} - 1 - \mathrm{i}(u \cdot x) \mathbf{I}_{\{|x| \leq 1\}} \right] \widetilde{\nu}_s(dx).$$

3. Example: A symmetric self-decomposable process

Suppose that Y is an additive process, considered in [2] as a model for the risky security, with no drift or Gaussian component and jump density

$$g_Y(t, y) = h_\nu(|y|/t^\gamma) \frac{\gamma}{\nu t^{\gamma+1}}, \quad \text{where} \quad h_\nu(y) = \frac{1}{\nu} \exp(-y/\nu) \mathbf{I}_{\{y > 0\}}.$$

Then in law the process Y is equal to a Brownian motion time-changed by an independent additive subordinator Z with $\beta \equiv 0$ and jump density

$$g(t, r) = a_t \mathrm{e}^{-r/b_t}, \quad \text{where} \quad a_t = \frac{\gamma}{\nu^3 t^{2\gamma+1}}, \quad b_t = 2\nu^2 t^{2\gamma}.$$

It is clear from Proposition 2.1 that $\widetilde{c}_t = \widetilde{Q}_t = 0$ for all $t \in \mathbb{R}_+$ and that the moment-generating functions of measures $\widetilde{\nu}_t(dx)$ and $g_Y(t, x) dx$ coincide

$$\int_{\mathbb{R} \setminus \{0\}} \mathrm{e}^{\lambda x} \, \widetilde{\nu}_t(dx) = \frac{2\gamma}{\nu t (1 - \lambda^2 \nu^2 t^{2\gamma})} = \int_{\mathbb{R} \setminus \{0\}} \mathrm{e}^{\lambda x} g_Y(t, x) dx$$

for $|\lambda| < 1/\nu t^\gamma$. This implies that the two additive processes coincide in law.

4. Proofs

4.1. *Proof of Proposition 2.1*

Let $\Psi_X(u)$ denote the characteristic exponent of the Lévy process X, i.e. $\mathbf{E}[\exp(\mathrm{i}u \cdot X_s)] = \exp(s\Psi_X(u))$ for any $u \in \mathbb{R}^n$. Since X and Z are independent processes with independnent increments, for any sequence of positive real numbers $0 \le t_0 < \ldots < t_m$ and vectors $u_1, \ldots, u_m \in \mathbb{R}^n$ it follows that

$$
\mathbf{E}\left[e^{i\sum_{i=1}^m u_i \cdot (Y_{t_i} - Y_{t_{i-1}})}\right] = \mathbf{E}\left[\mathbf{E}\left[\prod_{i=1}^m e^{iu_i \cdot (X_{Z_{t_i}} - X_{Z_{t_{i-1}}})}\,\Big|\, Z_{t_0}, \ldots, Z_{t_m}\right]\right]
$$

$$
= \mathbf{E}\left[\prod_{i=1}^m e^{(Z_{t_i} - Z_{t_{i-1}})\Psi_X(u_i)}\right]
$$

$$
= \prod_{i=1}^m \mathbf{E}\left[e^{iu_i \cdot (Y_{t_i} - Y_{t_{i-1}})}\right].
$$

Hence the process Y also has independent increments. Since Y is clearly càdlàg (as X and Z are), it is an additive process.

Finally, we have to determine the characteristic curve of Y. An argument similar to the one above implies that the characteristic function of Y_t equals

$$
\mathbf{E}[e^{iu \cdot Y_t}] = \mathbf{E}[e^{\Psi_X(u) Z_t}] = e^{\int_0^t \psi_s(-\Psi_X(u))\,ds} \quad \text{for any} \quad u \in \mathbb{R}^n.
$$

The last equality holds since $\Re(\Psi_X(u)) \le 0$ for all u and the integral in (1) is well-defined. It is not difficult to prove that for any Lévy process X started at 0 there exists a constant $C > 0$ such that the inequality holds

$$
\max\left\{\mathbf{P}(|X_r| > 1), |\mathbf{E}[X_r \mathbf{I}_{\{|X_r| \le 1\}}]|, \mathbf{E}[|X_r|^2 \mathbf{I}_{\{|X_r| \le 1\}}]\right\} \le C(r \wedge 1) \quad \forall r \in \mathbb{R}_+
$$

(see e.g. Lemma 30.3 in Sato [6]). Therefore, since $\int_0^\infty g(s,r)(r \wedge 1)dr < \infty$ by assumption, we have

$$
\int_0^\infty g(s,r) \max\left\{\mathbf{P}(|X_r| > 1), |\mathbf{E}[X_r \mathbf{I}_{\{|X_r| \le 1\}}]|, \mathbf{E}[|X_r|^2 \mathbf{I}_{\{|X_r| \le 1\}}]\right\} dr < \infty.
$$

We can thus define the measure $\widetilde{\nu}_s(dx)$, the vector $\widetilde{c}_s$, the matrix $\widetilde{Q}_s$ and the function $\Psi_s(u)$ by the formulae in Proposition 2.1. The Lévy-Khintchine representation

$$
\Psi_X(u) = iu \cdot c - \frac{1}{2}u \cdot Qu + \int_{\mathbb{R}^n \setminus \{0\}} \left[e^{iu \cdot x} - 1 - i(u \cdot x)\mathbf{I}_{\{|x| \le 1\}}\right] \nu(dx)
$$

and Fubini's theorem, which applies by the inequality above, yield the following calculation, whcih concludes the proof of the proposition:

$$
\psi_s(-\Psi_X(u)) = \beta(s)\Psi_X(u) + \int_0^\infty (\mathbf{E}[e^{iu \cdot X_r}] - 1)g(s,r)dr
$$

$$
= \beta(s)\Psi_X(u) + iu \cdot \int_0^\infty \mathbf{E}[X_r \mathbf{I}_{\{|X_r| \le 1\}}]g(s,r)dr
$$

$$
+ \int_0^\infty (\mathbf{E}[e^{iu \cdot X_r}] - 1 - iu \cdot \mathbf{E}[X_r \mathbf{I}_{\{|X_r| \le 1\}}])g(s,r)dr = \Psi_s(u).
$$

4.2. *Proof of Theorem 2.1*

Note first that the paths of the process (D, Y) are càdlàg. In what follows we prove that (D, Y) is a Markov process that satisfies the Feller property and find the generator of its semigroup.

1. Markov property. For any $g \in C_0(\mathbb{R}_+ \times \mathbb{R}^n)$ define

$$Q_t g(s, x) := \mathbf{E}[g(D_t, Y_t) | D_0 = s, Y_0 = x] = \mathbf{E}[g(s + t, X_{Z_{s+t}}) | X_{Z_s} = x].$$

Let $\lambda_{s,s+t}(\mathrm{d}r) := \mathbf{P}(Z_{s+t} - Z_s \in \mathrm{d}r)$ denote the law of the increment of Z which may have an atom at 0. Then, since X and Z are independent processes and the increments of Z are independent of the past, it follows from the definition that

$$Q_t g(s, x) = \int_{[0,\infty)} \mathbf{E}[g(s + t, X_{Z_s + r}) | X_{Z_s} = x] \lambda_{s,s+t}(\mathrm{d}r).$$

Define for any $v \in \mathbb{R}_+$ a σ-algebra $\mathcal{G}_v = \sigma(X_l : l \in [0, v])$. Then for a Borel set $A \in \mathcal{B}(\mathbb{R}^n)$ and any $X_0 = x_0 \in \mathbb{R}^n$ the Markov property of X yields

$$\mathbf{E}^{x_0}[g(t + s, X_{Z_s + r})\mathbf{I}_{\{X_{Z_s} \in A\}}] = \int_{[0,\infty)} \mathbf{E}^{x_0}[g(t + s, X_{v+r})\mathbf{I}_{\{X_v \in A\}}] \lambda_{0,s}(\mathrm{d}v)$$

$$= \int_{[0,\infty)} \mathbf{E}^{x_0}\left[\mathbf{E}[g(t + s, X_{v+r}) | \mathcal{G}_v]\mathbf{I}_{\{X_v \in A\}}\right] \lambda_{0,s}(\mathrm{d}v)$$

$$= \mathbf{E}^{x_0}\left[\mathbf{E}^{X_{Z_s}}[g(t + s, X_r)]\mathbf{I}_{\{X_{Z_s} \in A\}}\right].$$

Hence we get $\mathbf{E}[g(t+s, X_{Z_s+r}) | X_{Z_s}] = \mathbf{E}^{X_{Z_s}}[g(t+s, X_r)]$ a.s. for any $r \in \mathbb{R}_+$ and the following identity holds

$$Q_t g(s, x) = \int_{[0,\infty)} \mathbf{E}^x[g(s + t, X_r)] \lambda_{s,s+t}(\mathrm{d}r). \tag{2}$$

A similar argument and the monotone class theorem imply that, if $\mathcal{F}_s = \sigma(X_{Z_l} : l \in [0, s])$, then

$$\mathbf{E}[g(t + s, X_{r+Z_s}) | \mathcal{F}_s] = \mathbf{E}^{X_{Z_s}}[g(t + s, X_r)] \quad \text{a.s.}$$

The process (D, Y), started at $(0, x_0)$, satisfies

$$\mathbf{E}[g(D_{s+t}, Y_{s+t}) | \mathcal{F}_s] = \mathbf{E}[g(t + s, X_{Z_{s+t}}) | \mathcal{F}_s] = Q_t g(s, X_{Z_s}) = Q_t g(D_s, Y_s)$$

and is therefore Markov with the semigroup $(Q_t)_{t \geq 0}$.

2. Feller property. Since (D, Y) and Z are right-continuous, identity (2) implies that $\lim_{t \searrow 0} Q_t f(s, x) = f(s, x)$ for each $(s, x) \in \mathbb{R}_+ \times \mathbb{R}^n$. It is well-known that in this case pointwise convergence implies convergence in the Banach space $(C_0(\mathbb{R}_+ \times \mathbb{R}^n), \|\cdot\|_\infty)$. It also follows from representation (2), the dominated convergence theorem and the Feller property of X that a continuous function $(s,x) \mapsto Q_t g(s,x)$ tends to zero at infinity for any $g \in C_0(\mathbb{R}_+ \times \mathbb{R}^n)$. Hence (D,Y) is a Feller process.

3. Infinitesimal generator of the semigroup $(Q_t)_{t \geq 0}$. As before, let $\lambda_{s,s+t}$ be the law of the increment $Z_{s+t} - Z_s$ and let ψ_s be as in (1). Let $(t_n)_{n \in \mathbb{N}}$ be a decreasing sequence in $(0,\infty)$ that converges to zero. Denote by $\widehat{\mu}_n$ the Laplace transform of a compound Poisson process with Lévy measure $t_n^{-1} \lambda_{s,s+t_n}$. Hence we find for any $u \in \mathbb{C}$ that satisfies $\Re(u) \geq 0$

$$\widehat{\mu}_n(u) = \exp\left(\frac{1}{t_n} \int_0^\infty (e^{-ur} - 1)\lambda_{s,s+t_n}(dr)\right)$$
$$= \exp\left(t_n^{-1}(e^{\int_s^{s+t_n} \psi_v(u)dv} - 1)\right).$$

Since the function $t \mapsto \int_0^t \psi_{s+v}(u)dv$ is right-differentiable at zero with derivative $\psi_s(u)$, we get

$$\lim_{n \to \infty} \widehat{\mu}_n(u) = \exp(\psi_s(u)).$$

It is clear from (1) that $\exp(\psi_s(u))$ is a Laplace transform of an infinitely divisible distribution with Lévy measure $g(s,r)dr$. Therefore by Theorem 8.7 in [6] for every continuous bounded function $k : \mathbb{R} \to \mathbb{R}$ that vanishes on a neighbourhood of zero we get

$$\lim_{n \to \infty} t_n^{-1} \int_0^\infty k(r)\lambda_{s,s+t_n}(dr) = \int_0^\infty k(r)g(s,r)dr. \tag{3}$$

Furthermore the same theorem implies that for any continuous function h such that $h(r) = 1 + o(|r|)$ for $|r| \to 0$ and $h(r) = O(1/|r|)$ for $|r| \to \infty$ we have

$$\lim_{n \to \infty} t_n^{-1} \int_0^\infty rh(r)\lambda_{s,s+t_n}(dr) = \beta(s) + \int_0^\infty rh(r)g(s,r)dr. \tag{4}$$

A key observation is that (3) and (4) together imply that (3) holds for every continuous bounded function k that satisfies $k(r) = o(|r|)$ as $r \searrow 0$.

Claim. Let the function $f \in C_0(\mathbb{R}_+ \times \mathbb{R}^n)$ satisfy the assumptions of Theorem 2.1. Then for any $(s,x) \in \mathbb{R}_+ \times \mathbb{R}^n$ the limit holds

$$\lim_{t \searrow 0} t^{-1}(Q_t f - f)(s,x) = \frac{\partial f}{\partial s}(s,x) + \beta(s)\mathcal{L}f_s(x)$$
$$+ \int_0^\infty [P_r f_s(x) - f_s(x)]g(s,r)dr.$$

To prove this claim recall first that $(P_t)_{t \geq 0}$ is the semigroup of X and note that the identity holds

$$(Q_t f - f)(s,x) = \mathbf{E}^{s,x}[f(D_t, Y_t) - f(D_0, Y_t)] + \int_0^\infty [P_r f_s(x) - f_s(x)]\lambda_{s,s+t}(dr).$$

If we divide this expression by t and take the limit as $t \searrow 0$, the first term converges to the partial derivative $\frac{\partial f}{\partial s}(s, x)$ by the dominated convergence theorem (recall that the paths of Y are right-continuous).

Choose a function h as above, define $D(r) := P_r f_s(x) - f_s(x)$ and express the second term as

$$t^{-1} \int_0^\infty D(r)\lambda_{s,s+t}(\mathrm{d}r) = t^{-1} \int_0^\infty D(r)(1 - h(r))\lambda_{s,s+t}(\mathrm{d}r)$$
$$+ t^{-1} \int_0^\infty (D(r) - r\mathcal{L}f_s(x))h(r)\lambda_{s,s+t}(\mathrm{d}r)$$
$$+ \mathcal{L}f_s(x)t^{-1} \int_0^\infty rh(r)\lambda_{s,s+t}(\mathrm{d}r).$$

The first and second integrals on the right-hand side converge by (3) to

$$\int_0^\infty D(r)(1 - h(r))g(s,r)\mathrm{d}r \quad \text{and} \quad \int_0^\infty (D(r) - r\mathcal{L}f_s(x))h(r)g(s,r)(\mathrm{d}r)$$

respectively and the third integral converges by (4) to

$$\mathcal{L}f_s(x) \int_0^\infty rh(r)g(s,r)\mathrm{d}r.$$

This proves the claim.

Since $(Q_t)_{t\geq 0}$ is a strongly continuous contraction semigroup on the function space $C_0(\mathbb{R}_+ \times \mathbb{R}^n)$ with some generator $\mathcal{L}'$, if the pointwise limit in the claim exists and is in $C_0(\mathbb{R}_+ \times \mathbb{R}^n)$ for some continuous function f that vanishes at infinity, then f is in the domain of $\mathcal{L}'$ and $\mathcal{L}'f$ equals this limit (see e.g. Lemma 31.7 in [6]). This concludes the proof of the theorem.

References

1. R. Phillips, On the generation of semigroups of linear operators, *Pacific Journal of Mathematics* **2**, 343–369 (1952).
2. P. Carr, D. Madan, H. Geman and M. Yor, Self-decomposability and option pricing, *Mathematical Finance* **17**, 31–57 (2007).
3. A. Mijatović and M. Pistorius, Continuously monitored barrier options under Markov processes (2009), `http://arxiv.org/abs/0908.4028`
4. J. Jacod and A. Shiryaev, *Limit theorems for stochastic processes*, A Series of Comprehensive Studies in Mathematics, Vol. 288, 2nd edn. (Springer-Verlag, 2003).
5. S. Ethier and T. Kurtz, *Markov Processes: Characterization and Convergence* (Wiley, 2005).
6. K. Sato, *Lévy processes and infinitely divisible distributions*, Cambridge studies in advanced mathematics, Vol. 68 (CUP, 1999).

Statistical inference for differential equations driven by rough paths

A. Papavasiliou

*Departments of Statistics, Warwick University,
Coventry, CV4 7AL, UK
E-mail: a.papavasiliou@warwick.ac.uk*

We construct the "expected signature matching" estimator for differential equations driven by rough paths with polynomial vector field and we prove its consistency and asymptotic normality.

Keywords: Rough paths; Generalized Moment Matching; parameter estimation.

1. Introduction

Statistical inference for stochastic processes is a huge field, both in terms of research output and importance. In particular, a lot of work has been done in the context of diffusions.[2,3,12,18] Nevertheless, the problem of statistical inference for diffusions still poses many challenges, as for example constructing the Maximum Likelihood Estimator (MLE) for the general multi-dimensional diffusion. An alternative method in this case is that of the Generalized Moment Matching Estimator (GMME). While, in general, less efficient compared to the MLE, the GMME is usually easier to use, more flexible and has been successfully applied to general Markov processes.[6]

On the other hand, most methods for statistical inference in the context of non-Markovian continuous processes, such as stochastic delay equations, differential equations driven by fractional Brownian motion or, more generally, differential equations driven by Volterra type processes, are restricted to models that depend linearly on the parameter.[3,7,8,11,19]

The theory of rough paths provides a general framework for making sense of differential equations driven by any type of noise modelled as a rough path – this includes diffusions, differential equations driven by Volterra type processes, stochastic delay equations and even delay equation driven by fractional Brownian motion.[16] The basic ideas have been developed in the nineties.[15] In this paper, we discuss the problem of statistical inference for differential equations driven by rough paths.

The exact setting of the statistical problem considered here is the following: we observe many independent copies of specific iterated integrals of the response $\{Y_t,\ 0 < t < T\}$ of a differential equation

$$dY_t = f(Y_t; \theta) \cdot dX_t, \ Y_0 = y_0$$

driven by the *rough path* X. Two examples of interest are $X_t = (t, W_t)$ where W_t is Brownian motion and the differential equation is a diffusion and $X_t = (t, B_t)$ where B_t is fractional Brownian motion or other Volterra type process. The iterated integrals are observed at a fixed time T. However, if the response lives in more than one dimension, the iterated integrals will be functions of the whole path. For example, suppose that $Y_t = (Y_t^{(1)}, Y_t^{(2)})$ and we observe

$$\int\int_{0 < u_1 < u_2 < T} dY_{u_1}^{(2)} dY_{u_2}^{(1)}$$

for fixed time T. We further assume that the vector field $f(y; \theta)$ is polynomial in y and depends on the unknown parameter θ. Finally, we assume that we know the expected signature of the rough path X on the interval $[0, T]$, where by *signature of a path* we mean the set of all iterated integrals of X.

This setting is chosen for two reasons. The first is its simplicity: we develop here some basic tools for statistical inference of differential equation driven by rough paths. We expect that these can be generalized to other settings, such as observing one continuous path or for f more general.

The second reason was that such settings arise in the context of *equation-free* medelling of multiscale models.[9] Suppose that we have access to some code that simulates the dynamics of a complex system, such as molecular dynamics. We treat the code as a *black box*. We are interested in the global behavior of a function of our system that lives in the slow scale, i.e. in some limit its dynamics follow a diffusion, which is, however, unknown. The basic idea of equation-free modelling is to run the code for a *short time* and use the output to *locally estimate* the parameters of the differential equation. This process is repeated several times with carefullly chosen initial conditions, so as to get an estimate of the global dynamics. To summarize, in this problem:

(a) we observe many independent paths;
(b) time is short;

(c) we locally approximate the vector field by a polynomial.

Currently, the estimation is done using the MLE approach, pretending that the data comes from the diffusion rathen than the multiscale model.[4] However, for short time T we cannot expect the diffusion approximation to be a good one. We believe that in the scale of T, we can always approximate the dynamics by a rifferential equation driven by a rough path. This was our motivation for developing this method. The problem was first addressed in.[17] In this paper, we summarize the main results.

2. Setting

Our goal is to develop a method of statistical inference for differential equations driven by rough paths. However, the theory of rough paths is a deterministic theory. First, we put it in a probabilitic framework and describe the precise setting for the statistical inference problem we will discuss later.

Let $(\Omega, \mathcal{F}, \mathbb{P})$ be a probability space and $\mathbf{X} : \Omega \to G\Omega_p(R^n)$ a random variable, taking values in the space of geometric p-rough paths endowed with the p-variation topology. For each $\omega \in \Omega$, the rough path $\mathbf{X}(\omega)$ drives the following differential equation

$$dY_t(\omega) = f(Y_t(\omega); \theta) \cdot dX_t(\omega), Y_0 = y_0 \tag{1}$$

where $\theta \in \Theta \subseteq R^d$, Θ being the parameter space and for each $\theta \in \Theta$. As before, $f : R^m \times \Theta \to \mathrm{L}(R^n, R^m)$ and $f_\theta(y) := f(y; \theta)$ is a polynomial on y for each $\theta \in \Theta$. According to the Universal Limit Theorem,[15] we can think of equation (1) as a map

$$I_{f_\theta, y_0} : G\Omega_p(R^n) \to G\Omega_p(R^m), \tag{2}$$

sending a geometric p-rough path $\mathbf{X}$ to a geometric p-rough path $\mathbf{Y}$ and is continuous with respect to the p-variation topology. Consequently,

$$\mathbf{Y} := I_{f_\theta, y_0} \circ \mathbf{X} : \Omega \to G\Omega_p(R^m)$$

is also a random variable, taking values in $G\Omega_p(R^m)$ and if $\mathbb{P}^T$ is the distribution of $\mathbf{X}_{0,T}$, the distribution of $\mathbf{Y}_{0,T}$ will be $\mathbb{Q}_\theta^T = \mathbb{P}^T \circ I_{f_\theta, y_0}^{-1}$.

Suppose that the know the **expected signature** of $\mathbf{X}$ at $[0, T]$, i.e. we know

$$\mathbb{E}\left(\mathbf{X}_{0,T}^{(i_1, \ldots, i_k)}\right) := \mathbb{E}\left(\int \cdots \int_{0 < u_1 < \cdots < u_k < T} dX_{u_1}^{(i_1)} \cdots dX_{u_k}^{(i_k)}\right),$$

for all $i_j \in \{1, \ldots, n\}$ where $j = 1, \ldots, k$ and $k \geq 1$. Our goal will be to estimate θ, given several realizations of $\mathbf{Y}_{0,T}$, i.e. $\{\mathbf{Y}_{0,T}(\omega_i)\}_{i=1}^N$.

3. Method

In order to estimate θ, we are going to use a method that is similar to the "Method of Moments". The idea is simple: we will try to (partially) match the empirical expected signature of the observed p-rough path with the theoretical one, which is a function of the unknown parameters. Remember that the data we have available is several realizations of the p-rough path $\mathbf{Y}_{0,T}$ described in section 2. To make this more precise, let us introduce some notation: let

$$E^{\tau}(\theta) := \mathbb{E}_{\theta}(\mathbf{Y}_{0,T}^{\tau})$$

be the *theoretical expected signature* corresponding to parameter value θ and word τ and

$$M_N^{\tau} := \frac{1}{N} \sum_{i=1}^{N} \mathbf{Y}_{0,T}^{\tau}(\omega_i) \tag{3}$$

be the *empirical expected signature*, which is a Monte Carlo approximation of the actual one. The word τ is constructed from the alphabet $\{1, \ldots, m\}$, i.e. $\tau \in W_m$ where $W_m := \bigcup_{k \geq 0} \{1, \ldots, m\}^k$. The idea is to find $\hat{\theta}$ such that

$$E^{\tau}(\hat{\theta}) = M_N^{\tau}, \ \forall \tau \in V \subset W_m$$

for some choice of a set of words V. Then, $\hat{\theta}$ will be our estimate.

3.1. *Computing the Theoretical Expected Signature*

We need to get an analytic expression for the expected signature of the p-rough path $\mathbf{Y}$ at $(0, T)$, where $\mathbf{Y}$ is the solution of (1) in the sense described above. We are given the expected signature of the p-rough path $\mathbf{X}$ which is driving the equation, again at $(0, T)$. Unfortunately, we need to make one more approximation since the solution $\mathbf{Y}$ will not usually be available: we will approximate the solution by the rth Picard iteration $\mathbf{Y}(\mathbf{r})$, described in the Universal Limit Theorem.[15] Finally, we will approximate the expected signature of the solution corresponding to a word τ, $E^{\tau}(\theta)$, by the expected signature of the rth Picard iteration at τ, which we will denote by $E_r^{\tau}(\theta)$:

$$E_r^{\tau}(\theta) := \mathbb{E}_{\theta}(\mathbf{Y}(\mathbf{r})_{0,T}^{\tau}). \tag{4}$$

The good news is that when f_{θ} is a polynomial of degree q on y, for any $q \geq 0$, the rth Picard iteration of the solution is a linear combination of iterated integrals of the driving force $\mathbf{X}$. More specifically, for any realization ω and any time interval $(s, t) \in \Delta_T$, we can write:[17]

$$\mathbf{Y}(\mathbf{r})_{s,t}^{\tau} = \sum_{|\sigma| \leq |\tau| \frac{q^r - 1}{q-1}} \alpha_{r,\sigma}^{\tau}(y_0, s; \theta) \mathbf{X}_{s,t}^{\sigma}, \tag{5}$$

where $\alpha_{r,\sigma}^{\tau}(y; \theta)$ is a polynomial in y of degree $|\tau| q^r$ and $|\cdot|$ gives the length of a word. Thus,

$$E_r^{\tau}(\theta) = \sum_{|\sigma| \leq |\tau| \frac{q^r - 1}{q-1}} \alpha_{r,\sigma}^{\tau}(y_0, s; \theta) \mathbb{E}\left(\mathbf{X}_{s,t}^{\sigma}\right), \tag{6}$$

3.2. *The Expected Signature Matching Estimator*

We can now give a precise definition of the estimator, which we will formally call the **Expected Signature Matching Estimator** (ESME): suppose that we are in the setting of the problem described in section 2 and M_N^{τ} and $E_r^{\tau}(\theta)$ are defined as in (3) and (4) respectively, for every $\tau \in W_m$. Let $V \subset W_m$ be a set of d words constructed from the alphabet $\{1, \ldots, m\}$. For each such V, we define the ESME $\hat{\theta}_{r,N}^{V}$ as the solution to

$$E_r^{\tau}(\theta) = M_N^{\tau}, \ \forall \tau \in V. \tag{7}$$

This definition requires that (7) has a *unique* solution. This will not be true in general. Let $\mathcal{V}_r$ be the set of all V such that $E_r^{\tau}(\theta) = M, \ \forall \tau \in V$ has a unique solution for all $M \in S_{\tau} \subseteq R$ where S_{τ} is the set of all possible values of M_N^{τ}, for any $N \geq 1$. We will assume the following:

Assumption 3.1 (Observability). *The set $\mathcal{V}_r$ is non-empty and known (at least up to a non-empty subset).*

Then, $\hat{\theta}_{r,N}^{V}$ can be defined for every $V \in \mathcal{V}_r$.

It is very difficult to say anything about the solutions of system (7), as it is very general. However, if we assume that f is also a polynomial in θ, then (7) becomes a system of polynomial equations.

3.3. *Properties of the ESME*

It is possible to show that the ESME defined as the solution of (7) will converge to the true value of the parameter and will be asymptotically normal. More precisely, the following holds:[17]

Theorem 3.1. *Let $\hat{\theta}_{r,N}^{V}$ be the Expected Signature Matching Estimator for the system described in section 2 and $V \in \mathcal{V}_r$ with cardinality $|V| = d$. Assume that the expected signature of $\mathbf{Y}_{0,T}$ is finite and that $f(y; \theta)$ is a* *polynomial of degree q with respect to y and twice differentiable with respect to θ. Let θ_0 be the 'true' parameter value, meaning that the distribution of the observed signature $\mathbf{Y}_{0,T}$ is $\mathbb{Q}_{\theta_0}^T$. Set*

$$D_r^V(\theta)_{i,\tau} = \frac{\partial}{\partial \theta_i} E_r^\tau(\theta) \quad and \quad \Sigma_V(\theta_0)_{\tau,\tau'} = \mathrm{cov}\left(\mathbf{Y}_{0,T}^\tau, \mathbf{Y}_{0,T}^{\tau'}\right) \quad (8)$$

and assume that $\inf_{r>0,\theta\in\Theta} \|D_r^V(\theta)\| > 0$, i.e. $D_r^V(\theta)$ is uniformly nondegenerate with respect to r and θ. Then, for $r \propto \log N$ and T are sufficiently small,

$$\hat{\theta}_{r,N}^V \to \theta_0, \quad with\ probability\ 1, \quad (9)$$

and

$$\sqrt{N}\Phi_V(\theta_0)^{-1}\left(\hat{\theta}_{r,N}^V - \theta_0\right) \xrightarrow{\mathcal{L}} \mathcal{N}(0, I) \quad (10)$$

as $N \to \infty$, where

$$\Phi_V(\theta_0) = D_r^V(\theta_0)^{-1}\Sigma_V(\theta_0)^{1/2}. \quad (11)$$

References

1. Y. Aït-Sahalia and P. A. Mykland. An Analysis of Hansen-Scheinkman Moment Estimators for Discretely and Randomly Sampled Diffusions. *Journal of Econometrics* **144** 1–26, 2008.
2. A. Beskos, O. Papaspiliopoulos and G. Roberts. Monte Carlo maximum likelihood estimation for discretely observed diffusion processes. *Ann. Statist.* **37**(1): 223–245, 2009.
3. J. P. N. Bishwal. *Parameter Estimation in Stochastic Differential Equations.* Lecture Notes in Mathematics 1923, Springer-Verlag, Berlin 2008.
4. C.P. Calderon. Fitting effective diffusion models to data associated with a "glassy" potential: estimation, classical inference procedures, and some heuristics. *Multiscale Model. Simul.* 6(2): 656–687, 2007.
5. P. K. Friz and N. Victoir. *Multidimensional Stochastic Processes as Rough Paths.* Cambridge University Press (forthcoming).
6. L. P. Hansen and J. A. Scheinkman. Back to the Future: Generating Moment Implications for Continuous-Time Markov Processes. *Econometrica* **63**(4):767–804, 1995.
7. Y. Hu and D. Nualart. Parameter estimation for fractional Ornstein-Uhlenbeck processes. *Preprint:* `arxiv:0901.4925v1`.
8. H. Hult. Approximating some Volterra type stochastic integrals with applications to parameter estimation. *Stoch. Process. Appl.* 105: 1–32, 2003.
9. C.W. Gear and I.G. Kevrekidis, Projective methods for stiff differential equations: Problems with gaps in their eigenvalue spectrum, *SIAM J. Sci. Comput*, 24: 1091–1106, 2003.
10. P. E. Kloeden and E. Platen. *Numerical Solution of Stochastic Differential Equations.* Application of Mathematics 23. Springer-Verlag, Berlin 1999.

11. U. Küchler and M. Sørensen. *Exponential Families of Stochastic Processes.* Springer-Verlag, New York 1997.
12. Y. A. Kutoyants. *Statistical inference for ergodic diffusion processes.* Springer-Verlag, London, 2004.
13. W. E, D. Liu and E. Vanden-Eijnden. Analysis of multiscale methods for stochastic differential equations. *Comm. Pure Appl. Math.* **58**(11): 1544–1585, 2005.
14. T. J. Lyons, M. J. Caruna and T. Lévy. *Differential Equations Driven by Rough Paths.* In *Ecole d'été de Probabilités de Saint-Flour XXXIV* (Editor: J. Picard), Springer, Berlin 2007.
15. T. Lyons and Z. Qian. *System control and rough paths.* Oxford University Press, Oxford, 2002.
16. A. Neuenkirch, I. Nourdin and S. Tindel. Delay equations driven by rough paths. *Electronic Journal of Probability* 13, 2031–2068, 2008.
17. A. Papavasiliou and C. Ladroue. Parameter Estimation for Rough Differential Equations. *Preprint*: `arXiv:0812.3102`.
18. B. L. S. Prakasa Rao. *Statistical Inference for Diffusion Type Processes.* In *Kendall's Library of Statistics* 8, Oxford University Press, New York, 1999.
19. M. Reiss. Adaptive estimation for affine stochastic delay equations. *Bernoulli* 11(1): 67–102, 2005.

Constructing discrete approximations algorithms for financial calculus from weak convergence results

Radu S. Tunaru

City University London, 106 Bunhill Row, London EC1Y 8TZ, UK
E-mail: r.tunaru@city.ac.uk

A general method for generating approximation algorithms for integral calculations is proposed here, starting from weak convergence results and then adjusting the integral calculations such that the gaussian probability kernel appears inside the integral. While this technique can be applied in a wide applied mathematical context we focus here on European option pricing as a class of applications. We prove that the weak convergence characterizing condition can still be applied under some mild assumption on the payoff function of financial options. It is also shown that the approximation grid is a dense set in the set of real numbers.

Keywords: Weak convergence; local limit theorem; dense set; European option pricing.

1. Background and Notations

In [4] Duffie and Protter investigate the weak convergence of the financial gain process, showing how to deal with change from discrete to continuous time finance. Here we proceed somehow in reverse, using general known weak convergence results of carefully chosen probability measures to derive numerical exact form approximations schemes for calculations in finance. The results are proved with probability tools but are deterministic in nature, thus avoiding common pitfalls related to Monte Carlo simulations techniques, see [7] for background discussions. Without reduction of generality we focus on the risk-neutral valuation of the contingent claims under nonstochastic interest rates, similar to the framework in [5].

Given the complexity of the models and payoffs involved, the vast majority of financial derivatives are priced using numerical methods. An extensive review covering the literature on option pricing since its origins has been provided by Broadie and Detemple in [3]. Although Monte Carlo methods have become the most popular techniques in this applied area following the seminal work of Boyle, Evnine and Gibbs [2] in the late 80's, a series of pitfalls led to the emergence of a new type of numerical technique, broadly called *quasi-Monte Carlo*, that seem to be preferred recently for financial applications, see [1,6] for recent discussion. Our algorithms can be proved to have the topological density embedding of the latter while retaining the simplicity of the former.

Here we shall assume that we work with an asset process following a geometric Brownian motion process. At maturity T the random quantity to work with is $S_T = S_0 e^{(r-\sigma^2/2)T+\sigma\sqrt{T}Y}$, with $Y \sim N(0,1)$ a standard gaussian variable, and for option pricing or risk-management purposes one must calculate integrals such as

$$\int_{-\infty}^{\infty} \Pi(S_0 e^{(r-\sigma^2/2)T+\sigma\sqrt{T}x})\frac{1}{\sqrt{2\pi}}e^{-\frac{x^2}{2}}\,dx.$$

Denoting $\varphi(x) = \Pi(S_0 e^{(r-\sigma^2/2)T+\sigma\sqrt{T}x})\frac{1}{\sqrt{2\pi}}e^{-\frac{x^2}{2}}$ and $\psi(x) = \varphi(x)\sqrt{2\pi}e^{\frac{x^2}{2}}$, if μ is the standard gaussian measure it can be shown that

$$\int \psi(x)d\mu(x) = \int \varphi(x)dx.$$

It is evident that the test function ψ is given by the payoff function Π. Consider Y_n, Y real random variables defined on the same probability space $(\Omega, \mathcal{F}, \mathbb{P})$ and the probability measures $\mu_n = \mathbb{P} \circ Y_n^{\leftarrow}$ and $, \mu = \mathbb{P} \circ Y^{\leftarrow}$. By definition, Y_n converges in distribution to Y if and only if μ_n converges weakly to μ. This is equivalent to the condition

$$\lim_{n\to\infty} \int f\, d\mathbb{P} \circ Y_n^{\leftarrow} = \int f\, d\mathbb{P} \circ Y^{\leftarrow}. \tag{1}$$

for any function $f : \mathbb{R} \to \mathbb{R}$ that is bounded and continuous. As proved in [8, page 190] this condition can be relaxed to require the convergence to be satisfied only for functions f that are bounded, measurable and with $\mu(\{\omega \in \Omega : \text{f is discontinuous at } \omega\}) = 0$.

Let $\{X_n\}_{n\geq 1}$ be an iid sequence of $\mathbb{R}$-valued random variables defined on the probability space $(\Omega, \mathcal{F}, \mathbb{P})$ with finite mean and finite nonzero variance. If

$$Y_n = \frac{\sum_{k=1}^{n} X_k - n\mathrm{E}(X_1)}{\sqrt{n\,var(X_1)}} \tag{2}$$

then the central limit theorem implies that Y_n converges in distribution to Y, a standard gaussian random variable. Making various choices for the sequence of random variables one can develop an approximation scheme for the integrals representing the expectations of any derivative payoff with European exercise. The method can also be applied to other calculations in finance, it is not restricted to options pricing.

2. Approximation Algorithms

Our approximation scheme may start with sequences of random variables that are only independent and not necessarily identical distributed. For example, consider $\{X_n\}_{n\geq 1}$ a sequence of independent binary random variables such that

$$\mathbb{P}([\omega : X_j(\omega) = 1]) = p_j; \quad \mathbb{P}([\omega : X_j(\omega) = 0]) = 1 - p_j = q_j \qquad (3)$$

with $0 < p_j < 1$ for all $j \geq 1$. It is obvious that the random variable $X^{(n)} = X_1 + \ldots X_n$ takes only the values $0, 1, 2, \ldots, n$, and $\widetilde{P}_n^k = \mathbb{P}(X^{(n)} = k)$ is equal to the coefficient of t^k in the polynomial $(p_1 t + q_1)(p_2 t + q_2) \ldots (p_n t + q_n)$. Applying Liapounov's theorem one can prove easily the next result.

Theorem 2.1. *Let* $\{X^{(n)}\}_{n\geq 1}$ *be a sequence of random variables such that for each positive integer* n, $X^{(n)}$ *has the distribution given by* $\{\widetilde{P}_n^k\}_{0\leq k\leq n}$. *If* $\sum_{n\geq 1} p_n(1 - p_n)$ *is divergent then*

$$\lim_{n\to\infty} \mathbb{P} \circ \left(\frac{X^{(n)} - \sum_{i=1}^{i=n} p_i}{\sqrt{\sum_{i=1}^{i=n} p_i(1 - p_i)}} \right)^{\leftarrow} = \mathbf{N}(0, 1). \qquad (4)$$

As usual $\mathbf{N}(0, 1)$ denotes the probability measure corresponding to the gaussian distribution with mean 0 and variance 1. An example when the series $\sum_{n\geq 1} p_n(1 - p_n)$ is divergent is easily obtained for $p_n = 1/n$. Employing the condition (1) for

$$\mu_n = \mathbb{P} \circ \left(\frac{X^{(n)} - \sum_{i=1}^{i=n} p_i}{\sqrt{\sum_{i=1}^{i=n} p_i(1 - p_i)}} \right)^{\leftarrow} \quad \text{and} \quad \mu = \mathbf{N}(0, 1) \qquad (5)$$

leads to the next result proved in [9].

Theorem 2.2. *Let* $\varphi : [a, b] \to \mathbb{R}$ *be Riemann integrable and* $\{p_n\}_{n\geq 1}$ *a sequence of real numbers such that* $0 < p_n < 1$ *and the series* $\sum_{n\geq 1} p_n(1 - p_n)$ *is divergent. Then*

$$\int_a^b \varphi(x) \, dx = \lim_{n\to\infty} a_n, \qquad (6)$$

$$a_n = \sum_{k=0}^{k=n} \tilde{P}_n^k \sqrt{2\pi}\, \varphi_1 \left(\frac{k - \sum_{i=1}^{i=n} p_i}{\sqrt{\sum_{i=1}^{i=n} p_i(1 - p_i)}} \right) \exp\left[\frac{1}{2} \left(\frac{k - \sum_{i=1}^{i=n} p_i}{\sqrt{\sum_{i=1}^{i=n} p_i(1 - p_i)}} \right)^2 \right],$$

and $\varphi_1 : \mathbb{R} \to \mathbb{R}$, where $\varphi_1(x) = \varphi(x)$ if $x \in [a, b]$ and $\varphi_1(x) = 0$ otherwise.

If the integrand φ is continuous it is easy to see that the above result is still valid for $a = -\infty$ and $b = \infty$.

A simpler result is obtained if $p_i \equiv p$ for all i, in which case we fall back onto the binomial distribution and the classical central limit theorem. Applying the CLT for the sequence of variables $X^{(n)}$ with binomial distribution with parameters n and p, we get the approximation scheme

Proposition 2.1. *Let* $\varphi : [a, b] \to \mathbb{R}$ *be Riemann integrable. Then, for any* $p \in (0, 1)$

$$\int_a^b \varphi(x)\, dx = \lim_{n \to \infty} a_n \tag{7}$$

$$a_n = \sum_{k=0}^{k=n} P_n^k \sqrt{2\pi}\, \varphi_1 \left(\frac{k - np}{\sqrt{np(1 - p)}} \right) e^{\left[\frac{1}{2} \left(\frac{k - np}{\sqrt{np(1-p)}} \right)^2 \right]}$$

where $P_n^k = \binom{n}{k} p^k (1 - p)^{n-k}$ *and* $\varphi_1 : \mathbb{R} \to \mathbb{R}$, *where* $\varphi_1(x) = \varphi(x)$ *if* $x \in [a, b]$ *and* $\varphi_1(x) = 0$ *otherwise.*

It would be helpful to be able to apply the results given by formulas (6-7) to financial calculus such as option pricing. However, the convergence condition (1) cannot be used for unbounded payoffs. For the scheme derived from the binomial distribution we provide a proof that we can still make use of this condition under some mild regularity conditions on the payoff function. The same idea can be used for other discrete distributions.

2.1. *Adaptation to Infinite Payoff*

If one could prove the uniform integrability for the test function ψ associated with the payoff of the contract, then we can still apply the weak convergence condition. In other words if

$$\left. \begin{array}{l} \lim_{n \to \infty} \mu_n = \mu, \\ \lim_{\delta \to \infty} \sup_n \int_{B^c(0,\delta)} \psi d\mu_n = 0, \end{array} \right\} \Rightarrow \lim_{n \to \infty} \int \psi d\mu_n = \int \psi d\mu.$$

We shall demonstrate that

$$\lim_{\delta \to \infty} \sup_{n} \mu_n(B^c(0,\delta)) = 0, \quad \text{and}$$

$$\sup_{n} \left(\int \psi^v d\mu_n \right) < \infty$$

which together imply the uniform integrability. Consider the application $U(u) = \frac{u - np}{\sqrt{np(1-p)}}$ and make the notation $\alpha = S_0 e^{(r - \frac{\sigma^2}{2})T}$.

$$\int \psi d\mu_n = \int \psi d\mathbb{P} \circ (U(X^{(n)}))^{\leftarrow}$$

$$= \sum_{k=0}^{n} \psi(U(k))\mathbb{P}(X^{(n)} = k)$$

$$= \sum_{k=0}^{n} P_n^k \Pi(\alpha e^{\sigma \sqrt{T} \frac{k - np}{\sqrt{np(1-p)}}})$$

Let $B(0,\delta)$ be a ball of radius δ and let $B^c(0,\delta)$ be its complementary set. Now we shall apply Holder's inequality for positive v and s such that $\frac{1}{v} + \frac{1}{s} = 1$

$$\int_{B^c(0,\delta)} \psi d\mu_n = \int \psi 1_{B^c(0,\delta)} d\mu_n \tag{8}$$

$$\leq \left(\int \psi^v d\mu_n \right)^{\frac{1}{v}} (\mu_n(B^c(0,\delta))^s)^{\frac{1}{s}}$$

2.1.1. *Tightness condition*

Proposition 2.2. *Consider μ_n the probability measure corresponding to the normalized binomially distributed random variable with parameters n and p, and δ a positive number. Then*

$$\lim_{\delta \to \infty} \sup_{n} \mu_n(B^c(0,\delta)) = 0$$

Proof. $Y_n = \frac{X^{(n)} - np}{\sqrt{np(1-p)}}$ where $X^{(n)} \sim Binomial(n,p)$. Consider the continuous function $h_\delta(x) = \frac{x^2}{\delta^2}$, when $|x| > \delta$ and zero otherwise.

$$\mu_n(B^c(0,\delta)) = \int 1_{B^c(0,\delta)} d\mu_n$$

$$\leq \int h_\delta d\mu_n \leq \frac{1}{\delta^2} \int x^2 d\mu_n$$

Since $E(Y_n) = 0$ and $var(Y_n) = 1$ it follows that

$$\mu_n(B^c(0,\delta)) \leq \frac{1}{\delta^2} \implies 0 \leq \sup_n \mu_n(B^c(0,\delta)) \leq \frac{1}{\delta^2} \qquad \square$$

2.1.2. *Finite moments condition*

Proposition 2.3. *Assuming that there is $m \geq 1$ a real number such that $\Pi(u) \leq u^m$ for any $u \in [0,\infty)$ then*

$$\sup_n \left(\int \psi^v d\mu_n \right) < \infty.$$

Remark. The power parameter v is coming from the Holder inequality so it is a number greater than 1. The assumption $\Pi(u) \leq u^m$ is obviously satisfied for a the European call option with strike price K which has the payoff $\Pi(u) = \max(u - K, 0)$. For example, assuming without loss of generality that $K > 1$, one can take $m = 2$.

Proof. The integral represents the moment of order v with respect to a binomial mass density function.

$$\sup_n \left(\int \psi^v d\mu_n \right) = \sup_n \sum_{k=0}^{n} P_n^k \left[\Pi \left(\alpha e^{\sigma\sqrt{T}\frac{k-np}{\sqrt{np(1-p)}}} \right) \right]^v$$

$$\leq \sup_n \sum_{k=0}^{n} P_n^k \alpha^{mv} \left[e^{\sigma\sqrt{T}\frac{k-np}{\sqrt{np(1-p)}}} \right]^{mv}$$

$$= \sup_n \sum_{k=0}^{n} P_n^k \alpha^{mv} e^{mv\sigma\sqrt{T}\frac{k-np}{\sqrt{np(1-p)}}}.$$

Since $\alpha < \infty$ is fixed and denoting $w = mv\sigma\sqrt{T}$ it suffices to prove that

$$\sup_n \sum_{k=0}^{n} P_n^k e^{w\frac{k-np}{\sqrt{np(1-p)}}} < \infty.$$

After some algebra we get that

$$\sup_n \sum_{k=0}^{n} P_n^k e^{w\frac{k-np}{\sqrt{np(1-p)}}} = e^{w^2/2}. \qquad \square$$

3. A Topological Characteristic

For the binomial scheme, $p \in (0,1)$ is fixed and consider the grid

$$\mathcal{G} = \{\xi_k^n = \frac{k-np}{\sqrt{np(1-p)}} : 0 \leq k \leq n, k, n \in \mathbb{N}\}.$$

We shall prove that $\forall a < b, \quad \exists \, \xi_k^n \in \mathcal{G}$ such that $\xi_k^n \in [a, b]$. Consider the identity function $\varphi(x) = x$, that is continuous and bounded over the interval $[a, b]$. Applying our approximation result to $\varphi : [a, b] \to \mathbb{R}$ it follows

$$\varphi(b) - \varphi(a) = \int_a^b \varphi'(x)dx$$

$$= \lim_{n \to \infty} \sum_{k=0}^n \binom{n}{k} p^k (1 - p)^{n-k} \frac{1}{\rho(\xi_k^n)} 1_{[a,b]}(\xi_k^n)$$

because $\varphi'(x) \equiv 1$. Assuming that for any $k \le n$ it is true that $\xi_k^n \notin [a, b]$ implies that the series is zero, so $\varphi(b) - \varphi(a) = 0$, which is a contradiction.

4. Conclusions

The approximation algorithm identified here can be applied for any European option pricing traded in financial industry. The payoff functions in finance are quite smooth, with finite set of discontinuities and non-explosive growth. This important characteristic allowed us to prove the uniform integrability of the family of binomial probability measures defining our numerical scheme. Moreover, this grid is a dense set in the set of real numbers so our approximate calculations cover the entire targeted range.

References

1. Avramidis A.N., L'Ecuyer P., Efficient Monte Carlo and Quasi-Monte Carlo Option Pricing Under the Variance-Gamma Model, *Management Science*, **52**, 2006, 1930–1944
2. Boyle Ph.P., Evnine J., Gibbs S., Numerical Evaluation of Multivariate Contingent Claims, *The Review of Financial Studies*, **2**, 1989, 241–250
3. Broadie M., Detemple J., Option Pricing: Valuation Models and Applications, *Management Science*, **50**, 2004, 1145–1177
4. Duffie D., Protter Ph., From Discrete- To Continuous-Time Finance: Weak Convergence of the Financial Gain Process, *Mathematical Finance*, **2**, 1992, 1–15
5. Dufresne D., Laguerre Series for Asian and Other Options, *Mathematical Finance*, **10**,407–428
6. L'Ecuyer P., Quasi-Monte Carlo Methods with Applications in Finance, *Finance and Stochastics*, 2009, to appear
7. Glasserman P., *Monte Carlo Methods in Financial Engineering*, NewYork: Springer-Verlag, 2004
8. Loève M., *Probability Theory*,vol. I, 4th edition, NewYork:Springer-Verlag, 1977
9. Tunaru R. S., Approximating Riemann Integral Using Central Limit Theorem, *Mathematica Japonica*, **49**, 1999, 191–193

V.3. Coercivity and functional inequalities

Organisers: D. Bakry, B. Zegarlinski

Convexity along vector fields and applications to equations of Monge-Ampére type

M. Bardi

*Dipartimento di Matematica Pura e Applicata, Università di Padova,
Padova, 35121, Italy
E-mail: bardi@math.unipd.it*

F. Dragoni

*Department of Mathematics, Imperial College London,
London, SW7 2AZ, UK
E-mail: f.dragoni@imperial.ac.uk*

We introduce a notion of convexity in the geometry of vector fields and we prove a PDE-characterization for such notion and related properties.

Keywords: Covexity; vector fields; sub-Riemannian geometry; subelliptic PDE.

1. Introduction: The Euclidean case

Convex functions have many important properties which play a crucial role in the study of PDEs. We recall that a function $u : \mathbb{R}^n \to \mathbb{R}$ is convex whenever

$$u(tx+(1-t)y) \leq tu(x)+(1-t)u(y), \quad \text{for any } x, y \in \mathbb{R}^n \text{ and } t \in (0,1). \quad (1)$$

The previous definition makes sense in any convex domain $\Omega \subset \mathbb{R}^n$. If the domain is not convex, a function is convex in Ω if (1) holds for any $x, y \in \Omega$ and for any $t \in (T_1, T_2) \subset (0,1)$ such that $tx + (1 - t)y \in \Omega$ for any $t \in (T_1, T_2)$. In other words, for non-convex domains, one has to consider only the segments which are contained in the domain.

Moreover if $u \in C^2(\Omega)$, it is not difficult to show that convexity can be characterized by the sign of the second derivaties. To be more precise

$$u : \Omega \to \mathbb{R} \text{ is convex} \quad \Longleftrightarrow \quad -D^2u(x) \leq 0, \ \forall \, x \in \Omega. \quad (2)$$

We recall that the Hessian $D^2u(x)$ is a symmetric $n \times n$ matrix, therefore the previous inequality means $D^2u(x)$ is non-negative definite, that is

$y^t D^2 u(x)\, y \geq 0$, for all $y \in \mathbb{R}^n$ and it is equivalent to requiring that the minimum eigenvalue of $D^2 u(x)$ is non-negative.

If the function u is just continuous, in [1] Alvarez, Lasry and Lions proved that the characterization (2) is still true once you interpret the inequality for the Hessian in the viscosity sense, i.e. $D^2 \phi(x) \geq 0$, for all $\phi \in C^2(\Omega)$ such that $u - \phi$ has a local maximum point at x.

The aim is to introduce an intrinsic definition of convexity depending on the geometry of vector fields that can be characterized in a similar way via a viscosity inequality for intrinsic second-order derivatives and then investigate some related properties.

2. Convexity along vector fields

Let us look closer at the standard notion of convexity, recalled in (1). For the sake of simplicity, from now on we are going to consider the case $\Omega = \mathbb{R}^n$. A real-valued function (which we assume to be upper semicontinuous) u is convex in $\mathbb{R}^n$ if u is convex along all the lines, which means that the 1-variable function $u \circ x(t)$ is convex in $\mathbb{R}$, for any line $x(t)$. The idea is to give the same definition substituting the (Euclidean) lines with suitable curves related to the geometry of the vector fields. Lines are special curves in Euclidean spaces for two different reasons: they are geodesics and they have constant velocity. These two properties in more general geometries do not lead to the same class of curves. If we assume that our vector fields generate a metric structure on $\mathbb{R}^n$ (as it happens e.g. under the Hörmander condition), the geodesics are the curves minimizing the distance. One can say that a function is convex in the geometry induced by the vector fields if it is convex along any geodesic. This notion is called geodesical convexity and it has been extensively studied e.g. in Riemannian manifolds. Unfortunately in the model we have in mind (the sub-Riemannian manifolds), this does not lead to any interesting information on the function since there are too few geodesically convex sets. So our idea is to use the other property of Euclidean lines, which means to look at the curves whose velocity is constant w.r.t. to a given family of vector fields.

Definition 2.1. Let $\mathcal{X} = \{X_1(x), \ldots, X_m(x)\}$ be a family of C^2-vector fields on $\mathbb{R}^n$, u is convex along the vector fields (or briefly $\mathcal{X}$-*convex*) in $\mathbb{R}^n$ if $u \circ x(t)$ is convex in t, for any absolutely continuous curve $x : \mathbb{R} \to \mathbb{R}^n$ such that $\exists\, \alpha = (\alpha_1, \ldots, \alpha_m) \in \mathbb{R}^m$:

$$\dot{x}(t) = \sum_{j=1}^{m} \alpha_j X_j(x(t)), \quad \text{a.e. } t \in \mathbb{R}. \tag{3}$$

We call the previous curves $\mathcal{X}$-*lines*.

Let us remark that Definition 2.1 coincides with the standard notion of convexity whenever $\mathcal{X} = \{e_1, \ldots, e_m\}$ with e_i the i-th vector of the canonical basis in $\mathbb{R}^n$. Nevertheless, for general vector fields, the $\mathcal{X}$-convexity is not related to the Euclidean notion. In fact:

(1) $\mathcal{X}$-convexity does not imply the standard convexity.
    E.g. $X(x,y) = \partial_x$ and $u(x,y) = x^2 - y^2$.
(2) The standard convexity does not imply the $\mathcal{X}$-convexity.
    E.g. $X(x,y) = y\partial_x - x\partial_y$ and $u(x,y) = x + y$.

One of the main applications of our results is the case of sub-Riemannian geometries.

Example 2.1. [Hörmander vector fields and horizontal convexity.] Let us assume that $X_1(x), \ldots, X_m(x)$ satisfy the Hörmander condition, then they induce a sub-Riemannian structure on $\mathbb{R}^n$. In particular, if they are associated to a Carnot group, then Definition 2.1 is equivalent to the notion of horizontal convexity (see [5]), which was introduced first by Lu-Manfredi-Stroffolini [8] in the Heisenberg group and then extended to Carnot groups by Danielli-Garofalo-Nhieu [6] and later on studied by many other authors. Horizontal convexity is related to the algebra and until now, to our knowledge, it was never studied outside of Carnot groups. The notion of $\mathcal{X}$-convexity and the related results that we present here can be seen as a first step to generalize the theory of horizontal convexity to general sub-Riemannian manifolds.

3. $\mathcal{X}$-convexity and intrinsic second derivatives

We want now to establish a characterization for $\mathcal{X}$-convexity via the sign (in the viscosity sense) of intrinsic second-order derivatives.

Let $\mathcal{X} = \{X_1(x), \ldots, X_m(x)\}$, the *intrinsic Hessian* is defined as

$$D^2_{\mathcal{X}} u := \left(X_i(X_j u)\right)^m_{i,j=1} = \sigma^t D^2 u \, \sigma + \left(\nabla_{X_j} X_i \cdot Du\right)^m_{i,j=1},$$

where σ is the $m \times n$ matrix associated to $\mathcal{X}$, i.e. $\sigma(x) := [X_1(x), \ldots, X_m(x)]^T$.

We recall the following viscosity notion of convexity, introduced and studied in [8] in the Heisenberg group and then extended in [7] for general Carnot groups.

Definition 3.1. Let $u : \mathbb{R}^n \to \mathbb{R}$ be upper semicontinuous, we say that u is viscosity-convex (or briefly ν-*convex*) whenever

$$-D_{\mathcal{X}}^2 u(x) \leq 0, \quad \text{in the viscosity sense, in } \mathbb{R}^n. \tag{4}$$

The goal in [3] is to prove the following result, for a huge class of vector fields.

Theorem 3.1 (3). *Let* $\mathcal{X} = \{X_1(x), \ldots, X_m(x)\}$ *a family of* C^2-*vector fields and* $u : \mathbb{R}^n \to \mathbb{R}$ *upper semicontinuous, then* u *is* $\mathcal{X}$-*convex (Definition 2.1) if and only if* u *is* ν-*convex (Definition 3.1).*

Remark 3.1. The result is trivial in the smooth case. In fact, if $u \in C^2(\mathbb{R}^n)$ and $\dot{x}(t) = \sigma(x(t))\alpha = \sum_{i=1}^m \alpha_i X_i(x(t))$, then

$$\frac{d^2}{dt^2} u(x(t)) = \alpha^T D_{\mathcal{X}}^2 u \, \alpha.$$

Therefore $u \circ x(t)$ is convex for any $\mathcal{X}$-line $x(\cdot)$, if and only if, $D_{\mathcal{X}}^2 u \geq 0$.

To pass to the viscosity result in Theorem 3.1 is not trivial even in the Euclidean case. One implication is very easy while the other one is not. We first show the easy implication and then we give a sketch how to prove the other one.

Proof: $\mathcal{X}$-convexity $\Rightarrow$ ν-convexity.. Fix $x^0 \in \Omega$, $\alpha \in \mathbb{R}^m$ and let $x(\cdot)$ be a $\mathcal{X}$-line, i.e. $x(\cdot)$ solving (3). Let us consider $\varphi \in C^2$, test function for u at x^0, that means $u(x^0) = \varphi(x^0)$ and $u(x) \leq \varphi(x)$ for x near x^0. By Definition 2.1,

$$u(x^0) \leq \frac{u(x(t)) + u(x(-t))}{2} \leq \frac{\varphi(x(t)) + \varphi(x(-t))}{2}.$$

Taking the Taylor expansions around 0 for $\varphi \circ x$ at t and $-t$, one can deduce

$$u(x^0) \leq \varphi(x^0) + \frac{1}{2} D^2 \varphi(x(t))\big|_{t=0} t^2 + o(t^2) = u(x^0) + \frac{1}{2} \alpha^T D_{\mathcal{X}}^2 \varphi(x^0) \alpha t^2 + o(t^2).$$

Dividing by t^2 and taking the limit as $t \to 0$, we get $\alpha^T D_{\mathcal{X}}^2 \varphi(x^0)\alpha \geq 0$, which means $-D_{\mathcal{X}}^2 u \leq 0$ in the viscosity sense. $\qquad\square$

Sketch of proof: ν-convexity $\Rightarrow$ $\mathcal{X}$-convexity.. The idea comes from [2], where the authors use the same approach in order to prove that a function u is nondecreasing along a vector field X, if and only if, $Du \cdot X \geq 0$ in the viscosity sense. Rougly speaking, the approach consists in showing the result for 1-variable functions case and then, via a suitable diffeomorphism, to reduce the general case to the 1-dimensional one. Next we give a brief sketch of the proof.

Step 1 We first prove the result for functions defined on $\mathbb{R}$. Then we fix a point $x^0 \in \mathbb{R}^n$ and we look at a $\mathcal{X}$-line $x(\cdot)$ solving (3) for a some $\alpha \in \mathbb{R}^m$ and with $x(0) = x^0$. Let us assume $\sigma(x^0)\alpha \neq 0$. Otherwise the proof is trivial.

Step 2 Since $\sigma(x^0)\alpha \neq 0$ there is a diffeomorphism $\xi = \Phi(x)$ (depending on α) that brings locally the trajectories of $\dot{x} = \sigma(x)\alpha$ to the trajectories of $\dot{\xi} = e_1$ (Rectification Theorem for ODEs).

Step 3 Let us define $v(\xi) := u(\Phi^{-1}(\xi))$, then

$$-D_{\mathcal{X}}^2 u \leq 0 \quad \text{in } U \quad \Longleftrightarrow \quad -\frac{\partial^2 v}{\partial \xi_1^2} \leq 0 \quad \text{in } \Phi(U),$$

in the viscosity sense.

This step is the most complicate to prove. Even for smooth functions, the result is not trivial and needs several manipulations of formulas involving 3-tensors.

Step 4 Let $v^z(\xi_1) := v(\xi_1, z)$, with $z = (\xi_2, \ldots, \xi_n) \in \mathbb{R}^{n-1}$ frozen. We can show that

$$-\frac{\partial^2 v}{\partial \xi_1^2}(\xi_1, \ldots \xi_n) \leq 0 \quad \Longleftrightarrow \quad -(v^z)''(\xi_1) \leq 0,$$

in the viscosity sense.

Step 5 Assuming that $-D_{\mathcal{X}}^2 u \leq 0$ in the viscosity sense in a set U $\Rightarrow$ $-(v^z)'' \leq 0$ in the viscosity sense in a corresponding real interval. Hence, we can apply the result for functions of 1-variable, proved in Step 1, which implies (by Steps 3 and 4) that $v^z(\xi_1)$ is convex.

By the definition of Φ, $v^z(\xi_1)$ convex means exactly $u \circ x(t)$ convex.

Step 6 Applying the diffeomorphism for any fixed $\alpha \in \mathbb{R}^m$ (with $|\alpha| = 1$), we can conclude that u is $\mathcal{X}$-convex. $\qquad\square$

The ideas used in the above proof (and in particular Step 3) allow us to prove that the notion of $\mathcal{X}$-convexity is well-defined in any geometry, as the next remark tells in more details.

Remark 3.2. We show that the notion of ν-convexity depends just on the geometry induced by the vector fields and not on the particular vector fields chosen. Hence, by Theorem 3.1, the same holds for $\mathcal{X}$-convexity, too. In fact, if we consider two families of vector fields $\mathcal{X} = \{X_1(x), \ldots, X_m(x)\}$ and $\mathcal{Y} = \{Y_1(x), \ldots, Y_m(x)\}$ inducing the same geometry, i.e. there exists a diffeomorphism $F : \mathbb{R}^n \to \mathbb{R}^n$ which leads the trajectories along $\mathcal{X}$ into the trajectories along to $\mathcal{Y}$. Then, by the techniques used in Step 3, it is possible to prove that u is $\mathcal{X}$-convex if and only if $u \circ F^{-1}$ is $\mathcal{Y}$-convex.

The previous Lemma in particular means that in the Riemannian and sub-Riemannian manifolds $\mathcal{X}$-convexity is a well-defined geometric notion, independent on the chosen basis.

4. Gradient estimate and d-Lipschitz continuity

One important property of convex functions is that they have a priori bound on the gradient and so they are Lipschitz-continuous. We can prove that these results are still true for our more general notion of convexity.

Theorem 4.1. *[3] Let $u \in USC(\mathbb{R}^n)$ be a $\mathcal{X}$-convex and bounded function then, for any $K \subset\subset \mathbb{R}^n$, $\exists\, L$ such that*

(1) the intrinsic gradient is bounded in the viscosity sense, i.e.

$$|D_{\mathcal{X}} u(x)| := |\sigma^T(x) Du(x)| \leq L, \quad \text{in the viscosity sense, in } K, \quad (5)$$

(2) u is locally L-Lipschitz w.r.t. the metric $d(x,y)$ induced by the family of vector fields $\mathcal{X}$, i.e. $|u(x) - u(y)| \leq L\, d(x,y)$ in K, where

$$d(x,y) = \inf \left\{ l(\gamma) \,\big|\, \gamma \text{ admissible with} \gamma(0) = x, \gamma(T) = y \right\},$$

where "γ admissible curve" means that $\gamma : [0,T] \to \mathbb{R}^n$ is absolutely continuous and such that

$$\dot{\gamma}(t) = \sum_{i=1}^{m} \alpha_i(t) X_i(\gamma(t)), \quad a.e.\ t \in [0,T],$$

and

$$l(\gamma) := \int_0^T \left(\sum_{i=1}^{m} \alpha_i^2(t) \right)^{\frac{1}{2}} dt.$$

Moreover if d satisfies the following continuity property:

$$\forall\, K \subset\subset \Omega\ \exists\ \text{modulus } \omega_K : \quad d(x,y) \leq \omega_K(|x - y|), \quad (6)$$

then u is continuous with the same modulus of continuity. This implies that (5) holds a.e., too.

Remark 4.1. Whenever $\mathcal{X} = \{X_1(x), \ldots, X_m(x)\}$ satisfies the Hörmander condition, then (6) holds.

The previous properties have many applications, in particular to the study of nonlinear PDEs. We conclude giving an application to a family of subelliptic Monge-Ampère type equations.

5. Application to Monge-Ampère type equations

The classical Monge-Ampère equation is of the form:

$$- \det(D^2 u) = f(x).$$

More general Monge-Ampère type equations can be written as

$$- \det(D^2 u) + H(x, u, Du) = 0,$$

where H is a nonlinear first-order Hamiltonian. Monge-Ampère type equations are elliptic whenever $-D^2 u \leq 0$, which means excatly on convex functions.

Subelliptic Monge-Ampère type equations have been introduced and studied in the Heisenberg group and in more general Carnot groups, by Manfredi in 2003, Garofalo et collaborators 2003-2006, Gutierrez-Montanari 2004.

They are equations of the form:

$$- \det(D_{\mathcal{X}}^2 u) + H(x, u, D_{\mathcal{X}} u) = 0, \tag{7}$$

where $\mathcal{X}$ is a family of smooth vector fields satisfying the Hörmander condition.

Similarly to the classical Monge-Ampère setting, equations as in (7) are degenerate elliptic exactly on $\mathcal{X}$-convex functions, since there $-D_{\mathcal{X}}^2 u \leq 0$.

In [4], Bardi and Mannucci have recently proved Comparison Principles for subelliptic Monge-Ampère type equations in Carnot groups.

Theorem 5.1. *[4] Let* (7) *be a subelliptic Monge-Ampère type equation associated to a Carnot group. Suppose that the Hamiltonian* $H : \overline{\Omega} \times \mathbb{R} \times \mathbb{R}^n \to (0, +\infty)$ *is positive and continuous, and strictly increasing in the second variable and* $u \in USC(\overline{\Omega})$ *is a bounded, uniformly $\mathcal{X}$-convex viscosity subsolution of* (7), *and* $v \in LSC(\overline{\Omega})$ *is a bounded viscosity supersolution of* (7). *Assume also that H satisfies for some* $\lambda > 0$

$$H(x, r, q) - H(x, s, q) \geq \lambda(r - s), \quad \forall x \in \overline{\Omega},\ q \in \mathbb{R}^n,\ r, s \in [-M, M],$$

where $M = \max\{\|u\|_\infty, \|v\|_\infty\}$. *Then*

$$\sup_{\Omega}(u - v) \leq \max_{\partial \Omega}(u - v)^+.$$

Let us recall that a function is uniformly $\mathcal{X}$-convex if (4) holds as the strict inequality.

The proof of Theorem 5.1 relies on the geometry of vector fields and one key-step is an apriori bound on the intrinsic gradient for ν-convex functions, which so far was known just on Carnot groups.
Hence, by Theorem 4.1, this result can be extended to any family of C^2-vector fields satisfying the Hörmander condition.

References

1. O. Alvarez, J.-M. Lasry, P.-L. Lions, Convex viscosity solutions and state constraints. *J. Math. Pures Appl.* **76**, 265–288 (1997).
2. M. Bardi, I. Capuzzo Dolcetta, *Optimal Control and Viscosity Solutions of Hamilton-Jacobi-Bellman Equations*, (Birkhäuser, Boston, 1997).
3. M. Bardi, F. Dragoni, Convexity along vector fields and intrinsic Hessian, submitted paper.
4. M. Bardi, P. Mannucci, Comparison principles for equations of Monge-Ampère type associated to vector fields, submitted paper.
5. A. Bonfiglioli, E. Lanconelli, F. Uguzzoni, *Stratified Lie Groups and Potential Theory for Their Sub-Laplacians*, (Springer, Berlin 2007).
6. D. Danielli, N. Garofalo, D.-M. Nhieu, Notions of convexity in Carnot groups, *Comm. Anal. Geom.* **11**, 263–341 (2003).
7. P. Juutinen, G. Lu, J.-J. Manfredi, B. Stroffolini, Convex functions on Carnot groups. *Rev. Mat. Iberoam.* **23**, 191–200 (2007).
8. G. Lu, J.-J. Manfredi, B.Stroffolini, Convex functions in the Heisenberg group. *Calc. Var. Partial Differential Equations* **19**, 1–22 (2004).

Phi-entropy inequalities and Fokker-Planck equations

François Bolley and Ivan Gentil

Ceremade, Université Paris-Dauphine, UMR CNRS 7534
Place du Maréchal De Lattre De Tassigny, F-75775 Paris cedex 16, France
E-mails: bolley@ceremade.dauphine.fr,
gentil@ceremade.dauphine.fr
http://www.ceremade.dauphine.fr/~bolley
http://www.ceremade.dauphine.fr/~gentil

We present new Φ-entropy inequalities for diffusion semigroups under the curvature-dimension criterion. They include the isoperimetric function of the Gaussian measure. Applications to the long time behaviour of solutions to Fokker-Planck equations are given.

Keywords: Logarithmic Sobolev inequality; Poincaré inequality; Φ-entropies; Bakry-Emery criterion; diffusion semigroups; Fokker-Planck equation.

We consider a Markov semigroup $(\mathbf{P}_t)_{t\geq 0}$ on $\mathbb{R}^n$, acting on functions on $\mathbb{R}^n$ by $\mathbf{P}_t f(x) = \int_{\mathbb{R}^n} f(y)\, p_t(x, dy)$ for x in $\mathbb{R}^n$. The kernels $p_t(x, dy)$ are probability measures on $\mathbb{R}^n$ for all x and $t \geq 0$, called transition kernels. We assume that the Markov infinitesimal generator $L = \left.\dfrac{\partial \mathbf{P}_t}{\partial t}\right|_{t=0^+}$ is given by

$$Lf(x) = \sum_{i,j=1}^{n} D_{ij}(x) \frac{\partial^2 f}{\partial x_i \partial x_j}(x) - \sum_{i=1}^{n} a_i(x) \frac{\partial f}{\partial x_i}(x)$$

where $D(x) = (D_{ij}(x))_{1 \leq i,j \leq n}$ is a symmetric $n \times n$ matrix, nonnegative in the sense of quadratic forms on $\mathbb{R}^n$ and with smooth coefficients, and where the $a_i, 1 \leq i \leq n$, are smooth. Such a semigroup or generator is called a *diffusion*, and we refer to Refs. 5, 6, 13 for backgrounds on them.

If μ is a Borel probability measure on $\mathbb{R}^n$ and f a μ-integrable map on $\mathbb{R}^n$ we let $\mu(f) = \int_{\mathbb{R}^n} f(x)\, \mu(dx)$. If, moreover, Φ is a convex map on an interval I of $\mathbb{R}$ and f an I-valued map with f and $\Phi(f)$ μ-integrable, we let

$$\mathbf{Ent}_\mu^\Phi(f) = \mu(\Phi(f)) - \Phi(\mu(f))$$

be the Φ-entropy of f under μ (see Ref. 10 for instance). Two fundamental examples are $\Phi(x) = x^2$ on $\mathbb{R}$, for which $\mathbf{Ent}_\mu^\Phi(f)$ is the variance of f, and

$\Phi(x) = x \ln x$ on $]0, +\infty[$, for which $\mathbf{Ent}_\mu^\Phi(f)$ is the Boltzmann entropy of f. By Jensen's inequality, $\mathbf{Ent}_\mu^\Phi(f)$ is always nonnegative and, if Φ is strictly convex, it is positive unless f is a constant, equal to $\mu(f)$. The semigroup $(\mathbf{P_t})_{t\geq 0}$ is said μ-*ergodic* if $\mathbf{P_t}f$ tends to $\mu(f)$ as t tends to infinity in $L^2(\mu)$, for all f.

In Section 1 we shall derive bounds on $\mathbf{Ent}_\mu^\Phi(f)$ and $\mathbf{Ent}_{\mathbf{P_t}}^\Phi(f)(x)$ which will measure the convergence of $\mathbf{P_t}f$ to $\mu(f)$ in the ergodic setting. This is motivated by the study of the long time behaviour of solutions to Fokker-Planck equations, which will be discussed in Section 2.

Some results of this note with their proofs are detailled in Ref. 9.

1. Phi-entropy inequalities

Bounds on $\mathbf{Ent}_{\mathbf{P_t}}^\Phi(f)$ and assumptions on L will be given in terms of the *carré du champ* and Γ_2 operators associated to L, defined by

$$\Gamma(f,g) = \frac{1}{2}\Big(L(fg) - f\,Lg - g\,Lf\Big), \ \ \Gamma_2(f) = \frac{1}{2}\Big(L\Gamma(f) - 2\Gamma(f, Lf)\Big).$$

If ρ is a real number, we say that the semigroup $(\mathbf{P_t})_{t\geq 0}$ satisfies the $CD(\rho, \infty)$ *curvature-dimension* (or Bakry-Émery) *criterion* (see Ref. 7) if

$$\Gamma_2(f) \geq \rho\,\Gamma(f)$$

for all functions f, where $\Gamma(f) = \Gamma(f,f)$.

The carré du champ is explicitely given by

$$\Gamma(f,g)(x) = \ <\nabla f(x), D(x)\,\nabla g(x)> .$$

Expressing Γ_2 is more complex in the general case but, for instance, if D is constant, then L satisfies the $CD(\rho, \infty)$ criterion if and only if

$$\frac{1}{2}\Big(Ja(x)D + (Ja(x)D)^*\Big) \geq \rho\,D \tag{1}$$

for all x, as quadratic forms on $\mathbb{R}^n$, where Ja is the Jacobian matrix of a and M^* denotes the transposed matrix of a matrix M (see Ref. 2,4) .

Poincaré and logarithmic Sobolev inequalities for the semigroup $(\mathbf{P_t})_{t\geq 0}$ are known to be implied by the $CD(\rho, \infty)$ criterion. More generally, and following Ref. 6,7,10, let $\rho > 0$ and Φ be a C^4 strictly convex function on an interval I of $\mathbb{R}$ such that $-1/\Phi''$ is convex. If $(\mathbf{P_t})_{t\geq 0}$ is μ-ergodic and satisfies the $CD(\rho, \infty)$ criterion, then μ satisfies the Φ-entropy inequality

$$\mathbf{Ent}_\mu^\Phi(f) \leq \frac{1}{2\rho}\,\mu(\Phi''(f)\,\Gamma(f)) \tag{2}$$

for all I-valued functions f.

The main instances of such Φ's are the maps $x \mapsto x^2$ on $\mathbb{R}$ and $x \mapsto x \ln x$ on $]0, +\infty[$ or more generally, for $1 \leq p \leq 2$

$$\Phi_p(x) = \begin{cases} \frac{x^p - x}{p(p-1)}, & x > 0 \ \text{ if } p \in]1,2] \\ x \ln x, & x > 0 \ \text{ if } p = 1. \end{cases} \tag{3}$$

For this Φ_p with p in $]1, 2]$ the Φ-entropy inequality (2) becomes

$$\frac{\mu(g^2) - \mu(g^{2/p})^p}{p - 1} \leq \frac{2}{p\rho}\, \mu(\Gamma(g)) \tag{4}$$

for all positive functions g. For given g the map $p \mapsto \frac{\mu(g^2) - \mu(g^{2/p})^p}{p-1}$ is nonincreasing with respect to $p > 0, p \neq 1$. Moreover its limit for $p \to 1$ is $\mathbf{Ent}_\mu(g^2)$, so that the so-called *Beckner inequalities* (4) for p in $]1, 2]$ give a natural monotone interpolation between the weaker Poincaré inequality (for $p = 2$), and the stronger logarithmic Sobolev inequality (for $p \to 1$).

Long time behaviour of the semigroup

The Φ-entropy inequalities provide estimates on the long time behaviour of the associated diffusion semigroups. Indeed, let $(\mathbf{P_t})_{t \geq 0}$ be such a semigroup, ergodic for the measure μ. If Φ is a $\mathcal{C}^2$ function on an interval I, then

$$\frac{d}{dt}\mathbf{Ent}_\mu^\Phi(\mathbf{P_t}f) = -\mu(\Phi''(\mathbf{P_t}f)\,\Gamma(\mathbf{P_t}f)) \tag{5}$$

for all $t \geq 0$ and all I-valued functions f. As a consequence, if C is a positive number, then the semigroup converges in Φ-entropy with exponential rate:

$$\mathbf{Ent}_\mu^\Phi(\mathbf{P_t}f) \leq e^{-\frac{t}{C}}\,\mathbf{Ent}_\mu^\Phi(f) \tag{6}$$

for all $t \geq 0$ and all I-valued functions f, if and only if the measure μ satisfies the Φ-entropy inequality for all I-valued functions f,

$$\mathbf{Ent}_\mu^\Phi(f) \leq C\mu(\Phi''(f)\,\Gamma(f)). \tag{7}$$

1.1. *Refined Φ-entropy inequalities*

We now give and study improvements of (2) for the Φ_p maps given by (3):

Theorem 1.1 ([9]). *Let $\rho \in \mathbb{R}$ and $p \in]1,2[$. Then the following assertions are equivalent, with $\left(1 - e^{-2\rho t}\right)/\rho$ and $\left(e^{2\rho t} - 1\right)/\rho$ replaced by $2t$ if $\rho = 0$:*

(i) the semigroup $(\mathbf{P}_t)_{t \geq 0}$ satisfies the $CD(\rho, \infty)$ criterion;

(ii) $(\mathbf{P}_t)_{t\geq0}$ *satisfies the refined local Φ_p-entropy inequality*

$$\frac{1}{(p-1)^2}\left[\mathbf{P_t}(f^p)-\mathbf{P_t}(f)^p\left(\frac{\mathbf{P_t}(f^p)}{\mathbf{P_t}(f)^p}\right)^{\frac{2}{p}-1}\right]\leq\frac{1-e^{-2\rho t}}{\rho}\mathbf{P_t}\left(f^{p-2}\,\Gamma(f)\right)$$

for all positive t and all positive functions f;

(iii) $(\mathbf{P}_t)_{t\geq0}$ *satisfies the reverse refined local Φ_p-entropy inequality*

$$\frac{1}{(p-1)^2}\left[\mathbf{P_t}(f^p)-\mathbf{P_t}(f)^p\left(\frac{\mathbf{P_t}(f^p)}{\mathbf{P_t}(f)^p}\right)^{\frac{2}{p}-1}\right]$$

$$\geq\frac{e^{2\rho t}-1}{\rho}\left(\frac{(\mathbf{P_t}f)^p}{\mathbf{P_t}(f^p)}\right)^{\frac{2}{p}-1}(\mathbf{P_t}f)^{p-2}\,\Gamma(\mathbf{P_t}f)$$

for all positive t and all positive functions f.

If, moreover, $\rho>0$ and the measure μ is ergodic for the semigroup $(\mathbf{P_t})_{t\geq0}$, then μ satisfies the refined Φ_p-entropy inequality

$$\frac{p^2}{(p-1)^2}\left[\mu(g^2)-\mu(g^{2/p})^p\left(\frac{\mu(g^2)}{\mu(g^{2/p})^p}\right)^{\frac{2}{p}-1}\right]\leq\frac{4}{\rho}\mu(\Gamma(g)) \qquad (8)$$

for all positive maps g.

The bound (8) has been obtained in Ref. 3 for the generator L defined by $Lf=\mathrm{div}(D\nabla f)-<D\nabla V,\nabla f>$ with $D(x)$ a scalar matrix and for the ergodic measure $\mu=e^{-V}$, and under the corresponding $CD(\rho,\infty)$ criterion.

It improves on the Beckner inequality (4) since

$$\frac{\mu(g^2)-\mu(g^{2/p})^p}{p-1}\leq\frac{p}{2(p-1)^2}\left[\mu(g^2)-\mu(g^{2/p})^p\left(\frac{\mu(g^2)}{\mu(g^{2/p})^p}\right)^{\frac{2}{p}-1}\right]. \qquad (9)$$

We have noticed that for all g the map $p\mapsto\frac{\mu(g^2)-\mu(g^{2/p})^p}{p-1}$ is continuous and nonincreasing on $]0,+\infty[$, with values $\mathbf{Ent}_\mu(g^2)$ at $p=1$ and $\mathbf{Var}_\mu(g)$ at $p=2$. Similarly, for the larger functional introduced in (9), the map

$$p\mapsto\frac{p}{2(p-1)^2}\left[\mu(g^2)-\mu(g^{2/p})^p\left(\frac{\mu(g^2)}{\mu(g^{2/p})^p}\right)^{\frac{2}{p}-1}\right]$$

is nonincreasing on $]1,+\infty[$ (see [9, Prop. 11]). Moreover its value is $\mathbf{Var}_\mu(g)$ at $p=2$ and it tends to $\mathbf{Ent}_\mu(g^2)$ as $p\to1$, hence providing a new monotone interpolation between Poincaré and logarithmic Sobolev inequalities.

The pointwise $CD(\rho,\infty)$ criterion can be replaced by the integral criterion

$$\mu\left(g^{\frac{2-p}{p-1}}\Gamma_2(g)\right)\geq\rho\,\mu\left(g^{\frac{2-p}{p-1}}\Gamma(g)\right)$$

for all positive functions g, and one can still get the refined Φ_p-entropy inequality (2), even in the case of non-reversible semigroups (see [9, Prop. 14]).

Remark 1.1. For $\rho = 0$, and following Ref. 3, the convergence of $\mathbf{P_t}f$ towards $\mu(f)$ can be measured on $H(t) = \mathbf{Ent}_\mu^\Phi(\mathbf{P_t}f)$ as

$$|H'(t)| \le \frac{|H'(0)|}{1 + \alpha t}, \quad t \ge 0$$

where $\alpha = \frac{2-p}{p}|H'(0)|/H(0)$. This illustrates the improvement offered by (8) instead of (4), which does not give here any convergence rate.

1.2. *The case of the Gaussian isoperimetry function*

Let F be the distribution function of the one-dimensional standard Gaussian measure. The map $\mathcal{U} = F' \circ F^{-1}$, which is the isoperimetry function of the Gaussian distribution, satisfies $\mathcal{U}'' = -1/\mathcal{U}$ on the set $[0, 1]$, so that the map $\Phi = -\mathcal{U}$ is convex with $-1/\Phi''$ also convex on $[0, 1]$.

Theorem 1.2. *Let ρ be a real number. Then the following three assertions are equivalent, with $\left(1 - e^{-2\rho t}\right)/\rho$ and $\left(e^{2\rho t} - 1\right)/\rho$ replaced by $2t$ if $\rho = 0$:*

(i) the semigroup $(\mathbf{P_t})_{t\ge 0}$ satisfies the $CD(\rho, \infty)$ criterion;
(ii) the semigroup $(\mathbf{P_t})_{t\ge 0}$ satisfies the local Φ-entropy inequality

$$\mathbf{Ent}_{\mathbf{P_t}}^\Phi(f) \le \frac{1}{\Phi''(\mathbf{P_t}f)} \log\left(1 + \frac{1 - e^{-2\rho t}}{2\rho}\Phi''(\mathbf{P_t}f)\,\mathbf{P_t}(\Phi''(f)\Gamma(f))\right)$$

for all positive t and all $[0, 1]$-valued functions f;
(iii) the semigroup $(\mathbf{P_t})_{t\ge 0}$ satisfies the reverse local Φ-entropy inequality

$$\mathbf{Ent}_{\mathbf{P_t}}^\Phi(f) \ge \frac{1}{\Phi''(\mathbf{P_t}f)} \log\left(1 + \frac{e^{2\rho t} - 1}{2\rho}\Phi''(\mathbf{P_t}f)^2\Gamma(\mathbf{P_t}f))\right)$$

for all positive t and all $[0, 1]$-valued functions f.

If, moreover, $\rho > 0$ and the measure μ is ergodic for the semigroup $(\mathbf{P_t})_{t\ge 0}$, then μ satisfies the Φ-entropy inequality for all $[0, 1]$-valued functions f:

$$\mathbf{Ent}_\mu^\Phi(f) \le \frac{1}{\Phi''(\mu(f))} \log\left(1 + \frac{\Phi''(\mu(f))}{2\rho}\mu(\Phi''(f)\Gamma(f))\right).$$

The proof is based on [9, Lemma 4]. For $\Phi = -\mathcal{U}$ it improves on the general Φ-entropy inequality (2) since $\log(1+x) \le x$. Links with the isoperimetric bounds of Ref. 8 for instance will be addressed elsewhere.

2. Long time behaviour for Fokker-Planck equations

Let us consider the linear Fokker-Planck equation

$$\frac{\partial u_t}{\partial t} = \mathrm{div}[D(x)(\nabla u_t + u_t(\nabla V(x) + F(x)))], \quad t \geq 0,\, x \in \mathbb{R}^n \qquad (10)$$

where $D(x)$ is a positive symmetric $n \times n$ matrix and F satisfies

$$\mathrm{div}\left(e^{-V} DF\right) = 0. \qquad (11)$$

It is one of the purposes of Refs. 2 and 4 to rigorously study the asymptotic behaviour of solutions to (10)-(11). Let us formally rephrase the argument.

Assume that the Markov diffusion generator L defined by

$$Lf = \mathrm{div}(D\nabla f) - < D(\nabla V - F), \nabla f > \qquad (12)$$

satisfies the $CD(\rho, \infty)$ criterion with $\rho > 0$, that is (1) if D is constant, etc.

Then the semigroup $(\mathbf{P_t})_{t \geq 0}$ associated to L is μ-ergodic with $d\mu = e^{-V}/Zdx$ where Z is a normalization constant. Moreover, a Φ-entropy inequality (7) holds with $C = 1/(2\rho)$ by (2), so that the semigroup converges to μ according to (6). However, under (11), the solution to (10) for the initial datum u_0 is given by $u_t = e^{-V}\mathbf{P_t}(e^V u_0)$. Then we can deduce the convergence of the solution u_t towards the stationary state e^{-V} (up to a constant) from the convergence estimate (6) for the semigroup, in the form

$$\mathbf{Ent}_\mu^\Phi\left(\frac{u_t}{e^{-V}}\right) \leq e^{-2\rho t}\mathbf{Ent}_\mu^\Phi\left(\frac{u_0}{e^{-V}}\right), \quad t \geq 0. \qquad (13)$$

In fact such a result holds for the general Fokker-Planck equation

$$\frac{\partial u_t}{\partial t} = \mathrm{div}[D(x)(\nabla u_t + u_t a(x))], \quad t \geq 0,\, x \in \mathbb{R}^n \qquad (14)$$

where again $D(x)$ is a positive symmetric $n \times n$ matrix and $a(x) \in \mathbb{R}^n$. Its generator is the dual (for the Lebesgue measure) of the generator

$$Lf = \mathrm{div}(D\nabla f) - < Da, \nabla f >. \qquad (15)$$

Assume that the semigroup associated to L is ergodic and that its invariant probability measure μ satisfies a Φ-entropy inequality (7) with a constant $C \geq 0$: this holds for instance if L satisfies the $CD(1/(2C), \infty)$ criterion.

In this setting when $a(x)$ is not a gradient, the invariant measure μ is not explicit. Moreover the relation $u_t = e^{-V}\mathbf{P_t}(e^V u_0)$ between the solution of (14) and the semigroup associated to L does not hold, so that the asymptotic behaviour (13) for solutions to (14) can not be proved by using (6). However, this relation can be replaced by the following argument, for which the ergodic measure is only assumed to have a positive density u_∞ with respect to the Lebesgue measure.

Let u be a solution of (14) with initial datum u_0. Then, by [9, Lemma 7],

$$\frac{d}{dt}\mathbf{Ent}_\mu^\Phi\left(\frac{u_t}{u_\infty}\right)=\int\Phi'\left(\frac{u_t}{u_\infty}\right)L^*u_t\,dx=\int L\left[\Phi'\left(\frac{u_t}{u_\infty}\right)\right]\frac{u_t}{u_\infty}\,d\mu=-\int\Phi''\left(\frac{u_t}{u_\infty}\right)\Gamma\left(\frac{u_t}{u_\infty}\right)d\mu.$$

Then a Φ-Entropy inequality (7) for μ implies the exponential convergence:

Theorem 2.1. *With the above notation, assume that a Φ-entropy inequality* (7) *holds for μ and with a constant C. Then all solutions $u = (u_t)_{t\geq 0}$ to the Fokker-Planck equation* (14) *converge to u_∞ in Φ-entropy, with*

$$\mathbf{Ent}_\mu^\Phi\left(\frac{u_t}{u_\infty}\right) \leq e^{-t/C}\mathbf{Ent}_\mu^\Phi\left(\frac{u_0}{u_\infty}\right), \quad t \geq 0.$$

Acknowledgment

This work was presented during the 7th ISAAC conference held in Imperial College, London in July 2009. It is a pleasure to thank the organizers for giving us this opportunity.

References

1. C. Ané, S. Blachère, D. Chafaï, P. Fougères, I. Gentil, F. Malrieu, C. Roberto, and G. Scheffer. *Sur les inégalités de Sobolev logarithmiques*, S.M.F., Paris, 2000.
2. A. Arnold, A. Carlen, and Q. Ju. Comm. Stoch. Analysis, **2** (1) (2008), 153–175.
3. A. Arnold and J. Dolbeault. J. Funct. Anal., **225** (2) (2005), 337–351.
4. A. Arnold, P. Markowich, G. Toscani, and A. Unterreiter. Comm. Partial Diff. Equations, **26** (1-2) (2001), 43–100.
5. D. Bakry. *L'hypercontractivité et son utilisation en théorie des semigroupes.* Lecture Notes in Math. **1581** Springer, Berlin, 1994.
6. D. Bakry. *Functional inequalities for Markov semigroups.* In Probability measures on groups: recent directions and trends. Tata Inst., Mumbai, 2006.
7. D. Bakry and M. Émery. *Diffusions hypercontractives.* In Séminaire de probabilités, XIX, 1983/84, Lecture Notes in Math. **1123**. Springer, Berlin, 1985.
8. D. Bakry and M. Ledoux. Invent. math., **123** (1996), 259–281.
9. F. Bolley and I. Gentil. *Phi-entropy inequalities for diffusion semigroups. Preprint.*
10. D. Chafaï. J. Math. Kyoto Univ., **44** (2) (2004), 325–363.
11. D. Chafaï. ESAIM Proba. Stat., **10** (2006), 317–339.
12. B. Helffer. *Semiclassical analysis, Witten Laplacians, and statistical mechanics.* World Scientific Publishing Co. Inc., River Edge, 2002.
13. M. Ledoux. Ann. Fac. Sci. Toulouse Math., **9** (2) (2000) 305–366.

Isoperimetry for product of heavy tails distributions

Nathael Gozlan, Cyril Roberto and Paul-Marie Samson

Université Paris-Est Marne la Vallée - Laboratoire d'Analyse et de Mathématiques Appliquées UMR-8050, 5 bd Descartes, 77454 Marne la Vallée Cedex 2, France Emails: nathael.gozlan@univ-mlv.fr, cyril.roberto@univ-mlv.fr, paul-marie.samson@univ-mlv.fr

Extending an approach by Bobkov we obtain some isoperimetric inequalities for product of heavy tails distributions.

Keywords: Isoperimetric inequality; heavy tails distribution; Cheeger's inequality.

Consider a separable metric space (X, d) equipped with a probability measure μ which is not a Dirac mass at a point. In this note we study the following isoperimetric inequality

$$\mu_s(\partial A) \geq J(\mu(A)) \qquad A \subset X \text{ Borel} \tag{1}$$

where $J : [0, 1] \to \mathbb{R}^+$ is symmetric around $1/2$ and where the surface measure is defined by the Minkowski content $\mu_s(\partial A) = \liminf_{\varepsilon \to 0} \frac{\mu(A_\varepsilon \setminus A)}{\varepsilon}$ with $A_\varepsilon = \{x \in X : d(x, A) < \varepsilon\}$. For any function $f : X \to \mathbb{R}$ we define the modulus of the gradient of f by $|\nabla f|(x) = \limsup_{d(x,y) \to 0} \frac{|f(y) - f(x)|}{d(x,y)}$ with the convention that $|\nabla f|(x) = 0$ as soon as x is an isolated point of X. We define similarly on the product space X^n equipped with the distance $d_n(x, y) = \sqrt{\sum_{i=1}^n d(x_i, y_i)^2}$ and the n-fold product measure $\mu^n = \mu \otimes \cdots \otimes \mu$, the modulus of the gradient of $f : X^n \to \mathbb{R}$. A function is said to be "locally Lipschitz" if its Lipschitz constant is finite on every ball of X (or X^n). We assume that the product structure is of Euclidean type for the gradient, *i.e.*, for any locally Lipschitz function $f : X^n \to \mathbb{R}$, μ^n almost surely, $|\nabla f|^2(x) = \sum_{i=1}^n |\nabla_{x_i} f|^2(x)$ where $|\nabla_{x_i} f|$ is the modulus of the gradient of the function $X \ni x_i \mapsto f(x)$ with $(x_j)_{j \neq i}$ fixed. This is for example the case when $X = \mathbb{R}^k$ and μ is any absolutely continuous probability measure with respect to the Lebesgue measure.

Isoperimetric Inequalities are related to some Sobolev-type inequalities and to the concentration of measure phenomenon. Thus it has a lot of applications in high dimension (*e.g.* semi-group contraction properties, convergence to equilibrium of Markov processes etc.). It is therefore interesting to understand how Inequality (1) evolves on the product X^n. We refer to Refs. 1, 2, 3 and 4 for survey papers on the isoperimetric inequalities for probability measures and a more complete bibliography of the field and to Refs. 5 and 6 for an introduction to Sobolev-type inequalities and their applications.

When $J(s) = h_\mu \min(s, 1 - s)$ then (1) is the celebrated Cheeger's isoperimetric inequality: $\mu_s(\partial A) \geq h_\mu \min(\mu(A), 1 - \mu(A))$ and $h_\mu > 0$ is the Cheeger's constant. See also Refs. 7 and 8. Cheeger[9] proved that the constant h_μ is related to the spectral gap of the Laplacian on compact Riemannian manifolds. See Refs. 10, 11 and 12 for more references and related results. In this particular case, Bobkov and Houdré[13] proved that if the above Cheeger's isoperimetric inequality holds on X, then the same isoperimetric inequality holds on X^n with $h_{\mu^n} \geq h_\mu/(2\sqrt{6})$. In this work we extend their result in the following way:

Theorem 1. *Assume that for any Borel set $A \subset X$:*

$$\mu_s(\partial A) \geq J\left(\mu(A)\right)$$

for some $J : [0,1] \to \mathbb{R}^+$ symmetric around $1/2$. Assume that $s \mapsto J(s)/s$ is non-decreasing on $(0, \frac{1}{2})$. Then, for any integer $n \geq 1$,

$$\mu_s^n(\partial A) \geq \frac{n}{2} J\left(\frac{h}{4\sqrt{6}n} \min(\mu^n(A), 1 - \mu^n(A))\right) \qquad \forall A \subset X^n \text{ Borel.}$$

When $X = \mathbb{R}$ and $d\mu(x) = e^{-\Phi(|x|)}dx$ with Φ convex, the optimal function J in (1) is known[14] to be $I = F'_\mu \circ F_\mu$, with $F_\mu(x) := \mu(-\infty, x)$, and is concave. Since $J(0) = 0$, $s \mapsto J(s)/s$ is non-increasing. Hence our result does not apply to log-concave distributions. For results in this direction, see Refs. 15, 16, 17, 18, 19 and 20. When Φ is concave, it is known[21] that the optimal J in (1) is $J(t) = \min(I(t), 2I(\min(t, 1 - t)/2))$ (I as above). It follows easily[22] that $s \mapsto J(s)/s$ is non-decreasing on $(0, \frac{1}{2})$. Hence typical examples of application of our result are sub-exponential laws $d\mu_p(x) = e^{-|x|^p}/(2\Gamma(1 + (1/p)))$, $p \in (0, 1)$, Cauchy-type distributions $dm_\alpha(x) = \frac{\alpha}{2(1+|x|)^{1+\alpha}}$, $\alpha > 0$ and more generally κ-concave probability measures ($\kappa \leq 0$). In all these cases our result is optimal, see Refs. 22,23.

Note that the standard Cheeger's inequality, corresponding to $J(s) = h_\mu \min(s, 1 - s)$, enters the framework of Theorem 1 and leads to a weak version of Bobkov and Houdré's result. Namely, with the notation above, we get that $h_{\mu^n} \geq h_\mu/(8\sqrt{6})$. We are off by a factor 4, for technical reasons. This can anyway be improved to $h_{\mu^n} \geq h_\mu/(4\sqrt{6})$, see Remark 1 below.

Isoperimetric inequalities for product of heavy tails distributions are also obtained in Ref. 22. However our result is by nature very different from Ref. 22 and more intrinsic in the sense that we start with an isoperimetric inequality on X and derive from it an isoperimetric inequality on X^n. Also our approach, based on Bobkov's ideas,[24] is very elementary.

The proof of Theorem 1 relies on Sobolev-type inequalities which are known to be equivalent to isoperimetric inequalities. Indeed, it is easy to prove (see *e.g.* Ref. 25) that (1) is equivalent to the following weak Cheeger inequality: for any $f : X \to \mathbb{R}$ locally Lipschitz,

$$\int |f - m(f)|d\mu \leq \beta(s) \int |\nabla f|d\mu + s \, \mathrm{Osc}(f) \qquad \forall s \in (0, 1/2) \qquad (2)$$

where $m(f)$ is a the median of f under μ and $\mathrm{Osc}(f) = \sup f - \inf f$. More precisely (1) implies (2) with $\beta(s) = \sup_{s \leq t \leq \frac{1}{2}} \frac{t-s}{J(t)}$, and (2) implies (1) with $J(t) = \sup_{0 < s \leq t} \frac{t-s}{\beta(s)}$ for $t \in (0, \frac{1}{2})$ and $J(t) = J(1-t)$. Unfortunately, the weak Cheeger inequality (2) does not behave in a proper way on product spaces, due to the $\mathbb{L}^1$ norm of the gradient. Bobkov proposed an alternative functional form of (and equivalent to) the isoperimetric inequalities:

$$I\left(\int f d\mu\right) \leq \int \sqrt{I(f)^2 + C^2|\nabla f|^2}d\mu \qquad (3)$$

where $I : [0, 1] \to \mathbb{R}^+$ and $f : X \to [0, 1]$. Such inequalities enjoy the tensorisation property and was used by Bobkov[26] as an alternative proof of the Gaussian dimension free isoperimetric inequality of Sudakov and Tsirel'son[27] and Borell[28] and in Ref. 24 as shorter proof of Bobkov-Houdré's result mentioned above. See also Refs. 15, 29, 30, 31 and 32 for related results. Since our result are dimension dependent we shall prove a weak form of (3):

Theorem 2. *Let* $I(t) = 4t(1-t)$, $t \in [0,1]$. *Assume that for any Borel set* $A \subset X$, $\mu_s(\partial A) \geq J(\mu(A))$ *for some* $J : [0,1] \to \mathbb{R}^+$ *symmetric around* $1/2$. *Let* $C = 4\sqrt{6}$ *and* $\beta(s) = \sup_{s \leq t \leq \frac{1}{2}} \frac{t-s}{J(t)}$ *for* $s \in (0, \frac{1}{2})$. *Then, for any* $n \geq 1$, *any locally Lipschitz function* $f : X^n \to [0,1]$ *and any* $s \in (0, \frac{1}{2})$,

$$I\left(\int_{X^n} f d\mu^n\right) \leq \int_{X^n} \sqrt{I(f)^2 + 4C^2\beta^2(s)|\nabla f|^2}d\mu^n + Cns\,\mathrm{Osc}(f). \qquad (4)$$

Theorem 1 will easily follow from Theorem 2 by approximating indicator functions of sets by locally Lipschitz functions taking values in $[0, 1]$.

Our starting point is the following one dimensional functional inequality, derived from (2).

Lemma 1. *Let $\Phi(x) = \sqrt{1 + x^2} - 1$, $x \in \mathbb{R}$. Assume that for any Borel set $A \subset X$, $\mu_s(\partial A) \geq J\left(\mu(A)\right)$ for some $J : [0, 1] \to \mathbb{R}^+$ symmetric around $1/2$. Let $\beta(s) = \sup_{s \leq t \leq \frac{1}{2}} \frac{t-s}{J(t)}$ for $s \in (0, \frac{1}{2})$. Then for all locally Lipschitz functions $f : X \to \mathbb{R}$ with $m(f) = 0$,*

$$\int_X \Phi(f) d\mu \leq \int \Phi\left(4\beta(s)|\nabla f|\right) d\mu + 2s\operatorname{Osc}(f) \qquad \forall s \in (0, 1/2). \qquad (5)$$

Proof. By the above discussion (see (2)), the assumption $\mu_s(\partial A) \geq J\left(\mu(A)\right)$ implies that all locally Lipschitz functions $f : X \to \mathbb{R}$ with $m(f) = 0$ satisfy

$$\int_X |f| d\mu \leq \beta(s) \int_X |\nabla f| d\mu + s\operatorname{Osc}(f) \qquad \forall s \in (0, 1/2). \qquad (6)$$

Now consider a bounded function f, locally Lipschitz, with $m(f) = 0$. Set $f_+ = \max(f, 0)$ and $f_- = \max(-f, 0)$. Then $m(f_+) = m(f_-) = 0$ and thus $m(\Phi(f_+)) = m(\Phi(f_-)) = 0$. Hence, applying twice (6) to $\Phi(f_+)$ and $\Phi(f_-)$, we have for all $s \in (0, \frac{1}{2})$

$$\int_X \Phi(f_+) d\mu \leq \beta(s) \int_X \Phi'(|f|)|\nabla f|\chi_{\{f>0\}} d\mu + s\operatorname{Osc}(\Phi(f_+)),$$

$$\int_X \Phi(f_-) d\mu \leq \beta(s) \int_X \Phi'(|f|)|\nabla f|\chi_{\{f<0\}} d\mu + s\operatorname{Osc}(\Phi(f_-)).$$

Summing up we arrive at

$$\int_X \Phi(f) d\mu \leq \beta(s) \int_X \Phi'(|f|)|\nabla f| d\mu + s\left(\operatorname{Osc}(\Phi(f_+)) + \operatorname{Osc}(\Phi(f_-))\right)$$

$$\leq \beta(s) \int_X \Phi'(|f|)|\nabla f| d\mu + s\operatorname{Osc}(f) \qquad (7)$$

where in the last line we used the fact that $\Phi(x) \leq |x|$ and thus that

$$\operatorname{Osc}(\Phi(f_+)) + \operatorname{Osc}(\Phi(f_-)) = \Phi(\sup f) + \Phi(|\inf f|) \leq \sup f + |\inf f| = \operatorname{Osc}(f).$$

Now, using the Young inequality $xy \leq \Phi(2x) + \Phi^*(y/2)$ with $x = \beta(s)|\nabla f|/h$ and $y = \Phi'(|f|)$, where $\Phi^*(y) = \sup_u \{uy - \Phi(u)\}$, we have

$$\beta(s) \int_X \Phi'(|f|)|\nabla f| d\mu \leq \int_X \Phi\left(2\beta(s)|\nabla f|\right) d\mu + \int_X \Phi^*\left(\Phi'(|f|)/2\right) d\mu. \qquad (8)$$

A simple computation gives that $\Phi^*(y) = 1 - \sqrt{1 - y^2}$ for $|y| \leq 1$. Hence, since $|\Phi'| \leq 1$, we have $\Phi^*(\Phi'(x)) = 1 - \sqrt{1 - \frac{x^2}{1+x^2}} = \frac{\sqrt{1+x^2}-1}{\sqrt{1+x^2}} \leq \Phi(x)$ for any $x \in \mathbb{R}$. In turn, using the convexity of Φ^*, we get

$$\int_X \Phi^* \left(\frac{\Phi'(|f|)}{2} \right) d\mu \leq \frac{1}{2} \int_X \Phi^*(\Phi'(|f|)) d\mu \leq \frac{1}{2} \int_X \Phi(f) d\mu.$$

Plugging this bound into (8), it follows from (7), after simplifications, that

$$\int_X \Phi(f) d\mu \leq 2 \int_X \Phi\left(\frac{2\beta(s)}{h} |\nabla f| \right) d\mu + 2s \operatorname{Osc}(f).$$

This ends the proof since $2\Phi(x) \leq \Phi(2x)$ for $x \geq 0$, by convexity of Φ. $\square$

Proof of Theorem 2. Let $\Phi(x) = \sqrt{1 + x^2} - 1$, let $n = 1$ and consider a locally Lipschitz function $f : X \to [0,1]$. By Lemma 1 applied to $\sqrt{24}(f - m(f))$ we have for any $s \in (0, \frac{1}{2})$

$$\int_X \Phi(\sqrt{24}(f - m)) d\mu \leq \int_X \Phi\left(4\sqrt{24}\beta(s)|\nabla f| \right) d\mu + 2\sqrt{24}s \operatorname{Osc}(f).$$

Now observe that for $|t| \leq 1$, $\Phi(\sqrt{24}t) \geq 4t^2$. Hence,

$$\int_X \Phi(\sqrt{24}(f-m)) d\mu \geq 4 \int_X (f-m)^2 \geq 4 \operatorname{Var}_\mu(f) = I\left(\int_X f d\mu \right) - \int_X I(f) d\mu.$$

Let $C = 2\sqrt{24} = 4\sqrt{6}$. For any $s \in (0, \frac{1}{2})$ it follows that

$$
\begin{aligned}
I\left(\int_X f d\mu \right) &\leq \int_X \left[\Phi\left(2C\beta(s)|\nabla f| \right) + I(f) \right] d\mu + Cs \operatorname{Osc}(f) \\
&= \int_X \left[\sqrt{1 + (2C\beta(s)|\nabla f|)^2} - 1 + I(f) \right] d\mu + Cs \operatorname{Osc}(f) \\
&\leq \int_X \left[\sqrt{I(f)^2 + (2C\beta(s)|\nabla f|)^2} \right] d\mu + Cs \operatorname{Osc}(f)
\end{aligned}
$$

since $u = I(f) \leq 1$ and the function $u \mapsto \sqrt{u^2 + v^2} - u$ is non-increasing in $u \geq 0$ (for any v). Inequality (4) follows in dimension $n = 1$ and from Lemma 2 below, in any dimension. $\square$

Lemma 2. *Let $I : [0,1] \to \mathbb{R}^+$. Assume that for some constants $a, b > 0$ and all locally Lipschitz function $f : X \to [0,1]$ we have*

$$I\left(\int_X f d\mu \right) \leq \int_X \sqrt{I(f)^2 + a^2|\nabla f|^2} d\mu + b \operatorname{Osc}(f).$$

Then, for any $n \geq 1$ and any locally Lipschitz function $f : X^n \to [0,1]$

$$I\left(\int_{X^n} f d\mu^n \right) \leq \int_{X^n} \sqrt{I(f)^2 + a^2|\nabla f|^2} d\mu^n + bn \operatorname{Osc}(f).$$

Proof. The proof is by induction. Let $f : X^{n+1} \to [0,1]$. For simplicity we decompose any element of X^{n+1} as $(y,x) \in X^n \times X$. Let $g(x) = \int_{X^n} f(y,x)d\mu^n(y)$. Then we have,

$$I\left(\int_{X^{n+1}} f d\mu^{n+1}\right) = I\left(\int_X g(x)d\mu(x)\right)$$
$$\leq \int_X \sqrt{I(g(x))^2 + a^2|\nabla g|^2(x)}d\mu(x) + b\operatorname{Osc}(g).$$

Note that $\operatorname{Osc}(g) \leq \operatorname{Osc}(f)$. Also $|\nabla g|(x) \leq \int_{X^n} |\nabla_x f|(x,y)d\mu^n(y)$ for any $x \in X$, where $|\nabla_x f|$ is the modulus of the gradient of $x \mapsto f(y,x)$ with fixed $y \in X^n$. Furthermore,

$$I(g(x)) = I\left(\int_{X^n} f(y,x)d\mu^n(y)\right)$$
$$\leq \int_{X^n} \sqrt{I(f)^2(y,x) + a^2|\nabla_y f|^2(y,x)}d\mu^n(y) + bn\operatorname{Osc}(f)$$

where $|\nabla_y f|$ is the modulus of the gradient of $y \mapsto f(y,x)$ with fixed $x \in X$. Hence, using the following Hölder-Minkowski inequality $\sqrt{(\int u)^2 + (\int v)^2} \leq \int \sqrt{u^2 + v^2}$, where the integral is over X^n with respect to $d\mu^n(y)$, with $u = \sqrt{I(f)^2 + a^2|\nabla_y f|^2} + bn\operatorname{Osc}(f)$ and $v = C|\nabla_x f|$, we end up with

$$I\left(\int_{X^n+1} f d\mu^{n+1}\right)$$
$$\leq \int_X \int_{X^n} \sqrt{\left(\sqrt{I(f)^2 + a^2|\nabla_y f|^2} + bn\operatorname{Osc}(f)\right)^2 + a^2|\nabla_x f|^2 d\mu^n}\, d\mu + b\operatorname{Osc}(f)$$
$$\leq \int_{X^{n+1}} \sqrt{I(f)^2 + a^2\left(|\nabla_y f|^2 + |\nabla_x f|^2\right)}d\mu^{n+1} + b(n+1)\operatorname{Osc}(f)$$

where in the last line we used the following inequality, $\sqrt{(\alpha+\beta)^2 + \gamma^2} \leq \sqrt{\alpha^2 + \gamma^2} + \beta$, valid for any $\alpha, \beta, \gamma \geq 0$, that we applied to $\alpha = \sqrt{I(f)^2 + a^2|\nabla_y f|^2}$, $\beta = bn\operatorname{Osc}(f)$ and $\gamma = a|\nabla_x f|$. This ends the proof. $\square$

The next result is a more general version of Theorem 1 (without the assumption on $J(s)/s$).

Corollary 1. *Assume that for any Borel set $A \subset X$, $\mu_s(\partial A) \geq J(\mu(A))$ for some function $J : [0,1] \to \mathbb{R}^+$ symmetric around $1/2$. Let $\beta(s) = \sup_{s \leq t \leq \frac{1}{2}} \frac{t-s}{J(t)}$ for $s \in (0,\frac{1}{2})$ and $H(t) = \sup_{0 < s \leq t} \frac{t-s}{\beta(s)}$ for $t \in (0,\frac{1}{2})$. Then, for any $n \geq 1$ and any Borel set $A \subset X^n$,*

$$\mu_s^n(\partial A) \geq \frac{n}{2}H\left(\frac{1}{2\sqrt{6n}}\min(\mu^n(A), 1-\mu^n(A))\right).$$

Remark 1. When $J(s) = h_\mu s$, then $H(t) = h_\mu t$. In this case, using the notation of the introduction, we get that $h_{\mu^n} \geq h_\mu/(4\sqrt{6})$. This is Bobkov-Houdré's result[13] (see also Ref. 24) with a worst constant (off by a factor 2).

Proof. Fix $n \geq 1$ and a Borel set $A \subset X^n$. Thanks to Theorem 2 and the inequality $\sqrt{a^2 + b^2} \leq |a| + |b|$ we have for any locally Lipschitz function $f : X^n \to [0, 1]$ and all $s \in (0, 1/2)$,

$$I\left(\int_{X^n} f d\mu^n\right) \leq \int_{X^n} \sqrt{I(f)^2 + 4C^2\beta^2(s)|\nabla f|^2} d\mu^n + Cns\,\mathrm{Osc}(f)$$

$$\leq \int_{X^n} I(f)d\mu^n + 2C\beta(s)\int_{X^n} |\nabla f| d\mu^n + Cns\,\mathrm{Osc}(f),$$

where $I(t) = 4t(1-t)$, $t \in [0, 1]$ and $C = 2\sqrt{24}$. Approximating the indicator function χ_A of A by locally Lipschitz functions on X^n with values in $[0, 1]$ (see [13, Lemma 3.5]), we get that

$$4\mu^n(A)(1 - \mu^n(A)) \leq 2C\beta(s)\mu_s^n(\partial A) + Cns \qquad \forall s \in (0, 1/2).$$

Since $2t(1 - t) \geq \min(t, 1 - t)$, we end up with

$$\sup_{s \in (0,1/2)} \frac{2\min(\mu^n(A), 1 - \mu^n(A)) - Cns}{2C\beta(s)} \leq \mu_s^n(\partial A).$$

This leads to the expected result after some rearrangements. $\qquad\square$

Theorem 1 is a direct consequence of the previous corollary together with Lemma 2 below that compares H to J under the extra assumption that the function $s \mapsto J(s)/s$ is non-decreasing on $(0, 1/2)$ (we omit the proof).

Lemma 3. *Let J, β and H as in Corollary 1. Assume furthermore that the function $s \mapsto J(s)/s$ is non-decreasing on $(0, 1/2)$. Then, for all $t \in (0, \frac{1}{2})$, $H(t) \geq J\left(\frac{t}{2}\right)$.*

Proof. Since $s \mapsto J(s)/s$ is non-decreasing, we have

$$\beta(s) = \sup_{s \leq t \leq \frac{1}{2}} \frac{t - s}{J(t)} = \sup_{s \leq t \leq \frac{1}{2}} \frac{t}{J(t)} \frac{t - s}{t}$$

$$\leq \frac{s}{J(s)} \sup_{s \leq t \leq \frac{1}{2}} \frac{t - s}{t} = \frac{s(1 - 2s)}{J(s)} \leq \frac{s}{J(s)}.$$

Therefore, choosing $s = t/2$ we get

$$H(t) = \sup_{0 < s \leq t} \frac{t - s}{\beta(s)} \geq \sup_{0 < s \leq t} \frac{(t - s)J(s)}{s} \geq J\left(t/2\right). \qquad \square$$

Note that by similar reasoning it is also possible to prove (under the assumptions of Lemma 3) that $H(t) \leq 2J\left(2t\right)$ for $t \in (0, 1/4)$.

Acknowledgments

The second author warmly thank D. Bakry and B. Zegarlinski for their kind invitation to the 7th ISAAC congress, London, july 2009. This work was supported by the European Research Council through the "Advanced Grant" PTRELSS 228032 and by GDRE 224 GREFI-MEFI, CNRS-INdAM.

References

1. M. Ledoux, Isoperimetry and Gaussian analysis, in *Lectures on probability theory and statistics (Saint-Flour, 1994)*, , Lecture Notes in Math. Vol. 1648 (Springer, Berlin, 1996) pp. 165–294.
2. F. Barthe, Isoperimetric inequalities, probability measures and convex geometry, in *European Congress of Mathematics*, (Eur. Math. Soc., Zürich, 2005) pp. 811–826.
3. A. Ros, The isoperimetric problem, in *Global theory of minimal surfaces*, , Clay Math. Proc. Vol. 2 (Amer. Math. Soc., Providence, RI, 2005) pp. 175–209.
4. C. Roberto, Isoperimetry for product of probability measures: recent results, preprint, (2009).
5. C. Ané, S. Blachère, D. Chafaï, P. Fougères, I. Gentil, F. Malrieu, C. Roberto and G. Scheffer, *Sur les inégalités de Sobolev logarithmiques*, Panoramas et Synthèses, Vol. 10 (Société Mathématique de France, Paris, 2000).
6. F. Y. Wang, *Functional inequalities, Markov processes and Spectral theory* (Science Press, Beijing, 2005).
7. V. G. Maz'ja, *Soviet Math. Dokl.* **1**, 882 (1960).
8. V. G. Maz'ja, *Sobolev spaces*Springer Series in Soviet Mathematics, Springer Series in Soviet Mathematics (Springer-Verlag, Berlin, 1985). Translated from the Russian by T. O. Shaposhnikova.
9. J. Cheeger, A lower bound for the smallest eigenvalue of the Laplacian, in *Problems in analysis (Papers dedicated to Salomon Bochner, 1969)*, (Princeton Univ. Press, Princeton, N. J., 1970) pp. 195–199.
10. P. Buser, *Ann. Sci. École Norm. Sup. (4)* **15**, 213 (1982).
11. G. F. Lawler and A. D. Sokal, *Trans. Amer. Math. Soc.* **309**, 557 (1988).
12. M. Ledoux, Spectral gap, logarithmic Sobolev constant, and geometric bounds, in *Surveys in differential geometry. Vol. IX*, Surv. Differ. Geom., IX (Int. Press, Somerville, MA, 2004) pp. 219–240.

13. S. G. Bobkov and C. Houdré, *Ann. Probab.* **25**, 184 (1997).
14. S. Bobkov, *Ann. Probab.* **24**, 35 (1996).
15. D. Bakry and M. Ledoux, *Invent. Math.* **123**, 259 (1996).
16. F. Barthe, P. Cattiaux and C. Roberto, *Rev. Mat. Iberoam.* **22**, 993 (2006).
17. F. Barthe, P. Cattiaux and C. Roberto, *Electron. J. Probab.* **12**, no. 44, 1212 (2007).
18. E. Milman, On the role of convexity in isoperimetry, spectral gap and concentration, to appear in Inventiones Mathematicae, available at arxiv.0712.4092v5, (2007).
19. E. Milman, On the role of convexity in functional and isoperimetric inequalities, to appear in the Proceedings of the London Mathematical Society, available at arxiv.0804.0453, (2008).
20. E. Milman, Isoperimetric and concentration inequalities - part i: Equivalence under curvature lower bound, preprint available at arxiv.0902.1560, (2009).
21. S. G. Bobkov and C. Houdré, *Mem. Amer. Math. Soc.* **129**, viii+111 (1997).
22. P. Cattiaux, N. Gozlan, A. Guillin and C. Roberto, Functional inequalities for heavy tails distributions and application to isoperimetry, Preprint, (2008).
23. F. Barthe, P. Cattiaux and C. Roberto, *AMRX Appl. Math. Res. Express* , 39 (2005).
24. S. G. Bobkov, On the isoperimetric constants for product measures, Preprint, (2009).
25. S. G. Bobkov, *Electron. J. Probab.* **12**, 1072 (2007).
26. S. G. Bobkov, *Ann. Probab.* **25**, 206 (1997).
27. V. N. Sudakov and B. S. Tsirel'son, *Zap. Naučn. Sem. Leningrad. Otdel. Mat. Inst. Steklov. (LOMI)* **41**, 14 (1974), Problems in the theory of probability distributions, II.
28. C. Borell, *Invent. Math.* **30**, 207 (1975).
29. F. Barthe, *Geom. Funct. Anal.* **12**, 32 (2002).
30. F. Barthe and B. Maurey, *Ann. Inst. H. Poincaré Probab. Statist.* **36**, 419 (2000).
31. S. G. Bobkov and F. Götze, *J. Funct. Anal.* **163**, 1 (1999).
32. B. Zegarlinski, *Ann. Probab.* **29**, 802 (2001).

Hypoellipticity in infinite dimensions

Martin Hairer

Courant Institute, NYU and University of Warwick

We consider semilinear parabolic stochastic PDEs driven by additive noise. The question addressed in this note is that of the regularity of transition probabilities. If the equation satisfies a Hörmander 'bracket condition', then any finite-dimensional projection of the solution has a smooth density with respect to Lebesgue measure. One key ingredient in the argument is a bound on 'Wiener polynomials' that plays a role analogue to Norris' lemma.

1. Introduction

In this note, we report on recent results obtained in collaboration with J.C. Mattingly[1] regarding the behaviour of transition probabilities for a large class of semilinear stochastic PDEs. While these results were motivated mainly by the study of the long-time behaviour of solutions,[2] the aspect that we will focus on in this note is that of the regularity of transition probabilities which is an interesting (and still partially open) mathematical question in its own right.

The class of problems that we are able to treat are semilinear parabolic stochastic PDEs of the type

$$du = -Au\,dt + F(u)\,dt + Q\,dW(t)\,, \quad u_0 \in \mathcal{H}\,, \tag{1}$$

where $\mathcal{H}$ is a separable Hilbert space, A is a positive selfadjoint negative linear operator with compact resolvent, W is a finite-dimensional Wiener process taking values in $\mathbb{R}^d$ and $Q\colon \mathbb{R}^d \to \mathcal{H}$ with $Qe_i = q_i$. The nonlinearity $F\colon \mathcal{D}(A^\alpha) \to \mathcal{H}$ is assumed to be 'polynomial' in the sense that it can be written as a finite sum of multilinear terms. Prime examples of equations satisfying our assumptions are the 2D stochastic Navier-Stokes equations on a bounded regular domain, reaction-diffusion equations in dimension 3 or less, the stochastic Kuramoto-Sivashinsky equation, stochastic Burgers' equation, etc.

We address the question whether the solutions to Eq. (1) have 'smooth' transition probabilities. For finite-dimensional SDEs, the meaning of this question is clear: do the transition probabilities have a C^∞ density with respect to Lebesgue measure? In the infinite-dimensional case, it is much less clear what we mean by 'smooth' since there is no natural reference measure. In the special case where W is a cylindrical Wiener process on $\mathcal{H}$ and Q is 'large' (either with bounded inverse or comparable to some inverse power of A), it is often possible[3-5] to show that the transition probabilities have a density with respect to the Gaussian measure which is invariant for the linearised equation (i.e. the same equation with $F = 0$) and that this density is well-behaved. However, we are interested in the case where Q has finite-dimensional range, so that these arguments certainly do not apply. The next best notion of smoothness is then the following:

Question 1.1. Given $N > 0$ and an orthonormal projection $\Pi\colon \mathcal{H} \to \mathbb{R}^N$, does the law of $\Pi u(t)$ have a C^∞ density with respect to Lebesgue measure on $\mathbb{R}^N$?

In the finite-dimensional case, Question 1.1 was answered successfully by Hörmander's celebrated 'sums of squares' theorem:[6,7]

Theorem 1.1. *Consider a collection $\{V_i\}_{i=0}^d$ of smooth vector fields on an n-dimensional compact manifold $\mathcal{M}$ and consider the Stratonovich SDE*

$$dx = V_0(x)\, dt + \sum_{i=1}^d V_i(x) \circ dW_i(t)\,. \tag{2}$$

Then, if the Lie algebra generated by $\partial_t + V_0$ and $\{V_i\}_{i=1}^d$ spans the tangent space of $\mathbb{R} \times \mathcal{M}$ at every point, the law of the solutions to Eq. (2) has a C^∞ density with respect to the volume measure on $\mathcal{M}$.

It turns out that a similar result still holds in the context of Eq. (1). Of course, the first question that needs consideration is that of the definition of the 'Lie brackets' between the 'drift vector field' $-A + F$ and the 'diffusion vector fields' q_i. If G_1 is a symmetric k-multilinear map and G_2 is a symmetric ℓ-multilinear map, we can define a symmetric $k + \ell - 1$-multilinear map $[G_1, G_2]$ by

$$[G_1, G_2](u) = \ell G_2(u, \ldots, u, G_1(u)) - k G_1(u, \ldots, u, G_2(u))\,, \tag{3}$$

for every u such that the right hand side makes sense. (The case $k = 0$ where G_1 is equal to some constant element in $\mathcal{H}$ is included.) The problem of course is that since A and F are unbounded operators in general, the domain of definition of iterated Lie brackets can rapidly shrink. We therefore need to introduce a notion of an 'admissible' Lie bracket as being one which is still defined on $\mathcal{D}(A^\beta)$ for a sufficiently low value of β. The precise definition depends on the details of the equation and can be found in.[1]

Note that these definitions do indeed boil down to the usual Lie brackets between vector fields with polynomial coefficients when $\mathcal{H}$ is finite-dimensional. Note also that the k-th iterated Lie bracket of a k-multilinear map with constant elements is again a constant element. In particular, if we look at the iterated Lie brackets between $-A + F$ and the q_i, many will consist of constant elements of $\mathcal{H}$. With these notions in place, our main result can be formulated as:

Theorem 1.2. *In the context of Eq. (1), consider the collection $\Lambda \subset \mathcal{H}$ of all constant elements among the admissible iterated Lie brackets between the drift $-A + F$ and the constant elements q_i. (The drift is allowed to appear multiple times, so that Λ is countably infinite in general.) Then, if F is sufficiently regular and the solutions to Eq. (1) are well-behaved, Question 1.1 has a positive answer, provided that the linear span of Λ is dense in $\mathcal{H}$.*

Remark 1. It seems that the first result on the regularity of finite-dimensional projections for infinite-dimensional systems under Hörmander-type assumptions was obtained in the linear case by Ocone.[8] These techniques were later extended by Baudoin and Teichmann.[9] However, both of these works required the equation to generate a flow, rather than a semi-flow, thus excluding parabolic PDEs. The first regularity result of this type for parabolic SPDEs was obtained by Mattingly and Pardoux[10] for the particular setting of the 2D stochastic Navier-Stokes equations. Finally, the existence of densities (but not their regularity) was first obtained in a setting similar to ours by Bakhtin and Mattingly.[11]

Remark 2. One can deal with the case where the non-constant Lie brackets are also included in Λ. In this case, we obtain a collection $\Lambda(u)$ for every $u \in \mathcal{H}$ and we need some weak form of uniformity of this density. However, in most interesting examples of the form (1), it is known[12,13] that considering the non-constant Lie brackets does not provide more information.

The remainder of this note is devoted to a short explanation of some aspects of the proof of Theorem 1.2.

2. Some aspects of the proof

A natural line of attack is to try to mimic the probabilistic proof of Hörmaner's theorem, as obtained by Malliavin,[14] Kusuoka and Stroock,[15,16] Bismut,[17] etc. It is a fact that, under some growth restrictions, if a smooth function $\Phi\colon \mathbb{R}^M \to \mathbb{R}^N$ with $M \geq N$ is such that its derivative $D\Phi(x)$ is of maximal rank at every point, then the image of the standard Gaussian measure under g has a smooth density with respect to Lebesgue measure. The main insight of the probabilistic proof of Hörmander's theorem is that this fact still holds even if N is infinite, so that we can take for Φ the solution to Eq. (2) viewed as a map from Wiener space (an infinite-dimensional Gaussian space) to $\mathbb{R}^n$.

The problem then reduces to obtaining a moment bound for the inverse of the 'Malliavin matrix' $\mathcal{M}_t = D\Phi D\Phi^*$. Let us first recall in a nutshell how this is achieved in the finite-dimensional case. In the case of Eq. (2), the Malliavin matrix is given by

$$\langle \xi, \mathcal{M}_t \xi \rangle = \sum_{i=1}^{d} \int_0^t \langle \xi, J_{s,t} V_i(x(s)) \rangle^2 \, ds \, , \tag{4}$$

where $J_{s,t}$ denotes the derivative of the solution map at time t with respect to a change in its initial condition at time $s < t$. In the finite-dimensional case, a key step to the analysis is to use the fact that $J_{s,t} = J_{0,t} J_{0,s}^{-1}$, so that the invertibility of $\mathcal{M}_t$ is equivalent to the invertibility of the 'reduced Malliavin matrix'

$$\langle \xi, \mathcal{C}_t \xi \rangle = \sum_{i=1}^{d} \int_0^t \langle \xi, J_{0,s}^{-1} V_i(x(s)) \rangle^2 \, ds \, , \tag{5}$$

which in turn is equivalent to showing that $\mathbb{P}(\langle \xi, \mathcal{C}_t \xi \rangle \leq \varepsilon) = \mathcal{O}(\varepsilon^p)$ for every $p > 0$. The point is that the integrand in the definition of $\mathcal{C}_t$ is a semimartingale adapted to the filtration generated by the Wiener process driving the equation. The idea is then to repeatedly apply Itô's formula to the integrand, using the fact that if we set $Z_V(t) = \langle \xi, J_{0,s}^{-1} V(x(t)) \rangle$ for any smooth vector field V, one has

$$dZ_V(t) = Z_{[V_0,V]}(t)\, dt + \sum_{i=1}^{d} Z_{[V_i,V]}(t) \circ dW(t) \, . \tag{6}$$

Norris' lemma[18] (a quantitative version of the Doob-Meyer decomposition theorem) then ensures that if Z_V is small for some V, it must also be small for $\{[V, V_j]\}_{j=0}^{d}$. Since Hörmander's condition ensures that these quantities cannot all be small simultaneously, the claim follows.

The problem with the generalisation to parabolic SPDEs comes from the fact that $J_{0,s}$ is not invertible, so that we have to deal with the non-adapted integrand in Eq. (4). The trick is to use the fact that all the vector fields appearing in our case have a polynomial structure. This allows us to exploit the additive structure of our noise by setting $v = u - QW$, so that v satisfies the random PDE

$$\frac{dv}{dt} = -Av + F(v + QW) + AQW \ . \tag{7}$$

Since F is polynomial, the right hand side of this equation can be written as a sum over finitely many terms, each of them being multilinear in both v and W. A similar procedure to the finite-dimensional analysis then allows us to reduce the question to the following version of Norris' lemma. For a multiindex $\alpha = (\alpha_1, \dots, \alpha_\ell)$ write $W_\alpha(t) = W_{\alpha_1}(t) \cdot \dots \cdot W_{\alpha_\ell}(t)$ (with the convention that $W_\emptyset(t) = 1$) and consider a stochastic process of the form

$$Z(t) = \sum_{|\alpha| \leq m} A_\alpha(t) W_\alpha(t) \ , \tag{8}$$

where the A_α are stochastic processes that are not necessarily adapted to the Brownian filtration, but that are almost surely Lipschitz continuous in time. Then, one has:

Proposition 2.1. *In the above setting, there exists a universal family of events Ω_ε depending only on m such that $\mathbb{P}(\Omega_\varepsilon) = \mathcal{O}(\varepsilon^p)$ for every p and such that the implication*

$$\|Z\|_{L^\infty} \leq \varepsilon \quad \Longrightarrow \quad \begin{cases} either \ \sup_\alpha \|A_\alpha\|_{L^\infty} \leq \varepsilon^{3^{-m}} \\ or \ \sup_\alpha \|A_\alpha\|_{\mathrm{Lip}} \geq \varepsilon^{-3^{-(m+1)}} \end{cases} \tag{9}$$

holds for every $W \notin \Omega_\varepsilon$ and for every $\varepsilon \in (0,1]$. Here, the supremum norms are taken over the time interval $[0,1]$.

Remark 1. Note that for any given $W \notin \Omega_\varepsilon$, Eq. (9) is a *deterministic* implication that holds simultaneously for *all* processes of the form (8).

In order to be able to make use of this proposition, it is important to note that if G is any polynomial map (i.e. a sum of finitely many multilinear maps), then there exist $m > 0$ and finitely many polynomial maps G_α that all consist of iterated Lie brackets between G, $-A + F$, and the q_i such that the identity

$$\frac{d}{ds}\langle \xi, J_{s,t} G(v(s))\rangle = \sum_\alpha \langle \xi, J_{s,t} G_\alpha(v(s))\rangle W_\alpha(t) \tag{10}$$

holds. Using Proposition 2.1 repeatedly, we conclude that if $\langle \xi, \mathcal{M}_t \xi \rangle$ is small, then $\langle \xi, J_{s,t} G(v(s)) \rangle$ must be small for every admissible Lie bracket G constructed from the drift and the diffusion coefficients of our original SPDE. Since these span a dense linear subspace of $\mathcal{H}$, this cannot be true, so that any finite-dimensional projection of $\mathcal{M}_t$ must indeed be invertible, thus concluding the proof.

References

1. M. Hairer and J. C. Mattingly, A theory of hypoellipticity and unique ergodicity for semilinear stochastic PDEs, Preprint, (2009).
2. M. Hairer and J. C. Mattingly, *Ann. of Math. (2)* **164**, 993 (2006).
3. G. Da Prato and J. Zabczyk, *Ergodicity for Infinite Dimensional Systems*, London Mathematical Society Lecture Note Series, Vol. 229 (University Press, Cambridge, 1996).
4. F. Flandoli and B. Maslowski, *Commun. Math. Phys.* **172**, 119 (1995).
5. B. Goldys and B. Maslowski, *Ann. Probab.* **34**, 1451 (2006).
6. L. Hörmander, *Acta Math.* **119**, 147 (1967).
7. L. Hörmander, *The Analysis of Linear Partial Differential Operators I–IV* (Springer, New York, 1985).
8. D. Ocone, *J. Funct. Anal.* **79**, 288 (1988).
9. F. Baudoin and J. Teichmann, *Ann. Appl. Probab.* **15**, 1765 (2005).
10. J. C. Mattingly and É. Pardoux, *Comm. Pure Appl. Math.* **59**, 1742 (2006).
11. Y. Bakhtin and J. C. Mattingly, Malliavin calculus for infinite-dimensional systems with additive noise, Preprint, (2006).
12. M. Romito, *J. Statist. Phys.* **114**, 155 (2004).
13. W. E and J. C. Mattingly, *Comm. Pure Appl. Math.* **54**, 1386 (2001).
14. P. Malliavin, Stochastic calculus of variation and hypoelliptic operators, in *Proceedings of the International Symposium on Stochastic Differential Equations (Res. Inst. Math. Sci., Kyoto Univ., Kyoto, 1976)*, (Wiley, New York, 1978).
15. S. Kusuoka and D. Stroock, Applications of the Malliavin calculus. I, in *Stochastic analysis (Katata/Kyoto, 1982)*, , North-Holland Math. Library Vol. 32 (North-Holland, Amsterdam, 1984) pp. 271–306.
16. S. Kusuoka and D. Stroock, *J. Fac. Sci. Univ. Tokyo Sect. IA Math.* **32**, 1 (1985).
17. J.-M. Bismut, *Large Deviations and the Malliavin Calculus* (Birkhäuser Boston Inc., Boston, MA, 1984).
18. J. Norris, *Simplified Malliavin Calculus*, Lecture Notes in Mathematics, Vol. 1204 (Springer, New York, 1986).

Isoperimetry for spherically symmetric log-concave measures

Nolwen Huet

Institut de Mathématiques de Toulouse,
UMR CNRS 5219, Université de Toulouse,
31062 Toulouse, France
Email: nolwen.huet@math.univ-toulouse.fr

In this paper, we study the isoperimetric problem for probability measures μ on $\mathbb{R}^n$ with density proportional to $e^{-\phi(\lambda|x|)}$, where ϕ is a concave function and $|x|$ is the euclidean norm in $\mathbb{R}^n$. We choose $\lambda > 0$, such that μ is isotropic, i.e. such that its covariance is the identity. For such spherically symmetric measures, Bobkov[1] proved that μ satisfies a Poincaré inequality with a universal constant $C > 0$, in particular not depending on the dimension:

$$\forall f : \mathbb{R}^n \to \mathbb{R} \text{ smooth}, \quad C^2 \mathrm{Var}_\mu(f) \leq \mathrm{E}_\mu(|\nabla f|^2).$$

As soon as ϕ is non-decreasing, μ is a log-concave measure and then the Poincaré inequality implies a Cheeger inequality with the same constant C up to a universal multiplicative constant:[2–4] for every Borel sets $A \subset \mathbb{R}^n$,

$$\mu^+(\partial A) \geq C \min\{\mu(A), 1 - \mu(A)\}. \tag{1}$$

Here $\mu^+(\partial A)$ denotes the boundary measure of A defined by

$$\mu^+(\partial A) = \lim_{\varepsilon \to 0^+} \frac{\mu(A^\varepsilon) - \mu(A)}{\varepsilon},$$

and A^ε is the ε-neighborhood of A:

$$A^\varepsilon = \{x \in \mathbb{R}^n; \exists a \in A, |x - a| \leq \varepsilon\}.$$

This result means that the conjecture of Kannan, Lovász and Simonovits[5] is true in the particular case of spherically symmetric log-concave measures. This conjecture asserts that every log-concave isotropic measure satisfies a Cheeger inequality with a universal constant. We would like to improve the isoperimetric inequality (1) assuming more hypotheses on ϕ. Let us define the isoperimetric function of μ on $[0, 1]$ by

$$\mathrm{Is}_\mu(a) = \inf\{\mu^+(\partial A); \mu(A) = a\}.$$

With this notation, the Cheeger inequality is equivalent to the following inequality:

$$\forall a \in \left[0, \frac{1}{2}\right], \quad \mathrm{Is}_\mu(a) \geq C\, a.$$

In the following, assume that $\phi : \mathbb{R}^+ \to \mathbb{R}^+$ is convex, non-decreasing, of class $\mathcal{C}^2$, with $\phi(0) = 0$. For every $n \in \mathbb{N}$, choose $\lambda = \lambda_{n,\phi} > 0$ such that the corresponding measure μ defined as above is isotropic. We will make moreover two different assumptions:

(H1) $x \mapsto \phi(x)/x^2$ is non-increasing *(e.g. $\phi(x) = x^\alpha$ with $\alpha \in [1,2]$).*
(H2) $x \mapsto \phi(x)/x^2$ is non-decreasing and there exists $\alpha \geq 2$ such that $x \mapsto \phi(x)/x^\alpha$ is non-increasing *(e.g. $\phi(x) = x^\alpha$ with $\alpha \geq 2$).*

Theorem 1. *There exists a universal constant $C > 0$ and a universal $n_0 \geq 0$, such that, for every $n \geq n_0$,*

- *if ϕ satisfies (H1), then*

$$\forall a \in \left[0, \frac{1}{2}\right], \quad \mathrm{Is}_\mu(a) \geq C\, \phi^{-1}(1) \frac{a \log \frac{1}{a}}{\phi^{-1}\left(\log \frac{1}{a}\right)} \;;$$

- *if ϕ satisfies (H2), then*

$$\forall a \in \left[0, \frac{1}{2}\right], \quad \mathrm{Is}_\mu(a) \geq C\, a\sqrt{\log \frac{1}{a}} \;.$$

When applied to $\phi(x) = x^\alpha$, with $\alpha \geq 1$, this leads to

$$\mathrm{Is}_\mu(a) \geq C\, a \left(\log \frac{1}{a}\right)^{1 - 1/(\alpha \wedge 2)} .$$

Theorem 1 is optimal when looking for bounds not depending on the dimension, because of previous results known in dimension one[6] for the first case, and of the central limit theorem of Klartag[7] in the second case. We can improve our theorem when allowing the bounds on the isoperimetric function to depend on n. Actually, the best bound switches between the two profiles involved in Theorem 1 according to the values of a.

Theorem 2. *Assume that ϕ satisfies (H1) or (H2), and let $n \in \mathbb{N}^*$. Let $J_{n,\phi}$ be the continuous function on $[0,1]$ symmetric around $1/2$, defined on*

$[0, 1/2]$ *by*

$$
J_{n,\phi}(a) = \begin{cases} \dfrac{\phi^{-1}(n)}{\sqrt{n}} \dfrac{a \log \frac{1}{a}}{\phi^{-1}\left(\log \frac{1}{a}\right)} & \textit{if } a \le e^{-n}, \\[2em] a\sqrt{\log \dfrac{1}{a}} & \textit{if } a \ge e^{-n}. \end{cases}
$$

There exists a universal constant $C > 0$ and universal $n_0 \ge 0$, such that, for every $n \ge n_0$,

$$
\forall a \in [0, 1/2], \quad \mathrm{Is}_\mu(a) \ge C \, J_{n,\phi}(a).
$$

The latter theorem is almost optimal since we can exhibit almost isoperimetric sets, i.e. whose boundary measure is minimal, up to a multiplicative constant. When a is large, these are half-spaces which are the exact isoperimetric sets for the Gaussian measure. This is due to the fact that marginals of spherically symmetric log-concave measures are approximately Gaussian in high dimension. When a is small, these are the complementary of balls. This is quite surprising since it holds also for the Gaussian measure. These two facts are stated in the next lemmas.

Lemma 1. *Let $n \ge 4$. For every r such that*

$$
c_1 \le r \le c_2 n^{1/8},
$$

or every $a := \mu\big(\{x_1 \ge r\}\big)$ such that

$$
e^{-c_3 n^{1/16}} \le a \le e^{-c_4} < 1/2,
$$

then

$$
\mu^+\big(\partial\{x_1 \ge r\}\big) \le C \, a\sqrt{\log \frac{1}{a}},
$$

where $C, c_1, \ldots, c_4 > 0$ are universal constants.

Lemma 2. *Assume that there exists $\alpha \ge 1$ such that $x \mapsto \phi(x)/x^\alpha$ is non-increasing, and that ϕ is log-concave. Then there exists a universal constant $C > 0$ such that, for every $n \ge \alpha + 1$, whenever $a := \mu\{|x| \ge r\} \le e^{-3n} \wedge 1/2$,*

$$
\mu^+\big(\partial\{|x| \ge r\}\big) \le C\alpha \, \frac{\phi^{-1}(n)}{\sqrt{n}} \, \frac{a \log \frac{1}{a}}{\phi^{-1}\left(\log \frac{1}{a}\right)}.
$$

Let us now give a sketch of the proof of Theorems 1 and 2. The first idea is to use the following isoperimetric inequality of Bobkov[8] for log-concave measures: for each choice of $r > 0$ and $x_0 \in \mathbb{R}^n$,

$$\forall a \in [0, 1/2], \quad 2r \mathrm{Is}_\mu(a) \geq a \log \frac{1}{a}$$

$$+ (1 - a) \log \frac{1}{1-a} + \log \mu\{|x - x_0| \leq r\}. \quad (2)$$

We set $r = \phi^{-1}(K_n \log 1/a)$ and try to prove that the sum of the two last terms is non-negative to get another inequality. This requires estimates of probability of balls. Because of the symmetry of μ, we center the ball at 0. We prove the following bound thanks to an integration by parts:

$$\forall r \geq \phi^{-1}(2n), \quad \mu\{\lambda|x| \geq r\} \leq F_{n,\phi}(r) = \left(\frac{er}{\phi^{-1}(n)}\right)^n e^{-\phi(r)} \leq 1. \quad (3)$$

As this is true only for r large enough and that r is a decreasing function of a, this leads to a result valid only for a small enough. Actually, we obtain the case $a \leq e^{-n}$ of Theorem 2.

The second idea is to decompose μ into its radial measure ν, i.e. its projection onto the radius, and the uniform measure σ_{n-1} on the sphere $\mathbb{S}^{n-1}$. Indeed if R and θ are independent random variables respectively of law ν and σ_{n-1}, then $R\theta$ is of law μ. Now, (3) gives also a bound for the tails of ν, and thus ν satisfies the same isoperimetric inequality as μ when $a \leq e^{-n}$. Moreover, as ν is a real measure of density proportional to $r^{n-1}e^{-\phi(r)}$, we can use a lemma of Klartag[7] which gives estimates for the probability of balls centered at the point r_0 where the density of ν reaches its maximum:

$$\forall t \in [0, r_0], \quad \nu\{|r - r_0| \geq t\} \leq C e^{-cnt^2/r_0^2} \quad (4)$$

where $C > 1$ and $0 < c < 1$ are universal constants. One can compute r_0 and see that if μ is isotropic, then

$$r_0 \approx \sqrt{\mathrm{E}_\mu(|X|^2)} = \sqrt{n}.$$

As (4) is valid only for small r, this yields a Gaussian-type isoperimetric inequality for a large enough, once plugged into (2). Actually, we obtain

$$\forall a \in [0, 1/2], \quad \mathrm{Is}_\nu(a) \geq C \, J_{n,\phi}(a).$$

On the other hand, we know the exact solution to the isoperimetric problem on the sphere thanks to Lévy[9] and Schmidt.[10] In particular,

$$\forall a \in [0, 1/2], \quad \mathrm{Is}_{\sigma_{n-1}}(a) \geq C\sqrt{n} \, J_{n,\phi}(a).$$

So it remains to tensorize both isoperimetric inequalities to get the one for μ. It is easier to deal with functional inequalities. One can show that there exists $\kappa > 0$ such that, whenever a measure m on $\mathbb{R}^d$ satisfies

$$\mathrm{Is}_m \geq C J_{n,\phi},$$

then for all smooth functions $f : \mathbb{R}^d \to [0,1]$,

$$\kappa J_{n,\phi}\left(\int f\,dm\right) \leq \int J_{n,\phi}(f)\,dm + \frac{1}{C}\int |\nabla f|\,dm,$$

with the same constant $C > 0$. Let f be a smooth function on $\mathbb{R}^n$. We apply successively the functional inequality to $\theta \mapsto \int f(r\theta)\,d\nu(r)$ and σ_{n-1}, and then to $r \mapsto f(r\theta)$ and ν for each $\theta \in \mathbb{S}^{n-1}$, to obtain:

$$\kappa^2 J\left(\int f\,d\mu\right) \leq \int J(f)\,d\mu$$

$$+ \frac{1}{C}\int |\partial_r f|\,d\mu + \frac{\kappa}{C\sqrt{n}}\int |x|\,|\Pi_{\theta\perp}(\nabla f)|\,d\mu(x), \quad (5)$$

where $\partial_r f$ denotes the radial derivative of f and $\Pi_{\theta\perp}$ is the orthogonal projection onto the hyperplane orthogonal to θ. As $|\nabla f|^2 = |\partial_r f|^2 + |\Pi_{\theta\perp}(\nabla f)|^2$, we would like to get $|x|$ outside of the last integral in (5). Unfortunately $|x|$ is not bounded, but we can use a cut-off argument similar to the one used by Sodin[11] for the ℓ^p-balls. We replace f by fh where $h(r\theta) = h_1(r)$ with

$$h_1 = \begin{cases} 1 & \text{on } [0, r_1) \\ \dfrac{r_2 - r}{r_2 - r_1} & \text{on } [r_1, r_2] \\ 0 & \text{on } (r_2, +\infty) \end{cases}$$

Typically, r_1 and r_2 will be of the same order as $\sqrt{\mathrm{E}_\mu|X|^2} = \sqrt{n}$. As h and ∇h vanish beyond r_2, we can bound $|x|$ by r_2 and we obtain

$$\kappa^2 J\left(\int fh\,d\mu\right) - \int J(fh)\,d\mu - \frac{\|f\|_\infty \nu([r_1, r_2])}{C_\nu(r_2 - r_1)}$$

$$\leq \sqrt{2}\max\left(\frac{1}{C}, \frac{\kappa r_2}{C\sqrt{n}}\right)\int |\nabla f|\,d\mu.$$

We apply this to functions approximating $\mathbf{1}_A$, and using the continuity, concavity, and symmetry of $J_{n,\phi}$, we show that if

$$C J_{n,\phi}(1/2)(r_2 - r_1) \geq 1, \quad (6)$$

and

$$\kappa_1 \, \nu\{[r_1, +\infty)\} \le a \le \frac{1}{2},\tag{7}$$

then

$$\mathrm{Is}_\mu(a) \ge \kappa_2 \min\left(C, \frac{C\sqrt{n}}{r_2}\right) J_{n,\phi}(a).$$

Here κ_1 and κ_2 are positive constants depending only on κ. The condition (6) is a technical one not really restrictive. The condition (7) means that the set must be large enough so that the error made by cutting-off does not matter. We take then r_1 and r_2 of order $\sqrt{n}$ which gives us the isoperimetric inequality for a large enough by (7), namely $a \ge e^{-n}$ from the tail estimate (3). This achieves the proof since we already proved it for $a \le e^{-n}$.

For a complete proof and for the proofs of Lemmas 1 and 2, see our more detailed paper[12] or our PhD thesis.[13]

References

1. S. G. Bobkov, Spectral gap and concentration for some spherically symmetric probability measures, in *Geometric aspects of functional analysis*, , Lecture Notes in Math. Vol. 1807 (Springer, Berlin, 2003) pp. 37–43.
2. J. Cheeger, A lower bound for the smallest eigenvalue of the Laplacian, in *Problems in analysis (Papers dedicated to Salomon Bochner, 1969)*, (Princeton Univ. Press, Princeton, N. J., 1970) pp. 195–199.
3. V. G. Maz'ja, *Dokl. Akad. Nauk SSSR* **144**, 721 (1962).
4. V. G. Maz'ja, *Dokl. Akad. Nauk SSSR* **147**, 294 (1962).
5. R. Kannan, L. Lovász and M. Simonovits, *Discrete Comput. Geom.* **13**, 541 (1995).
6. F. Barthe, P. Cattiaux and C. Roberto, *Electron. J. Probab.* **12**, 1212 (2007).
7. B. Klartag, *Invent. Math.* **168**, 91 (2007).
8. S. G. Bobkov, *Ann. Probab.* **27**, 1903 (1999).
9. P. Lévy, *Problèmes concrets d'analyse fonctionnelle. Avec un complément sur les fonctionnelles analytiques par F. Pellegrino* (Gauthier-Villars, Paris, 1951). 2d ed.
10. E. Schmidt, *Math. Nachr.* **1**, 81 (1948).
11. S. Sodin, *Ann. Inst. H. Poincaré Probab. Statist.* **44**, 362 (2008).
12. N. Huet, Isoperimetry for spherically symmetric log-concave probability measures (2009), Preprint.
13. N. Huet, Inégalités géométriques pour des mesures log-concaves, PhD thesis, Université de Toulouse (2009).

Operators on the Heisenberg group with discrete spectra

J. Inglis

Department of Mathematics, Imperial College London, UK
E-mail: james.inglis06@imperial.ac.uk

We show that a certain class of hypoelliptic operators on the Heisenberg group have discrete spectra, using both a spectral representation of the Heisenberg Laplacian and methods based on functional inequalities.

Keywords: Heisenberg group; essential spectrum; super-Poincaré inequality.

1. Introduction

There are some well-known criteria that ensure classical Schrödinger operators defined on $\mathbb{R}^n$ have discrete spectra. For example, we can consider the Schrödinger operator

$$H = -\Delta + V,$$

where Δ is the standard Laplacian on $\mathbb{R}^n$ and $V \in L_1^{loc}(\mathbb{R}^n)$ is a potential which is bounded from below. If $V(x) \to \infty$ as $|x| \to \infty$, we can conclude that H has a purely discrete spectrum (see for example Theorem XIII.67 of Ref. 1). In this paper we consider a direct analogue of this class of operator, but now defined in the sub-Riemannian setting of the Heisenberg group, and prove some corresponding results.

We will work in the Heisenberg group $\mathbb{H}$, which can be realised as $\mathbb{R}^3$ equipped with the following group operation:

$$(x_1, x_2, x_3) \cdot (y_1, y_2, y_3) = \left(x_1 + y_1, x_2 + y_2, x_3 + y_3 + \frac{1}{2}(x_1 y_2 - x_2 y_1) \right)$$

for $(x_1, x_2, x_3), (y_1, y_2, y_3) \in \mathbb{R}^3$. $\mathbb{H}$ is a Lie group, and its Lie algebra can be identified with the space of left-invariant vector fields in the standard way. By direct computation, we can calculate that this space is spanned by the vector fields

$$X_1 = \partial_{x_1} + \frac{1}{2} x_2 \partial_{x_3}, \quad X_2 = \partial_{x_2} - \frac{1}{2} x_1 \partial_{x_3}, \quad X_3 = \partial_{x_3}.$$

Note that we can generate X_3 by taking the commutator of X_1 and X_2 i.e. $-[X_1, X_2] = X_3$. Thus the tangent space at every point in the group is spanned by X_1, X_2 and $[X_1, X_2]$, so that Hörmander's condition is satisfied. In view of this structure, we make the standard definitions of the *sub-gradient* $\nabla^{\mathbb{H}}$ and *sub-Laplacian* $\Delta^{\mathbb{H}}$ on $\mathbb{H}$ by setting

$$\nabla^{\mathbb{H}} := (X_1, X_2) \quad \text{and} \quad \Delta^{\mathbb{H}} := X_1^2 + X_2^2.$$

It is useful to note that we can write $\Delta^{\mathbb{H}} = -Y_1^2 - Y_2^2$ where

$$Y_1 = i\partial_{x_1} + \frac{1}{2}ix_2\partial_{x_3}, \quad Y_2 = i\partial_{x_2} - \frac{1}{2}ix_1\partial_{x_3}.$$

We also define the associated distance function on $\mathbb{H}$, the so-called *Carnot-Carathéodory distance* by

$$d(x, y) := \sup \left\{ |f(x) - f(y)| : |\nabla^{\mathbb{H}} f| \le 1 \right\},$$

and note that the Haar measure on $\mathbb{H}$ is the Lebesgue measure.

2. Generalisation of a classical result

Our first aim is to prove a direct analogue of the classical result for Schrödinger operators on $\mathbb{R}^n$ described above. We will utilise a useful representation of the sub-Laplacian $\Delta^{\mathbb{H}}$ on $\mathbb{H}$ given in Ref. 2. Indeed, denote by $\mathcal{F}_3$ the partial Fourier transform with respect to the third variable:

$$\mathcal{F}_3 u(x_1, x_2, \xi_3) := (2\pi)^{-1/2} \int_{-\infty}^{\infty} e^{-ix_3\xi_3} u(x_1, x_2, x_3) dx_3.$$

Then it can be shown that

$$\mathcal{F}_3(-\Delta^{\mathbb{H}} u)(x', \xi_3) = \left(i\nabla_{x'}^{\mathbb{H}} + \xi_3 \mathbf{A}(x')\right)^2 \mathcal{F}_3 u(x', \xi_3)$$

where $x' = (x_1, x_2)^T$ and $\mathbf{A}(x') = \frac{1}{2}(-x_2, x_1)^T$. For fixed ξ_3 the operator $\left(i\nabla_{x'}^{\mathbb{H}} + \xi_3 \mathbf{A}(x')\right)^2$ has been well studied, since it corresponds to the Hamiltonian of a particle moving in a magnetic field.[3] Indeed, we have the following spectral decomposition:

$$\mathcal{F}_3(-\Delta^{\mathbb{H}} u)(x', \xi_3) = \sum_{k=0}^{\infty} \lambda_k(\xi_3) P_k \mathcal{F}_3 u(x', \xi_3)$$

where $\lambda_k(\xi_3) := |\xi_3|(2k+1), k \in \{0, 1, \dots\}$, and P_k is the orthogonal eigenprojection given by

$$P_k u(x') = \int_{\mathbb{R}^2} u(y') \pi_k(x', y') dy',$$

where

$$\pi_k(x',y') = \frac{|\xi_3|}{2\pi} e^{-\frac{|\xi_3|}{2}i(x_1y_2-x_2y_1)-\frac{|\xi_3|}{4}|x'-y'|^2} L_k\left(\frac{|\xi_3|}{2}|x'-y'|^2\right)$$

and L_k is the k-th Laguerre polynomial. One can then calculate that for any smooth ψ

$$\int_{\mathbb{H}} \psi(x)(-\Delta^{\mathbb{H}}\psi)(x)dx = \sum_{k=0}^{\infty}\int_{\mathbb{R}} |\xi_3|(2k+1)\left|\hat{\psi}(\xi_3,k)\right|^2 d\xi_3 \qquad (1)$$

where $\hat{\psi}(\xi_3,k) := \|P_k\mathcal{F}_3\psi\|_{L^2(dx')}$. Moreover, we can also use the spectral theorem to see that for suitable functions α

$$\alpha(-\Delta^{\mathbb{H}}) = \alpha\left(\lambda_k(\xi_3)\right), \qquad (2)$$

where the right hand side represents the operator $\mathcal{F}_3^*\sum_k \alpha(\lambda_k(\xi_3))P_k\mathcal{F}_3$, and we have that

$$\int_{\mathbb{H}} \psi(x)\alpha(-\Delta^{\mathbb{H}})\psi(x)dx = \sum_{k=0}^{\infty}\int_{\mathbb{R}} \alpha\left(|\xi_3|(2k+1)\right)\left|\hat{\psi}(\xi_3,k)\right|^2 d\xi_3. \qquad (3)$$

Finally, for a self-adjoint operator H on $L^2(\mathbb{H},dx)$, let $\theta_m(H), m \in \{0,1,\dots\}$ be as in the Min-Max principle i.e. either $\theta_m(H)$ is the m-th eigenvalue below the bottom of $\sigma_{ess}(H)$, or $\theta_m(H)$ is the bottom of the essential spectrum and $\theta_m(H) = \theta_{m+1}(H) = \dots$ (see for example Theorem XIII.1 of Ref. 1).

Theorem 2.1. *Suppose V is in $L^1_{loc}(\mathbb{H},dx)$ and is bounded from below. Suppose also that $V(x) \to \infty$ as $d(x) \to \infty$. Then the operator $A = -\Delta^{\mathbb{H}} + V$ defined as the sum of quadratic forms is a semi-bounded, self-adjoint operator on $L^2(\mathbb{H},dx)$ with compact resolvent. In particular it has a purely discrete set of eigenvalues and a complete set of eigenfunctions.*

Proof.
   Suppose W is a bounded function, supported in a compact set $\Omega \subset \mathbb{R}^3$, so that $\sup_{x\in\Omega}|W(x)| \leq M$ for some $M \in \mathbb{R}$. For $\varepsilon > 0$ consider the operator

$$W\phi_\varepsilon(-\Delta^{\mathbb{H}})$$

where $\phi_\varepsilon(t) = \left(\varepsilon t^2 + t + 1\right)^{-1}$ for $t \in \mathbb{R}$. Now, using (2), we have that

$$\mathbf{Tr}(W^2 \phi_\varepsilon^2(-\Delta^{\mathbb{H}})) \leq M^2 \frac{1}{2\pi} \int_\Omega \int_{-\infty}^\infty \sum_{k=0}^\infty \phi_\varepsilon^2(\lambda_k(\xi_3)) \pi_k(x', x') d\xi_3 dx'$$

$$= M^2 \frac{|\Omega|}{4\pi^2} \sum_{k=0}^\infty \int_{-\infty}^\infty \phi_\varepsilon^2(\lambda_k(\xi_3)) |\xi_3| d\xi_3$$

$$= M^2 \frac{|\Omega|}{2\pi^2} \sum_{k=0}^\infty \int_0^\infty \frac{|\xi_3|}{\left(\varepsilon|\xi_3|^2(2k+1)^2 + |\xi_3|(2k+1) + 1\right)^2} d\xi_3$$

$$< \infty.$$

Since $W\phi_\varepsilon(-\Delta^{\mathbb{H}})$ is positive and self-adjoint, we therefore see that $W\phi_\varepsilon(-\Delta^{\mathbb{H}})$ is Hilbert-Schmidt. Moreover,

$$\left(\varepsilon\lambda_k(\xi_3)^2 + \lambda_k(\xi_3) + 1\right)^{-1} \to (\lambda_k(\xi_3) + 1)^{-1}$$

in $L^\infty(\mathbb{R}) \times l^\infty(\mathbb{N} \cup \{0\})$ as $\varepsilon \to 0$. Indeed

$$|(\phi_0 - \phi_\varepsilon)(\lambda_k(\xi_3))| = \left| \frac{1}{\lambda_k(\xi_3) + 1} - \frac{1}{\varepsilon\lambda_k(\xi_3)^2 + \lambda_k(\xi_3) + 1} \right|$$

$$\leq \varepsilon \frac{\lambda_k(\xi_3)^2}{(\lambda_k(\xi_3) + 1)^2} \leq \varepsilon.$$

Therefore,

$$\|W(\phi_0 - \phi_\varepsilon)(-\Delta^{\mathbb{H}})\psi\|_2^2 \leq M^2 \sum_{k=0}^\infty \int_\mathbb{R} (\phi_0 - \phi_\varepsilon)^2(\lambda_k(\xi_3)) \left|\hat{\psi}(\xi_3, k)\right|^2 d\xi_3$$

$$\leq \varepsilon^2 M^2 \sum_{k=0}^\infty \int_\mathbb{R} \left|\hat{\psi}(\xi_3, k)\right|^2 d\xi_3 = \varepsilon^2 M^2 \|\psi\|_2^2,$$

using equation (3), so that $W\phi_0(-\Delta^{\mathbb{H}})$ is the norm limit of Hilbert-Schmidt operators. We can thus conclude that $W\phi_0(-\Delta^{\mathbb{H}})$ is a compact operator.

Since $W\phi_0(-\Delta^{\mathbb{H}})$ is compact, by Weyl's Theorem,

$$\sigma_{ess}(-\Delta^{\mathbb{H}} + W) = \sigma_{ess}(-\Delta^{\mathbb{H}}) = [0, \infty).$$

Then, by the Min-Max principle, we can see that $\theta_m(-\Delta^{\mathbb{H}} + W) \geq -1$ for m sufficiently large, and all bounded functions W of compact support.

Now, given $a > 0$, define $V_a(x) := \min\{V(x), a + 1\} - a - 1$. Then V_a has compact support, since $V(x) \to \infty$ as $d(x) \to \infty$. Finally, since

$$\theta_m(A) \geq \theta_m\left(-\Delta^{\mathbb{H}} + V_a\right) + a + 1$$

we see that $\theta_m(A) \geq a$ for sufficiently large m, and because a is arbitrary we reach the desired conclusion. $\qquad\square$

Now consider a probability measure $\mu_U(dx) := Z_U^{-1} e^{-U(x)} dx$ on $\mathbb{H}$, where $Z_U = \int_{\mathbb{H}} e^{-U(x)} dx < \infty$ and $U(x)$ is twice differentiable almost everywhere.

Corollary 2.1. *Suppose that* $V = \frac{1}{4}|\nabla^{\mathbb{H}} U|^2 - \frac{1}{2}\Delta^{\mathbb{H}} U$ *is in* $L^1_{loc}(\mathbb{H})$*, is bounded from below, and is such that* $V(x) \to \infty$ *as* $d(x) \to \infty$*. Let* $A = -\Delta^{\mathbb{H}} + \nabla^{\mathbb{H}} U . \nabla^{\mathbb{H}}$*, so that* A *is a positive self-adjoint operator on* $L^2(\mu_U)$*. Then* $\sigma_{ess}(A) = \emptyset$*.*

Proof. Follows from Theorem 2.1 and the identity

$$\int_{\mathbb{H}} f(-\Delta^{\mathbb{H}} + \nabla^{\mathbb{H}} U . \nabla^{\mathbb{H}}) f \, d\mu = \int_{\mathbb{H}} g\left(-\Delta^{\mathbb{H}} + V\right) g \, dx, \quad g = f e^{-\frac{1}{2} U}. \qquad \square$$

Following a recent trend (see Ref. 4 and references therein), we are particularly interested in the case when $U(x) = -\alpha d^p(x)$ with $p \in (1, \infty)$ and $\alpha > 0$. In this case we can explicitly calculate that

$$\frac{1}{4}|\nabla^{\mathbb{H}} U|^2 - \frac{1}{2}\Delta^{\mathbb{H}} U = \frac{\alpha^2 p^2}{4} d^{2(p-1)} - \frac{\alpha p(p-1)}{2} d^{p-2} - \frac{\alpha p}{2} d^{p-1} \Delta^{\mathbb{H}} d. \quad (4)$$

Even though $\Delta^{\mathbb{H}} d$ has singularities on the x_3 axis, in Ref. 4 it was shown that there exists a constant K such that $|\Delta^{\mathbb{H}} d| \leq \frac{K}{d}$. Thus for $p \geq 2$, (4) is locally integrable, bounded from below, and $\to \infty$ as $d(x) \to \infty$, so that we can apply Corollary 2.1 to deduce that $-\Delta^{\mathbb{H}} + \nabla^{\mathbb{H}} U . \nabla^{\mathbb{H}}$ has a discrete spectrum in $L^2(\mu_U)$. However, when $p \in (1, 2)$, the expression is not locally integrable, and so we look for another approach.

3. Functional inequalities

An alternative approach to proving operators have discrete spectra is provided by the the strong relationship between the spectral properties of operators and functional inequalities involving their related Dirichlet forms. The relationship is beautifully illustrated by the work of Cipriani[5] and Wang[6] , in which functional inequalities are introduced which characterise the essential spectra of operators under some very general conditions. More specifically we introduce the following class of operators:

Definition 3.1. Let X be a Lusin space and μ a positive Radon measure on X having full topological support. A semi-bounded, self-adjoint operator $(A, \mathcal{D}(A))$ on $L^2(\mu)$, with associated closed quadratic from $(\mathcal{E}, \mathcal{D}(\mathcal{E}))$ defined by $\mathcal{E}(f, g) = (f, Ag)_{L^2(\mu)}$ for $f, g \in \mathcal{D}(A)$, is called a Persson's operator if

$$\inf \sigma_{ess}(A) = \sup\{\Sigma(K) : K \subset X \text{ is compact}\}$$

where

$$\Sigma(K) := \inf\left\{\frac{\mathcal{E}(f,f)}{\|f\|_2^2} : f \in \mathcal{D}(\mathcal{E}),\ \mathrm{supp}(f) \subset K^c\right\}.$$

The independent work of Cipriani and Wang leads to the following result.

Theorem 3.1. *Let A be a Persson's operator on $L^2(\mu)$. Then $\sigma_{ess}(A) = \emptyset$ if and only if the following super-Poincaré inequality (which we denote by (SP_2)) holds*

$$\mu(f^2) \leq r\mathcal{E}(f,f) + \beta(r)\left(\mu|f|\right)^2, \quad \forall r > 0 \qquad (SP_2)$$

for some decreasing $\beta : (0,\infty) \to (0,\infty)$.

We now aim to use this result to deal with the situation considered at the end of the previous section, i.e. when we have a probability measure $\mu_p(dx) := Z_p^{-1} e^{-\alpha d^p(x)} dx$ on $\mathbb{H}$, where $Z_p = \int_{\mathbb{H}} e^{-\alpha d^p(x)} dx < \infty$ with $p \in (1,\infty)$, $\alpha > 0$ and associated non-positive self-adjoint operator on $L^2(\mu_p)$ and Dirichlet form given by

$$A_p := -\Delta^{\mathbb{H}} + \nabla^{\mathbb{H}}(\alpha d^p).\nabla^{\mathbb{H}}, \quad \mathcal{E}_p(f,g) := \mu_p(fA_pg) = \int_{\mathbb{H}} \nabla^{\mathbb{H}} f.\nabla^{\mathbb{H}} g \, d\mu_p$$

respectively. We first need the following Lemma, which shows that Theorem 3.1 is applicable.

Lemma 3.1. *The operator A_p on $L^2(\mu_p)$ is a Persson's operator for all $p > 1$.*

Proof. This follows easily from Theorem 1 of Ref. 7. $\qquad\qquad\square$

We will actually prove a stronger version of the (SP_2) inequality, which we call an (SP_q) inequality:

Definition 3.2. *For $q \in (1,2]$, we say that a measure μ satisfies a q super-Poincaré or a (SP_q) inequality on $\mathbb{H}$ if*

$$\mu(|f|^q) \leq r\mu|\nabla^{\mathbb{H}} f|^q + \beta(r)\left(\mu|f|^{\frac{q}{2}}\right)^2, \qquad \forall r > 0 \qquad (5)$$

for all f for which the right-hand side is well defined.

Remark 1. We note here that (SP_q) is stronger than (SP_2) in the sense that $(SP_q) \Rightarrow (SP_2)$ for $q \in (1,2]$. This follows simply by applying (SP_q) to $f^{\frac{2}{q}}$.

Theorem 3.2.

(i) Suppose $p \geq 2$. Then μ_p satisfies (SP_q) where $\frac{1}{p} + \frac{1}{q} = 1$.
(ii) Suppose $p \in (1, 2)$. Then μ_p satisfies (SP_2).

Proof. In the case $p \geq 2$, our starting point is the q-logarithmic Sobolev inequality for the measure μ_p with $\frac{1}{p} + \frac{1}{q} = 1$, which is proved in Ref. 4. This inequality states that there exists a constant $C \in (0, \infty)$ such that

$$\mu_p\left(|f|^q \log \frac{|f|^q}{\mu_p|f|^q}\right) \leq C\mu_p|\nabla^{\mathbb{H}} f|^q.$$

We can then adapt the method described in Theorem 3.1 of Ref. 6 to conclude that (SP_q) holds. In the case $p \in (1, 2)$ the method is the same, but our starting point is the existence of constants $C, D \in (0, \infty)$ such that

$$\mu_p\left(f^2 (\log(1 + f^2))^{\frac{2(p-1)}{p}}\right) \leq C\mu_p|\nabla^{\mathbb{H}} f|^2 + D,$$

for all f such that $\mu_p(f^2) = 1$, which is also proved in Ref. 4. $\square$

We can thus conclude with the following Corollary, which follows by combining Theorem 3.1 with Lemma 3.1, Remark 1 and Theorem 3.2:

Corollary 3.1. *For any $p > 1$, the operator A_p has discrete spectrum in $L^2(\mu_p)$.*

Acknowledgements

The author wishes to thank Prof. L. Saloff-Coste for valuable discussions. The author was supported by EPSRC grant EP/D05379X/1.

References

1. M. Reed and B. Simon, *Analysis of Operators*, Methods of Modern Mathematical Physics, Vol. IV (Academic Press, 1978).
2. A. M. Hansson and A. Laptev, *Sharp spectral inequalities for the Heisenberg Laplacian*, in *Groups and Analysis*, ed. K. Trent, London Math. Soc. Lecture Note Ser., Vol. 354 (Cambridge Univ. Press, 2008), pp. 100–115.
3. L. D. Landau and E. M. Lifshitz, *Quantum Mechanics: Non-relativistic Theory* (Pergamon Press, 1958).
4. W. Hebisch and B. Zegarlinski, *arXiv:0905.1713v1* (2008).
5. F. Cirpriani, *J. Funct. Anal.* **177**, 89 (2000).
6. F. Y. Wang, *J. Funct. Anal.* **170**, 219 (2000).
7. G. Grillo, *Z. Anal. Anwendungen* **17**, 329 (1998).

498

Liggett-type inequalities and interacting particle systems: The Gaussian case

J. Inglis[§], M. Neklyudov[‡] and B. Zegarlinski[†*]

[§]*Department of Mathematics, Imperial College London, UK*
[‡]*Department of Mathematics, University of York, Heslington, UK*
[†]*CNRS, Toulouse, France*

We describe Liggett-type inequalities for certain degenerate infinite dimensional sub-elliptic generators and obtain estimates on the long-time behaviour of the corresponding Markov semigroups.

Keywords: Liggett inequality; interacting particle systems.

1. Introduction

In this contribution we present some results concerning the behaviour of the semigroup $(P_t)_{t\geq 0}$, defined by a generator of the following form

$$L = \frac{1}{4} \sum_{\mathbf{i}\in\mathbb{Z}^N} \sum_{\mathbf{j}\in\mathbb{Z}^N:|\mathbf{i}-\mathbf{j}|_1=1} (\partial_{\mathbf{i}} V(x)\partial_{\mathbf{j}} - \partial_{\mathbf{j}} V(x)\partial_{\mathbf{i}})^2.$$

Generators of a similar type appear in the study of dissipative dynamics in which certain quantities are preserved. For more information in this direction, in particular in connection with an effort of explaining the so-called Fourier law of heat conduction, we refer to a nice review Ref. 1 as well as Ref. 2 and the references therein. However, our main motivation stems from the works Refs. 3,4 and 9, where an attempt was made to understand infinite systems from the point of view of functional inequalities. Although formally similar to our approach, one can obtain a variety of different long-time behaviours depending on the underlying space.

The semigroup $(P_t)_{t\geq 0}$ that we consider is one of the simplest examples of a semigroup where the family of vector fields corresponding to the generator has a non-trivial Lie bracket which is degenerate (at point 0).

---

[*]On leave from Imperial College London

One motivation to study this particular $(P_t)_{t\geq 0}$ follows from the fact that, formally, we can easily deduce an exact formula for the invariant measure of the semigroup ("$e^{-V}dx$"). Indeed, we can notice that V is conserved under the action of P_t. As a result, we can formally conclude that the measures $d\mu_r = e^{-\frac{V}{r}}dx, r > 0$, are invariant. On the one hand, the semigroup $(P_t)_{t\geq 0}$ is quite simple, since we can calculate many quantities we are interested in directly. On the other hand, standard methods from interacting particle theory[5,8] do not help in this situation because they require some type of non-degeneracy condition such as Hörmander's condition, which is not satisfied in this case. Another difficulty stems from the intrinsic difference between the infinite dimensional case we consider, and the finite dimensional case i.e. the case when V depends on only a finite number of variables, and instead of the lattice we use its truncation with a periodic boundary condition. Indeed, in the finite dimensional case we can notice that V is a non-trivial fixed point for $P.$, and therefore the semigroup is not ergodic. This reasoning turns out to be incorrect in the infinite dimensional case. The situation here is more subtle because the function V is equal to infinity on the support set of the invariant measure.

We show that in the case of $V(x) = \sum_{i\in\mathbb{Z}^N} x_i^2$ the system is polynomially ergodic and derive Liggett-type inequalities. The proofs of the results presented here can be found in the forthcoming paper Ref. 7.

2. Framework

The Lattice: For $N \in \mathbb{N}$, let $\mathbb{Z}^N$ be the N-dimensional square lattice equipped with the l_1 lattice metric defined by

$$dist(\mathbf{i},\mathbf{j}) := |\mathbf{i} - \mathbf{j}|_1 \equiv \sum_{l=1}^{N} |i_l - j_l|$$

for $\mathbf{i} = (i_1,\ldots,i_N), \mathbf{j} = (j_1,\ldots,j_N) \in \mathbb{Z}^N$. For $\mathbf{i},\mathbf{j} \in \mathbb{Z}^N$, we define

$$\mathbf{i} \sim \mathbf{j} \Leftrightarrow dist(\mathbf{i},\mathbf{j}) = 1.$$

The Configuration Space: Let $\Omega \equiv (\mathbb{R})^{\mathbb{Z}^N}$. Define the following Hilbert spaces

$$E_\alpha = \left\{ x \in \Omega : \|x\|_E^2 := \sum_{\mathbf{i}\in\mathbb{Z}^N} x_\mathbf{i}^2 e^{-\alpha|\mathbf{i}|_1} < \infty \right\}$$

where $\alpha > 0$, equipped with inner product

$$\langle x, y \rangle_{E_\alpha} := \sum_{\mathbf{i}\in\mathbb{Z}^N} x_\mathbf{i} y_\mathbf{i} e^{-\alpha|\mathbf{i}|_1}$$

and

$$H = \left\{ (h^{(1)}, \ldots, h^{(N)}) \in (\Omega)^N : \|(h^{(1)}, \ldots, h^{(N)})\|_H^2 := \sum_{\mathbf{i} \in \mathbb{Z}^N} \sum_{k=1}^N \left(h_{\mathbf{i}}^{(k)} \right)^2 < \infty \right\}$$

with inner product

$$\langle (g^{(1)}, \ldots, g^{(N)}), (h^{(1)}, \ldots, h^{(N)}) \rangle_H := \sum_{\mathbf{i} \in \mathbb{Z}^N} \sum_{k=1}^N g_{\mathbf{i}}^{(k)} h_{\mathbf{i}}^{(k)}.$$

The Gibbs Measure: Let $\mu_{\mathbf{G}}$ be a Gaussian probability measure on $(E_\alpha, \mathcal{B}(E_\alpha))$ with mean zero and covariance $\mathbf{G}$. We assume that the inverse $\mathbf{G}^{-1}$ of the covariance is of finite range i.e.

$$\mathbf{M}_{\mathbf{i},\mathbf{j}} := \mathbf{G}_{\mathbf{i},\mathbf{j}}^{-1} = 0 \quad \text{if } dist(\mathbf{i},\mathbf{j}) > R,$$

with $\sup_{\mathbf{i},\mathbf{j} \in \mathbb{Z}^N} |\mathbf{M}_{\mathbf{i},\mathbf{j}}| \equiv M < \infty$. Later on such measures will be shown to be invariant with respect to a stochastic dynamics described as follows.

The System: Let $W = \{ (W^{(1)}, \ldots, W^{(N)}) \}$ be a cylindrical Wiener process in H. For $x \in E_\alpha$, $\mathbf{i} \in \mathbb{Z}^N$, define

$$V_{\mathbf{i}}(x) := \sum_{\mathbf{j} \in \mathbb{Z}^N} x_{\mathbf{i}} \mathbf{M}_{\mathbf{i},\mathbf{j}} x_{\mathbf{j}},$$

which is a finite sum since $\mathbf{M}_{\mathbf{i},\mathbf{j}} = 0$ if $dist(\mathbf{i},\mathbf{j}) > R$. Using formal notation set

$$V(x) := \frac{1}{2} \sum_{\mathbf{i} \in \mathbb{Z}^N} V_{\mathbf{i}}(x).$$

It will be convenient to simplify the notation for $\partial_{\mathbf{i}} V_{\mathbf{i}}$ as follows

$$\partial_{\mathbf{i}} V(x) = \frac{1}{2} \partial_{\mathbf{i}} \left(\sum_{\mathbf{j},\mathbf{l} \in \mathbb{Z}^N} x_{\mathbf{j}} \mathbf{M}_{\mathbf{j},\mathbf{l}} x_{\mathbf{l}} \right) \equiv \sum_{\mathbf{j}} \mathbf{M}_{\mathbf{i},\mathbf{j}} x_{\mathbf{j}} = \partial_{\mathbf{i}} V_{\mathbf{i}}.$$

We consider the following system of Stratonovich SDEs:

$$dX_{\mathbf{i}}(t) = \sum_{k=1}^N \left(\partial_{\mathbf{i}^-(k)} V(X(t)) \circ dW_{\mathbf{i}^-(k)}^{(k)}(t) - \partial_{\mathbf{i}^+(k)} V(X(t)) \circ dW_{\mathbf{i}}^{(k)}(t) \right)$$

$$(1)$$

where for $\mathbf{i} = (i_1, \ldots, i_N) \in \mathbb{Z}^N$ and $k \in \{1, \ldots, N\}$ we set

$$\mathbf{i}^{\pm}(k) := (i_1, \ldots, i_{k-1}, i_k \pm 1, i_{k+1}, \ldots, i_N).$$

For such systems one obtains the following existence result:

Proposition 2.1. *The stochastic evolution equation* (1) *has a mild solution* X *taking values in the Hilbert space* E_α, *unique up to equivalence among the processes satisfying*

$$\mathbb{P}\left(\int_0^T |X(s)|_{E_\alpha}^2\, ds < \infty\right) = 1.$$

Moreover, it has a continuous modification.

One can then show that the generator, $\mathcal{L}$, of the system (1) is given by

$$\mathcal{L} = \frac{1}{4}\sum_{\mathbf{i}\in\mathbb{Z}^N}\sum_{\mathbf{j}\in\mathbb{Z}^N:|\mathbf{i}-\mathbf{j}|_1=1} (\partial_{\mathbf{i}}V(x)\partial_{\mathbf{j}} - \partial_{\mathbf{j}}V(x)\partial_{\mathbf{i}})^2.$$

For $n \in \{0,1,\dots\}$, let $\mathcal{U}C_b^n(E)$ denote the set of all functions which are uniformly continuous and bounded, together with their Fréchet derivatives up to order n.

Corollary 2.1. *The semigroup* $(P_t)_{t\geq 0}$ *acting on* $\mathcal{U}C_b(E_\alpha), \alpha > 0$, *corresponding to the system* (1) *is Feller and can be represented by the formula*

$$P_t f(\cdot) = \mathbb{E}f\left(X_t(\cdot)\right), t \geq 0,$$

where $X_t(x)$ *is a mild solution to the system* (1) *with initial condition* $x \in E_\alpha$. *Furthermore,* $(P_t)_{t\geq 0}$ *satisfies Kolmogorov's backward equation, and solutions of the system are strong Markov processes.*

Example: Suppose

$$\mathbf{M}_{\mathbf{i},\mathbf{i}} = 1, \quad \mathbf{M}_{\mathbf{i},\mathbf{j}} = 0 \ if \ \mathbf{i} \neq \mathbf{j}.$$

Then $\partial_{\mathbf{i}}V(x) = x_{\mathbf{i}}$, and the system (1) becomes

$$dX_{\mathbf{i}}(t) = -\sum_{k=1}^N X_{\mathbf{i}}(t)dt + \sum_{k=1}^N \left(X_{\mathbf{i}-(k)}(t)dW_{\mathbf{i}-(k)}^k(t) - X_{\mathbf{i}+(k)}(t)dW_{\mathbf{i}}^k(t)\right),$$

which has generator

$$\mathcal{L} = \frac{1}{4}\sum_{\mathbf{i}\in\mathbb{Z}^N}\sum_{\mathbf{j}\in\mathbb{Z}^N:|\mathbf{i}-\mathbf{j}|_1=1} (x_{\mathbf{i}}\partial_{\mathbf{j}} - x_{\mathbf{j}}\partial_{\mathbf{i}})^2.$$

In this case the Gaussian measure $\mu_{\mathbf{G}}$ on E_α is the product measure.

3. Invariant measure

Let $\mu_{\mathbf{G}}, E_\alpha, H$ be as in section 2. First of all, one can show that $\mu_{\mathbf{G}}$ is reversible for the system (1):

Theorem 3.1. *The semigroup $(P_t)_{t\geq 0}$ acting on $\mathcal{U}C_b(E_\alpha)$ can be extended to $L^p(\mu_{\mathbf{G}})$ for any $p \geq 1$. Moreover, for any $f, g \in L^2(\mu_{\mathbf{G}})$ and $t \geq 0$, we have*

$$\mu_{\mathbf{G}}(fP_tg) = \mu_{\mathbf{G}}(gP_tf).$$

4. Weak continuity

We say that the semigroup of bounded linear operators $(S_t)_{t\geq 0}$ defined on $\mathcal{U}C_b(E)$ is weakly continuous if Definition 2.1, p. 350 of Ref. 6 is satisfied.

Theorem 4.1. *There exists $\beta > 0$ such that the semigroup $(S_t)_{t\geq 0} = \left(e^{-\beta t}P_t\right)_{t\geq 0}$ is weakly continuous in $\mathcal{U}C_b(E_\alpha)$. Hence the generator $\mathcal{L}$ is closed and the semigroup $(P_t)_{t\geq 0}$ is strongly continuous on $L^2(E_\alpha, d\mu_{\mathbf{G}})$.*

5. Ergodicity

Everywhere below we consider the example of the second section i.e when $\mathbf{M} = \mathbf{Id}$ and

$$\mathcal{L} = \frac{1}{4} \sum_{\mathbf{i}\in\mathbb{Z}^N} \sum_{\mathbf{j}\in\mathbb{Z}^N : |\mathbf{i}-\mathbf{j}|_1 = 1} (x_{\mathbf{i}}\partial_{\mathbf{j}} - x_{\mathbf{j}}\partial_{\mathbf{i}})^2,$$

and we show that the corresponding semigroup is ergodic with a polynomial rate of convergence. Let us introduce the Hilbert space

$$\mathbb{X} = \{f \in L^2(E_\alpha, d\mu_{\mathbf{G}}) : |f|_{\mathbb{X}}^2 = |f|_{L^2(d\mu_{\mathbf{G}})}^2 + \mathcal{A}^2(f) < \infty\} \subset L^2(d\mu_{\mathbf{G}}),$$

where

$$\mathcal{A}^2(f) \equiv \sum_{\mathbf{i}\in\mathbb{Z}^N} \mu_{\mathbf{G}}|\partial_{\mathbf{i}}f|^2.$$

The operator $\mathcal{L}$ is symmetric and non-positive in $\mathbb{X}$, and its self-adjoint extension, denoted by the same symbol $\mathcal{L}$, generates a strongly continuous contraction semigroup $T_t = e^{t\mathcal{L}} : \mathbb{X} \to \mathbb{X}$. Moreover, $T_t = P_t|_{\mathbb{X}}$.

First of all one can deduce the following estimate.

Lemma 5.1. *There exists a constant C such that for any $f \in \mathbb{X}$, $\mathbf{i} \in \mathbb{Z}^N$, and $t > 0$,*

$$\mu_{\mathbf{G}}|\partial_{\mathbf{i}}(P_tf)|^2 \leq \frac{C\mathcal{A}^2(f)}{t^{\frac{N}{2}}}. \tag{2}$$

This allows us to show that the system is ergodic with polynomial rate of convergence:

Corollary 5.1. *For $f \in \mathbb{X}$ such that $\mathcal{B}(f) \equiv \sum\limits_{\mathbf{i} \in \mathbb{Z}^N} \left(\mu_{\mathbf{G}} |\partial_{\mathbf{i}} f|^2 \right)^{\frac{1}{2}} < \infty$ and $t > 0$ we have*

$$\mathcal{A}^2(P_t f) \leq C \frac{\mathcal{A}(f)\mathcal{B}(f)}{t^{\frac{N}{4}}}. \tag{3}$$

Furthermore,

$$\mu_{\mathbf{G}}(P_t f - \mu_{\mathbf{G}}(f))^2 \leq C \frac{\mathcal{A}(f)\mathcal{B}(f)}{t^{\frac{N}{4}}}. \tag{4}$$

Remark 1. The convergence in Lemma 5.1 cannot be improved and the rate of convergence in Corollary 5.1 is optimal. Indeed, for $W(\mathbf{k}, t) = P_t(x_{\mathbf{k}}^2), t \geq 0, \mathbf{k} \in \mathbb{Z}^N$, we can deduce that $W(t) = e^{t \triangle_{\mathbb{Z}^N}} W(0)$, with $\triangle_{\mathbb{Z}^N}$ denoting the lattice Laplacian, for which the convergence in Lemma 5.1 is precise.

Remark 2. As a consequence we have that following functional (which is not just an L_p norm)

$$\widetilde{\mathcal{A}}(f) \equiv \sup_{s>1} \mathcal{A}(P_s f)$$

is monotone in the sense that

$$\widetilde{\mathcal{A}}(P_t f) \leq \widetilde{\mathcal{A}}(f).$$

Let $D_{\mathbb{X}}(\mathcal{L})$ denote the domain of generator $\mathcal{L}$ in $\mathbb{X}$. The following result shows that the class of functions for which system is ergodic is larger than the one considered in Corollary 5.1.

Proposition 5.1. *For any $f \in L^2(d\mu_{\mathbf{G}})$ semigroup $(P_t)_{t \geq 0}$, is ergodic i.e.*
$\mu_{\mathbf{G}}(P_t f - \mu_{\mathbf{G}} f)^2 \to 0$, *as $t \to \infty$.*
Furthermore, for all $f \in \mathbb{X}$, $|P_t f - \mu_{\mathbf{G}} f|_{\mathbb{X}} \to 0$, as $t \to \infty$.

6. Liggett-type inequalities

From the results of the previous section we can deduce the following Liggett-type inequalities.

Theorem 6.1.
(i) For $f \in \mathbb{X} \cap D(\mathcal{L})$ such that $\mathcal{B}(f) < \infty$, we have

$$\mu_{\mathbf{G}}(f - \mu_{\mathbf{G}}(f))^2 \leq C \langle -\mathcal{L}f, f \rangle_{L^2(d\mu_{\mathbf{G}})}^{\frac{N}{N+4}} (\mathcal{A}(f)\mathcal{B}(f))^{\frac{4}{N+4}}.$$

(ii) Let $\gamma \equiv \{\gamma_{\mathbf{i}}\}_{\mathbf{i} \in \mathbb{Z}^N}$ be a sequence such that $\gamma_{\mathbf{i}} > 0$, $\mathbf{i} \in \mathbb{Z}^N$, and $\sum_{\mathbf{i} \in \mathbb{Z}^N} \frac{1}{\gamma_{\mathbf{i}}} < \infty$. Then for all $f \in D_{\mathbb{X}}(\mathcal{L})$, we have

$$\mathcal{A}^2(f) \leq C(N,\gamma) \left[\sum_{\mathbf{i} \in \mathbb{Z}^N} \gamma_{\mathbf{i}} |\partial_{\mathbf{i}} f|^2_{L^2(d\mu_{\mathbf{G}})} \right]^{\frac{2}{N+2}} \left[\sum_{\mathbf{i} \in \mathbb{Z}^N} \int_{E_\alpha} \partial_{\mathbf{i}} f \partial_{\mathbf{i}} (-\mathcal{L}f) d\mu_{\mathbf{G}} \right]^{\frac{N}{N+2}}.$$

Acknowledgements

We would like to thank Z. Brzeźniak and B. Gołdys for useful remarks and attention to the work. The authors also would like to thank organisers of the ISAAC congress for their hospitality.

References

1. F. Bonetto, J.L. Lebowitz, L. Rey-Bellet, in Mathematical Physics 2000, A. Fokas et al., (Imperial College Press, London, 2000), pp. 128-150.
2. F. Bonetto, J. L. Lebowitz, J. Lukkarinen, and S. Olla, *J. Stat. Phys.* **134** (2009), 5-6, 1097-1119.
3. L. Bertini and B. Zegarliński, *J. Funct. Anal.* **162** (1999), 2 , 257-286.
4. L. Bertini and B. Zegarliński, *Markov Processes and Rel. Fields* **5** (1999), 2, 125-162.
5. P. Ługiewicz, B. Zegarliński, *J. Funct. Anal.* **247** (2007), 2, 438-476.
6. S. Cerrai, *Semigroup Forum* **49** (1994), 3, 349-367.
7. J. Inglis, M. Neklyudov, B. Zegarliński, in preparation
8. T. Liggett, *Interacting particle systems*, Springer, 2004.
9. G. Kondrat, S. Peszat and B. Zegarliński, *J. Phys. A* **33** (2000), 5901-5912.

Enlarging the functional space of decay estimates on semigroups

C. Mouhot

CNRS & DMA, ÉNS Paris,
45, rue d'Ulm, F-75230 Paris cedex 05, France
E-mail: Clement.Mouhot@ens.fr

This note briefly presents a new method for enlarging the functional space of a "spectral-gap-like" estimate of exponential decay on a semigroup. A particular case of the method was first devised in Ref. 1 for the spatially homogeneous Boltzmann equation, and a variant was used in Ref. 2 in the same context for inelastic collisions. We present a generalized abstract version of it, a short proof of the algebraic core of the method, and a new application to the Fokker-Planck equation. More details and other applications shall be found in the work in preparation Ref. 3 (another application to quantum kinetic theory can be found in the work in preparation Ref. 4).

Keywords: Spectral gap; exponential decay; semigroup; Fokker-Planck equation; Poincaré inequality.

1. The "space enlargement" issue

Consider a Hilbert space $\mathcal{H}$, a (possibly unbounded) linear operator $\mathcal{T}$ on $\mathcal{H}$ which generates a strongly continuous semigroup $e^{t\mathcal{T}}$ with spectrum $\Sigma(\mathcal{T})$. Assume that for some Hilbert *subspace* $H \subset \mathcal{H}$ the restricted operator $T := \mathcal{T}\big|_H$ generates a strongly continuous semigroup e^{tT} with spectrum $\Sigma(T)$ in H.

Assume some "spectral-gap-like" information on $\Sigma(T)$, typically when T is self-adjoint assume

$$\forall f \in H, \ f \perp \mathrm{Null}(T), \quad \left\| e^{tT} f \right\|_H \le e^{\lambda t} \|f\|_H, \quad \lambda < 0.$$

An important class of applications is the following: $\mathcal{T}$ is a partial differential operator (acting on a large class of function on $\mathbb{R}^d$, say L^1), with equilibrium μ and detailed spectral information available in a much smaller space $H = L^2(\mu^{-1})$ where it is symmetric. The latter space is much smaller than $\mathcal{H}$ in the sense that it requires a stronger decay condition, e.g. when

μ is a gaussian in statistical mechanics.

The question addressed here is: can one deduce from the spectral-gap information in the space H some spectral-gap information in the larger space $\mathcal{H}$, and if possible in a quantitative way? More explicitly, does $e^{t\mathcal{T}}$ have the same decay property as e^{tT} above?

We give a positive answer for a class of operators $\mathcal{T}$ which split into a part $\mathcal{A}$ "regularizing" $\mathcal{H}$ into H and a coercive part $\mathcal{B}$. We then show that, under some assumption on the potential force, the Fokker-Planck equation belongs to this class and, as a consequence, we prove that its spectral gap property can be extended from the linearization space (with gaussian decay) to larger L^2 spaces with, say, polynomial weights.

2. The abstract result

Let us start with an almost equivalent condition for the decay of the semi-group in terms of a uniform bound on a vertical line for the resolvent. We omit the proof to keep this note short. It can be found in Ref. 3 and it mainly relies on a careful use of the Parseval identity between the resolvent operator and the semigroup.

For some closed densely defined unbounded operator T in a Hilbert space E, denote by by $R(z) = (T - z)^{-1}$, $z \notin \Sigma(T)$ its resolvent operator, and $\mathcal{L}(E)$ the space of bounded linear operators on E. Finally for any $a \in \mathbb{R}$, define the half complex plane $\Delta_a := \{z \in \mathbb{C},\ \Re e\, z > a\}$.

Theorem 2.1. *Assume for the operator T in the Hilbert space E:*

(H1) Localization of the spectrum: $\Sigma(T) \subset (\overline{\Delta_a})^c \cup \{\xi_1, \ldots, \xi_k\}$ *with* $a \in \mathbb{R}$, *and* $\xi_j \in \Delta_a$, $1 \le j \le k$ *some discrete eigenvalues;*

(H2) Control on the resolvent operators:

$$\exists\, K > 0, \quad \forall y \in \mathbb{R}, \quad \|R(a + i\, y)\|_{\mathcal{L}(E)} \le K.$$

(H3) Weak control on the semigroup: *There exist* $b, C_b \ge 0$ *such that*

$$\forall\, t \ge 0 \quad \|e^{tT}\|_{\mathcal{L}(E)} \le C_b\, e^{bt}.$$

Then, for any $\lambda > a$, there exists C_λ explicit from a, b, C_b, K such that

$$\forall\, t \ge 0, \quad \left\| e^{tT} - \sum_{i=1}^{k} e^{\xi_i t}\, \Pi_i \right\|_{\mathcal{L}(E)} \le C_\lambda\, e^{\lambda t} \tag{1}$$

for the spectral projectors Π_i of eigenvalues ξ_i.

We also have the following converse result: assume

$$\forall\, t \geq 0, \quad \left\| e^{t\,T} - \sum_{j=1}^{k} e^{\xi_j\, t}\, \Pi_j \right\|_{\mathcal{L}(E)} \leq C_a\, e^{a\, t} \tag{2}$$

for some constants $a \in \mathbb{R}$, $C_a \in (0, \infty)$, some complex numbers $\xi_j \in \Delta_a$ and some operators Π_j which all commute with $e^{t\,T}$. Then T satisfies (H1), (H2), (H3).

Remark 1. Assumption **(H3)** is required in this theorem in order to obtain quantitative constants in the rate of decay. Therefore, under assumptions **(H1)** and **(H3)**, assertions **(H2)** and Eq. (1) are equivalent in a quantitative way.

The following theorem is the core of the method:

Theorem 2.2. *Assume that $\mathcal{T}$ is a closed unbounded densely defined operator in a Hilbert space $\mathcal{H}$, and that $T := \mathcal{T}\big|_H$ is a closed unbounded densely defined operator in a Hilbert subspace $H \subset \mathcal{H}$ which satisfies (H1) and (H2) (with $E = H$). Assume moreover that $\mathcal{T}$ satisfies:*

(H4) **Decomposition:** *$\mathcal{T} = \mathcal{A} + \mathcal{B}$ where $\mathcal{A}$ and $\mathcal{B}$ are closed unbounded densely defined operators with domains included in the one of $\mathcal{T}$ such that*

- *for some $r > 0$, the operator $\mathcal{B} - \xi$ is invertible with uniform bound for any $\xi \in \Delta_a \backslash (\cup_{i=1}^{k} B(\xi_i, r))$ (where every balls $B(\xi_i, r)$ are strictly included in Δ_a);*
- *$B = \mathcal{B}_{|H}$ is well-defined as a closed unbounded densely defined operator with domain included in the one of T, and $B - \xi$ is invertible for any $\xi \in \Delta_a \backslash (\cup_{i=1}^{k} B(\xi_i, r))$;*
- *$\mathcal{A}\,(\mathcal{B} - \xi)^{-1} : \mathcal{H} \to H$ and $(\mathcal{B} - \xi)^{-1}\,\mathcal{A} : \mathcal{H} \to H$ are bounded for any $\xi \in \Delta_a \backslash (\cup_{i=1}^{k} B(\xi_i, r))$.*

Then $\mathcal{T}$ satisfies (H2) in the space $E = \mathcal{H}$ (with constructive bounds in terms of the above assumptions).

The proof of the following corollary is immediate by combining Theorem 2.2 and Theorems 2.1.

Corollary 2.1. *Assume that T satisfies (H1) and (H2) in the space H and $\mathcal{T}$ satisfies (H3) and (H4) in the space $\mathcal{H}$. Assume moreover that the eigenvalues of $\mathcal{T}$ in Δ_a are the same as those of T, that is $\{\xi_1, \ldots, \xi_k\}$.*

Then the conclusion of Theorem 2.1 holds in the space $\mathcal{H}$: for any $\lambda > a$, there exists C_λ explicit from a, b, C_b, K such that

$$\forall t \geq 0, \quad \left\| e^{t\,T} - \sum_{i=1}^{k} e^{\xi_i\,t}\,\Pi_i \right\|_{\mathcal{L}(\mathcal{H})} \leq C_\lambda\,e^{\lambda t}.$$

Remark 2. If r can be taken as small as wanted **(H4)** (for some decompositions depending on r), it can be proved that the eigenvalues of T in Δ_a are the same as those of T in Δ_a (that is $\{\xi_1, \ldots, \xi_k\}$) and this assumption can be relaxed.

Remark 3. Thanks to the reciprocal part of Theorem 2.1, assumption **(H2)** on T can be replaced by assuming a decay on the semigroup:

$$\forall t \geq 0, \quad \left\| e^{t\,T} - \sum_{j=1}^{k} e^{\xi_j\,t}\,\Pi_j \right\|_{L(H)} \leq C_\lambda\,e^{\lambda t}.$$

Proof of Theorem 2.2. Assume that $k = 1$ and $\xi_1 = 0$ for the sake of simplicity, the proof being similar in the general case.

Take $\xi \notin \Delta_a \backslash B(0, r)$ and define

$$U(\xi) := \mathcal{B}(\xi)^{-1} - R(\xi)\,\mathcal{A}\,\mathcal{B}(\xi)^{-1},$$

where $R(\xi)$ is the resolvent of T in H and $\mathcal{B}(\xi) = \mathcal{B} - \xi$. Since by assumption $\mathcal{B}(\xi)^{-1} : H \to H$, $\mathcal{A}\,\mathcal{B}(\xi)^{-1} : \mathcal{H} \to H$ and $R(\xi) : H \to H$ are bounded operators, $U(\xi) : H \to H$ is well-defined and bounded from $\mathcal{H}$ to $\mathcal{H}$. Then,

$$\begin{aligned}
(T - \xi)\,U(\xi) &= (\mathcal{A} + \mathcal{B}(\xi))\,\mathcal{B}(\xi)^{-1} - (T - \xi)\,R(\xi)\,\mathcal{A}\,\mathcal{B}(\xi)^{-1} \\
&= \mathcal{A}\,\mathcal{B}(\xi)^{-1} + \mathrm{Id}_{\mathcal{H}} - (T - \xi)\,R(\xi)\,\mathcal{A}\,\mathcal{B}(\xi)^{-1} \\
&= \mathcal{A}\,\mathcal{B}(\xi)^{-1} + \mathrm{Id}_{\mathcal{H}} - \mathcal{A}\,\mathcal{B}(\xi)^{-1} = \mathrm{Id}_{\mathcal{H}}.
\end{aligned}$$

To be more precise, introduce the canonical injection $J : H \to \mathcal{H}$ and use that $R = J\,R$, $A = J\,\mathcal{A}$, $T\,J = J\,T$ to write:

$$\begin{aligned}
(T - \xi)\,R(\xi)\,\mathcal{A}\,\mathcal{B}(\xi)^{-1} &= (T - \xi)\,J\,R(\xi)\,\mathcal{A}\,\mathcal{B}(\xi)^{-1} = J\,(T - \xi)\,R(\xi)\,\mathcal{A}\,\mathcal{B}(\xi)^{-1} \\
&= J\,\mathrm{Id}_H\,\mathcal{A}\,\mathcal{B}(\xi)^{-1} = J\,\mathcal{A}\,\mathcal{B}(\xi)^{-1} = \mathcal{A}\,\mathcal{B}(\xi)^{-1}.
\end{aligned}$$

The operator $T - \xi$ is also one-to-one. Indeed, if g satisfies

$$g \in \mathrm{Dom}(T), \qquad (T - \xi)\,g = 0,$$

the decomposition **(H4)** yields

$$\mathcal{B}(\xi)\,g = -\mathcal{A}\,g \in H,$$

and therefore $g \in \mathrm{Domain}(B) \subset \mathrm{Domain}(T) \subset H$ because $B(\xi) = \mathcal{B}(\xi)_{|H}$ is invertible on H. We conclude that $g = 0$ since $T - \xi$ is one-to-one. As a conclusion, $U(\xi)$ is the inverse of $T - \xi$ which in turn implies that $\xi \notin \Sigma(T)$ and $\mathcal{R}(\xi) = U(\xi)$ satisfies the announced estimate. This concludes the proof.

3. Application to the Fokker-Planck equation

In this section we are concerned with the Fokker-Planck equation

$$\partial_t f = T f := \mathrm{div}(\nabla f + E f)$$

for the real valued density function $f = f(t,x)$, $t \geq 0$, $x \in \mathbb{R}^d$. In this equation $E = E(x) \in \mathbb{R}^d$ is a given force field, written as

$$E = \nabla U + F \tag{3}$$

where the potential $U : \mathbb{R}^d \to \mathbb{R}$ is such that $\mu(dx) = e^{-U(x)}\,dx$ is a probability measure satisfying the "Poincaré inequality condition": there exists $\lambda_P < 0$ such that

$$-\int |\nabla u|^2 \, \mu(dv) \leq \lambda_P \int u^2 \, \mu(dv) \qquad \forall\, u \in H^1(\mathbb{R}^d), \quad \int u\,d\mu = 0, \quad (4)$$

(*cf.* for instance Refs. 5–7 and the references therein). The additionnal force field F satisfies

$$\nabla \cdot F = 0, \qquad \nabla U \cdot F = 0, \qquad |F| \leq C(1 + |\nabla U|). \tag{5}$$

Thanks to that structural assumptions we can split T between a symmetric term and a skew-symmetric term:

$$T = T^s + T^{as}, \quad T^s f = \mathrm{div}(\nabla f + \nabla U\, f), \quad T^{as} f = \mathrm{div}\,(F\, f).$$

The operator T^s is symmetric in $H = L^2(\mu^{-1})$

$$\langle T^s f, g\rangle_H = -\int \nabla(f/\mu) \cdot \nabla(g/\mu)\,\mu = \langle f, T^s g\rangle_H,$$

while the operator T^{as} is anti-symmetric in $\mathcal{H} = L^2(m^{-1})$ for any weight function $m^{-1}(v) = \theta(U(x))$, with $\theta : \mathbb{R}_+ \to \mathbb{R}_+$:

$$(T^{as} f, g)_{\mathcal{H}} = \int \left[(\nabla \cdot F) f + F \cdot \nabla f \right] g\, m^{-1} = -\int f \nabla \cdot (F\, g\, m^{-1})$$

$$= -\int f \left[(\nabla \cdot F)\, g\, m^{-1} + (F \cdot \nabla g)\, m^{-1} + g\, \theta'(U)\, (\nabla U \cdot F) \right]$$

$$= -\int \left[(\nabla \cdot F)\, g + F \cdot \nabla g \right] f\, m^{-1} = -\langle f, T^{as} g\rangle_{\mathcal{H}}.$$

As an important consequence, we have

$$\langle T^{as} f, f \rangle_{\mathcal{H}} = \int \nabla \cdot (F\,f)\, f\, \theta(U) = 0.$$

In $H := L^2(\mu^{-1})$ the restricted operator $T := \mathcal{T}\big|_H$ is non-positive, its first eigenvalue is 0 associated to the eigenspace $\mathbb{R}\mu$, and it has a spectral gap thanks to the Poincaré inequality:

$$\int f\,(Tf)\,\mu^{-1} = -\int \mu\, |\nabla(f/\mu)|^2 \le \lambda_2\, \|f - \langle f \rangle\|_H^2.$$

A natural question to ask is whether it is possible to obtain an exponential decay on the semigroup in a space larger than H. The following theorem gives an answer in L^2 spaces with polynomial or "stretched" exponential weights. The proof follows from the application of the abstract method and some careful computations on the Dirichlet form in the larger space.

Theorem 3.1. *Let $\mu = e^{-U}$ with $U(v) = (1 + |x|^2)^{s/2}$, $s \ge 1$ (so that Poincaré inequality holds for μ).*

Let $m \in C^2(\mathbb{R}^d)$ be a weight function such that $m^{-1}(x) = \theta(U(x))$ with $\theta(x) = (1 + |x|^2)^{k/2}$ with $k > d$ or $\theta(x) = e^{(1+|x|^2)^{k/2}}$ with $k \in (0,1)$. Let us define $\mathcal{H} := L^2(m^{-1})$.

Then there exist explicit $\lambda \in (-\lambda_P, 0)$ and $C_\lambda \in [1, \infty)$ such that

$$\forall\, f_0 \in \mathcal{H}, \qquad \forall\, t \ge 0 \quad \|f_t - \langle f_0 \rangle\,\mu\|_{\mathcal{H}} \le C_\lambda\, e^{\lambda t}\, \|f_0 - \langle f_0 \rangle\,\mu\|_{\mathcal{H}}.$$

Remark 1. In this theorem $C_\lambda > 1$ is allowed, which means that we do not prove that the Dirichlet form of $\mathcal{T}$ has a sign.

Remark 2. The smoothness assumption on U and m at the origin can be relaxed.

Acknowledgements. The author thanks Stéphane Mischler for useful discussion during the preparation of this note.

References

1. C. Mouhot, *Comm. Math. Phys.* **261**, 629 (2006).
2. S. Mischler and C. Mouhot, *Comm. Math. Phys.* **288**, 431 (2009).
3. M. Gualdani, S. Mischler and C. Mouhot, Spectral gap in small and large functional spaces, work in progress.
4. A. Arnold, I. Gambda, M. Gualdani, S. Mischler, C. Mouhot and C. Sparber, The Wigner-Fokker-Planck equation: stationary states and large time behavior, work in progress.

5. J.-D. Deuschel and D. W. Stroock, *J. Funct. Anal.* **92**, 30 (1990).

6. M. Ledoux, *The concentration of measure phenomenon* (Amer. Math. Soc., 2001).

7. D. Bakry and M. Émery, Propaganda for Γ_2, in *From local times to global geometry, control and physics (Coventry, 1984/85)*, , Pitman Res. Notes Math. Ser. Vol. 150 (Longman Sci. Tech., Harlow, 1986) pp. 39–46.

The q-logarithmic Sobolev inequality in infinite dimensions

I. Papageorgiou

*Department of Mathematics, Imperial College,
London, SW7 2AZ, UK
E-mail: ioannis.papageorgiou05@imperial.ac.uk*

The q Logarithmic Sobolev inequality in infinite dimensions is investigated for both quadratic and non quadratic interactions. We assume that the one site measure satisfies the q Log-Sobolev inequality and we determine conditions under which the (LSq) can be extended to the infinite volume Gibbs measure.

Keywords: Log-Sobolev inequality; Gibbs measure; infinite dimensions.

1. Introduction

We are interested in the q Log-Sobolev Inequality for measures related to systems of spins on the lattice with nearest neighbour interactions, with values on an unbounded space $\mathbb{M}$. In particular, the cases of $\mathbb{M} = \mathbb{R}$ and $\mathbb{M} = \mathbb{H}$ the Heisenberg group (see [I-P]) will be considered. Let $\Omega = \mathbb{M}^{\mathbb{Z}^N}$. For any subset $\Lambda \subset\subset \mathbb{Z}^N$ and $\omega \in \Omega$ we define the probability measure

$$d\mathbb{E}^{\Lambda,\omega}(x_\Lambda) = \frac{e^{-H^{\Lambda,\omega}}dx_\Lambda}{Z^{\Lambda,\omega}} \tag{1}$$

where

- $x_\Lambda = (x_i)_{i\in\Lambda}$ and $dx_\Lambda = \prod_{i\in\Lambda} dx_i$
- $i \sim j$ means that the nodes i and j on $\mathbb{Z}^N$ are neighbours, i.e. $|i - j| = 1$
- $Z^{\Lambda,\omega} = \int e^{-H^{\Lambda,\omega}}dx_\Lambda$
- $H^{\Lambda,\omega}(x_\Lambda) = \sum_{i\in\Lambda} \phi(x_i) + \sum_{i\in\Lambda, j\sim i} J_{ij}V(x_i, z_j)$
- $z_j = (x_\Lambda \circ \omega_{\Lambda^c})_j = \begin{cases} x_j & ,j\in\Lambda \\ \omega_j & ,j \notin \Lambda \end{cases}$

We call ϕ the phase and V the potential of the interaction. We assume that $|J_{ij}| \in [0, J_0]$ for some $J_0 > 0$. A Gibbs measure ν for the local specification

$\{\mathbb{E}^{\Lambda,\omega}\}_{\Lambda\subset\subset\mathbb{Z}^N,\omega\in\Omega}$ is defined as a probability measure which solves the DLR equation

$$\nu\mathbb{E}^{\Lambda,\cdot}f \equiv \int_\Omega \mathbb{E}^{\Lambda,\omega}(f)\nu(d\omega) = \nu f$$

for all finite sets $\Lambda \subset\subset \mathbb{Z}^N$ and bounded measurable functions f on Ω. For any subset $\Lambda \subset \mathbb{Z}^N$ we define the gradient

$$|\nabla_\Lambda f|^q = \sum_{i\in\Lambda} |\nabla_i f|^q$$

for any compactly supported cylindrical smooth function f. When a variable x_i takes values in $\mathbb{M} = \mathbb{R}$, then $\nabla_i = \frac{\partial}{\partial x_i}$, while in the case where $\mathbb{M} = \mathbb{H}$, ∇_i and Δ_i are the sub-gradient and the sub-Laplacian on $\mathbb{H}$ respectively (see [I-P]). When $\Lambda = \mathbb{Z}^N$ we will simply write $\nabla := \nabla_{\mathbb{Z}^N}$. If we denote

$$\mathbb{E}^{\Lambda,\omega}f = \int f\mathbb{E}^{\Lambda,\omega}(dx_\Lambda)$$

we can define the q Log-Sobolev Inequality

$$\mathbb{E}^{\Lambda,\omega}|f|^q log\frac{|f|^q}{\mathbb{E}^{\Lambda,\omega}|f|^q} \leq C_{LS}\mathbb{E}^{\Lambda,\omega}|\nabla_\Lambda f|^q \quad (LSq)$$

2. The Logarithmic Sobolev Inequality for Quadratic Interactions

In this section we describe conditions which allow to pass from the uniform (LS_q) inequality for the single site measures $\mathbb{E}^{i,\omega}$ to the (LS_q) inequality for the corresponding Gibbs measure ν when the interactions are quadratic. In the Euclidean model, this problem has been extensively studied in the case $q = 2$, for example in [B-H], [G-Z], [L], [Ma], [Z1] and [Z2] and more recently in [O-R]. The case $1 < q < 2$ was looked at in [B-Z] for Gibbs measures with super Gaussian tails. The following argument is strongly related to these methods, though it is based on the work contained in [Z1] and [Z2]. In particular, we extend the results obtained in [O-R], [M] and [Z2] for $q = 2$, to the more general case $q \in (1,2]$. Furthermore, our result can be applied beyond the Euclidean case, like for instance to spins on the Heisenberg group.

We will work with the following hypothesis:

(H0): The one dimensional single site measures $\mathbb{E}^{i,\omega}$ satisfy (LS_q) with a constant c which is independent of the boundary conditions ω.

(H1′): The interaction V is such that

$$\|\nabla_i\nabla_j V(x_i, x_j)\|_\infty < \infty.$$

The main theorem of the section follows ([I-P]):

Theorem 2.1. *Suppose the local specification* $\{\mathbb{E}^{\Lambda,\omega}\}_{\Lambda\subset\subset\mathbb{Z}^N,\omega\in\Omega}$ *satisfies* **(H0)** *and* **(H1′)**. *Then, for sufficiently small* J_0, *the corresponding infinite dimensional Gibbs measure* ν *is unique and satisfies the* (LS_q) *inequality*

$$\nu\left(|f|^q \log \frac{|f|^q}{\nu|f|^q}\right) \le C\nu(|\nabla f|^q) \tag{2}$$

for some positive constant C.

3. The Log-Sobolev Inequality on the Heisenberg Group I

Concerning non-elliptic Hörmander generators, the $CD(\rho,\infty)$ condition ([B], [B-E]) does not hold (see [B-B-B-C]). Thus, the usual tools of obtaining the logarithmic Sobolev inequality do not apply here. In this section we present the main results from [I-P] were it was proven that a certain class of nontrivial Gibbs measures with quadratic interaction potential on an infinite product of Heisenberg groups satisfy the logarithmic Sobolev inequality.

Let $(\tilde{\mathbb{E}}^{\Lambda,\omega})_{\Lambda\subset\subset\mathbb{Z}^N,\omega\in\Omega}$ be the local specification defined by

$$\tilde{\mathbb{E}}^{\Lambda,\omega}(dx_\Lambda) = \frac{e^{-\tilde{H}^{\Lambda,\omega}(x_\Lambda)}}{\int e^{-\tilde{H}^{\Lambda,\omega}(x_\Lambda)}dx_\Lambda}dx_\Lambda \tag{3}$$

where dx_Λ is the Lebesque product measure on $\mathbb{H}^\Lambda$ and

$$\tilde{H}^{\Lambda,\omega}(x_\Lambda) = \alpha \sum_{i\in\Lambda} d^p(x_i)$$

$$+\varepsilon \sum_{i\in\Lambda} \sum_{j:j\sim i} (\mathcal{I}_{\{d(x_i)\geq\eta,\}}d(x_i) + \rho\mathcal{I}_{\{d(\omega_j)\geq\eta,\}}d(\omega_j))^2 + \theta U^{\Lambda,\omega}(x_\Lambda) \tag{4}$$

for $d(.)$ the Carnot-Carathéodory distance (see [I-P]), $\alpha > 0$, $\varepsilon, \rho, \theta \in \mathbb{R}$, and $p \geq 2$, while $U^{\Lambda,\omega}(x_\Lambda) = \sum_{\{i,j\}\cap\,\Lambda,j\sim i} \Phi_{\{i,j\}}(x_i, x_j)$ with $\left\|\Phi_{\{i,j\}}\right\|_\infty \leq M$, $\left\|\nabla_i\nabla_j\Phi_{\{i,j\}}\right\|_\infty \leq M$ and $x_i = \omega_i$ for $i \notin \Lambda$. In the case when $p = 2$, we must have that $\varepsilon > -\frac{\alpha}{2N}$ to ensure that $\int e^{-\tilde{H}_\Lambda}dx_\Lambda < \infty$.

The main result for the infinite volume Gibbs measure ν is the following:

Theorem 3.1. *Let* ν *be a Gibbs measure corresponding to the local specification defined by (3) and (4). Let* q *be dual to* p *i.e.* $\frac{1}{p} + \frac{1}{q} = 1$ *and suppose* $\varepsilon\rho > 0$, *with* $\varepsilon > -\frac{\alpha}{2N}$ *if* $p = 2$. *Then there exists* $\varepsilon_0, \theta_0 > 0$ *such that for* $|\varepsilon| < \varepsilon_0$ *and* $|\theta| < \theta_0$, ν *is unique and satisfies the* (LS_q) *inequality (2).*

The work is strongly related to the methods of Hebisch and Zegarlinski described in [H-Z]. A key result to get Theorem 3.1 is the following theorem on the one site measure:

Theorem 3.2. *Let $\frac{1}{q} + \frac{1}{p} = 1$, and $\varepsilon\rho > 0$ with $\varepsilon > -\frac{\alpha}{2N}$ if $p = 2$ for the local specification defined by* (3) *and* (4). *Then there exists a constant c, independent of the boundary conditions $\omega \in \Omega$ such that*

$$\tilde{\mathbb{E}}^{i,\omega}\left(|f|^q \log \frac{|f|^q}{\tilde{\mathbb{E}}^{i,\omega}|f|^q}\right) \leq c\tilde{\mathbb{E}}^{i,\omega}(|\nabla_i f|^q)$$

for all smooth $f : \mathbb{M}_i \to \mathbb{R}$, where $\mathbb{M}_i$ is a copy of $\mathbb{M}$ corresponding to site i.

4. The Logarithmic Sobolev Inequality with non Quadratic Interactions

We are interested in the q Logarithmic Sobolev Inequality for measures related to systems of unbounded spins on the one dimensional Lattice with nearest neighbour interactions with possible unbounded two sites derivatives. Suppose that the Log-Sobolev Inequality is true for the single site measure with a constant not depending on the boundary conditions. The aim of this section is to present a criterion under which the inequality can be extended to the infinite volume Gibbs measure. More specifically, for measures $\mathbb{E}^{\Lambda,\omega}$ as in (1), we extend the already known results by replacing the boundness of $\nabla_x \nabla_y V(x,y)$ by some uniform exponential integrability (w.r.t the Gibbs measure) of the potential and its first derivative around a single site. Part of the results presented in the current section can be found in [P1].

Hypothesis We consider the following three main hypothesis:

(H0): The one site measures $\mathbb{E}^{i,\omega}$ satisfy the (LSq) with a constant uniformly bounded with respect to the boundary conditions ω.

(H1): There exists $\epsilon > 0$ and $K > 0$, such that for any $i \in \mathbb{Z}$ and $r, s \in \{i-2, i-1, i, i+1, i+2\}$

$$\nu_{\Lambda(i)} e^{\epsilon V(x_r, x_s)} \leq e^K \quad \text{and} \quad \nu_{\Lambda(i)} e^{\epsilon|\nabla_r V(x_r, x_s)|^q} \leq e^K$$

(H2): The coefficients $J_{i,j}$ are such that $|J_{i,j}| \in [0, J]$ for some $J < 1$ sufficiently small.

The main theorem of the section follows.

Theorem 4.1. *If hypothesis (H0)-(H2) are satisfied, then the infinite dimensional Gibbs measure ν for the local specification $\{\mathbb{E}^{\Lambda,\omega}\}_{\Lambda \subset\subset \mathbb{Z}, \omega \in \Omega}$ satisfies the (LS_q) inequality* (2).

Consider now the following condition

(H3): The restriction $\nu_{\Lambda(k)}$ of the Gibbs measure ν to the $\sigma-$algebra $\Sigma_{\Lambda(k)}$,

$$\Lambda(k) = \{k-2, k-1, k, k+1, k+2\}$$

satisfies the (LSq) with a constant independent of k.

A key step on the proof of Theorem 4.1 is the following theorem (see [P1]).

Theorem 4.2. *If hypothesis (H0)-(H3) are satisfied, then the infinite dimensional Gibbs measure ν for the local specification $\{\mathbb{E}^{\Lambda,\omega}\}_{\Lambda\subset\subset\mathbb{Z},\omega\in\Omega}$ satisfies the (LS_q) inequality (2).*

The following proposition allows to pass from Theorem 4.2 to Theorem 4.1.

Proposition 4.1. *Hypothesis (H0)-(H2) imply (H3).*

5. A Perturbation Result for the Logarithmic Sobolev Inequality in Infinite Dimensions with non Quadratic Interactions

In the current section we want to relax the main hypothesis of section 4 to the case where the Log-Sobolev Inequality is true for the single site boundary-free measure. The results presented in this section can be found in [P2]. Consider the one dimensional measure

$$\mathbb{E}^{i,\omega}(dx_i) = \frac{e^{-\phi(x_i)-\sum_{j\sim i} J_{ij} V(x_i-\omega_j)}dx_i}{Z^{i,\omega}} \quad \text{with} \quad \|\nabla_i\nabla_j V(x_i, x_j)\|_\infty = \infty$$

and assume that

$$\mu(dx_i) = \frac{e^{-\phi(x_i)}dx_i}{\int e^{-\phi(x_i)}dx_i}$$

satisfies the Log-Sobolev q inequality. We want to investigate under which conditions the infinite volume Gibbs measure ν for the local specification $\{\mathbb{E}^{\Lambda,\omega}\}_{\Lambda\subset\subset\mathbb{Z},\omega\in\Omega}$ satisfies the Log-Sobolev inequality.

Hypothesis In addition to hypothesis (H1) - (H3) as in the previous section, we consider the following hypothesis:

(H0′): The one dimensional measures $\mu(dx_i) = \frac{e^{-\phi(x_i)}dx_i}{\int e^{-\phi(x_i)}dx_i}$ satisfies the Log-Sobolev q Inequality.

(H4): $\nu e^{\epsilon d(x_i)} < e^K$

The main theorem follows ([P2]).

Theorem 5.1. *If conditions (H0′) and (H1)-(H4) are satisfied, then the infinite dimensional Gibbs measure ν for the local specification $\{\mathbb{E}^{\Lambda,\omega}\}_{\Lambda\subset\subset\mathbb{Z},\omega\in\Omega}$ satisfies the Log-Sobolev q inequality (2).*

The last result is in particular interesting in the case of spins with values on the Heisenberg group, since in [H-Z] the authors showed that a related class of measures on $\mathbb{H}$ satisfy (LS_q) inequalities (see Theorem 5.2 below).

Theorem 5.2. *[H-Z] Let μ_p be the probability measure on $\mathbb{H}$ given by*

$$\mu_p(dx) = \frac{e^{-\beta d^p(x)}}{\int_{\mathbb{H}} e^{-\beta d^p(x)}\,dx}\,dx$$

where $p \geq 2$, $\beta > 0$, dx is the Lebesgue measure on $\mathbb{R}^3$ and $d(x)$ is the Carnot-Carathéodory distance. Then μ_p satisfies an (LS_q) inequality, where $\frac{1}{p} + \frac{1}{q} = 1$.

6. The Log-Sobolev on the Heisenberg Group II

The U-bound inequalities introduced in [H-Z] are an essential tool in proving the Spectral Gap and the Log-Sobolev inequality, under the framework of the Heisenberg group. Furthermore, in [I-P], following the [H-Z] approach, U-bound inequalities were also used to show the inequality for the infinite volume Gibbs measure on $(\mathbb{H})^{\mathbb{Z}^N}$. In this [section we investigate cases were weaker U-bound inequalities (see (H0″) below) hold for

$$d\mathbb{E}^{\{\sim i\},\omega} = \frac{e^{-H^{\{\sim i\},\omega}}d\mu_0}{\int e^{-H^{\{\sim i\},\omega}}d\mu_0} \tag{5}$$

where $\{\sim i\} \equiv \{j : j \sim i\}$ and μ_0 a measure on $\mathbb{H}^{\{\sim i\}}$.

Hypothesis In addition to (H1) - (H4) as before, we consider the following hypothesis for the measure $\mathbb{E}^{\{\sim i\},\omega}$ as in (5):

(H0″): Non uniform U-bound: There exists $\hat{C}$ such that for any $i \in \mathbb{Z}$ and any $\omega \in \Omega$, there exists $\hat{D}_i(\omega)$ such that

$$\mathbb{E}^{\{\sim i\},\omega}\,|f|^q\,(|\nabla_{\{\sim i\}}H^{\{\sim i\},\omega}|^q + H^{\{\sim i\},\omega}) \leq$$

$$\hat{C}\mathbb{E}^{\{\sim i\},\omega}\,|\nabla_{\{\sim i\}}f|^q + \hat{D}_{\{\sim i\}}(\omega)\mathbb{E}^{\{\sim i\},\omega}\,|f|^q$$

for any f compactly supported function. We require that $\hat{D}_i(\omega)$ satisfies $\nu e^{\epsilon\hat{D}_{\{\sim i\}}(\omega)} \leq e^{K_1}$ and that $\hat{D}_{\{\sim i\}}(\omega)$ only depends on ω_{i-2}, ω_i and ω_{i+2}.

(H*) There exist constants $B_*(L), B^*(L) \in (0,\infty)$ such that

$$\int e^{-H^{\{\sim i\},\omega}}d\mu_0 \geq \frac{1}{B^*(L)} \quad \text{and} \quad e^{-H^{\{\sim i\},\omega}} \geq \frac{1}{B_*(L)}$$

for $\omega_{i-2}, x_{i-1}, \omega_i, x_{i+1}, \omega_{i+2} \in B_L$, where $B_R = \{x \in \mathbb{H} : d(x) \leq R\}$

$(\mathbf{H^{**}})$ The measure μ_0 satisfies the Classical-Sobolev inequality

$$\left(\int |f|^{q+\epsilon} d\mu_0 \right)^{\frac{q}{q+\epsilon}} \leq \alpha \int |\nabla f|^q d\mu_0 + \beta \int |f|^q d\mu_0$$

for some constants $\alpha, \beta \in [0, \infty)$ independent of f.

The theorem of the section follows:

Theorem 6.1. *If conditions (H0′), (H1)-(H4) and (H*)-(H**) are satisfied, then the infinite dimensional Gibbs measure ν satisfies the (LS_q) inequality (2).*

References

B. D. BAKRY , *L'hypercontractivité et son utilisation en théorie des semi-groupes* ,Séminaire de Probabilités XIX, Lecture Notes in Math., 1581, Springer, New York, 1-144 (1994)

B-B-B-C. D. BAKRY, F. BAUDOIN, M. BONNEFONT AND D. CHAFFAÏ, *On gradient bounds for the heat kernel on the Heisenberg group* , J of Funct Analysis, 255, 1905-1938 (2008)

B-E. D. BAKRY AND M. EMERY, *Diffusions hypercontractives* , Séminaire de Probabilités XIX, Springer Lecture Notes in Math., 1123, 177-206 (1985)

B-Z. S.G. BOBKOV AND B. ZEGARLINSKI, *Entropy Bounds and Isoperimetry.* Memoirs of the American Mathematical Society, 176, 1 - 69 (2005)

B-H. T. BODINEAU AND B. HELFFER, *Log-Sobolev inequality for unbounded spin systems* , J of Funct Anal., 166, 168-178 (1999)

G-Z. A.GUIONNET AND B.ZEGARLINSKI, *Lectures on Logarithmic Sobolev Inequalities*, IHP Course 98, Séminare de Probabilités XXVI, Lecture Notes in Mathematics 1801, Springer, 1-134 (2003).

H-Z. W.HEBISCH AND B.ZEGARLINSKI, Coercive inequalities on metric measure spaces. J of Funct Anal. (to appear)

I-P. J. INGLIS AND I. PAPAGEORGIOU, *Logarithmic Sobolev Inequalities for Infinite Dimensional Hörmander Type Generators on the Heisenberg Group*. Potential Anal., 31, 79-102 (2009)

L. M. LEDOUX, *Concentration of measure and logarithmic Sobolev inequalities*, Seminaire de Probabilites, XXXIII, Lecture Notes in Math. 1709, Springer-Verlag, 120-216. (1999)

Ma. K.MARTON, *Logarithmic Sobolev Inequality for Weakly Dependent Random Variables.* (preprint)

O-R. F. OTTO AND M. REZNIKOFF, *A new criterion for the Logarithmic Sobolev Inequality and two Applications* , J. Func. Anal., 243, 121-157 (2007).

P1. I.PAPAGEORGIOU, *The Logarithmic Sobolev Inequality in Infinite dimensions for Unbounded Spin Systems on the Lattice with non Quadratic Interactions.* (preprint) arXiv:0901.1765

P2.   I.PAPAGEORGIOU, *A Perturbation result for the Logarithmic Sobolev Inequality with non Quadratic Interactions.* (preprint) arXiv:0901.1482

Z1.   B. ZEGARLINSKI, *On log-Sobolev Inequalities for Infinite Lattice Systems*, Lett. Math. Phys. 20, 173-182 (1990)

Z2.   B. ZEGARLINSKI, *The strong decay to equilibrium for the stochastic dynamics of unbounded spin systems on a lattice*, Comm. Math. Phys. 175, 401-432 (1996)

V.4. Dynamical systems

Organisers: J. Lamb, S. Luzzatto

Dynamical systems aim to provide the mathematical tools to describe and model deterministic systems that arise in the study of ordinary differential equations and in the iteration of maps on smooth manifolds. A variety of ideas and techniques from analysis and other areas come together to provide existence and classification results regarding the dynamical properties of systems from geometrical, topological, and probabilistic points of view.

The dynamical systems section organized during the 7th ISAAC Congress provided a forum for high level researchers working mainly in bifurcation theory and ergodic theory to present their recent research and to discuss open problems and technical issues. The session comprised 19 talks on a variety of topics, including: hyperbolic, partially hyperbolic and nonuniformly hyperbolic dynamics, holomorphic dynamics, tilings and quasicrystals, entropy, statistical properties of deterministic systems, limit cycles, bifurcations, and dynamical systems arising in algebraic logic. The following is a list of speakers for this session: M. Abate, F. Abdenur, K. Burns, P. Berger, Y.-C. Chen, B.-S. Du, M. Field, J. Freitas, G. Forni, V. Gaiko, J.C. van der Meer, M. Nicol, A. Niknam, G. Panti, C.C. Peng, F. Sadyrbaev, J. Schmeling, M. Todd, Q. Wang.

Homogeneous vector fields and meromorphic connections

Marco Abate

*Dipartimento di Matematica, Università di Pisa,
Largo Pontecorvo 5, 56127 Pisa, Italy
E-mail: abate@dm.unipi.it
http://www.dm.unipi.it/~abate*

This note describes how to use geodesics of meromorphic connections to study real integral curves of homogeneous vector fields in $\mathbb{C}^2$.

Keywords: Homogeneous vector fields; meromorphic connections; maps tangent to the identity; holomorphic dynamics; Poincaré-Bendixson theorem.

1. Introduction

In this short note we shall summarize a method recently introduced[1] to study the *real* dynamics of *complex* homogeneous vector fields. Besides its intrinsic interest, this is an useful problem to study because the discrete dynamics of the time 1-map is encoded in the real integral curves of the vector field, and time 1-maps of homogeneous vector fields are prototypical examples of holomorphic maps tangent to the identity at the origin (that is, of holomorphic self-maps $f\colon \mathbb{C}^n \to \mathbb{C}^n$ with $f(O) = O$ and $df_O = \mathrm{id}$). Indeed, Camacho[2,3] has proved that every (germ of a) holomorphic self-map tangent to the identity in $\mathbb{C}$ is locally topologically conjugated to the time-1 map of a homogeneous vector field, and it is natural to conjecture that such a statement should hold for generic holomorphic self-maps in several variables too; so understanding the real dynamics of complex homogeneous vector fields will go a long way toward the understanding of the dynamics of holomorphic self-maps tangent to the identity in a full neighborhood of the origin, one of the main open problems in contemporary local dynamics in several complex variables.

The main idea is that, roughly speaking, integral curves for homogeneous vector fields are geodesics for a meromorphic connection on a projective space. To understand the dynamics of geodesics of meromorphic connections is another very interesting problem, and it naturally splits in two parts: study of the global dynamics of geodesics (e.g., recurrence properties and Poincaré-Bendixson-like theorems), and study of the local dynamics nearby the poles of the connection (via normal forms and local conjugacies).

Due to space limitations, we shall describe our results in dimension 2 only; but part of the construction can be extended to any dimension, up to replace meromorphic connections by partial meromorphic connections. See Refs. 1,4 for details and proofs.

2. The construction

A *homogeneous vector field* of *degree* $\nu + 1 \geq 2$ in $\mathbb{C}^2$ is a vector field of the form

$$Q = Q^1 \frac{\partial}{\partial z^1} + Q^2 \frac{\partial}{\partial z^2} ,$$

where Q^1, Q^2 are homogeneous polynomials of degree $\nu + 1$ in two complex variables. A homogeneous vector field Q is *dicritical* if it is of the form

$$Q = P_\nu(z) \left(z^1 \frac{\partial}{\partial z^1} + z^2 \frac{\partial}{\partial z^2} \right)$$

where P_ν is a homogeneous polynomial of degree ν; *non-dicritical* otherwise.

Let $[\cdot]\colon \mathbb{C}^n \setminus \{O\} \to \mathbb{P}^1(\mathbb{C})$ be the canonical projection. A direction $[v] \in \mathbb{P}^1(\mathbb{C})$ is a *characteristic direction* for a homogeneous vector field Q if the complex line $L_{[v]} = \mathbb{C}v$ is Q-invariant (and then $L_{[v]}$ is a *characteristic line* of Q). We shall moreover say that a characteristic direction $[v]$ is *degenerate* if $Q|_{L_{[v]}} \equiv O$, and *non-degenerate* otherwise. It is easy to check that all directions are characteristic if Q is dicritical, and that a non-dicritical homogeneous vector field only has a finite number of characteristic directions. The dynamics on a characteristic line is one-dimensional, and very easy to study; so from now on we shall deal with non-dicritical homogeneous vector fields only, and we shall mainly be interested in the dynamics outside characteristic lines.

Let $\pi\colon M \to \mathbb{C}^2$ be the blow-up of the origin in $\mathbb{C}^2$, with exceptional divisor $E = \mathbb{P}^1(\mathbb{C})$. Let $p\colon N_E^{\otimes \nu} \to E$ be the ν-th tensor power of the normal bundle of E into M. There exists a natural ν-to-one holomorphic covering map $\chi_\nu\colon \mathbb{C}^2 \setminus \{O\} \to N_E^{\otimes \nu} \setminus E$ generalizing the usual biholomorphism between $\mathbb{C}^2 \setminus \{O\}$ and $N_E \setminus E = M \setminus E$: in the coordinates (ζ, v) induced by the canonical chart of M in $\pi^{-1}(z^1 \neq 0)$ the map χ_ν is given by $\zeta(z) = z^2/z^1$ and $v(z) = (z^1)^\nu$, where ζ is the coordinate on $E = \mathbb{P}^1(\mathbb{C})$

and v is the coordinate on the fiber of $N_E^{\otimes \nu}$; in particular, $p \circ \chi_\nu(z) = [z]$ for all $z \in \mathbb{C}^2 \setminus \{O\}$.

The homogeneity of Q has a first important consequence: whereas the push-forward of a vector field in general is not a vector field, the push-forward $d\chi_\nu(Q)$ of Q by χ_ν is a global holomorphic vector field G defined on the total space of $N_E^{\otimes \nu}$, and vanishing only on the zero section and on the fibers over the degenerate characteristic directions.

The point is that G is, in a suitable sense, the geodesic field of a meromorphic connection. To explain why, we need two more objects. First of all, using Q it is possible to define a global morphism $X_Q \colon N_E^{\otimes \nu} \to TE$ vanishing only over the characteristic directions of Q (and hence it gives an isomorphism between $N_S^{\otimes \nu}$ and TS, where $S \subset E$ is the complement in E of the characteristic directions). If we denote by ∂_1 the local generator of N_E in the canonical chart of M in $\pi^{-1}(\{z^1 \neq 0\})$, and by $\partial/\partial\zeta$ the local generator of TE, the local expression of X_Q is

$$X_Q(\partial_1^{\otimes \nu}) = [Q^2(1,\zeta) - \zeta Q^1(1,\zeta)] \frac{\partial}{\partial \zeta} \ .$$

Notice that $[1 : \zeta]$ is a characteristic direction of Q if and only if $Q^2(1,\zeta) - \zeta Q^1(1,\zeta) = 0$.

Furthermore, we can also define a meromorphic connection ∇ on $N_E^{\otimes \nu}$. The global definition of ∇ is a bit involved;[1,4] in the usual coordinates is locally expressed by

$$\nabla_{\partial/\partial\zeta} \partial_1^{\otimes \nu} = -\frac{\nu Q^1(1,\zeta)}{Q^2(1,\zeta) - \zeta Q^1(1,\zeta)} \partial_1^{\otimes \nu} \ .$$

Notice that ∇ is actually holomorphic on S; its poles are contained in the characteristic directions.

Mixing X_Q and ∇ we can get a linear connection, that is a connection ∇^o defined on the tangent space of S; it suffices to set

$$\nabla^o_v w = \nabla_v X_Q^{-1}(w)$$

for any tangent vector fields v and w on S. It is clear that ∇^o is a holomorphic linear connection on S (and a meromorphic linear connection on E); we can then use it to define the notion of geodesic in this context. A smooth curve $\sigma \colon I \to S$, where $I \subseteq \mathbb{R}$ is an interval, is a *geodesic* for ∇^o if σ' is ∇^o-parallel, that is $\nabla^o_{\sigma'}\sigma' \equiv O$. In local coordinates, this equation is equivalent to the clearly geodesic-looking equation

$$\sigma'' + (k \circ \sigma)(\sigma')^2 = 0 \ ,$$

where k is the meromorphic function defined by $\nabla^o_{\partial/\partial\zeta}(\partial/\partial\zeta) = k(\partial/\partial\zeta)$. Again, the poles of ∇^o are contained in the set of characteristic directions of Q. Furthermore, to each pole p of ∇^o is associated a *residue* $\mathrm{Res}_p(\nabla^o) \in \mathbb{C}$, locally defined as the residue of the meromorphic function k just introduced (but again the definition is independent of the local coordinates). Similarly, one can define the residue $\mathrm{Res}_p(\nabla) \in \mathbb{C}$ of ∇ at a pole $p \in E$; the difference between the two residues is given by the order of vanishing of X_Q.

The relations between integral curves of Q, integral curves of G and geodesics of ∇^o is summarized by the following:

Proposition 2.1. *Let Q be a non-dicritical homogeneous vector field in $\mathbb{C}^2$, and let $\hat{S}_Q$ be the complement in $\mathbb{C}^2$ of the characteristic lines of Q. Then for a real curve $\gamma\colon I \to \hat{S}_Q$ the following are equivalent:*

(i) γ is an integral curve in $\mathbb{C}^2$ of Q;
(ii) $\chi_\nu \circ \gamma$ is an integral curve in $N_S^{\otimes\nu}$ of the geodesic field G;
(iii) $[\gamma]$ is a geodesic in S for the induced connection ∇^o.

The big advantage of this approach is that we can now bring into play the differential geometry machinery developed to study geodesics of connections. It is true that the connection ∇^o in general is *not* globally induced by a metric, and thus the theory of our geodesics is subtly different from the usual theory of metric geodesics. However, ∇^o is locally induced by a conformal family of *flat* metrics, and the flatness enables the use of global results like the Gauss-Bonnet theorem. Furthermore, we can use the residues of ∇^o to express the relations between the holomorphic structure and the behavior of geodesics. All of this yields a fairly complete description of the recurrence properties of the geodesics, that is a Poincaré-Bendixson theorem for meromorphic connections:

Theorem 2.1. *Let $\sigma\colon [0,\varepsilon_0) \to S$ be a maximal geodesic for a meromorphic connection ∇^o on $\mathbb{P}^1(\mathbb{C})$, where $S = \mathbb{P}^1(\mathbb{C}) \setminus \{p_0, \ldots, p_r\}$ and $p_0, \ldots, p_r$ are the poles of ∇^o. Then either*

(i) $\sigma(t)$ tends to a pole of ∇^o as $t \to \varepsilon_0$; or

(ii) σ is closed, and then surrounds poles $p_1, \ldots, p_g$ with
$$\sum_{j=1}^{g} \mathrm{Re}\,\mathrm{Res}_{p_j}(\nabla^o) = -1;\ or$$

(iii) the ω-limit set of σ in $\mathbb{P}^1(\mathbb{C})$ is given by the support of a closed geodesic surrounding poles $p_1, \ldots, p_g$ with $\displaystyle\sum_{j=1}^{g} \mathrm{Re}\,\mathrm{Res}_{p_j}(\nabla^o) = -1;\ or$

(iv) the ω-limit set of σ in $\mathbb{P}^1(\mathbb{C})$ is a simple cycle of saddle connections (see below) surrounding poles $p_1, \ldots, p_g$ with $\sum_{j=1}^{g} \operatorname{Re} \operatorname{Res}_{p_j}(\nabla^o) = -1$; or

(v) σ intersects itself infinitely many times, and in this case every simple loop of σ surrounds a set of poles whose sum of residues has real part belonging to $(-3/2, -1) \cup (-1, -1/2)$.

In particular, a recurrent geodesic either intersects itself infinitely many times or is closed.

In this statement, a *saddle connection* is a geodesic connecting two (not necessarily distinct) poles of ∇^o; and a *simple cycle of saddle connections* is a Jordan curve composed of saddle connections. Notice furthermore that a closed geodesic is not necessarily periodic: it is if and only if the sum of the imaginary parts of the residues at the poles it surrounds is zero.

As a consequence, we get a Poincaré-Bendixson theorem for homogeneous vector fields:

Theorem 2.2. *Let Q be a homogeneous holomorphic vector field on $\mathbb{C}^2$ of degree $\nu + 1 \geq 2$, and let $\gamma \colon [0, \varepsilon_0) \to \mathbb{C}^2$ be a recurrent maximal integral curve of Q. Then γ is periodic or $[\gamma] \colon [0, \varepsilon_0) \to \mathbb{P}^1(\mathbb{C})$ intersect itself infinitely many times.*

Proposition 2.1 and Theorem 2.1 are very helpful in describing the global behavior of integral curves away from the characteristic lines; to complete the picture we need to know what happens nearby the characteristic lines. It turns out that the best way of solving this problem is by studying the integral curves of G nearby the fibers over the characteristic directions; the advantage here is that G extends holomorphically everywhere, and this makes the local study easier.

The characteristic directions can be subdivided in three classes: the *apparent singularities*, which are the characteristic directions which are not poles of ∇, the *Fuchsian singularities*, which are poles of ∇ of order 1, and the *irregular singularities*, which are poles of ∇ of order greater than 1. Fuchsian singularities are generic; and non-degenerate characteristic directions are Fuchsian singularities. We have a complete formal description of all kinds of singularities, and a complete holomorphic description of Fuchsian and apparent singularities. For instance, the holomorphic classification of Fuchsian singularities, revealing in particular the existence of resonance phenomena, is the following

Theorem 2.3. *Let $z_0 \in \mathbb{P}^1(\mathbb{C})$ be a Fuchsian pole of ∇, that is assume that in local coordinates (U_α, z_α) centered at z_0 we can write*

$$G = z_\alpha^\mu(a_0 + a_1 z_\alpha + \cdots)\partial_\alpha - z_\alpha^{\mu-1}(b_0 + b_1 z_\alpha + \cdots)\frac{\partial}{\partial v_\alpha} ,$$

with $\mu \geq 1$ and $a_0,\, b_0 \neq 0$. Put $\rho = b_0/a_0$. Then μ and ρ are (formal and) holomorphic invariants, and we can find a chart (U, z) centered in p_0 in which G is given by

$$z^{\mu-1}\left(zv\partial - \rho v^2\frac{\partial}{\partial v}\right)$$

if $\mu - 1 - \rho \notin \mathbb{N}^*$, or by

$$z^{\mu-1}\left(zv\partial - \rho(1 + az^n)v^2\frac{\partial}{\partial v}\right)$$

for a suitable $a \in \mathbb{C}$ (another formal and holomorphic invariant) if $n = \mu - 1 - \rho \in \mathbb{N}^*$.

Putting together all previous results (and several similar results proved in Ref. 1) one gets a fairly complete description of the dynamics of a large class of homogeneous vector fields. An example of statement we are able to prove is the following:

Theorem 2.4. *Let Q be a non-dicritical homogeneous vector field on $\mathbb{C}^2$ of degree $\nu+1 \geq 2$. Assume that all characteristic directions of Q are Fuchsian singularities of order 1 (this is the generic case). Assume moreover that for no set of characteristic directions the real part of the sum of the residues of ∇^o is equal to -1. Let $\gamma\colon [0, \varepsilon_0) \to \mathbb{C}^2$ be a maximal integral curve of Q. Then:*

(a) If $\gamma(0)$ belongs to a characteristic line L then the image of γ is contained in L. Moreover, either $\gamma(t) \to O$ (and this happens for a Zariski open dense set of initial conditions), or $\|\gamma(t)\| \to +\infty$.

(b) If $\gamma(0)$ does not belong to a characteristic line, then either

(i) γ converges to the origin tangentially to a characteristic direction whose residue with respect to ∇ has negative real part; or

(ii) $\|\gamma(t)\| \to +\infty$ tangentially to a characteristic direction whose residue with respect to ∇ has positive real part; or

(iii) $[\gamma]\colon [0, \varepsilon_0) \to \mathbb{P}^1(\mathbb{C})$ intersects itself infinitely many times.

Furthermore, if (iii) never occurs then (i) holds for a Zariski open dense set of initial conditions.

In this theorem, the assumption on the sum of the residues of ∇^o is used just to exclude closed geodesics or simple cycles of saddle connections with the aim of simplifying the statement, but we have a fairly good understanding of the dynamics in those cases too. For instance, we have examples of homogeneous vector fields with periodic integral curves of arbitrarily high period accumulating the origin — and thus of holomorphic self-maps tangent to the identity with periodic orbits of arbitrarily high period accumulating the origin, an unexpected phenomenon the cannot happen in one variable. Furthermore, since the only constraint on the residues of ∇^o is that their sum must be -2, using Theorem 2.1.(v) it is easy to construct a large class of homogeneous vector fields with only Fuchsian singularities of order 1 where the case (b.iii) in Theorem 2.4 cannot occur; and so for this large class of homogeneous vector fields we have a complete description of the dynamics. To have a complete description of the dynamics of all homogeneos vector fields in $\mathbb{C}^2$ it remains to understand better what happens for irregular singularities and when there are geodesics intersecting themselves infinitely often; and we plan to attack these problems in future papers.

References

1. M. Abate, F. Tovena, Poincaré-Bendixson theorems for meromorphic connections and homogeneous vector fields. Preprint, arXiv:0903.3485 (2009).
2. C. Camacho, *Astérisque* **59–60**, 83 (1978).
3. A.A. Shcherbakov, *Moscow Univ. Math. Bull.* **37**, 60 (1982).
4. M. Abate, F. Bracci, F. Tovena, *Ann. of Math.* **159**, 819 (2004).

Period annuli and positive solutions of nonlinear boundary value problems

S. Atslega

Daugavpils University,
Dept. Nat. Sciences and Mathematics,
Parades Str. 1, Daugavpils, LV-5400, Latvia
E-mail: svetlana_og@inbox.lv
www.du.lv/en

F. Sadyrbaev

University of Latvia,
Institute of Mathematics and Computer Science,
Rainis boul. 29, Riga, LV-1459, Latvia
E-mail: felix@latnet.lv

The existence of multiple positive solutions of the Dirichlet problem for an equation $x'' + g(x) = 0$ is discussed. It is shown that under suitable profile of the graph of the primitive $G(x)$ multiple period annuli exist containing trajectories of solutions to the problem.

Keywords: Multiple positive solutions; period annuli; phase portrait; nonlinear boundary value problem.

1. Introduction

In this article we consider the actual problem of existence of multiple positive solutions to nonlinear boundary value problems. It was shown in the work[1] that multiple positive solutions of the problem $u'' + a_\mu(t)u^{\gamma+1} = 0$, $u(0) = 0$, $u(1) = 0$, where $\gamma > 0$ and $a_\mu(t)$ is a sign changing weight function, can exist. The example was constructed showing seven positive solutions with diverse graphs. The work,[1] in turn, was motivated by the paper.[2]

We show that the same is possible for autonomous equations of the form

$$x'' + g(x) = 0, \tag{1}$$

provided that the primitive $G(x) = \int_0^x g(s)\, ds$ satisfies certain conditions.

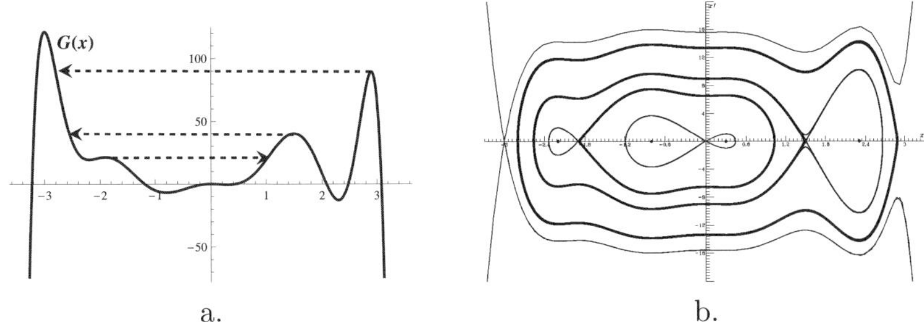

Fig. 1.   **a.** Function $G(x)$; **b.** Phase portrait for equation $x'' + g(x) = 0$.

The positive solutions we are constructing are contained in the sets of solutions which constitute the so called period annuli. So first we give definitions and discuss how period annuli emerge. Statements on the existence of positive solutions follow. The example is analyzed in details.

2. Period annuli

Consider the system

$$x' = y, \quad y' = -g(x). \tag{2}$$

Recall that a critical point O of (2) is *a center* if it has a punctured neighborhood covered with nontrivial cycles.

The largest connected region covered with cycles surrounding O is called *central region.*

Every connected region covered with nontrivial concentric cycles is usually called *a period annulus.*

We will call a period annulus associated with a central region by *a trivial period annulus.* Periodic trajectories of a trivial period annulus encircle exactly one critical point of the type center.

Respectively period annuli enclosing several (more than one) critical points will be called *nontrivial period annuli.*

There are four central regions and three nontrivial period annuli shown in Fig. 1b.

Theorem 2.1 (^{3}). *Let M_1 and M_2 ($M_1 < M_2$) be non-neighbouring points of maxima of the primitive function $G(x)$. Suppose that*

$$G(x) < min\{G(M_1); G(M_2)\}$$

at any point of maximum $x \in (M_1; M_2)$.

Then there exists a nontrivial period annulus associated with the pair M_1 and M_2.

Intersection of a period annulus $\mathbb{P}$ with the positive half-plane (x, x'), $x > 0$, will be called *a positive part of* $\mathbb{P}$ and denoted $\mathbb{P}_+$.

The parts of the inner and outer boundary of a period annulus $\mathbb{P}$ which lie in the right half-plane $\{x \geq 0\}$ will be denoted $\partial\mathbb{P}_{+i}$ and $\partial\mathbb{P}_{+o}$ respectively.

We call a period annulus *proper* if $\mathbb{P}_+$ is not empty and both components $\partial\mathbb{P}_{+i}$ and $\partial\mathbb{P}_{+o}$ have critical points of the system (2).

We call a period annulus *semi-proper* if $\mathbb{P}_+$ is not empty and exactly one of the components $\partial\mathbb{P}_{+i}$ or $\partial\mathbb{P}_{+o}$ has critical points of the system (2).

3. Positive solutions

Consider the problem

$$x'' + \lambda g(x) = 0, \quad x(0) = x(1) = 0 \tag{3}$$

where g is a C^1-function with simple zeros (if $g(z) = 0$ then $g'(z) \neq 0$) only, $\lambda > 0$ is a parameter.

Suppose we are interested in the number of positive solutions to the problem (3). Consider the Cauchy problem

$$x'' + g(x) = 0, \quad x(0) = 0, \ x'(0) = \gamma, \ \gamma > 0. \tag{4}$$

Recall that the first zero $t(\gamma)$ of a solution $x(t; \gamma)$ of the problem (4) is called *time map*.

The number of solutions to the equation $t(\gamma) = 1, \quad \gamma > 0$, gives the number of positive solutions to the problem (3) for $\lambda = 1$. The unity in the right side refers to the length of the interval.

Let $T(\gamma, \lambda)$ be the first zero function (the time map) for the Cauchy problem

$$x'' + \lambda g(x) = 0, \quad x(0) = 0, \ x'(0) = \gamma. \tag{5}$$

The rescaling arguments give us the relation

$$T(\gamma, \lambda) = \frac{1}{\sqrt{\lambda}} t\left(\frac{\gamma}{\sqrt{\lambda}}\right). \tag{6}$$

between the functions $T(\gamma, \lambda)$ and $t(\gamma) = T(\gamma, 1)$. Of course, the relation $T(\gamma, \lambda) = 1$ gives the number of positive solutions to the problem (3) for fixed λ.

Suppose that the primitive function $G(x)$ is such that there exist multiple period annuli with non-empty positive parts $\mathbb{P}_+$. Notice that critical points of the system (2) which can belong to either $\partial\mathbb{P}_{+i}$ or $\partial\mathbb{P}_{+o}$ are necessarily the saddle points.

Theorem 3.1. *Suppose there are m_p proper and m_s semi-proper period annuli in system (2). Then there exists $\lambda_0 > 0$ such that the problem (3) with $\lambda = \lambda_0$ has at least $2m_p + m_s$ positive solutions.*

Proof. Consider a proper period annulus $\mathbb{P}$, $\lambda = 1$. Notice that any period annulus is symmetric with respect to the x-axis. Consider $\Gamma_+ := \{x' > 0\} \cap \mathbb{P}_+$ and let $\gamma \in \Gamma_+$. Consider the upper component $[\gamma_1, \gamma_2]$ of Γ_+. The time map $t(\gamma) = T(\gamma, 1)$ tends to infinity as γ goes to γ_1 or to γ_2 since $\partial\mathbb{P}_{+i}$ and $\partial\mathbb{P}_{+o}$ contain saddle points of the equivalent system. Therefore $T(\gamma, 1)$ is a U-shaped function and has at least one positive minimum.

For λ going to infinity period annuli $\mathbb{P}(\lambda)$ enlarge and the upper component of $\Gamma_+(\lambda)$ is now $[\sqrt{\lambda}\gamma_1, \sqrt{\lambda}\gamma_2]$. One has that for fixed λ

$$\min_{[\sqrt{\lambda}\gamma_1, \sqrt{\lambda}\gamma_2]} T(\gamma, \lambda) = \min_{[\gamma_1, \gamma_2]} \frac{1}{\sqrt{\lambda}} T\left(\frac{\sqrt{\lambda}\gamma}{\sqrt{\lambda}}, 1\right) = \frac{1}{\sqrt{\lambda}} \min_{[\gamma_1, \gamma_2]} T(\gamma, 1).$$

Therefore for λ fixed $T(\gamma, \lambda)$ is also a U-shaped function and has at least one positive minimum. The increasing of λ unboundedly increases the velocity of movement of a point along a trajectory of the system equivalent to equation in (3). Due to the above formula minimum of $T(\gamma, \lambda)$ tends to zero as $\lambda \to +\infty$. Thus λ can be chosen large enough that $\min T(\gamma, \lambda) < 1$, where 1 refers to the length of the interval in (3). Hence at least two positive solutions of the problem (3) for large enough λ.

In case of a semi-proper period annulus one has only that $T(\gamma, \lambda) \to \infty$ as $\gamma \to \Gamma_+ \cap \partial\mathbb{P}_{+i}$ or $\gamma \to \Gamma_+ \cap \partial\mathbb{P}_{+o}$. The value of $T(\gamma, \lambda)$ at the opposite end point of the interval Γ_+ is finite. One may assert only that at least one positive solution exists in a semi-proper period annulus for large enough λ.

Since the total number of the proper and semi-proper period annuli is $m_p + m_s$ by assumption, a common λ exists and this completes the proof.
$\square$

4. Example

Consider equation (1), where

$$g(x) = -x(x+3)(x+2.2)(x+1.9)(x+0.8)(x-0.3)(x-1.5)(x-2.3)(x-2.9). \tag{7}$$

So $\lambda = 1$. The graph of the primitive function G is depicted in Fig. 1a.

The first period annulus is depicted in Fig. 2. The x'-axis goes through the center of "figure eight". The function $T_+(\gamma)$ and the respective two positive solutions are shown in Fig. 3.

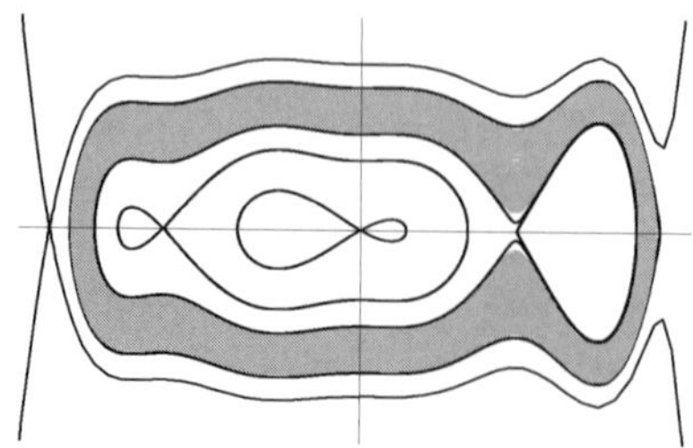

Fig. 2.   The outer (first) nontrivial period annulus

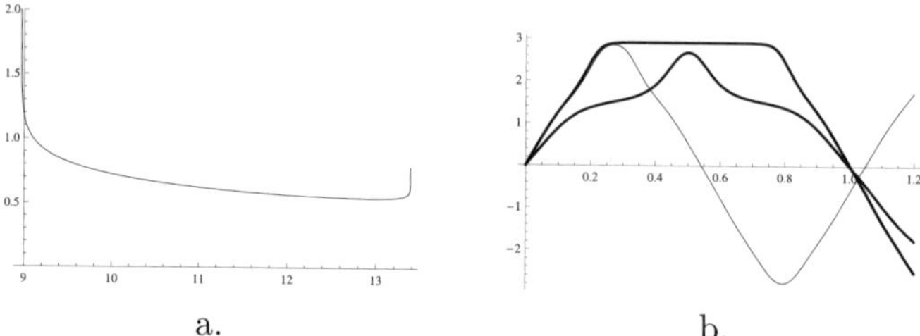

a.   b.

Fig. 3.   **a.** Time T_+, $T_{+min} = 0.436609$; **b.** Solutions (bold), solution with T_{+min}.

The second (middle) period annulus is depicted in Fig. 4. The graph of $T_+(\gamma)$ function is given in Fig. 5a. The graph of the unique positive solution is given in Fig. 5b.

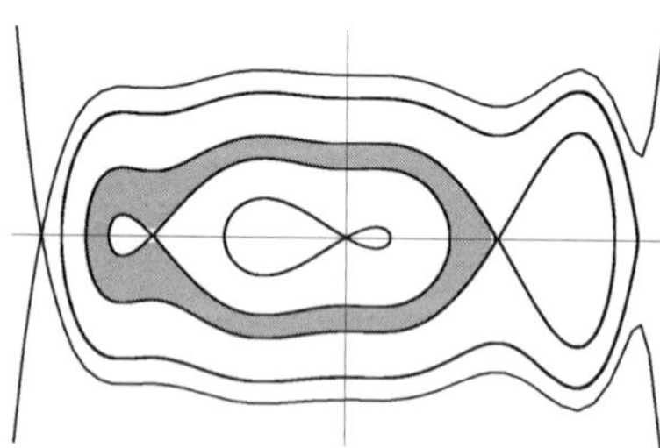

Fig. 4.   The second (middle) nontrivial period annulus

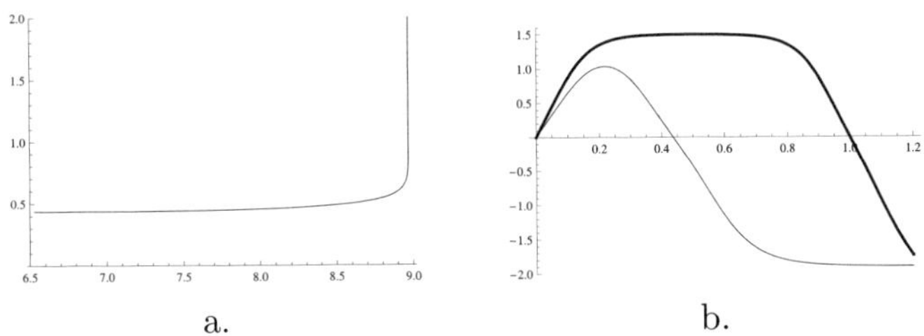

a. b.

Fig. 5. **a.** Time T_+, $T_{+min} \approx 0.42$; **b.** Solution (bold), solution with T_{+min}.

The third (inner) period annulus and the respective graphs of T_+ and a positive solution are depicted in Fig. 6 and Fig. 7 respectively.

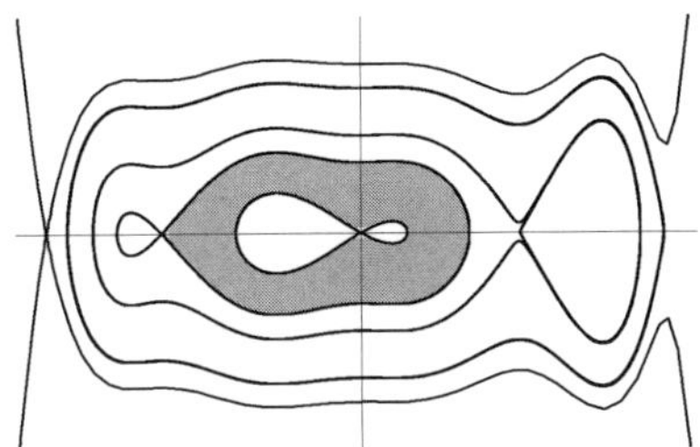

Fig. 6. The second (middle) nontrivial period annulus

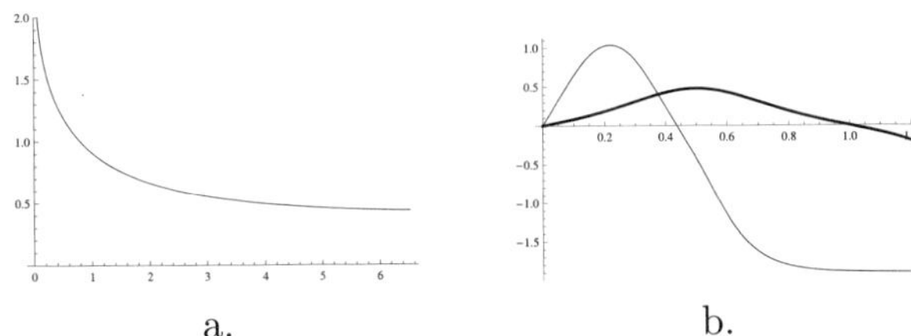

a. b.

Fig. 7. **a.** Time T_+, $T_{+min} \approx 0.5$; **b.** Solution (bold), solution with T_{+min}.

References

1. M. Gaudenzi, P. Habets and F. Zanolin, A Seven-Positive-Solutions Theorem for a Superlinear Problem. *Advanced Nonlinear Studies*, **4**, 149 - 164 (2004).
2. G.J. Butler, Rapid oscillation, nonextendability, and the existence of periodic solutions to second order nonlinear differential equations. *J. Differential Equations*, **22**, 467 - 477 (1976).

3. S.Atslega and F. Sadyrbaev. Multiple solutions of the second order nonlinear Neumann BVP. *Dynamics of Continuous, Discrete and Impulsive Systems (Series A)*. DCDIS A Supplement dedicated to the 6th International Conference on Differential Equations and Dynamical Systems held in Baltimore, U.S.A., May 22 - 26, Watam Press, 2009, 100-103.

4. J. Smoller and A. Wasserman, Global bifurcation of steady-state solutions. *J. Diff. Equations*, **39**, 269–290 (1981).

Global limit cycle bifurcations in a biomedical model*

V.A. Gaiko

*United Institute of Informatics Problems,
National Academy of Sciences of Belarus,
L. Beda Str. 6-4, Minsk 220040, Belarus
E-mail: valery.gaiko@yahoo.com*

In this paper, using the global bifurcation theory, we establish the qualitative analysis of a quartic dynamical system which can be used as a biomedical model. In particular, we prove that such a system can have at most two limit cycles.

Keywords: Biomedical dynamical system; field rotation parameter; bifurcation; limit cycle; Wintner–Perko termination principle.

1. Introduction

In this paper, we continue studying biomedical models of "predator–prey" type which have been considered in [2]. For example, in immunology models of the tumour growth dynamics in view of the interaction of two types of cells determining the untitumour organism resistance are very interesting. There are two dynamic variables in such models: the concentration of tumour cells and the concentration of specific resistance cells. Such a model looks like

$$\dot{x} = \nu - \alpha\,x - \beta\,xy + \frac{\gamma\,xy}{1+y},$$
$$\dot{y} = y - xy - \frac{\delta\,y}{1+y}, \qquad (1.1)$$

where x is the concentration of cells of a specific resistance subsystem (effective cells) and y is the concentration of tumour cells. In the framework of the model, the following processes are taken into account:

- the growth rate of the effective cells is equal to ν;

---

*This work was supported by the Netherlands Organization for Scientific Research.

- the natural death relative rate of the effective cells is equal to α;
- the death of the effective and tumour cells is a result of their interaction (see the members proportional to xy in both equations);
- the quantitative growth of the effective cells is a result of the growth of previous ones; the rate of this process depends on the concentration of the tumour cells, and it reaches a value equal to γ under their excess;
- the reproduction of tumour cells is also taken into account.

System (1.1) can be reduced to a cubic dynamical system, and it has been studied in [2]. We will study a more general (quartic) dynamical system which models the dynamics of the populations of predators and their prey that use the group defense strategy in a given biomedical system and which is a variation on the classical Lotka–Volterra system:

$$
\begin{aligned}
\dot{x} &= x((1 - \lambda x)(\alpha x^2 + \beta x + 1) - y) \equiv P, \\
\dot{y} &= -y((\delta + \mu y)(\alpha x^2 + \beta x + 1) - x) \equiv Q,
\end{aligned}
\tag{1.2}
$$

where $\alpha \geq 0$, $\delta > 0$, $\lambda > 0$, $\mu \geq 0$ and $\beta > -2\sqrt{\alpha}$ are parameters. Such a quartic dynamical model was studied earlier, for instance, in [4], [10]. However, the qualitative analysis was incomplete, since the global bifurcations of limit cycles could not be studied properly by means of the methods and techniques which were used earlier in the qualitative theory of dynamical systems.

Together with (1.2), we will also consider an auxiliary system [1], [5]

$$
\dot{x} = P - \gamma Q, \qquad \dot{y} = Q + \gamma P,
\tag{1.3}
$$

applying to these systems new bifurcation methods and geometric approaches developed in [3], [5]–[8] and completing the qualitative analysis of system (1.2).

2. Bifurcations of Limit Cycles

Let us first formulate the Wintner–Perko termination principle [5], [9] for the polynomial system

$$
\dot{\boldsymbol{x}} = \boldsymbol{f}(\boldsymbol{x}, \boldsymbol{\mu}),
\tag{$2.1_{\boldsymbol{\mu}}$}
$$

where $\boldsymbol{x} \in \mathbf{R}^2$; $\boldsymbol{\mu} \in \mathbf{R}^n$; $\boldsymbol{f} \in \mathbf{R}^2$ ($\boldsymbol{f}$ is a polynomial vector function).

Assume that system ($2.1_{\boldsymbol{\mu}}$) has a limit cycle

$$
L_0 : \boldsymbol{x} = \boldsymbol{\varphi}_0(t)
$$

of minimal period T_0 at some parameter value $\boldsymbol{\mu} = \boldsymbol{\mu}_0 \in \mathbf{R}^n$. Let l be the straight line normal to L_0 at the point $\boldsymbol{p}_0 = \boldsymbol{\varphi}_0(0)$ and s be the coordinate along l with s positive exterior of L_0. It then follows from the implicit function theorem [5], [9] that there is a $\delta > 0$ such that the Poincaré map $h(s, \boldsymbol{\mu})$ is defined and analytic for $|s| < \delta$ and $\|\boldsymbol{\mu} - \boldsymbol{\mu}_0\| < \delta$. Besides, the displacement function for system $(2.1_{\boldsymbol{\mu}})$ along the normal line l to L_0 is defined as the function

$$d(s, \boldsymbol{\mu}) = h(s, \boldsymbol{\mu}) - s.$$

In terms of the displacement function, a multiple limit cycle can be defined as follows [5], [9].

Definition 2.1. A limit cycle L_0 of $(2.1_{\boldsymbol{\mu}})$ is a *multiple limit cycle* iff $d(0, \boldsymbol{\mu}_0) = d_r(0, \boldsymbol{\mu}_0) = 0$ and it is a *simple limit cycle* if it is not a multiple limit cycle; furthermore, L_0 is a limit cycle of multiplicity m iff

$$d(0, \boldsymbol{\mu}_0) = d_r(0, \boldsymbol{\mu}_0) = \ldots = d_r^{(m-1)}(0, \boldsymbol{\mu}_0) = 0, \quad d_r^{(m)}(0, \boldsymbol{\mu}_0) \neq 0.$$

Theorem 2.1 (Wintner–Perko termination principle). *Any one-parameter family of multiplicity-m limit cycles of relatively prime polynomial system $(2.1_{\boldsymbol{\mu}})$ can be extended in a unique way to a maximal one-parameter family of multiplicity-m limit cycles of $(2.1_{\boldsymbol{\mu}})$ which is either open or cyclic.*

If it is open, then it terminates either as the parameter or the limit cycles become unbounded; or, the family terminates either at a singular point of $(2.1_{\boldsymbol{\mu}})$, which is typically a fine focus of multiplicity m, or on a (compound) separatrix cycle of $(2.1_{\boldsymbol{\mu}})$, which is also typically of multiplicity m.

The proof of this principle for general polynomial system $(2.1_{\boldsymbol{\mu}})$ with a vector parameter $\boldsymbol{\mu} \in \mathbf{R}^n$ parallels the proof of the planar termination principle for the system

$$\dot{x} = P(x, y, \lambda), \quad \dot{y} = Q(x, y, \lambda) \tag{2.1_λ}$$

with a single parameter $\lambda \in \mathbf{R}$ [5], [9], since there is no loss of generality in assuming that system $(2.1_{\boldsymbol{\mu}})$ is parameterized by a single parameter λ; i.e., we can assume that there exists an analytic mapping $\boldsymbol{\mu}(\lambda)$ of $\mathbf{R}$ into $\mathbf{R}^n$ such that $(2.1_{\boldsymbol{\mu}})$ can be written as $(2.1_{\boldsymbol{\mu}(\lambda)})$ or even (2.1_λ) and then we can repeat everything, what had been done for system (2.1_λ) in [9]. In particular, if λ is a field rotation parameter of (2.1_λ), the following Perko's theorem on monotonic families of limit cycles is valid [9].

Theorem 2.2. *If L_0 is a nonsingular multiple limit cycle of (2.1_0), then L_0 belongs to a one-parameter family of limit cycles of (2.1_λ); furthermore:*

1) *if the multiplicity of L_0 is odd, then the family either expands or contracts monotonically as λ increases through λ_0;*

2) *if the multiplicity of L_0 is even, then L_0 bifurcates into a stable and an unstable limit cycle as λ varies from λ_0 in one sense and L_0 disappears as λ varies from λ_0 in the opposite sense; i. e., there is a fold bifurcation at λ_0.*

Applying the definition of a field rotation parameter [1], [5], [9], i. e., a parameter which rotates the field in one direction, to system (1.2), let us calculate the corresponding determinants for the parameters α and β:

$$\Delta_\alpha = PQ'_\alpha - QP'_\alpha = x^3 y(y(\delta + \mu y) - x(1 - \lambda x)), \tag{2.2}$$

$$\Delta_\beta = PQ'_\beta - QP'_\beta = x^2 y(y(\delta + \mu y) - x(1 - \lambda x)). \tag{2.3}$$

It follows from (2.2) and (2.3) that on increasing α or β the vector field of (1.2) in the first quadrant is rotated in positive direction (counterclockwise) only on the outside of the ellipse

$$y(\delta + \mu y) - x(1 - \lambda x) = 0. \tag{2.4}$$

Therefore, to study limit cycle bifurcations of system (1.2), it makes sense together with (1.2) to consider also an auxiliary system (1.3) with a field rotation parameter γ:

$$\Delta_\gamma = P^2 + Q^2 \geq 0. \tag{2.5}$$

Using system (1.3) and applying Perko's results, we will prove the following theorem.

Theorem 2.3. *System (1.2) can have at most two limit cycles.*

Proof. First let us prove that system (1.2) can have at least two limit cycles.

Let the parameters α, β vanish and consider first the quadratic system

$$\begin{aligned}
\dot{x} &= \ \ x(1 - \lambda x - y), \\
\dot{y} &= -y(\delta + \mu y - x).
\end{aligned} \tag{2.6}$$

It is clear that such a system, with two invariant straight lines, cannot have limit cycles at all [5].

Inputting a negative parameter β into this system, the vector field of the cubic system

$$\begin{aligned}
\dot{x} &= \ \ x((1 - \lambda x)(\beta x + 1) - y), \\
\dot{y} &= -y((\delta + \mu y)(\beta x + 1) - x)
\end{aligned} \tag{2.7}$$

will be rotated in negative direction (clockwise) at infinity, the structure and the character of stability of infinite singularities will be changed, and an unstable limit, Γ_1, will appear immediately from infinity in this case [1], [5]. This cycle will surround a stable antisaddle (a node or a focus), A_1, which is in the first quadrant of system (2.7).

Inputting a positive parameter α into system (2.7), the vector field of quartic system (1.2) will be rotated in positive direction (counterclockwise) at infinity, the structure and the character of stability of infinite singularities will be changed again, and a stable limit, Γ_2, surrounding Γ_1 will appear immediately from infinity in this case [1], [5]. On further increasing the parameter α, the limit cycles Γ_1 and Γ_2 combine a semi-stable limit, Γ_{12}, which then disappears in a "trajectory concentration" [1], [5].

On further increasing α, two other singular points, a saddle S and an antisaddle A_2, will appear in the first quadrant in system (1.2). We can fix the parameter α, fixing simultaneously the positions of the finite singularities A_1, S, A_2, and consider system (1.3) with a positive parameter γ which acts like a positive parameter α of (1.2), but on the whole phase plane.

So, consider system (1.3) with a positive parameter γ. On increasing this parameter, the stable nodes A_1 and A_2 becomes first stable foci, then they change the character of their stability, becoming unstable foci. At these Andronov–Hopf bifurcations [1], [5], stable limit cycles will appear from the foci A_1 and A_2. On further increasing γ, the limit cycles will expand and will disappear in small separatrix loops of the saddle S. If these loops are formed simultaneously, we will have a so-called eight-loop separatrix cycle. In this case, a big stable limit surrounding three singular points, A_1, S, and A_2, will appear from the eight-loop separatrix cycle after its destruction, expanding to infinity on increasing γ. If a small loop is formed earlier, for example, around the point A_1 (A_2), then, on increasing γ, a big loop formed by two lower (upper) adjoining separatrices of the saddle S and surrounding the points A_1 and A_2 will appear. After its destruction, we will have simultaneously a big limit cycle surrounding three singular points, A_1, S, A_2, and a small limit cycle surrounding the point A_2 (A_1). Thus, we have proved that (1.2) can have at least two limit cycles [4], [10].

Let us prove now that this system can have at most two limit cycles. The proof is carried out by contradiction applying Catastrophe Theory, see [5], [9]. Consider system (1.3) with three parameters: α, β, and γ (the parameters δ, λ, and μ can be fixed, since they do not generate limit cycles). Suppose that (1.3) has three limit cycles surrounding the only point, A_1, in the first quadrant. Then we get into some domain of the parameters α, β, and γ being restricted by definite conditions on three other parameters, δ, λ, and μ. This domain is bounded by two fold bifurcation surfaces forming a cusp bifurcation surface of multiplicity-three limit cycles in the space of the parameters α, β, and γ [5].

The corresponding maximal one-parameter family of multiplicity-three limit cycles cannot be cyclic, otherwise there will be at least one point corresponding to the limit cycle of multiplicity four (or even higher) in the parameter space. Extending the bifurcation curve of multiplicity-four limit cycles through this point and parameterizing the corresponding maximal one-parameter family of multiplicity-four limit cycles by the field rotation parameter, γ, according to Theorem 2.2, we will obtain two monotonic curves of multiplicity-three and one, respectively, which, by the Wintner–Perko termination principle (Theorem 2.1), terminate either at the point A_1 or on a separatrix cycle surrounding this point. Since we know at least the cyclicity of the singular point which is equal to two [4, 10], we have got a contradiction with the termination principle stating that the multiplicity of limit cycles cannot be higher than the multiplicity (cyclicity) of the singular point in which they terminate. If the maximal one-parameter family of multiplicity-three limit cycles is not cyclic, using the same principle (Theorem 2.1), this again contradicts the cyclicity of A_1 [4], [10] not admitting the multiplicity of limit cycles to be higher than two. This contradiction completes the proof in the case of one singular point in the first quadrant.

Suppose that system (1.3) with three finite singularities, A_1, S, and A_2, has two small limit cycles around, for example, the point A_1 (the case when limit cycles surround the point A_2 is considered in a similar way). Then we get into some domain in the space of the parameters α, β, and γ which is bounded by a fold bifurcation surface of multiplicity-two limit cycles [5].

The corresponding maximal one-parameter family of multiplicity-two limit cycles cannot be cyclic, otherwise there will be at least one point corresponding to the limit cycle of multiplicity three (or even higher) in the parameter space. Extending the bifurcation curve of multiplicity-three limit cycles through this point and parameterizing the corresponding maximal one-parameter family of multiplicity-three limit cycles by the field rotation parameter, γ, according to Theorem 2.2, we will obtain a monotonic curve which, by the Wintner–Perko termination principle (Theorem 2.1), terminates either at the point A_1 or on some separatrix cycle surrounding this point. Since we know at least the cyclicity of the singular point which is equal to one in this case [4], [10], we have got a contradiction with the termination principle (Theorem 2.1). If the maximal one-parameter family of multiplicity-two limit cycles is not cyclic, using the same principle (Theorem 2.1), this again contradicts the cyclicity of A_1 [4, 10] not admitting the multiplicity of limit cycles higher than one. Moreover, it also follows from the termination principle that either an ordinary (small) separatrix loop or a big loop, or an eight-loop cannot have the multiplicity (cyclicity) higher than one in this case. Therefore, according to the same principle, there are no more than one limit cycle in the exterior domain surrounding all three finite singularities, A_1, S, and A_2.

Thus, taking into account all other possibilities for limit cycle bifurcations [4], [10], we conclude that system (1.2) cannot have either a multiplicity-three limit cycle or more than two limit cycles in any configuration. The theorem is proved. $\qquad\square$

References

1. N. N. Bautin and E. A. Leontovich, *Methods and Ways of the Qualitative Analysis of Dynamical Systems in a Plane* (Nauka, Moscow, 1990, in Russian).
2. A. D. Bazykin, *Nonlinear Dynamics of Interacting Populations* (World Scientific, Singapore, 1998).
3. F. Botelho and V. A. Gaiko, Global analysis of planar neural networks, *Nonlinear Anal.* **64**, 1002 (2006).
4. H. W. Broer, V. Naudot, R. Roussarie, K. Saleh and F. O. O. Wagener, Organizing centers in the semi-global analysis of dynamical systems, *Int. J. Appl. Math. Stat.* **12**, 7 (2007).
5. V. A. Gaiko, *Global Bifurcation Theory and Hilbert's Sixteenth Problem* (Kluwer, Boston, 2003).
6. V. A. Gaiko, Limit cycles of quadratic systems, *Nonlinear Anal.* **69**, 2150 (2008).
7. V. A. Gaiko, Limit cycles of Liénard-type dynamical systems, *CUBO Math. J.* **10**, 115 (2008).
8. V. A. Gaiko and W. T. van Horssen, Global bifurcations of limit and separatrix cycles in a generalized Liénard system, *Nonlinear Anal.* **59**, 189 (2004).
9. L. Perko, *Differential Equations and Dynamical Systems* (Springer, New York, 2002).
10. H. Zhu, S. A. Campbell and G. S. K. Wolkowicz, Bifurcation analysis of a predator-prey system with nonmonotonic functional response, *SIAM J. Appl. Math.* **63**, 636 (2002).

V.5. Functional differential and difference equations

Organisers: L. Berezansky, J. Diblik, A. Zafer

It is a well-established principle to model the evolution of physical, biological and economic systems using ordinary differential equations, in which the response of the system depends purely on the current state of the system. However, in many applications the response of the system can be delayed, or depend on the past history of the system in a more complicated way. Dynamical systems which respond in this way are called Functional Differential Equations or Delay Difference Equations. They include: Delay Differential Equations, Integro-Differential Equations, Differential Equations with Distributed delays, Neutral Differential Equations, Dynamic Equations on times scales and corresponding classes of difference equations. The Functional Differential and Difference Equations section organized during the 7th ISAAC Congress in the period 15-17 July 2008 attracted many well-known experts in the field of FDE and its applications.

The session comprised 30 talks on a variety of topics which include: stability and oscillation of FDE, qualitative properties of impulsive and stochastic differential and difference equations, control problems for FDE, boundary value problems for FDE, applications to population dynamics, gene regulatory networks, and models arising in hydrodynamics, equations with state-dependent delays and Volterra equations, inverse problems of the calculus of variations for FDE, invariant manifold for FDE, bifurcation problems, existence and nonexistence of asymptotically periodic solutions.

The following is the list of speakers to the session: Ağacık Zafer, Abdullah Özbekler, István Györi, Mihaly Pituk, Ferenc Hartung, Alexandra Rodkina, Conall Kelly, Zeynep Kayar, Leonid Berezansky, David Reynolds,

Mehmet Ünal, Benzion Shklyar, Irena Rachůnková, Svatoslav Staněk, Milan Tvrdý, Alexander Domoshnitsky, Andrei Shindiapin, Miroslava Růžičková, Martina Langerová, Aleksandr Boichuk, Josef Diblík, Jaromír Baštinec, Ewa Schmeidel, Yakov Goltser, Marcia Federson, Malgorzata Migda, Gabor Kiss, Vladimir Savchin, Andrejs Reinfelds.

Asymptotic analysis of gene regulatory networks with delay effects

Yuriy Nepomnyashchikh[2], Arcady Ponosov[1],

Andrei Shindiapin[2] and Irina Shlykova[1,*]

[1] *Department of Mathematical Sciences and Technology,*
Norwegian University of Life Sciences,
N-1430 Ås, Norway
** E-mail: irinsh@umb.no*

[2] *Department of Mathematics and Informatics,*
Eduardo Mondlane University,
C.P. 257 Maputo, Mozambique

A method of formalizing the analysis of asymptotic properties of solutions to systems of differential equations with distributed time-delays and Boolean-type nonlinearities is offered. Such objects arise in many applications, but of most importance are systems coming from gene regulatory networks (GRN). The dynamics of GRN are governed by sigmoid-type nonlinearities which are close to the step functions. This is due to the fact that genes are only activated if certain concentrations are close to the respective threshold values. The delay effects arise from the time required to complete transcription, translation and diffusion to the place of action of a protein.

We describe an algorithm of localizing stationary points in the presence of delays as well as stability analysis around such points. The basic technical tool consists in replacing step functions with the so-called "logoid functions", combined with a special modification of the well-known "linear chain trick", and investigating the smooth systems thus obtained.

A significant part of this framework is based on asymptotic analysis of singularly perturbed matrices, where we apply Mathematica to be able to derive exact stability criteria. This work is a brief review of the results presented in Ref. 1.

Keywords: Gene regulation; delay equations; stability.

1. The main object

In this paper we study asymptotically stable steady states (stationary points) of the delay system

$$\dot{x}_i = F_i(Z_1, ..., Z_m) - G_i(Z_1, ..., Z_m)x_i$$
$$Z_k = \Sigma(y_{i(k)}, \theta_k, q_k) \tag{1}$$
$$y_i(t) = (\Re_i x_i)(t) \quad (t \geq 0), \quad (i = 1, ..., n; \ k = 1, ..., m).$$

This system describes a gene regulatory network with autoregulation,[2-5] where changes in one or more genes happen slower than in the others, which causes delay effects in some of the variables.

The functions F_i, G_i, which are affine in each Z_k and satisfy

$$F_i(Z_1, ..., Z_m) \geq 0, \quad G_i(Z_1, ..., Z_m) > 0$$

for $0 \leq Z_k \leq 1$, $k = 1, ..., m$, stand for the production rate and the relative degradation rate of the product of gene i, respectively, and x_i denoting the gene product concentration. The input variables y_i endow Eqs. (1) with feedbacks which, in general, are described by nonlinear Volterra ("delay") operators $\Re_i$ depending on the gene concentrations $x_i(t)$.

In this work we assume $\Re_i$ to be integral operators of the form

$$(\Re_i x_i)(t) = {}^0c x_i(t) + \int_{-\infty}^{t} K_i(t - s)x_i(s)ds, \qquad t \geq 0, \quad i = 1, ..., n, \tag{2}$$

where

$$K_i(u) = \sum_{\nu=1}^{p} {}^{\nu}c_i \cdot {}^{\nu}K_i(u),$$
$$ {}^{\nu}K_i(u) = \frac{\alpha_i^{\nu} \cdot u^{\nu-1}}{(\nu-1)!} e^{-\alpha_i u} \quad (i = 1, ..., n).$$

The coefficients ${}^{\nu}c_i$ ($\nu = 0, ..., p$, $i = 1, ..., n$) are real nonnegative numbers satisfying $\sum_{\nu=0}^{p} {}^{\nu}c_i = 1$ for any $i = 1, ..., n$. It is also assumed that $\alpha_i > 0$ for all $i = 1, ..., n$.

The response functions Z_k express the effect of the different transcription factors regulating the expression of the gene. Each $Z_k = Z_k(y_{i(k)})$ ($0 \leq Z_k \leq 1$ for $y_{i(k)} \geq 0$) is a smooth function depending on exactly one input variable $y_{i(k)}$ and on two other parameters: the threshold value θ_k and the steepness value $q_k \geq 0$. A gene may have more than one, or no thresholds. This is expressed in the dependence $i = i(k)$. If different k correspond to the same i, then gene $i(k)$ has more than one threshold. If some i does not correspond to any k, then gene $i(k)$ has no threshold.

In the vicinity of the threshold value θ_k the response function Z_k is increasing almost instantaneously from 0 to 1, i.e. gene $i(k)$ becomes activated very quickly. Thus, the response function is rather close to the step function that has the unit jump at the threshold $y_i = \theta_i$. There are many ways to model response functions. In this paper we will use approximations which were introduced in Ref. 4 and which are based on the so-called "tempered nonlinearities" or logoids.

The logoid function is given by

$$\Sigma(y, \theta, q) := L\left(0.5 + \frac{y - \max\{\theta, \sigma(q)\}}{2\sigma(q)}, \frac{1}{q}\right), \quad (\theta > 0, \ 0 < q < 1),$$

where

$$L(u, p) = \begin{cases} 0 & \text{if} \quad u < 0 \\ 1 & \text{if} \quad u > 1 \\ \frac{u^p}{u^p + (1-u)^p} & \text{if} \quad 0 \leq u \leq 1 \end{cases}$$

and $\sigma(q) \to +0$ if $q \to +0$.

The function coincides with the step function outside a narrow interval around the threshold, and rises continuously from 0 to 1 inside it. Using this kind of function simplifies significantly the stability analysis of the steady states belonging to the discontinuity set of the system in the non-delay model.[2,5] And as we will see, the logoid approach is also efficient in the delay case.

A method to study Eqs. (1) is well-known in the literature, and it is usually called "the linear chain trick" (see Ref. 6). However, a direct application of this "trick" in its standard form is not suitable for our purposes, because we want any Z_i depend on y_i, only. Modifying the linear chain trick we can remove this drawback of the method. Therefore we introduce the new modified variables

$$^1v_i = {}^0c_i x_i + \sum_{\nu=1}^{p} {}^p c_i \cdot {}^p w_i, \quad \text{and} \quad {}^\nu v_i = \sum_{j=1}^{p-\nu+1} {}^{j+\nu-1}c_i \cdot {}^j w_i \quad (\nu = 2, ..., p)$$

based on the variables

$$^\nu w_i(t) = \int_{-\infty}^{t} {}^\nu K_i(t - s) x_i(s)\, ds \quad (\nu = 1, \ldots, p, \ i = 1, \ldots, n)$$

which used in the ordinary LCT method.

Then Eqs. (1) is equivalent to the following system of ordinary differential equations:

$$\begin{aligned} \dot{x}_i(t) &= F_i(Z_1, ..., Z_m) - G_i(Z_1, ..., Z_m) x_i(t) \\ \dot{v}_i(t) &= A_i v_i(t) + \Pi_i(x_i(t)) \quad t > 0 \\ Z_k &= \Sigma(y_{i(k)}, \theta_k, q_k), \quad y_i = {}^1v_i \quad (i = 1, \ldots, n), \end{aligned} \tag{3}$$

where

$$A_i = \begin{pmatrix} -\alpha_i & \alpha_i & 0 & \ldots & 0 \\ 0 & -\alpha_i & \alpha_i & \ldots & 0 \\ 0 & 0 & -\alpha_i & \ldots & 0 \\ \vdots & \vdots & \ddots & \ddots & \vdots \\ 0 & 0 & \ldots & 0 & -\alpha_i \end{pmatrix}, \; v_i = \begin{pmatrix} {}^1v_i \\ {}^2v_i \\ \vdots \\ {}^pv_i \end{pmatrix}$$

and

$$\Pi_i(x_i) := \alpha_i x_i \pi_i + {}^0c_i f_i(Z, x_i)$$

with

$$\pi_i := \begin{pmatrix} {}^0c_i + {}^1c_i \\ {}^2c_i \\ \vdots \\ {}^pc_i \end{pmatrix}, \quad f_i(Z, x_i) := \begin{pmatrix} F_i(Z) - G_i(Z)x_i \\ 0 \\ \vdots \\ 0 \end{pmatrix}.$$

2. Stability analysis and the localization principle

It is easy to define stationary points for Eqs. (3) if $Z_k = \Sigma(y_{i(k)}, \theta_k, q_k)$ are all smooth ($q_k > 0$). However, in this case the stability analysis and computer simulations may be cumbersome and time-consuming. To simplify the model, one uses the step functions $Z_k = \Sigma(y_{i(k)}, \theta_k, 0)$ and the corresponding limit system.

Let us give some general notations and definitions related to geometric properties of Eqs. (3) in the limit case ($q_k = 0, \; k = 1, ..., m$). Assume that

- $M := \{1, ..., m\}, \quad J := \{1, ..., j\}, \quad N := \{1, ..., n\}, \quad n \leq j, \; m \leq j$, (i. e. $N \subset J, M \subset J$);
- $R := M - S$ for a given $S \subset M$;
- $a_R := (a_r)_{r \in R}, \; a_S := (a_s)_{s \in S}$.

Suppose that all $q_k = 0$. Then the right-hand side of Eqs. (3) can be discontinuous, namely, if one or several $y_{i(k)}$ assume their threshold values, i.e. $y_{i(k)} = \theta_k$. In this case we will call $y_{i(k)}$ a singular variable. We associate a Boolean variable B_k to each Z_k by $B_k = 0$ if $y_{i(k)} < \theta_k$ and $B_k = 1$ if $y_{i(k)} > \theta_k$, the variable $y_{i(k)}$ is called regular.

Suppose that given a subset $S \subset M, S \neq \emptyset$ and a Boolean vector $B_R \in \{0, 1\}^R$, where $R = M - S$, the set $SD(\theta_S, B_R)$, which consists of all $(x_i, v_i) \in R^J$, where $B_r = Z_r(y_{i(r)}) \; (r \in R)$ and $y_{i(s)} = \theta_s \; (s \in S)$, is called a singular domain.

The model with the step functions $Z_k = \Sigma(y_{i(k)}, \theta_k, 0)$ considerably simplifies the analysis of the original system with $Z_k = \Sigma(y_{i(k)}, \theta_k, q_k)$ $(q_k > 0)$, which is smooth. However, if a potential stationary point in the limit model belongs to the discontinuity set, then we still need the smooth system to define the very notion of a stationary point.

Definition 2.1. A point $\hat{P}$ is called a stationary point for Eqs. (3) with $Z_k = \Sigma(y_{i(k)}, \theta_k, 0)$ $(k \in M)$ if for any set of functions $Z_k = \Sigma(y_{i(k)}, \theta_k, q_k)$ $(k \in M)$ there exist a number $\varepsilon > 0$ and points $P(q)$, $q = (q_1, ..., q_m)$, $q_k \in (0, \varepsilon)$ $(k \in M)$ such that

- $P(q)$ is a stationary point for System (3) with $Z_k = \Sigma(y_{i(k)}, \theta_k, q_k)$ $(k \in M)$;
- $P(q) \to \hat{P}$ as $q \to 0$ (i.e. to the zero vector).

If the limit point $\hat{P}$ does not belong to the discontinuity set of Eqs. (3), i.e. if $y_{i(k)} \neq \theta_k$ $(k \in M)$, then $\hat{P}$ simply becomes a conventional stationary point for the limit system and called a regular stationary point.

The case where some of the coordinates coincide with the respective thresholds is more involved. The crucial role in this case will be played by the Jacoby matrix $\frac{\partial}{\partial Z_S} F_S(Z) - \frac{\partial}{\partial Z_S} G_S(Z) y_{i(S)}$. The entry in the s-th row and the σ-th column of this matrix amounts $\frac{\partial}{\partial Z_\sigma} F_{i(s)}(Z) - \frac{\partial}{\partial Z_\sigma} G_{i(s)}(Z) y_{i(s)}$. In other words,

$$
\begin{aligned}
J_S(Z_S, B_R, y_{i(S)}) &= \frac{\partial}{\partial Z_S} F_S(Z_S, B_R) - \frac{\partial}{\partial Z_S} G_S(Z_S, B_R) y_{i(S)} \\
&= \left[\frac{\partial}{\partial Z_\sigma} F_{i(s)}(Z_S, B_R) - \frac{\partial}{\partial Z_\sigma} G_{i(s)}(Z_S, B_R) y_{i(s)} \right]_{s,\sigma \in S}.
\end{aligned}
\tag{4}
$$

Below we provide a sufficient condition for $\hat{P}$ to be a singular stationary point.[7]

Theorem 2.1. *Assume that for some $S \subset M$ the system of algebraic equations*

$$
\begin{aligned}
F_{i(S)}(Z_S, B_R) - G_{i(S)}(Z_S, B_R)\theta_{i(S)} &= 0, \\
F_{i(R)}(Z_S, B_R) - G_{i(R)}(Z_S, B_R)y_{i(R)} &= 0
\end{aligned}
$$

with the constraints $0 < Z_s < 1$ $(s \in S)$ and $Z_r(y_{i(r)}) = B_r$ $(r \in R)$ has a solution $\hat{Z}_S := (\hat{Z}_s)_{s \in S}$, $\hat{y}_{i(R)} := (\hat{y}_{i(r)})_{r \in R}$, which, in addition, satisfies $\det J_S(\hat{Z}_S, B_R, \theta_S) \neq 0$, where $J_S(\hat{Z}_S, B_R, \theta_S)$ is given by (4).

Then there exists a stationary point $\hat{P} \in SD(\theta_S, B_R)$ for Eqs. (3). This point is independent of the choice of the delay operators $\Re_i$ given by (2).

In the non-delay case any regular stationary point is always asymptotically stable as soon as it exists. This is due to the assumptions $G_i > 0$. Stability of the matrix $J_S(Z_S, B_R, \theta_S)$ provides the asymptotic stability of singular stationary points (see e.g. Ref. 2 for delays).

Example 2.1. We consider the equation

$$
\begin{aligned}
\dot{x}(t) &= F(Z) - G(Z)x(t) \\
Z &= \Sigma(y, \theta, q) \\
y(t) &= (\Re x)(t) \quad (t \geq 0)
\end{aligned}
$$

with the delay operator given by

$$
(\Re x)(t) = {}^0 c x(t) + \int_{-\infty}^{t} K(t-s)x(s)ds, \qquad t \geq 0,
$$

where $K(u) = {}^1 c \cdot {}^1 K(u) + {}^2 c \cdot {}^2 K(u)$, ${}^1 K(u) = \alpha e^{-\alpha u}$ and ${}^2 K(u) = \alpha^2 e^{-\alpha u}$, $(\alpha > 0)$, ${}^\nu c \geq 0$ $(\nu = 0, 1, 2)$, ${}^0 c + {}^1 c + {}^2 c = 1$ and $Z = \Sigma(y, \theta, q)$ is the logoid function.

Applying the generalized linear chain trick, we arrive at the system

$$
\begin{aligned}
\dot{x} &= F(Z) - G(Z)x \\
{}^1\dot{v} &= {}^0 c\,(F(Z) - G(Z)x) + \alpha x({}^0 c + {}^1 c) - \alpha \cdot {}^1 v + \alpha \cdot {}^2 v \\
{}^2\dot{v} &= \alpha \cdot {}^2 c\, x - \alpha \cdot {}^2 v,
\end{aligned}
$$

where $Z = \Sigma(y, \theta, q)$. Assume that the equation $F(Z) - G(Z)\theta = 0$ have a solution $\hat{Z}$, $(0 < \hat{Z} < 1)$ then

A. For ${}^0 c > 0$ we have that the point $\hat{P}(\hat{x}, {}^1\hat{v}, {}^2\hat{v})$, where $\hat{x} = {}^1\hat{v} = \theta$ and ${}^2\hat{v} = {}^2 c\,\theta$, will be asymptotically stable if $J < 0$, and unstable if $J > 0$.

B. If ${}^0 c = 0$ then the point $\hat{P}(\hat{x}, {}^1\hat{v}, {}^2\hat{v})$, where $\hat{x} = {}^1\hat{v} = \theta$ and ${}^2\hat{v} = {}^2 c\,\theta$, has the following properties

(1) If $J > 0$, then $\hat{P}$ is unstable.
(2) If $J < 0$, ${}^1 c = 0$, then $\hat{P}$ is unstable.
(3) If $J < 0$, ${}^1 c \neq 0$ and $G(\hat{Z}) < \alpha({}^1 c)^{-1}(1 - 2\,{}^1 c)$, then $\hat{P}$ is unstable.
(4) If $J < 0$, ${}^1 c \neq 0$ and $G(\hat{Z}) > \alpha({}^1 c)^{-1}(1 - 2\,{}^1 c)$, then $\hat{P}$ is asymptotically stable.

Here $J = F'(Z) - G'(Z)\theta$ is independent of Z (as both F and G are affine).

Corollary 2.1. *The analytical formulas for more sophisticated delay operators can be obtained with the help of software Mathematica.*

A very important advantage of the logoid nonlinearities is the localization principle. Roughly speaking we may remove all regular variables in the stability analysis, because they did not influence local properties of solutions around stationary points. This principle is of particular importance for delay systems (which are non-local). On the other hand, the localization principle helps to simplify both notation and proofs.

Let $S \subset M$ and B_R be fixed. We are looking for singular stationary points in the singular domain $SD(\theta_S, B_R)$. Consider the reduced system

$$\begin{aligned}
\dot{\mathbf{x}}_s &= \mathbf{F}_s(\mathbf{Z}_s) - \mathbf{G}_s(\mathbf{Z}_s)\mathbf{x}_s \\
\mathbf{Z}_s &= \Sigma(\mathbf{y}_s, \theta_s, q_s) \\
\mathbf{y}_s(t) &= (\Re_s x_s)(t), \quad (s \in S),
\end{aligned} \tag{5}$$

where $\mathbf{F}_s(\mathbf{Z}_s) = F_{i(s)}(Z_s, B_R)$, $\mathbf{G}_s(\mathbf{Z}_s) = G_{i(s)}(Z_s, B_R)$.

Theorem 2.2. *(localization principle) Suppose that the conditions of Theorem 2.1 are fulfilled. Then Eqs. (5) has an isolated stationary point $\hat{\mathbf{P}}$. The point $\hat{\mathbf{P}}$ is asymptotically stable (unstable) iff $\hat{P}$ is asymptotically stable (unstable) for Eqs. (1).*

Acknowledgments

The present study was partially supported by the National Programme for Research for Functional Genomics in Norway (FUGE) in the Research Council of Norway, by the Norwegian Council of Universities' Committee for Development Research and Education (NUFU), grant no. PRO 06/02.

References

1. I. SHLYKOVA, A. PONOSOV, A. SHINDIAPIN A., AND YU. NEPOMNYASHCHIKH, *A general framework for stability analysis of gene regulatory networks with delay*, Electron. J. Diff. Eqns., Vol. 2008(2008), No. 104, 1-36.
2. E. PLAHTE, T. MESTL, AND S. W. OMHOLT, *A methodological basis for description and analysis of systems with complex switch-like interactions*, J. Math. Biol., v. 36 (1998), 321-348.
3. A. PONOSOV, *Gene regulatory networks and delay differential equations.* Special issue of Electronic J. Diff. Eq., v. 12 (2004), pp. 117-141.
4. E. PLAHTE, T. MESTL, AND S. W. OMHOLT, *Global analysis of steady points for systems of differential equations with sigmoid interactions*, Dynam. Stabil. Syst., v. 9, no. 4 (1994), 275-291.
5. T. MESTL, E. PLAHTE, AND S. W. OMHOLT, *A mathematical framework for describing and analysing gene regulatory networks*, J. Theor. Biol., v. 176 (1995), 291-300.
6. N. MCDONALD. *Time lags in biological models*, Lect. Notes in Biomathematics, Springer-Verlag, Berlin-Heidelberg-New York, 1978, 217 p.
7. L. GLASS AND S. A. KAUFFMAN . *The logical analysis of continuous, nonlinear biochemical control networks*, J. Theor. Biol., v. 39 (1973), 103-129.

On zero controllability of evolution equations by scalar control

B. Shklyar

*Department of Applied Mathematics, Holon Institute of Technology,
Holon, 58102, Israel
E-mail: shk_b@hit.ac.il
www.hit.ac.il/staff/benzionS*

The exact controllability to the origin for linear evolution control equation is considered. The problem is investigated by its transformation to infinite linear moment problem. Some examples are considered.

Keywords: Null controllability; evolution equation; linear moment problem.

1. Introduction and problem statement

Let X, U be complex Hilbert spaces, and let A be infinitesimal generator of strongly continuous C_0-semigroups $S(t)$ in X.[6,7] Consider the abstract evolution control equation[6,7]

$$\dot{x}(t) = Ax(t) + Bu(t), x(0) = x_0, \ 0 \le t < +\infty, \tag{1}$$

where $x(t), \ x_0 \in X, u(t) \in U, \ B : U \to X$ is a linear possibly unbounded operator, $W \subset X \subset V$ are Hilbert spaces with continuous dense injections, where $W = D(A)$ equipped with graphic norm, $V = W^*$, the operator B is a bounded operator from U to V (see more details in[8,9]).

It is well-known that[8,9]), etc. :

• for each $t \ge 0$ the operator $S(t)$ has an unique continuous extension $\mathcal{S}(t)$ on the space V and the family of operators $\mathcal{S}(t) : V \to V$ is the semigroup in the class C_0 with respect to the norm of V and the corresponding infinitesimal generator $\mathcal{A}$ of the semigroup $\mathcal{S}(t)$ is the closed dense extension of the operator A on the space V with domain $D(\mathcal{A}) = X$;

• the sets of eigenvalues and of generalized eigenvectors of operators $\mathcal{A}, \mathcal{A}^*$ and A, A^* are the same.

Let $x(t, x_0, u(\cdot))$ be a mild solution of equation (1) with initial condition $x(0) = x_0$.

Definition 1.1. Equation (1) is said to be exact null-controllable on $[0, t_1]$ by controls vanishing after time moment t_2, if for each $x_0 \in X$ there exists a control $u(\cdot) \in L_2([0, t_2], U), u(t) = 0$ a.e. on $[t_2, +\infty)$ such that $x(t_1, x_0, u(\cdot)) = 0$.

The goal of this paper is to establish necessary and sufficient conditions of exact null-controllability for linear evolution control equations with unbounded input operator by transformation of exact null-controllability problem (controllability to the origin) to linear infinite moment problem [a].

2. Main results

The assumptions on A are listed below.

(1) The operators A has purely point spectrum σ with no finite limit points. All the eigenvalues of A have finite multiplicities.
(2) There exists $T \geq 0$ such that all mild solutions of the equation $\dot{x}(t) = Ax(t)$ are expanded in a series of generalized eigenvectors of the operator A converging uniformly for any $t \in [T_1, T_2], T < T_1 < T_2$.

For the sake of simplicity we consider the following:

(1) The operator A has all the eigenvalues with multiplicity 1.
(2) $U = \mathbb{R}$ (one input case). It means that the possibly unbounded operator $B : U \to \mathbb{R}$ is defined by an element $b \in V$, i.e. equation (1) can be written in the form

$$\dot{x}(t) = Ax(t) + bu(t), x(0) = x_0, b \in V, \ 0 \leq t < +\infty. \qquad (2)$$

Let the eigenvalues $\lambda_j \in \sigma, j = 1, 2, \ldots$ of the operator A be enumerated in the order of non-decreasing of their absolute values, and let $\varphi_j, \psi_j, j = 1, 2, \ldots$, be eigenvectors of the operator A and the adjoint operator A^* respectively. It is well-known, that $(\varphi_k, \psi_j) = \delta_{kj}, \ j, k = 1, 2 \ldots,$ where $\delta_{kj}, \ j, k = 1, 2 \ldots$ is the Kroneker delta.
Denote: $x_j(t) = (x(t, x_0, u(\cdot)), \psi_j), \ x_{0j} = (x_0, \psi_j), \ b_j = (b, \psi_j), \ j = 1, 2, \ldots,$. All scalar products here are correctly defined, because $\psi_j \in W, \ b \in V = W^*$.

Theorem 2.1. *For equation (1) to be exact null-controllable on* $[0, t_1],$ *$t_1 > T$, by controls vanishing after time moment $t_1 - T$, it is necessary and*

---

[a]Proofs of results presented in the paper are omitted because of restrictions of the size of a paper for Proceedings. They will be published in the full version of the paper.

sufficient that for any $x_0 \in X$ *the infinite moment problem*

$$x_{0j} = -\int_0^{t_1-T} e^{-\lambda_j \tau} b_j u\left(\tau\right) d\tau, \ \ j = 1, 2, ... \tag{3}$$

has a solution $u\left(\cdot\right) \in L_2\left[0, t_1 - T\right]$.

2.1. *Solution of moment problem (3)*

The solvability of moment problem (3) for each $x_0 \in X$ essentially depends on properties of the eigenvalues λ_j, $\ j = 1, 2, ...,$. If the sequence of exponents $\left\{e^{-\lambda_j t} b_j, j = 1, 2, ..., \right\}$ forms a Riesz basic in their linear span, then the moment problem

$$c_j = -\int_0^{t_1-T} e^{-\lambda_j \tau} b_j u\left(\tau\right) d\tau, \ \ j = 1, 2, ... \tag{4}$$

is solvable if and only if $\sum_{j=1}^{\infty} |c_j|^2 < \infty$.

There are very large number of papers and books devoted to conditions for sequence of exponents to be a Riesz basic. All these conditions can be used for conditions of null-controllability of equation (1). They are very useful for the investigation of the null-controllability of hyperbolic partial control equations and functional differential control systems of neutral type.

However moment problem (4) may also be solvable when the sequence $\left\{e^{-\lambda_n t} b_n, n = 1, 2, ..., \right\}$ doesn't form a Riesz basic in $L_2\left[0, t_1 - T\right]$. Below we will try to find more extended controllability conditions which are applicable for cases when the sequence $\left\{e^{-\lambda_n t} b_n, n = 1, 2, ..., \right\}$ doesn't form a Riesz basic in $L_2\left[0, t_1 - T\right]$.

Definition 2.1. The sequence $\{x_j \in X, j = 1, 2, ..., \}$ is said to be minimal, if there no element of the sequence belonging to the closure of the linear span of others. By other words,

$$x_j \notin \overline{\mathrm{span}}\left\{x_k \in X, k = 1, 2, ..., k \neq j\right\}.$$

The investigation of the controllability problem defined above is based on the following result of Boas[4] (see also[2,11]).

Theorem 2.2 (Boas, 1941). *Let* $x_j \in X, j = 1, 2, ...,$. *The linear moment problem* $c_j = (x_j, g), j = 1, 2, ...$ *has a solution* $g \in X$ *for each square summable sequence* $\{c_j, j = 1, 2, ...\}$ *if and only if there exists a positive constant* γ *such that all the inequalities*

$$\gamma \sum_{k=1}^{n} |c_k|^2 \leq \left\|\sum_{j=1}^{n} c_j x_j\right\|^2, n = 1, 2, ..., \tag{5}$$

are valid.

Let $\{x_j \in X, j = 1, 2, ..., \}$ be a sequence of elements of X, and let $G_n = \{(x_i, x_j), i, j = 1, 2, ..., n\}$ be the Gram matrix of n first elements $\{x_1, ..., x_n\}$ of above sequence. Denote by $\gamma_n^{\min}$ the minimal eigenvalue of the $n \times n$-matrix G_n. Each minimal sequence $\{x_j \in X, j = 1, 2, ..., \}$ is linear independent, so any first n elements $\{x_1, ..., x_n\}, n = 1, 2, ...,$ of this sequence are linear independent. Hence $\gamma_n^{\min} > 0, \forall n = 1, 2, ...,..$ It is easily to show that the sequence $\{\gamma_n^{\min}, n = 1, 2, ..., \}$ decreases , so there exists $\lim_{n \to \infty} \gamma_n^{\min} \geq 0$.

Definition 2.2. The sequence $\{x_j \in X, j = 1, 2, ..., \}$ is said to be strongly minimal, if $\gamma^{\min} = \lim_{n \to \infty} \gamma_n^{\min} > 0$.

It is well-known that for Hermitian $n \times n$-matrix $G_n = \{(x_j, x_k), j, k = 1, 2, ..., n\}$

$$\gamma_n^{\min} \sum_{k=1}^{n} |c_k|^2 \leq \sum_{j=1}^{n} \sum_{k=1}^{n} c_j (x_j, x_k) \overline{c_k}, n = 1, 2, ..., . \tag{6}$$

From the well-known formula $\sum_{j=1}^{m} \sum_{k=1}^{m} c_j (x_j, x_k) \overline{c_k} = \left\| \sum_{j=1}^{m} c_j x_j \right\|^2$, formula (6) and the inequality $\gamma_n^{\min} \geq \gamma^{\min} > 0$ it follows that

$$\gamma^{\min} \sum_{k=1}^{n} |c_k|^2 \leq \left\| \sum_{j=1}^{n} c_j x_j \right\|^2, n = 1, 2, ..., . \tag{7}$$

Hence the above result of Boas[4] can be reformulated as follows.

Theorem 2.3. *The linear moment problem $c_j = (x_j, g), j = 1, 2, ...$ has a solution $g \in X$ for each square summable sequence $\{c_j, j = 1, 2, ...\}$ if and only if the sequence $\{x_j, j = 1, 2, ..., \}$ is strongly minimal.*

3. Solution of the exact null-controllability problem

Theorem 3.1. *For equation (1) to be exact null-controllable on $[0, t_1]$, $t_1 > T$, by controls vanishing after time moment $t_1 - T$, it is necessary, that the sequence $\{e^{-\lambda_j \tau} b_j, t \in [0, t_1 - T], j = 1, 2, ..., \}$ is minimal, and sufficient , that:*

- *the sequence $\{e^{-\lambda_j \tau} b_j, t \in [0, t_1 - T], j = 1, 2, ...\}$ is strongly minimal;*
- $\sum_{j=1}^{\infty} |(x_0, \psi_j)|^2 < +\infty, \forall x_0 \in X.$

If the sequence of eigenvectors of operator A forms a Riesz basic in their linear span, then

for equation (1) to be exact null-controllable on $[0, t_1], t_1 > T$, by controls vanishing after time moment $t_1 - T$, it is necessary and sufficient, that the sequence sequence $\left\{ e^{-\lambda_j t} b_j, t \in [0, t_1 - T], \ j = 1, 2, ... \right\}$ is strongly minimal.

Obviously, the condition $b_j \neq 0, j = 1, 2, ...,$ is the necessary for the solvability of the moment problem (2).

Lemma 3.1. *If the sequence*

$$\left\{ e^{-\lambda_j t}, t \in [0, t_1 - T], \ j = 1, 2, ... \right\} \tag{8}$$

is strongly minimal and

$$\inf_{n \in \mathbb{N}} |b_n| = \beta > 0 \tag{9}$$

holds, then the sequence $\left\{ e^{-\lambda_j t} b_j, t \in [0, t_1 - T], \ j = 1, 2, ... \right\}$ is also strongly minimal.

The problem of expansion into a series of eigenvectors of the operator A producing a Riesz basic in their linear span is widely investigated in the literature In this case one can set $T = 0$, so Theorem 3.1 and Lemma 3.1 can be proven with $T = 0$.

Examples of strongly minimal sequences. Of course all sequences producing Riesz basic in their linear span are strongly minimal.

One can prove[b] that the sequence $\left\{ e^{n^2 \pi^2 t}, n = 1, 2, ..., t \in [0, t_1] \right\}$ is strongly minimal for any $t_1 > 0$, but in accordance with results of[1] this sequence is not Riesz basic in its linear span.

4. Approximation Theorems

The condition $\lim_{n \to \infty} \lambda_n^{\min} > 0$ can be checked by numerical methods. The problem appears to be rather difficult in general for theoretical puprposes. However there are sequences for which the validity of above inequality can be easily established. For example, every orthonormal sequence is strongly minimal. Below we will show that if the sequence $\{ y_j \in X, j = 1, 2, ... \}$ can be approximated in the some sense by strongly minimal sequence $\{ x_j \in X, j = 1, 2, ... \}$, then it is also strongly minimal.

---

[b]The proof is omitted here.

Theorem 4.1. *If the sequence $\{x_j \in X, j = 1, 2, ...\}$ is strongly minimal, and the sequence $\{y_j \in X, j = 1, 2, ...\}$ is such that*

$$\left\| \sum_{j=1}^{n} c_j \left(y_j - x_j \right) \right\| \leq q \left\| \sum_{j=1}^{n} c_j x_j, \right\| , n = 1, 2, ... , \tag{10}$$

for any sequence $\{c_j, j = 1, 2, ...\}$ of complex numbers, where q is a constant, $0 < q < 1$, then the sequence $\{y_j \in X, j = 1, 2, ...\}$ also is strongly minimal.

4.1. *Example*

Let $X = l_2$ be the Hilbert space of square summable sequences. Consider the evolution system

$$\begin{cases} \dot{x}_k \left(t \right) = \lambda_k x_k \left(t \right) + u \left(t \right), \ k = 1, 2, ..., 0 < t < t_1, \\ \quad x_k \left(0 \right) = x_{k0}, \qquad k = 1, 2, ..., \end{cases} \tag{11}$$

where $u\left(t \right), 0 < t < t_1$ is a scalar control function,
$$x \left(t \right) = \{x_k \left(t \right), k = 1, 2, ..., \}, \{x_{k0}, k = 1, 2, ..., \} \in l^2,$$
$$\lambda_k \in \{z \in \mathbb{C} : |\operatorname{Re} z| \leq \gamma\}, k = 1, 2, ..., .$$

Definition 4.1. Equation (11) is said to be exact null-controllable on $[0, t_1]$ by controls vanishing after time moment t_2, if for each
$x_0 \left(\cdot \right) = \{x_{k0}, k = 1, 2, ..., \} \in l_2$
there exists a control $u\left(\cdot \right) \in L_2 \left[0, t_2 \right], u\left(t \right) = 0$ a.e. on $[t_2, +\infty)$ such that
$x_k \left(t \right) \equiv 0, \ k = 1, 2, ..., \forall t \geq t_1.$

System (11) can be written in the form of (1), where $x\left(t \right) = \{x_k \left(t \right), k = 1, 2, ..., \} \in l^2, u\left(\cdot \right) \in L_2 \left[0, t_1 \right]$;
the self-adjoint operator $A : l_2 \to l_2$ is defined for $x = \{x_k, k = 1, 2, ..., \} \in l_2$ by $Ax = \{\lambda_k x_k, k = 1, 2, ...\}$ with domain $D\left(A \right) = \{x \in l_2 : Ax \in l_2\}$, and the unbounded operator B is defined by $Bu = bu, u \in \mathbb{R}$, where $b = \{1, 1, ..., 1, ...\} \notin l_2$. One can show that all the assumptions imposed on equation (1) are fulfilled for equation (11) with $T = 0$.

Obviously, the numbers $\lambda_k, k = 1, 2, ...$, are eigenvalues of the operator A defined above; the sequences $e_k = \left\{ \underbrace{0, ..., 0, 1, 0, ..., 0, ...,}_{1 \text{ on } k\text{-th place}} \right\}$ are correspond-

ing eigenvectors, producing the Riesz basic of l_2, so $b_j = 1, j = 1, 2, ..., .$

Together with system (11) consider the other evolution system

$$\begin{cases} \dot{x}_k \left(t \right) = \mu_k x_k \left(t \right) + b_k u \left(t \right), \ n = 1, 2, ..., 0 < t < t_1, \\ \quad x_k \left(0 \right) = x_{k0}, k = 1, 2, ..., \quad n = 1, 2, ..., \end{cases} \tag{12}$$

where $\mu_k = \lambda_k + O\left(\frac{1}{k}\right), k = 1, 2, ..., \inf\limits_{k=1,2,...} \{|b_k|\} > 0$.

Proposition 4.1. *If system (11) is exact null-controllable on $[0, t_1]$ by controls vanishing after time moment t_2, then the same is valid for system (12).*

This proposition and results of J. Burns, T. Herdmann and H. Stech[5] provides a possibility to use asymptotical formulas for zeros of quasipolynomials[3] in order to prove null-controllability conditions for linear time-invariant functional-differential neutral control systems having only neutral chains of eigenvalues.

References

1. S. Avdonin and A. Ivanov, it Families of exponentials. The method of Moments in Controllability Problems for Distributed parameter Systems (Cambridge University Press,Cambridge, UK, 1995).
2. N. Bari, *Uchen. Zap. Mosk. Univ.*, **148**, Nat, **4** (1951).
3. R. Bellmann and K. Cooke, *Differential-Difference Equations,* (New York Academic Press London, 1963).
4. R. Boas, *Amer. J. Math.*, **63** (1941).
5. J. Burns, T. Herdmann and H. Stech, *SIAM J. Math. Anal.*, **14** (1983).
6. E. Hille and R. Philips, *Functional Analysis and Semi-Groups* (AMS, 1957).
7. M. Krein, *Linear Differential Equations in Banach Spaces* (Moscow, Nauka Publisher, 1967 (in Russian)).
8. D. Salamon, *Trans. Amer. Math. Soc.*, **300** (1987).
9. G. Weiss, *SIAM J. Contr. and Optimiz.*, **27** (1989).
10. R. Young, *An Introduction to Nonharmonic Analysis* (Academic Press, New York, 1980).
11. R. Young, *Proceedings of AMS,* **126** (1998).

V.6. Mathematical biology

Organisers: R. Gilbert

New computer technologies for construction and numerical analysis of mathematical models of molecular genetic systems

I.R. Akberdin[2], S.I. Fadeev[1], I.A. Gainova[1]*, F.V. Kazantsev[2], V.K. Korolev[1],

V.A. Likhoshvai[2] and A.E. Medvedev[3]

[1] *Sobolev Institute of Mathematics,*
Siberian Branch of the Russian Academy of Sciences,
Koptyug Avenue, 4, Novosibirsk, 630090, Russia
E-mails: gajnova@math.nsc.ru; fadeev@math.nsc.ru; korolev@math.nsc.ru

[2] *Institute of Cytology and Genetics,*
Siberian Branch of the Russian Academy of Sciences,
Lavrentyev Avenue, 10, Novosibirsk, 630090, Russia
E-mails: akberdin@bionet.nsc.ru; kazfdr@bionet.nsc.ru; likho@bionet.nsc.ru

[3] *Khristianovich Institute of Theoretical and Applied Mechanics,*
Siberian Branch of the Russian Academy of Sciences,
Institutskaya Street, 4/1, Novosibirsk, 630090, Russia
E-mail: medvedev@itam.nsc.ru

We have represented a new computer system for generation and analysis of mathematical models describing the dynamics of the molecular genetic systems functioning in pro- and eukaryotes. The system consists of two program modules: MGSgenerator and STEP+.

Keywords: Molecular-genetic systems (MGS); numerical analysis; mathematical model; nonlinear system; program module.

1. Introduction

Mathematical models of molecular-genetic systems are based on the information about the structural and functional organization of gene networks and their dynamic properties disseminated over hundreds and thousands of scientific papers. The problem arises of data comparison and analysis of non-uniformed experimental data, analysis of cause-and-effect relations between molecular structure, dynamics and phenotypic features of MGS, and software development for automatic generation of mathematical models, storage of creating models in the database and their numerical analysis.

In the context of solving some of the above mentioned problems, a group of researchers from the Institute of Cytology and Genetics and Sobolev Institute of Mathematics is developing an integrated computer system that does not only render automatically the process of reconstruction of mathematical models based on the structural and functional organization of gene networks but also implements original approaches and algorithms to modeling and studying molecular-genetic systems. Part of this system is the computer system consisting of two interlinked program modules (MGSgenerator [1] and STEP+ [2]), presented in the given paper.

2. Description of the MGSgenerator program module

The MGSgenerator program module generates automatically a mathematical model from the structural model of a gene network extracted from the GeneNet database [3] and delivers it in an appropriate computer format. The model can be exported to SiBML [4], SBML [5], and Step+ formats.

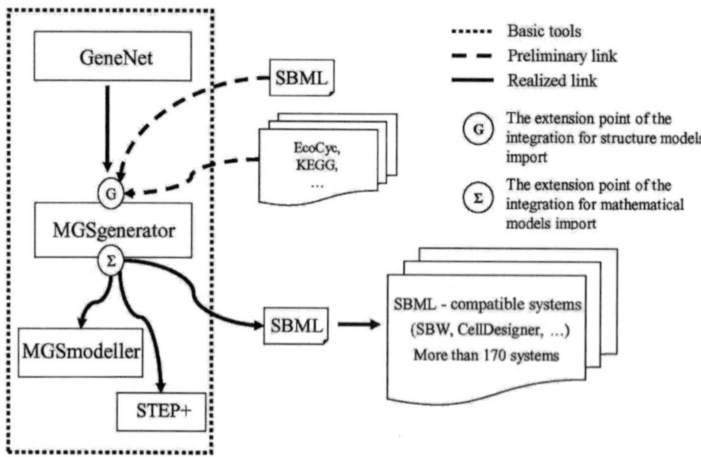

Fig. 1.   Diagram of the MGSgenerator software.

The construction of a mathematical model is based on the generalized chemical-kinetic method [6] that allows describing the molecular-genetic processes linking the components of the gene network, by a nonlinear system of differential equations. The MGSgenerator matches every biological process to some template function and, based on this information, generates a mathematical model. To describe the biological processes, template models of three various levels of complexity are used that form the functions determining the right-hand sides of the system of ordinary differential equations.

So, the processes of replication, transcription, and translation with ignored regulatory interactions are described by the simple equation

$$V = xk_0 \,,$$

where V is the rate of the process, x is the number of genes (concentrations), mRNA in a cell, etc., and k_0 is the reaction rate constant.

If molecular-genetic regulation of these processes is taken into account, another template model is used:

$$V = xk_0 \left[\frac{\delta + \sum_i \left(\frac{a_i}{k_{1,ai}} \right)^{h_i}}{1 + \sum_i \left(\frac{a_i}{k_{2,ai}} \right)^{hd_i} + \sum_i \left(\frac{r_i}{k_{ri}} \right)^{g_i}} \right] .$$

Here x is a chosen element (gene/RNA/enzyme), a_i is the activator concentration, r_i is the inhibitor concentration, k_0 is the reaction rate constant, k_{ai} is the activator constant, k_{ri} is the inhibitor constant, δ is the basal activity, and h_i, g_i and hd_i are the Hill coefficients. By default, the interaction of regulators with each other is assumed to be competitive, and their interaction with the chosen element (x) is assumed to be non-competitive.

The mathematical description of enzymatic synthesis with allowance for molecular-genetic regulation and reaction reversibility is even more complicated:

$$V = x \frac{k_{s,0} \prod_i \frac{S_i}{k_{mS,i}} - k_{p,0} \prod_i \frac{P_i}{k_{mP,i}}}{\prod_i \left(1 + \frac{S_i}{k_{mS,i}} + \frac{P_i}{k_{mP,i}} \right)} \times \frac{\delta + \sum_i \left(\frac{a_i}{k_{1,ai}} \right)^{h_i}}{1 + \sum_i \left(\frac{a_i}{k_{2,ai}} \right)^{hd_i} + \sum_i \left(\frac{r_i}{k_{ri}} \right)^{g_i}} .$$

Here x is the enzyme (regulator), S is the substrate (input), P is the product (output), a_i is the activator concentration, r_i is the inhibitor concentration, k_0 is the reversal constant, $k_{s,0}$ is the constant of the direct reaction, $k_{p,0}$ is the constant of the reverse reaction, k_{mS} is the Michaelis constant for the substrate, k_{mP} is the Michaelis constant for the product, k_{ai} is the activator constant, k_{ri} is the inhibitor constant, δ is the basal activity, and h_i, g_i and hd_i are the Hill coefficients.

The MGSgenerator software was created with the use of plug-in technology and has a module architecture. Such a structure of the generator allows for the attachment of new data sources, export models from other databases, such as KEGG [7], EcoCyc [8], etc., change or add to existing program modules, and integrate the given module into other computer systems.

3. Description of the STEP+ program module

The STEP+ program module allows a comprehensive study of the mathematical model generated by the MGSgenerator in the STEP+ input format in the form of an autonomous system composed of n differential equations with the vector of parameters p:

$$\dot{x} = f(x, p). \tag{1}$$

Here $f(x, p)$ is a sufficiently smooth vector-function of the vector arguments $x \in R^n$ and $p \in R^m$ in the domain of its definition.

The STEP+ program module includes algorithms for numerical analysis of the solution of the autonomous system, depending on the model parameters: multistep Gear method with a variable order of accuracy for integration of stiff systems [9]; method of solution continuation with respect to a parameter for constructing the stationary solutions diagram of the autonomous system (1) [10, 11]; Godunov-Bulgakov numerical criterion for determining the guaranteed asymptotic stability of stationary solutions, depending on the model parameter $\alpha \in p$, [12].

3.1. *Continuation method and parametrization*

The continuation method with respect to a parameter is also used to study the dependence of the solution of a system of nonlinear equations (not related to the autonomous system)

$$f(x, \alpha) = 0, \tag{2}$$

on a scalar parameter α, $\alpha \in p \in R^m$. The plots of one or several components of the vector-function $x = x(\alpha)$ which is the solution of Eq. (2) will be referred to as the diagram of stationary solutions. The method is based on the implicit function theorem. According to this theorem, the plot of the solution of Eq. (2) in the $(n + 1)$-dimensional space will be a smooth space curve if in a neighborhood of the space curve the rank of matrix A ($A = [f_x, f_\alpha]$) of the derivatives of the right-hand sides is always equal to n, regardless of α. Notice that the smooth space curve may intersect the hyperplane $\alpha = \alpha^*$, $\alpha^* \in [\alpha_0, \alpha_1]$ several times, meaning that there is a multiplicity of solutions of (2) when $\alpha = \alpha^*$.

Let the solution of Eq. (2) be known for a certain value of α and A be the matrix formed by this solution. As the rank of A is n, there is a nondegenerate square matrix $B = [f_z]$ that is obtained from A by deleting some k-th column. Denote by z the vector whose components are equal to those of the vector (x, α) obtained by deleting the component with the index k, which will be denoted by μ. By defining the derivative vector z_μ as a solution of the system with the matrix B and the right-hand side f_μ, let us find the maximal by module component of vector $(z_\mu, 1)$ with the index j. The vector consisting of the components of vector (x, α) after deleting the j-th component is denoted by u, while the component deleted is denoted by λ.

Vector u can be considered as a solution of Eq. (2) at the same value of α, but with parameter λ. In this case, the derivative vector of the solution with respect to λ can be found by normalization of the components of vector $(z_\mu, 1)$, using its maximal by module component for the normalization. As a result, this maximal by module component of the derivative vector will be equal to 1. This means that λ can be seen as a parameter of Eq. (2) in the neighborhood of the solution under consideration.

The procedure of determining λ will be referred to as parametrization, and λ itself as a current parameter of Eq. (2) as it is defined regularly for one $\Delta\lambda$ step in the continuation method. The role of the current parameter can be equitably taken by any component of vector (x, α). To sum, the solution continuation with respect to the parameter, together with the parametrization and adaptation of the current step, allows for plotting a smooth space curve defined by Eq. (2) that may have turning points and, consequently, have several solutions within some ranges of α.

3.2. *Godunov-Bulgakov method*

The STEP+ program module contains a numeric criterion of the guaranteed asymptotic stability of the stationary solution developed under the guidance of academician S.K.Godunov. The method, which does not require calculations of eigenvalues of the Jacobi matrix ($J = [f_x]$) of Eq. (1), is based on estimating the norm of the solution H of the Lyapunov matrix equation

$$J^* H + H J = -E \,,$$

where E is a unit matrix, J^* is a matrix adjoint to J. Matrix J is a Hurwitz matrix (i.e. the considered stationary solution is asymptotically stable) if H is a Hermitian positive-definite matrix. In this case, H has an integral presentation in which the matrix exponents of J and J^* are used:

$$H = \int_0^\infty \exp(tJ^T)\exp(tJ)\,dt\,.$$

The numeric characteristic $\kappa(J)$ of asymptotic stability when J is a Hurwitz matrix has the form

$$\kappa(J) = 2\|J\|\ \sup_{v(0)\neq 0}\ \frac{\int_0^\infty \|v(t)\|^2\,dt}{\|v(0)\|^2}\ ,\ \ v(t) = [\exp(tJ)]v(0)\,.$$

At each iteration in the process of calculating $\kappa(J)$, the verification for the inequality is carried out: $\bar\kappa(J) < k_0$, where $\bar\kappa(J)$ is the approximation of $\kappa(J)$ at the iteration, and k_0 is a constant whose value depends only on the precision of number representation by the computer. The asymptotic stability is guaranteed if the limiting value $\kappa(J)$ also satisfy this inequality. If, at some iteration, the inequality is not satisfied, then the guaranteed conclusion about 'practical' instability of the considered stationary solution is given.

The interface of STEP+ module automatically constructs the Jacobi matrix and the matrix of partial derivatives of the right-hand sides of the system with respect to the parameters. The analytic representation of the matrix elements allows us to perform a numerical study of the system solution depending on an arbitrary model parameter of interest.

4. Conclusion

The system has been tested on the gene network that regulates auxin metabolism in the meristem of shoots of *Arabidopsis thaliana* L [13,14]. Not only has testing the system of mathematical models automatic generation with MGSgenerator module and numerical analysis of the model by STEP+ module demonstrated its validity, but it has also led to interesting biological results on the dynamics of auxin metabolism functioning as well as the evolution of its transformations.

Acknowledgments

The work has been conducted during the Special Semester on Quantitative Biology Analyzed by Mathematical Methods, October 1st, 2007 – January 27th, 2008, organized by RICAM, Austrian Academy of Sciences.

This work has been partially supported by the Russian Foundation for Basic Research (Grant Nos. 09-01-08109, 08-01-91204-YaF_a), by the Grant NSh-2447.2008.4, and by the SB RAS Interdisciplinary project Nos. 91, 107, and 119, and also by the RAS projects 21.26, 22.8, 23.29.

References

1. F. V. Kazantsev, I. R. Akberdin, K. D. Bezmaternykh, and V. A. Likhoshvai, *System of automated generation of mathematical models of gene networks*, VOGiS Herald journal, **13** 1, (2009).

2. S. I. Fadeev, V. K. Korolev, I. A. Gainova, A. E. Medvedev, *The package Step+ for numerical study of autonomous systems arising when modeling dynamics of genetic-molecular systems*, in *Proc. of the 6th Intern. Conf. on Bioinformatics of Genome Regulation and Structure (BGRS-06)*, **2** (2006).

3. E. A. Ananko, N. L. Podkolodny, I. L. Stepanenko, O. A. Podkolodnaya, D. A. Rasskazov, D. S. Miginsky, V. A. Likhoshvai, A. V. Ratushny, N. N. Podkolodnaya, and N. A. Kolchanov, *GeneNet in 2005*, Nucleic Acids Res. Vol., **33** (2005).

4. V. A. Likhoshvai, Yu. G. Matushkin, A. V. Ratushny, E. A. Ananko, E. V. Ignatyeva, O. V. Podkolodnaya, *Generalized chemical-kinetic approach for modeling gene networks*, Molecular biology, **3**, 6 (2001).

5. M. Hucka, A. Finney, H. Sauro, H. Bolouri, J. Doyle, H. Kitano, et al., *The systems biology markup language (SBML): a medium for representation and exchange of biochemical network models*, Bioinformatics. March (2003).

6. V. Likhoshvai and A. Ratushny, *Generalized Hill function method for modeling molecular processes*, J. of Bioinformatics and Comp. Biology, **5**, 2 (2007).

7. M. Kanehisa and S. Goto, *KEGG: Kyoto Encyclopedia of Genes and Genomes*, Nucleic Acids Res., **28** (2000).

8. I. Keseler, J. Collado-Vides, S. Gama-Castro, J. Ingraham, S. Paley, I. Paulsen, M. Peralta-Gil and P. Karp, *EcoCyc: a comprehensive database resource for Escherichia coli*, Nucleic Acids Res. Vol., **33** (2005).

9. G. W. Gear, *The automatic integration of ordinary differential equations*, Comm. ACM.V., **14**, 3 (1971).

10. S. I. Fadeev, S. A. Pokrovskaya, A. Yu. Berezin, and I. A. Gainova, STEP program package for a numerical study of systems of nonlinear equations and autonomous systems of the general form. Description of STEP package operation by examples of the tasks from the educational course "Engineering chemistry of catalytic processes". Tutorial. (Izd. Novosib. Gos. Univ., Novosibirsk, Russia, 1998).

11. S. I. Fadeev and V. V. Kogai, *Using parameter continuation based on the multiple shooting method for numerical research of nonlinear boundary value problems*, Int. J. of Pure and Applied Mathematics, **14**, 4 (2004).

12. S. K. Godunov, Ordinary Differential Equations with Constant Coefficients. (Izd. Novosib. Gos. Univ., Novosibirsk, 1994).

13. I. R. Akberdin, N. A. Omelyanchuk, S. I. Fadeev, V. M. Efimov, I. A. Gainova and V. A. Likhoshvai, *Mathematical model of auxin metabolism in shoots of Arabidopsis thaliana L*, Proc. of the 6th Intern. Conf. on Bioinformatics of Genome Regulation and Structure (BGRS'2008), (2008).

14. I. R. Akberdin, F. V. Kazantsev, N. A. Omeliyanchuk and V. A. Likhoshvai, *Mathematical model of auxin metabolism in meristem cells of plant shoots*, VOGiS Herald journal, **13**, 1 (2009).

Investigation of the acoustic properties of the cancellous bone

R.P. Gilbert*

*Department of Mathematical Sciences,
University of Delaware, Newark, Delaware 19716, USA
E-mail: gilbert@math.udel.edu*

K. Hackl and S. Ilic

*Institute of Mechanics, Ruhr-University Bochum,
Bochum, 44780, Germany
E-mails: sandra.ilic@rub.de,klaus.hackl@rub.de*

The current paper is concerned with simulating the behavior of cancellous bone by using the multiscale finite element method. This approach belongs to the group of homogenization methods and is special in so far as it solves boundary value problems at two scales by using the finite element method. In order to activate the viscous effects in the fluid phase of the cancellous bone, a dynamic investigation and an analysis in the complex domain are necessary.

Keywords: Multiscale FEM; cancellous bone; homogenization; wave propagation.

1. Introduction

Cancellous bone, also known as spongy bone (lat.: substantia spongiosa) is one of two main types of osseous tissues. In contrast to the cortical bone which is dense and forms the surface of the bones, the cancellous bone has a mesh structure and constitutes the interior of the majority of bones, including the vertebrae. The mesh typical for this tissue is build from the thin trabeculae whose interspace is filled by the fluid marrow.

The microstructure of a cortical bone is influenced by many factors, such as the age of the bone or the load acting on it. These factors can sometimes lead to an extreme resorption of the solid phase typical for the pathological phenomenon known as osteoporosis. At the beginning of this process, the thickness of trabeculae decreases gradually, while in later stages, a severe

---

*R.P. Gilbert was partially supported by the NSF Research Grant DMS-0920850.

loss of the strength of the material is caused by the complete disappearance of individual solid elements.

Within recent years, there has been an increasing interest in the investigation of the cancellous bone,[1-3] and in this work, the multiscale finite element method (FEM) is applied for this purpose.[4-6] The mentioned method is a numerical approach resulting from combining the homogenization theory with the FE technique. Its application is illustrated by examples simulating the wave propagation through the studied tissue and calculating its effective attenuation coefficient.

The paper is structured as follows. In Sec. 2, the idea of the multiscale FEM is explained. Emphasis is placed on the connection of the macroscopic and microscopic quantities as well as on the conditions for the transition between the scales. Sec. 3 deals with the material modeling of particular phases while Sec. 4 is concerned with the approximation of the microarchitecture. The paper closes with a section devoted to numerical examples (5) and brief conclusions (6).

2. Concept of the multiscale FEM

The multiscale FEM belongs to the group of homogenization methods and is applied for statistically uniform composites. This group of materials is characterized by the fact that a representative volume element (RVE) can be defined for them. The characteristic size of this RVE must be much smaller than the characteristic size of the modeled body and if this condition is met, the following definitions can be used for the macroscopic quantities[5,6]

$$\bar{\epsilon} = \frac{1}{V} \left[\int_{\mathcal{B}} \epsilon \, dV - \int_{\mathcal{L}} \mathbf{x} \otimes \mathbf{n} \, dA \right] = \frac{1}{V} \int_{\partial \mathcal{B}} \mathbf{x} \otimes \mathbf{n} \, dA, \qquad (1)$$

$$\bar{\sigma} = \frac{1}{V} \int_{\mathcal{B}} \sigma \, dV = \frac{1}{V} \int_{\partial \mathcal{B}} \mathbf{t} \otimes \mathbf{x} \, dA. \qquad (2)$$

In the previous expressions, the standard notation is used; $\mathbf{x}$ is the position vector, $\mathbf{n}$ is the normal vector to the surface, $\epsilon = \frac{1}{2}(\nabla \mathbf{u} + \mathbf{u}\nabla)$ is the strain tensor defined depending on the displacement vector $\mathbf{u}$, σ the stress tensor and $\mathbf{t}$ the traction. The averaging is performed over the volume V of the RVE $\mathcal{B}$ with the boundary $\partial \mathcal{B}$ and the boundary of the voids inside the RVE $\mathcal{L}$. The overbar symbol is introduced to distinguish macro- from microscopic quantities. The definitions satisfy the Hill postulate that the macroquantities have to be defined depending on the microquantities acting on the boundary of the RVE. In addition, the connection of the scales requires an energetic balance, i.e., the equality of the macrowork and the volume average of the microwork which can be written

$$\bar{\sigma} : \bar{\epsilon} = \frac{1}{V} \int_{\mathcal{B}} \sigma : \epsilon \, dV \quad \Leftrightarrow \quad \frac{1}{V} \int_{\partial \mathcal{B}} (\mathbf{t} - \bar{\sigma}\mathbf{n}) \cdot (\mathbf{x} - \bar{\epsilon}\mathbf{x}) \, dA = 0. \quad (3)$$

The previous expression is the so called Hill-Mandel macrohomogeneity condition and it is used to define possible boundary conditions at the microlevel. Two of them, the static and the kinematic one, can be seen directly from the second expression in (3)

$$\mathbf{t} = \bar{\sigma} \cdot \mathbf{n} \quad \text{on } \partial \mathcal{B} \qquad \text{- static b.c.} \quad (4)$$

$$\mathbf{x} = \bar{\epsilon} \cdot \mathbf{x} \quad \text{on } \partial \mathcal{B} \qquad \text{- kinematic b.c.} \quad (5)$$

However, for the purposes of this contribution, the type known as periodic boundary conditions is used. In that case, the deformation takes a form which depends on the macrostrain $\bar{\epsilon}$ and the microfluctuations $\tilde{\mathbf{w}}$

$$\mathbf{x} = \bar{\epsilon}\mathbf{x} + \tilde{\mathbf{w}}. \quad (6)$$

The microfluctuations $\tilde{\mathbf{w}}$ must be periodic and the tractions $\mathbf{t}$ antiperiodic on the periodic boundary of the RVE

$$\tilde{\mathbf{w}}^+ = \tilde{\mathbf{w}}^- \qquad \text{and} \qquad \mathbf{t}^+ = -\mathbf{t}^- \qquad \text{on } \partial \mathcal{B}. \quad (7)$$

Assumption (6) for the deformation leads to the additive decomposition of the microstrain tensor

$$\epsilon = \bar{\epsilon} + \tilde{\epsilon} \quad (8)$$

where the definition $\tilde{\epsilon} = \frac{1}{2}(\nabla \tilde{\mathbf{w}} + \tilde{\mathbf{w}}\nabla)$ is applied.

Note that a more general formulation related to the finite deformation is also available,[4-6] but for the purpose of bone modeling the use of the linear elasticity is more suitable.

3. Material behavior of phases

In order to activate the viscous phenomena in the fluid phase of the RVE, the behavior of a sample under sound excitation is considered. For this particular kind of load, displacements are a harmonic function

$$\mathbf{u} = \mathbf{u}(\mathbf{x})e^{i\omega t}, \quad (9)$$

such that the position-dependent part $\mathbf{u}(\mathbf{x})$ might be complex and ω represents the angular frequency.

The behavior of each of the phases composing the considered tissue is described by a set of two equations, i.e. the equation of motion and the material specific constitutive law. By using the time derivatives of expression (9)

$$\ddot{\mathbf{u}} = -\omega^2 \mathbf{u}, \qquad \dot{\mathbf{u}} = i\omega \mathbf{u} \tag{10}$$

the equations describing the solid phase become

$$-\omega^2 \rho_s \mathbf{u} - \nabla \cdot \boldsymbol{\sigma}_s = \rho_s \mathbf{b}(\mathbf{x}), \tag{11}$$

$$\boldsymbol{\sigma}_s = \mathbf{C} : \boldsymbol{\epsilon}. \tag{12}$$

Here, the index s denotes the solid phase, ρ is the density, $\mathbf{b}(\mathbf{x})$ represents the body forces and $\mathbf{C}$ is the complex elasticity tensor calculated in a standard manner for linear elasticity, but using the complex bulk modulus $K = K^{\mathrm{R}} + iK^{\mathrm{I}}$ and the complex shear modulus $\mu = \mu^{\mathrm{R}} + i\mu^{\mathrm{I}}$. Imaginary parts of these parameters are defined by the expressions $K^{\mathrm{I}} = \frac{\delta}{\pi} K^{\mathrm{R}}$ and $\mu^{\mathrm{I}} = \frac{\delta}{\pi}\mu^{\mathrm{R}}$ depending on the logarithmic decrement δ.

The state of deformations and stresses in the fluid phase (subscript f) is determined by the equations

$$-\omega^2 \rho_f \mathbf{u} - \nabla \cdot \boldsymbol{\sigma}_f = \rho_f \mathbf{b}(\mathbf{x}), \tag{13}$$

$$\boldsymbol{\sigma}_f = c^2 \rho_f \, \nabla \cdot \mathbf{u}\,\mathbf{I} + 2i\omega\eta\,\boldsymbol{\epsilon} + i\omega\xi\,\nabla \cdot \mathbf{u}\,\mathbf{I}, \tag{14}$$

where the constitutive law for the barotropic fluid (14) has been chosen. It depends on the velocity of the wave in the fluid c, the viscosity parameters η and ξ , the identity tensor $\mathbf{I}$ and the imaginary unit $i = \sqrt{-1}$.

Finally, the continuity of the displacements perpendicular to the contact surface ($\mathbf{u}_s^{\perp}$, $\mathbf{u}_f^{\perp}$) is required as a coupling condition between the phases

$$\mathbf{u}_s^{\perp} = \mathbf{u}_f^{\perp} \quad \text{on} \quad \Gamma = \Omega_s \cap \Omega_f. \tag{15}$$

Here Ω_s and Ω_f denote domains occupied by the solid and the fluid phase respectively. Expression (15) is just one possible assumption for the contact condition which might have a significant influence on the effective values.

4. Modeling the RVE

Apart from the material behavior, the geometry of the microstructure is an additional property which is important in modeling the RVE. This property particularly influences the number and the type of elements needed for FE simulations. Within the model presented here, eight-node brick elements are used for modeling both phases. The formulation of this type of element is based on the minimization of the Lagrangian

$$L = \frac{1}{2} \int_{\Omega^e} \rho\, \dot{\mathbf{u}}^{\mathrm{T}} \cdot \dot{\mathbf{u}}\, dv - \frac{1}{2} \int_{\Omega^e} \boldsymbol{\epsilon} \cdot \mathbf{C} \cdot \boldsymbol{\epsilon}\, dv - \Pi^{e,\mathrm{ext}} \quad , \tag{16}$$

where $\Pi^{e,\mathrm{ext}}$ denotes the power of external forces and the superscript e indicates that a finite element is considered. The expression can be applied to both phases, and the distinction between them is achieved by the choice of the density ρ and the elasticity tensor $\mathbf{C}$. Finally, $\mathbf{u}$ is the vector of the complex displacements which, in an extended form, is expressed

$$\mathbf{u} = \left\{ \mathbf{u}^{\mathrm{R}}\ i\mathbf{u}^{\mathrm{I}} \right\}^{\mathrm{T}} = \left\{ u^{\mathrm{R}}\quad v^{\mathrm{R}}\quad w^{\mathrm{R}}\quad i\,u^{\mathrm{I}}\quad i\,v^{\mathrm{I}}\quad i\,w^{\mathrm{I}} \right\}^{\mathrm{T}}.$$

The FE approximation of the previous vector has the following form

$$\mathbf{u} = \mathbf{N} \cdot \hat{\mathbf{u}}^{e}, \qquad \hat{\mathbf{u}}^{e} = \left\{\ (\hat{\mathbf{u}}^{e\mathrm{R}})^{\mathrm{T}}\quad i(\hat{\mathbf{u}}^{e\mathrm{I}})^{\mathrm{T}}\ \right\}^{\mathrm{T}}, \tag{18}$$

$$\hat{\mathbf{u}}_{i}^{\mathrm{R}} = \left\{\hat{u}^{\mathrm{R}}\quad \hat{v}^{\mathrm{R}}\quad \hat{w}^{\mathrm{R}}\right\}^{\mathrm{T}}, \qquad \hat{\mathbf{u}}_{i}^{\mathrm{I}} = \left\{\hat{u}^{\mathrm{I}}\quad \hat{v}^{\mathrm{I}}\quad \hat{w}^{\mathrm{I}}\right\}^{\mathrm{T}}. \tag{19}$$

Here $\mathbf{N}$ is a matrix with terms which are dependent on the shape functions for an eight-node brick element while $\hat{\mathbf{u}}^{e}$ is the vector of the element DOFs consisting of the vectors of the real and the imaginary nodal DOFs $\hat{\mathbf{u}}_{i}^{\mathrm{R}}$ and $\hat{\mathbf{u}}_{i}^{\mathrm{I}}$. The implementation of $(18)_1$ into (16) yields the discrete equation of motion

$$(-\omega^2 \mathbf{M}^{e} + \mathbf{K}^{e}) \cdot \hat{\mathbf{u}}^{e} = \mathbf{f}^{e} \tag{20}$$

where $\mathbf{M}^{e}$ is the mass matrix, $\mathbf{K}^{e}$ the stiffness matrix and $\mathbf{f}^{e}$ the vector of the nodal forces, all of them being complex quantities.[4,7]

5. Numerical results

The numerical examples presented in this section deal with the simulation of the wave propagation through the bone. To this end, the RVE shown in Fig. 1 is assumed at the microlevel. The side length a amounts to 1 mm and the trabeculae thickness b is 0.125 mm. The porosity of the chosen sample amounts to 84.4% and the density to 1108 kg/m^3. These values lie in the ranges characteristic for a cancellous bone, i.e. 72-95% for the porosity and 1100-1200 kg/m^3 for the density.

At the macrolevel, a bone sample with a thickness of 30 mm and harmonic wave acting with a pressure of 4 kN/mm^2 on its surface are chosen. Fig. 2 shows the amplitudes of the oscillations for two waves with different excitation frequencies, 700 and 900 kHz. The dashed line represents the

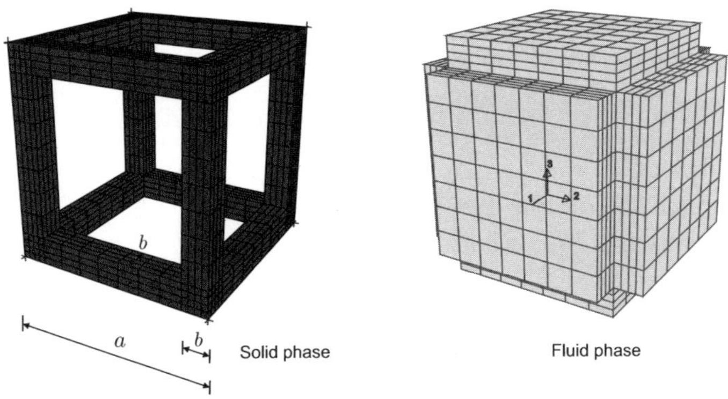

Fig. 1. Geometry of the RVE.

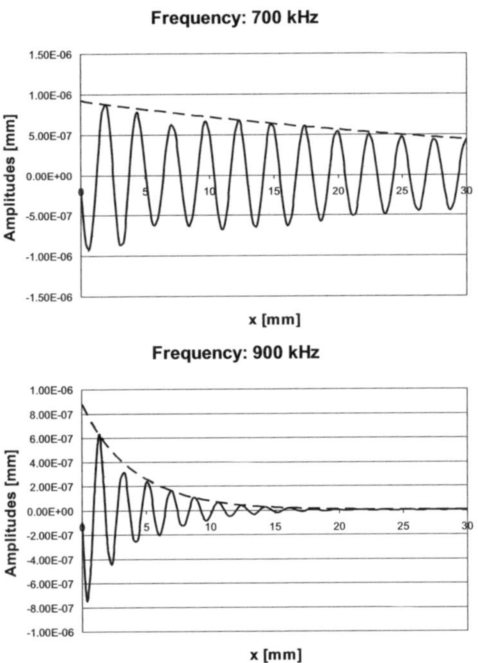

Fig. 2. Amplitudes of the particle oscillations for waves with different excitation frequencies. The material parameters used in these simulations are: $\rho_\mathrm{f}{=}950\,\mathrm{kg\,m}^{-3}$, $K_\mathrm{f}{=}2.00{\times}10^9\,\mathrm{Pa}$, $\eta{=}1.5\,\mathrm{Ns\,m}^{-2}$, $\xi{=}0$, $\rho_\mathrm{s}{=}1960\,\mathrm{kg\,m}^{-3}$, $K_\mathrm{s}^\mathrm{R}{=}2.04{\times}10^{10}\,\mathrm{Pa}$, $\mu_\mathrm{s}^\mathrm{R}{=}0.833{\times}10^{10}\,\mathrm{Pa}$, $\delta = 0.1$.

envelope $y(x) = A^* \exp(\alpha x)$ where A^* is the first maximum amplitude and α the attenuation coefficient calculated according to the expression

$$\alpha = \frac{\ln \Delta A}{\Delta d}. \tag{21}$$

This coefficient depends on ΔA which represents the ratio of the amplitudes and Δd which denotes the distance between the points where the amplitudes A are measured. Δd must be the multiple value of the wave length. For the cases presented in Fig. 2, the attenuation coefficients amount to 0.242 and 2.454 Neper/cm. However, these values show a significant discrepancy in comparison to experimental results which supports the observation that the consideration of the pure internal viscosity of the phases is not sufficient to explain the high attenuation coefficient of the cancellous bone.

6. Conclusions

Within the scope of this contribution, the results obtained by using the multiscale FEM for modeling the cancellous bone are presented. To this end, a cubic RVE consisting of the solid frame and the fluid core is proposed. Special emphasis is placed on the modeling of viscous effects requiring a formulation in the complex domain. The chosen numerical examples are concerned with a wave propagation through the considered tissue. However, a comparison to experimental results suggests that a further development of the RVE is necessary. Two envisaged modifications are the consideration of a complex RVE consisting of a greater number of individual cells and an alternative modeling of the coupling conditions at the contact surface of the phases.

References

1. J. Buchanan, R. Gilbert and K. Khashanah, *J. Comput. Acoust.* **12(2)**, 99 (2004).
2. J. Buchanan and R. Gilbert, *Math. Comput. Model.* **45(3-4)**, 281 (2006).
3. S. Ilic, K. Hackl and R. Gilbert, *Biomechan. Model. Mechanobiol.* **9(1)**, 87 (2010).
4. S. Ilic, *Application of the multiscale FEM to the modeling of composite materials* 2008. Ph.D. Thesis, Ruhr University Bochum, Germany.
5. S. Ilic and K. Hackl, *J. Theor. Appl. Mech.* **47**, 537 (2009).
6. J. Schröder, *Homogenisierungsmethoden der nichtlinearen Kontinuumsmechanik unter Beachtung von Stabilitätsproblemen* 2000. Habilitationsschrift, Universität Stuttgart, Deutschland.
7. O. Zienkiewicz and R. Taylor, *The finite element method* (Butterworth-Heinemann, 2000).

VI. Others

Organisers: Local Organising Committee

The relationship between Bezoutian matrix and Newton's matrix of divided differences

R.G. Airapetyan

Department of Mathematics, Kettering University,
Flint, Michigan, USA
E-mail: rhayrape@kettering.edu

Let $x_1, \ldots, x_n$ be real numbers, $P(x) = p_n(x - x_1) \cdots (x - x_n)$, and $Q(x)$ be a polynomial of order less than or equal to n. Denote by $\Delta(Q)$ the matrix of generalized divided differences of $Q(x)$ with nodes $x_1, \ldots, x_n$ and by $B(P, Q)$ the Bezoutian matrix of P and Q. A relationship between the corresponding principal minors of the matrices $B(P, Q)$ and $\Delta(Q)$ counted from the right lower corner is established. It implies that if the principal minors of the matrix of divided differences of a function $g(x)$ are positive or have alternating signs then the roots of the Newton's interpolation polynomial of g are real and separated by the nodes of interpolation.

Keywords: Bezoutian matrix; Newton's matrix of generalized divided differences; Newton's interpolation polynomial.

1. Introduction

In this paper a relationship between two well known matrices is established. The first one is a Bezoutian matrix (Bezoutiant) B playing an important role in the theory of separation of polynomial roots. The second one is Newton's matrix of divided differences Δ or, in the case of multiple nodes, Hermite's matrix of generalized divided differences, playing an important role in numerical analysis and approximation theory. Despite the fact that these matrices are intensively used, the fact that there is a simple relationship between them seems to be unknown. In this paper we establish the following relationship between these two matrices. We show that the corresponding principal minors of B and Δ counted from the right lower corner are related by a simple formula (are equal when $p_n = 1$). It is well known that such minors of Bezoutiants play are important for the theory of polynomial root separation. However, the importance of similar minors of Newton's matrix, which, in general, is not a symmetric matrix, was not known. As a simple application of the relationship between B and Δ, a theorem about locations of the roots of interpolation polynomials in terms of the principal minors of Δ is established.

2. Main Results

With the polynomials $P(x) = \sum_{j=0}^{n} p_j x^j$ and $Q(x) = \sum_{j=0}^{n} q_j x^j$ let us associate the bilinear form

$$\sum_{i,j=1}^{n} b_{ij} x^{i-1} y^{j-1} = \frac{P(x)Q(y) - P(y)Q(x)}{x - y}, \tag{1}$$

which Sylvester[1] named "Bezoutiant". If the order of Q is less than the order of P, that is $Q(x) = \sum_{j=0}^{m} q_j x^j$, $m < n$, then one adds zero coefficients $q_{m+1}, \ldots, q_n$ to Q. In what follows we assume that $m \leq n$. Bezoutiants have been studied in papers of many mathematicians such as J. Sylvester[1], C. Hermite[2], A. Cayley[3], A. Hurwitz[4], and later M.G. Krein and M.A. Naimark[5], A.S. Householder,[6] F. I. Lander[7], P.A. Fahrmann and B.N. Datta[8], A. Olshevsky and V. Olshevsky[9].

It has been shown[7] that

$$B(P,Q) = ||b_{ij}||_{i,j=1,\ldots,n} = \begin{pmatrix} p_1 & p_2 & \cdots & p_n \\ p_2 & & \cdots & 0 \\ & \cdot & & \cdot \\ & \cdot & \cdot & \\ & \cdot & & \cdot \\ p_n & 0 & \cdots & 0 \end{pmatrix} \begin{pmatrix} q_0 & \cdots & q_{n-2} & q_{n-1} \\ 0 & \cdots & \cdot & q_{n-2} \\ \cdot & & & \cdot \\ \cdot & & \cdot & \cdot \\ \cdot & \cdot & q_0 & \cdot \\ 0 & \cdots & 0 & q_0 \end{pmatrix}$$

$$- \begin{pmatrix} q_1 & q_2 & \cdots & q_n \\ q_2 & & \cdots & 0 \\ & \cdot & & \cdot \\ & \cdot & \cdot & \\ & \cdot & & \cdot \\ q_n & 0 & \cdots & 0 \end{pmatrix} \begin{pmatrix} p_0 & \cdots & p_{n-2} & p_{n-1} \\ 0 & \cdots & \cdot & p_{n-2} \\ \cdot & & & \cdot \\ \cdot & & \cdot & \cdot \\ \cdot & \cdot & p_0 & \cdot \\ 0 & \cdots & 0 & p_0 \end{pmatrix}. \tag{2}$$

The main properties of the Bezoutian matrix are[5,7-9]:

- The defect of the Bezoutian matrix equals the degree of the greatest common divisor of the polynomials P and Q.
- The rank of the Bezoutian matrix equals the order of the last principal minor of the matrix $B = ||b_{i,j}||_{i,j=1,\ldots,n}$ which does not vanish if, in

constructing the consecutive major minors, one starts from the lower right-hand corner.

- If the Bezoutian matrix is positive definite then both polynomials $P(x)$ and $Q(x)$ have real, distinct roots. Moreover, the roots of $P(x)$ and $Q(x)$ interlace.
- If all consecutive principal minors starting from the lower right-hand corner are positive or have alternating signs, then the roots of $P(x)$ and $Q(x)$ are real, distinct, and interlace.

Since principal minors of Bezoutiants play so important a role, it seems interesting to find explicit formulas for them. If the roots $x_1, x_2, \ldots, x_n$ of $P(x)$ are simple, such formulas were established in[10].

Theorem 2.1. *Let $|b_{ij}|_{i,j=k+1}^{n}$ be the principal minors counted from the lower right corner of the Bezoutian matrix $B(P,Q)$ of polynomials $P(x) = p_n(x - x_1) \ldots (x - x_n)$ and $Q(x)$. Then,*

$$|b_{i,j}|_{i,j=k+1}^{n} = p_n^{2(n-k)} \sum_{\substack{(i_1, \ldots, i_{n-k}) \subset (1, \ldots, n) \\ i_1 < i_2 < \cdots < i_{n-k}}} \frac{Q(x_{i_1}) \cdots Q(x_{i_{n-k}})}{P'(x_{i_1}) \cdots P'(x_{i_{n-k}})}$$

$$\times \prod_{\substack{(j_1, j_2) \subset (i_1, \ldots, i_{n-k}) \\ j_1 < j_2}} (x_{j_1} - x_{j_2})^2. \tag{3}$$

Remark 1. If $k = n - 1$ then formula (3) becomes

$$b_{n,n} = p_n^2 \sum_{i=1}^{n} \frac{Q(x_i)}{P'(x_i)}. \tag{4}$$

Remark 2. Since $|b_{ij}|_{i,j=k+1}^{m}$ are continuous functions of $x_1, \ldots, x_m$ in case of multiple roots one has to find the corresponding limit which is technically difficult and leads to complicated expressions.

In order to consider the case of $x_1, \ldots, x_n$ which are not necessarily different, let us introduce the following generalized divided differences.

Definition 2.1. (see[11]):

$$g[x_i] := g(x_i), \quad i = 1, \ldots, n,$$

$$g[x_{i_1}, \ldots, x_{i_k}] := \begin{cases} \dfrac{g[x_{i_2}, \ldots, x_{i_k}] - g[x_{i_1}, \ldots, x_{i_{k-1}}]}{x_{i_k} - x_{i_1}}, & \text{if } x_{i_1} \neq x_{i_k} \\[2ex] \lim\limits_{x \to x_{i_1}} \dfrac{\partial}{\partial x} g[x, x_{i_2}, \ldots, x_{i_{k-1}}], & \text{if } x_{i_1} = x_{i_k}. \end{cases} \tag{5}$$

Remark 3. This definition of generalized divided differences is equivalent to the definition given in[11] when $x_1 \leq x_2 \leq \cdots \leq x_n$.

Consider the following triangular matrix of the generalized divided differences:

$$\Delta(g) = \|\Delta_{ij}\|_{i,j=1,\ldots,n}, \quad \Delta_{ij} = \begin{cases} 0, & \text{if } i+j < n+1 \\ g[x_{n-i+1}, \ldots, x_j], & \text{if } i+j \geq n+1, \end{cases} \tag{6}$$

that is

$$\Delta(g) = \begin{pmatrix} & & & 0 & \Delta_n \\ & & 0 & \cdot\ \Delta_{n-1}\ \Delta_{n-1,n} \\ & & \cdot\ \cdot & \cdot & \cdot \\ & & \cdot\ \cdot & \cdot & \cdot \\ 0 & \cdot & & \cdot & \cdot \\ \Delta_1\ \Delta_{1,2} & \ldots & \Delta_{1,n-1} & \Delta_{1,n} \end{pmatrix}. \tag{7}$$

As it is well known, Newton-Hermite's interpolation polynomial for n nodes $\{x_1, \ldots, x_n\}$ is:

$$\Delta_1 + \Delta_{1,2}(x - x_1) + \cdots + \Delta_{1,n}(x - x_1) \ldots (x - x_{n-1}). \tag{8}$$

Denote by $|\Delta_{i,j}|_{i,j=k+1}^n$ the principal minors of the matrix Δ counted from the lower right corner. The following theorem establishes a relationship between principal minors of the Bezoutian and Newton's matrices.

Theorem 2.2. *Let $|b_{i,j}|_{i,j=k+1}^n$ and $|\Delta_{i,j}|_{i,j=k+1}^n$ be the principal minors of the matrices $B(P,Q)$ and $\Delta(Q)$ counted from the lower right corner. Then*

$$|b_{i,j}|_{i,j=k+1}^n = p_n^{n-k} |\Delta_{i,j}|_{i,j=k+1}^n, \quad k = 0, 1, \ldots, n-1. \tag{9}$$

The proof of this theorem is rather technical and will be published in a different paper.

The relationship between $B(P,Q)$ and $\Delta(Q)$ established in this theorem is surprising taking into account that these matrices are of very different type, the Bezoutiant is a symmetric matrix and Newton's matrix is a triangular matrix. Two simple examples below show these matrices for some polynomials of order three.

Example 2.1. Let us consider polynomials $P(x) = x^3 - 4x^2 - x + 4 = (x+1)(x-1)(x-4)$ and $Q(x) = x^3 - 6x^2 + 11x - 6 = (x-1)(x-2)(x-3)$. Then

$$B(P,Q) = \begin{pmatrix} -38 & 48 & -10 \\ 48 & -60 & 12 \\ -10 & 12 & -2 \end{pmatrix}, \quad \Delta(Q) = \begin{pmatrix} 0 & 0 & 6 \\ 0 & 0 & 2 \\ -24 & 12 & -2 \end{pmatrix}.$$

Since $p_3 = 1$ the corresponding principal minors of these two matrices counted from the lower right-hand corner are equal, they are $-2, -24, 0$.

Example 2.2. Consider polynomials $P(x) = x^3 - 12x^2 + 44x - 48 = (x-2)(x-4)(x-6)$ and $Q(x) = x^3 - 9x^2 + 23x - 15 = (x-1)(x-3)(x-5)$. Then

$$B(P,Q) = \begin{pmatrix} 444 & -252 & 33 \\ -252 & 153 & -21 \\ 33 & -21 & 3 \end{pmatrix}, \quad \Delta(Q) = \begin{pmatrix} 0 & 0 & 15 \\ 0 & -3 & 9 \\ 3 & -3 & 3 \end{pmatrix}.$$

Principal minors counted from the lower right-hand corner are $3, 18, 135$.

Theorem 2.2 and the properties of the Bezoutian matrix described above imply the following theorem.

Theorem 2.3. *If, for some function $g(x)$, all consecutive principal minors of the matrix of divided differences (see (7)) starting from the lower right-hand corner are positive or have alternating signs, then the roots of Newton's interpolation polynomial of $g(x)$ are real, distinct, and interlace with the nodes of interpolation.*

The author thanks the referee for the useful suggestions.

References

1. J. Sylvester, *On a Theory of the Syzygetic relations of two rational integral functions, comprising an application to the theory of Sturm's Functions, and that of the greatest Algebraic Common Measure*, Philos. Trans. Roy. Soc. London **143** (1853), 407-548.
2. C. Hermite, *Extrait d'une lettre de Mr. Ch. Hermite de Paris à Mr. Borchardt de Berlin, sur le nombre des racines d'une équation algébrique comprises entre des limites données*, J. Reine Angew. Math. **52** (1856), 39-51.

3. A. Cayley, *Note sur la méthode d'élimination de Bezout*, J. Reine Angew. Math. **53** (1857), 366-376.

4. A. Hurwitz, Ueber die Bedingungen unter welchen eine Gleichung nur Wurzeln mit negativen reellen Teilen besitzt, Math. Ann. **46** (1895), 273-284.

5. M.G. Krein and M.A. Naimark, The method of symmetric and Hermitian forms in the theory of the separation of the roots of algebraic equations, Linear and multilinear algebra, 10 (1981), 265-308 (The paper was originally published in Kharkov in 1936).

6. A.S. Householder, *Bezoutiants, elimination and localization*, SIAM Review, **12** (1970), N0. 1, 73-78.

7. F. I. Lander, Bezoutiant and inversion of hankel and toplitz matrices, Matematicheskie Issledovaniya, 9, N2 (32), (1974), 69-87.

8. P.A. Fahrmann and B.N. Datta, *On Bezoutians, Van der Monde matrices, and the Lienard-Chipart stability criterion,* Linear Algebra and its applications, **120** (1989), 23-37.

9. A. Olshevsky and V. Olshevsky, Kharitonovis theorem and Bezoutians, Linear Algebra Appl., **399**, (2005), 285–297.

10. R.G. Airapetyan, *On the reduction of the Cauchy problem for a hyperbolic equation to symmetric systems,* Soviet Journal of Contemporary Mathematical Analysis, **21** (1986), N 1.

11. J. Stoer and R. Bulirsch, Introduction to Numerical Analysis, 3rd ed., Texts in Applied Mathematics, Springer, New York, 2002.

Interpolation beyond the interval of convergence:
An extension of Erdos-Turan Theorem

H. Al-Attas and M.A. Bokhari

*Department of Mathematics and Statistics,
King Fahd University of Petroleum and Minerals,
Dhahran, 31261, Saudi Arabia
E-mails: halattas@kfupm.edu.sa,mbokhari@kfupm.edu.sa*

An elegant result due to Erdos and Turan states that the sequence of Lagrange interpolants to a continuous function f at the zeros of orthogonal polynomials over an interval $[c, d]$ converges to f in mean square. We introduce certain sequences of polynomials which preserve both interpolation as well as convergence properties of Erdos-Turan Theorem. In addition, they interpolate f at a finite number of pre-assigned points lying outside $[c, d]$. We shall introduce a method to construct the suggested polynomials and also investigate their properties. Some computational aspects are also discussed.

Keywords: Fundamental polynomials; Lagrange and Hermite interpolants; orthogonal zero-interpolants; convergence in L_2-norm.

1. Introduction

Over the past few decades, the orthogonal polynomials have been extensively used for the approximation of functions. Most of the approaches have their roots in a result due to Erdos and Turan which is as follows:

Theorem 1.1.[1] *Let P_0, P_1, P_2, ... be a system of polynomials which are orthogonal on a finite closed interval $[c, d]$ with respect to a weight function ω. For a real valued function f continuous on $[c,d]$, let $L_n f$ denote the polynomial of degree n which interpolates to f at the zeros of P_{n+1}. Then*

$$\lim_{n\to\infty} \int_c^d (L_n f(x) - f(x))^2 \, \omega(x)dx = 0. \tag{1}$$

In Theorem 1.1, the interpolating nodes being the zeros of orthogonal polynomial are always distinct and located within the interval (c, d).[2] In some applications, approximating polynomials are required to interpolate a given function f at the end-point(s) of $[c, d]$ or sometimes at the points lying outside (c, d). With this aim, we expand the set of nodes considered in Theorem 1 by including a finite number of points that lie outside (c, d). To achieve this objective, we shall modify the interpolants without affecting their interpolation and convergence characteristics as discussed in Theorem 1.1.

Our first result deals with the construction of a sequence of interpolants that include both end points of $[c, d]$ and yet orthogonal and convergent in the sense of Theorem 1.1. This result is further generalized by including a finite number of multiple nodes which lie outside (c, d). The structure of both extensions is based on the zeros of interpolating orthogonal polynomials which we also term as orthogonal 0-interpolants.[3]

2. First Extension of Erdos-Turan Theorem

We shall require the following notations to state our first result:

(a) $\omega(x) :=$ non - negative weight function continuous on $[c, d]$

(b) $\langle f, g \rangle_{\omega, [c,d]} := \int_{c}^{d} f(x)g(x)\omega(x)dx$

(c) $\|f\|_{\omega, [c,d]} := \sqrt{\langle f, f \rangle_{\omega, [c,d]}}$

(d) $\pi_n :=$ Class of all polynomials of degree $\leq n$

(e) $W(x) := (x - c)(x - d)$

(f) $\pi_n(W) := \langle W, xW, \cdots, x^n W \rangle$, a linear space with basis $W, xW, \cdots, x^n W$

(g) $C[c, d] :=$ Class of all continuous real valued functions on $[c, d]$

(h) $K_\delta[c, d] :=$ All functions in $C[c, d]$ which are differentiable on $[c, c + \delta]$ and $[d - \delta, d]$ for some $\delta > 0$

(i) $L_n(., A, g) := n^{th}$ degree Lagrange interpolant to g at the points of set A

(j) $f_U(x) := f(x) - L_1(x, U, f)$ where $U = \{c, d\}$

(k) $f_{U,W}(x) := \begin{cases} \frac{f_U(x)}{W(x)}, & x \notin U \\ \lim_{t \to x} \frac{f_U(t)}{W(t)}, & x \in U \end{cases}$

Remark 1. $\pi_{n+1}(W)$(cf. (f)) is an $(n+2)$ dimensional space and has a basis comprised of monic orthogonal polynomials with respect to weight function $\omega(x)$ over $[c, d]$.[4] Having denoted this basis by $\{\phi_j\}_{j=0}^{n+1}$, we observe that each ϕ_j has a unique decomposition (cf. (e))$\phi_j(x) = p_j(x) W(x)$ for some $p_j \in \pi_j$.

Remark 2. In the notation of (b) it follows that $0 = \langle p_i, p_j \rangle_{\omega W^2}$, for $i \neq j$. Therefore, p_j, $j = 0, 1, 2, \ldots, n+1$, are monic orthogonal polynomials with respect to the weight function $\omega(x) W^2(x)$ over $[c, d]$, and thus, each p_j has j real distinct zeros lying in the open interval (c, d).[2] Letting Z_{n+1} as the set of all zeros of the orthogonal polynomial p_{n+1}, we set (cf. (e),(i),(f))

$$\boldsymbol{L}_n(x, U, f) := L_1(x, U, f) + W(x) L_n(x, Z_{n+1}, f_{U,W}) \tag{2}$$

and state our first result as follows:

Theorem 2.1. *If $f \in K_\delta[c, d]$ (cf. (h)) with $\delta > 0$, then the polynomial $\boldsymbol{L}_n(x, U, f)$ given in 2 interpolates f at $n+3$ points of $U \cup Z_{n+1}$. In addition, (cf. (c)), we have $\lim\limits_{n \to \infty} \|\boldsymbol{L}_n(., U, f) - f\|_{\omega, [c,d]} = 0.$*

Remark 3 The polynomial $\boldsymbol{L}_n(., U, f)$ reduces to $L_n(., Z_{n+1}, f)$ if interpolation conditions at c and d are dropped. In such case, Theorem 2.1 reproduces Theorem 1.1.

3. Second Extension of Erdos-Turan Theorem

We further generalize Theorem 1.1 by constructing the interpolants which include a finite number of simple or multiple nodes lying outside the interval (c, d) . To formulate our result, we fix a finite set $U_k := \{u_1, u_2, \ldots, u_k\}$ outside the interval (c, d) where each node u_i has a multiplicity m_i. Consider the interval $[a, b]$ with $a := \inf(U_k \cup [c, d])$ and $b := \sup(U_k \cup [c, d])$, let $K_{\delta, U_k}[c, d]$ be the class of all functions from $C[c, d]$ which are m_i -times differentiable on the intervals of the form

$$I_i = \begin{cases} [u_i, u_i + \delta] & \text{if} \quad u_i = a \\ [u_i - \delta, u_i] & \text{if} \quad u_i = b \\ (u_i - \delta, u_i + \delta) & , \text{ otherwise} \end{cases} \tag{3}$$

for some $\delta > 0$. With the notations $s(k) := \{m_i\}_{i=1}^{k}$, $S(k) := \sum\limits_{i=1}^{k} m_i$ and $m^* := \left(\max\limits_{1 \leq i \leq k} m_i \right) - 1$ let $H_{S(k)-1}(., U_k, f)$ be the polynomial of degree $\leq S(k) - 1$ that satisfies the conditions

$$H_{S(m)-1}^{(j)}(u_i, U_k, f) = f^{(j)}(u_i), \quad i = 1, 2, \ldots k; \ j = 0, 1, \ldots m_i \tag{4}$$

Now we initiate the construction of 0-orthogonal interpolants in the sense of Hermite by setting $W_{s(k)}(x) := \prod\limits_{j=1}^{k} (x - u_j)^{m_j}$ and then considering the $(n + 2)$ dimensional linear space $\pi_{n+1}(W_{S(k)}) :=$

$\langle W_{S(k)}, x W_{S(k)}, \dots, x^{n+1} W_{S(k)} \rangle$. Each member of the orthogonal basis, say $\{\phi_{j,s(k)}\}_{j=0}^{n+1}$, of this space will be referred to as 0-orthogonal interpolants.

Remark 4. On the lines of preceding section, we note that

(i) $\phi_{j,s(k)}(x) := p_{j,s(k)}(x) W_{s(k)}(x)$ for some $p_{j,s(k)} \in \pi_j$

(ii) $p_{j,s(k)}$, $j = 0, 1, \dots$, are orthogonal polynomials with respect to weight function $\omega(x) W_{s(k)}^2(x)$ over $[c, d]$,

(iii) $Z_{n+1,s(k)}$, the set of all zeros of $p_{n+1,s(k)}$, is comprised of exactly $n+1$ 'free orthogonal zeros' which are distinct and lie inside the interval (c, d),

(iv) every $\phi_{j,s(k)}$, $j = 1, 2, \dots$, has k *fixed orthogonal zeros* u_i, $i = 1, 2, \dots, k$, each with multiplicity m_i (cf. (4)). Because of this property we call every $\phi_{j,s(k)}$ an orthogonal 0-interpolant at the points u_i, $i = 1, 2, \dots, k$.

For a given $f \in K_{\delta, U_k}[c, d]$, we set

$$f_{U_k, s(k)}(x) := f(x) - H_{S(k)-1}(x, U_k, f). \tag{5}$$

Then with the choice of a and b as given above,

$$f_{U_k, W_{S(k)}}(x) := \begin{cases} \dfrac{f_{U_k, S(k)}(x)}{W_{S(k)}(x)}, & x \notin U_k \\ \lim\limits_{t \to x} \dfrac{f_{U_k, S(k)}(t)}{W_{S(k)}(t)}, & x \in U_k \end{cases} \tag{6}$$

is well-defined on $[a, b]$. Note that the polynomial

$$\boldsymbol{H}_{S(k)-1}(x, U_k, f) := H_{S(k)-1}(x, U_k, f) + W_{S(k)}(x) L_n(x, Z_{n+1,s(k)}, f_{U_k, W_{S(k)}}) \tag{7}$$

interpolates the function f m_i-times at the points u_i in the sense of Hermite and at the $n+1$ points of $Z_{n+1,s(k)}$ in the sense of Lagrange. With this observation, we have the second generalization of Theorem 1.1 as follows:

Theorem 3.1. *If* $f \in K_{\delta, U_k}[c, d]$ *with* $\delta > 0$, *then the polynomial* $\boldsymbol{H}_{S(k)-1}(., U_k, f)$ *given in (7) interpolates* f *on the set* $U_k \cup Z_{n+1,s(k)}$. *The interpolation at the points of* U_k *is in the sense of Hermite. In addition, we have* $\lim\limits_{n \to \infty} \left\| \boldsymbol{H}_{S(k)-1}(., U_k, f) - f \right\|_{\omega, [c, d]} = 0$.

4. Computational Aspects

In this section we discuss application of Theorems 1.1 and 3.1 to Runge function $f(x) = 1/(1 + x^2)$ for which a high degree Lagrange interpolant at evenly spaced points on an interval $[-a, a]$ with $a > 3$ leads to extreme *wiggle* near the interval's end-points.[2] We have computed the maximum point-wise error (M-Er) and root mean squared error (RMS-Er) between this function and its various interpolants constructed in Theorems 1.1 and 3.1. With the choice of weight function $\omega(x) \equiv 1$, we subdivide our study into three parts. In each of the figures given below, the solid and the dashed curves respectively correspond to the function and its interpolant. The markers * and o are used to represent the interpolation data related to the *free orthogonal zeros* (in case of Theorems 1.1 and 3.1) and the *pre-assigned points* (in case of Theorem 3.1) respectively. The degree of interpoants in all cases is kept constant at 19 for the sake of logical comparison.

Fig. 1 and 2 deal with Theorem 1.1 where in both cases the interpolant is based on 20 orthogonal zeros over the interval [-3,3]. However, the errors are determined on [-3,3] in case of fig. 1 and on the extended interval [-5,5] in case of fig. 2. These are as follows: Fig. 1: M-Er =0.0024, RMS-Er = 8.1155×10^{-4} and Fig. 2: M-Er = 2.5249×10^5 and RMS-Er = 4.0967×10^4.

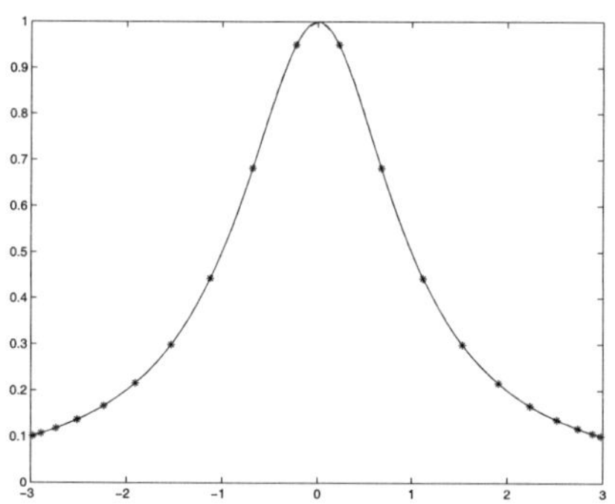 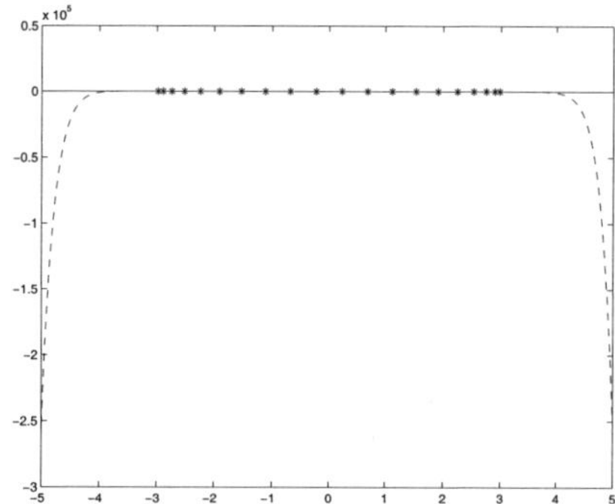

Fig. 1. Interpolation without pre-assigned nodes (Theorem 1.1)

Fig. 2. Interpolation without pre-assigned nodes on [-5,5] (Theorem 1.1)

Theorem 3.1 in case of simple *pre-assigned zeros* $\pm 3.5, \pm 4.5, \pm 5$ outside the interval of convergence along with 14 *free orthogonal zeros* within (-3,3) leads to M-Er = 0.0148 and RMS-Er = 0.0041 over [-3,3] (see fig. 3) and M-Er =11.4783 and RMS-Er = 2.4455 over [-5,5] (see fig. 4).

While considering $\pm 3.5, \pm 4.5, \pm 5$ as *double zeros* outside the interval of convergence, i.e., (-3,3) along with 8 *free orthogonal zeros* within (-3,3), we observe that M-Er =0.0346 and RMS-Er = 0.0110 over [-3,3] (see fig. 5) and M-Er =0.0349 and RMS-Er =0.0103 over [-5,5] (see fig. 6).

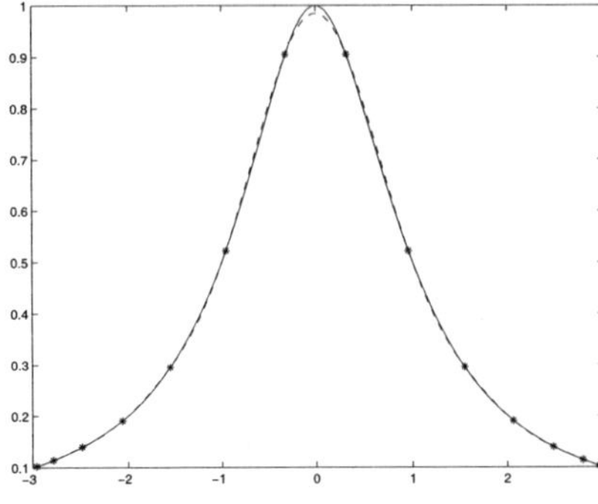

Fig. 3.   Interpolation with simple pre-assigned nodes (Theorem 3.1)

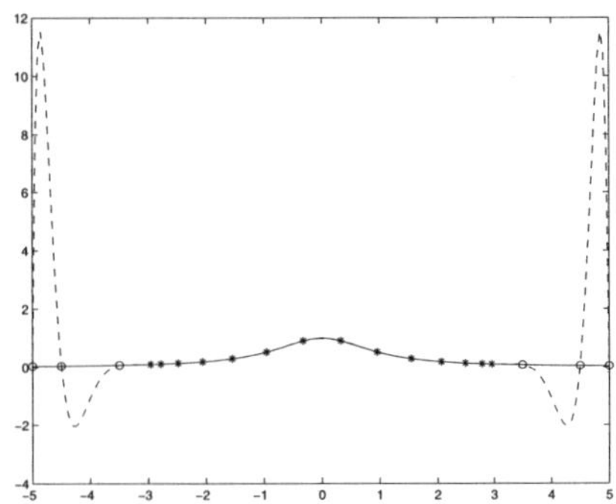

Fig. 4.   Interpolation with simple pre-assigned nodes on [-5,5] (Theorem 3.1)

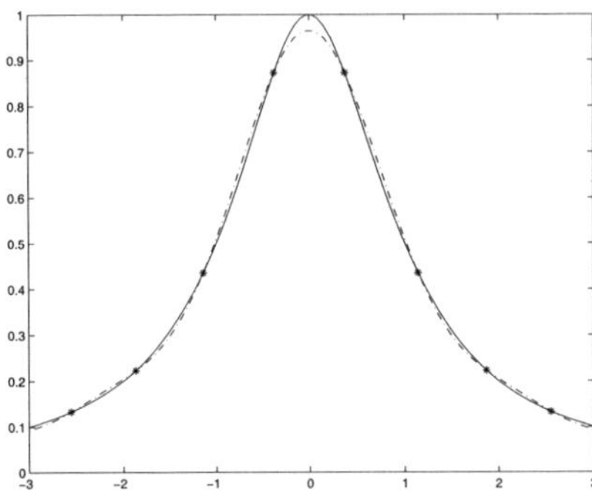

Fig. 5.   Interpolation with double pre-assigned nodes (Theorem 3.1)

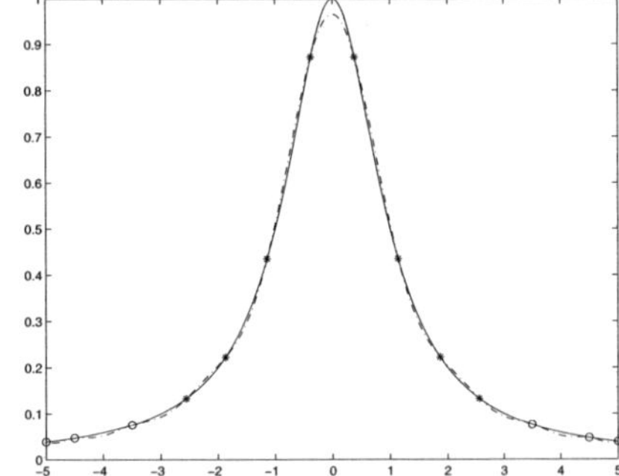

Fig. 6.   Interpolation with double pre-assigned nodes on [-5,5] (Theorem 3.1)

5. Conclusion

Some sequences of polynomials are constructed that interpolate a function f at a triangular array of the zeros of certain orthogonal polynomials over $[c, d]$. In addition, each polynomial interpolates f at a finite number of pre-assigned simple or multiple points outside the interval $[c, d]$. These sequences converge to f in the L_2-sense on $[c, d]$ under suitable conditions. Our work is based on the construction of orthogonal polynomials, each having an identical factor with simple or multiple zeros at the pre-assigned nodes outside $[c, d]$. Our result generalizes a result of Erdos and Turan (cf. Theorem 1.1).

Acknowledgments

The work carried out in this paper is a part of Research Project No. FT-080008 funded by King Fahd University of Petroleum and Minerals, Saudi Arabia.

References

1. P. Erdos and P. Turan, *AM* **38**, 142 (1937).
2. J. Stoer and R. Bulirsch, *An Introduction to Numerical Analysis*, 2nd edn. (Springer, 1993).
3. M. A. Bokhari, *Dynamical Systems and applications* **16**, 203 (2007).
4. M. A. Bokhari and M. Iqbal, *J. Comp. Appl Math* **70**, 201 (1996).

To solutions of the one non-divergent type parabolic equation with double non-linearity

M. Aripov

National University of Uzbekistan,
Tashkent 100174, Uzbekistan
E-mail: mirsaidaripov@mail.ru
www.mpamit09.uz/aripov

S.A. Sadullaeva

National University of Uzbekistan,
Tashkent 100174, Uzbekistan

The properties of the weak solution of problem Cauchy for one parabolic equation with double nonlinearity and with lower members are investigated. The researched equation is best combination of forms of the equation of nonlinear diffusion, fast diffusion, the equation to very fast diffusion and p-Laplace heat conductivity equation by action of convective transfer the velocity of which depend from time.

Keywords: Weak solution; double nonlinear equation.

1. Introduction

The properties of the weak solution of the following Cauchy problem in the domain $Q = \{(t, x) : t > 0, x \in R^N\}$ for non divergent type double nonlinear parabolic equation with lower members are studied

$$u_t = u^k div\left(u^{m-1}\left|\nabla u\right|^{p-2}\nabla u\right) - div(v(t)u) + \varepsilon\gamma(t,x)u^\beta \qquad (1)$$

$$u\left(0, x\right) = u_0\left(x\right) \geqslant 0, \ x \in R^N, \qquad (2)$$

where $m \geqslant 1, p \geqslant 2, \beta \geqslant 1$ are given positive numbers and functions $0 < \gamma(t, x) \in C(Q), 0 < v(t) \in C(0, \infty), \varepsilon = \pm 1, \nabla = grad_x$, the function $u_0\left(x\right)$ is not identically equal to zero and messupp $u_0\left(x\right) < \infty$.

The problem (1)-(2) is basic for describing various physical, biological, chemical and other processes [1-4] and it was a subject of the numerous researches in a particular case of the equation (1) [1-4].

In particular values of parameters of the equation (1) ($k = 0, p = 2$ $\gamma(t) = 1, v(t) = 0$) was shown [1] the phenomenon of finite speed of distribution of perturbation: there exist a continuous function $l(t) > 0$ such that $u(t, x) \equiv 0$ when $|x| \geqslant l(t)$, $t > 0$. It is shown (see [1]) by the construction of self similar solution, the localization of the solution i.e. $u(t, x) \equiv 0$ $|x| \geqslant l(t) < +\infty$ takes place (see [1) and references in it]. The reason of this situation is a presence of an absorption in the equation (1). When $\gamma(t, x) = \gamma(t)$, $p = 2, k = 0$, $\beta = 1$, it was proved that takes place the phenomena of a "wall" [4] if carrying out of the certain relations between functions $v(t)$ and $\gamma(t)$. The global solvability and non solvability (J. L. Lyons's problem) for the problem (1)-(2) when $k = 0$, $v(t) \equiv 0$, $\gamma(t) = 1$; $k = 0$, $v(t) \equiv 0$, $\gamma(t) = 1$, $p = 2$ or $m = 1$, also the critical value of parameters at which the behavior of the solution of the problem (1) - (2) is chanced were established (see [1] and references in it). The surface $|x| = l(t)$ is called a free boundary or a front of perturbation.

Since the equation (1) in the domain where $u = 0$ is degenerate, then in the domain of degeneration it may can't have the classical solution. Therefore we consider the weak solution of the equation (1), having physical sense and satisfying to the equation (1) in sense of distribution.

2. Method of the nonlinear splitting for the problem

Below one manner of nonlinear splitting (decomposition) [2,4] for construction of the self similar, an approximately self similar equation to the equation (1) which is relatively easier for investigating of the qualitative properties of the solution of the problem (1)-(2) is offered. For this goal at first we solve the simple equation

$$\frac{d\bar{u}}{dt} = \varepsilon\gamma(t)\bar{u}^{\beta}. \tag{3}$$

After we will search solution of the equation (1) in the form

$$u(t, x) = \bar{u}(t)w(\tau, x), \tag{4}$$

where the function $\bar{u}(t)$ is the solution of the equation (3) and $w(\tau, x)$ is the solution of the equation (1) without lower members.

But instead of a variable t is used a new variable

$$\tau(t) = \int_0^t [\bar{u}(t)]^{k+m+p-3} \, dt. \tag{5}$$

Therefore substituting (4) into the equation (1) one gets the following approximately self similar equation

$$f^p \eta^{1-N} \frac{d}{d\eta}(\eta^{N-1} f^{m-1}|\frac{df}{d\eta}|^{p-2}\frac{df}{d\eta}) + \frac{\eta}{p}\frac{df}{d\eta}$$
$$+ \gamma(t)\tau(t)\overline{u}^{\beta-(k+p+m-2)}\varepsilon(-f+f^\beta) = 0 \quad (6)$$

where

$$\eta = \frac{|\xi|}{[\tau(t)]^{\frac{1}{p}}}, \quad \xi_i = x_i - \int_0^t v_i(t)dt, \quad i = 1,...N, \quad \xi = (\xi_1, \xi_2...\xi_N). \quad (7)$$

3. Condition of localization of the Cauchy problem

Now consider the Cauchy problem for (1), (2) in the case when $\beta = 1$, $\gamma(t) \neq 0$, $u_0(x) = q\delta(x)$, where q is the power of instantaneously source, $\delta(x)$ is Dirac function. We will show the condition of localization of solution of the problem (1)-(2).

Using the method of nonlinear splitting and the self similar presentation of the solution (1) in the form (6) it is easy prove that the problem (1), (2) have the following exact solution

$$u(t, x) = \tau(t)^{-\frac{N}{2+(k+p+m-3)N}} \overline{f}(\eta), \quad (8)$$

$$\overline{f}(\eta) = \begin{cases} l\left(a - b|\eta|^{p/(p-1)}\right)_+^{(p-1)/(p+m-3)}, & \text{if } p + m - 3 \neq 0; \\ \exp(-(\frac{\eta}{p})^p), & \text{if } p + m - 3 = 0, \end{cases}$$

where $b = (k+p+m-3)(\frac{1}{p})^{\frac{p}{p-1}}$, the notation $(a)_+$ means $(a)_+ = \max(0, a)$.
$\eta = |\xi|/\tau^{\frac{1}{2+(p+m-3)N}}$, if $k + p + m - 3 > 0$
and $\eta = |\xi|/t^{1/p}$, if $k + m + p - 3 = 0, T > 0$ is constant and the constant à will be found from the condition

$$\int_{R^N} u(t, x)dx = q.$$

For the case $u_0(x) \neq q\delta(x)$ the solution of the equation (1) is

$$u(t, x) = (T + \tau(t))^{-\frac{N}{2+(k+p+m-3)N}} \overline{f}(\eta),$$

where $\eta = |\xi|/(T + \tau(t))^{\frac{1}{2+(k+p+m-3)N}}$, if $k + p + m - 3 > 0$ and $\eta = |\xi|/(T + t)^{1/p}$, if $k + m + p - 3 = 0, T > 0$.

In fact in order to construct the solution (6) we introduce in (1) the replacement $u(t,x) = \bar{u}(t)w(\tau(t),x)$, where

$$\tau(t) = \int_0^t [\bar{u}(t)]^{k+p+m-3}dt, \quad \bar{u}(t) = \exp(\varepsilon \int_0^t \gamma(\eta)d\eta)$$

Putting in the equation $(1)w(\tau(t),\xi) = w(\tau(t),|\xi| = r)$, it is reduced to the radial symmetrical form

$$\frac{\partial w}{\partial t} = r^{1-N}\frac{\partial}{\partial r}(r^{N-1}w^{m-1}|\frac{\partial w}{\partial r}|^{p-2}\frac{\partial w}{\partial r}), \tag{9}$$

Theorem 1. Let $u(t,x)$ be the weak solution of the problem (9),(2) and $u_0(x) \leqslant u_+(0,x)$, where $u_+(t,x) = \bar{f}(\eta)$. Then in Q for weak solution $u(t,x)$ of the problem (1), (2) and for the free boundary $x_i(t)$ $i = 1,...N$ the estimates $u(t,x) \leqslant u_+(t,x)$,

$$\sum_{i=1}^{N}(x_i - \int_0^t v_i(t)dt) \leqslant (\frac{a}{b})^{\frac{p-1}{p}}\tau(t)^{\frac{1}{p}}$$

are valid.

Below on the basis of self similar and approximately the self similar approach and of a method of the standard equations the following condition of global solvability of the problem (1), (2) is proved.

4. Global solvability of solution, estimate of the free boundary

Theorem 2. Let the following conditions: $u_0(x) \leqslant u_+(0,x)$, $x \in R^N$, $\varepsilon = -1$, $\gamma(t)\tau(t)[\bar{u}(t)]^{\beta-(m+p-2)} \leqslant N/p$, $t > 0$ be valid. Then there exists a global solution $u(t,x)$ of the Cauchy problem (1), (2) for which in Q the estimate

$$u(t,x) \leqslant z_+ (t,x) = \bar{u}(t)\,\bar{f}(\eta)$$

and the estimate for the free boundary

$$\sum_{i=1}^{N}(x_i - \int_0^t v_i(t)dt) \leqslant (a/b)^{(p-1)/p}\tau^{1/p} \tag{10}$$

are valid.

Theorem 3. Let the following conditions: $u_0(x) \leqslant u_+(0,x)$, $x \in R^N$, $\varepsilon = +1$, $\gamma(t)\tau(t)[\bar{u}(t)]^{\beta-(k+m+p-2)} < N/p$, $t > 0$ be valid. Then there exists a global solution $u(t, x)$ of the Cauchy problem (1), (2) for sufficiently small $u_0(x)$, for which in Q the estimate

$$u(t, x) \leqslant u_+ (t, x)$$

and the estimate for the free boundary (10) are valid.

It is easy to check that when $\gamma(t, x) = 1$ in the equation (1), the condition of a global solvability is $\beta > k + m + p - 2 + p/N$, which generalize previously known results by H. Fujite, A.A Samarskii, V.A Galaktionov, S.P Kurdyomov, A.P. Mikhailov (see [1]) in particular value of parameters.

5. Results of numerical experiments and visualization

At the numerical solution of a problem the primary equation was approximated on a grid under the implicit circuit of variable directions (for a multidimensional case) in combination to the method of balance. Iterative processes were under construction on the method of Picard, Newton and a special method. Results of computational experiments show that all listed iterative methods are effective for the solution of nonlinear problems and leads to nonlinear effects if we will use as initial approximation the solutions of self similar equations constructed by a method of nonlinear splitting and by the method of standard equation [4].

References

1. A.A Samarskii, V.A Galaktionov, S.P Kurdyomov, A.P. Mikhailov, Blow-up in quasilinear parabolic equations. Berlin, 4, Walter de Grueter, 1995, 535 p.
2. M. Aripov, Asymptotic of the solutions of the non-Newton polytrophic filtration equation. ZAMM, vol. 80, Sup. 3, (2000), 767-768
3. I. Kombe, Double nonlinear parabolic equations with singular lower order term, Nonlinear Analysis 2004, 56, 185-199
4. M. Aripov, Methods of the standard Equation for Solutions of the Nonlinear Problems. 1988 Tashkent, FAN. pp.137.

Long-time behavior of periodic Navier-Stokes equations in critical spaces

Jamel Benameur

*Departement of Mathematics, College of Science, King Saud University,
Riadh 11451, Kingdom of Saudi Arabia
E-mail: jbenameur@ksu.edu.sa
www.ksu.edu.sa*

Ridha Selmi

*Département de Mathématiques, Faculté des Sciences de Gabès,
Cité Erriadh 6072, Tunisia
E-mail: ridha.selmi@isi.rnu.tn
www.fsg.rnu.tn*

We establish smoothing effects for the 3-D incompressible periodic Navier–Stokes equations for initial data in the critical Sobolev space $H^{1/2}(\mathbb{T}^3)$. We use this to prove that any global solution to the 3D periodic NSE decays exponentially fast to zero as time tends to infinity, as soon as the data (hence the solution) is mean free. Otherwise, the difference to the average do so.

Keywords: Periodic incompressible Navier–Stokes Equation; critical spaces; energy estimate; smoothing effects; long time behavior.

1. Introduction

We deal with the 3-D incompressible periodic Navier-Stokes equations:

$$(NS_\nu) \qquad \begin{cases} \partial_t u - \nu \Delta u + (u.\nabla)u = -\nabla p, & \text{in } \mathbb{R}^+ \times \mathbb{T}^3 \\ \text{div } u = 0 & \text{in } \mathbb{R}^+ \times \mathbb{T}^3 \\ u_{|t=0} = u^0 & \text{in } \mathbb{T}^3, \end{cases}$$

where $\nu > 0$ is the viscosity of the fluid, $u = u(t,x) = (u_1, u_2, u_3)$ and $p = p(t,x)$ denote respectively the unknown velocity and the unknown pressure of the fluid at the point $(t,x) \in \mathbb{R}^+ \times \mathbb{T}^3$. Here, the periodic box $\mathbb{T}^3 = [0, 2\pi)^3$ and $u^0 = (u_1^0, u_2^0, u_3^0)$ is a given initial velocity.

Some existence, uniqueness and regularity in time and space variables can be found in [4,5,9] and references therein. About long time behavior, author established in [13] the decay of high homogeneous Sobolev norm for the solutions to NSE in the 2D case. That is where the global existence is actually known. Also, in the case of the whole space [7], without giving any convergence rate, authors proved that for any time-continuous solution, the homogeneous Sobolev norm $\dot{H}^{\frac{1}{2}}$ have zero limit as time goes to infinity. This result can be extended to the periodic case. In [2], the following Theorem was proved.

Theorem 1.1. *Let $u^0 \in H^{1/2}(\mathbb{R}^3)$ be a divergence-free vector field. There exist a time $T > 0$ and a unique solution u to (NS_ν) satisfying:*

$$\forall 0 \le t \le T, \quad \int_0^t \int_{\mathbb{R}^3} |\xi|^2 e^{(\nu\tau)^{1/2}|\xi|} |\hat{u}(\tau, \xi)|^2 d\tau d\xi \le 4\|\nabla e^{\tau\frac{\nu}{2}\Delta} u^0\|^2_{L^2([0,t]\times\mathbb{R}^3)}.$$

Moreover, a constant c exist such that, if u^0 satisfy $\|u^0\|_{\dot{H}^{1/2}} < c\nu$, then

$$\int_{\mathbb{R}^+\times\mathbb{R}^3} |\xi|^2 e^{(\nu t)^{1/2}|\xi|} |\hat{u}(t, \xi)|^2 dt d\xi \le \frac{4}{\nu}\|u^0\|^2_{L^2(\mathbb{R}^3)}.$$

This result is closely related to our work. We note that in [2] the author do not gave any strong time decay result in the whole space $\mathbb{R}^3$. This is due to the difficulties to handle the low frequency part. Although, in the case of the torus, we are able to do so thanks to the possibility of representing those frequencies in the only point $k = 0$, by one hand, and the use of homogenous Sobolev spaces, in the other hand.

Our main results states that any global solution to the 3D periodic NSE decays exponentially fast to zero as time tends to infinity, as soon as the data (hence the solution) is mean free. Otherwise, the difference to the average do so. Explicitly, we have

Theorem 1.2. *If $u \in \mathcal{C}(\mathbb{R}^+, H^{1/2}(\mathbb{T}^3)) \cap L^2(\mathbb{R}^+, \dot{H}^{3/2}(\mathbb{T}^3))$ is a solution of (NS_ν) such that $\int_{\mathbb{T}^3} u(0, x)dx = 0$, then $u \in L^\infty(\mathbb{R}^+, H^{1/2})$, and for all $a \in (1/2, 1)$, there exist $t_a > 0$, such that*

$$\|u(t, .)\|_{H^{1/2}} \le c(1 - a)e^{-\nu a(t-t_a)}, \quad \forall t > t_a. \tag{1}$$

Note that the zero-mean value of the solution is a required hypothesis. In fact, $u(t, x) = (1, 1, 1)$ is the unique solution of (NS_ν) system, associated to the data $u^0 = (1, 1, 1)$ and $\|u(t, .)\|_{H^{1/2}} = 1 \nrightarrow 0$, when $t \to \infty$.

Theorem 1.3. *If $u \in \mathcal{C}(\mathbb{R}^+, H^{1/2}(\mathbb{T}^3)) \cap L^2(\mathbb{R}^+, \dot{H}^{3/2}(\mathbb{T}^3))$ is a solution of (NS_ν), then*

- $\forall t > 0$, $\displaystyle\int_{\mathbb{T}^3} u(t,x)dx = \int_{\mathbb{T}^3} u(0,x)dx,$ $\hspace{2cm}$ (2)

- $u \in L^\infty(\mathbb{R}^+, H^{1/2}),$ $\hspace{4cm}$ (3)

- $\forall a \in (0,1), \exists t_a > 0, \ \forall t > t_a$

$$\left\| u(t,.) - \int_{\mathbb{T}^3} u(0,x)dx \right\|_{H^{1/2}} \leq c(1-a)e^{-\nu a(t-t_a)}. \quad (4)$$

To prove these results, our main idea is to use a change of function that depends explicitly on time. This change of function leads to a non classical energy estimate that allows to derive the decay rate as time goes to infinity. In [13], such decay was inferred by means of ideas from [8,14] and the "Fourier spitting method" first developed in [12,13]. In [7], the zero limit was obtained using decomposition into high/low frequencies [1] and the "small solution theory" [10]. Combining methods from [7], and decay properties of the heat operator in a periodic box, we can reobtain by a simple argument our main results.

In the following, we give some notations and definitions. Then, we prove Theorem 1.2. The proof is based on Friedrich methods, classical product laws, compactness methods and technical results. Finally, the proof of Theorem 1.3 is inspired from the one of Theorem 1.2.

2. Notations

- For $k \in \mathbb{Z}^3$ and $x \in \mathbb{T}^3$, $\mathcal{F}(f)(k) = \hat{f}(k) = \displaystyle\int_{\mathbb{T}^3} \exp(-ix.k)f(x)dx$ denotes the Fourier transform and the inverse Fourier formula is defined by $\mathcal{F}^{-1}\left((a_k)_{k\in\mathbb{Z}^3}\right)(x) = \displaystyle\sum_{k\in\mathbb{Z}^3} \exp(ik.x)a_k.$

- For $s \in \mathbb{R}$, $H^s(\mathbb{T}^3)$ denotes the usual non homogeneous Sobolev space on $\mathbb{T}^3$ and $< .,. >_{H^s(\mathbb{T}^3)}$ is the associated scalar product. The homogeneous Sobolev space is defined by:

$$\dot{H}^s = \{f \in \mathcal{S}'(\mathbb{T}^3); \ \sum_{k\in\mathbb{Z}^3} |k|^{2s}|\hat{f}(k)|^2 < \infty\}.$$

- We denote by $\mathcal{C}_b(X)$ the space of bounded and continuous functions on the space X.

- If $f = (f_1, f_2, f_3)$ and $g = (g_1, g_2, g_3)$ are two vector fields, we set

$$f \otimes g := (g_1 f, g_2 f, g_3 f),$$

and

$$\mathrm{div}\,(f \otimes g) := (\mathrm{div}\,(g_1 f), \mathrm{div}\,(g_2 f), \mathrm{div}\,(g_3 f)).$$

3. Proof of Theorem 1.2

We begin by recalling the following product laws stated in [3].

Lemma 3.1. *Let s, s' tow reals numbers such that $s < 3/2$ and $s + s' > 0$. There exists a positive constant $C := C(s, s')$, such that for all $f \in \dot{H}^s(\mathbb{T}^3)$ and $g \in \dot{H}^{s'}(\mathbb{T}^3)$,*

$$\|fg\|_{\dot{H}^{s+s'-\frac{3}{2}}(\mathbb{T}^3)} \leq C\Big(\|f\|_{\dot{H}^s(\mathbb{T}^3)}\|g\|_{\dot{H}^{s'}(\mathbb{T}^3)} + \|f\|_{\dot{H}^{s'}(\mathbb{T}^3)}\|g\|_{\dot{H}^s(\mathbb{T}^3)}\Big).$$

If $s, s' < 3/2$ and $s + s' > 0$, there exists a constant $c = c(s, s')$,

$$\|fg\|_{\dot{H}^{s+s'-\frac{3}{2}}(\mathbb{T}^3)} \leq c\|f\|_{\dot{H}^s(\mathbb{T}^3)}\|g\|_{\dot{H}^{s'}(\mathbb{T}^3)}.$$

In a first step, we prove the proposition below.

Proposition 3.1. *Let $a \in (1/2, 1)$. Let $u^0 \in H^{1/2}(\mathbb{T}^3)$ a divergence-free vector field, such that $\|u^0\|_{\dot{H}^{1/2}} \leq (1-a)c\nu$ and $\int_{\mathbb{T}^3} u^0 = 0$, then there exists a unique solution $u_\nu \in \mathcal{C}_b(\mathbb{R}^+, H^{1/2}(\mathbb{T}^3)) \cap L^2(\mathbb{R}^+, \dot{H}^{3/2}(\mathbb{R}^3))$. Moreover,*

$$\int_{\mathbb{T}^3} u_\nu(t, x)dx = 0, \quad \forall t \geq 0, \tag{5}$$

and

$$\|u_\nu(t)\|_{\dot{H}^{1/2}} \leq c\nu(1 - a)e^{-\nu at}, \ \forall t \geq 0. \tag{6}$$

Proof. For a strictly positive integer n, the Friedrich's operator J_n is

$$J_n(f) := \mathcal{F}^{-1}\Big(\mathbf{1}_{\{|k|<n\}}\mathcal{F}(f)(k)\Big).$$

Using Friedrich's method, we consider the following approximate Navier–Stokes system $(NS_{n,\nu})$ in $\mathbb{R}_+ \times \mathbb{T}^3$,

$$\begin{cases} \partial_t u - \nu\Delta J_n u + J_n\mathrm{div}\,(J_n u \otimes J_n u) = \nabla\Delta^{-1}J_n\mathrm{div}\,(J_n u \otimes J_n u), \\ u_{|t=0} = J_n u_0. \end{cases}$$

By the ordinary differential equations theory, the system $(NS_{n,\nu})$ has a unique maximal solution $u_{n,\nu}$ in the space $\mathcal{C}^1([0, T_n^*), L^2(\mathbb{T}^3))$. Using uniqueness and the fact that J_n is a projector we obtain

$$\begin{cases} \partial_t u_{n,\nu} - \nu\Delta u_{n,\nu} + J_n\mathrm{div}\,(u_{n,\nu} \otimes u_{n,\nu}) = \nabla\Delta^{-1}J_n\mathrm{div}\,(u_{n,\nu} \otimes u_{n,\nu}) \\ u_{n,\nu|t=0} = J_n u_0. \end{cases}$$

Taking the scalar product in $L^2(\mathbb{T}^3)$, we obtain that for $t \in [0, T_n^*)$,

$$\partial_t \|u_{n,\nu}\|_{L^2}^2 + 2\nu \|\nabla u_{n,\nu}\|_{L^2}^2 \leq 0.$$

It follows that for all $t \in [0, T_n^*)$, $\|u_{n,\nu}(t)\|_{L^2}^2 \leq \|u^0\|_{L^2}^2$, which implies $T_n^* = +\infty$, and the following estimate holds for all $t \geq 0$:

$$\|u_{n,\nu}(t)\|_{L^2}^2 + 2\nu \|\nabla u_{n,\nu}\|_{L_t^2(L^2)}^2 \leq \|u^0\|_{L^2}^2. \tag{7}$$

Let $v_{n,\nu,a} := \mathcal{F}^{-1}\left(e^{a\nu t|k|}\, \hat{u}_{n,\nu}\right)$, by Fourier transform one has (I_1):

$$(I_1) \quad \partial_t \hat{v}_{n,\nu,a} + \nu|k|(|k| - a)\, \hat{v}_{n,\nu,a} + e^{a\nu t|k|}\mathcal{F}\Big(J_n(u_{n,\nu}.\nabla u_{n,\nu})\Big)$$
$$= e^{a\nu t|k|}\mathcal{F}\Big(\nabla \Delta^{-1} J_n(u_{n,\nu}.\nabla u_{n,\nu})\Big)$$

Taking $\overline{I}_1.\hat{v}_{n,\nu,a} + I_1.\overline{\hat{v}}_{n,\nu,a}$, one obtains

$$\partial_t |\hat{v}_{n,\nu,a}|^2 + 2\nu|k|(|k| - a)|\hat{v}_{n,\nu,a}|^2$$
$$= -2\Re\Big(e^{a\nu t|k|}\mathcal{F}\big(J_n(u_{n,\nu}.\nabla u_{n,\nu})\big)\overline{\hat{v}}_{n,\nu,a}\Big).$$

Using the inequalities:

$$e^{a|k|} \leq e^{a|k-p|}e^{a|p|} \quad \forall k, p \in \mathbb{Z}^3$$

and

$$(1 - a)|k|^2 \leq |k|(|k| - a) \quad \forall k, p \in \mathbb{Z}^3,$$

one obtains (I_2):

$$(I_2): \partial_t |\hat{v}_{n,\nu,a}|^2 + 2\nu(1 - a)|k|^2|\hat{v}_{n,\nu,a}|^2 \leq 2|\hat{v}_{n,\nu,a}| * |\hat{v}_{n,\nu,a}|.|\widehat{\nabla v}_{n,\nu,a}|,$$

where "$*$" the convolution product. Here, we note that the second elementary inequality does not hold in the whole space because of the rang of low frequencies ξ such that $|\xi| < 1$, so that our technics are suitable only for the periodic case.

Using lemma 3.1, identity (I_2) leads to:

$$\partial_t \|v_{n,\nu,a}\|_{\dot{H}^{1/2}}^2 + 2\nu(1-a)\|\nabla v_{n,\nu,a}\|_{\dot{H}^{1/2}}^2 \leq C\|v_{n,\nu,a}\|_{\dot{H}^{1/2}}\|\nabla v_{n,\nu,a}\|_{\dot{H}^{1/2}}^2. \tag{8}$$

Let

$$T_{n,\nu,\varepsilon} := \sup\{t \geq 0,\ \|v_{n,\nu,a}\|_{L_t^\infty(\dot{H}^{1/2})} < 2(1 - a)c\nu\}.$$

By a classical energy argument, $T_{n,\nu,\varepsilon} = +\infty$ and for all $t \geq 0$,

$$\|v_{n,\nu,a}(t)\|_{\dot{H}^{1/2}}^2 + \nu(1 - a)\int_0^t \|\nabla v_{n,\nu,a}\|_{\dot{H}^{1/2}}^2 \leq \|u^0\|_{\dot{H}^{1/2}}^2. \tag{9}$$

Finally, a standard compactness argument gives the global existence result: $u_\nu \in \mathcal{C}_b(\mathbb{R}^+, H^{1/2}(\mathbb{T}^3)) \cap L^2(\mathbb{R}^+, \dot{H}^{3/2}(\mathbb{T}^3))$ of (NS_ν), such that for all $t \geq 0$,

$$\sum_k e^{2a\nu t|k|}|k|.|\widehat{u_\nu}(t,k)|^2 d + \nu(1-a)\int_0^t \sum_k e^{2a\nu t|k|}|k|^3|\widehat{u_\nu}(\tau,k)|^2 d\tau$$
$$\leq 2\|u^0\|_{\dot{H}^{1/2}}^2.$$
(10)

Multiplying by $e^{-2a\nu t}$, one deduces inequality (6). To obtain equality (5), one integrates the first equation of (NS_ν) over the torus $\mathbb{T}^3$.

In a **second step**, we consider a solution u of (NS_ν) in $\mathcal{C}_b(\mathbb{R}^+, H^{1/2}(\mathbb{T}^3)) \cap L^2(\mathbb{R}^+, \dot{H}^{3/2}(\mathbb{T}^3))$, such that $\int_{\mathbb{T}^3} u = 0$. Moreover, let a subset A of $\mathbb{R}$ defined by

$$A := \{t > 0 : \|u(t)\|_{H^{1/2}} \geq c\nu(1-a)\}.$$

Denote by $\mathbf{1}_A$ its characteristic function and $\lambda(A)$ its Lebesgue measure. Using the fact that $\int_{\mathbb{T}^3} u = 0$, we obtain that for all $t > 0$,

$$\left[(1-a)c\nu\mathbf{1}_A(t)\right]^2 \leq \|u(t)\|_{H^{1/2}}^2 \leq \|\nabla u(t)\|_{H^{1/2}}^2.$$

So that

$$\lambda(A) \leq \frac{\|u\|_{L^2(\mathbb{R}^+,H^{3/2})}^2}{(1-a)^2 c^2 \nu^2} := T_a.$$

Since $t \to \|u(t)\|_{H^{1/2}}$ is continuous on $\mathbb{R}^+$ then it is Lebesgue measurable. The measure of the set A is strictly inferior then the one of interval $(0, T)$, so for all $T > T_a$, there exists $t_a \in (0, T)$ where t_a do not belong to A. Then, t_a is such that $\|u(t_a)\|_{H^{1/2}} < c\nu(1-a)$. Then by proposition (3.1), the following system

$$\begin{cases} \partial_t v - \nu\Delta v + (v.\nabla)v = -\nabla p, & \text{on } \mathbb{R}^+ \times \mathbb{T}^3, \\ div\,(v) = 0 & \text{on } \mathbb{R}^+ \times \mathbb{T}^3, \\ v_{|t=0} = u(t_a) & \text{on } \mathbb{T}^3, \end{cases}$$

has a unique solution $u \in \mathcal{C}_b(\mathbb{R}^+, H^{1/2}(\mathbb{T}^3)) \cap L^2(\mathbb{R}^+, \dot{H}^{3/2}(\mathbb{T}^3))$ satisfying (6). By an existence result due to [6], we have $v_\nu(t) = u(t-t_a)$, $\forall t \geq t_a$ which establish (1). Since the $H^{1/2}$ norm of u is time continuous on the compact $[0, t_a]$ and bounded on $(t_a, +\infty)$, then $u \in L^\infty(\mathbb{R}^+, H^{1/2})$.

4. Proof of Theorem 1.3

To obtain equality (2), one integrates the first equation of (NS_ν) over $\mathbb{T}^3$. Inequality (4) is a direct consequence of Theorem (1.2). To prove assertion (3), we note that we have

$$\int_0^t \|u\|_{\dot{H}^{1/2}}^4 d\tau \leq \int_0^t \|u\|_{L^2}^2 \|\nabla u\|_{L^2}^2 d\tau \leq \frac{1}{2\nu}\|u^0\|_{L^2}^4,$$

then, $u \in L^4(\mathbb{R}^+, \dot{H}^{1/2})$. Let, $a \in (1/2, 1)$ and the set of time t such that $\|u(t)\|_{H^{1/2}} < c\nu(1-a)$ and we proceed as in the proof of theorem 1.2.

References

1. C. P. Calderón, *Existence of Weak Solutions for the Navier-Stokes Equations with Initial Data in L^p*, Tran. Amer. Math. Soc., 318 (1), pages 179–200, (1990).

2. J.-Y. Chemin, *Le système de Navier–Stokes incompressible soixante dix ans après Jean Leray*, Société Mathématique de France 2004, Séminaire et Congrès 9, pages 99–123, 2004.

3. J.-Y. Chemin, *About Navier–Stokes equations*, Publication du Laboratoire Jaques-Louis Lions, Université de Paris VI, R96023, 1996.

4. J.-Y. Chemin and I. Gallagher, *On the global wellposedness of the 3-D Navier–Stokes equation with large initial data*, Annales de l'école Normale supérieure, 39, pages 679–698, 2006.

5. H. Fujita and T. Kato, *On the Navier-Stokes initial value problem I*, Archive for rational mechanics and analysis 16, pages 269–315, 1964.

6. I. Gallagher, *Application of Schochet's Methods to Parabolic Equation*, Journal des Mathématiques Pures et Appliquées, 77, pages 989–1054, 1998.

7. I. Gallagher, D. Iftimie, F. Planchon, *Non-Explosion en Temps Grand et Stabilité de Solutions Globales des Équations de Navier-Stokes,* C. R.Acad. Sci. Paris, Ser. I 334, pages 289–292, (2002).

8. T. Kato, *Strong L^p Solutions of the Navier-Stokes Equations in $\mathbb{R}^n$ with Applications to Weak Solutions*, Math. Zeit., 187, pages 471-480, (1984).

9. J. Leray, *Essai sur le mouvement d'un liquide visqueux emplissant l'espace*, Acta mathematica, 63, pages 22–25, 1933.

10. F. Planchon, *Global Strong Solutions in Sobolev or Lebesgue Spaces to the Incompressible Navier-Stokes Equations in $\mathbb{R}^3$*, Ann. Inst. H. Poicaré Anal. Non Linéaire, 13, Pages 319-336, 1996.

11. M. Schonbek, *Large Time Behaviour of Solutions to the Navier-Stokes Equations in H^m Spaces*, Comm. Part. Diff. Eq, 20, Pages 103-117, 1995.

12. M. Schonbek, *L^2 Decay for Weak Solutions to the Navier-Stokes Equation*, Ardi. Rat. Mech. Anal. Pages 209-222, 1985.

13. M. Schonbek, *Decay of Parabolic Conservation Laws*, Communications in Part. Diff. Eq, 7, Pages 449-473, 1980.

14. M. Wiegner, *Decay Results for Weak Solutions of the Navier-Stokes Equations in $\mathbb{R}^n$*, J. London Math. Soc. (2), 35, Pages 363-413, 1987.

Regularity theory in asymptotic extensions of topological modules and algebras

Maximilian F. Hasler

*Laboratoire AOC & CEREGMIA, Université des Antilles et de la Guyane,
B.P. 7209, 97275 Schoelcher, Martinique, France
E-mail: mhasler @ univ-ag.fr*

Based on a previously established framework of asymptotic extensions of topological modules and algebras, we use results concerning sheaf theoretic properties and functoriality of the construction, to develop some tools for microlocal algebraic analysis in this setting. In particular, we introduce the notion of (ϕ, F)–singular spectrum of sections $\mathbf{f} \in \mathcal{G}_M(E)$, and give results concerning its linear and non-linear properties.

Keywords: Asymptotic analysis; regularity theory; topological module or algebra; ultrametric topology; sequence spaces; algebra of generalized functions.

1. Introduction

Results presented in this paper extend our construction of asymptotic extensions of topological modules and algebras.[1] This framework can be seen as a generalisation and unification of several approaches to algebras of New Generalized Functions in the spirit of Colombeau,[2] Delcroix and Scarpalézos' asymptotic algebras,[3,4] Marti's $(\mathcal{C}, \mathcal{E}, \mathcal{P})$–algebras,[5] and the more general sequence space approach of Delcroix, Hasler, Pilipović and Valmorin reviewed in,[6] which also allows for inductive limit spaces arising in theories of ultradistributions.[7]

The present approach avoids the limitation to locally convex topological vector spaces, while preserving relatively simple definitions. Allowing for topological modules already as an *input*, we are able to iterate the construction, which is not possible otherwise. As we show in the examples, it is nonetheless immediate to get back the known situation of Colombeau algebras based on topological vector spaces. Here we extend our previous work by elements of regularity theory, which will be developed further elsewhere.

2. The M–extended topology on $(E^\Lambda, +)$

Let R denote a topological ring, and Λ an arbitrary set of indices with a given filter base $\mathcal{B}_\Lambda$. The first and most fundamental step in our construction is to define a topology on the sequence space E^Λ, by "stretching" neighborhoods of zero in the topological R–module E by sequences $(r_\lambda) \in M \subset R^\Lambda$ defining the growth scale. Here we don't give proofs of results established elsewhere.[1]

Definition 2.1. For a given subset $M \subset R^\Lambda$ and a topological R–module E, let
$$\mathcal{B}_M(E) = \left\{ B_r^V \; ; \; r \in M, \ V \in \mathcal{V}(o_E) \right\}$$

where $\mathcal{V}(o_E)$ denotes the set (or a basis) of neighborhoods of o_E, and
$$B_r^V = \left\{ f \in E^\Lambda \mid \exists \omega \in \mathcal{V}(o_R) : \omega f \underset{(\Lambda)}{\subset} r V \right\} .$$

The subscript (Λ) means that the relation holds "asymptotically with respect to the given filter base $\mathcal{B}_\Lambda$ on the index set Λ", *i.e.*,
$$\omega f \underset{(\Lambda)}{\subset} r V \iff \exists \Lambda' \in \mathcal{B}_\Lambda \ \forall \lambda \in \Lambda' : \omega f_\lambda \subset r_\lambda V.$$

In the sequel it shall be tacitely understood that any relation involving objects indexed by $\lambda \in \Lambda$ is considered to hold only on some $\Lambda' \in \mathcal{B}_\Lambda$.

Example 2.1. If E is a locally convex vector space and the seminorm p is the gauge of the absolutely convex set $V = \{u \in E \mid p(u) < 1\}$, then
$$B_r^V = \left\{ f \in E^\Lambda \mid p(f) = O(r) \right\} .$$

Theorem 2.1. *The set $\mathcal{B}_M(E)$ is a basis of neighbourhoods of zero on $(E^\Lambda, +)$, if $M \subset R^\Lambda$ is nonvoid and $\forall r, s \in M \ \exists t \in M : t = o(r) \wedge t = o(s)$.*

Here we write, for $r, s \in R^\Lambda$: $r = o(s) \iff \forall \omega \in \mathcal{V}(o_R) : r \subset s\,\omega$, using again the above convention $(r \in s\omega \iff \exists \Lambda' \in \mathcal{B}_\Lambda, \ \forall \lambda \in \Lambda', \ r_\lambda \subset s_\lambda \omega)$.

Definition 2.2. Let $\tau_M(E)$ denote the topology induced on $(E^\Lambda, +)$ by the basis $\mathcal{B}_M(E)$ of neighbourhoods of zero. Let
$$E_M^\circ = \overline{\{o_{E^\Lambda}\}}^{\,\tau_M(E)} = \bigcap_{B \in \mathcal{B}_M(E)} B$$

be the closure of zero with respect to $\tau_M(E)$. Then, given a subspace F of $(E^\Lambda, +)$, the factor space F/E_M° is the associated Haussdorff space and will be denoted by $\widetilde{F}$ in the sequel.

Example 2.2. In Colombeau theory this closure of zero is called the ideal of "negligible sequences". If $(E, \mathcal{P})$ is a semi-normed vector space and $M = \{\mathbf{a}_m\}_{m \in \mathbb{Z}} \subset (\mathbb{R}_+^*)^\Lambda$ is an asymptotic scale in the sense of Delcroix and Scarpalézos,[3] then $E_M^\circ = \{f \in E^\Lambda \mid \forall p \in \mathcal{P} \ \forall m \in \mathbb{Z} : p(f) = o(\mathbf{a}_m)\}$.

Remark 1. The non-archimedean nature of $\tau_M(E)$ is seen from the fact that for any $B \in \mathcal{B}_M(E)$ and $f \in B$, we have $\alpha f \in B$ and $n f \in B$ for any $\alpha \in R$ and any $n \in \mathbb{N}$.

Remark 2. It should be clear that all of the above applies in particular to $E = R$, *i.e.*, we have the topology $\tau_M(R)$ on the space of nets of scalars $(R^\Lambda, +)$, and the associated Hausdorff space $\widetilde{R^\Lambda} = R^\Lambda / R_M^\circ$.

3. The M–extension of E

Definition 3.1. We call M–bounded the nets in
$$E_M = \{f \in E^\Lambda \mid \forall B \in \mathcal{B}_M(E) \ \exists \beta \in \mathcal{B}_M(R) : \beta f \in B\}$$
and define the M–extension of E as the associated Hausdorff space
$$\mathcal{G}_M(E) := \widetilde{E}_M = E_M / E_M^\circ .$$
Elements of $\mathcal{G}_M(R) = \widetilde{R}_M$ are called M–generalized numbers.

Example 3.1. In Colombeau theory, the M–bounded nets are called "moderate". For $M = \{(\varepsilon^m)_{\varepsilon \in (0,1)} ; m \in \mathbb{N}\}$ and a semi-normed vector space $(E, \mathcal{P})$, $E_M = \{f \in E^{(0,1)} \mid \forall p \in \mathcal{P} \ \exists m \in \mathbb{N} : p(f_\varepsilon) = O(\varepsilon^{-m})\}$.

Prop 3.1. We assume throughout this paper that M contains a sequence with zero limit. Then the canonical ("constant") embedding
$$\mathbf{i}_0 : E \to \widetilde{E}_M ; \ f \mapsto (f)_\lambda + E_M^\circ$$
is indeed an injection and allows to consider E as a subspace of $\widetilde{E}_M$. However, $\mathbf{i}_0$ is not continuous for the respective topologies, since in view of the ultrametric topology $\tau_M(E)$, any constant sequence remains "far" from $o \in \widetilde{E}_M$.

Remark 1. The condition defining the set E_M is exactly the condition for continuity of (component-wise) multiplication $R^\Lambda \times E^\Lambda \to E^\Lambda$ in the point (o, f), with respect to the topologies $\tau_M(R)$ and $\tau_M(E)$. It turns out that on the space of M–bounded nets, multiplication will also be continuous at the origin and in a point $(x, o) \in R_M \times E_M$:

Theorem 3.1. *With the preceding conventions, $\mathcal{G}_M(R) = \widetilde{R}_M$ is a topological ring and $\mathcal{G}_M(E) = \widetilde{E}_M$ is a topological $\widetilde{R}_M$–module.*

4. Functoriality and sheaf theory

It is natural to expect the following result, proved in:[1]

Lemma 4.1. *Continuous R–linear maps $\varphi \in \mathcal{L}(E, F)$ canonically extend to continuous $\widetilde{R}_M$–linear maps*

$$\mathcal{G}_M(\varphi) \in \mathcal{L}(\mathcal{G}_M(E), \mathcal{G}_M(F)) \ : \ (f_\lambda)_\lambda + E_M^0 \mapsto (\varphi(f_\lambda))_\lambda + F_M^0 \ .$$

Indeed, linearity ensures that the image of a representative will again respect the bounds defining F_M, and together with continuity implies the independence of the representative.

Given that continuous linear maps are the morphisms in the category of topological modules and that we already defined $\mathcal{G}_M(E)$ for any topological R–module E, we have:

Corollary 4.1. $\mathcal{G}_M$ *is a functor* $\mathbf{ModTop}_R \to \mathbf{ModTop}_{\mathcal{G}_M(R)}$, *where* $\mathbf{ModTop}_R$ *denotes the category of topological R–modules.*

Prop 4.1. *If E is a topological A–algebra (e.g., for $A = R$ or $A = E$), then polynomial maps*

$$\varphi : E \to E \ ; \ f \mapsto \sum a_k \, f^k \quad (a_i \in A)$$

canonically extend to continuous maps $\mathcal{G}_M(\varphi)$ on the A–algebra $\widetilde{E}_M$.

We now proceed to generalize the functor $\mathcal{G}_M$ to *sheaves* of R–modules or algebras, which will be essential for the following development of regularity theory.

Theorem 4.1. *If E is a presheaf of topological R–algebras over a topological space X, then*

$$\mathcal{G}_M(E) : \ \Omega \mapsto \mathcal{G}_M(E)(\Omega) := \mathcal{G}_M(E(\Omega))$$

is a presheaf of topological $\widetilde{R}_M$–algebras over X.
Moreover, if E is a fine sheaf, then $\mathcal{G}_M(E)$ is a fine sheaf.

Since a fine sheaf is defined in terms of a locally finite family of sheaf morphisms summing up to the identity morphism, the latter statement is a direct corollary of the Lemma 4.1 about canonical extensions of such morphisms.

This opens the way to the development of most of the usual machinery of sheaf theory applied to local analysis, *e.g.*, the following standard

Definition 4.1. For a sub(pre)sheaf $\mathcal{F}$ of $\widetilde{E}_M$, the $\mathcal{F}$–("singular") support of a section $\mathbf{f} \in \widetilde{E}_M$ is the closed set

$$\mathop{\mathrm{supp}}_{\mathcal{F}}(\mathbf{f}) = X \setminus \{x \in X \mid \exists V \in \mathcal{V}(x) : \mathbf{f}|_V \in \mathcal{F}(V)\} \ .$$

In particular, taking the subsheaf $\mathcal{F} = \{o\}$ we get the support $\mathrm{supp}(\mathbf{f})$ in the usual sense, and for $\mathcal{F} = E$, we get the "singular support" $\mathrm{sing\,supp}(\mathbf{f})$.

More generally, if E is contained in a larger space, then we associate a subsheaf of $\widetilde{E}_M$ to this space as follows:

Definition 4.2. If E is a sub(pre)sheaf of F, then we denote by $\mathcal{F}_F$ the sub(pre)sheaf of $\widetilde{E}_M$ defined by

$$\mathcal{F}_F : \Omega \to \left\{ \mathbf{f} \in \widetilde{E}_M(\Omega) \mid \lim f_\lambda \in F(\Omega) \right\} \ .$$

With this definition, we have two different notions of E–singular support: The presheaf $E = \mathbf{i}_0(E)$ is obviously a subpresheaf of $\mathcal{F}_E$, and thus

$$\forall \mathbf{f} \in \widetilde{E}_M(\Omega) : \quad \mathop{\mathrm{supp}}_{E}(\mathbf{f}) \supset \mathop{\mathrm{supp}}_{\mathcal{F}_E}(\mathbf{f}) \ .$$

It turns out that for the purpose of regularity theory as developed in the next chapter, the space $\mathcal{F}_E$ will be more appropriate than the space $\mathbf{i}_0(E)$, even if the latter seems more natural to consider.

5. Regularity theory

The singular spectrum we introduce below specifies the points, but also the degree of irregularity of $\mathbf{f} \in \widetilde{E}_M$ with respect to a subsheaf $\mathcal{F}$ of regular elements. This subsheaf $\mathcal{F}$ will usually be an R–module, but not an $\widetilde{R}_M$–module. Then we measure how small a generalized number $\phi \in \widetilde{R}_M$ must be in order to have $\phi \mathbf{f} \in \mathcal{F}$ in a neighborhood of a given point.

Definition 5.1. Consider a probing function $\phi : \Gamma \to \widetilde{R}_M$ defined on a partially ordered monoid $(\Gamma, +, \preceq)$ with minimal element 0, such that

$$\phi(0) = 1_{\widetilde{R}_M} \ ; \ \forall m, m' \in \Gamma \setminus \{0\} : \phi(m) = o(1), \ \phi(m+m') = O(\phi(m)\phi(m'))$$

Then we define the $(\phi, \mathcal{F})$–(singular) spectrum of $\mathbf{f} \in \widetilde{E}_M(\Omega)$ as

$$\mathop{\mathrm{spec}}_{(\phi,\mathcal{F})}(\mathbf{f}) = \{ (x,m) \in \Omega \times \Gamma \mid m \in \Gamma \setminus N_x(\mathbf{f}) \}$$

where $N_x(\mathbf{f}) = \{ m \in \Gamma \mid \exists V \in \mathcal{V}(x) : \phi(m)\,\mathbf{f}|_V \in \mathcal{F}(V) \} \ .$

Example 5.1. In the setting of Colombeau's algebra (Ex. 3.1), take $\Gamma = \mathbb{R}_+$ and $\phi(m) = [(\varepsilon^m)_\varepsilon]$. Then $\mathrm{spec}_{(\phi,\mathcal{F})}(\mathbf{f})$ is the parametric spectrum of $\mathbf{f}$ as considered by Marti.[8] This can be generalized to multi-parameter algebras, say $\Lambda = (0,1)^k$, with $\Gamma = \mathbb{R}_+^k$ and $\phi(m) = \left[(\lambda_1^{m_1} \cdots \lambda_k^{m_k})_\lambda\right]_{R_M^\circ}$. Usually, ϕ should respect the scale M used to define $\widetilde{E}_M$. When the scale is generated by a set $B = \left\{b^{(i)}; i \in I\right\}$, then a natural way to achieve this is to take powers of the families $b^{(i)}$, *i.e.*, $\Gamma = \mathbb{N}^{(I)}$ and $\phi(m) = \left[\prod_{i \in I}(b^{(i)})^{m_i}\right]_{R_M^\circ}$.

Remark 1. From the properties of ϕ, we see that $m \in N_x(\mathbf{f})$ implies that the whole cone $m + \Gamma \subset N_x(\mathbf{f})$, and thus $N_x(\mathbf{f}) = \bigcup_{m \in N_x(\mathbf{f})} (m + \Gamma)$.

Remark 2. The set $\Sigma_x(\mathbf{f}) = \Gamma \setminus N_x(\mathbf{f})$ appearing in the definition of $\mathrm{spec}_{(\phi,\mathcal{F})}\mathbf{f}$ could be said "the fibre of badness" of $\mathbf{f}$ at $x \in \Omega$. It is empty iff $0 \in N_x(\mathbf{f})$, *i.e.*, iff $\lim (f_\lambda)_{\lambda \in \Lambda}$ exists in $\mathcal{F}$. This can also be stated as

Corollary 5.1. *The projection of the* $(\phi,\mathcal{F})$*–singular spectrum of* $\mathbf{f} \in \widetilde{E}_M(\Omega)$ *on the first component,* $x \in \Omega$*, is the* $\mathcal{F}$*–singular support* $\mathrm{supp}_{\mathcal{F}}(\mathbf{f})$*.*

5.1. *Linear properties of the singular spectrum*

Obviously we have for any subsheaf $\mathcal{F}$ of $\widetilde{E}_M$ and any probing function ϕ the inclusion $\mathrm{spec}_{(\phi,\mathcal{F})}(\mathbf{f} + \mathbf{g}) \subset \mathrm{spec}_{(\phi,\mathcal{F})}(\mathbf{f}) \cup \mathrm{spec}_{(\phi,\mathcal{F})}(\mathbf{g})$, for all $\mathbf{f}, \mathbf{g} \in \widetilde{E}_M$.

Prop 5.1. Let $L \in \mathcal{L}(F)$ be a continuous linear operator on a presheaf F which contains E as a subpresheaf, and denote $\mathbf{L} = \mathcal{G}_M(L)$ the canonical extension of L (restricted to E) to $\widetilde{E}_M$. Then,

$$\forall \mathbf{f} \in \widetilde{E}_M : \quad \mathrm{spec}_{(\phi,\mathcal{F}_F)}(\mathbf{L}\mathbf{f}) \subset \mathrm{spec}_{(\phi,\mathcal{F}_F)}(\mathbf{f}).$$

Proof. If $\lim \phi_\lambda u_\lambda|_V \in F(V)$ for some open set V, then, using continuity and linearity, we have $\lim \phi_\lambda L(u_\lambda)|_V = L(\lim \phi_\lambda u_\lambda|_V) \in L(F(V)) \subset F(V)$. Therefore, $m \in N_x(\mathbf{f}) \implies m \in N_x(\mathbf{L}\mathbf{f})$ at any point $x \in \Omega$, and taking complements gives the desired result. $\square$

Corollary 5.2. *If F is a presheaf of differential R–modules, E a subpresheaf of (differential) algebras, and $P(D) = \sum \xi_\alpha D^\alpha$ a multivariate polynomial in the derivatives D, with coefficients $\xi_\alpha \in E$, then*

$$\mathrm{spec}_{(\phi,\mathcal{F}_F)}(P(D)\mathbf{f}) \subset \mathrm{spec}_{(\phi,\mathcal{F}_F)}(\mathbf{f}).$$

Example 5.2. In the usual Colombeau algebra, we can take $F = \mathcal{D}'$.

5.2. *Nonlinear properties*

To study nonlinear properties of the singular spectrum, we assume that E is a topological algebra, such that $\widetilde{E}_M$ is also an algebra.

Theorem 5.1. *If $\mathcal{F}$ is a presheaf of algebras, then for* $\mathbf{f}, \mathbf{g} \in \widetilde{E}_M$,

$$\operatorname*{spec}_{(\phi,\mathcal{F})}(\mathbf{f}\,\mathbf{g}) \subset S(\mathbf{f},\mathbf{g}) \cup S(\mathbf{g},\mathbf{f}) \cup T(\mathbf{f},\mathbf{g})$$

where $S(\mathbf{f},\mathbf{g}) = \{(x,m) \in (\operatorname*{supp}_{\mathcal{F}}(\mathbf{f}) \setminus \operatorname*{supp}_{\mathcal{F}}(\mathbf{g})) \times \Gamma \mid m \in \Gamma \setminus N_x(\mathbf{f})\}$ *and*

$$T(\mathbf{f},\mathbf{g}) = \{(x,m) \in (\operatorname*{supp}_{\mathcal{F}}(\mathbf{f}) \cap \operatorname*{supp}_{\mathcal{F}}(\mathbf{g})) \times \Gamma \mid m \in \Gamma \setminus (N_x(\mathbf{f}) + N_x(\mathbf{g}))\} \ .$$

Remark 3. Note that $N_x(\mathbf{f}) + N_x(\mathbf{g}) = \emptyset$ if $N_x(\mathbf{f}) = \emptyset$ or $N_x(\mathbf{g}) = \emptyset$; otherwise, the sum is *smaller* than either of the sets on its own, *cf.* Rem. 1.

Proof. If $m \in N_x(\mathbf{f})$ and $m' \in N_x(\mathbf{g})$, then $m + m' \in N_x(\mathbf{f}\,\mathbf{g})$, from the last property of ϕ in Def. 5.1 and the fact that $\mathcal{F}(X)$ is an algebra. Thus, $\Gamma \setminus N_x(\mathbf{f}\,\mathbf{g}) \subset \Gamma \setminus (N_x(\mathbf{f}) + N_x(\mathbf{g}))$, *i.e.*, $\operatorname*{spec}_{(\phi,\mathcal{F})}(\mathbf{f}\,\mathbf{g}) \subset \{(x,m) \mid m \notin N_x(\mathbf{f}) + N_x(\mathbf{g})\}$. The result follows by distinguishing the cases $x \in \operatorname*{supp}_F(\mathbf{f})$ and/or $x \in \operatorname*{supp}_F(\mathbf{g})$ and using Remark 1. $\square$

Corollary 5.3. *If P is a polynomial of degree k with coefficients in E, then*

$$\operatorname*{spec}_{(\phi,\mathcal{F})}(P(\mathbf{f})) \subset \{(x,m) \in \operatorname*{supp}_{\mathcal{F}}(\mathbf{f}) \times \Gamma \mid m \notin k\,N_x(\mathbf{f})\}$$

where $kN = \{n_1 + ... + n_k \ ; \ n_i \in N\} \supset \{k\,n; n \in N\}$.

Proof. It is sufficient to consider the case $P = X^k$, *i.e.*, $\operatorname*{spec}_{(\phi,\mathcal{F})}(\mathbf{f}^k)$. We proceed as in the previous proof with $\mathbf{g} = \mathbf{f}^{k-1}$. Since $\mathcal{F}$ is an algebra, $\operatorname*{supp}_{\mathcal{F}}(\mathbf{f}^{k-1}) \subset \operatorname*{supp}_{\mathcal{F}}(\mathbf{f})$. The result follows by induction over k. $\square$

References

1. M F HASLER, Asymptotic extension of topological modules and algebras. *Int. Trans. Spec. Func.* **20**, 291 (2009).
2. J-F COLOMBEAU, New Generalized Functions and Multiplication of Distributions. (North-Holland, Amsterdam, 1984).
3. A DELCROIX, D SCARPALÉZOS, Asymptotic scales, asymptotic algebras. *Int. Trans. Spec. Func.* **6**, 157 (1997).
4. A DELCROIX, D SCARPALÉZOS, Topology on Asymptotic Algebras of Generalized Functions and Applications. *Mh Math* **129**, 1 (2000).

5. J-A MARTI, $(\mathcal{C}, \mathcal{E}, \mathcal{P})$ -sheaf structures and applications. in: Nonlinear theory of generalized functions. (Grosser & al., eds.), (Res. Notes Math., Chapman & Hall/CRC, 1999), 175–186.

6. A DELCROIX, M F HASLER, S PILIPOVIĆ, V VALMORIN, Sequence spaces with exponent weights – Realizations of Colombeau type algebras. *Diss. Math.* **447**, 1 (2007).

7. M. HASUMI, Note on the n-dimensional tempered ultra-distributions *Tohoku Math. J.* **13**, 94 (1961).

8. J-A MARTI, Regularity, Local and Microlocal Analysis in Theories of Generalized Functions *Acta Appl. Math.* **105**, 267-302 (2008).

612

Some fixed point theorems on the cone Banach spaces

Erdal Karapinar

Department of Mathematics, Atılım University,
Ankara, 06836, Turkey
E-mail:ekarapinar@atilim.edu.tr

In this paper, existence of fixed points for operators T defined on a subset C of a cone Banach space X and belonging to a class $D_c(a,b)$ with $0 \le a, b \le 1$ are considered.

Keywords: Cone normed spaces, fixed point theorem.

1. Introduction and Preliminaries

In 1987, Shy-Der Lin [6] introduced the notion of K-metric spaces by replacing real numbers with a cone K in the metric function, that is $d : X \times X \to K$. Later, without mentioning the paper of Shy-Der Lin, in 2007, Huang ad Zhang [5] introduced the notion of cone metric spaces (CMS) by replacing real numbers with an ordering Banach space. In that paper, they also discussed some properties of convergence of sequences and proved the fixed point theorems of contractive mapping for cone metric spaces: any mapping T of a complete cone metric space X into itself that satisfies, for some $0 \le k < 1$, the inequality

$$d(Tx, Ty) \le kd(x, y) \tag{1}$$

for all $x, y \in X$, has a unique point.

Recently, many results on fixed point theorems have been extended to cone metric spaces (see e.g.[1,2,4,5,8–11]) but none of them recall the paper of Shy-Der Lin [6] which is the initiatory paper in this direction. One of the aim of this paper is to write and emphasize this historical remark. Nova [6] considered the operators T defined on a subset C of a Banach space X that satisfy the inequality:

$$\|Tx - Ty\| \le a\|x - y\| + b(\|x - Tx\| + \|y - Ty\|), \tag{2}$$

for all $x, y \in C$, where $0 \le a, b \le 1$. Any operator T satisfying the condition (2) is said to the belong to class $D(a, b)$, in short $T \in D(a, b)$. Note that a contraction operator is in the class $D(k, 0)$ with $0 \le k < 1$. Note also that although condition (1) yields the continuity of the operator T, condition (2) may hold even if T is not continuous. Indeed, the triangle inequality implies that any operator T belongs to class $D(1, 1)$: $\|Tx - Ty\| \le \|Tx - x\| + \|x - y\| + \|y - Ty\|$.

In this manuscript, first, a cone Banach space X is defined and then existence of fixed points for operators $T \in D(a, b)$ with $0 \le a, b \le 1$ defined on a subset C of a cone Banach space X is discussed.

Throughout this paper $E := (E, \| \cdot \|)$ stands for real Banach space. Let $P := P_E$ always be a closed non-empty subset of E. P is called *cone* if $ax + by \in P$ for all $x, y \in P$ and non-negative real numbers a, b where $P \cap (-P) = \{0\}$ and $P \ne \{0\}$.

For a given cone P, one can define a partial ordering (denoted by $\le$: or $\le_P$) with respect to P by $x \le y$ if and only if $y - x \in P$. The notation $x < y$ indicates that $x \le y$ and $x \ne y$ while $x << y$ means that $y - x \in \mathrm{int}P$, where $\mathrm{int}P$ denotes the interior of P. From now on, it is assumed that $\mathrm{int}P \ne \emptyset$.

The cone P is called

(N) *normal* if there is a number $K \ge 1$ such that for all $x, y \in E$:

$$0 \le x \le y \Rightarrow \|x\| \le K\|y\|. \tag{3}$$

(R) *regular* if every increasing sequence which is bounded from above is convergent. That is, if $\{x_n\}_{n\ge 1}$ is a sequence such that $x_1 \le x_2 \le ... \le y$ for some $y \in E$, then there is $x \in E$ such that $\lim_{n\to\infty} \|x_n - x\| = 0$.

In (N), the least positive integer K, satisfying the equation (3), is called the normal constant of P. Note that, in [4] and [8], the normal constant K is considered a positive real number, $(K > 0)$, although it is proved that there is no normal cone for $K < 1$ in ([8], Lemma 2.1).

Lemma 1.1.

(i) *Every regular cone is normal.*

(ii) *For each $k > 1$, there is a normal cone with normal constant $K > k$.*

(iii) *The cone P is regular if every decreasing sequence which is bounded from below is convergent.*

Proof of (i) and (ii) are given in [8] and the last one follows from definition.

Definition 1.1. Let X be a non-empty set. Suppose the mapping $d : X \times X \to E$ satisfies:

$(M1)$ $0 \le d(x,y)$ for all $x, y \in X$,
$(M2)$ $d(x,y) = 0$ if and only if $x = y$,
$(M3)$ $d(x,y) \le d(x,z) + d(z,y)$, for all $x, y \in X$.
$(M4)$ $d(x,y) = d(y,x)$ for all $x, y \in X$

then d is called cone metric on X, and the pair (X,d) is called a cone metric space (CMS).

Example 1.1. Let $E = \mathbb{R}^3$ and $P = \{(x,y,z) \in E : x, y, z \ge 0\}$ and $X = \mathbb{R}$. Define $d : X \times X \to E$ by $d(x,\tilde{x}) = (\alpha|x - \tilde{x}|, \beta|x - \tilde{x}|, \gamma|x - \tilde{x}|)$, where α, β, γ are positive constants. Then (X,d) is a CMS. Note that the cone P is normal with the normal constant $K = 1$.

Analogously we can define Cone Normed Spaces (CNS) as follow:

Definition 1.2. (See e.g.[1,5]) Let X be a vector space over $\mathbb{R}$. Suppose the mapping $\| \cdot \|_c : X \to E$ satisfies:

$(N1)$ $\|x\|_c > 0$ for all $x \in X$,
$(N2)$ $\|x\|_c = 0$ if and only if $x = 0$,
$(N3)$ $\|x + y\|_c \le \|x\|_c + \|y\|_c$, for all $x, y \in X$.
$(N4)$ $\|kx\|_c = |k|\|x\|_c$ for all $k \in \mathbb{R}$,

then $\| \cdot \|_c$ is called cone norm on X, and the pair $(X, \| \cdot \|_c)$ is called a cone normed space (CNS).

Note that each CNS is CMS. Indeed, $d(x,y) = \|x - y\|_c$. Complete cone normed spaces will be called cone Banach spaces.

Definition 1.3. Let (X,d) be a CMS, $x \in X$ and $\{x_n\}_{n \ge 1}$ a sequence in X. Then

(i) $\{x_n\}_{n \ge 1}$ *converges to* x whenever for every $c \in E$ with $0 << c$ there is a natural number N, such that $d(x_n, x) << c$ for all $n \ge N$. It is denoted by $\lim_{n \to \infty} x_n = x$ or $x_n \to x$.

(ii) $\{x_n\}_{n \ge 1}$ is a *Cauchy sequence* whenever for every $c \in E$ with $0 << c$ there is a natural number N, such that $d(x_n, x_m) << c$ for all $n, m \ge N$.

(iii) (X, d) is a complete cone metric space if every Cauchy sequence is convergent.

Lemma 1.2. *(See e.g. [5]) Let (X, d) be a CMS, P be a normal cone with normal constant K, and $\{x_n\}$ be a sequence in X. Then,*

- (i) *the sequence $\{x_n\}$ converges to x if and only if $d(x_n, x) \to 0$ (or equivalently $\|d(x_n, x)\| \to 0$),*
- (ii) *the sequence $\{x_n\}$ Cauchy if and only if $d(x_n, x_m) \to 0$ (or equivalently $\|d(x_n, x_m)\| \to 0$).*
- (iii) *the sequence $\{x_n\}$ converges to x and the sequence $\{y_n\}$ converges to y then $d(x_n, y_n) \to d(x, y)$.*

Lemma 1.3. *Let (X, d) be a CMS over a cone P in E. Then (1) $Int(P) + Int(P) \subseteq Int(P)$ and $\lambda Int(P) \subseteq Int(P), \lambda > 0$. (2) If $c >> 0$ then there exists $\delta > 0$ such that $\|b\| < \delta$ implies $b << c$. (3) For any given $c >> 0$ and $c_0 >> 0$ there exists $n_0 \in \mathbb{N}$ such that $\frac{c_0}{n_0} << c$. (4) If a_n, b_n are sequences in E such that $a_n \to a$, $b_n \to b$ and $a_n \leq b_n, \forall n$ then $a \leq b$.*

The proof of the first two parts is followed from the definition of $Int(P)$. The third part is obtained by the second part. Namely, if $c >> 0$ is given then find $\delta > 0$ such that $\|b\| < \delta$ implies $b << c$. Then find n_0 such that $\frac{1}{n_0} < \frac{\delta}{\|c_0\|}$ and hence $\frac{c_0}{n_0} << c$. Since P is closed, the proof of fourth part is achieved.

Definition 1.4. (See e.g. [3]) P is called minihedral cone if $\sup\{x, y\}$ exists for all $x, y \in E$; and strongly minihedral if every subset of E which is bounded from above has a supremum.

Lemma 1.4. *Every strongly minihedral normal cone is regular.*

Proof. Let $P \subset E$ be strongly minihedral normal cone with normal constant K and $a_1 \leq a_2 \leq a_3 \leq \ldots$ increasing and bounded above in E. Since P is strongly minihedral, one can find $\sup\{a_1, a_2, a_3, \ldots\}$, say a. Claim: $a_n \to a$ in E. Let $\varepsilon > 0$ be given. Choose $c >> 0$ such that $\|c\| \leq K\varepsilon$. Now, $a - c << a$, hence find M such that

$$a - c << a_M << a.$$

Yet, then $0 < a - a_n << a - a_M << c, \quad \forall n \geq M$ since a_n is increasing. Thus,

$$\|a - a_n\| \leq K\|c\| < \varepsilon, \quad \forall n \geq M \Rightarrow \lim_{n \to \infty} a_n = a. \qquad \square$$

Example 1.2. Let $E = C[0,1]$ with the supremum norm and $P = \{f \in E : f \geq 0\}$. Then P is a cone with normal constant $M = 1$ which is not regular. This is clear, since the sequence x^n is monotonicly decreasing, but not uniformly convergent to 0. This cone, by Lemma 1.4, is not strongly minihedral. However, it is easy to see that the cone mentioned in Example 1.1 is strongly minihedral.

Definition 1.5. (See e.g. [5]) Let (X,d) be a CMS and $A \subset X$. A is said to be sequentially compact if for any sequence $\{x_n\}$ in A; there is a subsequence $\{x_{n_k}\}$ of $\{x_n\}$ such that $\{x_{n_k}\}$ is convergent in A.

Lemma 1.5. *(See e.g. [9]) Every cone metric space (X,d) is topological space which is denoted by (X, τ_c). Moreover, every sequentially compact subset $A \subset X$ is compact (each cover of A by subsets from τ_c can be reduced to finite subcover).*

2. Main Results

From now on, let X be a Banach space, P be a strongly minihedral normal cone, T be self-mapping operator defined on subset C of X. We write $T \in D_c(a,b)$ if we replace usual norm with $\|\cdot\|$ with cone norm $\|\cdot\|_c$ in (2).

Proposition 2.1. *Let $T \in D_c(a,b)$, $a,b \geq 0$, $a < 1$ and $F_T := \{x \in C : Tx = x\}$. If F_T is not empty, then F_T consists of a single point.*

Proof. Let x,y be fixed points of $T \in D_c(a,b)$, that is, $\|x - y\|_c = \|Tx - Ty\|_c$ and $\|Tx - Ty\|_c \leq a\|x - y\|_c + b(\|x - Tx\|_c + \|y - Ty\|_c)$ which implies that $x = y$. $\qquad\square$

Let $T : X \to X$ and $x \in X$. The mapping T is *asymptotically regular in* x if $\|T^{n+1}x - T^n x\|_c \to 0$ as $n \to \infty$.

Proposition 2.2. *Let $T \in D_c(a,b)$, $a+2b < 1$. Then $\inf_{x \in C} \|x - Tx\|_c = 0$ and T is asymptotically regular at any point.*

Proof. Note that existence of infimum is due to strongly minihedrality of the cone P. Define $x_n = T^n x_0$ where $T \in D_c(a,b)$ and x_0 is an arbitrary. Then $\|x_n - x_{n+1}\|_c = \|Tx_{n-1} - Tx_n\|_c \leq a\|x_{n-1} - x_n\|_c + b(\|x_{n-1} - x_n\|_c + \|Tx_{n-1} - Tx_n\|_c)$ which implies that $(1 - b)\|(1 - T)x_n\|_c \leq (a + b)\|(1 - T)x_{n-1}\|_c$. Thus, $\|(1 - T)x_n\|_c \leq (\frac{a+b}{1-b})^n\|(1 - T)x_0\|_c$. So, $\|x_n - Tx_n\|_c \to 0$ as $n \to \infty$ since $\frac{a+b}{1-b} < 1$. Hence, $\inf_{x \in C} \|x - Tx\|_c = 0$.

Observe that, $T^{n+1}x_0 - T^n x_0 = x_n - x_{n+1} = (1-T)x_n \to 0$ as $n \to \infty$. Thus T is asymptotically regular. $\qquad\square$

Theorem 2.1. *Let $T \in D_c(a,b)$, $0 \le a,b < 1$. Then the sequence $\{x_n\}$ contained in the set C, and satisfies*

$$\lim_{n \to \infty} (x_n - Tx_n) = 0 \qquad (4)$$

if and only if the sequence converges to unique fixed point of T.

Proof. Necessary part is easy because $\|Tx_n - Tx_m\|_c \le a\|x_n - x_m\|_c + b(\|x_n - Tx_n\|_c + \|x_m - Tx_m\|_c)$ regarding the triangular inequality yields that $\|Tx_n - Tx_m\|_c \le \frac{a+b}{1-a}(\|x_n - Tx_n\|_c + \|x_m - Tx_m\|_c)$. From the hypothesis, $\{Tx_n\}$ is a Cauchy Sequence. Since C is closed and X is complete, there exists $z \in C$ such that $\lim_{n \to \infty} Tx_n = z$. Since $x_n - Tx_n \to 0$ as $n \to \infty$, then $x_n \to z$ as $n \to \infty$. By triangular inequality and $T \in D_c(a,b)$, $0 \le a,b < 1$, one can obtain that

$$\|z - Tz\|_c \le \frac{a+b}{1-a}\|z - x_n\|_c + \frac{a+b}{1-a}\|x_n - Tx_n\|_c.$$

Since $x_n - Tx_n \to 0$ and $x_n \to z$ as $n \to \infty$ implies that z is a fixed point of T. Uniqueness follow from Proposition 2.1.

Sufficiency. Assume there exists $z \in C$ such that $z = Tz$ and $\lim_{n \to \infty} x_n = z$. Regarding $T \in D_c(a,b)$, $0 \le a,b < 1$ and triangular inequality, one can observe that

$$\|Tx_n - x_n\|_c - \|x_n - z\|_c \le \|Tx_n - z\|_c \le a\|x_n - z\|_c + b\|x_n - Tx_n\|_c.$$

Hence, $(1-b)\|Tx_n - x_n\|_c \le (1+a)\|x_n - z\|_c$ and thus $Tx_n - x_n \to 0$ as $n \to \infty$ $\qquad\square$

References

1. Abdeljawad, T.: Completion of cone metric spaces, *Hacet. J. Math. Stat.*, (to appear).
2. Abdeljawad, T. and Karapınar, E.: Quasi-Cone Metric Spaces and Generalizations of Caristi Kirk's Theorem. *Fixed Point Theory Appl.* (2009), Article ID 574387, 9 pages, doi:10.1155/2009/574387 (2009).
3. Deimling, K.: Nonlinear Functional Analysis, Springer-Verlage, (1985).
4. Huang Long-Guang, Zhang X.: Cone metric spaces and fixed point theorems of contractive mappings, *J. Math. Anal. Appl.*, **332**,1468–1476 (2007).
5. Karapınar, E.: Fixed Point Theorems in Cone Banach Spaces *Fixed Point Theory Appl.* (to appear).
6. Shy-Der L.: A common fixed point theorem in abstract spaces, *Indian J. Pure Appl. Math.*, **18**(8),685–690 (1987).

7. Nova, L.: Fixed point theorems for some discontinuous operators, *Pasific J. Math.,* **123**(1),189–196 (1986).

8. Rezapour, Sh., Hamlbarani, R.: Some notes on the paper "Cone metric spaces and fixed point theorems of contractive mappings, *J. Math.Anal. Appl.,* **347**,719–724 (2008).

9. Turkoglu, D., Abuloha, M.: Cone Metric Spaces and Fixed Point Theorems in Diametrically Contractive Mappings, *Acta Mathematica Sinica,* English Series (to appear) .

10. Turkoglu, D., Abuloha, M., Abdeljawad, T.: KKM mappings in cone metric spaces and some fixed point theorems *Nonlinear Analysis: Theory, Methods & Applications,* doi: 10.1016.na.2009.6.58 (2009).

11. Turkoglu, D., Abuloha, M., Abdeljawad, T.: Some theorems and examples of cone metric spaces, *J. Comput. Anal. Appl.,* (to appear).

A boundary value problem for 3D elliptic equations with singular coefficients

E.T. Karimov

Institute of Mathematics and Information Technologies,
Uzbek Academy of Sciences, Tashkent, 100125, Uzbekistan
E-mail: erkinjon@gmail.com
www.karimovet.narod.ru

In the present paper a boundary-value problem with Neumann condition for three-dimensional elliptic equation with singular coefficients is studied. The uniqueness theorem for considered problem is proven by energy integral's method. A solution of studied problem is found in an explicit form using method of Green's functions.

Keywords: Elliptic equation with singular coefficient; Neumann condition; Green's function; Lauricella's hypergeometric function.

1. Introduction

Known that theory of degenerate equations or equations with singular coefficients is one of the central parts of modern theory of partial differential equations, which has many applications in aero and gas dynamics [1]. Detailed bibliography and description of main boundary problems for various type degenerate equations, in particular, for elliptic equations with singular coefficients, one can find in monographes [2-6]. Omitting huge amount of works on studying local and non-local problems for elliptic equation with singular coefficients, note some works, which are close to the present work. In the work [7] fundamental solutions for bi-axially-symmetric Helmholtz equation is constructed and in the work [8] the unique solvability of the Dirichlet-Neumann problem for aforementioned equation is proved. In the work [9], fundamental solutions for a class of many-dimensional degenerate elliptic equations with spectral parameter were constructed.

Note also that by many authors, such as A.Altyn, A.J.Fryant, R.Gilbert, P.Henrici, P.Kumar, P.A.McCoy, R.J.Weinacht, axially-symmetric potentials were investigated.

Fundamental solution for equation

$$H_{\alpha,\beta,\gamma}\left(u\right) = u_{xx} + u_{yy} + u_{zz} + \frac{2\alpha}{x}u_x + \frac{2\beta}{y}u_y + \frac{2\gamma}{z}u_z = 0 \qquad (1)$$

is found in the work [10] and further one of them will be used at solving a boundary problem for equation (1). Here $0 < 2\alpha, 2\beta, 2\gamma < 1$, $\alpha, \beta, \gamma = const.$

2. Formulation of the problem and the uniqueness theorem

Let $D \subset R_3^+$ be a domain, bounded by planes
$D_1 = \{(x, y, z): \ x = 0, \ 0 < y < b, \ 0 < z < c\}$,
$D_2 = \{(x, y, z): \ y = 0, \ 0 < x < a, \ 0 < z < c\}$,
$D_3 = \{(x, y, z): \ z = 0, \ 0 < x < a, \ 0 < y < b\}$
and by a surface D_4, which intersects with domains $D_i \left(i = \overline{1,3}\right)$. Lines of intersections designate through $\Gamma_1 = D_4 \cap D_1$, $\Gamma_2 = D_4 \cap D_2$, $\Gamma_3 = D_4 \cap D_3$, respectively. Here $a, b, c = const > 0$.

Problem N. To find a regular solution $u\left(x, y, z\right)$ of the equation (1) from a class of functions $C\left(\overline{D}\right) \cap C^2\left(D\right)$, satisfying conditions:

$$x^{2\alpha} u_x\left(x, y, z\right)\big|_{x=0} = \nu_1\left(y, z\right), \quad (y, z) \in D_1, \qquad (2)$$

$$y^{2\beta} u_y\left(x, y, z\right)\big|_{y=0} = \nu_2\left(x, z\right), \quad (x, z) \in D_2, \qquad (3)$$

$$z^{2\gamma} u_z\left(x, y, z\right)\big|_{z=0} = \nu_3\left(x, y\right), \quad (x, y) \in D_3, \qquad (4)$$

$$u\left(x, y, z\right) = \varphi\left(x, y, z\right), \quad (x, y, z) \in \overline{D_4}, \qquad (5)$$

where $\nu_1\left(y, z\right), \nu_2\left(x, z\right), \nu_3\left(x, y\right), \varphi\left(x, y, z\right)$ are given functions, moreover $\nu_1\left(y, z\right), \nu_2\left(x, z\right), \nu_3\left(x, y\right)$ can reduce to infinity of the order less than $1 - 2\alpha;\ 1 - 2\beta;\ 1 - 2\gamma$ on points $O,\ B,\ C;\ O,\ A, C;\ O,\ A,\ B$, respectively. Here $A\left(a, 0, 0\right)$, $B\left(0, b, 0\right)$, $C\left(0, 0, c\right)$, $O\left(0, 0, 0\right)$.

Consider an identity

$$x^{2\alpha} y^{2\beta} z^{2\gamma}\left[u H_{\alpha,\beta,\gamma}\left(w\right) - w H_{\alpha,\beta,\gamma}\left(u\right)\right] = \left(x^{2\alpha} y^{2\beta} z^{2\gamma}\left(u w_x - w u_x\right)\right)_x +$$
$$+ \left(x^{2\alpha} y^{2\beta} z^{2\gamma}\left(u w_y - w u_y\right)\right)_y + \left(x^{2\alpha} y^{2\beta} z^{2\gamma}\left(u w_z - w u_z\right)\right)_z.$$

Integrating the both side of this identity along the domain D and using formula of Gauss-Ostrogradsky [11], obtain:

$$\iiint\limits_{D} x^{2\alpha} y^{2\beta} z^{2\gamma}\left[u H_{\alpha,\beta\gamma}(w) - w H_{\alpha,\beta\gamma}(u)\right] dx dy dz =$$
$$= \iint\limits_{\partial D} x^{2\alpha} y^{2\beta} z^{2\gamma}\left[(u w_x - w u_x)\cos(n, x)+ \right. \qquad (6)$$
$$\left. + (u w_y - w u_y)\cos(n, y) + (u w_z - w u_z)\cos(n, z)\right] dS.$$

Here $\partial D = D_1 \cup D_2 \cup D_3 \cup D_4$, $\cos(n,x)\,dS = dydz$, $\cos(n,y)\,dS = dxdz$, $\cos(n,z)\,dS = dxdy$, n is outer normal to ∂D.

The following equality is directly checkable:

$$\iiint_D x^{2\alpha} y^{2\beta} z^{2\gamma} u H_{\alpha,\beta\gamma}(u) dxdydz = \iiint_D x^{2\alpha} y^{2\beta} z^{2\gamma} \left[u_x^2 + u_y^2 + u_z^2 \right] dxdydz +$$

$$+ \iiint_D \left[\left(x^{2\alpha} y^{2\beta} z^{2\gamma} u u_x \right)_x + \left(x^{2\alpha} y^{2\beta} z^{2\gamma} u u_y \right)_y + \left(x^{2\alpha} y^{2\beta} z^{2\gamma} u u_z \right)_z \right] dxdydz.$$

After the application of Gauss-Ostrogradsky's formula, obtain

$$\iiint_D x^{2\alpha} y^{2\beta} z^{2\gamma} \left[u_x^2 + u_y^2 + u_z^2 \right] dxdydz = \iint_{D_1} y^{2\beta} z^{2\gamma} u\nu_1 dydz +$$

$$+ \iint_{D_2} x^{2\alpha} z^{2\gamma} u\nu_2 dxdz + \iint_{D_3} x^{2\alpha} y^{2\beta} u\nu_3 dxdy - \iint_{D_4} x^{2\alpha} y^{2\beta} z^{2\gamma} \varphi \frac{\partial u}{\partial n} dS. \qquad (7)$$

If consider homogeneous problem N, then from (7) one can get that

$$\iiint_D x^{2\alpha} y^{2\beta} z^{2\gamma} \left[u_x^2 + u_y^2 + u_z^2 \right] dxdydz = 0.$$

Hence, follows that $u(x,y,z) = 0$ in $\overline{D}$. Therefore the following uniqueness theorem is true:

Theorem 2.1. *If the problem N has a solution, then it is unique.*

3. The existence of the solution

The existence of the solution will be proved by method of Green's functions. For this aim suppose that $a = b = c$ and D_4 is $1/8$ part of sphere with origin on point $O(0,0,0)$, and with radius $R = a$.

Definition 3.1. Green's function of the problem N for the equation (1) is a function $G_1(x,y,z;x_0,y_0,z_0)$, satisfying conditions:

(1) this function is a regular solution of (1) in the domain D, except the point (x_0,y_0,z_0);

(2) it satisfies boundary conditions

$$G_1(x,y,z;x_0,y_0,z_0)|_{D_4} = 0, \quad x^{2\alpha}\frac{\partial}{\partial x}G_1(x,y,z;x_0,y_0,z_0)\Big|_{x=0} = 0,$$

$$y^{2\beta}\frac{\partial}{\partial y}G_1(x,y,z;x_0,y_0,z_0)\Big|_{y=0} = 0, \quad z^{2\gamma}\frac{\partial}{\partial z}G_1(x,y,z;x_0,y_0,z_0)\Big|_{z=0} = 0;$$

(3) it can be represented as

$$G_1(x,y,z;x_0,y_0,z_0) = q_1(x,y,z;x_0,y_0,z_0) + q_1^*(x,y,z;x_0,y_0,z_0),$$

$$(8)$$

where

$$q_1\left(x,y,z;x_0,y_0,z_0\right) = k_1\left(r^2\right)^{-\alpha-\beta-\gamma-\frac{1}{2}} \times$$
$$\times F_A^{(3)}\left(\alpha+\beta+\gamma+\tfrac{1}{2};\alpha,\beta,\gamma;2\alpha,2\beta,2\gamma;\xi,\eta,\zeta\right)$$

is a fundamental solution [10], and function

$$q_2^*\left(x,y,z;x_0,y_0,z_0\right) = -\left(\frac{a}{R_0}\right)^{2\alpha+2\beta+2\gamma+1} q_1\left(x,y,z;\overline{x_0},\overline{y_0},\overline{z_0}\right)$$

is a regular solution of (1) in the domain D. Here
$\xi = \frac{r^2-r_1^2}{r^2},\ \eta = \frac{r^2-r_2^2}{r^2},\ \zeta = \frac{r^2-r_3^2}{r^2}, r^2 = \left(x-x_0\right)^2+\left(y-y_0\right)^2+\left(z-z_0\right)^2,$
$r_1^2 = \left(x+x_0\right)^2+\left(y-y_0\right)^2+\left(z-z_0\right)^2,\ r_2^2 = \left(x-x_0\right)^2+\left(y+y_0\right)^2+\left(z-z_0\right)^2,$
$r_3^2 = \left(x-x_0\right)^2+\left(y-y_0\right)^2+\left(z+z_0\right)^2,$
$\overline{x_0} = \frac{a^2}{R_0^2}x_0,\ \overline{y_0} = \frac{a^2}{R_0^2}y_0,\ \overline{z_0} = \frac{a^2}{R_0^2}z_0,\ R_0^2 = x_0^2+y_0^2+z_0^2,$
$F_A^{(3)}\left(a;b_1,b_2,b_3;c_1,c_2,c_3;x,y,z\right)$ is Lauricella's hypergeometric function
from three variables [12, p. 117], k_1 is a constant, which will be defined
later.

Let $\left(x_0,y_0,z_0\right) \in D$. Excise a small ball with a center on $\left(x_0,y_0,z_0\right)$
and with radius ε, from the domain D. Remaining part of D denote as D_ε,
and through C_ε designate a sphere of excised ball. Applying formula (6),
obtain

$$\iint\limits_{C_\varepsilon} x^{2\alpha}y^{2\beta}z^{2\gamma}\left[u\frac{\partial G_1}{\partial n} - G_1\frac{\partial u}{\partial n}\right]dS =$$
$$= -\iint\limits_{D_1} y^{2\beta}z^{2\gamma}\nu_1\left(y,z\right)G_1\left(0,y,z;x_0,y_0,z_0\right)dydz-$$
$$-\iint\limits_{D_2} x^{2\alpha}z^{2\gamma}\nu_2\left(x,z\right)G_1\left(x,0,z;x_0,y_0,z_0\right)dxdz- \qquad (9)$$
$$-\iint\limits_{D_3} x^{2\alpha}y^{2\beta}\nu_3\left(x,y\right)G_1\left(x,y,0;x_0,y_0,z_0\right)dxdy-$$
$$-\iint\limits_{D_4} x^{2\alpha}y^{2\beta}z^{2\gamma}\varphi\left(S\right)\frac{\partial G_1(x,y,z;x_0,y_0,z_0)}{\partial n}dS.$$

Considering (8) and

$$F_A^{(3)}\left(a;b_1,b_2,b_3;c_1,c_2,c_3;0,y,z\right) = F_2\left(a;b_2,b_3;c_2,c_3;y,z\right),$$
$$F_A^{(3)}\left(a;b_1,b_2,b_3;c_1,c_2,c_3;x,0,z\right) = F_2\left(a;b_1,b_3;c_1,c_3;x,z\right),$$
$$F_A^{(3)}\left(a;b_1,b_2,b_3;c_1,c_2,c_3;x,y,0\right) = F_2\left(a;b_1,b_2;c_1,c_2;x,y\right),$$

find

$$G_1\left(0, y, z; x_0, y_0, z_0\right) = k_1 \frac{F_2\left(\alpha+\beta+\gamma+\frac{1}{2};\beta,\gamma;2\beta,2\gamma;\eta_{01}^{(x)},\zeta_{01}^{(x)}\right)}{\left[x_0^2+(y-y_0)^2+(z-z_0)^2\right]^{\alpha+\beta+\gamma+\frac{1}{2}}} -$$

$$-k_1 \frac{F_2\left(\alpha+\beta+\gamma+\frac{1}{2};\beta,\gamma;2\beta,2\gamma;\eta_{02}^{(x)},\zeta_{02}^{(x)}\right)}{\left[\left(a-\frac{yy_0}{a}\right)^2+\left(a-\frac{zz_0}{a}\right)^2+\frac{x_0^2+z_0^2}{a^2}y^2+\frac{x_0^2+y_0^2}{a^2}z^2-a^2\right]^{\alpha+\beta+\gamma+\frac{1}{2}}},$$

$$G_1\left(x, 0, z; x_0, y_0, z_0\right) = k_1 \frac{F_2\left(\alpha+\beta+\gamma+\frac{1}{2};\alpha,\gamma;2\alpha,2\gamma;\xi_{01}^{(y)},\zeta_{01}^{(y)}\right)}{\left[(x-x_0)^2+y_0^2+(z-z_0)^2\right]^{\alpha+\beta+\gamma+\frac{1}{2}}} -$$

$$-k_1 \frac{F_2\left(\alpha+\beta+\gamma+\frac{1}{2};\alpha,\gamma;2\alpha,2\gamma;\xi_{02}^{(y)},\zeta_{02}^{(y)}\right)}{\left[\left(a-\frac{xx_0}{a}\right)^2+\left(a-\frac{zz_0}{a}\right)^2+\frac{y_0^2+z_0^2}{a^2}x^2+\frac{x_0^2+y_0^2}{a^2}z^2-a^2\right]^{\alpha+\beta+\gamma+\frac{1}{2}}}, \tag{10}$$

$$G_1\left(x, y, 0; x_0, y_0, z_0\right) = k_1 \frac{F_2\left(\alpha+\beta+\gamma+\frac{1}{2};\alpha,\beta;2\alpha,2\beta;\xi_{01}^{(z)},\eta_{01}^{(z)}\right)}{\left[(x-x_0)^2+(y-y_0)^2+z_0^2\right]^{\alpha+\beta+\gamma+\frac{1}{2}}} -$$

$$-k_1 \frac{F_2\left(\alpha+\beta+\gamma+\frac{1}{2};\alpha,\beta;2\alpha,2\beta;\xi_{02}^{(z)},\eta_{02}^{(z)}\right)}{\left[\left(a-\frac{xx_0}{a}\right)^2+\left(a-\frac{yy_0}{a}\right)^2+\frac{y_0^2+z_0^2}{a^2}x^2+\frac{x_0^2+z_0^2}{a^2}y^2-a^2\right]^{\alpha+\beta+\gamma+\frac{1}{2}}},$$

where

$$\xi_{0i}^{(y)} = \xi_i \,|_{y=0}\ , \ \ \xi_{0i}^{(z)} = \xi_i \,|_{z=0}\ , \ \ \eta_{0i}^{(x)} = \eta_i \,|_{x=0}\ , \ \ \eta_{0i}^{(z)} = \eta_i \,|_{z=0}\ ,$$

$$\zeta_{0i}^{(x)} = \zeta_i \,|_{x=0}\ , \ \ \zeta_{0i}^{(y)} = \zeta_i \,|_{y=0}\ ,$$

$$\xi_i = \sigma_i x x_0,\ \ \eta_i = \sigma_i y y_0,\ \ \zeta_i = \sigma_i z z_0,\ \ (i=1,2),\ \ \sigma_1 = -\frac{4}{r^2},$$

$$\sigma_2 = -\frac{4a^2}{R_0^2}\frac{1}{\left(a-\frac{xx_0}{a}\right)^2+\left(a-\frac{yy_0}{a}\right)^2+\left(a-\frac{zz_0}{a}\right)^2+\frac{y_0^2+z_0^2}{a^2}x^2+\frac{x_0^2+z_0^2}{a^2}y^2+\frac{x_0^2+y_0^2}{a^2}z^2-2a^2},$$

$$k_1 = \frac{1}{2\pi}\frac{\Gamma(\alpha)\Gamma(\beta)\Gamma(\gamma)\Gamma(2\alpha+2\beta+2\gamma)}{\Gamma(2\alpha)\Gamma(2\beta)\Gamma(2\gamma)\Gamma(\alpha+\beta+\gamma)}. \tag{11}$$

Using formula of differentiation [12]

$$\frac{\partial^{i+j+k}F_A^{(3)}}{\partial x^i y^j z^k} = \frac{(a)_{i+j+k}(b_1)_i(b_2)_j(b_3)_k}{(c_1)_i(c_2)_j(c_3)_k} \times$$

$$\times F_A^{(3)}\left(a+i+j+k, b_1+i, b_2+j, b_3+k; c_1+i, c_2+j, c_3+k; x, y, z\right)$$

and considering adjacent relation [12]

$$\frac{b_1}{c_1}x F_A^{(3)}\left(a+1; b_1+1, b_2, b_3; c_1+1, c_2, c_3; x, y, z\right)+$$

$$+\frac{b_2}{c_2}y F_A^{(3)}\left(a+1; b_1, b_2+1, b_3; c_1, c_2+1, c_3; x, y, z\right)+$$

$$+\frac{b_3}{c_3}z F_A^{(3)}\left(a+1; b_1, b_2, b_3+1; c_1, c_2, c_3+1; x, y, z\right)=$$

$$= F_A^{(3)}\left(a+1; b_1, b_2, b_3; c_1, c_2, c_3; x, y, z\right)-$$

$$-F_A^{(3)}\left(a; b_1, b_2, b_3; c_1, c_2, c_3; x, y, z\right),$$

obtain

$$\frac{\partial}{\partial n} q_1\left(x, y, z; x_0, y_0, z_0\right) =$$
$$= \frac{\partial}{\partial x} q_1 \cdot \cos\left(n, x\right) + \frac{\partial}{\partial y} q_1 \cdot \cos\left(n, y\right) + \frac{\partial}{\partial z} q_1 \cdot \cos\left(n, z\right) =$$
$$= -\left(\alpha + \beta + \gamma + \tfrac{1}{2}\right) k_1\left(r^2\right)^{-\alpha - \beta - \gamma - \frac{1}{2}} \times$$
$$\times F_A^{(3)}\left(\alpha + \beta + \gamma + \tfrac{3}{2}; \alpha, \beta, \gamma; 2\alpha, 2\beta, 2\gamma; \xi, \eta, \zeta\right) \frac{\partial}{\partial n}\left[\ln r^2\right] -$$
$$-2\left(\alpha + \beta + \gamma + \tfrac{1}{2}\right) k_1\left(r^2\right)^{-\alpha - \beta - \gamma - \frac{3}{2}} x_0 \times$$
$$\times F_A^{(3)}\left(\alpha + \beta + \gamma + \tfrac{3}{2}; \alpha + 1, \beta, \gamma; 2\alpha + 1, 2\beta, 2\gamma; \xi, \eta, \zeta\right) \frac{dydz}{dS} -$$
$$-2\left(\alpha + \beta + \gamma + \tfrac{1}{2}\right) k_1\left(r^2\right)^{-\alpha - \beta - \gamma - \frac{3}{2}} y_0 \times$$
$$\times F_A^{(3)}\left(\alpha + \beta + \gamma + \tfrac{3}{2}; \alpha, \beta + 1, \gamma; 2\alpha, 2\beta + 1, 2\gamma; \xi, \eta, \zeta\right) \frac{dxdz}{dS} -$$
$$-2\left(\alpha + \beta + \gamma + \tfrac{1}{2}\right) k_1\left(r^2\right)^{-\alpha - \beta - \gamma - \frac{3}{2}} z_0 \times$$
$$\times F_A^{(3)}\left(\alpha + \beta + \gamma + \tfrac{3}{2}; \alpha, \beta, \gamma + 1; 2\alpha, 2\beta, 2\gamma + 1; \xi, \eta, \zeta\right) \frac{dxdy}{dS}. \tag{12}$$

Left part of the equality (9) divide into two parts:

$$\iint_{C_\varepsilon} x^{2\alpha} y^{2\beta} z^{2\gamma} u \frac{\partial G_1}{\partial n} dS = \iint_{C_\varepsilon} x^{2\alpha} y^{2\beta} z^{2\gamma} u \frac{\partial q_1(x, y, z, x_0, y_0, z_0)}{\partial n} dS -$$
$$- \left(\frac{a}{R_0}\right)^{2\alpha + 2\beta + 2\gamma + 1} \iint_{C_\varepsilon} x^{2\alpha} y^{2\beta} z^{2\gamma} u \frac{\partial q_1(x, y, z, \overline{x_0}, \overline{y_0}, \overline{z_0})}{\partial n} dS \tag{13}$$

Substitute (12) into (13) and pass to the spherical system of coordinates:

$$x = x_0 + \varepsilon \sin\theta \cos\varphi, \; y = y_0 + \varepsilon \sin\theta \sin\varphi, \; z = z_0 + \varepsilon \cos\theta,$$
$$0 < \theta < \pi, \, 0 < \varphi < 2\pi, \, 0 < \varepsilon < \infty.$$

Further, using some formulas for Lauricella's hypergeometric function such a decomposition formula [13]

$$F_A^{(3)}\left(\alpha; \beta_1, \beta_2, \beta_3; \gamma_1, \gamma_2, \gamma_3; x, y, z\right) =$$
$$= \sum_{i,j,k=0}^{\infty} \frac{(\alpha)_{i+j+k} (\beta_1)_{i+j} (\beta_2)_{i+k} (\beta_3)_{j+k}}{(\gamma_1)_{i+j} (\gamma_2)_{i+k} (\gamma_3)_{j+k} i! j! k!} x^{i+j} y^{i+k} z^{j+k} \times$$
$$\times F\left(\alpha + i + j, \beta_1 + i + j; \gamma_1 + i + j; x\right) \times$$
$$\times F\left(\alpha + i + j + k, \beta_2 + i + k; \gamma_2 + i + k; y\right) \times$$
$$\times F\left(\alpha + i + j + k, \beta_3 + j + k; \gamma_3 + j + k; z\right),$$

auto-transformation formula [14] and considering properties of Gamma-functions [14], after complicated evaluations we find a solution of the stated problem in the following form:

$$u\left(x_0, y_0, z_0\right) = -\iint_{D_1} y^{2\beta} z^{2\gamma} \nu_1(y, z) G_1\left(0, y, z; x_0, y_0, z_0\right) dydz -$$
$$- \iint_{D_2} x^{2\alpha} z^{2\gamma} \nu_2(x, z) G_1\left(x, 0, z; x_0, y_0, z_0\right) dxdz -$$
$$- \iint_{D_3} x^{2\alpha} y^{2\beta} \nu_3(x, y) G_1\left(x, y, 0; x_0, y_0, z_0\right) dxdy - \tag{14}$$
$$- \iint_{D_4} x^{2\alpha} y^{2\beta} z^{2\gamma} \varphi(S) \frac{\partial G_1}{\partial n} dS,$$

where

$$G_1\left(0, y, z; x_0, y_0, z_0\right),\ G_1\left(x, 0, z; x_0, y_0, z_0\right),\ G_1\left(x, y, 0; x_0, y_0, z_0\right)$$

are defined by (10).

Theorem 3.1. *The unique solution of the problem N exists and defined by formula (14).*

Acknowledgments

The author is grateful to Organizers of ISAAC 2009, especially to Professor M.Ruzhansky, for given possibility via financial support, to discuss results of the present paper with specialists, during the ISAAC congress in London.

References

1. L. Bers, *Mathematical aspects of subsonic and transonic gas dynamics* (New York, London, 1958).
2. A.V.Bitsadze, *Some classes of partial differential equations* (Moscow, 1981).
3. R.Gilbert, *Function Theoretic Methods in Partial Differential Equations* (New York, London: Academic Press, 1969).
4. M.M.Smyrnov, *Degenerate elliptic and hyperbolic equations* (Moscow, 1966).
5. M.S.Salakhitdinov, *Mixed-composite type equations* (Tashkent: Fan, 1974).
6. M.S.Salakhitdinov and M.Mirsaburov, *Non-local problems for mixed type equations with singular coefficients* (Tashkent: Universitet, 2005).
7. A. Hasanov, *Compl.Var. and Ell.Eq.* **52(8)**, pp.673-683 (2007).
8. M.S.Salakhitdinov and A. Hasanov, *Compl.Var. and Ell.Eq.* **53(4)**, pp.355-364 (2008).
9. A.K.Urinov, *Reports of FSU* **1**, pp.5-11 (2006).
10. A.Hasanov and E.T.Karimov, *Appl. Math. Lett.* **22(12)** pp.1828-1832 (2009).
11. J.M.Rassias, *Lecture Notes on Mixed Type Partial Differential Equations* (World Scientific, 1990).
12. P. Appell and J. Kampe de Feriet, *Fonctions Hypergeometriques et Hyperspheriques; Polynomes d'Hermite* (Gauthier - Villars. Paris, 1926).
13. A. Hasanov and H.M.Srivastava, *Appl. Math. Lett.* **19(2)** pp.113-121 (2006).
14. A.Erdelyi and G.Bateman, *Higher transcendental functions*, vol. 1. (Moscow: Nauka, 1973). (Russian translation).

Numerical modelling of three-dimensional turbulent stream of reacting gas, implying from rectangle form nozzle, on base of $k - \varepsilon$ model of turbulence

S. Khodjiev

Bukhara State University, Bukhara, Uzbekistan
E-mail: safarkhodjiev@mail.ru

In the present work a method and efficient algorithm of the calculation of reacting gas, implying from rectangle form nozzle and spreading in oxidizer's stream is discussed. Three-dimensional partially parabolized systems of Navier-Stocks for multi-componential chemical reacting gas mixtures is used for description of flows.

Keywords: Numerical modeling; reacting gas; "$k - \varepsilon$" model; turbulence.

1. Formulation of a problem

Consider a reacting stream, implying from rectangle form nozzle and spreading in entangled air flow. As an origin of system of coordinates we choose a center of initial cross-section of the stream: axe OX directed along the stream and axes OY, OZ are parallel to sides of nozzle with length $2a$ and $2b$, respectively. Assume, flow is symmetric regarding to the axe OX and planes YOX, ZOX, which formate boundary of integrating domain, moreover they allow us to consider only one quarter of rectangle form stream.

This kind of flow is described by the following parabolized system of equations:

$$\frac{\partial \rho u}{\partial x} + \frac{1}{L}\frac{\partial \rho v}{\partial y} + \frac{\partial \rho w}{\partial z} = 0, \tag{1}$$

$$\rho u \frac{\partial u}{\partial x} + \rho v \frac{\partial u}{\partial y} + \rho w \frac{\partial u}{\partial z} = -\frac{\partial P}{\partial x} + \frac{\partial}{\partial y}\left(\mu_{\mathrm{T}}\frac{\partial u}{\partial y}\right) + \frac{\partial}{\partial z}\left(\mu_{\mathrm{T}}\frac{\partial u}{\partial z}\right), \tag{2}$$

$$\rho u \frac{\partial v}{\partial x} + \rho v \frac{\partial v}{\partial y} + \rho w \frac{\partial v}{\partial z} =$$
$$-\frac{\partial P}{\partial y} + \frac{4}{3}\frac{\partial}{\partial y}\left(\mu_T \frac{\partial v}{\partial y}\right) + \frac{\partial}{\partial z}\left(\mu_T \frac{\partial v}{\partial z}\right) - \frac{2}{3}\frac{\partial}{\partial y}\left(\mu_T \frac{\partial w}{\partial z}\right) + \frac{\partial}{\partial z}\left(\mu_T \frac{\partial w}{\partial y}\right), \tag{3}$$

$$\rho u \frac{\partial w}{\partial x} + \rho v \frac{\partial w}{\partial y} + \rho w \frac{\partial w}{\partial z} =$$
$$-\frac{\partial P}{\partial z} + \frac{4}{3}\frac{\partial}{\partial z}\left(\mu_T \frac{\partial w}{\partial z}\right) + \frac{\partial}{\partial y}\left(\mu_T \frac{\partial w}{\partial y}\right) + \frac{\partial}{\partial y}\left(\mu_T \frac{\partial v}{\partial z}\right) - \frac{2}{3}\frac{\partial}{\partial z}\left(\mu_T \frac{\partial v}{\partial y}\right), \tag{4}$$

$$\rho u \frac{\partial H}{\partial x} + \rho v \frac{\partial H}{\partial y} + \rho w \frac{\partial H}{\partial z} =$$
$$\frac{1}{\mathrm{Pr}_T}\frac{\partial}{\partial y}\left(\mu_T \frac{\partial H}{\partial y}\right) + \frac{1}{\mathrm{Pr}_T}\frac{\partial}{\partial z}\left(\mu_T \frac{\partial H}{\partial z}\right) + \left(1 - \frac{1}{\mathrm{Pr}_T}\right)\left[\frac{\partial}{\partial y}\left(\mu_T u \frac{\partial u}{\partial y}\right) + \right.$$
$$\left. + \frac{\partial}{\partial z}\left(\mu_T u \frac{\partial u}{\partial z}\right) + \frac{\partial}{\partial z}\left(\mu_T v \frac{\partial v}{\partial z}\right) + \frac{\partial}{\partial y}\left(\mu_T w \frac{\partial w}{\partial y}\right)\right] +$$
$$\left(\frac{4}{3} - \frac{1}{\mathrm{Pr}_T}\right)\left[\frac{\partial}{\partial y}\left(\mu_T v \frac{\partial v}{\partial y}\right) + \frac{\partial}{\partial z}\left(\mu_T w \frac{\partial w}{\partial z}\right)\right] -$$
$$-\frac{\partial}{\partial y}\left(\frac{2}{3}\mu_T v \frac{\partial w}{\partial z}\right) + \frac{\partial}{\partial z}\left(\mu_T v \frac{\partial w}{\partial y}\right) + \frac{\partial}{\partial y}\left(\mu_T w \frac{\partial v}{\partial z}\right) - \frac{\partial}{\partial z}\left(\frac{2}{3}\mu_T w \frac{\partial v}{\partial y}\right), \tag{5}$$

$$\rho u \frac{\partial C_i}{\partial x} + \rho v \frac{\partial C_i}{\partial y} + \rho w \frac{\partial C_i}{\partial z} =$$
$$\frac{1}{Sc_T}\frac{\partial}{\partial y}\left(\mu_T \frac{\partial C_i}{\partial y}\right) + \frac{1}{Sc_T}\frac{\partial}{\partial z}\left(\mu_T \frac{\partial C_i}{\partial z}\right) + \overset{\bullet}{w}_i, \tag{6}$$

$$\rho u \frac{\partial k}{\partial x} + \rho v \frac{\partial k}{\partial y} + \rho w \frac{\partial k}{\partial z} =$$
$$\frac{\partial}{\partial y}\left(\frac{\mu_T}{\sigma_k}\frac{\partial k}{\partial y}\right) + \frac{\partial}{\partial z}\left(\frac{\mu_T}{\sigma_k}\frac{\partial k}{\partial z}\right) + G - \rho\varepsilon, \tag{7}$$

$$\rho u \frac{\partial \varepsilon}{\partial x} + \rho v \frac{\partial \varepsilon}{\partial y} + \rho w \frac{\partial \varepsilon}{\partial z} =$$
$$\frac{\partial}{\partial y}\left(\frac{\mu_T}{\sigma_\varepsilon}\frac{\partial \varepsilon}{\partial y}\right) + \frac{\partial}{\partial z}\left(\frac{\mu_T}{\sigma_\varepsilon}\frac{\partial \varepsilon}{\partial z}\right) + (C_{\varepsilon_1} G - C_{\varepsilon_2} \rho\varepsilon)\frac{\varepsilon}{k}, \tag{8}$$

where

$$G = \mu_T \left[\left(\frac{\partial u}{\partial y}\right)^2 + \left(\frac{\partial u}{\partial z}\right)^2\right], \tag{9}$$

$$H = c_p T + \frac{u^2 + v^2 + w^2}{2} + \sum_{i=1}^{N_k} C_i h_i^*, \tag{10}$$

$$P = \rho T R \sum_{i=1}^{N_k} \frac{c_i}{m_i}. \tag{11}$$

In equations (1)–(11) x, y, z are coordinates, ρ is a density, u, v, w are components of velocity, T is temperature, H is full enthalpy, P is pressure, k is kinetic energy of turbulence, ε is dissipate of the kinetic energy of turbulence, C_i is mass concentration of i-th component, N_k is a number of mixture's components, c_p is a thermal capacity at constant pressure, Pr_T, Sc_T are turbulence number of Prandtle and Schmidt, h_i^* is heat formation of i-th component, R is universal gas constant, m_i is molecular mass of i-th component, $\dot{w}_i$, is a speed of formation of i-th component, $C_{\varepsilon_1}, C_{\varepsilon_2}, C_\mu, \sigma_k, \sigma_\varepsilon$ are empiric constants of "$k - \varepsilon$" model of turbulence.

Boundary conditions: I. $x = 0$: 1) $0 \leq y \leq a$, $0 \leq z \leq b$;

$$u = u_2,\ v = 0,\ w = 0,\ H = H_2,\ P = P_2,\ C_i = (C_i)_2, i = \overline{1, N_k},\ k = k_2,\ \varepsilon = \varepsilon_2$$

2) $a < y < y_{+\infty},\ b < z < z_{+\infty}$;

$$u = u_1,\ v = 0,\ w = 0,\ H = H_1,\ P = P_1,\ C_i = (C_i)_1, i = \overline{1, N_k},\ k = k_1,\ \varepsilon = \varepsilon_1. \tag{12}$$

II. $x > 0$: 1) $z = 0$; $0 < y <_{+\infty}$; $w = 0$, $\frac{\partial f}{\partial z} = 0$, $(f = u, v, H, C_i, k, \varepsilon)$.
2) $y = 0$, $0 < z < z_{+\infty}$; $v = 0$, $\frac{\partial f}{\partial y} = 0$, $(f = u, w, H, C_i, k, \varepsilon)$. 3) $z \to z_\infty$, $y \to y_\infty$, $u = u_1$, $v = 0$, $w = 0$, $H = H_1$, $P = P_1$, $C_i = (C_i)_1$, $k = k_1$, $\varepsilon = \varepsilon_1$.

For numerical integration of the system of equations (1)–(11) with boundary conditions (12) we use double-layer ten-point non-explicit finite difference scheme of variable directions [1].

2. Numerical method

In this work an effective method similar to SIMPLE, solving Poisson equation for defining corrections to velocities by direct method is used. In contrast to [2] corrections reduced by three components of velocities. Found solutions u, v, w in new iteration represented as rated (u_p, v_p, w_p) and plus corrections (u_c, v_c, w_c). They are defined from the equation of continuity by introducing potential $Q, \rho u_c = \frac{\partial Q}{\partial x}, \rho v_c = \frac{\partial Q}{L \partial y}, \rho w_c = \frac{\partial Q}{\partial z}$,, which is a solution of Poisson equation:

$$\frac{\partial^2 Q}{\partial x^2} + \frac{\partial^2 Q}{L^2 \partial y^2} + \frac{\partial^2 Q}{\partial z^2} = Q_p, \tag{13}$$

where Q_p is source term.

Difference equation (13) can be written for potential Q on every point of mesh crosswise to flow in a plane along i (numeration i along an axe OX, j along OY and k along OZ) and use three-diagonal system of equations with the following assumptions:

1) $Q_{i-1,j,k} = 0$, $Q_{i,j,k-1} = 0$ mean that corrections to the velocity in the plane $(i-1)$ and on cross-section $(k-1)$, in which mass storage is already guaranteed.

2) $Q_{i+1,j,k} = 0$, $Q_{i,j,k+1} = 0$ mean that corrections to the velocity will equal to zero in the plane $(i+1)$ and on cross-section $(k+1)$, when their convergence are achieved in plane and on cross-section,respectively.

Below we shortly give an **algorithm** of solving for aforementioned problem:

(1) Solving the difference equation (2) find u_p;
(2) Solving the difference equation (3) find v_p using a value of u_p.
(3) Solving the difference equation (4) find w_p using values of u_p, v_p.
(4) Solve the difference equation (13) with considering assumptions 1). 2) and make corrections of velocities as $u = u_p + u_c$, $v = v_p + v_c$, $w = w_p + w_c$.
(5) Solve an equation of energy and equation of concentration regarding to surplus concentrations with the help of corrected velocities. Further calculate individual components of concentrations.
(6) Calculating difference equations (7), (8) find k and ε, and then calculate turbulence viscosity μ_T by formula (11).

Test rates.

As a test problem for described method, combustion of propane-butane mixture in air is investigated:

$$C_3H_8 + 5O_2 = 3CO_2 + 4H_2O, \quad C_4H_{10} + 6,5O_2 = 4CO_2 + 5H_2O$$

at the following outputs of oxidizer and fuel:

$$u_1 = 0;\; u_2 = 61 m/s; T_1 = 300K; T_2 = 1200K; (C_2)_2 = 0,12'$$

$$(C_4)_2 = 0,88; (C_1)_1 = 0,232; (C_4)_1 = 0,768; P_1 = P_2 = 1atm;$$

$$Pr_T = Sc_T = 0,65; h_2^* = 11490 KKal/kg.$$

References

1. S. Khodjiev, *Uzb.J. Mech.Probl.* **2**, 1993, pp.28-33.
2. S.V.Patankar and D.B.Spolding, *Heat and mass transfer in boundary layers*, (Morgan-Grampion, London, 1967).

Sequence spaces of invariant means and some matrix transformations

M. Mursaleen

Department of Mathematics, Aligarh Muslim University,
Aligarh 202002, India
Email: mursaleenm@gmail.com

In this paper we characterize the matrix classes $(l_p, V_\sigma(\lambda))$ and $(l_p, V_\sigma^\infty(\lambda))$, where the sequence space $V_\sigma^\infty(\lambda)$ has recently been introduced by Mursaleen.[6] This space is related to the concept of σ-mean and the sequence $\lambda = (\lambda_n)$.

Keywords: Double sequences; sequence spaces; invariant means; matrix transformations.

1. Introduction and Preliminaries

We shall write w for the set of all complex sequences $x = (x_k)_{k=0}^\infty$. Let ϕ, l_∞, c and c_0 denote the sets of all finite, bounded, convergent and null sequences respectively. We write $l_p := \{x \in w : \sum_{k=0}^\infty | x_k |^p < \infty\}$ for $1 \leq p < \infty$. We denote the sequences $e = (1, 1, 1, ...)$ and $e^{(n)} = (0, 0, ..., 1(\text{nth place}), 0, ...)$. For any sequence $x = (x_k)_{k=0}^\infty$, we denote the *n-section* by $x^{[n]} = \sum_{k=0}^n x_k e^{(k)}$.

Note that c_0, c, and l_∞ are Banach spaces with the sup-norm $\|x\|_\infty = \sup_k |x_k|$, and $l^p (1 \leq p < \infty)$ are Banach spaces with the norm $\|x\|_p = (\sum |x_k|^p)^{1/p}$; while ϕ is not a Banach space with respect to any norm.

Let X be subset of w. Then the *β-dual* of X denoted by X^β is defined as the set $\{a = (a_k) \in \omega : \sum_k a_k x_k$ converges for all $x = (x_k) \in X\}$. Note that $c_0^\beta = c^\beta = l_\infty^\beta = l_1$, $l_1^\beta = l_\infty$ and $l_p^\beta = l_q$, where $p^{-1} + q^{-1} = 1$.

A sequence $(b^{(n)})_{n=0}^\infty$ in a linear metric space X is called *Schauder basis* if for every $x \in X$, there is a unique sequence $(\beta_n)_{n=0}^\infty$ of scalars such that $x = \sum_{n=0}^\infty \beta_n b^{(n)}$. A sequence space X with a linear topology is called a *K-space* if each of the maps $p_i : X \to \mathbb{C}$ defined by $p_i(x) = x_i$ is continuous for all $i \in \mathbb{N}$. A *K*-space is called an *FK-space* if X is a complete linear metric space; a *BK-space* is a normed *FK*-space. An *FK*-space $X \supset \phi$ is said to have property AK if every sequence $x = (x_k)_{k=0}^{\infty} \in X$ has a unique representation $x = \sum_{k=0}^{\infty} x_k e^{(k)}$, that is, $x = \lim_{n \to \infty} x^{[n]}$. We used here standard notations as used by Maddox[3] and Mursaleen.[5]

Let σ be a one-to-one mapping from the set $\mathbb{N}$ of natural numbers into itself. A continuous linear functional ϕ on the space l_{∞} is said to be an *invariant mean* or a *σ-mean* if and only if (i) $\phi(x) \geq 0$ when the sequence $x = (x_k)$ has $x_k \geq 0$ for all k, (ii) $\phi(e) = 1$, where $e = (1, 1, 1, \cdots)$, and (iii) $\phi(x) = \phi((x_{\sigma(k)}))$ for all $x \in \ell_{\infty}$. Throughout this paper we consider the mapping σ which has no finite orbits, that is, $\sigma^p(k) \neq k$ for all integer $k \geq 0$ and $p \geq 1$, where $\sigma^p(k)$ denotes the *pth* iterate of σ at k. Note that, a σ-mean extends the limit functional on the space c in the sense that $\phi(x) = \lim x$ for all $x \in c$, Mursaleen.[4] Consequently, $c \subset V_\sigma$ the set of bounded sequences all of whose σ-means are equal. We say that a sequence $x = (x_k)$ is *σ-convergent* if and only if $x \in V_\sigma$, where

$$V_\sigma := \{x \in l_\infty : \lim_{p \to \infty} t_{pn}(x) = L \ \text{ uniformly in } \ n; \ L = \sigma\text{-}\lim x\}, \ \text{ where}$$

$$t_{pn}(x) = \frac{1}{p+1} \sum_{m=0}^{p} x_{\sigma^m(n)}.$$

Using this concept, Schaefer[8] defined and characterized the σ-conservative, σ-regular and σ-coercive matrices. If σ is translation then σ-mean is often called a Banach limit[1] and the set V_σ is reduced to the set f of almost convergent sequences studied by Lorentz.[2]

The following space $V_\sigma^\infty(\lambda)$ was introduced by Mursaleen.[6]

Let $\lambda = (\lambda_n)$ be a non-decreasing sequence of positive numbers tending to ∞ such that $\lambda_{n+1} \leq \lambda_n + 1$, $\lambda_1 = 0$. Then

$$V_\sigma^\infty(\lambda) := \{x \in l_\infty : \sup_{m,n} | \ \tau_{mn}(x) \ |< \infty \ \},$$

where

$$\tau_{mn}(x) =: \frac{1}{\lambda_m} \sum_{j \in I_m} x_{\sigma^j(n)}.$$

If $\lambda_m = m + 1$ and $\sigma(n) = n + 1$ then $V_\sigma^\infty(\lambda)$ is reduced to the set f_∞ defined by Nanda.[7] Note that $c \subset V_\sigma(\lambda) \subset V_\sigma^\infty(\lambda) \subset l_\infty$, where

$$V_\sigma(\lambda) := \{x \in l_\infty : \lim_{m \to \infty} \tau_{mn}(x) = L \ \text{ uniformly in } \ n; \ L = (\sigma, \lambda)\text{-}\lim x\}.$$

The spaces $V_\sigma(\lambda)$ and $V_\sigma^\infty(\lambda)$ are BK spaces with the norm $\|x\| = \sup_{m,n} |\tau_{mn}(x)|$.

Recently, Mursaleen[6] has characterized the matrix classes $(l_1, V_\sigma^\infty(\lambda))$ and $(l_\infty, V_\sigma^\infty(\lambda))$. In this paper, we characterize the matrix classes $(l_p, V_\sigma^\infty(\lambda))$ and $(l_p, V_\sigma(\lambda))$.

2. Main Results

Let X and Y be two sequence spaces, $A = (a_{nk})_{n;k=1}^\infty$ be an infinite matrix of real or complex numbers and $A_n = (a_{nk})_{k=1}^\infty$ be the sequence in the n-th row of A. We write $Ax = (A_n(x))$, where $A_n(x) = \sum_k a_{nk} x_k$ provided that the series on the right converges for each n. If $x = (x_k) \in X$ implies that $Ax \in Y$, then we say that A defines a matrix transformation from X into Y and by (X, Y) we denote the class of such matrices, that is, $A \in (X, Y)$ if and only if $A_n \in X^\beta$ for all n and $Ax \in Y$ for all $x \in X$.

Let Ax be defined. Then, for all r, n, we write

$$\tau_{mn}(Ax) = \sum_{k=1}^\infty t(n, k, m) x_k,$$

where

$$t(n, k, m) = \frac{1}{\lambda_m} \sum_{j \in I_m} a(\sigma^j(n), k),$$

and $a(n, k)$ denotes the element a_{nk} of the matrix A.

Our first theorem is a generalization of both the theorems of Mursaleen.[6]

Theorem 2.1. $A \in (l_p, V_\sigma^\infty(\lambda))$ *if and only if*

(i) $M = \sup_{m,n} \sum_k |t(n, k, m)|^q < \infty, (1 < p \le \infty)$,
(ii) $\sup_{m,n,k} |t(n, k, m)|^p < \infty, (0 < p \le 1)$;

where $p^{-1} + q^{-1} = 1$.

Proof. We will prove the case *(i)* $1 < p < \infty$. The case *(ii)* $0 < p \le 1$ can be proved similarly.

Necessity. Let $A \in (l_p, V_\sigma^\infty(\lambda))$. Write $Q_n(x) = \sup_r |\tau_{mn}(Ax)|$. It is easy to see that Q_n is a continuous seminorm on l_p, since for $x \in l_p$

$$|Q_n(x)| \le M \, ||x||_p. \tag{1}$$

Suppose *(i)* is not true. Then there exists $x \in l_p$ with $\sup_n Q_n(x) = \infty$ (and hence $M = 0$). By the principle of condensation of singularities (Yosida[9]), the set $\{x \in l_p : \sup_n Q_n(x) = \infty\}$ is of second category in l_p and hence nonempty, that is, there is $x \in l_p$ with $\sup_n Q_n(x) = \infty$. But this contradicts the fact that Q_n is pointwise bounded on l_p. Now define $x = (x_k)$ by

$$
x_k = \begin{cases}
sgn\ t(n,k,m) & ; \text{ for each } m, n\ (1 \le k \le k_0), \\[2ex]
0 & ; \text{ for } k > k_0.
\end{cases}
$$

Then $x \in l_p$. Applying this sequence to Eq. (1), we get *(i)*.

Sufficiency. Suppose that $x = (x_k) \in l_p$. By using Hölder's inequality, we have

$$
|\tau_{mn}(Ax)| \le \sum_k |t(n,k,m)x_k|
$$

$$
\le \left(\sum_k |t(n,k,m)|^q\right)^{1/q}\left(\sum_k |x_k|^p\right)^{1/p}.
$$

Taking the supremum over n, m on both sides and using Eq. (1), we get $Ax \in V_\sigma^\infty(\lambda)$ for $x \in l_p$.

This completes the proof of the theorem. $\qquad\qquad\square$

Next we characterize the class $(l_p, V_\sigma(\lambda))$.

Theorem 2.2. *$A \in (l_p, V_\sigma(\lambda))$ if and only if*

$$
(i) \quad \begin{cases}
\sup_m \sum_k |t(n,k,m)|^q < \infty, (1 < p < \infty), \\[2ex]
\sup_{m,k} |t(n,k,m)|^p < \infty, (0 < p \le 1),
\end{cases}
$$

for every n, where $p^{-1} + q^{-1} = 1$; and
(ii) $a_{(k)} = \{a_{nk}\}_{n=1}^\infty \in V_\sigma(\lambda)$ for each k, i.e. $\lim_m t(n,k,m) = u_k$ uniformly in n, for each k.

Proof. We consider the case $1 < p < \infty$, the case $0 < p \le 1$ can be proved similarly.

Necessity. Let $A \in (l_p, V_\sigma(\lambda))$. Since $e_k \in l_p$, the condition *(ii)* holds. Put $f_{mn}(x) = \tau_{mn}(Ax)$, since $\tau_{mn}(Ax)$ exists for each m and $x \in l_p$. Therefore $\{f_{mn}(x)\}_m$ is a sequence of continuous real functionals on l_p and further we have $\sup_m \mid f_{mn}(x) \mid < \infty$ on l_p, for every n. Now condition *(i)* follows by the same argument as in the proof of necessity of Theorem 2.1.

Sufficiency. Suppose that the conditions *(i)* and *(ii)* hold and $x \in l_p$. Now for every $r \geq 1$, we have

$$\sum_{k=1}^{r} |t(n,k,m)|^q \leq \sup_m \sum_k |t(n,k,m)|^q.$$

Therefore

$$\sum_k |u_k|^q = \lim_r \lim_m \sum_{k=1}^{r} |t(n,k,m)|^q$$

$$\leq \sup_m \sum_k |t(n,k,m)|^q < \infty.$$

Thus the series $\sum_k t(n,k,m)x_k$ and $\sum_k u_k x_k$ converge for each m and $x \in l_p$. For a given $\varepsilon > 0$ and $x \in l_p$, choose k_0 such that

$$\sum_{k=k_0+1}^{\infty} |x_k|^p < \varepsilon.$$

Since *(ii)* holds, therefore there exists m_0 such that

$$\left| \sum_{k=1}^{k_0} (t(n,k,m) - u_k) \right| < \varepsilon \text{ for every} m > m_0.$$

Hence by the condition *(ii)* it follows that $\left| \sum_{k=k_0+1}^{\infty} (t(n,k,m) - u_k) \right|$ is arbitrary small and we have

$$\lim_m \sum_k t(n,k,m)x_k = \sum_k u_k x_k \text{ uniformly in } n.$$

This completes the proof of the theorem. $\qquad \square$

3. Corollaries

If we take $p=1$ and $p = \infty$ in our Theorem 2.1, we get the following corollaries which are Theorem 3.1 and 3.2 of Mursaleen,[6] respectively.

Corollary 3.1. *$A \in (l_1, V_\sigma^\infty(\lambda))$ if and only if*

$$\sup_{n,k,m} |t(n,k,m)| < \infty.$$

Corollary 3.2. $A \in (l_\infty, V_\sigma^\infty(\theta))$ *if and only if*

$$\sup_{n,m} \sum_k |t(n,k,m)| < \infty.$$

If we take $p=1$ in our Theorem 2.2, we get the following corollary.

Corollary 3.3. $A \in (l_1, V_\sigma(\lambda))$ *if and only if*

(i) $\sup_{k,m} |t(n,k,m)| < \infty$ *for every n,*
(i) $\lim_m t(n,k,m) = u_k$ *uniformly in n, for each k.*

Note that the case $p = \infty$ is not covered by Theorem 2.2 while we can get the following directly from Theorem 3 of Schaefer.[8]

Corollary 3.4. $A \in (l_\infty, V_\sigma(\lambda))$ *if and only if*

(i) $\| A \| < \infty$,
(ii) $a_{(k)} \in V_\sigma(\lambda)$ *for each k; and*
(iii) $\lim_m \sum_k | t(n,k,m) - u_k |$ *uniformly in n.*

References

1. S. Banach, *Théorie des operations linéaires* (Warsaw, 1932).
2. G. G. Lorentz, A contribution to the theory of divergent sequences, *Acta Math.*, **80**(1948) 167-190.
3. I. J. Maddox, *Elements of Function Analysis* (Camb. Univ. Press, 1970).
4. M. Mursaleen, Some new invariant matrix methods of summability, *Quart. J. Math. Oxford*(2), **34**(1983) 77-86.
5. M. Mursaleen, *Elements of Metric Spaces* (Anamaya Publ., New Delhi, 2005).
6. M. Mursaleen, A new sequence space of invariant mean and some matrix transformations, *J. Orissa Math. Soc.*, **27**(2008) 13-20.
7. S. Nanda, Matrix transformations and almost boundedness, *Glasnik Matematicki*, **14(34)**(1979) 99-107.
8. P. Schaefer, Infinite matrices and invariant means, *Proc. Amer. Math. Soc.*, **36**(1972) 104-110.
9. K. Yosida, *Functional Analysis* (Springer-Verlag, Berlin Heidelberg New York, 1966).

636

Motion stabilization of a solid body with fixed point

Zaure Rakisheva

Department of Mechanics and Mathematics,
Al-Farabi Kazakh National University,
Almaty, 050012, Kazakstan
E-mail: Zaure.Rakisheva@kaznu.kz
www.kaznu.kz

The problem of solid body dynamics in the central Newton field of forces is considered. This motion is described by the well-known system of Euler-Poisson equations. By original change of variables the system is reduced to the normal form with the first integral of norm type. The solution of this system is considered as the non-perturbed motion and its stability is investigated. The procedure is outlined for obtaining the steady motion in general case. The controlling force nature is defined. The methodology is applied to three cases with special restrictions on the body's inertia moments, so-called generalized classical cases of Euler, Lagrange and Kovalevskaya.

Keywords: A solid body with one fixed point; Newton field of forces; motion stabilization; asymptotically steady motion; controlling moments; first integrals.

1. Introduction

A solid body with one fixed point is the mathematical model of many real mechanisms, machines, technical devices, natural and artificial celestial bodies, such as gyroscopes, planets, satellites and so on. Such body motion is one of the most important problems of theoretical mechanics. Dynamics of the solid body with one fixed point in a gravity field of forces was first investigated by Euler.[1] He had derived the motion equations and proved that to solve the problem will be enough to find only one additive first integral besides the known three ones. Many scientists tried to find analytical solution of the motion equations, but it could be found only in three special cases with restrictions on body mass distribution. Kozlov[2] had proved that new analytical first integral could not exist in general case.

2. Problem definition and theory

In this paper the problem of dynamics of solid body with one fixed point in the central Newton field of forces is studied. This problem is considered as generalization of the classic problem about motion under force of gravity of the solid body around the fixed point, i.e. it is provided that the distance between the gravitating center O^* and the fixed point O is much more than the body size. As is well known, e.g. see Ref. 3, there is no general solution of the problem in analytical form. In this connection the question on studying of the qualitative features and control possibilities of motion appears.

Let the body's motionless point O is fixed on the distance R from the centre of gravity O^* (fig. 1).

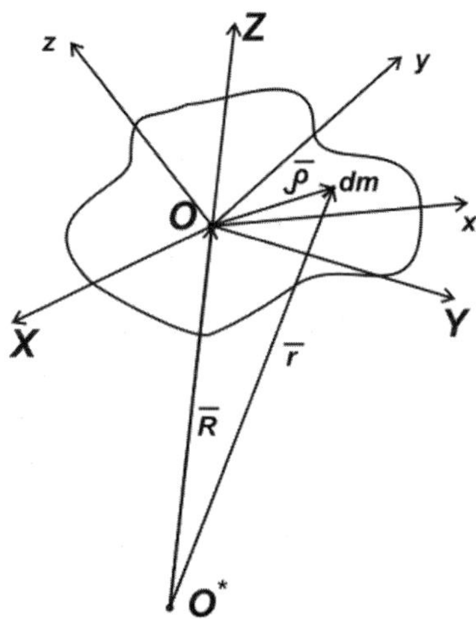

Fig. 1.

Let's combine with the body fixed point the beginnings of two co-ordinate systems: the immovable coordinate system $OXYZ$ with the axis OZ, directed along the vector $\overline{R} = \overline{OO^*}$, and the mobile coordinate system $Oxyz$ with the axes, directed along the body's main axes of inertia. Let α, β, γ are direct cosines of the mobile axes with axis OZ. It is known that the force function U of the Newton field of forces, acting on the solid body, is given by the formula:

$$U = \int_V \varphi^*(r)\,dm,$$
$$r^2 = R^2 + 2R(x\alpha + y\beta + z\gamma) + \rho^2,$$
$$\rho^2 = x^2 + y^2 + z^2,$$

where $\varphi^*(r)$ is the force function, acting on the body element dm and depending only on the distance from the body's point to the centre of gravity, ρ is the distance from the body's point to the fixed point O; the integral is taken on all the body volume.

The solid body motion around the fixed point O is described by the well-known system of the Euler dynamic equations and the Poisson equations, e.g. see Ref. 3:

$$A\dot{p} + (C - B)qr = \gamma\frac{\partial U}{\partial \beta} - \beta\frac{\partial U}{\partial \gamma},$$
$$B\dot{q} + (A - C)rp = \alpha\frac{\partial U}{\partial \gamma} - \gamma\frac{\partial U}{\partial \alpha}, \tag{1}$$
$$C\dot{r} + (B - A)pq = \beta\frac{\partial U}{\partial \alpha} - \alpha\frac{\partial U}{\partial \beta},$$

$$\dot{\alpha} = r\beta - q\gamma;$$
$$\dot{\beta} = p\gamma - r\alpha; \tag{2}$$
$$\dot{\gamma} = q\alpha - p\beta,$$

where p, q, r are projections of an angular speed of the body rotation on the mobile axes; A, B, C are the body's main moments of inertia.

If assume that the distance R is great in comparison with the body sizes, it is possible to present the function U in the form:

$$U(\alpha, \beta, \gamma) = -\frac{\lambda M}{R^2}(a\alpha + b\beta + c\gamma) - \frac{3}{2}\frac{\lambda}{R^3}(A\alpha^2 + B\beta^2 + C\gamma^2) + \ldots, \tag{3}$$

where M is the body mass; a, b, c are coordinates of the body centre of mass concerning the mobile coordinate system.

Substituting the expression (3) in the equations (1), we will obtain the motion equations in the following form:

$$A\frac{dp}{dt} + (C - B)qr = Mg(c\beta - b\gamma) + 3\frac{g}{R}(C - B)\beta\gamma,$$
$$B\frac{dq}{dt} + (A - C)rp = Mg(a\gamma - c\alpha) + 3\frac{g}{R}(A - C)\gamma\alpha, \tag{4}$$
$$C\frac{dr}{dt} + (B - A)pq = Mg(b\alpha - a\beta) + 3\frac{g}{R}(B - A)\alpha\beta.$$

where $g = \lambda/R^2$ is an acceleration of the gravitation force on distance R from the centre of gravity. Now the body motion is described by the differential equations (2) and (4).

3. Results

3.1. *Studying of qualitative features of the motion*

Let's apply the linear change of variables
$(p, q, r, \alpha, \beta, \gamma) \to (u_1, u_2, u_3, u_4, u_5, u_6)$ via transformation (see Ref. 4):

$$p = \frac{1}{\sqrt{lA}} u_1, \qquad q = \frac{1}{\sqrt{lB}} u_2, \qquad r = \frac{1}{\sqrt{lC}} u_3,$$

$$\alpha = \frac{1}{\sqrt{1 + \frac{3gl}{R} A}} \left(u_4 - \frac{Mgal}{\sqrt{1 + \frac{3gl}{R} A}} \right),$$

$$\beta = \frac{1}{\sqrt{1 + \frac{3gl}{R} B}} \left(u_5 - \frac{Mgbl}{\sqrt{1 + \frac{3gl}{R} B}} \right),$$

$$\gamma = \frac{1}{\sqrt{1 + \frac{3gl}{R} C}} \left(u_6 - \frac{Mgcl}{\sqrt{1 + \frac{3gl}{R} C}} \right),$$

where l is the unit, which has the following dimension: $[l] = s^2/(kg \cdot m^2)$.

With this transformation, equations (4) and (2) are reduced to the system of the non-dimensional equations with explicitly expressed linear parts:

$$\dot{u}_1 = \sqrt{\tfrac{l}{A}} Mgc \cdot \frac{\sqrt{1 + \frac{3gl}{R} B}}{1 + \frac{3gl}{R} C} u_5 - \sqrt{\tfrac{l}{A}} Mgb \cdot \frac{\sqrt{1 + \frac{3gl}{R} C}}{1 + \frac{3gl}{R} B} u_6 -$$
$$- \frac{C - B}{\sqrt{lABC}} u_2 u_3 + \sqrt{\tfrac{l}{A}} \cdot \frac{3g(C - B)}{\sqrt{(R + 3glB)(R + 3glC)}} u_5 u_6,$$

$$\dot{u}_4 = -\sqrt{\tfrac{l}{C}} Mgb \cdot \frac{\sqrt{1 + \frac{3gl}{R} A}}{1 + \frac{3gl}{R} B} u_3 + \sqrt{\tfrac{l}{B}} Mgc \cdot \frac{\sqrt{1 + \frac{3gl}{R} A}}{1 + \frac{3gl}{R} C} u_2 +$$
$$+ \frac{1}{\sqrt{lC}} \cdot \sqrt{\frac{R + 3glA}{R + 3glB}} u_3 u_5 - \frac{1}{\sqrt{lB}} \cdot \sqrt{\frac{R + 3glB}{R + 3glA}} u_2 u_6 \tag{5}$$

Here only the first and forth equations are given, the others can be obtained by the circle permutation of the quantities (A, B, C), (a, b, c), (u_1, u_2, u_3) and (u_4, u_5, u_6).

Let's consider the system (5) on stability on the first approximation. The characteristic equation of the system of the first approximation is:

$$\lambda^6 + N\lambda^4 + P^2\lambda^2 = 0, \tag{6}$$

With constants N and P^2, we get:

$$N = M^2 g^2 l \left\{ \frac{a^2\left[B\left(1+\frac{3gl}{R}B\right)+C\left(1+\frac{3gl}{R}C\right)\right]}{BC\left(1+\frac{3gl}{R}A\right)^2} + \frac{b^2\left[C\left(1+\frac{3gl}{R}C\right)+A\left(1+\frac{3gl}{R}A\right)\right]}{CA\left(1+\frac{3gl}{R}B\right)^2} + \right.$$
$$\left. + \frac{c^2\left[A\left(1+\frac{3gl}{R}A\right)+B\left(1+\frac{3gl}{R}B\right)\right]}{AB\left(1+\frac{3gl}{R}C\right)^2} \right\},$$

$$P^2 = M^4 g^4 l^2 \left[\frac{a^2}{BC\left(1+\frac{3gl}{R}A\right)^2} + \frac{b^2}{CA\left(1+\frac{3gl}{R}B\right)^2} + \frac{c^2}{AB\left(1+\frac{3gl}{R}C\right)^2} \right] \cdot$$

$$\cdot \left[\frac{a^2\left(1+\frac{3gl}{R}B\right)\left(1+\frac{3gl}{R}C\right)}{\left(1+\frac{3gl}{R}A\right)^2} + \frac{b^2\left(1+\frac{3gl}{R}C\right)\left(1+\frac{3gl}{R}A\right)}{\left(1+\frac{3gl}{R}B\right)^2} + \frac{c^2\left(1+\frac{3gl}{R}A\right)\left(1+\frac{3gl}{R}B\right)}{\left(1+\frac{3gl}{R}C\right)^2} \right],$$

The equation (6) gives immediately two zero characteristic roots and the biquadrate equation:

$$\lambda_1 = \lambda_2 = 0,$$
$$\lambda^4 + N\lambda^2 + P^2 = 0. \tag{7}$$

The roots of the equation (7) depend on sign of its discriminant. If discriminant D is positive, all the characteristic roots have zero real parts. In these cases, according to the theorem of Lyapunov[5] about motion stability on the first approximation we cannot determine the motion stability. If $D < 0$, the equation (7) decomposes into two quadratic equations:

$$\lambda^4 + N\lambda^2 + P^2 = (\lambda^2 + \sqrt{2P-N}\lambda + P)(\lambda^2 - \sqrt{2P-N}\lambda + P) = 0,$$
$$\lambda_{3,4} = \frac{-\sqrt{2P-N} \pm \sqrt{2P+N}i}{2}, \tag{8}$$
$$\lambda_{5,6} = \frac{\sqrt{2P-N} \pm \sqrt{2P+N}i}{2}. \tag{9}$$

It may be estimated that expression $2P - N$ is more than zero. This implies that the characteristic roots $\lambda_{5,6}$ have positive real parts. Therefore, according to the theorem of Lyapunov the unperturbed motion is unstable in this case.

3.2. *Motion stabilization*

Now we will try to make unstable motion in the last case the stable one with a help of some additive moments of forces. Let's add to the right part of the equations of the system (5) the terms of the form $\mu K u_i$, $i = 1, \ldots, 6$, where $\mu < 1$, $K = -\sqrt{2P-N}$.

The characteristic equation for such controlled system will be:

$$(\mu K - \lambda)^2 \left[(\mu K - \lambda)^4 + N(\mu K - \lambda)^2 + P^2 \right] = 0, \tag{10}$$

The roots λ_i of this equation are:

$$\lambda_1 = \lambda_2 = \mu K,$$

$$\lambda_{3,4} = \frac{-\sqrt{2P - N}(2\mu - 1) \pm \sqrt{2P + N}i}{2},$$

$$\lambda_{5,6} = \frac{-\sqrt{2P - N}(2\mu + 1) \pm \sqrt{2P + N}i}{2}.$$

Taking into account that $K < 0$ and constraining μ within the limits $0,5 < \mu < 1$, we shall obtain the negative real parts for all the roots of the characteristic equation. Hence, the Lyapunov's theorem about asymptotical stability of the unperturbed motion of the controlled system is true.

3.3. *Nature of the controlling forces and integral of the norm type*

In order to define the nature of controlling forces let's realize reverse change of the variables. In the right part of first three equations of the controlled system in comparison with initial system (4) the small additional moment with the vector, opposite to the vector of kinetic moment $\overline{K}_O = Ap\bar{i} + Bq\bar{j} + Cr\bar{k}$, is observed.

It is easy to prove that the forces, creating the controlling moment, are the dissipative ones.

Small additional members in the right part of last three equations of the controlled system characterize a small deviation of the direction of the acting force from the axis OZ.

The known first integrals of the system (2), (4) in new variables give us the integral of the norm type:

$$u_1^2 + u_2^2 + u_3^2 + u_4^2 + u_5^2 + u_6^2 = C_1. \tag{11}$$

The same integral can be obtained from the motion equations.

3.4. *Research of the generalized cases of Euler, Lagrange, Kovalevskaya*

The proposed method was applied to three cases with special restrictions on the body's inertia moments, so-called generalized classical cases of Euler, Lagrange and Kovalevskaya.

The Euler case is characterized by the fact that the moment of the enclosed force concerning the fixed point is equal to zero, i.e. $a = b = c = 0$. In the Lagrange case the motion of dynamically symmetric solid body is considered ($A = B$) and the body's centre of gravity lays on an axis of dynamic symmetry ($a = b = 0$). The Kovalevskaya case is characterized by the certain ratio between the main moments of inertia ($A = B = 2C$) and the gravity centre arrangement in an equatorial plane of the ellipsoid of inertia ($c = 0$).

In all these cases the discriminant D is positive, but the proposed method can be applied. If we constrain μ within the limits $-1 < \mu < 0$, we will obtain the negative real parts for all the roots of the characteristic equations in all three cases. Hence, the Lyapunov's theorem about asymptotical stability of the unperturbed motion of the controlled system by the above-stated restrictions is true.

4. Conclusions

The linear change of the variables, which reduced the initial system to the system of the nondimensional equations with explicitly expressed linear part, had been suggested. The first integral of the norm type was found. The transformed system of the equations was investigated on stability on the first approximation.

The control problem was defined for the case of instable by Lyapunov solution. It was shown that we can find the small additive forces, which provide asymptotically steady motion of the body.

Change of the first integrals of the controlled system was studied. Nature of the controlling forces was determined and their influence upon the body motion was revealed.

The motion stabilization was realized in the generalized cases of Euler, Lagrange, Kovalevskaya.

References

1. Euler L., *Theoria motus corporum solidorum seu rigidorum ex primis nostrae cognitionis principiis stabilita et ad omnes motus, qui in huiusmodi corpora cadere possunt, accomodata.* (Greifswald, Rose, 1790).
2. Kozlov V.V., *Non-existing of an addition analytical integral in the problem of non-symmetric heavy solid body motion round the fixed point*, Vestnik MGU, series of mathematics and mechanics, No 1, (1975), pp. 105-110.
3. Arkhangelsky Ju.A., *Analytical dynamics of the solid body.* (Nauka, Moscow, 1977).
4. Rakisheva Z.B., *The problem of the solid body dynamics and its first integrals*, Vestnik KazGU, series of mathematics, mechanics, informatics, No 1(29), (2002), pp. 189-192.
5. Malkin I.G., *Theory of the motion stability.* (Nauka, Moscow, 1966).

Session Organisers / Subeditors

Aliyev, T., 1

Barsegian, G., 121
Begehr, H., 65
Berezansky, L., 545
Berlinet, A., 247
Bucci, F., 355
Burenkov, V., 305

Crisan, D., 379
Csordas, G., 121

Dai, D.-Q., 65
Diblik, J., 545
Du, J., 65

Gilbert, R.P., 561
Gürlebeck, K.G., 203

Hirosawa, F., 323

Kilbas, A., 287
Kisil, V., 203
Kurylev, Y., 363

Lamb, J., 521
Lanza de Cristoforis, M., 1
Lasiecka, I., 355
Luzzatto, S., 521
Lyons, T., 379

Plaksa, S., 1

Reissig, M., 323

Sabadini, I., 171
Saitoh, S., 247, 287
Samko, S., 305
Sommen, F., 171
Sprössig, W., 203

Tamrazov, P., 1

Yamamoto, M., 363

Zafer, A., 545
Zegarlinski, B., 453

Author Index

Abate, M., 523
Abdous, B., 249
Airaptyan, R.G., 579
Akberdin, I.R, 563
Akhalaia, G., 67
Al-Attas, H., 585
Andreian Cazacu, C., 3
Arena, O., 357
Aripov, M., 592
Atslega, S., 530

Ballantine, C., 13
Bardi, M., 455
Barsegian, G., 123, 129
Begehr, H., 74
Benameur, J., 597
Berglez, P., 81
Berlinet, A., 249
Bokhari, M.A., 585
Bolley, F., 463
Bory Reyes, J., 173
Brändén, P., 136
Brody, D.C., 381

Cardon, D.A., 143
Charalambides, M., 150
Chen, X., 398
Costache, M.-R., 74
Crisan, D., 389

D'Abbico, M., 325
Dalla Riva, M., 23, 31
Davie, A.M., 405
Dovbush, P.V., 39
Dragoni, F., 455

Fadeev, S.I., 563

Faustino, N., 205
Fei, M., 180
Feichtinger, H.G., 257
Fernandez Arias, A., 157
Fujita, K., 266
Fujiwara, H., 289

Gaiko, V.A., 537
Gainova, I.A., 563
Galstyan, A., 332
Gentil, I., 463
Georgiev, S.G., 212
Ghisa, D., 3, 13
Gilbert, R.P., 570
Giorgadze, G.K., 89
Golberg, A., 45
Gozlan, N., 470

Hackl, K., 570
Hairer, M., 410, 479
Hasler, M.F., 604
Herrmann, T., 339
Higgins, J.R., 273
Huet, N., 485
Hughston, L.P., 417

Ilic, S., 570
Inglis, J., 491, 498

Kaasalainen, M., 365
Karapinar, E., 612
Karimov, E.T., 619
Karupu, O.W., 53
Kats, B.A., 59
Kazantsev, F.V., 563
Kelbert, M., 424
Khaybullin, A.G., 105

Kheyfits, A.I., 114
Khodjiev, S., 626
Kinoshita, T., 332
Kisil, V., 219
Korolev, V.K., 563

Léandre, R., 226
Lanza de Cristoforis, M., 31
Li, S., 180
Li, X.-M., 398
Likhoshvai, V.A., 563
Littman, W., 357
Lucente, S., 325

Macrina, A., 417
Makatsaria, G., 67
Malonek, H.R., 173
Manjavidze, N., 67
Manolarakis, K., 389
Manson, C., 410
Medvedev, A.E., 563
Mijatović, A., 431
Mouhot, C., 505
Moura Santos, A., 307
Mursaleen, M., 630

Neklyudov, M., 498
Nepomnyashchikh, Y., 547

Onchiş, D.M., 257
Opic, B., 315

Papageorgiou, I., 512
Papavasiliou, A., 438
Peña Peña, D., 173
Perez Alvares, J., 157
Perotti, A., 188
Pinotsis, D.A., 240
Pistorius, M., 431
Pivetta, M., 346
Plaksa, S.A., 232

Poirier, A., 164
Ponosov, A., 547

Qiao, Y., 195

Rajabova, L., 96
Rakisheva, Z., 636
Rappoport, J.M., 296
Reissig, M., 339
Roberto, C., 470

Sadullaeva, S.A., 592
Sadyrbaev, F., 530
Saitoh, S., 372
Saks, R.S., 105
Samson, P.-M., 470
Sargsyan, A., 129
Selmi, R., 597
Sharma, A.K., 280
Sharma, S.D., 280
Shindiapin, A., 547
Shklyar, B., 554
Shlykova, I., 547
Shpakivskyi, V.S., 232
Sommen, F., 173

Taglialatela, G., 325
Tappert, S., 74
Tunaru, R.S., 445

Vaitekhovic, T., 74

Wu, B., 398

Xie, Y., 195

Yagdjian, K., 332
Yamada, M, 372
Yang, H., 195

Zegarlinski, B., 498